Practical Clinical Chemistry

WITHDRAWN
UTSA Libraries

Practical Clinical Chemistry

Gelson Toro, Ph.D.
Director of Biochemistry,
Missouri Clinical and Biochemical Laboratory;
Director of Research Laboratories,
Reproductive Biology Research Foundation,
St. Louis, Missouri

Philip G. Ackermann, Ph.D.
Biochemist, DePaul Hospital;
Consulting Biochemist to Faith Hospital
West and Homer G. Phillips Hospital;
Research Associate,
Reproductive Biology Research Foundation,
St. Louis, Missouri

Little, Brown and Company   **Boston**

Preface

Practical Clinical Chemistry is an outgrowth of our previous contributions of sections on clinical chemistry to several editions of a book on clinical laboratory methods. While working in the clinical laboratory, during professional consultations, and in the process of research and teaching medical technology students, we frequently felt the need of a book that would treat the subject of clinical chemistry in greater detail and emphasize at the same time some of the more recent developments in clinical laboratory techniques, such as automation and radioimmunoassay methods. As none of the clinical laboratory texts currently available seemed to fit these needs, the present book was developed.

The first chapters are concerned with instrumentation, quality control, and normal values. The textual discussion of procedures for determining normal values is, we believe, more complete than that found in most other treatments of this subject. The chapters dealing with routine chemical determinations are quite complete, with alternative methods given for all the more important determinations.

To keep the book to a reasonable size, discussion of the historical background of various laboratory determinations is minimal. Formerly popular methods are mentioned but not discussed in detail. Similarly, information on the abnormal values for the constituents of body fluids is not given in great detail. The more common causes for elevated or depressed levels of body fluid constituents are discussed, but unusual or uncommon causes for variations from the established clinical norms have been omitted. Our attention is focused primarily on the chemical methods themselves.

The chapter on enzymes is much more extensive than that found in most books in the field. Detailed determination procedures are given for practically all the enzymes usually reported by clinical laboratories in the United States. Some details of laboratory determinations also are described for a considerable number of other enzymes. The Enzyme Commission names and numbers are included with the revised recommendations of The Commission on Enzymes of the International Union of Biochemistry.

Methods for the chemical determination of a number of hormones in blood and urine are included. The discussion of gas-liquid chromatographic methods has been purposely limited. Proper discussion of this field would require a chapter in itself. This discussion was not presented because radioimmunoassay methods mentioned in this text will probably in time supersede other methods such as gas-liquid chromatography.

In view of the increasing importance of the new radioimmunoassay (RIA) methods for the determination of blood serum levels of many hormones and drugs, the theory of a methodology for such determinations is treated in some detail in a separate chapter. Details of the evaluation techniques for a number of important hormones and drugs are given. Since most of these methods are best performed with kits or at least with labeled antigens from commercial suppliers, a list of current sources of supply is given in this chapter.

The chapter on automation is an extensive review of the types of automated and partially automated equipment available in this country at the time of writing (1974). General information is given as to the type of operation, number of different tests available, speed of operation, and other pertinent facts. Since the type of instrument best suited for a given laboratory depends so much on the particular clinical requirements of that laboratory, only general information can be given, but it should be sufficient to enable concerned professionals to decide which particular instruments should be investigated further for possible use in a particular laboratory.

We feel that this book will be of value to virtually all clinical chemistry laboratories. The smaller ones will find methods for the more common analyses, which are satisfactory on a small scale, and the larger laboratories will find good methods for the less commonly performed tests. We endeavor to make the book as complete as possible without being unduly lengthy. With rapid advances in the field, it is difficult to keep up with new developments. We include as many newer methods as appear to give satisfactory results with a saving in time or reagents compared to older methods, although not all of them have been extensively evaluated in our laboratories.

We are very appreciative of the great assistance and advice given by our director, William H. Masters, M.D., and by Fred Belliveau, David Rollow, and other members of the editorial staff of Little, Brown; we are also grateful to the illustrator Donna Werber from the Washington University Medical School, Department of Illustration, St. Louis, Missouri, and to Joan Bauman of the Reproductive Biology Research Foundation, St. Louis, Missouri, for her contribution to the RIA chapter. We are grateful for the advice, assistance, and moral support given by our wives Elena Toro and Momoyo Ackermann, without whose assistance we would not have reason to be grateful to anyone.

G. T.
P. G. A.

St. Louis, Missouri

Contents

1. Laboratory Instruments Using Light Measurement

Most tests in the clinical chemistry laboratory are performed with the aid of electronic instrumentation: spectrophotometers, flame photometers, atomic absorption spectrophotometers, fluorometers, and many other types of automated instruments. In order to use these instruments properly, the laboratory worker must understand the principles of their operation. The basic electronic principles for many of these instruments have been given by Ackermann [1], and many of the operational details have beeen described by Lee [2] and by White et al. [3]. Each manufacturer's instrument has its own special features and the instruction manual should be studied thoroughly; only the common basic principles will be given here. In almost all the instruments the net result is the measurement of some type of electrical signal, the magnitude of which is related in some way to the concentration of the substance being analyzed.

Photometry and Spectrophotometry

Photometers and Spectrophotometers

PRINCIPLE AND DESIGN

The basic principle of photometers and spectrophotometers is the measurement of the absorption of light by a solution, the concentration in solution of the absorbing substance being related to the amount of light absorbed [4–7]. The absorption of light in the visible range by a solution will cause the solution to appear colored to the eye, and the intensity of the color is related to the concentration of the absorbing substance. Thus, the early instruments were called *colorimeters*. Today the colorimetric measurements are rarely made by visual comparison of colors, and the instruments are more properly called *photometers*. However, the general field of measurement of concentration by means of light absorption is still often termed *colorimetry*.

Visible light is but a small portion of the entire electromagnetic spectrum. The various portions of the spectrum can be identified by means of the wavelength. At one end are the radiowaves with wavelengths that range from about 10^6 cm for long radiowaves to about 10^{-1} cm for very short radiowaves. Infrared light includes wavelengths from about 10^{-2} cm to about 7×10^{-5} cm. Visible light ranges rom 7×10^{-5} cm (red) through orange, yellow, green, blue, and violet with a lower limit of about 4×10^{-5} cm. The ultraviolet extends from this point down to wavelengths of about 10^{-6} cm. At shorter wavelengths are x-rays and finally gamma and cosmic rays. The various portions may also be designated by frequency (velocity/wavelength, where the velocity of electromagnetic radiation is approximately 3×10^{10} cm/sec) or, in some applications, by wave

Figure 1-1. Basic components of a photometer. L, light source; F, filter or grating; C, cuvette for sample; P, photocell; M, meter or other measuring device. The double lines represent the light path; the single line, the electrical path.

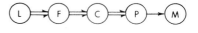

number (reciprocal of wavelength). In the visible and near ultraviolet light the preferred wavelength unit is the *nanometer* (nm) (1 nm = 10^{-9} meter). The *millimicron* (mμ), which is numerically equal to the nanometer, has also been used, as well as the *Angstrom* unit (Å) (1 nm = 10 Å).

The basic elements of a photometer are illustrated in Figure 1-1. The light source (L) for many instruments is simply a tungsten lamp. Although only a small proportion of the radiant energy from the lamp is in the visible spectrum, the tungsten lamp is a convenient source of light. It may be used for wavelengths down to about 330 nm. At shorter wavelengths not only is the relative amount of radiant energy small, but the glass envelope will absorb the short ultraviolet. For ultraviolet light the deuterium discharge tube may be used. Other special lamps such as the xenon-halogen lamp are available which give an intense light over a considerable range of wavelengths. The element labeled F in the illustration is an optical filter or other means of isolating a particular wavelength of light. When an optical filter is used, the instrument may be termed a *filter photometer* or merely a photometer. When a prism or grating is used to produce a spectrum from which the desired wavelength may be isolated, the instrument is called a *spectrophotometer.*

The cuvette (C) is the container for the solution to be measured. This will be discussed in greater detail later. The photocell (P) is a device for converting light energy into electrical energy. There are a number of different types of photocells. The simpler barrier layer photocells produce sufficient current to be measured directly with a sensitive meter. The photoemissive cells are much more sensitive, but usually require some electrical amplification of the output. In any type of photocell an electrical signal is produced which, over a certain range, is almost directly proportional to the intensity of light falling upon it. Most photocells are not equally sensitive to all wavelengths of light and some instruments employ different photocells for different portions of the spectrum.

In the simpler instruments the measuring device (M) is a meter in which the electrical signal is read directly by a pointer moving over a dial. The more sophisticated instruments may have additional circuitry so that the readout is displayed in digital form.

ABSORBANCE AND TRANSMITTANCE

The basic relation between the absorption of light by a solution and the concentration of the absorbing substance is known as Beer's law (more correctly, the Beer-Bouguer law). This is usually written in the form

$$\frac{I}{I_o} = e^{-kdc} \tag{1}$$

where I_o is the intensity of light entering the solution, I the intensity of the light that has passed through the solution, c the concentration of the absorbing substance, d the distance traveled by the light beam through the solution, and k a constant whose value depends on the units used for c and d. Taking the logarithm of both sides, the relation may be written:

$$\log \frac{I}{I_o} = -k'dc \tag{2}$$

where k' is a different constant. The ratio I/I_o is usually expressed as a percentage ($I < I_o$ if any light is absorbed) and termed *percent transmittance*.

$$\%T = 100 \, I/I_o$$

With this substitution we have

$$2 - \log(\%T) = k'dc \tag{3}$$

The expression on the left in this equation was formerly termed *optical density* (O.D.), but now the preferred term is *absorbance*. Thus, when Beer's law holds,

$$\text{absorbance} = k'dc \tag{4}$$

If the light path is held constant (as is usually the case),

$$\text{absorbance} = kc \tag{5}$$

Thus, the absorbance (A) is directly proportional to the concentration of the absorbing substance. In comparing the absorption of solutions of different substances, the results may be expressed in terms of the *absorptivity, a,* which is defined as: $a = A/cd$. Provided that Beer's law holds, this value a is independent

Figure 1-2. Double-beam spectrophotometer. L, light source; F, optical filter; R, beam splitter; C_1 and C_2, cuvettes; P_1 and P_2, photocells; M, meter. For further explanation see text.

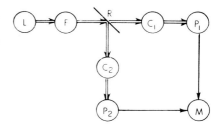

of the values of c and d used in the measurements. Thus different measurements can be more readily compared, provided the same units are always used for c and d. The preferred units are grams per liter for concentration and centimeters for the light path distance. The molar absorptivity, E, is the absorptivity when the concentration is expressed in moles per liter rather than grams per liter. If a or E were accurately known, one could in theory calculate the proper value of k' to insert in equation 4 without making any measurements (see discussion of NADH absorption in Chapter 14).

In the usual method for the determination of I/I_o, first a cuvette containing water or a blank solution is placed in the instrument and the reading is set to $100\%\,T$ (or zero A). Then the cuvette is replaced by one containing the solution to be measured. The intensity of light entering the cuvette must be the same in both instances (I_o constant). This means that the light from the lamp must be constant. In the ordinary laboratory the line voltage can vary substantially when other equipment such as centrifuges are switched off and on. Most photometers have a voltage regulator circuit to keep the voltage applied to the lamp constant. Voltage fluctuation can also be avoided by the use of the double-beam spectrophotometer (Fig. 1-2). The light from the source L passes through the filter and then to a beam splitter R which diverts a definite fraction of the light through cuvette 1 and the rest through cuvette 2. The ratio of the outputs from the two photocells is compared by the meter, M. Cuvette C_2 may contain a blank or a dummy cuvette. When the test solutions are placed in C_1 and compared with C_2, fluctuations in the light intensity should have little effect. If, for example, the output from the lamp increased by 5 percent, the amount of light entering each photocell would also be increased by 5 percent, but the ratio would remain unchanged.

If two solutions are compared in a photometer, either the same cuvette must be used for both measurements or the different cuvettes must have exactly the same light path; otherwise the comparison will not be valid. With accurate

Figure 1-3. Plot of concentration against percent transmission and absorbance.

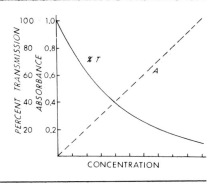

spectrophotometers, cuvettes can be obtained which agree in light path to less than 1 percent. With the round cuvettes used in the simpler photometers, there may be more variation. The cuvettes should be compared by measuring their transmission in relation to one arbitrarily taken as the standard. The comparison should be made not only with water in the cuvettes but also with some stable absorbing solution such as the cobalt solution mentioned later in this section.

As noted, when Beer's law is followed, the absorbance is directly proportional to the concentration, but the percent transmission is not. Thus, when the concentration is plotted against the absorbance or percent transmission, curves like those in Figure 1-3 are obtained. Often it is desirable to construct a calibration curve, plotting the results obtained with a number of different standards against the percent transmission or absorbance. Since it is much easier to plot a straight line, the use of absorbance is preferred. One can plot the concentration against the logarithm of the percent transmittance or use semilog paper but this is not so convenient. Furthermore, when Beer's law holds, the calculations as derived from the equations are simple:

$$\frac{\text{absorbance of sample}}{\text{absorbance of standard}} \times \text{conc. of standard} = \text{conc. of sample}$$

Since the output of the photocell usually varies directly with the light intensity, the scale will be linear in I/I_o or percent transmission. The simpler instruments may also have an absorbance scale but this will be compressed at the upper end making it difficult to read. It is usually more convenient to read in $\%T$, then convert to absorbance by means of a table such as Table 1-1.

Table 1-1. Percent Transmission ($\%T$)–Absorbance (A) Table

$\%T$	A	$\%T$	A	$\%T$	A	$\%T$	A
1.0	2.000	26.0	.585	51.0	.292	76.0	.119
1.5	1.824	26.5	.577	51.5	.288	76.5	.116
2.0	1.699	27.0	.569	52.0	.284	77.0	.114
2.5	1.602	27.5	.561	52.5	.280	77.5	.111
3.0	1.523	28.0	.553	53.0	.276	78.0	.108
3.5	1.456	28.5	.545	53.5	.272	78.5	.105
4.0	1.398	29.0	.538	54.0	.268	79.0	.102
4.5	1.347	29.5	.530	54.5	.264	79.5	.100
5.0	1.301	30.0	.523	55.0	.260	80.0	.097
5.5	1.260	30.5	.516	55.5	.256	80.5	.094
6.0	1.222	31.0	.509	56.0	.252	81.0	.092
6.5	1.187	31.5	.502	56.5	.248	81.5	.089
7.0	1.155	32.0	.495	57.0	.244	82.0	.086
7.5	1.125	32.5	.488	57.5	.240	82.5	.084
8.0	1.097	33.0	.482	58.0	.237	83.0	.081
8.5	1.071	33.5	.475	58.5	.233	83.5	.078
9.0	1.046	34.0	.469	59.0	.229	84.0	.076
9.5	1.022	34.5	.462	59.5	.226	84.5	.073
10.0	1.000	35.0	.456	60.0	.222	85.0	.071
10.5	.979	35.5	.450	60.5	.218	85.5	.068
11.0	.959	36.0	.444	61.0	.215	86.0	.066
11.5	.939	36.5	.438	61.5	.211	86.5	.063
12.0	.921	37.0	.432	62.0	.208	87.0	.061
12.5	.903	37.5	.426	62.5	.204	87.5	.058
13.0	.886	38.0	.420	63.0	.201	88.0	.056
13.5	.870	38.5	.415	63.5	.197	88.5	.053
14.0	.854	39.0	.409	64.0	.194	89.0	.051
14.5	.839	39.5	.403	64.5	.191	89.5	.048
15.0	.824	40.0	.398	65.0	.187	90.0	.046
15.5	.810	40.5	.392	65.5	.184	90.5	.043
16.0	.796	41.0	.387	66.0	.181	91.0	.041
16.5	.782	41.5	.382	66.5	.177	91.5	.039
17.0	.770	42.0	.377	67.0	.174	92.0	.036
17.5	.757	42.5	.372	67.5	.171	92.5	.034
18.0	.745	43.0	.367	68.0	.168	93.0	.032
18.5	.733	43.5	.362	68.5	.164	93.5	.029

Table 1-1 (*continued*)

%T	A	%T	A	%T	A	%T	A
19.0	.721	44.0	.357	69.0	.161	94.0	.027
19.5	.710	44.5	.352	69.5	.158	94.5	.025
20.0	.699	45.0	.347	70.0	.155	95.0	.022
20.5	.688	45.5	.342	70.5	.152	95.5	.020
21.0	.678	46.0	.337	71.0	.149	96.0	.018
21.5	.668	46.5	.333	71.5	.146	96.5	.016
22.0	.658	47.0	.328	72.0	.143	97.0	.013
22.5	.648	47.5	.323	72.5	.140	97.5	.011
23.0	.638	48.0	.319	73.0	.137	98.0	.009
23.5	.629	48.5	.314	73.5	.134	98.5	.007
24.0	.620	49.0	.310	74.0	.131	99.0	.004
24.5	.611	49.5	.305	74.5	.128	99.5	.002
25.0	.602	50.0	.301	75.0	.125	100.0	.000
25.5	.594	50.5	.297	75.5	.122		

PROCEDURE WHEN BEER'S LAW DOES NOT HOLD

When Beer's law does not hold over the entire range of interest, a curve similar to that of A of Figure 1-4 is obtained. In some exceptional instances the curve may be concave downward. Ordinarily the absorbance at higher concentrations is less than would be expected from Beer's law. In Figure 1-4 B represents the curve which would have been obtained if Beer's law held. (For accurate calculations in such cases, one either constructs a calibration curve such as A in the figure with each set of samples or at least checks the previous calibration curve at several points and then reads the sample values from the curve.) A method that is often used is to run three or four standards at different levels and then use for the calculation of each sample the standard whose absorbance is closest to the sample. This is illustrated in Figure 1-4. Here point a represents a standard having a concentration of 10 units and an absorbance of 0.900 (note that if Beer's law were followed, this would have had an absorbance of 1.00). Point b represents a sample with an absorbance of 0.84. If this falls on the calibration curve A, it corresponds to a concentration of 9.0 units. If it were assumed that Beer's law held, and calculating from a 5-unit standard, one would have obtained a value of 8.4 units (point d). If we calculate the value from the 10-unit standard, this is the same as assuming that the sample is read from curve C, which is a portion of the straight line passing through point a and the origin. This would yield a value of approximately 9.2 units (point c). Thus we see that if

Figure 1-4. Curve illustrating errors involved when Beer's law does not hold. See text.

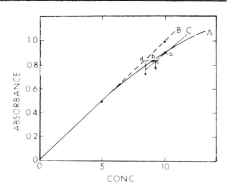

Beer's law does not hold, the only accurate method involves construction of a complete calibration curve. One should know through what range of absorbance Beer's law will be followed for a given photometer. This point is discussed later under Calibration of Spectrophotometers.

Development of color

In most instances the substance being determined is not colored or the light absorbance is too small for accurate measurement in the presence of other absorbing materials. The appropriate color is developed by means of a chemical reaction or a series of reactions involving the substance being determined. The amount of final color is taken as a measure of the amount of the substance present. The exact amount of color produced may depend upon the particular conditions under which the reaction is carried out. These may vary somewhat between different runs. For this reason the color developed by the samples is usually compared with that formed from standards run at the same time under identical conditions. The simple comparison of the absorbance of samples with that of previously observed standards (calibration curve) or calculations using a previously obtained absorptivity may not be accurate unless it is known definitely that the amount of color produced is constant under varying conditions.

Absorptivity comparison

The absorptivity value is helpful in comparing different methods which may have been developed with different instruments. If it is known, for example, that

Figure 1-5. Light paths in square (A) and round (B) cuvettes.

one method is stated to yield an absorptivity of **27,000** for a given substance, and another method an absorptivity of only **14,000**, one can safely conclude that for a given amount of substance, the former method would yield approximately twice the absorbance of the latter, and is to be preferred.

Another point in the comparison of absorptivities is that the length of the light path must be accurately known. In square or rectangular cuvettes in which the parallel light enters and leaves through parallel faces, the light path is constant throughout the beam (Fig. 1-5). In round cuvettes containing a liquid, the cuvette and liquid will act as a lens refracting the light. In this instance the average light path through the solution may not be the same as the internal diameter of the cuvette. Further, the amount of refraction will vary with the refractive index of the liquid and may be quite different for water and for an organic liquid such as chloroform. For aqueous solutions the effective light path can be calculated using solutions of known absorptivity. This is discussed later in this chapter under Calibration of Spectrophotometers.

Wavelength range

Theoretically Beer's law holds exactly only for monochromatic light (light of a single wavelength). Also, most substances absorb light in different amounts in different portions of the spectrum. This is illustrated by Figure 1-6 which gives the absorption as a function of wavelength for three different substances. Curve Ia shows a very sharp absorption peak at about 403 nm; curve II shows a broad absorption band with a maximum at 512 nm; curve III shows two absorption bands with maxima at 542 and 578 nm. This last compound, oxyhemoglobin, also has an absorption peak at 415 nm which is not shown in the figure since the absorption at 415 nm is about nine times that at 542 nm.

Thus the various photometers must have a means of isolating the particular wavelength desired. In the simpler filter photometers, optical filters of glass or other colored material are used. These transmit light in a small range of wavelengths and absorb other wavelengths. Strictly monochromatic light is ideal; most instruments will furnish light in a certain range of wavelengths.

Figure 1-6. Absorption curves for some compounds. I_a, coproporphyrin in $1.5N$ HCl; II, cobalt sulfate in $0.01N$ H_2SO_4; III oxyhemoglobin in dilute NH_3; I_b, coproporphyrin absorption measured with wide-band instrument.

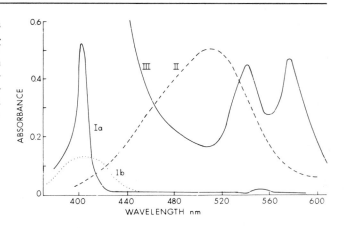

In discussing the range of wavelengths transmitted, the concept of the *band half width* is used (Fig. 1-7). Usually the desired wavelength will be transmitted to the greatest extent, with decreasing amounts of longer and shorter wavelengths. The band half width is defined by the two wavelengths at which the light is transmitted at half the maximum intensity. Most simple glass filters will have a band half width of 30 to 50 nm. Some special filters (interference filters) may have band widths of 5 to 10 nm but they are much more expensive. With a filter photometer, a separate filter is needed for each wavelength desired. Spectrophotometers use prisms or diffraction gratings to break up the white light into the spectrum; the desired portion then is isolated by means of slits. The wavelength desired is obtained by turning a dial to the appropriate position. Depending upon the instrument, the spectrophotometer may give light with a band width of from 35 nm down to less than 1 nm. The simpler, inexpensive instruments usually have wider band widths. In general, the smaller the band width, the greater the range of absorbance over which Beer's law will be followed.

For some purposes a narrow band width is not necessary. It is related to the width of the absorption band in the substance being determined. Referring again to Figure 1-7, note that curve I has an equivalent half width of about 45 nm and curve II a half width of about 15 nm. Generally speaking, for accurate measurements the half width of the light band used should be less than the equivalent half width of the absorption band of the substance being measured. Thus an instrument yielding a light beam with a half width of 20 nm would be satisfactory for a substance having an absorption curve like that of II. It would

Figure 1-7. Band half width. Width of band at one-half (15%) the maximum (30%) transmission. Curve I has half width from A to D; curve II, from B to C.

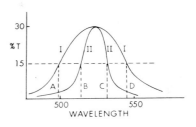

not be sufficiently narrow for the analysis of a substance having an absorption curve like I.

The curves of Figure 1-6 were obtained experimentally with a good spectro-photometer having a band half width of less than 0.5 nm and with a stray light of less than 0.1% T. When the absorption curve for the cobalt solution (II) was measured with a spectrophotometer having a nominal band width of 20 nm, an almost identical curve was obtained. This indicates that the latter spectrophoto-meter would be satisfactory for measuring the absorption of substances having absorption curves similar to that of the cobalt solution. When, however, the coproporphyrin absorption (Ia) was measured with the wide-band instrument, curve Ib was obtained. This clearly indicates that this instrument would not be satisfactory for determining coproporphyrin.

Usually the absorbance of a compound being determined is measured at a wavelength corresponding to the maximum (or one of the maxima) of the ab-sorption curve. Measurement at the point of maximum absorbance gives the greatest sensitivity (absorbance per unit concentration), and Beer's law will usually be valid over a larger range of absorbance when a diluted solution is used. Oxyhemoglobin is usually measured at 541 or 578 nm although this com-pound has a much larger absorption peak at 415 nm, since in this instance adequate sensitivity can be attained at 541 nm using a 1:250 dilution of whole blood. For reading at 415 nm, a dilution of about 1:2000 would be required, which would not be practical. Also, at 415 nm, with the simpler photometers, the problem of stray light would be greater than at the longer wavelength and with a more concentrated solution.

There are a few other instances in which the absorbance is measured at other than near the maximum. In many reactions for developing a color, the reagents themselves may have some absorbance at the wavelength used or they may form some color even when none of the substance being analyzed is present. In these instances, a reagent blank is usually run and the absorbance of the blank

Figure 1-8. Absorbance of solutions in determination of creatinine by the Jaffé reaction. Curve A is net absorbance of color formed from creatinine; curve B is absorbance of the blank.

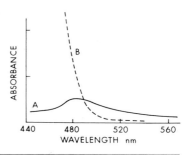

compensated for either by reading the blank and samples against water and subtracting the absorbance of the blank from that of the standards and samples, or by setting the photometer to zero absorbance with the reagent blank in the cuvette and reading the other solutions against this.

Sometimes the absorbance of the blank may be so great that it is difficult to set the instrument to zero absorbance with the blank. This is illustrated by Figure 1-8 which gives the absorbance of blank (B) and the net absorbance of the colored compound formed (A) for the determination of creatinine by the Jaffé reaction. Note that at the maximum absorbance of the color formed (in curve A), the absorbance of the blank (B) is very high. It is actually much more accurate to read at a higher wavelength such as 510 or 520 nm. At this wavelength, although the net absorbance of the sample is somewhat less than the maximum, the absorbance of the blank is greatly reduced so that any error due to slight variations in the blank is much less.

Another instance of a procedure in which the maximum absorbance is not used is illustrated in Figure 1-9 which represents the blue color produced in the copper reduction–molybdate methods for glucose determination. Although the color produced has a maximum absorbance at around 680 nm, a wavelength of around 500 nm is often used. Because of the intensity of the color produced, sufficient sensitivity can be obtained at the shorter wavelength and it has been found that absorbance at the shorter wavelength is more constant with time than at the peak. The absorbance at 680 nm will gradually increase over a considerable period of time, whereas the absorbance around 500 nm reaches a constant value in a short time and thus is more reliable and convenient to use.

Spectrophotometric Analysis of Multicomponent Systems [8]

If one had a mixture of two substances that had absorption peaks at distinctly different wavelengths with neither having appreciable absorption at the maxi-

Figure 1-9. Absorbance curve for colored compound formed in the reduction of phosphomolybdate by cuprous oxide in the determination of glucose.

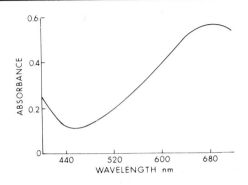

Figure 1-10. Two compounds with separate, nonoverlapping absorption peaks.

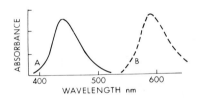

mum absorption of the other—such as the compounds having the absorption peaks illustrated in Figure 1-10—one could readily determine the concentration of one in the presence of the other by a single measurement at the appropriate wavelength (about 440 nm for A and 590 nm for B in the figure). If, however, the absorption curves for the two compounds resembled those in Figure 1-11, a single reading would not be sufficient to determine the concentration of either; measurements at two wavelengths are required. The relation between the concentrations of the two substances and the two wavelength readings is readily derived.

One reading is made at 510 nm (dashed line a) where compound A has a much greater absorbance and a second reading at 620 nm (dashed line b) where compound B has a greater absorbance. At 510 nm the absorption due to compound A would be $A_{1a} = K_{1a} \cdot C_1$, where A_{1a} is the absorbance of compound 1 at wavelength a (510 nm), C_1 is the concentration of compound A, and K_{1a} is the appropriate constant assuming Beer's law to hold. Similarly the absorbance of

Figure 1-11. Two compounds with overlapping absorption peaks.

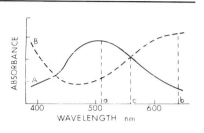

compound B at this wavelength would be $A_{2a} = K_{2a} \cdot C_2$, where the symbols have similar meanings.

At wavelength b (620 nm) the similar equations would be

$$A_{1b} = K_{1b} \cdot C_1$$
$$A_{2b} = K_{2b} \cdot C_2$$

(The two concentrations remain the same as at wavelength a). The total measured absorbance at 510 nm is then

$$A_{510} = A_{1a} + A_{2a} = K_{1a} \cdot C_1 + K_{2a} \cdot C_2$$

The measured absorbance at 620 nm is

$$A_{620} = A_{1b} + A_{2b} = K_{1b} \cdot C_1 + K_{2b} \cdot C_2$$

Solving these two equations for the unknowns, C_1 and C_2, we obtain:

$$C_1 = \frac{A_{510} \cdot K_{2b} - A_{620} \cdot K_{2a}}{K_{1a} \cdot K_{2b} - K_{1b} \cdot K_{2a}}$$

$$C_2 = \frac{A_{620} \cdot K_{1a} - A_{510} \cdot K_{1b}}{K_{1a} \cdot K_{2b} - K_{1b} \cdot K_{2a}}$$

where the Ks may be determined by measuring at the two wavelengths, two separate solutions containing known amounts of only compound A or compound B. This equation appears rather complicated, but after the various Ks have been reduced to the simpler form, we obtain:

$$C_1 = A_{510} \cdot K_1 - A_{620} \cdot K_2 \text{ and } C_2 = A_{510} \cdot K_3 - A_{620} \cdot K_4$$

where the new Ks may be readily derived from the above equations. A further simplification can be made if one of the chosen wavelengths (560 nm in the illustration) is one at which the same concentration of the two compounds have the same absorbance (*isobestic point*). If in the above equations we let the subscript b refer to 560 rather than 620 nm and remember that at 560 nm, $K_{1b} = K_{2b}$ (since the two compounds have the same absorbance at the same concentration at this wavelength), then the equations can be reduced to

$$C_1 = \frac{A_{510} - A_{560} \cdot K_{2a}}{K_{1a} - K_{2a}}$$

$$C_2 = \frac{A_{560} \cdot K_{1a} - A_{510}}{K_{1a} - K_{2a}}$$

If we are interested in the relative amounts of the two compounds A and B, it can be shown that the above equations can be expressed in the form

$$\frac{C_1}{C_1 + C_2} \times 100 = \text{percent } C_1 = R \cdot a + b$$

where R is the ratio of the absorbance at 510 to that at 560 nm, $R = \dfrac{A_{510}}{A_{560}}$, $a = (K_{1a} - K_{2a})$, and $b = \dfrac{K_{2a}}{(K_{1a} - K_{2a})}$.

From the above derivation we also have

$$C_1 + C_2 = K_{1b} \times A_{560}$$

The same procedure can theoretically be applied to the measurement of three components in a mixture by measuring at three different wavelengths. The calculations become somewhat more complicated but can be readily made with a desk calculator, particularly if the wavelengths can be chosen at the isobestic points for the three pairs of absorption curves. The Instrumentation Laboratory's CO-oximeter* uses this principle to measure carboxyhemoglobin, oxyhemoglobin, and reduced hemoglobin in a blood sample.

* Instrumentation Laboratory, Lexington, Mass. 02173.

Calibration of Spectrophotometers

In the calibration and checking of photometers and spectrophotometers, the main points to be considered are wavelength calibration, absolute absorbance, and linearity [9–12]. With filter photometers the problem of wavelength calibration does not arise. One must assume that the peak of the transmission band for the filter is approximately as stated. With wide-band filters the exact wavelength is not so important. Usually only a relatively few filters are furnished with the instrument, so that the exact preferred wavelength cannot always be chosen. Using a stable filter, the measurements will always be made with the same spectral band. It is not advisable to transfer a calibration curve from one filter photometer to another even of the same type since the filters can be slightly different. With spectrophotometers having a relatively wide band width, the wavelength calibration is particularly important in assuring that the actual peak of the wavelength band is the same as that showing in the wavelength scale, especially after changing the exciter lamp. The narrow-band didymium and holmium oxide glass filters which are useful for accurate spectrophotometers (see below) cannot be used with the wide-band instruments (see previous discussion of the porphyrin absorption band). Some manufacturers furnish special filters for use in calibrating their instruments. For approximate use some solutions that are relatively stable and have a fairly wide absorption band can be used for checking the wavelength. These include oxyhemoglobin with peaks at 415, 542, and 578 nm, *p*-nitrophenol in alkaline solution at 405 nm, and the previously mentioned cobalt ammonium sulfate solution with a maximum absorbance at 512 nm.

With precise, narrow-band instruments the wavelength calibration is more important since these instruments are often used for precise measurements of absorption curves and absorptivities for reference purposes. There are a number of ways of checking wavelengths accurately. Mercury has a number of narrow emission lines. If a mercury lamp can be used in place of the regular exciter lamp, the exact position of these emission lines can be noted in reference to the wavelength scale. Although a small penlight is available for insertion into the instrument, it is often impossible to place it in the correct position in some instruments. If a mercury lamp can be used, it will have emission peaks at 313 nm, 365 nm (with a secondary peak at 366 nm), 405 nm (with a secondary peak at 408 nm), 436 nm, and 546 nm. Many instruments have a deuterium lamp for use in the ultraviolet. In addition to the continuous spectrum this lamp emits two sharp peaks at 486 and 656 nm which can be used for checking the wavelength. Holmium and didymium oxide glasses are available which have a number of rather sharp absorption peaks. The principal peaks for holmium glass are at 279, 334,

361, 418, 453, 536, and 637 nm and for didymium glass at 573, 586, and 685 nm. Absorption curves are usually furnished with the glass filter when purchased.

In the sensitive narrow-band instruments with which actual absorptivities may be measured or used in calculations, not only must the wavelength scale be exact but the absorbance scale must also be correct. Two solutions whose absorptivities are accurately known are those of potassium dichromate and cobalt ammonium sulfate. A solution containing 0.0500 g of reagent grade potassium dichromate in $0.01N$ sulfuric acid to make exactly 1 liter should have an absorptivity of 10.7 at 350 nm and 14.4 at 257 nm. The former wavelength is useful for checking instruments when measurements are to be made at 340 nm. A solution of cobalt ammonium sulfate (28.962 g in water containing 10 ml of concentrated sulfuric acid and diluted to exactly 1 liter, $0.0735M$) will have a molar absorptivity of 4.71 at 510 nm (absorbance $= 0.346$ with 1 cm light path).

For most applications it is not as important that the absorbance be correct in absolute value, as that over the range of values used it should be directly proportional to the concentration. Thus it is advisable to check the linearity of the absorbance scale. With the inexpensive wide-beam instruments, the absorbance may be linear with the concentration only up to an absorbance value of 0.8 or even less. Solutions of cobalt sulfate, oxyhemoglobin, cyanmethemoglobin, or p-nitrophenol can be used for checking linearity. A relatively large quantity is prepared of a solution having an absorbance in the instrument being tested of about 1.0. Five to ten different dilutions of this solution are made, using precision pipets and volumetric flasks. If, for example, 10 ml of the solution is diluted to 100 ml and the original solution was made up in $0.01N$ sulfuric acid, this same solution is used as a diluent and as a blank in reading the dilutions. The dilutions are preferably made up in duplicate. The same cuvette should be used for all measurements and the process repeated on a second day. A plot of the absorbance values against the dilution will show the extent of linearity. If the instrument is found to be linear up to, for example, 0.8 absorbance, then in later spot checks it will be necessary to use only two solutions having absorbances of about 0.4 and 0.8.

If the instrument is found by these tests to be linear up to, for example, 1.0 absorbance, this does not mean that for all actual determinations the results will be linear in that range. If the colored compound whose absorbance is measured is developed by a chemical reaction, this reaction may not proceed in a strict stoichiometric proportion. For example, at higher concentrations the reaction may not proceed as completely as it should. This would mean a less than theoretical absorbance at higher concentrations. In some methods it will be definitely stated that the reaction does not follow Beer's law irrespective of the linearity of the instrument. In such cases several standards of different concentration must be run with each set of samples. In addition to this factor and the use of a very wide band width, as mentioned earlier, another cause of deviations from linearity is the presence of stray light.

Stray light may be defined as light of wavelengths other than that of the desired band which enters the cuvette and photocell. The effect is similar to that of an excessively wide band width. It is usually particularly noticeable at the ends of the visible spectrum (around 400 and 700 nm). Most spectrophotometers using a grating or prism have extra optical filters which are inserted in the light beam just in front of the cuvette or photocell to reduce stray light. For narrow-band instruments, filters are available for checking stray light. The Corning filter No. 3060 has a very low transmittance below 340 nm, and filter No. 3389 has a very low transmittance below 400 nm. Using these filters the transmittance should be less than 1% just below the wavelength given. A simple way of checking for stray light is to use a solution of $0.5M$ nickel sulfate in $0.01N$ sulfuric acid (13.14 g of $NiSO_4 \cdot 6H_2O$ and 1 ml of $1N$ sulfuric acid dissolved in water to make 100 ml). This will have a transmission of less than 1% in 1 cm cuvettes at 390 to 400 cm. A transmission of over 1.5% indicates that some stray light is present. With larger cuvettes the transmission should be less, and with 19 mm round cuvettes, the transmission should be not more than 1%, although these desirable criteria may not be reached with the simpler instruments. Stray light is most commonly found in the violet end of the spectrum. The nickel sulfate solution has a transmission of under 10% in the range of 700 to 740 nm when measured with a narrow-band instrument in 1 cm cuvettes. This is hardly low enough to use for an accurate estimation of stray light in this region, but if a photometer with 19 mm round cuvettes is used, the transmission should be less than 3% at 720 nm.

Flame Photometers and Atomic Absorption Spectrophotometers

Both of these types of instruments involve aspiration of a solution containing the test substance into a flame (except the so-called flameless atomic absorption apparatus to be mentioned later) [13–19].

FLAME PHOTOMETERS

In the flame photometer the element is aspirated into the flame. Some of the atoms are excited by the thermal energy of the flame and emit light of certain wavelengths characteristic of the element. Thus sodium atoms when excited by the flame will emit light of a wavelength about 590 nm (actually two lines of 589.0 and 589.6 nm, but these are usually not separated). If the light from the flame is passed through an optical filter that has a narrow transmission peak near 590 nm, only light of this wavelength will pass through and be measured. The intensity of the light emitted by the flame will depend upon the number of atoms of the element in the flame and the temperature of the flame. If the temperature and size of the flame and the rate of aspiration of a solution into the flame are all kept constant, then the amount of light emitted will be proportional to the concentration of the element in the solution being analyzed. Since it is difficult to keep these factors constant for extended periods of time, the solutions are always analyzed by running standards along with each batch of unknowns under supposedly identical conditions. One way to reduce the variability is by the use of an internal standard. A substance normally not present in the biological fluids being analyzed is introduced in a constant amount in all standards and samples. If the internal standard emits light of a different wavelength from that of the substance being analyzed, then by the use of two filters and photocells the light from the internal standard can be compared with that from the element under analysis. In this case any small fluctuations in flame temperature and aspiration rate will affect both internal standard and sample to the same extent and their ratio would be relatively constant. For sodium and potassium determinations lithium has been used as an internal standard since it is ordinarily not found in biological fluids. Recently this element has been used in the treatment of some psychiatric disorders, but with most instruments and methods for determining sodium and potassium by flame photometry with an internal lithium standard, the amount of lithium introduced into the serum by therapy is not great enough to cause an appreciable error. Obviously serum lithium cannot be determined with an internal lithium standard. Some instruments have an arrangement by which lithium may be determined using potassium as an internal standard. Lithium may also be determined in a flame photometer that does not use an internal standard or by atomic absorption spectrophotometry.

In flame photometry the presence of one element such as sodium may influence the emission of another element such as potassium. This effect is decreased by a greater dilution of the samples. The use of a relatively high concentration of lithium as an internal standard also decreases the interference or makes it more constant between different samples. Little error is introduced for plasma or serum determinations of sodium and potassium by their mutual interference as long as the dilution is 1:100 or more. An internal standard is used and the standards contain sodium and potassium in about the ratio normally found in blood. For urine, where the ratio of sodium to potassium is different, there may be a slight error unless the standards are made up correspondingly. However, the error will not be significant since for most clinical purposes it is not necessary to know the urinary concentration as accurately as the serum concentration.

Calcium may also be determined by flame photometry but atomic absorption is preferred. In calcium determinations the presence of phosphate will interfere with the calcium emission. Lanthanum is usually added to combine with the phosphate and reduce its interference with the calcium. Sodium also has a slight effect on the calcium emission and either a small correction must be applied based on the determined sodium concentration or an excess of sodium is added to samples and standards alike to make the effect relatively constant.

ATOMIC ABSORPTION SPECTROPHOTOMETERS

In flame photometry a few of the atoms in the flame are excited to a higher energy level by the thermal energy of the flame. In returning to their normal state these atoms emit the excitation energy in the form of the characteristic radiation. Usually only a few of the atoms in the flame are excited at any one time; most of them remain in the lowest energy level unless a very hot flame is used. In atomic absorption methods, use is made of this fact. It is a basic principle of physics that if an atom in an excited state emits light of a definite wavelength in returning to the normal state, then atoms in the normal state can be excited to the higher energy state by absorbing light of exactly this same wavelength. For purposes of illustration we will discuss this in terms of a particular atomic species, lithium, although the same principles apply to other atoms as well.

Excited lithium atoms emit the characteristic radiation of light of a wavelength of 670.8 nm in returning to the normal state. Consequently the normal atoms in the flame will absorb light of this wavelength and be changed to the excited state. (The excited atoms will later return to the normal state, emitting this energy in the form of the characteristic radiation.) The amount of light absorbed by the atoms in the flame will depend upon the number of absorbing atoms in the flight path through the flame. If a solution containing the lithium is aspirated into the flame at a constant rate, the amount of absorption will be proportional to the concentration of lithium in the solution. Only light whose wavelength is very close to that emitted by the lithium atoms at 671 nm will be absorbed. In order to attain a sufficient intensity of light of the proper wavelength, the light source is obtained by emission from excited lithium atoms. A second flame is not used for this but rather a more constant source, a gas discharge tube (similar in principle to a neon light) containing lithium. The lithium atoms in the tube are excited by the gas discharge and emit the characteristic radiation. Because of their construction these tubes are generally known as hollow cathode tubes. For each element to be determined by atomic absorption a separate lamp is required containing atoms of the element being determined to give the characteristic light for that element. Lamps containing several different elements are sometimes used but these generally have a shorter life and give a less intense light than do the single-element tubes. Also in some instances there may be interference between the atoms of the different elements. Usually single-element lamps are preferred.

Referring again to the lithium determination, not only is some of the light from the lamp source absorbed by the atoms in the flame but the excited atoms in the flame also emit the lithium radiation. The problem is to separate the unabsorbed light from the lamp we wish to measure from the other light of exactly the same wavelength emitted by the atoms in the flame. This is accomplished by mechanically or electrically interrupting the light from the lamp many times per second before it reaches the flame. Thus the light from the lamp which passes through the flame and reaches the phototube consists of many short pulses of light (modulated) in contrast to the light from the flame itself which is more or less continuous. The light from the source will thus produce a modulated current from the photocell which is distinguishable by electrical means from the more constant current produced by the light from the flame itself. After passing through the flame, the light is directed to a prism or diffraction-grating monochromator to isolate the desired wavelength (since the flame will also produce light of many other wavelengths as well).

Since this is an absorption process, the equivalent of Beer's law will hold here also, and the readings must be converted to absorbance units before calculation. In atomic absorption the readings are often made in percent absorption ($\%A$) rather than percent transmission as in photometry. Since $\%T = 100 - \%A$, the absorbance is then given as $A = 2 - \log (100 - \%A)$. In atomic absorption the wavelength scale need not be accurately calibrated, since one sets the dial to the approximately correct position and then adjusts the wavelength to give the maximum light intensity. As with flame photometry, the rate of aspiration of the solution into the flame must be kept constant and standards run with each set of unknowns. Some instruments have an arrangement so that an internal standard can be used (e.g., strontium is used as the internal standard for determination of calcium), but this adds to the complexity of the instrument.

As with flame photometry, there may be interferences between different elements in the solution. In the determination of calcium, lanthanum is usually added to reduce the interference by phosphate. In the determination of serum lithium it may be helpful to add an equivalent amount of sodium to the standards to equal the amount normally found in serum. In some cases the biological fluid (e.g., serum, urine) can be merely diluted with water or other appropriate solution and aspirated directly into the flame, or the preparation of a protein-free filtrate (usually with trichloracetic acid) may be necessary. In some cases the metallic element can be extracted into an organic solvent by means of a chelating agent and the extract aspirated into the flame. This usually results in some increase in sensitivity. In order to obtain a long light path for absorption, a long narrow burner is often used with the light passing along the length of the burner.

For increased sensitivity and the use of smaller samples some modifications have been recently introduced. One of these is the *sample cup method:* A tube of a high-melting metal such as tantalum is placed over the flame with the light passing down the axis of the tube as illustrated in Figure 1-12. The tube is usually closer to the flame than illustrated and is heated very hot by the flame. The sample is placed in a small sample cup which may also be of tantalum. After some preliminary treatment to evaporate any solvent and oxidize some of the organic material present, the cup containing the sample is introduced into the flame directly under a hole in the bottom of the tube as illustrated. The cup holder is constructed so as to position the cup accurately just below the hole. The hot flame vaporizes the element in the cup. The vapors rise into the tube and diffuse out the ends so that for a short time there is a relatively high concentration of absorbing atoms in the light path. Since the vapor remains in the tube for only a short time, it is necessary to have a recorder connected to the instrument to measure the short absorption peaks. If the material in the cup contains some organic matter when introduced into the flame, smoke from this will cause a nonspecific absorption peak. Usually this occurs before the peak due to the absorption by the element being determined and the two peaks can be distinguished on the recorder chart. If this is not possible, a more complicated method of background correction is necessary.

Another modification is the *flameless atomic absorption apparatus.* The metal tube of Figure 1-12 is replaced by a hollow carbon cylinder with the light passing down the long axis. The carbon tube may be heated to a high temperature by the passage of a heavy electric current through it from electrodes attached to the two ends. The sample is introduced into the center of the tube by a small pipet inserted through a hole in the top or side, or a solid sample may be introduced through the end of the tube by means of a special sampling spoon. The rod is first heated to a low temperature to volatilize any solvent, then heated at a somewhat higher temperature to decompose any organic matter present without volatilizing any of the substance being determined. Finally, the recorder

Figure 1-12. Simplified diagram of sample cup method of atomic absorption spectrophotometry.

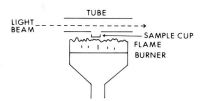

is turned on and the rod is heated to a very high temperature for a short time to volatilize the material into the path of the light beam. The atoms remain in the tube long enough to obtain an adequate recorder trace. The times and temperatures for the various steps are automatically controlled so that one needs only to press one switch to start the cycle. A current of an inert gas such as nitrogen or argon is passed through and around the tube to prevent oxidation of the carbon, and the tube is surrounded by a water jacket to prevent excess heating of the instrument and to cool down the carbon tube more rapidly so that successive samples can be introduced in a shorter time. Although the accessories required for this method add considerably to the cost of the instrument, the sensitivity may be increased by as much as 100-fold.

One element, mercury, is sufficiently volatile at room temperature so that it can be determined by flameless atomic absorption without the application of heat. The sample containing the mercury is treated to destroy most of the organic matter and a reducing agent added to reduce the mercury compounds to the metallic metal. A stream of air is then bubbled through this solution. The air then passes through a closed tube with its long axis in the path of the light beam from a mercury lamp. Sufficient mercury vapor is carried along with the air to enable one to measure the atomic absorption. A small commercial instrument, the MSA-50,† based on this principle is available for the determination of mercury.

Fluorometry

Another method that uses light measurement as a means of analysis is fluorometry [20, 21]. A number of compounds when illuminated by light will absorb some of the light and reemit some of the energy in the form of light of a longer wavelength. This is the phenomenon of *fluorescence*. The energy of a light quantum (photon) increases with decreasing wavelengths (increasing frequency).

† Coleman Instruments Division, Perkin Elmer Corporation, Maywood, Ill., 60153.

Figure 1-13. Fluorometric spectrum distribution. Curve A is the wavelength distribution of fluorescent light when excited at optimum exciting wavelength. Curve B is the relative intensity of fluorescent light at 520 nm when excited by the different wavelengths.

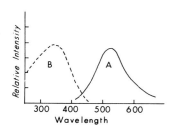

Since some of the energy is lost in the process of absorption and reemission, the emitted light must always be of longer wavelength (lower energy) than the exciting light. Usually only a certain wavelength range is effective in exciting fluorescence and the fluorescent light is usually confined to another wavelength range of longer wavelength. This is illustrated in Figure 1-13, where curve A represents the relative intensity of the fluorescent light at various wavelengths when excited at the optimum exciting wavelength, and curve B represents the relative intensity of the fluorescent light when excited by light of the different wavelengths shown. Usually the light from the exciting lamp is passed through an optical filter so that only light that is effective in exciting fluorescence will reach the fluorescent substance (about 300 to 380 nm in the figure). The fluorescent light is measured at right angles to the exciting beam after passage through a filter which removes scattered exciting light (below about 450 nm in the figure).

Under proper conditions the amount of fluorescent light emitted will be proportional to the intensity of the exciting light and to the concentration of the fluorescent material. If the former is kept constant, the amount of fluorescent light emitted will be a measure of the concentration of the fluorescent material. Since it is difficult to keep the intensity of the exciting light constant for a long period of time and to duplicate conditions exactly from day to day, standards are always run along with each group of samples. It is usually necessary also to run a reagent blank to compensate for any residual fluorescence in the reagents.

One difficulty often encountered in flourescent measurements is that other substances in the sample may tend to decrease the fluorescence of the substance being determined in comparison with that in a pure solution. Since it is difficult to determine the exact effect of these unknown impurities (as, for example, when the sample is an extract from a biological fluid), an internal standard method is used. Two identical samples are obtained and to one of them is added a known amount of the standard. The two samples are then carried through the procedure in the usual manner. If it is assumed that the effect of the impurities is the same in the two samples, the true concentration of the substance in the sample may be determined.

Fluorometry is essentially more sensitive than colorimetry, by a factor which may be as much as 100 or greater. As with colorimetry, if the substance being determined is not itself fluorescent, it may be changed into a fluorescent compound by a chemical reaction. Not all compounds can be made fluorescent so the method is somewhat limited in this respect. Fluorometry may not be as simple in procedure as colorimetry, but its higher sensitivity is often a great advantage. The simple fluorometers use optical filters for isolating the wavelengths for the exciting and fluorescent light. If the instrument uses prisms or diffraction gratings for this purpose, it is known as a *spectrofluorometer*.

Turbidimetry and Nephelometry [22]

Methods involving the measurement of turbidity are occasionally used in the clinical laboratory. A reagent may be added to the solution of the substance being determined so that it is precipitated in such a finely divided form that the particles remain as a stable suspension. If such a turbid solution is placed in a colorimeter cuvette and the light absorption determined, the absorbance found will be a measure of the concentration of the substance in the original solution. Some of the incident light may be absorbed by the suspended particles but usually a larger part of the light not passing through the cuvette is scattered out of the light path by the particles. Because of the nature of the light absorbance, this method is not as dependent upon wavelength as are colorimetric methods, and Beer's law is usually not obeyed, so a calibration curve must be constructed with each determination. The light scattering will vary somewhat with the wavelength used and the size of the particles. The latter factor constitutes one of the disadvantages of the method. The size of the particles formed may depend upon the exact method of formation, the order in which the reagents are added, temperature of the solution, presence of other salts, and other conditions. Thus it is more difficult to secure reproducible results than with colorimetric methods. The particles may also tend to coalesce or dissolve on standing, so that the turbidity may change with time. With too concentrated solutions the particles formed may be too large and will settle out too rapidly for good measurements. The method is not of greatest accuracy but is used for a few tests.

In nephelometry the light actually scattered out of the path of the incident beam is measured at right angles to the beam. Some fluorometers have been adapted for use as nephelometers, but this method is rarely used in the clinical laboratory.

References

1. Ackermann, P. G. *Electronic Instrumentation in the Clinical Laboratory.* Boston: Little, Brown, 1972.
2. Lee, L. W. *Elementary Principles of Laboratory Instruments,* 3rd ed. St. Louis: Mosby, 1974.
3. White, W. L., Erickson, M. M., and Stevens, S. C. *Practical Automation for the Clinical Laboratory,* 2nd ed. St. Louis: Mosby, 1972.
4. Delory, D. E. *Photoelectric Colorimetry in Clinical Biochemistry.* London: Hilger & Watts, 1966.
5. Snell, F. D., and Snell, C. T. *Colorimetric Methods of Analysis, Including Photometric Methods,* vol. 4A. New York: Van Nostrand-Reinhold, 1971.
6. Meloan, C. E. *Instrumental Analysis Using Spectroscopy,* vol. 1. Philadelphia: Lea & Febiger, 1968.
7. Mellon, M. G. *Analytical Absorption Spectroscopy.* New York: Wiley, 1950.
8. Van Assendelft, O. W. *Spectrophotometry of Hemoglobin Derivatives.* Springfield, Ill.: Thomas, 1970.
9. Rand, R. N. *Clin. Chem.* 15:839, 1969.
10. Frings, C. S. *Clin. Chem.* 17:568, 1971.
11. Martinek, R. G., Jacobs, S. L., and Hemmer, F. E. *Clin. Chim. Acta* 36:75, 1972.
12. Edisbury, J. R. *Practical Hints on Absorption Spectrometry.* New York: Plenum, 1967.
13. Dean, J. A. *Flame Photometry.* New York: McGraw-Hill, 1960.
14. Dvorak, J., Rubeska, I., and Rezac, Z. *Flame Photometry: Laboratory Practice.* Cleveland: Chemical Rubber, 1971.
15. Margoshes, M., and Vallee, B. L. *Methods Biochem. Anal.* 3:353, 1956.
16. Slavin, W. *Atomic Absorption Spectroscopy.* New York: Wiley, 1968.
17. Christian, G. D., and Feldman, F. J. *Atomic Absorption Spectroscopy: Applications in Agriculture, Biology and Medicine.* New York: Wiley-Interscience, 1970.
18. Willis, J. B. *Methods Biochem. Anal.* 11:1, 1963.
19. Robinson, J. W. *Atomic Absorption Spectroscopy.* New York: Dekker, 1966.
20. Udenfriend, S. *Fluorescent Assay in Biology and Medicine.* New York: Academic. Vol. 1, 1962; vol. 2, 1968.
21. Elevitch, F. R. *Fluorometric Techniques in Clinical Chemistry.* Boston: Little, Brown, 1973.
22. Oser, B. L. *Hawk's Physiological Chemistry,* 14th ed. New York: Blakiston–McGraw-Hill, 1965. P. 1010.

2. Physical and Other Miscellaneous Methods

Specific Ion Electrodes

The use of these electrodes is an electrical method for measuring the concentration of certain ions in solution [1–4]. If, for example, a metal M is dipped into a solution containing some M^+ ions (assuming the metal is univalent), an equilibrium will be set up between the metal and the solution:

$$M \rightleftharpoons M^+ + e^-$$

(e^- is the symbol for the negative electron, which will actually not exist as such in the solution.) A few metal atoms will dissolve or plate out on the metal (depending upon conditions) and an electrical potential will be developed between the metal and the solution. The magnitude of this potential will be related to the concentration of M^+ ions in solution. If we could determine this potential, we would have a method of measuring the concentration of the metallic ion in the solution.

To measure a potential difference, two electrodes are needed. One electrode will be the metal itself. As a connection to the solution, we insert an indifferent electrode X. A potential will also be developed between this electrode and the solution, but if the indifferent electrode does not react in any way with the solution we may assume that this latter potential will be constant or at least independent of the concentration of M^+ ions. The measured potential will then be the algebraic difference between the potential of the metal electrode E_M and that of the indifferent electrode E_X. The latter is assumed constant and the former will be found to be proportional to the logarithm of the concentration of M^+ ions, C_{M+}. Thus $E = E_X - K \log (C_{M+})$. By making measurements under similar conditions but with different known concentrations of the metallic ion, we can relate the measured potential to the concentration of the ion. The K in the above equation actually includes as one of its factors the reciprocal of the absolute temperature, so that the temperature must be included in the constant factors. The factor in the equation that we have labeled E_X will also include any other potentials that might develop in the complete measuring circuit. These may vary from time to time so the electrode system must be calibrated with known concentrations each time it is used.

Theoretically, instead of the concentration of the ionic species we should use the thermodynamic activity a. This will be equal to the concentration in very dilute solutions but will be somewhat less in other solutions. It may be considered to be the effective concentration of the ion in regard to the electrode reaction. Also, the standards may be made up to have definite concentrations of total metal rather than a concentration of the metallic ions which may not be exactly known. However, in many instances the conditions may be adjusted so that when the electrode is calibrated in terms of total metal concentration, the results on the samples will also give results close to the true total metal concentration. This will be discussed further in considering the electrodes for specific ions.

The most common indifferent electrodes used are the calomel electrode and the silver–silver chloride electrode. These electrodes are designed to give a constant reference potential. The essential part of a calomel electrode is an electrode of metallic mercury in contact with a saturated solution of mercurous chloride (calomel). Mercury was used because it was found that with many metals the exact electrode potential varied somewhat with the physical state of the metal (size of crystals, previous mechanical stress, etc.), but mercury, being in a liquid state at room temperature, is not subject to these factors. The mercury is in contact with a saturated solution of the mercury salt, since with an excess of the salt present, the concentration of mercurous ions in the solution would always be the same at a given temperature. Contact with the solution to be measured is usually made by means of a bridge of saturated potassium chloride solution. The silver electrode consists of a silver wire coated with solid silver chloride and in contact with a saturated solution of silver chloride in KCl.

Hydrogen Electrode

The ion most commonly determined by an ion electrode is the hydrogen ion. The original hydrogen electrode was made by bubbling hydrogen gas over an electrode coated with platinum black. The absorbed hydrogen functioned as a metallic hydrogen electrode. This system is rather inconvenient and most electrodes for hydrogen ions are now composed of a thin membrane of a special glass. It was found that an electrical potential was developed across this membrane when it separated two solutions containing different concentrations of hydrogen ions. A common form is a small bulb of this glass containing a known concentration of hydrogen ions and to which one electrical connection is made. This is dipped into the solution containing the unknown hydrogen ion concentration. The second connection to the solution is made through a calomel or silver chloride electrode. As in the example given earlier, calibration is made by measuring the actual potential developed with different solutions of known hydrogen ion concentration. Most instruments give a direct reading in pH (which is related to the logarithm of the hydrogen ion concentration; see discussion of buffers and pH in Chapter 20). As mentioned above, the potential developed is actually proportional to the logarithm of the activity of the hydrogen ion rather than to its concentration. Most measurements of hydrogen ion concentration in relation to biological systems are made at such low hydrogen ion concentrations (less than $0.001M$) that the difference between the activity and the concentration is small. The only exception is in the use of the electrode to measure the hydrogen ion concentration in gastric juice which may contain a relatively high concentration of hydrogen ions. The corrections that must be applied in this case are discussed in the section on gastric acidity in Chapter 17.

Sodium and Potassium Electrodes

The sodium electrode uses a special glass that is sensitive to changes in the sodium ion concentration. The potassium electrode has a special membrane utilizing the organic compound, valinomycin, producing a membrane sensitive to changes in the potassium ion concentration. These electrodes have been used to measure the concentration of serum sodium and potassium. The electrodes actually measure the thermodynamic activities of the sodium and potassium ions rather than the total concentration of ions or the total concentration of sodium or potassium in the solution. This is overcome by diluting the samples and standards alike in a fairly concentrated buffer (which, of course, contains no sodium or potassium), so that the ionic activity remains proportional to the total amount of sodium or potassium and the samples are thus compared with standards made up on the basis of total sodium or potassium. The buffer also serves to keep the pH constant as electrodes may be slightly sensitive to hydrogen ions. The electrodes are not entirely specific—the sodium electrode is slightly sensitive to potassium ions and vice versa—but by making up the standards to contain sodium and potassium in approximately the ratio found in the samples, this interference is minimized. Commercial instruments are available for determining sodium and potassium in blood serum with the ion electrodes

Chloride Electrode

The chloride electrode has been used for the determination of this element. The electrode has been used for some time for the determination of sweat chloride in the diagnosis of cystic fibrosis and more recently for the determination of serum chloride.

Calcium Electrode

One electrode that is used for the determination of the actual concentration of the ionic species is the calcium electrode for the determination of ionic calcium in serum.

P_{CO_2} and P_{O_2} Electrodes

Two other electrodes have some use in clinical chemistry. These are the P_{CO_2} electrode for measuring the partial pressure of CO_2 in blood or serum and the P_{O_2} electrode for measuring the partial pressure of oxygen in whole blood.

In the P_{CO_2} electrode the actual measurement is of pH (hydrogen ion concentration). The small hydrogen ion electrode assembly is immersed in a bicarbonate buffer system. This is separated from the solution being measured by a membrane of a plastic permeable to gaseous CO_2 but not permeable to dissolved ions. As the CO_2 diffuses through the membrane, it reacts with the buffer system to change the pH (see discussion of buffers and pH, Chapter 20). The electrode measurement is usually calibrated to read directly in P_{CO_2}.

The oxygen electrode is based on a different principle. It is representative of a polarographic method. These methods have been used to some extent in clinical chemistry in Europe but are seldom used in this country. The basic equation of the reaction may be written as

$$H_2O + \tfrac{1}{2} O_2 + 2 E^- \rightarrow 2 OH^-$$

This is the reverse of the reaction that occurs in the electrolysis of water, in which the OH^- ions attracted to the anode combine to form water and oxygen. At a potential of about $0.6V$, less than that required to decompose water, the reaction as presented will occur in the presence of free oxygen. The reaction will result in a small current whose magnitude will be proportional to the amount of oxygen in the solution. In the actual assembly, the electrodes in a buffer are separated from the solution being analyzed by a membrane permeable to gaseous oxygen but not to ions. As the oxygen diffuses across the membrane into the buffer, a small current is developed which is proportional to the partial pressure of oxygen. Since the reaction uses oxygen, the electrodes and current are made small so that the oxygen will not be used up faster than it will diffuse in. The small current formed can be readily amplified by electronic means. The instrument is usually made to give a direct reading in terms of partial pressure of oxygen, P_{O_2}. Both the oxygen and the carbon dioxide electrodes are calibrated by using gases containing known concentrations of O_2 and CO_2.

Other Specific Ion Electrodes

A number of other specific ion electrodes have been developed; some that might have use in clinical chemistry or toxicology are those sensitive to cupric, fluoride, iodide, lead, mercury, and cadmium ions. The ammonia electrode has some special applications. Theoretically it could be used for the determination of ammonia in micro-Kjeldahl determinations and in other reactions that liberate or form ammonia. For example, the enzyme reaction

adenosine $+ H_2O \rightleftharpoons$ inosine $+ NH_3$

which is catalyzed by the enzyme, adenosine deaminase, could be followed with an ammonia electrode. Since in such a measurement only the change in potential would be of interest, the other interfering substances present might have a smaller effect since they would remain constant during the reaction. One could easily calibrate the electrode by adding to the substrate a known amount of ammonia instead of serum or other fluid containing the enzyme. Another suggestion is the urea electrode. An ammonia electrode is surrounded by a gel containing the enzyme urease. When this is dipped into a urea solution, the urea diffusing into the gel is decomposed to ammonia and CO_2 by the enzyme. The concentration of ammonia as measured by the electrode could, under proper conditions, be related to the concentration of urea in the solution.

Coulometric Methods

In coulometric methods one of the reacting substances or reagents is generated in situ by the passage of an electric current through the solution [5, 6]. The substance thus formed reacts with the compound being determined. When all the latter has reacted, the excess reagent present is detected by some means, usually electrical, and the production of the reagent stopped by turning off the electric current. The amount of reagent formed to react completely with the substance being determined is thus a measure of the latter. The amount of reagent formed is proportional to the amount of current that has passed through the solution. The latter can be measured by electrical means, or, if the current is kept constant, the rate of generation of the reagent is constant and the time taken to form sufficient reagent to react completely with the sample is a measure of the amount of the substance in the sample.

The most common determination made by coulometric methods in the clinical laboratory is the analysis for chloride by some modification of the chloridometer originally developed by Cotlove. In this instrument the passage of an electric current generates silver ions from a silver electrode:

$Ag \rightleftharpoons Ag^+ + e^-$

The silver ions formed react with the chloride in the sample to produce insoluble silver chloride. When all the chloride ions have been precipitated, the excess Ag

Figure 2-1. Polarographic curve. For explanation see text.

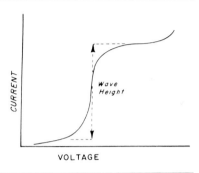

ions are detected by electrical means (chloride electrode) and the formation of the ions is stopped. Usually the current is kept constant and the time taken for generation of the silver ions is a measure of the chloride present. The reaction is usually carried out in a solution containing approximately $0.17M$ acetic acid, a small amount of nitric acid, and an organic substance such as gelatin to give a better reaction. In the new models there is a direct readout in milliequivalents per liter (mEq/liter). Also, the rate of Ag ions formation can be varied so that different concentrations of chloride can be titrated readily.

Few other coulometric determinations are in actual use in the clinical laboratory. One that has been suggested is the determination of ammonia (which may be formed from urea by the action of urease) by the reaction with hypobromite:

$$2\,NH_3 + 3\,NaOBr \rightleftharpoons N_2 + 3NaBr + 3\,H_2O$$

A similar reaction had been used earlier for the determination of urea, with the resulting nitrogen gas being determined by the gasometric method. In the present coulometric method the hypobromite is generated in the solution by the passage of an electric current. The endpoint is detected electrometrically.

Polarography

In this method the substance being determined reacts as an electrode when a current is passed through the solution [7, 8]. Usually the reaction is a reduction of the compound by the electric current. The applied voltage is increased linearly and the current produced is recorded on a strip chart recorder. The result will be a trace similar to that of Figure 2-1. By comparison of the known rate of voltage change and the speed of the chart movement, the horizontal axis can be calibrated

Figure 2-2. Enlarged trace of left-hand portion of curve of Fig. 2-1.

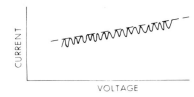

in terms of voltage. The increase in current is measured by extrapolation as shown by the dashed lines in the figure. The amount of increase will be proportional to the amount of substance in solution, and the voltage at which this increase occurs will be characteristic of the particular compound being determined. The solution usually contains another electrolyte that does not take part in the reaction (supporting electrolyte) to carry the current since the substance being determined may be present in very small amounts. When the reaction involved is a reduction, oxygen must be absent. The solution may be freed of oxygen by bubbling a stream of oxygen-free nitrogen through it for some time. In order to prevent undesired reactions at the reducing electrode, a dropping mercury electrode is used. Drops of mercury are slowly formed at the end of a capillary from a mercury reservoir, which constitutes the electrode. The mercury drops eventually fall into the bottom of the container. In this way there is always a fresh mercury surface at the electrode. Because of this periodic change in the electrode, the curve formed is not the smooth line indicated in Figure 2-1 but actually contains small fluctuations as shown in Figure 2-2. These fluctuations offer no difficulty in the measurement of the wave height since a line can be drawn along the top (or bottom) of them for extrapolation as indicated.

Polarography is rarely used in this country in the clinical laboratory, though it may be more common in Europe where the method originated. About the only procedure mentioned in this book which might be considered a polarographic one is the determination of Po_2 by the oxygen electrode. Note that there the oxygen is reduced to OH^- ions.

Refractometry

The refractometer measures the *refractive index* (ratio of the velocity of light in the solution to that in vacuo) of a solution in comparison with that of pure water [9–11]. When almost any substance is dissolved in water, the refractive index of the solution is increased over that of the pure water. In serum the largest contributor to the increase in refractive index is protein; the other dissolved material contributes only to a small extent. Thus the refractive index of the serum will be a measure of the protein content. The relation is established experimentally. The refractive index of a number of serums whose protein content has been accurately measured by chemical means is measured and an empirical formula derived relating the refractive index to the protein content of the serum. Figure 2-3 is derived from the data given with the usual hand refractometer. Note that because of the other substances present in the serum, the curve does not pass through the origin. The procedure is not completely specific for protein, but is accurate to within 0.1 g/dl for usual serum proteins in the range of approximately 2 to 10 g/dl. The error may be somewhat greater in very abnormal serum protein ratios.

The instrument may be calibrated directly in grams of protein per deciliter, and may also have a refractive index scale. Occasionally one may wish to check the calibration. For this purpose use either several different serums whose protein content has been accurately determined by a chemical method or standard solutions of sodium chloride. Since the effect on the refractive index is slightly different for different compounds, a solution containing, for example, 8.00 g of NaCl/dl will not have the same refractive index as one containing 8.00 g of serum protein/dl. It has been found that a solution containing exactly 7.50 g of NaCl/dl will have the same refractive index as a solution containing 5.15 g of serum protein/dl and accordingly this NaCl solution should give a reading of 5.15 on the protein scale. Similarly a solution containing 10.00 g of NaCl/dl should give a reading on the protein scale of 7.15 since the NaCl solution will have the same refractive index as a solution containing 7.15 g of serum protein/dl.

Specific Gravity

The measurement of specific gravity of an aqueous solution as a means of determining the amount of dissolved solute is rarely used (except for the use of urine specific gravity, discussed below) [12, 13]. The specific gravity is defined as the ratio of the weight of a given volume of a solution to the weight of an equal volume of pure water. The International Union of Pure and Applied Chemistry has suggested the use of the term *relative density* rather than *specific gravity*, but we will use the more common term. (The IUPAC recommendation has some value since in other terms in physics, such as specific conductivity and specific refractivity, the adjective has a somewhat different connotation from that in specific gravity.) Like the refractive index mentioned earlier, the specific gravity of a solution increases with increasing amounts of dissolved solute, and the rate of increase is different for different solutes. Thus the specific gravity is only a relative measure of the amount of total dissolved material in a solution.

Figure 2-3. Relation between protein content of serum and refractive index. N, difference in refractive index between serum and water.

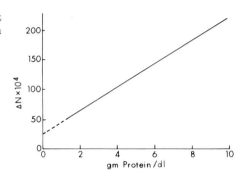

However, like the refractive index, the specific gravity has been used as a method of determining the amount of serum protein. One equation that has been used is:

serum protein (g/dl) = 348 (sp.g. − 1.0073)

where sp.g. is the specific gravity of the serum. For this determination the specific gravity was usually determined by the falling drop method. This will not be discussed in detail here; a good explanation is found in Natelson [37]. The principle of the falling drop method is based on the fact that the rate of fall of a uniform drop through an immiscible liquid of slightly lower specific gravity than the drop will depend upon the difference in densities of the drop and the liquid. By using drops of liquids of known specific gravity for comparison, that of the serum can be readily estimated. The automated urinalysis instrument developed by Ames* uses the rate of fall of urine droplets to determine the specific gravity. The refractometer method is much simpler and requires only a small amount of liquid, so that the falling drop method is now rarely used.

Practically the only clinical use of specific gravity determinations is in urinalysis. Although this may not strictly fall in clinical chemistry, a short explanation may be useful since there appears to be some confusion about the determination of urine specific gravity using the refractometer. The specific gravity of urine is merely a measure of the amount of dissolved substances. This in turn is a measure of the ability of the kidneys to concentrate urine and has

* Clinilab, Ames Company, Elkhart, Ind. 46514.

been used as such for many years by physicians. In the past the specific gravity has been usually determined by a simple hydrometer.

The refractometer has been used as a measure of the dissolved material in urine. The change in refractive index is as valid a measure of dissolved material as is the change in specific gravity, but because physicians are not familiar with the former, an empirical relation was derived, relating the two. A large number of urines were measured for both refractive index and specific gravity so that the refractometer could be calibrated in terms of specific gravity. The refractive index and specific gravity are different properties of a solution (though both are related to the amount of dissolved solutes) and there is no reason why solutions of different solutes and having the same refractive index will necessarily have the same specific gravity. Thus in the calibration of a refractometer that reads directly in *urine* specific gravity, one cannot use a solution that has an *actual* specific gravity of say, 1.015, but must use a solution that has the same refractive index as a urine having a specific gravity of 1.015. We find that a solution containing 7.50 g of NaCl/dl of solution, with a specific gravity of 1.051, has the same refractive index as an average urine with a specific gravity of 1.033; thus this NaCl solution should read 1.033 on the urine specific gravity scale. Similarly NaCl solutions containing 2.50 and 5.00 g/dl would give readings of 1.011 and 1.022, respectively, on the urine specific gravity scale.

Conductivity

Conductivity measures the total concentration of those substances in solution that are ionic and thus conduct an electric current [14–16]. Before flame photometers were available, the conductivity method was used to measure the total base in serum, which was taken to be a measure of the amount of electrolytes in solution. This was used for following electrolyte or acid-base therapy since the chemical determinations of serum sodium or potassium were lengthy. In essence, the method measures the resistance of the solution to an electric current between two electrodes placed in the solution. The measurements were made with a Wheatstone bridge arrangement using platinized electrodes and an alternating current (1,000 Hz). Conventionally the results were expressed as conductivity (the reciprocal of the resistance). Since the actual resistance in solution varied with the size of the electrodes and their distance apart, these were calibrated by using a solution of known conductivity such as $0.1N$ KCl solution. Using the known value for the KCl solution, the results for the samples are calculated in terms of the specific conductivity (the conductivity that would be found for the solution with two electrodes, each of 1 cm^2 area and exactly 1 cm apart). One suggested formula is then total base $= 10.2\,K + 35.16 - 72.67/P$

where K is the specific conductivity multiplied by 1,000, and P is the total protein in grams per deciliter.

Osmolality

The measurement of osmolality gives some information concerning the amount of dissolved substances in biological fluids [17–21]. The measurement of osmotic pressure assays a more basic physiological phenomenon than estimated by refractive index or conductivity. If an aqueous solution is separated from pure water by a membrane permeable to the water molecules but not to those of the solute, water tends to diffuse into the solution and dilute it. If the hypothetical membrane were capable of withstanding pressure, one would find that by exerting pressure on the solution one could slow down or even reverse the flow of water into the solution. That pressure which is just sufficient to prevent any diffusion of water into the solution is the *osmotic pressure* of the solution. Since the osmotic pressure is the measure of the tendency of water to diffuse through a membrane from a lower to a higher solute concentration, its application to biological phenomena and membrane diffusion should be apparent. The direct method of measuring osmotic pressure as implied in the definition above is difficult and this property of solutions is measured in other ways.

The osmotic pressure is one of a number of *colligative* properties of solutions which depend upon the *number* of particles in the solution and not on their charge, size, weight, or chemical composition. Thus a glucose molecule, a sodium ion, and an albumin molecule will all have the same effect on these properties. When one mole of sodium chloride is dissolved in a given volume of water, it will have twice the osmotic effect of one mole of glucose, since each molecule of sodium chloride will yield one sodium and one chloride ion in solution (complete ionization is assumed), whereas a molecule of glucose will give only one particle in solution. Thus the concentration of solutions in regard to colligative properties are usually expressed in osmoles (or milliosmoles). One *osmole* of any substance is that amount that will give the same osmotic pressure as one mole of a compound that is neither associated nor disassociated in solution but gives one dissolved particle per molecule (e.g., glucose). Concentrations in osmoles are usually expressed as per kilogram of solvent (water) rather than per liter of solution. Thus the concentrations are correctly expressed as osmolalities and osmolarities (in chemical terminology molar solutions are in weight per 1,000 ml of solution and molal solutions are in weight per 1,000 g of solvent).

Direct Vapor Pressure Method

The decrease in vapor pressure with the amount of dissolved (nonvolatile) solute is also a colligative property (depending only on the number of particles in solution). (In the following discussion we will assume the solvent to be water although the theoretical considerations hold for any solvent.) Thus the osmotic properties of the solution could be determined by measuring the change in vapor pressure. There are instruments available[△] for measuring the change in vapor pressure directly. In these, small measured drops of the solution (sample or standard) are placed on minute, sensitive temperature-sensing probes (thermistors) in a chamber of controlled humidity. If the partial pressure of water vapor in the chamber is less than the vapor pressure of the solution in the drop, water will tend to evaporate from the latter. The adiabatic evaporation of the water will produce a slight cooling effect which is measured. By comparison of standards and samples, the relative osmolality of the sample as compared with the standard may be determined. Although this is a sensitive method, it requires a more careful technique than the freezing point method mentioned below.

Boiling Point and Freezing Point Methods

It can be shown that a decrease in vapor pressure in a solution means an increase in the boiling point and a decrease in the freezing point. The boiling point is by definition that temperature at which the vapor pressure of the solution becomes equal to the external (atmospheric) pressure. Thus at 100°C the vapor pressure of water has increased to exactly 760 mm Hg. If the water contains some dissolved nonvolatile substances, its vapor pressure will be lower and a further increase in temperature will be necessary to bring it up to the required 760 mm Hg.

Although the osmolalities can be measured through the increase in boling point of the solution, this is obviously not a satisfactory way for the determination of the osmolality of blood serum, for example. The osmolality is similarly related to the decrease in freezing point, which may be defined as the point at which the solid and liquid phases are in equilibrium and hence have the same vapor pressure. The presence of dissolved material will lower the vapor pressure and consequently the freezing point. This is illustrated in Figure 2-4. The solid line in the graph to the right of the zero point on the temperature scale is the vapor pressure curve of the liquid phase (this curve has been drawn for illustrative purposes only and does not reproduce the actual numerical conditions).

△ Wescor, Inc. Logan, Utah **87321**.

Figure 2-4. Vapor pressure curves for (A) water, (C) ice, and (B) a solution.

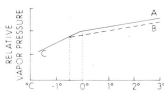

The solid line to the left of zero is the vapor pressure curve for the solid phase (ice). The curves intersect at zero as they must. The dashed line is the vapor pressure curve for a solution containing some dissolved material. This has a lower vapor pressure at all temperatures. The vapor pressure of this solution does not become equal to that of ice until it reaches a somewhat lower temperature (in the figure, about $-0.5°C$). If it is assumed that the dissolved material does not dissolve in the ice, the above interpretation will be correct and the solution will have a lower freezing point than pure water. Thus the freezing point lowering can be used to measure the osmolality. In fact, it is found that a solution containing 1 osmole/kg of water will have a freezing point of $-1.86°C$.

The procedure for measuring the freezing point is relatively simple. The solution is slowly cooled in a bath to a temperature slightly below its freezing point. It is then violently stirred for a few seconds when the water will begin to freeze and the temperature will rise to the freezing point. The temperature is measured with a sensitive detector since it must be measured to within a few thousandths of a degree. Comparison is made with standards treated similarly. Since the temperature of freezing varies proportionally with the osmolality (at least over the range encountered in biological samples), the instrument may be calibrated to read directly in milliosmoles per kilogram.

Radioactivity Measurements

Radioactive atoms are used as tracers in many determinations in clinical chemistry [22]. The advantage of their use is that they can be determined rather simply without as much separation from interfering material as is required for chemical determinations and in many instances smaller amounts can be easily determined. Radioactive atoms are those that spontaneously decompose into other atomic species with the liberation of energy. The decomposition of each individual atom results in the release of a definite pulse of energy which can be measured electronically. The number of pulses per unit time is then a measure of the amount of radioactivity present.

In the decomposition of radioactive atoms several different types of energy may be liberated. In those most commonly used in the clinical laboratory, the energy is in the form of gamma rays (similar to high-voltage x-rays) since these are most easily detected. Other atoms may emit beta rays (high-velocity electrons) but these are more difficult to measure; a few types of radioactive atoms may emit alpha rays (doubly charged helium nuclei), which are still more difficult to measure and are not used in ordinary applications.

Each particular radioactive atomic species decomposes at a constant characteristic rate. In any given amount of radioactive material the rate of decomposition is proportional to the amount of radioactivity present and is virtually independent of external conditions (such as temperature, pressure, chemical combination of the atoms, or presence of other atomic species). The rate of decomposition is usually specified in terms of the *half-life,* defined as the time required for one-half of the atoms present to decompose. Thus if a particular atomic species has a half-life of 8 days, this means that after 8 days one-half of the radioactive atoms in a given sample will remain undecomposed and after 32 days only one-sixteenth will remain undecomposed. When a radioactive element decomposes with the liberation of energy, it is usually changed into a stable (nonradioactive) element. The relation between the half-life and the number of atoms remaining at any time is given by the equation:

$$\frac{N_c}{N_0} = e^{-0.693 \, t/\lambda}$$

where N_c is the number of atoms remaining at time t compared with N_0, the number present at zero time, and λ is the half-life. The two quantities t and λ must be measured in the same units such as hours or days. For computational purposes this equation may be written in the form:

$$\log (\% \text{ remaining}) = 2 - (0.301 \, t/\lambda)$$

A given atomic species such as, for example, iodine exists in several different types of atoms which differ slightly in their mass (and radioactivity), but all will be—in the chemical sense—iodine atoms. These different *isotopes* may be designated as $^{125}_{53}\text{I}$, $^{127}_{53}\text{I}$, $^{131}_{53}\text{I}$, and so on. The nucleus of an atom may be considered for our purposes as composed of protons (positively charged particles) and neutrons (neutral particles of approximately the same weight as the protons). In the designations above, the superscript designates the total number of neutrons and protons in the nucleus (mass number) and the subscript designates the number of protons in the nucleus (number of positive charges). Often the subscript is omitted since any atom that has a nucleus with 53 positive charges is, chemically speaking, iodine.

The use of radioactive elements is based on the fact that the radioactive atoms (until they have decomposed) act chemically in a nearly identical manner as the usual nonradioactive atoms of the same element.

Thus if some iodine-containing molecules, such as thyroxine, are treated so that a small percentage of the normal iodine atoms are randomly replaced by radioactive iodine atoms, the radioactivity can be used to follow the thyroxine through a series of reactions. Suppose, for instance, that we measure the initial amount of radioactivity in our sample containing the labeled thyroxine, then we subject it to a chromatographic separation. If in one fraction we find 25 percent of the radioactivity, this means that 25 percent of all the thyroxine molecules are in this fraction. Radioactivity might be measured very simply by merely placing the container with the fraction in the measuring device without having to carry through any lengthy chemical procedure.

Gamma rays are usually detected with a crystal scintillation counter. This consists of a large crystal of sodium iodide containing a small amount of thallium iodide or other activator placed against the window of a photomultiplier tube (a very sensitive phototube for measuring small amounts of light). The crystal and tube are enclosed in a light-tight chamber. When a gamma ray enters the chamber and strikes the crystal, it may be absorbed with the emission of a minute flash of light. This flash is detected by the phototube and converted to a small electrical impulse which is then amplified and counted. Each gamma ray entering the chamber may represent the decomposition of one radioactive atom. Since the radioactive atoms may emit the gamma ray in any direction, one tries to surround the sample as much as possible with the crystal to detect as many of the gamma rays as possible. This is usually done by having a well in the crystal into which the container with the sample is placed so that the sample is more nearly surrounded by the crystal.

The energies of the gamma rays or other energy particles emitted by radioactive atoms are usually expressed in million (or mega) electron volts (mev). One electron volt (ev) is the energy received by an electron in falling through a potential of 1 volt (v). In some radioactive decompositions, several different energies of gamma rays may be emitted, but usually most of the energy is concentrated in one or two of the rays.

Two of the radioactive species that may be used in the laboratory do not emit gamma rays, but only relatively weak beta rays (electrons). These species are ^3H (tritium) and ^{14}C. Both emit only weak beta rays which would not even penetrate the thin glass of a tube in which the sample is placed. A different method known as *liquid scintillation counting* is used. The material containing the ^3H or ^{14}C (in combination as organic compounds) is placed in a vial in a special solution usually containing toluene, naphthalene, and other substances such that whenever the radioactive atoms in the solution decompose, the liberated electrons are rather quickly absorbed by the liquid but in doing so a minute flash of light is emitted. The container is placed in a light-tight compartment near the faces of one or more photomultiplier tubes so that the light flashes can be detected and counted.

No matter which method of counting is used, there will always be some spurious counts due to the natural radioactivity present everywhere and to cosmic rays. To eliminate this as much as possible, the scintillation crystal or solution and the phototube are usually surrounded by several inches of lead to filter out extraneous rays. Even then it is usually necessary to make a background count without any sample. This background count which corresponds somewhat to a reagent blank is subtracted from all readings for the samples. In some instruments this is done automatically.

Each pulse of light detected by the phototube and the resultant electrical pulse in theory represents one atomic disintegration. The pulses are counted electronically. The results may be expressed in a number of ways such as counts per minute, or seconds required to make 10,000 counts, or other units. The results may be displayed on a series of neon lights, by a meter pointer, or by a digital readout. Most methods using gamma ray emitters usually specify sufficient radioactivity so that the individual measurements can be made in 10 min or less. With the methods using the weaker beta rays (^3H or ^{14}C), much longer measuring times are required and automatic sample changers and counters are used.

Although on the average the disintegration rate will follow the equation given above exactly, the process is statistically random. In successive counts on the same sample, one would not obtain the same count each time. In the counting of discrete events the *standard deviation* (a measure of the random error, see Chapter 4) is equal to the square root of the number of events counted. Thus if the "true" activity were 10,000 counts in 5 min, or 2,000/min, the successive 5-min counts would have a standard deviation of 100 counts ($\sqrt{10,000}$), and 95 percent of the counts would be within two standard deviations of the mean, or between 9,800 and 10,200 counts, a difference from the true value of ± 200 counts. Thus, if possible, 10,000 counts should be made for each sample to give a relative standard deviation of ± 1 percent. Note that the standard deviation depends upon the total number of counts made and not on the calculated counts per minute or other unit of time.

The fact that the theoretical activity would have decreased slightly in the interval between the first and last of a series of counts is of little importance. It can be calculated that with a radioactive substance having a half-life of 10 days, the theoretical decrease in activity over 30 min would be only 0.15 percent or less. Of course, in comparing a count with one made 24 hr earlier or later, the difference in activity must be taken into account.

In crystal scintillation counting not all the liberated gamma rays will reach the crystal. The percentage reaching the crystal to be counted will vary with the relative positions of the sample and crystal. Thus, in making a number of counts on different samples, one must take care that all the samples are in the same relative position. This usually includes using the same volume of solution for each count and similar-sized tubes or other containers.

Another possible source of error is in coincident losses. When high activities are measured, the light flashes in the crystal and the electrical pulses from the phototube may occasionally be so close together that the electrical circuitry cannot distinguish between them and only one count will be registered instead of the actual two or three. This will result in a count lower than the theoretical value. The amount of error will increase with the counting rate. With the activities ordinarily used in the clinical laboratory, this is rarely a serious problem. The maximum rate for counting with less than 1 percent error will depend upon the particular electrical circuits involved. The instruction manuals for many instruments give some information on this. Theoretically the coincidence error could be determined by counting a solution of very high activity and then making accurate dilutions of this for further counting, in a manner similar to checking the linearity of a photometer.

In order to obtain most radioactivity materials for use, a license issued by the Atomic Energy Commission is required. The suppliers of the radioactive materials can inform the laboratory if this is required and of the procedure for obtaining it.

Competitive Protein Binding and Radioimmunoassay

These two methods of analysis are coming more into use for the measurement of minute quantities of a number of substances of biological interest including some hormones [23–29]. The main requirements are that there must be available some of the substance to be determined which has been labeled with radioactive atoms (commonly ^{125}I, ^{14}C, or ^{3}H), some of the pure material to be used as a standard, an additional substance that will quantitatively bind or otherwise combine with the material being determined, and a method for separating the material that has been bound from that which remains free.

Standard known amounts of the material to be measured (X) and samples containing unknown amounts of X are incubated along with a known, constant amount of X that has been made radioactive (X^*), and a small quantity of the binding substance. The nonradioactive X and the radioactive X^* compete for binding sites on the molecules of the binding substance. The number of X^* molecules bound after incubation is then inversely related to the number of X molecules present in the system. The X^* molecules that are bound are then separated from those that are not bound, and the radioactivity in either the bound or free portion can be measured. A curve is drawn, plotting the percentage of bound (or free) radioactivity in each of the standard tubes vs. the concentration of X known to have been added to that tube. The concentration of X in the unknown samples may then be determined from the standard curve.

In competitive protein binding, the binding substance is a specific protein usually derived from serum which binds the X molecules strongly. Since the binding agent is usually a relatively large protein molecule, the complex formed will have a much higher molecular weight and can be separated by means of Sephadex filtration or by the use of dextran-coated charcoal. Alternatively the protein complex may be precipitated by the use of ammonium sulfate or other precipitating reagents.

In radioimmunoassay the binding substance is an antibody, a molecule produced by the immunological system of the body in response to a foreign substance, or antigen, which is then bound specifically by the antibody. The antigen in radioimmunoassay is the substance to be measured. The complex formed is the antigen-antibody combination which is a much larger molecule and can be separated by one of the methods mentioned above, or in some cases merely by centrifugation. In the double antibody method, the separation is made by adding in the final step a second antibody that combines with the first antibody. The resulting complex is a still larger molecule and may precipitate out directly or otherwise be more readily separated.

The antibody-antigen reaction is, in general, much more specific than the protein binding, so that when this method is used, less preliminary separation of interfering material may be required.

These methods are among the most sensitive available. They are usually the preferred (sometimes the only practicable) methods for the assay of many hormones and other substances that occur in biological fluids in nanogram quantities. Some problems associated with these methods are discussed briefly here; more specific information is given in the sections dealing with the particular assays elsewhere in this volume. The general problems include the availability of the standards, the radioactive material, and the binding substance. For some substances such as thyroxine, the steroid hormones, and insulin, purified standards are readily available. For other substances, particularly hormones such as the parathyroid and the various pituitary hormones, good standards may be difficult to obtain.

Although it is possible for the ordinary clinical laboratory to prepare some substances labeled with ^{125}I, it is not practical to do so. It would not ordinarily be possible for a laboratory to prepare compounds labeled with ^{3}H or ^{14}C unless the services of a chemist skilled in organic synthesis were available. The measurement of the radioactivity from ^{125}I offers no difficulty and can be readily done in any laboratory, but the measurement of the much weaker radiation from ^{14}C and ^{3}H requires a liquid scintillation counter, preferably one with automatic counting. The problems involved in liquid scintillation counting are much greater than those involved in counting of gamma radiation by a crystal scintillation counter. The ordinary laboratory would be advised to limit itself to those assays that can be done with ^{125}I-labeled material.

The preparation of the binding substance (B) or (Ab) may also be far from simple. For a few assays by competitive protein binding where serum proteins are used, a simple dilution of the serum obtained from the proper source may be satisfactory. Certainly the preparation of the antibodies to specific antigens is beyond the scope of most laboratories, except those connected directly with a research project in this field. The antigen must be prepared in pure form, the animals injected with the antigen over a period of time, and the final serum tested for the amount of antibodies present. Although the antibodies as purchased are rather expensive, usually only small amounts are used, and the authors would not advise anyone without considerable experience in this field to attempt to prepare them.

Reagents for most competitive protein binding assays and radioimmunoassays that would be of interest to the clinical laboratory are commercially available, many in convenient kit form including standards, radioactive label, binding substance, and detailed instructions.

Microchemical Methods and Equipment

In the earlier days of clinical chemistry, most determinations on blood or serum were made with samples of 1 or 2 ml in size. Any method using from 0.05 to 0.1 ml was generally considered a micromethod. Today many common methods, including a number found in this volume, use samples of 0.02 to 0.10 ml. These methods usually use ordinary glassware and apparatus (except for micropipets) and volumes of reagents sufficient to give a final volume of 3 ml or more. The final colorimetric readings can thus be made in regular spectrophotometer cuvettes. A true micromethod might be interpreted by some as being a method that uses not only samples of the order of 5 to 20 μl in size but also smaller amounts of reagents in special microapparatus. This type of microtechnique will not be treated here. In general, these microsystems require considerable and constant experience for the user to obtain satisfactory precision. For further information on systems of this type, the reader is referred to the references cited [30–32].

The following discussion will deal with micropipets that can be used with the various procedures, including a number of those mentioned in this volume, that utilize 100 μl or less of sample. There are generally a number of manufacturers or distributors for each type of pipet and it is not possible to note them all. A few of each type are mentioned to illustrate the features.

The simplest pipets are merely fine capillary tubes of uniform bore. Some of these will have a graduation at one point.* The tube is filled to slightly above this point by suction or capillary action. The outside of the tube is wiped off and the sample brought down to the graduation by gently touching the tip to an absorbent material. The sample is then blown into the reagent and the tube rinsed out. Some other types of capillary pipets are completely filled, usually by capillary action. In wiping this type off, care must be taken not to remove any liquid from the bore. In some applications, the filled capillary tube is merely dropped into the reagent solution and the reaction carried out. When the mixture is transferred to a cuvette for colorimetric reading, the capillary will remain behind. Or, if the reaction tube is also used as a cuvette, the capillary can be made to adhere to one wall of the reaction tube without interfering with the light path. Other pipets of this type which are completely filled are attached to a holder so that the liquid can be forced out of the capillary tube and the latter rinsed out.† These types are inexpensive and are made to be discarded after one use. With careful technique they can deliver samples with a precision of 1 percent.

Another type of pipet has a plunger to control the volume of liquid drawn up into a disposable plastic tip. The Eppendorf‡ and the MLA§ pipets are examples of this type. The MLA is illustrated in Figure 2-5. One attaches the tip, de-

* Corning Glass Works, Corning, N.Y. 14830; Clay-Adams Company, Parsippany, N.J. 07054.
† Unopette—Becton & Dickinson, Rutherford, N.J. 07070; Eskalab Pipettes—Smith Kline Instruments, Inc., Palo Alto, Calif. 94306.
‡ Brinckman Instruments, Westbury, N.Y. 11590.
§ Medical Laboratory Automation, Inc., Mt. Vernon, N.Y. 10550.

Figure 2-5. MLA micropipet.

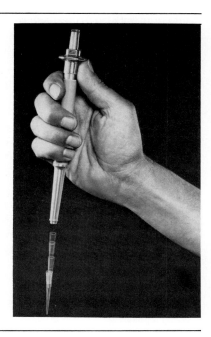

presses the plunger to expel the air, and then inserts the tip in the sample and slowly releases the plunger to draw up the required sample. The sample liquid is delivered by depressing the plunger. It is usually recommended that the tip touch the side of the tube when delivering the sample. Some pipets of this type have a second stop so that when the plunger is depressed to this stop, a small amount of air is expelled to complete the delivery of the sample. Since the disposable tips are made of plastic which is not wetted by aqueous solutions, complete delivery of the sample is possible. Some types have adjustable stops so that several different volumes can be delivered, but usually a separate pipet is required for each different volume. With care and experience these pipets can give excellent precision and are rapid in use.

Another type of micropipet is the glass pipets exemplified by the Accupettes|| and Technipettes.# These are similar in construction to the well-known Sahli pipets except that the larger sizes have a bulb in the pipet bore to increase the precision. They are fairly easy to use, but require some care when pipetting

|| Scientific Products Division, American Hospital Supply Corp., McGaw Park, Ill. 60085.
Curtin Scientific Company, Houston, Tex. 77001.

dilute aqueous solutions since it is sometimes difficult to adjust the meniscus properly to the graduation. This difficulty is not encountered with more viscous solutions such as blood serum. Since these types are of glass, they require washing after use and are therefore not as convenient as the disposable types.

Other glass micropipets of the Kirk and Levy-Lang type have been used in microchemical work for many years. These may be used in some of the true microchemical procedures but are much more expensive and fragile than the above-mentioned glass types. The latter are sufficiently accurate for most clinical work. Another device that might be classified in this category is the Hamilton syringe.** This is a syringe with a fine bore and needle which is made to be airtight. It is used chiefly for injecting samples into a gas chromatograph, but it can be used for other purposes for delivery of samples of the order of a few microliters such as for spotting exact amounts on chromatographic plates.

Many of the automatic dilutors and pipettors mentioned in Chapter 19 on automation can be used in micro work. Most of them will deliver samples of 10 to 20 μl with good accuracy. There are some dilutors which can measure quite accurately 5 μl or less but these are much more expensive and are not required for most purposes.

Gasometric Methods

These are methods in which a gas is formed in a chemical reaction and the amount of reacting substance determined by measuring the amount of gas formed [33–37]. In some instances one of the components of a gas mixture may be determined. The fundamental equation for these methods is usually stated in the form

$$PV = nRT$$

where P is the pressure exerted by the gas, V its volume, n the number of moles of gas present, T the absolute temperature (°C + 273), and R a universal constant whose actual value depends upon the units used for P and V. When the pressure, volume, and temperature are known or can be measured, the number of moles of gas (or the amount of gas in any other units) can be calculated. In the older volumetric methods the volume of gas was measured under the ambient atmospheric pressure. These methods are now rarely used in clinical chemistry but are still useful in the analysis of gas mixtures (using the Haldane or similar apparatus) as, for example, in determining the percentage of CO_2 in a gas

** Hamilton Company, Reno, Nev. 89502.

mixture used for calibration of the P_{CO_2} electrode. The gas is admitted to the apparatus to give a definite volume at atmospheric pressure. The CO_2 is absorbed by NaOH solution and the volume measured again. After appropriate corrections the volume percent of CO_2 can be easily calculated.

When the gas must be liberated from a solution, the manometric apparatus of Van Slyke or the microadaptation of Natelson is used. The essential parts of the apparatus are the extraction chamber in which the sample is treated to liberate the gas, the mercury manometer to measure the pressure exerted by the gas, the leveling bulb or other means of changing the mercury level in the chamber, and the stopcock closing the capillary through which the sample and reagents are admitted. Before accurate photometers were generally available, Van Slyke and his associates had developed a large number of procedures for the clinical laboratory all based on the measurement of liberated gas in the Van Slyke apparatus. These methods are rarely used today, and gasometric methods are used almost exclusively for the determination of gases dissolved or combined in blood—O_2, CO_2, and sometimes CO. The micro type of apparatus developed by Natelson is simpler to operate and requires a smaller sample. Although other methods are being widely used, the Natelson can still be used for the determination of carbon dioxide in serum.

The sample and reagents are admitted to the reaction chamber which is then sealed and a partial vacuum created by lowering the mercury level. The liberated gas is then extracted from the solution by shaking or with magnetic stirring. When the reaction is complete, the gas is compressed to a definite volume and the pressure exerted by the gas is measured. This can be used to calculate the amount of gas present. A number of corrections must be applied. The reagents and sample as introduced may contain small amounts of dissolved air which would be extracted and measured. This may be corrected by running a blank with all the reagents and using water instead of the serum or other sample. The pressure exerted by the gases extracted from the blank is subtracted from the pressure of the sample to give the corrected pressure for calculations. Or, after measurement, a reagent can be added to absorb the desired gas (e.g., NaOH to absorb CO_2). The difference in pressure before and after absorption is used to calculate the amount of gas. Another correction is based on the fact that the pressure in the chamber is not only that of the extracted gases but also of the water vapor present. This can vary somewhat with the amount of dissolved substances in solution and thus may not be exactly the same in the sample and blank or before and after absorption of the gas.

In addition, the gases may have a definite solubility in water and may not be completely extracted. They may also dissolve somewhat when the gas is compressed for measurement. This is particularly true of CO_2, which is much more soluble in water than the other gases usually determined. All these factors are also affected by the temperature so that when the difference between two readings is used, the readings must both have been made at the same temperature. These various corrections are usually included in a factor (which varies with the temperature) such that for a given volume of reagents and sample at a given temperature, the pressure difference multiplied by the factor gives the amount of gas present.

References

1. Durst, R. A. [Ed.]. *Ion Selective Electrodes*. National Bureau of Standards Special Publication 314. Washington, D.C.: Government Printing Office, 1969.
2. Eisenman, G. [Ed.]. *Glass Electrodes for Hydrogen and Other Cations: Principles and Practice*. New York: Dekker, 1967.
3. Belcher, D.   Hydrogen Ion Determinations. In O. Glaser [Ed.], *Medical Physics*, vol. 11. Chicago: Year Book, 1964.
4. Guilbault, G. G. *Enzymatic Methods of Analysis*. New York: Pergamon, 1970. Chap. 7.
5. Cotlove, E., Trantham, H. V., and Bowman, R. L.   *J. Lab. Clin. Med.* 51:461, 1958.
6. Cotlove, E.   *Stand. Methods Clin. Chem.* 3:81, 1961.
7. Kolthoff, I. M., and Lingane, J. J.   *Polarigraphy*, 2nd ed. New York: Interscience, 1952.
8. Meites, L.   *Polarigraphic Techniques*, 2nd ed. New York: Interscience, 1965.
9. Rubini, M. E., and Wolf, A. V.   *J. Biol. Chem.* 225:896, 1957.
10. Barry, K. G., McLaurin, A. W., and Parnell, B. L.   *J. Lab. Clin. Med.* 55:803, 1960.
11. Lines, J. G., and Raine, D. N.   *Ann. Clin. Biochem.* 7:1, 1970.
12. Moore, N. S., and Van Slyke, D. D.   *J. Clin. Invest.* 8:337, 1930.
13. Lowry, O. H., and Hunter, T. J.   *J. Biol. Chem.* 159:465, 1945.
14. Sunderman, F. W.   *Am. J. Clin. Pathol.* 46:679, 1966.
15. Sunderman, F. W.   *Am. J. Clin. Pathol.* 15:467, 1945.
16. Sunderman, F. W.   *J. Biol. Chem.* 143:185, 1942.
17. Haraway, A. W., and Becker, E. L.   *J.A.M.A.* 205:56, 1968.
18. Abele, J. E.   *Am. J. Med. Electron.* 2:14, 1963.
19. Warhol, R. M., Eichnholz, A., and Mulhausen, R. O.   *Arch. Intern. Med.* 116:743, 1965.
20. Johnson, R. B., Jr., and Hoch, H.   *Stand. Methods Clin. Chem.* 5:159, 1965.
21. Hendry, E. P.   *Clin. Chem.* 8:246, 1962.
22. Early, P. J., Rozzek, M. A., and Sodes, D. V.   *Textbook of Nuclear Medical Technology*. St. Louis: Mosby, 1969.
23. Hawker, C. D.   *Anal. Chem.* 45:11, 1973.
24. Kirkhan, K. E., and Hunter, W. M. [Eds.]. *Radioimmunoassay Methods: European Workshop. 1970*. Edinburgh, London: Churchill, 1971.
25. Murphy, B. E. P.   *Nature* (Lond.) 201:679, 1964.

26. O'Dell, W. D., and Daughaday, W. H. [Eds.]. *Principles of Competitive Protein Binding Assays*. Philadelphia: Lippincott, 1971.
27. Peron, F. G., and Caldwell, B. V. [Eds.]. *Immunologic Methods in Steroid Determination*. New York: Appleton-Century-Crofts, 1970.
28. Skelley, D. S., Brown, L. P., and Besch, P. K. *Clin. Chem.* 19:146, 1973.
29. Yalow, R. S., and Benson, S. A. *Nature* (Lond.) 184 (Suppl. 21):1648, 1959.
30. Mattenheimer, H. *Micromethods for the Clinical and Biochemical Laboratory*. Ann Arbor: Ann Arbor Science Publishers, 1970.
31. Knights, E. M., Jr., MacDonald, R. P., and Ploonpuu, J. *Ultramicro Methods for Clinical Laboratories*, 2nd ed. New York: Grune & Stratton, 1962.
32. O'Brien, D., Ibbot, F. A., and Rodgerson, D. O. *Laboratory Manual of Pediatric Micro and Ultramicro Biochemical Techniques*, 4th ed. New York: Harper & Row, 1968.
33. Peters, J. P., and Van Slyke, D. D. Quantitative Clinical Chemistry. Vol. 2, *Methods*. Baltimore: Williams & Wilkins, 1932.
34. Reiner, M. *Stand. Methods Clin. Chem.* 1:23, 1953.
35. Natelson, S. *Am. J. Clin. Pathol.* 21:1153, 1951.
36. Natelson, S., and Manning, C. W. *Clin. Chem.* 1:165, 1955.
37. Natelson, S. *Microtechniques of Clinical Chemistry*. Springfield, Ill.: Thomas, 1957.

3. Chromatography and Electrophoresis

Chromatography

Most methods of chromatography are technically methods of separating a number of similar components in a mixture from each other so that these can be determined with a minimum of interference [1, 2]. In some methods the end result of the separation is a number of separate fractions, each containing one of the original components in fairly pure form. These are then analyzed by conventional methods, or the individual components may be presented as portions of a stream to be analyzed by a continuous-flow method. In other types of chromatography the individual components are separated on some type of supporting medium and determined in situ by the application of different reagents. Often the actual determination by chemical or physical means is so intimately connected with the process of separation that the two are considered together as a method of chromatographic determination.

Column Chromatography

The first method used was that of column chromatography (Fig. 3-1). A column is filled with a slurry of an absorbing material in an appropriate solvent. When the absorbing material has been packed down and the excess solvent allowed to drain off, a mixture of the compounds to be separated is applied to the top of the column in a solvent. The solvent is allowed to flow through the column. This may be merely under the influence of gravity or in some cases a positive pressure may be applied to the top of the column. As the solvent moves through the column, it tends to carry the dissolved substances along with it. The column material will tend, by absorption or other means, to slow down the rate of passage of the solutes through the column. If the effect of the column packing on the different components of the mixture is slightly different, they will be slowed down at different rates and eventually be separated as illustrated at B in the figure. If the action of the column packing is by absorption of the substances on the column material, the process is known as *absorption chromatography*. If the slowing down of the solute molecules is due to an exchange of these molecules between the main body of the solvent and a layer of solvent absorbed on the column particles, the process is known as *partition chromatography*. Actually both processes are acting in varying degrees in most types of chromatography.

As the flow of solvent is continued, the different components move down the column (at different rates) and eventually the condition represented at C in the figure is reached. With further solvent flow, substance 3 will be eluted from the column. Further flow of solvent will then elute substance 2 and finally

Figure 3-1. Principle of column chromatography. Material applied at top of column (A) is separated into three components (B and C).

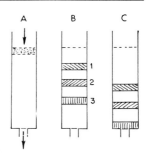

substance 1. Thus the different substances in the original mixture may be separated.

Often a number of slightly different solvents may be added in succession to give a better separation. With aqueous solutions the variations in solvents are usually changes in the pH or buffer concentration. With organic solvents the variations are usually in the polarity of the solvent, e.g., different proportions of a nonpolar solvent like benzene and a polar solvent like ethanol.

Often the separation is not as sharp as might be inferred from Figure 3-1. If a large number of small successive fractions of elutate were collected and analyzed, one might obtain instead of a clear separation as indicated by the dashed columns in Figure 3-2, a result similar to the solid curve of the figure. There is no clear-cut separation but usually the material represented between the points a, b, c, and d will give a good estimation of the compounds 3, 2, and 1, respectively. Considerable experimentation is usually necessary to find the best solvent and rate of flow for the optimum separation in a given instance. Usually the material on the column is not actually visible as might be inferred from Figure 3-1. The originator of the method, the Russian botanist Tswett, was working with the pigments extracted from plants (chlorophylls, xanthophylls, etc.) and he could see definite colored bands on his columns. From this was derived the name that is still used—chromatography.

Column chromatography may be used to separate several substances, each of which is to be determined, or to separate one substance to be determined from interfering materials. In the illustration given, one might be interested in component 3, the other components being impurities that would interfere with the determination of the desired constituent. Then only the desired component would be eluted from the column. In another instance one might be interested only in component 1, in which case the other components would be eluted from the

Figure 3-2. Theoretical distribution of fractions from columns (dashed columns in figure) and actual distribution (solid curve).

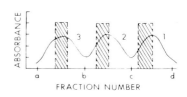

column first and discarded, and then the desired component eluted, usually by a change in solvent.

In addition to columns using absorbing substances such as alumina, silica gel, and powdered cellulose, two other types of columns may be used. One type uses an *ion exchange resin*. As the name implies, this column is based on the exchange of ionic substances between the solution and the column material. It is used chiefly for amino acids and amino acid derivatives which will exist in solution in ionized form at the proper pH. The other type of column uses *gel filtration*. The gels used (Sephadex*) are polymerized dextrans in the form of particles having many fine pores. The separation by gel filtration is usually on the basis of molecular weight or size. As a mixture of large and small particles in a solvent passes through the gel column, the small particles tend to become trapped in the pores of the gel while the large particles which cannot enter the pores travel around the particles with the solvent and are eluted first. Particles with different average pore sizes may be used to separate components of different sizes.

Paper and Thin Layer Chromatography [3–7]

METHOD

The separation is carried out on a supporting medium such as paper (an absorbent type similar to thick filter paper), or a thin layer (0.1 to 0.5 mm) of the absorbing substance on a supporting material such as glass plates, plastic, aluminum foil, or other inert material, or on glass fiber impregnated with the absorbent. A binder such as calcium sulfate or starch may be used to help keep the absorbing material in place, and a fluorescing substance may be added to help in the detection of substances by viewing under ultraviolet light.

In general the mixture of substances to be separated is applied as a spot or streak near one end of the sheet. (We will refer to supporting medium and absorbent as a sheet although it may be in any of the forms mentioned.) The sheet

* Pharmacia Fine Chemicals, Piscatawany, N.J. 08854.

Figure 3-3. Paper chromatography. A, ascending chromatography. The substances are applied as spots at the bottom (1, 2, 3). The paper dips into the solvent (S) at the bottom and the solvent ascends by capillary action to e. The letters b, c, and d represent different substances carried along by the solvent at different rates.

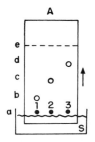

is then supported, with the end nearest the point of application dipping into the solvent. Usually the arrangement is such that the point of application is a half centimeter or so above the surface of the solvent. The solvent rises through the absorbing material on the strip by capillary action. This flow of solvent carries the applied substances along just like the flow of solvent through a column. Since the solvents used are often rather volatile and evaporation from the strip would interfere with the chromatography, the sheets are placed in an enclosed chamber and the space saturated with the solvent vapor before the sheet is introduced. A number of different samples along with several standards containing known amounts of the compounds under analysis may be applied to one sheet. When the solvent has risen the required distance, the sheet is removed and the solvent evaporated with the aid of heat if necessary. When paper chromatography is used with aqueous solvents, the time required may be several hours or more. With thin layer chromatography—thin layers of the absorbing media applied to an inert backing—the time is often less than 1 hr.

ANALYSIS

The rise of the solvent will cause the different components in the mixture to be distributed along the sheet at different heights similar to the distribution in the column in Figure 3-1. This is illustrated in Figure 3-3, where three different substances applied at different places have risen to different heights in the sheet. The dashed line e represents the height to which the solvent had risen when the sheet was removed. As with column chromatography, the different components are usually not visible on the sheet. In some instances their position may be visualized by viewing under ultraviolet light. The compounds may be fluorescent or they may quench the fluorescence of the inorganic phosphor originally added to the absorbent and the spots will appear dark against a fluorescent background.

Figure 3-4. Absorbance curve for three different spots on a chromatographic sheet.

In some instances the absorbent material containing the spot as located by ultraviolet light may be carefully scraped from the sheet and transferred to a test tube. It is then eluted with a solvent and the material determined by chemical means. If the sheet can be cut with scissors, portions containing the different substances can be cut and the substances eluted and analyzed by chemical means.

Usually the sheet is dipped in or sprayed with a reagent or series of reagents to develop colors with the substances on the sheet. The amount of substance in the various spots is estimated by comparing the size of the spot and the intensity of color with spots produced by known amounts of standards that have been carried through the same procedure, preferably on the same sheet. In some applications the relative amounts of the different substances in the different spots may be estimated by scanning the spots with a densitometer. This consists of a light source, optical filter, slit, and photocell in an arrangement similar to a photometer. As the spots on the sheet are moved along past the slit, the absorbance of the colored spots on the sheet is measured. A recorder is connected to the photocell output. The sheet is moved along mechanically at a constant rate and the changes in absorbance recorded.

For three spots of different intensity the recorder trace could be something like that shown in Figure 3-4. As a rough estimate the height of the peaks could be used to approximate the relative amounts of the three substances. In the figure the relative amounts in peaks A, B, and C would be approximately in the ratio of 3:1:2. For more accurate measurements the areas under the respective peaks can be determined. If the curve is drawn on fairly uniform paper, a simple way is to cut out the various peaks with scissors and weigh them. The relative weights would give a measure of the relative areas. Cutting up the chart paper on which the trace is made to obtain the relative weights of the individual fractions destroys the trace for future reference. This can be overcome by cutting up a Xerox copy of the trace. This also gives somewhat more accurate results as the Xerox paper is apparently more uniform than chart paper. The areas could also be measured by the use of a planimeter, but this is rather tedious.

Most commonly this is done by means of a mechanical or electrical integrator attached to the recorder, which simplifies the computations. This is treated in more detail in the section on electrophoresis later in this chapter. This refinement is often not used in paper or thin layer chromatography, where the results are expressed only semiquantitatively.

In paper and thin layer chromatography it is important to know which of a number of possible substances are present, as well as their relative amounts. Standards or known substances may be added at different points on the same sheet for comparison. A numerical figure that is helpful in identifying the various compounds forming spots is the Rf value. This is the ratio of the distance traveled by the substance in question to the distance traveled by the solvent front on the same sheet. When the sheet is first removed from the solvent after development of the chromatogram, the position of the solvent front is marked. Then, after the spots have been visualized, their positions are marked. Two measurements are made: the distance from the middle of the spot to the point of application, and the distance between the point of application and the solvent front. The ratio of these two distances is the Rf value for that compound using that particular solvent and absorbent. If, for example, the spot had traveled a distance of 8.1 cm and the solvent front was a distance of 12.0 cm from the starting position, the Rf value would be 8.1/12.0 = 0.68.

The Rf values for known compounds under specific conditions may be obtained experimentally or found in the literature. Even under the same conditions the Rf value may vary somewhat, but the ratio of two Rf values will be more constant. Thus we might find an Rf value of 0.82 for the standard compound and 0.65 for another compound. In a later experiment we find, say, 0.61 as the Rf value of a particular spot. Was this the same compound for which we had previously found an Rf value of 0.65? If in our second experiment we found that our standard had an Rf value of 0.77, we could be reasonably certain that the second compound was the same in both instances since the ratio of its Rf value to that of the standard was 0.79 and 0.80 in the two cases—a good agreement.

When a number of compounds are present, they may not all be completely separated by chromatography with a given solvent. Using a different solvent may give a different partial separation (Fig. 3-5). With solvent I, we see that compounds A and F are not separated, nor are B and G, nor D and H. With solvent II, most of the compounds move at different rates from those with solvent I; this time compounds F and G are not separated, nor B and H, nor C and D. Using what is known as two-dimensional chromatography, a more complete separation can be made. As illustrated in III, the mixture is applied at a spot in one corner of the sheet. The sheet is then dipped in solvent I so that the sol-

Figure 3-5. One- and two-dimensional chromatography. (I) Linear chromatography with one solvent. (II) Chromatography with a different solvent. (III) Chromatography with the two solvents in succession applied at right angles. The letters A to H represent different substances in the mixture that was originally applied at the spot.

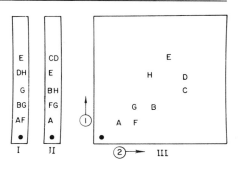

vent rises in the direction indicated by arrow 1. After chromatographic development, the compounds will be distributed along the left edge in the same manner as in strip I. The sheet is removed and the solvent evaporated. It is then dipped into the second solvent with what was the left edge now at the bottom. The solvent then rises in the sheet in the direction indicated by arrow 2. In this way the final distribution of the compounds will be indicated on the large sheet. The positions of the spots must then be compared with those obtained when a mixture of known compounds is treated similarly.

Gas-Liquid Chromatography [8–11]

METHOD

In gas-liquid chromatography, often called merely gas chromatography, the substances are separated in the gaseous phase. The "solvent" which carries the vapors along is an inert gas such as helium or sometimes nitrogen. The column absorbent is usually a high-boiling-point liquid which may be absorbed on the surface of an inert carrier such as crushed firebrick. This is placed in a column which is usually about 0.5 cm in diameter. The columns may be several meters long and are often coiled to conserve space. In some applications the column is merely a fine capillary with the high-boiling liquid absorbed on the walls. The material to be separated is usually dissolved in a volatile solvent and a few microliters injected with a syringe and fine needle through a septum near the top of the column. The material is carried along by the flow of the inert gas. The different substances are absorbed in varying degrees by the liquid and a separation is obtained similar to that of column chromatography. As the gas flow emerges from the column, it is analyzed by one of the methods discussed below.

One of the requirements of gas chromatography is that the substances have a sufficient vapor pressure to give readily detectable amounts of the material in the gaseous phase. Gases such as O_2, CO_2, and CO can be extracted from blood and analyzed directly. Other substances such as alcohols, acetone, and ether have a sufficient vapor pressure so that they can be extracted from biological specimens and chromatographed at room temperature or slightly above.

In most applications the columns are heated in a temperature-controlled oven to increase the vapor pressure of the substances. The temperature of the columns may be as high as 250°C or more. Although high temperatures may allow the analysis of higher-boiling materials, they possess some disadvantages. Too high a temperature may cause decomposition of the material being analyzed and cause some "bleeding" of the high-boiling absorbent from the column which will interfere with the separation. Often the compounds to be separated are not sufficiently volatile and are chemically converted to more volatile derivatives. Steroids, for example, are often converted into trimethyl silyl esters or similar compounds. When it is desired to separate components with very different degrees of volatility, the temperature of the column may be gradually increased during the run. For example, the column may be held at 150°C for 10 min after the sample has been injected, then gradually increased to 225°C during the next 15 min, then held at this temperature to the end of the run. This temperature programming may be done automatically by the instrument to give reproducible results. If the column were held at the higher temperature from the start, the more volatile components might come off the column so rapidly that they would not be separated.

ANALYSIS

There are three general methods for the analysis of the column effluent. One method uses changes in the thermal conductivity of the gas. The thermal conductivity of the carrier gas, helium, is high compared with most other vapors. The effluent gas is passed through a chamber containing a filament heated to a temperature somewhat above that of the gas. If the rate of flow of the gas is constant, the decrease in temperature of the filament due to cooling will depend upon the thermal conductivity of the gas. As the gas flow contains more or less of the organic vapors eluted from the column, the temperature of the filament— and hence its resistance—will vary. This is monitored electronically and the resulting current amplified to drive a recorder. The result may be a trace similar to that of Figure 3-6. The high peak at the beginning (on the left) is due to the solvent in which the sample was dissolved for injection. The other peaks represent varying amounts of the different components of the sample. This method is not as sensitive as the other analysis methods and is much more influenced by

Figure 3-6. Recorder trace of a gas-liquid chromatogram. High peak at the left (beginning of trace) is due to the solvent used to dissolve the substances being analyzed.

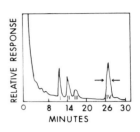

traces of water vapor in the sample (water is sometimes a difficult contaminant to remove completely).

A much more sensitive method uses a hydrogen flame detector. The column effluent is mixed with additional hydrogen and burned in a small flame. The flame is placed between two electrodes to which an electrical potential is applied. In the process of burning, some ions may be produced in the flame. These ions are attracted to the electrodes and constitute an electrical current. When pure hydrogen is burned, only a few ions are formed. As the organic substances eluted from the column enter the flame, the number of ions formed will increase proportional to the amount of substances present. The increased current is amplified to drive a recorder and again a trace similar to that of Figure 3-6 is produced.

The third method of analysis uses an electron capture detector. Some radioactive material or a glow discharge is used as a source of electrons, producing a current through the gas between two electrodes. When the organic molecules from the column effluent enter the detector, some constituents, particularly the halogens, tend to capture some of the electrons:

$$Cl + e^- \rightleftharpoons Cl^-$$

The chloride ions are so much heavier than the electrons that they move much more slowly and may be swept out by the gas flow before reaching the electrodes. The result is a decrease in the number of negative charges reaching the electrode and thus in the current between them. The amplified changes in current again are used to produce a recorder trace. This method is particularly applicable to halogenated compounds. Other compounds can be determined by forming halogenated derivatives before addition to the column. Since the method measures a decrease in current, it does not have the dynamic range of the hydrogen flame method.

The various compounds in the recorder trace are identified, as with other methods, by running known compounds through the procedure. They may be identified by their *retention time*, the time from the injection of the sample until the particular peak appears on the recorder trace. Like the Rf value, the retention times will vary somewhat with the exact conditions but under the same operating conditions the relative retention times are fairly constant.

In gas chromatography of a complex mixture, a known amount of an internal standard may be added. This will be a compound that is chemically similar to the compounds being separated but known not to be present in the type of biological sample used. The retention times of the other substances may be compared with the known retention time of the standard. If a complex extraction or other procedure is involved in the preparation of the sample for analysis, the internal standard is added as soon as possible in the procedure to compensate for losses during the manipulations. The relative amounts of the different components are estimated from the sizes of the peaks produced on the recorder trace. The simplest procedure is to use the relative heights of the peaks above the baseline. If all the peaks are similar in form, this may be satisfactory, but it is not very accurate in comparing sharp and broad peaks, since the area under the peaks is the factor that is more nearly proportional to the amount of the substance.

A second rather simple procedure is to multiply the height of the peak, in any convenient unit, by the width at a point halfway between the peak and the baseline. This is illustrated for peak IV in Figure 3-6 where the arrows indicate the width at half height. This is usually more satisfactory than using the peak height alone and does not require very much computation.

The best method is to have a mechanical or electrical integrator attached to the recorder for calculation of the areas under the peaks. If a known amount of the internal standards has been added, this may be used to calculate the absolute amounts of the various constituents in the sample. Suppose, in the figure, that peak IV corresponds to the internal standard and that exactly 50 μg of this had been added. If the area under this peak corresponded to 126 units and the area under peak II was 63 units, then the amount of the substance corresponding to peak II would be $63/126 \times 50 = 25$ μg. This assumes that the same amount of each component will give exactly the same peak area, but this may not necessarily be true. It would have to be determined experimentally by running known amounts of the different components and the standard through the column. If it were found, for example, that the compound represented by peak II gave only 80 percent of the response by the same amount of standard, then the true amount of this component in the above example would be $25/0.80 = 31$ μg.

The gas chromatographic method has been found to be valuable for the detection of small amounts of a number of substances such as the steroid hormones (e.g., estrogens and adrenocortical hormones), drugs (e.g., barbiturates), and many other compounds. Some modifications not generally used in the clinical laboratory are (1) the pyrolysis technique, in which the compounds are deliberately decomposed by heat just before entering the column and the different compounds identified by means of the peaks produced by the results of pyrolysis, and (2) a technique in which a mass spectrograph is attached to the output of the column for identification of the compounds. Another technique which may prove of value is identification of different microorganisms by gas chromatography of the more volatile components produced by the growth of the microorganism, each microorganism producing a somewhat different pattern of peaks.

Electrophoresis

Electrophoresis is a method of separation of different substances based on their different rates of migration in an electric field [12–19]. The substances, however, must be capable of forming charged particles in solution. When a voltage is applied to two electrodes in a conducting solution, an electrical potential is set up between the electrodes. Charged particles in the solution will then move under the influence of the electric field, the negative particles moving toward the positive electrode and the positively charged particles toward the negative electrode. The rate of migration will depend on the size and charge on the particles as well as the strength of the electric field. When the latter is held constant, different charged particles, which might be initially all at one position in a solution, will move at different rates, and after a time they will be found at different distances from the starting point. If the different portions of the solutions could be removed separately without disturbing the others and analyzed, the amounts of the different substances would be found. It is very difficult to separate the different portions of the solution, and in the original procedures for this "free" electrophoresis the analysis was done by optical means without disturbing the solution. Although this method is very useful for research purposes, it is too cumbersome for ordinary laboratory use.

The usual clinical laboratory electrophoresis methods involve the use of some sort of supporting medium which will absorb a certain amount of the buffer solution that is needed to carry the electric current. The mixture to be separated is added as a streak near one end of the medium, connections are made for an electric current at the two ends, and the potential applied. The substances move along the strip of supporting medium at different rates, depending upon their size and charge. After a suitable time the current is removed and the strip taken from the apparatus. The strip is then usually stained with some reagent that will give color with the material being separated. The strips of medium can be manipulated considerably without changing the relative positions of the separated substances.

One of the first supporting media used was an absorbent paper similar to heavy filter paper. This is still used in some applications, but for serum proteins, the chief application of electrophoresis in the clinical laboratory, cellulose acetate is now used almost exclusively. Other supporting media sometimes used are gels of agar, agarose, or polyacrylamide.

After the positions of the various components have been located by staining, they may be quantitated in a number of ways. For some purposes it is only necessary to estimate the amounts semiquantitatively by noting the relative sizes and intensity of the separate stained portions, or the strip may be cut into sections and the stained material in the separate sections eluted and the amount determined photometrically. Usually the different portions are scanned in a recording densitometer.

The substances usually determined by electrophoresis are some types of proteins such as serum proteins, lipoproteins, and isoenzymes which are proteins or associated with proteins. Some simpler compounds can be determined by high-voltage electrophoresis. In ordinary protein electrophoresis the applied voltage is usually in the range of 150 to 350 v. With high-voltage electrophoresis the applied voltage may be from 1,000 to 2,000 v. Because of the higher voltage some cooling of the electrophoresis strip is required. Urinary amino acids and other simple compounds such as vanillylmandelic acid (VMA) and its metabolites have been determined by electrophoresis. Urinary sugars have also been separated by this procedure. Sugars are not usually considered to be ionic substances, but in borate buffer they form ionic complexes with the borate and can be separated by electrophoresis.

The quantitation of densitometer scans by the use of a mechanical or electrical integrator attached to the recorder has been mentioned earlier as used in some paper or thin layer chromatography and in gas-liquid chromatography, but the chief use in the ordinary clinical laboratory is in the quantitation of the results of protein electrophoresis, not only of serum proteins but also of lipoproteins and isoenzymes. For that reason this will be discussed in more detail here. There are two different types of mechanical integrator traces. One type with the sawtooth curve is illustrated in Figure 3-7. An explanation of the mechanical principles involved in the production of this type of trace will be found in Ackermann [20].

The integrator trace in the figure is difficult to follow due to the reduction in size

65

Figure 3-7. Protein electrophoresis scan showing sawtooth type of integrator trace. (Beckman Analytrol, Beckman Instruments, Fullerton, Calif.)

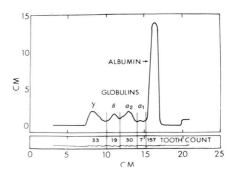

Figure 3-8. Enlargement of portions of the integrator trace. For explanation see text.

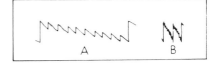

in the illustration; an enlargement is shown in Figure 3-8. The sawteeth or
pips are divided into groups of ten for ease in counting. Trace A is an enlargement of a part of the trace such as is found under the gamma globulin peak.
Each point represents an area of 0.1 cm^2. When the area is large, as under the
albumin peak, the individual teeth are not distinguishable, as shown at trace B,
and each of the peaks here represents ten teeth. As indicated on Figure 3-7, lines
are drawn separating the peaks and extending downward to intersect the integrator trace. Sometimes it is difficult to decide the proper position for the low
point of the valley and some experience is helpful. The total number of small
peaks between each of the vertical lines is counted. This represents an area
proportional to the amount of the particular protein fraction. In this particular
apparatus the integrator is not correct when the peak goes over the 14 cm mark
on the chart, and a correction must be made. With other apparatus this may not
be required. It could be avoided by reducing the sensitivity somewhat so that
the peak would not be quite so high. The total tooth count for each fraction is
determined and these added to obtain a grand total. The percentage of total
area or tooth count for each fraction represents the percentage of that component in the mixture.

In the other type of integrator trace (Fig. 3-9) the pen makes repeated

Figure 3-9. Serum protein electrophoresis scan showing another type of integrator trace. (Helena Quick-Scan, Helena Laboratories, Beaumont, Tex.)

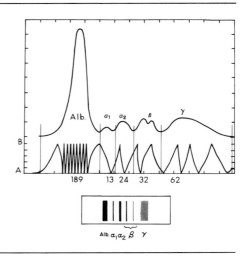

excursions between two lines. A traverse from line A to line B represents ten units of area (some scales have ten divisions between A and B for ease in calculation). Again lines are drawn downward from the correct positions in the valleys between the peaks. The gamma fraction, for example, has six complete traverses (60 units), plus approximately one unit at each end, for a total of 62. The beta fraction has two complete traverses plus a traverse of about nine units on the right and about three on the left, for a total of 32 units. Again the total units for each fraction are added and the relative proportion for each fraction calculated. These traces are sometimes easier to use than the sawtooth type.

Some of the newer electronic scanners may be equipped with an electronic integrator and calculator such that the relative amounts in the different fractions are automatically calculated from the curves and the results given in printout form.

It may be noted that in Figure 3-7 the albumin peak is on the right and in Figure 3-9 it is on the left. This sometimes is merely a matter of convenience, depending upon the particular apparatus used. The densitometer measures the light absorbance by the substances in the strip. These usually have been made colored (light absorbing) by some type of staining or other chemical reaction. The above calculations assume that Beer's law holds for the absorbance and that all the substances in the strip produce the same amount of color per unit of material. Neither of these assumptions need be true. This will be discussed further in the procedures for the electrophoresis of the different types of compounds (serum proteins, isoenzymes, etc.). A discussion of how some deviations from Beer's law can be corrected in the apparatus can be found in reference 1.

Immunodiffusion and Immunoelectrophoresis

These methods are based principally on the diffusion of an antigen (protein) through a gel (usually agar) from the position where it is placed in a small well in the gel. The antigen combines with the appropriate precipitating antibody which is either included in the gel or diffuses toward the antigen from a second well. Where the antibody and antigen meet, a visible arc or ring of precipitate is formed. The formation of the precipitate indicates the presence of the antigen that reacts with the particular antibody present. The amount of precipitate may be used as an indication of the amount of antigen present. In the Ouchterlony technique, the antigen is placed in the center well in an agar plate [21–26]. In other wells around the center are placed various antibodies to the antigens expected to be present or whose presence is to be detected. Where the antibody and antigen diffusing outward from their respective wells meet, an arc of precipitate forms (Fig. 3-10). Actually, when viewed with oblique or reflected light, the precipitate is seen as a white arc against a darker background. The different antibodies in outer wells 2 to 7 in the figure produce different degrees of precipitation in the arcs, depending upon the amount of the particular antigen in the sample placed in the center well. In order to detect a particular antigen, the appropriate antibody must be placed in one of the outer wells.

In *immunoelectrophoresis* the various protein antigens are first separated by electrophoresis before the diffusion is carried out [12, 24–26] (Fig. 3-11). The samples are placed in the agar gel at the points indicated by 1 and 3 in the figure. In position 1 might be placed the unknown serum sample and in position 3 a normal serum for comparison, or a mixture of known specific serum proteins. The gel slab is then placed in an electrophoresis chamber and the voltage applied for the required time. As with the other electrophoresis methods, the proteins are separated. In the illustration the migration is in the direction of the arrow. For purposes of illustration, we have assumed that the sample was separated into four fractions as illustrated by the small open circles. Actually with serum there would be more fractions and the separation would not be as regular as shown. After the electrophoresis has been carried out, a polyvalent antiserum, containing antibodies to all the constituents likely to be present in the sample, is placed in the trough, 2 in the figure, and the diffusion allowed to proceed for the required time which may be 48 hr or more. The various antigens diffusing out would meet the antibodies diffusing from the trough and form arcs of precipitate as indicated. The relative density of the precipitation indicates the relative amount

Figure 3-10. Results of immunodiffusion by Ouchterlony technique. For explanation see text.

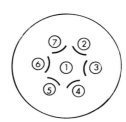

Figure 3-11. Immunoelectrophoresis of antigens.

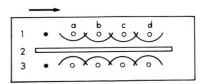

of antigen present. When human serum is used, a large number of different bands have been identified.

A more quantitative method is *radial immunodiffusion* [27–29]; however, it can be used with only one antigen at a time. The antibody is incorporated in the agar gel in the plate. A number of small wells are made and into them are placed samples or different concentrations of the antigen being determined (Fig. 3-12). As the antigen diffuses out into the gel with antibody, a ring of precipitate will be formed around each well. The diameter of the ring can be used as a measure of the amount of antigen present by comparing the sizes of the rings produced by different concentrations of the standards with those produced by the samples. Usually one plots the diameter of the rings against the logarithm of the concentration (most conveniently done with two-cycle semilog graph paper). A smooth curve drawn through the points is used to calculate the concentration of the antigen in the samples. Occasionally one may obtain a better curve by using the square root of the concentration instead of the logarithm. Accurate measurement of the ring diameters is critical, and some type of magnifying comparator is needed for this.

Another variation is *electroimmunodiffusion* (EID) introduced by Laurell [30–32]. The electrophoresis is carried out in agar gel containing the antibody to the substance being determined or on Mylar-backed cellulose acetate strips impregnated with buffer containing the antibody. The antigen is added at

Figure 3-12. Radial immunodiffusion. Diameter of the rings formed by precipitate is a measure of the concentration of the antigens in the wells.

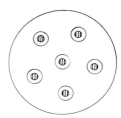

Figure 3-13. Typical results obtained with electroimmuno-diffusion (EID). The distance traveled by the precipitin arcs is a measure of the concentration of the added antigen.

positions across the strip and electrophoresis carried out. With agar gel the resulting precipitin arcs may be measured directly or they may be stained as is necessary with the cellulose acetate. Typical results are shown in Figure **3-13**. The distance traveled by the tip of the arc is carefully measured. Under controlled conditions this will be a measure of the amount of added antigen. The distance traveled will also be proportional to the duration of electrophoresis and this must be standardized for each run. By comparison of several standards with the sample, the amount of antigen in the sample can be calculated. This procedure is sometimes called the *rocket method* because of the characteristic shape of the precipitin arcs.

References

1. Snyder, L. R. *Methods Med. Res.* 12:2, 1970.
2. Sweig, G., and Serma, G. [Eds.]. *Handbook of Chromatography*. Cleveland: Chemical Rubber, 1972.
3. Block, R. J., Durrum, E. L., and Zweig, G. *Paper Chromatography and Paper Electrophoresis*, 2nd ed. New York: Academic, 1958.
4. Kirchner, J. C. *Thin Layer Chromatography*. New York: Wiley-Interscience, 1967.
5. Scott, R. M. *Clinical Analysis by Thin-Layer Chromatography Techniques*. Ann Arbor, Mich.: Ann Arbor-Humphrey, 1969.
6. Haer, F. C. *Introduction to Chromatography on Glass-Impregnated Fiber*. Ann Arbor, Mich.: Ann Arbor-Humphrey, 1969.
7. Randerath, K. *Dunschicht-Chromatographie*. Weinheim/Bergst.: Verlag Chemie, 1965.

8. Porter, R. [Ed.]. *Gas Chromatography in Biology and Medicine*. London: Churchill, 1969.
9. Szymanski, H. E. [Ed.]. *Biomedical Application of Gas-Liquid Chromatography*, vol. 2. New York: Plenum, 1968.
10. Purnell, H. *Gas Chromatography*. New York: Wiley, 1962.
11. Wotiz, H. H., and Clark, S. J. *Gas Chromatography in the Analysis of Steroid Hormones*. New York: Plenum, 1966.
12. Cawley, L. P. *Electrophoresis and Immunoelectrophoresis*. Boston: Little, Brown, 1969.
13. Cawley, L. P., Penn, G. M., Itano, M., Bell, H. E., and Minard, B. *Basic Electrophoresis, Immunoelectrophoresis and Immunochemistry*. Chicago: American Society of Clinical Pathologists, 1972.
14. Nerenberg, S. T. *Electrophoresis—A Practical Manual*. Philadelphia: Davis, 1966.
15. Chin, H. P. *Cellulose Acetate Electrophoresis, Techniques and Applications*. Ann Arbor, Mich.: Ann Arbor-Humphrey, 1970.
16. Zweig, G., and Whitaker, J. E. Paper Chromatography and Paper Electrophoresis. Vol. 1, *Electrophoresis in Stabilizing Media*. New York: Academic, 1967.
17. Mabry, C. C., and Karam, E. A. *Am. J. Clin. Pathol.* 42:421, 1964.
18. Mabry, C. C., and Todd, W. R. *J. Lab. Clin. Med.* 61:146, 1963.
19. Sachett, D. *J. Lab. Clin. Med.* 63:306, 1964.
20. Ackermann, P. G. *Electronic Instrumentation in the Clinical Laboratory*. Boston: Little, Brown, 1970.
21. Ouchterlony, O. *Acta Pathol. Microbiol. Scand.* 26:507, 1949.
22. Ouchterlony, O. *Handbook of Immunodiffusion and Immunoelectrophoresis*. Ann Arbor, Mich.: Ann Arbor Science Publishers, 1968.
23. Weiner, L. M., and Zak, B. *Stand. Methods Clin. Chem.* 7:305, 1972.
24. Iammarino, R. M. *Stand. Methods Clin. Chem.* 7:163, 1972.
25. Fuller, J. B. *Selected Topics in Clinical Chemistry*. Chicago: American Society of Clinical Pathologists, 1972.
26. Grabar, P. *Methods Biochem. Anal.* 7:1, 1959.
27. Mancini, G., Carbonara, A. O., and Heremans, J. F. *Immunochemistry* 2:235, 1965.
28. Fahey, J. L., and McKelvey, E. M. *J. Immunol.* 94:84, 1965.
29. Stoeriko, N. *Blut* 16:200, 1968.
30. Laurell, C. B. *Anal. Biochem.* 10:357, 1965.
31. Laurell, C. B. *Anal. Biochem.* 15:45, 1966.
32. Merrill, D., Hartley, T. F., and Claman, H. N. *J. Lab. Clin. Med.* 69:151, 1967.

4. Normal Values, Quality Control, and Units

Normal Values [1]

Clinical laboratory tests are usually performed for diagnostic or prognostic purposes. They furnish information that might indicate whether a patient has a certain disease or condition which might affect the level of the constituent being determined, or information about the results of therapy. In the following discussion we will refer to the levels of substances in blood serum, but the same arguments hold for levels in other biological fluids as well. The results may be correlated with the normal values or, more properly, the *range* of normal values. However, even the concept of an absolutely definite and completely established "normal range" for each blood constituent, with all subjects having levels outside this range being in some way abnormal, is entirely erroneous. One can speak only of the estimate of the normal range under certain specified conditions [2, 3].

Although a knowledge of the normal range is important to the physician, the problem of determining what is the best estimate of this range offers more difficulties than is often realized. This discussion will deal with some aspects of the problem. To determine the normal range of values for fasting blood sugar, for example, one might attempt to measure the fasting levels in a fairly large group of normal subjects. "Normal" here might mean those believed to be free of any disease or other condition that might affect carbohydrate metabolism. If we should take as one of the criteria for selecting the subjects, the condition that they have a fasting blood glucose within certain limits (as might be done in other circumstances), we would be involved in an erroneous circular argument. As stated, the error should be obvious. In some instances in which supposedly normal values have been established, similar errors have been made which were not so obvious as in our example. Some of the subjects could still theoretically have a latent condition that could influence their carbohydrate metabolism but that might not be detected by our other tests. Strictly speaking, such subjects should not be included in the normal group, but practically we might have no way of detecting them.

Suppose for the moment that we do have a group of normal individuals on whom we wish to measure the serum level of some constituent. If we make measurements on these individuals, we will generally find that the results are distributed over a range of values, and that there are more results closer to the average value than near the upper and lower limits. This is illustrated by the histogram of Figure 4-1, which is a plot of the number of measurements on different individuals having the integral values of 11, 12, 13, . . . , 30 units of concentration. (The actual units are irrelevant to the present discussion.) It is

71

Figure 4-1. Frequency histogram for hypothetical experimental results on the determination of a normal range. The dashed curve is the gaussian curve calculated for all the points, with a mean of 19.02 and a standard deviation of 4.22. The dotted curve represents the gaussian curve calculated for the range B, with a mean of 17.97 and a standard deviation of 3.11.

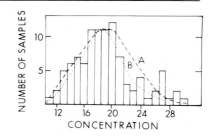

Figure 4-2. Two typical gaussian curves. Curve A has twice the standard deviation (S.D.) of curve B. The ranges of 1 and 2 S.D. are shown for each curve.

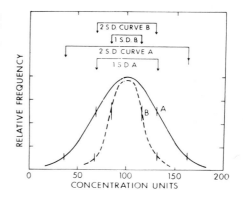

often assumed that the distribution of values follows what is known as the *gaussian error curve*. This curve is symmetrical about the mean value and may be defined in terms of two parameters, the mean and the standard deviation (S.D.). Two such curves are drawn in Figure 4-2. Along the abscissa are plotted the results of the determinations and along the ordinate the frequency with which the different results are obtained. The curves are drawn so that curve A has twice the S.D. of curve B. Note that the S.D. is a measure of the spread of the results. A property of this curve is that **68.3%** of all the values should fall within 1 S.D. from the mean (which is at the peak of the curve), and **95.5%** of all values within 2 S.D. from the mean. These two ranges are shown for each curve in the figure. The method for the calculation of the S.D. will be given later. The range of **2 S.D.** from the mean is sometimes taken as the range of normal limits. As will be seen, one must always choose some more or less arbitrary limits and 2 S.D. is a convenient one.

The gaussian error curve is often (sometimes erroneously) used in the treatment of experimental data. It is favored by statisticians as a number of mathematically interesting results can be derived from it. Actually even if we suppose that our data do follow a gaussian curve, the assumption is not (or should not be) that the determinations on a relatively small number of subjects (say, 20 or 30) will give a true normal range, but rather that if one could make measurements on a much larger number of subjects (of the order of 500 or more), the results of such measurements would follow a gaussian curve closely and the true normal range would be found. The relatively small number of determinations actually made is assumed to represent a true random sample from this hypothetical parent population. The results from such smaller groups will on the average follow the true gaussian curve more or less closely. The average and S.D. obtained from the smaller group are themselves only estimates of the true average and S.D. and are also subject to statistical variation.

The gaussian curve derived from the calculated mean and S.D. of the data plotted in Figure 4-1 is given as the dashed curve. Note that this curve does not follow the histogram very closely. From the inspection of the histogram, one might wonder whether the subjects giving values higher than 25 units were not actually an admixture of "abnormal" subjects into the normal group and that the true curve for the normal group should be limited to the values below 26 units. The gaussian curve corresponding to this distribution is shown as the dotted curve. It seems to give a better fit. However, we are not justified in discarding some data merely because it does not look right. This is discussed more fully in the mathematical section following. If the data shown were actual experimental results, they might lead one to examine the criteria used in the original selection of the subjects to determine if some abnormal subjects had actually been used.

Any normal range based on the determination of the S.D., such as $\pm$ 2 S.D. from the mean, will be strictly valid only if the experimental data actually follow a gaussian curve. How can we be certain of this? There is no a priori reason why they should follow the curve. In fact, some statisticians doubt if any results obtained from biological measurements will follow a gaussian curve. We note that the data of Figure 4-1 do not agree too well with the theoretical curve. In the mathematical section following, we will discuss some ways of determining whether the data do follow the theoretical curve and some coordinate transformations that can be used to give a more normal distribution. (Mathematicians sometimes speak of the gaussian curve as the normal error curve; this has nothing to do with the use of "normal" in the clinical sense.) The gaussian curve

Figure 4-3. Hypothetical frequency distribution curves for I (solid line), a group of normal individuals; II (dashed line), a group of abnormals; and III (dotted line), composite curve for the two groups. For the significance of points A to D see text.

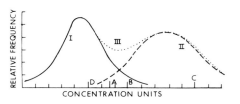

was originally derived as a means of estimating the precision of a number of measurements on the same sample or on a number of supposedly identical samples. As such it is satisfactory for determining the precision of a method or for use in quality control where the results may be more justifiably assumed to follow this distribution.

In any event we make our measurements on normal subjects and derive a normal range by some mathematical means. Whatever our method, it will exclude a few values on supposedly normal subjects. This is illustrated in Figure 4-3 where curve I is a frequency distribution of the results obtained on normal subjects and curve II represents measurements on abnormal or diseased subjects. These curves may not necessarily be gaussian in shape. In drawing them, no attempt was made to make them so. Points A and B represent two different upper limits of normal that might have been chosen. If point A were used, a certain percentage of the normal individuals would have levels of the constituent above the normal range. Such a result may be termed a *false positive* test: The result indicates that the individual may have a disease or condition that increases the level of this constituent, but actually he is normal. If point B were chosen, there would be a smaller number of false positives, but there would be a larger number of *false negatives*, that is, patients who were actually in the abnormal group but who have levels within the normal limits. The occurrence of false negatives may be less definite than that of false positives, as the exact shape of curve II is usually less well known than that of curve I.

The question may arise whether in any given test it is better to use a limit that gives a low percentage of false positives but a higher percentage of false negatives, or vice versa. The definition of "better" in this connection is purely a clinical problem having little to do with the mathematical analysis of the results. In some screening tests it may be advisable to include a higher percentage of false positives which can be later ruled out by further tests rather than to miss any subjects who actually have the disease. Of course, the final diagnosis will usually not be made on the basis of a single test, but a provisional one might be, and a decision made as to what further tests should be done. If the confirmatory tests are time consuming or expensive or can result in considerable discomfort to the patient, the decision on the tests becomes important. It is true that if the result on a subject were very high—as, for example, at point C in the figure—one would not be too concerned as to whether this might be a false positive.

The use of a normal range has been stated to be such as to enable us to estimate the percentage of false positives that would result by adopting certain cutoff points and applying them to other subjects like those who provided the standard series. The first part of this statement has been discussed. The latter part requires further comment. The qualification of applying the range to other subjects like those who provided the standard series is important. In other words, if our patient was healthy, would he have been included in the group used to provide the normal range on the basis of other requirements such as age, sex, etc.? It is well known that growing children have physiologically higher alkaline phosphatase levels than mature adults. It seems fairly well established that women have lower average uric acid levels in serum than men (at least in the 20 to 50 age group), and that endocrine function decreases with age. The glucose tolerance of normal individuals is affected by the previous dietary carbohydrate intake. There is evidence that individuals living in different geographical locations or under different climatic conditions may have somewhat different normal ranges for a number of blood constituents. Even normal individuals under a hospital regime can have normal levels for a number of substances that differ from levels they have when they are active. These factors are often not considered in determination of normal range but they are obviously of importance [4].

One can also make a distinction between a statistical normal and a clinically desirable normal. For example, the average serum cholesterol level in healthy males in the U.S. population is higher at age 40 than at age 20. If one believes that increased cholesterol levels play at least some part in the etiology of heart disease, then one might feel that the average level in the older group is too high and that this value should not be considered as a clinically desirable normal. This is a clinical judgment and is entirely separate from the statistical normal, although it may be influenced by the latter.

We will not discuss here the effect that different methods of chemical analysis will have on the normal, though these are obviously of importance. It is certainly impractical to keep in mind a whole series of normal ranges for a given constituent, subdivided as to age group, sex, etc., but these differences must be kept in mind when interpreting the results. It has often been suggested that each laboratory should establish its own normal ranges. This is often taken as referring to variations due to differences in chemical methods of analysis, but could also include some of the other factors mentioned, including geographical location and type of hospital patients.

One method that has been suggested for determining the normal range is to use the analysis of the admission screening tests on a large number of hospital patients. Theoretically this method should give normal values applicable to the type of patients entering the hospital. If, for example, one determines the serum urea nitrogen (still commonly labeled BUN) on all admission patients, a certain percentage of them will have diseases that may cause an elevation in the BUN. Others will have different conditions that may have little or no effect on the BUN. A plot of the frequency distribution of all the analytical results might result in a curve similar to the dotted curve of Figure 4-3, which represents a composite curve of I (normals) and II (abnormals). The points on the curve to the left of D, for example, will be influenced very little by the abnormal patients whose values would follow curve II. Thus the left-hand portion of the curve may be supposed to be due to the normals alone. If it is assumed that the curve is a gaussian one, the whole normal curve can be calculated from this portion and the range obtained. The same reasoning would apply to other serum levels such as those for calcium or protein. Some patients would have abnormal values (not necessarily the same ones as those who had abnormal BUN) and some would be normal for this constituent. With proper computer programming it would be simple to calculate the normal curve, assuming it to be gaussian [5, 6].

This method involves several assumptions. The first is that the curve follows a gaussian distribution. Unless the exact mathematical form of the curve is known, it is difficult to determine the complete curve from only one portion. As will be seen in the next section, given a simple curve (such as I of the figure), there are methods to determine whether the curve follows the gaussian distribution, and some coordinate transformations that can be used to give a curve following the gaussian distribution more closely. It is difficult to apply these methods to a composite curve like III of Figure 4-3.

If there were three types of curves making up the composite one—normal subjects, those having elevated levels, and those having levels below normal (this could readily occur with serum electrolytes, for example)—the manual method would be more complicated and probably inaccurate. With a proper computer set-up it should be possible to resolve a composite curve into any number of component gaussian curves, but such methods are not generally available. For these reasons the determination of normal ranges by these methods is not recommended by many investigators [3].

This rather lengthy discussion of normal values may seem confusing to the average technician or clinician. Its purpose is to emphasize that the determination of a normal range is not so straightforward as is often assumed. A normal range is not something that is definitely fixed, once and for all, but is empirically determined and depends upon many factors. The normal range is very helpful in diagnosis, but one must always be aware of its limitations.

It may be calculated that if the normal ranges will give 5 percent of false positives on normal individuals for each test, then with 12 test profiles the probability of obtaining at least one positive is $1 - (0.95)^{12} = 0.46$, and for 18 tests the probability is $1 - (0.95)^{18} = 0.60$. In other words, if a battery of 18 tests were run on normal individuals, 60 percent of them would have at least one value falling outside the normal range. This is actually somewhat of an exaggeration, since the 18 tests are not completely independent [7]. The albumin level, for example, does not vary completely independent of the total protein, and the chloride level is not completely independent of the sodium level.

Calculation of Standard Deviation and Estimation of Normal Range

The gaussian error curve has the mathematical form:

$$f(x)dx = \frac{1}{\sigma\sqrt{2\pi}}\, e^{-\frac{(\mu - x)^2}{2\sigma^2}}\, dx \tag{1}$$

where $f(x)\,dx$ is the probability that the measurement has a value lying between x and $(x + dx)$. The curve is defined by two parameters: μ, the mean value, and σ, the standard deviation. The theoretical curve of equation 1 corresponds to the results of a very large number of determinations on what is termed the *parent population*. The calculations using the actual measurements are then an attempt to estimate the actual values of μ and σ from a relatively small, assumed-random sampling from the parent population.

It turns out that the best estimate of μ is

$$\mu = \bar{x} = \frac{\Sigma x_i}{N} \tag{2}$$

where the expression on the right means the summation over all values of x, divided by N, the number of samples or measurements. This expression is what is usually called the average or mean value.

The best estimate of σ is given by

$$\sigma = \sqrt{\frac{\Sigma\,(\bar{x} - x_i)^2}{N - 1}} \tag{3}$$

In words: Find the difference between each value and the mean, square each difference, and add the results. Divide the sum by one less than the number of measurements and extract the square root.

It is usually more convenient to use a slightly different formula which is readily derived from equation 3. This is:

$$\text{S.D.} = \sqrt{\frac{\Sigma\,(x_i{}^2) - (\Sigma x_i)^2/N}{N - 1}} \tag{4}$$

An example of an S.D. calculation using equation 3 is given in Table 4-1, columns 1 to 3. Column 1 lists x_i, the results of the individual determinations; their sum is 1084. Dividing the sum by N, which is 12, gives the average $\bar{x} = 90.33$. Column 2 gives the values for $(\bar{x} - x_i)$, the difference between each result and the average; column 3 gives the squares of these differences. Note that since only the squares of the differences in column 2 are used, the sign of the differences need not be recorded. Summing the values in column 3 gives 214.6668. Dividing this by $N - 1 = 11$ and extracting the square root, one obtains S.D. $= 4.42$.

Column 4 gives the values of $(x_i)^2$ for the calculation of S.D. by formula 4. From the sum of column 4 we subtract $(\Sigma x_i)^2/N = (1084)^2/12$ (which equals 97921.333) and obtain 214.667, the same as the sum of column 3. Again we divide by $N - 1$ and extract the square root.

Columns 5 and 6 illustrate a variation of the method. From each number in column 1 we subtract the same convenient number (80 in this illustration) to give the smaller numbers x' of column 5; the squares of these numbers are in column 6. We find $\bar{x}' = 124/12 = 10.33$, and the true average is then $10.33 + 80 = 90.33$, where 80 is the number originally subtracted from the numbers of column 1. Using equation 4 we calculate $\Sigma(x_i'^2) - (\Sigma x_i')^2/N = 1496 - (124)^2/12 = 214.667$, the same as obtained previously. The smaller numbers of columns 5 and 6 are easier to work with than the larger numbers of columns 1 and 4.

It is often simpler to work with integers rather than decimals as there is then no chance of a decimal point being misplaced in the calculations. Suppose the numbers of column 1 were 8.7, 9.5, etc., instead of 87, 95, etc. The simplest

Table 4-1. Calculation of Standard Deviation

1 x_i	2 $\bar{x} - x_i$	3 $(\bar{x} - x_i)^2$	4 $(x_i)^2$	5 x'_i	6 $(x'_i)^2$
87	3.33	11.0889	7569	7	49
95	4.67	21.8089	9025	15	225
88	2.33	5.4289	7744	8	64
97	6.67	44.4889	9409	17	289
91	0.67	0.4489	8281	11	121
83	7.33	53.7289	6889	3	9
94	3.67	13.4689	8836	14	196
87	3.33	11.0889	7569	7	49
96	5.67	32.1489	9216	16	256
91	0.67	0.4489	8281	11	121
86	4.33	18.7489	7396	6	36
89	1.33	1.7689	7921	9	81
1084		214.6668	98136	124	1496

Using equation 3:

$$\bar{x} = \frac{\Sigma x_i}{N} = \frac{1084}{12} = 90.33$$

$$\sigma = \sqrt{\frac{\Sigma \, (\bar{x} - x_i)^2}{N - 1}} = \sqrt{\frac{214.6668}{11}} = 4.42$$

Using equation 4:

$$\text{S.D.} = \sqrt{\frac{\Sigma \, (x_i^2) - (\Sigma x_i)^2/N}{N - 1}}$$

$$= \sqrt{\frac{98136 - (1084)^2/12}{11}} = \sqrt{\frac{98136 - 97921.333}{11}}$$

$$= \sqrt{\frac{214.667}{11}} = 4.42$$

Variation using equation 4:

$$\bar{x}'_i = 124/12 = 10.33 \qquad \bar{x} = 10.33 + 80 = 90.33$$

$$\text{S.D.} = \sqrt{\frac{1496 - (124)^2/12}{11}} = \sqrt{\frac{214.667}{11}} = 4.42$$

way to do the calculations is mentally to multiply each number by 10, then subtract 80, obtaining the same figures as shown in column 5. Then after calculating the average and the standard deviation as 90.33 ± 4.42, one divides by 10 to obtain the correct result, 9.033 ± 0.442.

As mentioned earlier, the data from biological samples need not necessarily follow a gaussian curve. There is a method which will suggest whether or not the data actually follow such a curve. The data plotted in Figure 4-1 may be tabulated as shown in Table 4-2. This indicates that there was one determination giving the value 11, 2 determinations giving the value 12, 5 giving the value 13, and so on. The third column gives the cumulative total of all values up to a given level. Thus there are 14 determinations giving a value of 14 or less. The next column gives the percentage this cumulative value is of the total number of determinations, in this case, 100. As mentioned earlier, the shape of the curve indicates that possibly the last 11 values do not belong to the group; thus we have included in the last column the cumulative percentage for the first 89 determinations to see if possibly these would follow a gaussian curve more closely than all the 100 values. The cumulative percentages are then plotted on probability paper (K&E No. 46-8000) with the values along the horizontal axis and the probability scale along the vertical axis. If the data follow a gaussian curve, all the points should fall on a straight line. Strictly speaking, the scale used should not be the cumulative probability but a slightly different probit

Table 4-2. Data from Figure 4-1

Value	Number	Cumulative Total	Cumulative % of 100	Cumulative % of First 89
11	1	1	1.0	1.1
12	2	3	3.0	3.4
13	5	8	8.0	9.0
14	6	14	14.0	15.7
. . . .				
23	2	84	84.0	94.5
24	4	88	88.0	99.0
25	1	89	89.0	100.0
26	2	91	91.0	—
27	5	96	96.0	—
28	1	97	97.0	—
29	2	99	99.0	—
30	1	100	100.0	—

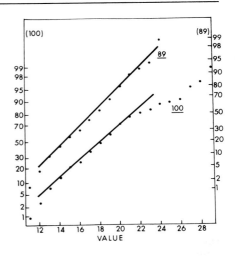

Figure 4-4. Probability graph for same data as presented in Fig. 4-1. For details see text.

scale. The probability paper is much more readily available and the differences are appreciable only at very small or large cumulative probabilities. Thus when plotted on probability paper a true gaussian curve may give a slight curvature at the two ends of the curve, but this is of little practical importance. In Figure 4-4 we have plotted the results for both sets of values (100 and 89) in the same figure but with different vertical scales to prevent the curves from overlapping. Note that the points for the full set of 100 do not fall on a straight line over the entire distance, but that the set of 89 gives a somewhat better fit.

When the data do not appear to follow a gaussian curve very well, i.e., do not give a straight line on the probability curve, one may plot the curves using for the horizontal axis either the square root of the values or the logarithm of the values. The square root transformation usually results in less change in the shape of the curve than does the logarithmic one. In Figures 4-5 and 4-6 we have plotted the same data as used in Figures 4-1 and 4-4 on a square root scale. Comparing Figures 4-6 and 4-4 and Figures 4-1 and 4-5, it is apparent that the transformation did not give any improvement in the fit [8].

The other transformation is to plot the frequencies or probabilities against the logarithm of the concentration. The same data as used previously are plotted on this basis in Figures 4-7 and 4-8. Comparing Figure 4-8 with Figures 4-4 and 4-6 one notes that the data now appear to follow a straight line somewhat more closely. When the data as plotted in Figure 4-1 give a skewed curve, the loga-

Figure 4-5. The same data as used in Fig. 4-1, plotted in terms of the square root of the concentration. Points are used instead of a histogram. The solid line corresponds to the calculated gaussian curve for all points (range A), and the dashed line the calculated gaussian curve for the B range. The averages and standard deviations for the two ranges are 4.223 ± 0.371 and 4.335 ± 0.477, respectively, in square root units. These correspond to means of 17.83 and 18.80 in regular units.

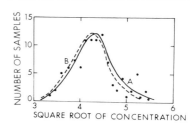

Figure 4-6. Probability graph for same data as used in Fig. 4-1 when plotted vs. the square root of the concentration values.

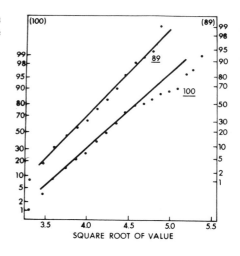

rithmic transformation often gives a curve following the gaussian one more closely. Methods [9, 10] have been suggested for plotting on the horizontal axis in the probability chart more general functions of the values such as $\log (x_i + C)$ or $x_i + C$, where the x_i are the various determined values and C is an empirically determined constant (not necessarily positive). Using one of these formulas one can often find the proper constant so that the data plotted in this manner follow a gaussian curve quite closely. However, in order to determine the constant C, a fairly extensive computer program is necessary, so that the method is not generally available. The examples given are not very satisfactory

Figure 4-7. The same data as used in Fig. 4-1, plotted in terms of the logarithm of the concentration. Points are used instead of a histogram. The solid curve again corresponds to the gaussian curve calculated from all the data with a mean and standard deviation in logarithmic units of 1.2689 ± 0.0954. The dashed curve corresponds to the gaussian curve calculated from the narrower range B with a mean and standard deviation of 1.2478 ± 0.0729. These two means correspond to values in regular units of 18.57 and 17.68, respectively.

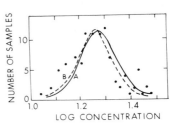

Figure 4-8. Probability graph for same data as used previously when plotted vs. the logarithm of the concentration values.

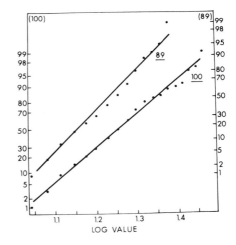

in that they do not show any very marked deviations from the gaussian curve on the linear scale. In many determinations in the laboratory, particularly with the normal levels of the enzymes, one may find markedly skewed curves and the simple transformations given here should be tried.

Note that the different ways of plotting and of calculating the means and S.D. yield slightly different results as indicated in Table 4-3 where the means and the range of 2 S.D. are given for the three methods and for the set of 89 and of 100 values. The various differences are not great, and if we are taking our range of normals to be ± 2 S.D., the various ranges are not sufficiently different to cause any serious errors in interpretation. However, as mentioned

Table 4-3. Means and Ranges of 2 S.D. for Various Methods of Computation
of Data in Figure 4-1

Method	Set of 89 Values		Set of 100 Values	
	Mean	Range	Mean	Range
Linear	18.0	11.7–24.2	19.0	10.6–27.5
Square root	17.8	11.6–26.7	18.8	11.4–28.0
Logarithm	17.7	11.0–28.8	18.6	11.6–26.7

earlier, the S.D. and hence the range is valid only if the data actually follow the gaussian curve.

In our discussion of the normal range, we have taken this to be 2 S.D. from the mean. In a given clinical situation, one might be interested in only one part of the range—for example, only values near the upper limit, low values not being of interest as is the case with some enzymes. In that case there would be at the most, 2.3% of false positives, and one might feel that this would lead to excluding too many patients who might actually have the disease (too many false negatives) and that a narrower range should be used. If one chooses a range with a 90% tolerance interval (that is, 90% of the normals would fall within the range with 5% above and 5% below), this would correspond to a range of ± 1.645 times the S.D. In the data given for Figure 4-1 this would correspond to limits of 12.9 to 23.0 for the set of 89 values and 12.0 to 26.0 for the set of 100.

If we knew the mean and S.D. exactly (i.e., knew the exact values of σ and $\bar{x}$), we could say with confidence that in any group of normals, 95% of them would have values falling within our 95% tolerance interval. However, our calculated values of the mean and S.D. are only estimates of the true values and are themselves subject to statistical variation. We can say only that x% of the time, 95% of the subjects will fall in the 95% tolerance interval. There are methods for calculating ranges such that x would have any desired value but these will not be given here.

There is a simple nonparametric method for estimating the normal range from the experimental data. The method is so called because it does not depend upon the form of the curve or the determination of any such parameter as the S.D. This is illustrated by the data given in Table 4-4. The data used are those plotted in Figures 4-1 and 4-7, using only the figures for the entire set of 100 values, for illustrative purposes only. The numerical values are first arranged in order of rank with the lowest value given the rank 1, the next lowest the rank 2, and so on to the highest value which is given the highest rank. In the table we have given only the first and last ten, which are sufficient for the present, but

Table 4-4. Calculation of Ranges by Percentile Method

Rank	Value	Log	Rank	Value	Log
1	11	1.041	91	26	1.415
2	12	1.079	92	27	1.431
3	12		93	27	
4	13	1.114	94	27	
5	13		95	27	
6	13		96	27	
7	13		97	28	1.447
8	13		98	29	1.462
9	14	1.146	99	29	
10	14		100	30	1.477

5% limit	95% limit
$0.05 \times 101 = 5.05$	$0.95 \times 101 = 95.95$

rank 4	13 (first)	rank 92	27 (first)
rank 9	14 (first)	rank 97	28

$$\frac{(5.05 - 4)}{(9 - 4)} \times (14 - 13) = 0.21$$

$$\frac{(95.95 - 92)}{(97 - 92)} \times (28 - 27) = 0.79$$

$$\text{limit} = 13 + 0.21 = 13.2$$

$$\text{limit} = 27 + 0.79 = 27.8$$

Logarithmic basis

rank 4	1.114	rank 92	1.431
rank 9	1.146	rank 97	1.447

$$\frac{(5.05 - 4)}{(9 - 4)} + (1.146 - 1.114) = 0.007$$

$$\frac{(95.95 - 92)}{(97 - 92)} \times (1.447 - 1.431) = 0.013$$

limit	$1.114 + 0.007 = 1.121$	limit	$1.431 + 0.013 = 1.444$

This corresponds to a value of 13.2   This corresponds to a value of 27.8

the intermediate values must be known in order to determine the rankings. Note that the determinations on five subjects have the same value of 13. These have been assigned the ranks 4 to 8 inclusive. These identical values can complicate the calculations. This can often be avoided by using values which have been recorded to an extra decimal. In the example given, the values as actually determined may have been rounded off to the nearest integer, as conventionally the results for this determination are reported in this fashion. If the actual results before rounding had been 10.8, 11.8, 12.4, 12.7, 12.9, 13.1, and so on, and these figures were used, the ranking would be unambiguous. Another point should be mentioned briefly: If the results are conventionally reported to the nearest integer only, is it meaningful to give a normal range of 20.5 to 37.3 instead of the rounded figures of 21 to 37? We would certainly be justified in reporting the average and standard deviation as 29.1 ± 4.3, and in all the calculations we should continue to use the extra figure. As a final report, the range should be reported as 21 to 37.

To return to the calculations: We have chosen to obtain the range for a 90% tolerance interval, excluding the lowest and highest 5%. We then multiply $N + 1$ (101 in this example) by 0.05 and 0.95 to give the numbers 5.05 and 95.95 shown in the table. We now find the interpolated values corresponding to these rankings. For the 5% limit we note that the first of the samples having the value of 13 has the rank of 4 and the first having the value of 14 has the rank of 9. To find the value corresponding to the rank of 5.05 we interpolate linearly as shown (this may be considered as a linear equation, $ax + b$, with the known value of 13 when $x = 4$ and value of 14 when $x = 9$; then we find the value of 13.2 when $x = 5.05$). A similar calculation is carried out for the 95% limit which gives an upper limit of 27.8. We have also made similar calculations using the logarithms of the values. These give substantially the same limits indicating that the range as calculated by the percentile method does not depend upon the shape of the frequency curve. The limits of 13.2 to 27.8 derived by this method may be compared with those calculated from an assumed gaussian or log gaussian curve for mean ± 1.645 S.D.; the limits are 13.1 to 25.9 for the gaussian curve (Fig. 4-1) and 13.6 to 25.2 for the log gaussian curve (Fig. 4-3). Also by extrapolating curve I of Figure 4-3 we would obtain a 5 to 95% range of 12.3 to 23.7, and from dashed curve II, a range of 12.6 to 25.1. This illustrates that a number of slightly different ranges can be arrived at, depending upon the method of calculation. In the present illustration we would probably settle for a range of something like 13 to 26.

A complicating factor should be mentioned. This applies to all methods of determining the range, but particularly to the percentile method which is more sensitive to this factor. The problem arises when a few values obtained from apparently normal individuals seem to be abnormal in that these values are higher (or lower) than the other values in the group. Suppose, for example, that the highest value in the series of determinations illustrated in Figure 4-1 and Table 4-4 was 38 instead of 30, all other values remaining as given. Should this high value be included in our calculations for determining the range? We may be tempted to discard this value as being actually abnormal. One should not discard a measurement merely because "it doesn't look right." The possible causes for the discrepancy should be investigated as far as possible. The abnormal result might have been the result of a clerical error in transcription. The chemical determination may have been in error for some unknown cause. When running determinations to obtain a normal range, it is best to save the specimens and rerun the few highest and lowest values to check the possibility of error. This may not always be possible. The constituent, such as an enzyme, may not be sufficiently stable on storage, or too little of a sample may have been obtained. Another possibility is that the subject from whom the questioned data was obtained may actually be abnormal but this was not recognized in the preliminary screening. If possible, the subject's history and data from other tests should be reviewed.

Suppose that these factors are checked as far as possible but offer no clue. There are mathematical checks that may be used. For the percentile method, one may calculate the ratio of the difference between the outlier (abnormally high or low result) and the next highest (or lowest) value to the difference between the highest and lowest values (including the outlier). Mathematically for a high outlier this would be

$$R = \frac{x_n - x_{n-1}}{x_n - x_1}$$

where the values are arranged in order of magnitude, as is done in the percentile method, with x_1 the lowest value and x_n the highest. For a low outlier, the numerator of the fraction would be $x_2 - x_1$. If R is greater than $\frac{1}{3}$ ($= 0.333$), the outlier is rejected from the calculations. In the example given above, if the value 30 of the original set is replaced by 38, the formula gives

$$R = \frac{38 - 29}{38 - 11} = \frac{9}{27} = \frac{1}{3}$$

In this case R equals exactly one-third and the answer is not definite. In any event the method serves merely as a useful guide and not as a hard and fast rule. If a gaussian curve is assumed, one may calculate

$$R = \frac{x_n - \bar{x}}{s}$$

where x_n is the value for the outlier, $\bar{x}$ the mean value, and s the S.D. calculated using all the values including the outlier. Using the same set of determinations as given above, one would find

$$R = \frac{38 - 19.1}{4.51} = 4.2$$

(The replacement of the value 30 by 38 gives a slightly higher mean and S.D. than shown in Fig. 4-1.) If one were to reject any outliers giving a value of R greater than 3, the above formula would indicate that the outlier value 38 should be rejected. We saw in the discussion of Figure 4-5 that the data used do not follow a gaussian curve very closely so that the rejection criterion for a gaussian curve might not be reliable. Since the curves of Figure 4-7 indicated that the data follow a log gaussian curve, one may calculate R on this basis, again assuming the highest value 30 to be replaced by 38. In logarithmic units,

$$R = \frac{1.580 - 1.275}{0.089} = 3.4$$

(Here also the use of a different value for the highest figure gives a slightly different mean and S.D. than given under Fig. 4-3.)

Quality Control [11–18]

It is common experience in the laboratory that if a given determination is re-peated a number of times on aliquots of the same sample, the results will not be identical. These differences can be due to slight fluctuations or errors in the pipetting of the samples or reagents, variations in the heating bath temperature or time of heating (if used), fluctuations in the colorimeter lamp or other elec-trical circuits, errors in colorimeter readings, or other causes including slight inhomogeneities in the sample itself. The differences will usually be smaller with automated methods but they can never be entirely eliminated. If the sample aliquots are not all run in one batch but on different days, the variation might be greater (even assuming no change in the sample itself). If the determination is actually made only once (as is usually the case), the result obtained will not necessarily be the same as the average value that would have been obtained from many determinations. Since the results of the test are to be used for diagnostic purposes, it is desirable that the result be known as accurately as possible, that the fluctuations be held to a minimum. Thus the basic purpose of quality control is to predict and control the variation in any technique in order to insure the reliability of the reported data. Later we will deal with the mathematical means for characterizing the variations in results. At present we will discuss it only in general terms.

Precision and Accuracy

The amount of variation in the results of the tests may be defined in terms of precision and accuracy. Although in common language these terms may be almost synonymous, in quality control they have distinctly different meanings. The *precision* of a method is the degree to which the determinations vary from the average value. A method in which nine-tenths of the determinations on a sample are within 5 percent of the average is more precise than one in which the same proportions were within 10 percent. The precision may be characterized in terms of the S.D., as will be shown later.

The *accuracy* of a method is the degree to which the determinations (and the average) differ from the true value of the concentration of the substance being determined. This is more difficult to measure since in a complex biological mixture the actual true value may not be exactly known, possibly due to interferences from other substances present. The true value is usually determined by analyzing the sample by two or three different methods that have been carefully studied and found to produce precise results. These reference methods may be too long or complicated for routine laboratory use, but they are useful in estimating the true concentrations. For some substances the true value may not be exactly known but this may not be as important as it may seem. If the normal range for a given constituent is determined by a given method, then the results of a determination on a patient's serum by the same method can be correlated with the normal range to give the physician useful information. Formerly, for example, the Folin-Wu method was widely used for blood glucose although it was soon shown that the method determined some other reducing substances in blood in addition to glucose. The method served clinical purposes well, in that the physician could always relate the patient's sugar level to the normal levels for this method. Such a correlation would always be valid if the Folin-Wu method always gave values, say, 15 mg/dl higher than the true glucose or was always higher than the true glucose by, say, 5 percent. However, it was found that the difference between the Folin-Wu and the true glucose method was not always exactly the same either absolutely or relatively, and depended upon the other substances actually present in the blood. Thus, particularly near the upper limits of normal, the comparison between the patient's value and the normal range was not as accurate as it would be if a true glucose method were used. In general, this is the basis for preferring as accurate a method as possible (within the limits of practicability, that is).

Quality Control Samples

In quality control the emphasis is more on precision than on accuracy. The former is more easily measured, and an imprecise method is of little value no matter what its accuracy. One usually analyzes one or more aliquots from one or more different samples at various times during each day's run and repeats this on successive days. This requires a large sample that is stable and of constant composition. Two types of such samples have been used. One is a *pooled serum sample*. This is prepared by pooling the excess serum from the sample determinations (omitting those that are grossly hemolyzed) and storing the serum in a refrigerator. When a sufficient amount has been collected, it is filtered or centrifuged to remove particulate matter and fibrin clots. It is preferable to have one quality control sample with levels of the various constituents approximating those found in normal serum and one with abnormal values. This is not always easy to obtain. The serum may be concentrated somewhat by partial freezing and pouring off the serum from the ice crystals. For the inorganic constituents and others such as glucose, urea, and uric acid, additional amounts can be added to bring the concentrations to approximately the desired levels. This may also be done with some other constituents but it is extremely difficult to do for cholesterol. Some enzymes can be purchased as fairly pure material and added to enhance the serum pool. Here it is sometimes difficult to add the right amount and the enzymes added will probably not be of human origin and thus react differently from the enzymes in the serum samples. When the pool has been prepared with desired levels of constituents, it may be refiltered and then dispensed into small vials, each holding the amount estimated to be needed for one day's run. The vials are tightly stoppered and stored in a freezer, preferably at $-30°C$. When a vial is needed it is removed from the freezer, thawed, and carefully mixed before use. The serum pools are not as stable as might be desired, particularly for glucose and the enzymes.

Many laboratories prefer to use the commercially available *lyophilized serums*. These are stable in the lyophilized state for a year or more when kept in the refrigerator. If a laboratory orders these serums on a contract basis for monthly or bimonthly delivery, most suppliers will set aside a sufficient amount so that the same lot number will be delivered over a year's time. The material is available in several different sizes of vials and in a normal and an abnormal range. It can be obtained assayed—with values stated for the different constituents, the values having usually been obtained by analysis by several different independent reference laboratories. The material can also be obtained unassayed—the exact concentrations of the constituents are not given, though they are usually stated to be within a certain range. For small laboratories the assayed form is more convenient, particularly when some tests are not done every day. One can then measure not only the precision of the analysis, but also how closely the results approach the labeled values. Some laboratories use these serums for standards, particularly for enzymes or for tests that are run only occasionally. It is best to use one lot number for a standard and a different one as a control serum. Although these serums are generally uniform, occasionally one will find a vial in which the values found differ markedly from the labeled values. In such instances the results should be checked by using a different vial.

For large laboratories with automated equipment it is satisfactory to use the unassayed serum for a control. When a new lot is first obtained, it is carefully compared with the old lot and with the standards run together in a similar series. When large numbers of samples are run, the control serums are randomly placed in the series of samples and, of course, treated similarly. If the control serums are to be used to judge the precision of the regular analyses, these serums should not be given any preferential treatment.

The lyophilized serums are obtained from pools of human serums. The serums may be dialyzed to remove most diffusible substances and then definite weight amounts of these added. Other serums may not be dialyzed but definite amounts of some substances may be added in order to bring their concentrations within the desired ranges. Usually it is necessary to add enzymes to bring their activities to the desired level. The enzymes added are usually not of human origin and may not react exactly the same as those found in human serum. Thus some care must be taken when the method used for the analysis of the enzymes is different from that used for its standardization. Another minor disadvantage of the serums is that they are generally much more turbid after reconstitution than most human serum samples (except for hyperlipemic ones). This may cause a slight interference when methods that do not involve dialysis or protein precipitation are used. It has been noted that in the determination of iron binding capacity by precipitation of the excess iron with magnesium carbonate, the removal of iron is much more dependent upon proper pH for reconstituted lyophilized serum than it is for regular serum samples [11]. This can lead to erroneous control values. It has also been found that the apparent alkaline phosphatase activity of some reconstituted lyophilized serums tends to increase during the first several hours [12]. There may also be other ways in which the

Figure 4-9. Quality control chart with average value 91; 2 standard deviations from the average is 9, and 3 standard deviations from the average is 13.5.

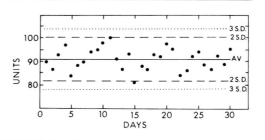

process of lyophilization and reconstitution may change the serum constituents, particularly the enzymes, the protein fractions present in small amounts, and other substances that are usually combined with proteins. Such changes have not yet been documented but are certainly possible. The lyophilized serums are nonetheless the best source of control material and for standards that must be made up in serum.

Quality Control Charts

Practically all laboratories that have quality control programs make use of some sort of quality control chart. Usually the results of the daily analysis of the control serum are plotted in a manner similar to that shown in Figure 4-9. The solid line through the center is the average value found (91 units in the illustration and since the units of concentration are irrelevant to the discussion we will merely call them units). The two dashed lines are the usual control limits of ± 2 S.D. calculated either from the data given or from previous measurements (see earlier discussion in this chapter on the S.D. and its calculation). In the illustration the S.D. is assumed to be 4.5 units so the two lines are at 91 ± 9, or 82 and 100 units. Sometimes a third set of lines is used at ± 3 S.D. (dotted lines in figure).

   The plotting of the results in this manner enables one to note easily the variation in the results of the analysis. The daily points will be scattered above and below the average. The points should fall within the 2-S.D. lines (actually, in theory, one point in every 21 will fall slightly outside the lines). Thus a point such as shown for day 15 which is just outside the limits should be investigated for possible causes of error, but a single point just outside the limits as shown is no cause for great alarm. If, however, two points are found outside the limits within a few days or if a point is found to lie outside the 3-S.D. limits, the procedure should be investigated for possible causes of error such as faulty reagents or standards, reading at the wrong wavelength, etc. The analysis should be repeated immediately using preferably a new vial of the control serum. Although the commercial serums are generally reliable, occasionally a vial or box of vials will on analysis give values for one or more constituents quite different from the expected ones. If a number of constituents are found to be in error in a given vial, the vial of serum itself is more likely to be at fault than if just one constituent is wrong. If a new vial also gives apparently erroneous results, the trouble lies somewhere in the procedure itself. The various steps should be checked until the difficulty is eliminated. If the fault is found in the procedure, the analysis of the samples should be repeated.

The quality control chart will often enable one to detect difficulties before the results fall outside the range. In the figure, we note that the values for days 5 to 11 show a continued upward trend. This would indicate that possibly a reagent is deteriorating or some other factor is changing. This should be investigated even before day 11. The figure indicates that at a later time (days 15 to 21) a similar upward trend is found. Such plots suggest a recurrent problem and the reagents should be carefully investigated for instability as this is the usual cause for such changes.

Often more than one aliquot of a given control serum is run each day, particularly when a large number of tests are run on automated equipment. When two aliquots are run, one near the beginning of the series (A) and one near the end (B), it is helpful to plot the algebraic difference (A − B), as shown in Figure 4-10. Theoretically the points should fall at random, equally above and below the zero line (no difference between the two aliquots). In the example shown, there are many more negative values (B greater than A) than positive values, particularly at the later days. This would indicate that the later analyses are on the average higher than the earlier ones with a drift in results. Although the average difference in this example is not large (−0.6 units), it could indicate the beginning of a more serious problem, especially as the last seven values are all negative. If the values obtained on the second analysis were consistently lower rather than higher than the first, one might suspect some deterioration in the control serum, though such changes are not likely to occur within a few hours.

When a large number of samples are to be run, it would be best to distribute aliquots of a pool throughout the run. This may be expensive to do with commercial lyophilized serum, but since the purpose of this is to determine whether the same result is obtained on the same serum pool, serum pooled in the laboratory could be used. The best arrangement of the repetitive samples and treatment of the data is given by Heggen et al. [13].

Figure 4-10. Chart showing the algebraic difference (A — B) between two aliquots of control serum run on the same day.

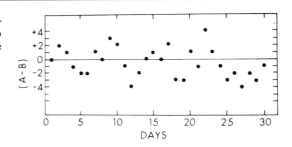

Normal Average Control

A supplementary method for quality control without the use of control serum has been suggested [14]. This method uses the average value of all those analytical results on a given constituent for the day's run that fall within the normal range for this constituent. If, for example, the BUN is determined on a relatively large number of hospital patients including all admissions, some of the patients will have conditions that cause an increase in BUN and others will have diseases that have no effect on the BUN and will have apparently normal values. The average of all the normal BUNs run on a given day should be quite constant from day to day, since the hospital admissions are usually drawn from the same type of population over a period of time. The actual computations for this method would be lengthy when done manually but if the results of the laboratory tests are entered into a computer, it could be easily programmed to print out the result in a very short time. Thus it might be an additional check that could be obtained without any great effort (once the program has been written) but it has not yet proved to be a substitute for the usual lyophilized serum controls [15].

Coefficient of Variation

In Figure 4-9 we have used ± 2 S.D. as the limits for the quality control. There is a certain ambiguity here. Any method, no matter how imprecise, will give approximately 95 percent of the results within ± 2 S.D. The procedure to be used must be such that the S.D. is sufficiently small to give acceptable results. Although most standard methods in common use will be sufficiently precise under optimum conditions, they may not always be so in practice. It is often convenient to use instead of the S.D., the *relative standard deviation* (RSD). This is the

S.D. calculated as a percentage of the average value, often termed the *coefficient of variation.*

One suggestion is that for a given determination the quality control limits as 2 RSD should not be more than one-fourth of the normal range expressed as a percentage of the average value for the range [16]. When this calculated figure is more than 10%, only the value of 10% is used. Thus if the normal range for serum sodium is taken as 135 to 145 mEq/liter, then the limits for the quality control chart (and of 2 RSD for the method) should be equal to 100 × ¼ (145 − 135)/140 = 1.8%. This suggestion has not been adopted very widely, but it is at least a useful criterion in checking the precision of a method. It seems a reasonable one, at least in the normal range, since the narrower the normal range, the more precise the method must be to detect accurately patients with borderline values. When the value for a normal control serum constituent is numerically a small figure, as, for example, in bilirubin levels where the normal range is about 0.3 to 1.3 mg/dl, it may not be possible to attain a precision of even 10%. Since the levels of bilirubin are customarily reported to only one decimal place, for a serum having an actual level of 1.0 mg/dl, adopting the limits of ± 10% would mean that all values should fall within 0.9 and 1.1 mg/dl. A value of 1.2 mg/dl would thus be 20% from the average and be out of control. Actually this is not an alarming difference in terms of actual measurement since the error in photometric measurements is greater at low absorbance readings and the difference between 1.1 mg/dl and 1.2 mg/dl is not of great clinical significance. Thus at this range a greater RSD should be allowable.

Even when samples in the normal range can be determined more accurately, one may find a greater precision at higher levels. Thus for glucose one might have a RSD of 3.5% at a level of 100 mg/dl and a RSD of only 2.5% at a level of 280 mg/dl, although this would mean a numerically wider range at the higher level. It is helpful in judging whether the laboratory results are in satisfactory control if one has information concerning the RSD obtained in other laboratories for the same method. Some of the surveys (see below) may give this information.

Youden Plot

Another type of plot that may be helpful in the laboratory when the control serums are run at two different levels is the Youden plot (Fig. 4-11) [17]. The scale for one control serum is marked on one axis (in the illustration, L = lower value) and the scale for the other serum is marked on the other axis (H = higher level). The two scales must be adjusted so that the S.D. for the serums are the same. In the illustration the low serum is taken to have an average of 99.5 mg/dl

Figure 4-11. Youden plot. For explanation see text.

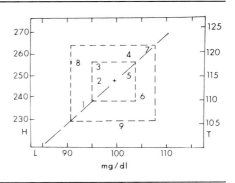

with a S.D. of ± 4.2 mg/dl and the other an average of 247.5 mg/dl with a S.D. of ± 9.1 mg/dl. This means that in the two scales, the actual distances corresponding to, say, 10 mg/dl must be in the ratio of 9.1:4.2, or 2.2:1. Thus if the L scale was marked off in convenient units of 40 mm, the H scale must be such that 10 mg/dl corresponds to a distance of 40/2.2 = 18.2 mm. Thus to draw this scale, one marks a convenient point near the middle of the scale as 250 mg/dl, and the points corresponding to 260 and 240 mg/dl are 18.2 mm above and below this point, and so on. This makes it convenient to plot values on the H scale.

A simpler method of plotting is illustrated on the right-hand scale (T = transposed scale). Since 250/2.2 = 115 (approximately), one marks a convenient point in the middle of the scale which is labeled 115, then the other points 105, 110, 120, and 125 are marked using the same scale distance as for the L scale (in the illustration, 10 mg/dl = 40 mm). To plot the high values one divides them by 2.2 (slide rule accuracy is sufficient) and plots them on the T scale. The average H value, 247.5, is then plotted at the point equal to 247.5/2.2 = 113 on the T scale. The cross (+) in the center of the plot thus corresponds to the two averages, L = 99.5 and H = 247.5. The daily points are then plotted in this manner, number 1 corresponding to a low value of 94 for the low serum and 237 for the high serum, and number 2 to values of 97 and 248, respectively. The successive values obtained on successive days are designated on the plot by the numbers 3 to 8. All the points falling within the inner square are within 1 S.D. of the mean for both values, the sides of this square corresponding to ± 1 S.D. from the mean and the sides of the larger square to ± 2 S.D. from the mean. The point 4, for example, is within 1 S.D. for L values but somewhat over 1 S.D. for the H value. For points falling on the diagonal dashed line, the ratio of the H and L values is the same as for the mean. This could indicate that the error could be in the standard or calibration as this would make all values high or low by proportional amounts. For points far from the diagonal, the indication is that the two values obtained differ from the mean in different ways. For example, the value obtained for the low serum was only slightly above the average, whereas the value obtained for the high serum was more than 2 S.D. below the average. This type of plot is not quite so convenient for comparing daily results, but it gives a somewhat different type of information than the other figures illustrated earlier. This type of plot is often used in comparing the survey results from a number of different laboratories. Barnett further discusses the various points mentioned in this section [18].

Survey Programs

A number of different organizations offer survey programs in which one or more lyophilized serums are sent to each laboratory on a regular basis. These are analyzed in the routine manner and the results returned to the supplier. The laboratory will then receive a computer printout comparing the results obtained by the laboratory with those obtained by other participating laboratories as well as, in some cases, the results obtained by some reference laboratories. In this way the precision and accuracy of one laboratory can be compared with others. This is very helpful in monitoring laboratory results. Participation in such a program will probably soon be mandatory for all clinical laboratories. The service offered by the College of American Pathologists* may be one of the best known, but similar programs are offered by the American Association of Bioanalysts† as well as the required program in proficiency testing from the Center for Disease Control.‡ Some commercial suppliers of lyophilized control serum offer a somewhat similar service which may be used as daily control serums or periodic sampling.

The College of American Pathologists also offers a quality assurance service which is essentially a computer treatment of the laboratory's quality control program. The results of the regular quality control analyses are entered on special cards and sent to the computer service each week. On a monthly basis a computer analysis of the results is received by the laboratory giving the S.D., control limits, and other information. This eliminates the tedious task of calculating these values each month. If the laboratory has its own computer, it may prefer to use it for such purposes. In all the survey programs the samples

* College of American Pathologists, 230 N. Michigan Ave., Chicago, Ill. 60601.
† American Association of Bioanalysts, 411 N. 7th St., St. Louis, Mo. 63101.
‡ Center for Disease Control, U.S. Dept. of Health, Education & Welfare, Atlanta, Ga. 30333.

received should be analyzed just the same as the routine samples from the patients. Many times the lyophilized material is given to the technician for reconstitution and analysis. This informs her that the material is a survey sample and she thus, consciously or not, gives it preferential treatment. Preferably the laboratory worker should receive the sample after reconstitution as a completely blind sample in order to eliminate bias.

Standards

The use of proper standards is another problem closely connected with quality control. In almost all determinations in the clinical laboratory the results are obtained by comparison with standards containing known amounts of the substance being determined. The accuracy of the method can thus be no better than the accuracy with which the standards are prepared. Not only must the standard substances be weighed out accurately and carefully made up in accurate volumetric flasks, but the material itself must be as pure as possible. The National Bureau of Standards§ provides a large number of standard substances. Those of interest for the clinical laboratory include glucose, cholesterol, urea, uric acid, bilirubin, potassium chloride, creatinine, and calcium carbonate. Also obtainable are potassium acid phthalate and tris-(hydroxymethyl)aminomethane for the standardization of acid and base solutions, and potassium dihydrogen phosphate and disodium phosphate for the preparation of standard pH solutions for blood pH determination. These substances are rather expensive for routine use, but should be used to check the daily standards which may be either made up from reagent grade chemicals or purchased in solution form. The National Committee on Clinical Laboratory Standards|| will in time have a number of standards available for the laboratory. The first one available is a bovine albumin for use as a protein standard.

When the laboratory method uses an aqueous solution of the standard (or a solution in another solvent, as for cholesterol), it is sufficient to compare the regularly used secondary standard with the primary standard by carefully analyzing the two solutions at several different concentration levels by the usual method. The results are used to correct the value of the secondary standard if necessary. If the method uses a protein-free filtrate of the serum (or other fluid), the usual practice is to neglect the correction due to the volume occupied by the precipitate; that is, one assumes that if 1 ml of serum is added to a total of 9 ml of water and precipitating reagents, the filtrate (or supernatant if

§ National Bureau of Standards, Washington, D.C. 20054.
|| National Committee on Clinical Laboratory Standards, 2525 W. 8th Ave., Los Angeles, Calif. 90057.

centrifugation is used) is exactly a 1:10 dilution of the serum. This can introduce a small error (usually not more than a few percent) if the filtrate is compared with a dilution of the standard. This is usually neglected since all tests, including the determination of the normal range, are done in the same manner.

When the procedure (as with many automated methods) requires a standard in serum, the verification of this standard may require elaborate testing. If a protein-free filtrate is made from the serum standard as well as the samples, comparison of the filtrate from the serum standard with an appropriate dilution of the primary standard is satisfactory. If dialysis is used or the serum is used directly, the results may not be quite the same as when an aqueous standard is carried through the procedure. In this case the serum is analyzed by a good reference method known to give accurate results when compared with the aqueous standard. The value so obtained for the serum is used as the standard value. Some exceptions to this are discussed under the procedures for the particular constituents involved. The standardization of enzymes is discussed in Chapter 14.

Quantities and Units

The results of quantitative measurements in clinical chemistry (or any other scientific discipline) are always expressed in some type of unit. In clinical chemistry these are usually concentration units (amount of substance per unit volume or weight) although the measurements may be made in some other units. Various types of units have been used for reporting the results of different determinations, and different units for the same determination are used in different localities. In the interest of uniformity, the Commission on Clinical Chemistry of the International Union of Pure and Applied Chemistry (IUPAC) and the International Federation of Clinical Chemistry (IFCC) have recommended the adoption of a consistent system of units (Système International d'Unites, or SI) based on the MKS (meter-kilogram-second) system adopted by many other scientific disciplines [19, 20]. Not all of these recommendations have been universally adopted (nor will they be in the near future), particularly in this country, though the acceptance has been greater in Europe. It is expected that many of the recommendations will be gradually adopted in this country. *Clinical Chemistry*, the official publication of the American Association of Clinical Chemists, has adopted many of the recommendations as requirements in papers submitted for publication [21]. One difficulty is that the results of the analyses are primarily used by physicians who may not be familiar with the new units. On the other hand, many clinical chemists take part in developing new procedures for known constituents or previously undetermined substances. In such instances the new system of units can be more readily introduced. Recently the International Committee on Standardization in Hematology and the World Association of (Anatomic and Clinical) Pathological Societies together with the IFCC have agreed to recommend to physicians and other health workers the use of the SI system for clinical laboratory results [22]. The following presentation gives some information concerning these recommendations.

Although the basic units in the MKS system are the meter, kilogram, and second, in clinical chemistry the working unit of volume is the liter (now defined to be exactly 0.001 cubic meters) and the working unit of mass is the gram. Some of the prefixes used for multiples and submultiples of the basic units are familiar but those more recently introduced may not be. The most commonly used prefixes in clinical chemistry are given here:

Prefix	Abbreviation	Factor
kilo-	k	10^3
deci-	d	10^{-1}
centi-	c	10^{-2}
milli-	m	10^{-3}
micro-	μ	10^{-6}
nano-	n	10^{-9}
pico-	p	10^{-12}

Others not so common in clinical chemistry are mega, M, 10^6, and femto, f, 10^{-15}. In general, the multiples and submultiples should be used in intervals of 10^3 and the proper interval chosen so that the numerical value is between 1 and 1,000, e.g., 75 μg is preferred to 0.075 mg.

The recommendations suggest that mass concentrations be expressed in grams (or multiples or submultiples) per liter, but it is realized that in this country the expression in terms of per 100 ml or per deciliter (dl) will continue to be used for some time. Such archaic expressions as mg% or mcg% should definitely not be used, and the use of percent is not recommended for describing the concentrations of solutions, as it is often ambiguous. The concentrations of reagents and similar solutions should be expressed in moles per liter (molarity) (M). The recommendations of the committee are that eventually all results of clinical tests should be expressed in moles per liter when possible. One mole is defined as that amount of substance that contains as many formula units as there are carbon atoms in exactly 0.012 kg of the carbon isotope ^{12}C. This corresponds to the usual molecular weight in grams, since the atomic weights are now all based on $^{12}C =$

12.0000. It was also recommended that the concentrations of inorganic ions be expressed as moles per liter rather than in equivalents per liter as is now done. Note that for the common ions Na, K, and Cl, as well as bicarbonate, the equivalent and molar weights are the same, so this makes no difference in the numerical value. The advantages of reporting the serum electrolytes in moles per liter rather than milligrams per deciliter has been recognized for some time. Thus in patients with normal electrolyte balance the sum of the Na^+ and K^+ concentrations in moles per liter might be found to be 15 ± 5 mols/liter greater than the sum of Cl^- and HCO_3^-; variations from this would indicate electrolyte imbalance. This would not be so easily recognized if the sodium and potassium were reported in milligrams per deciliter and the bicarbonate as volume % CO_2 as was once done. (In this and the following illustrations the numerical figures are given for illustrative purposes only and may not be the accepted normal ranges for a given method.) Physiologically it can be expected that the significance of the amount of a given substance is more related to the number of molecules of that substance than to the weight concentration. Theoretically it would be of little significance how the concentration of a substance is expressed if this is totally unrelated to the concentrations of any other substance in the body fluid analyzed, but the concentration of the different substances in serum, for example, are apparently correlated with each other [23]. The expression of the concentrations in moles per liter may aid in discovering the exact nature of these relationships which may not always be as apparent as for the electrolytes.

Another SI unit, which is used only minimally in clinical chemistry, is that for pressure. This is designated as the pascal (Pa) which is equivalent to newtons per square meter (N/m^2). The relationship between the pascal and the common unit of millimeters of mercury is: 1 mm Hg = 0.133 kPa. The pressure unit is used chiefly for recording the measurements of the partial pressures of CO_2 and O_2 (P_{CO_2} and P_{O_2}). Since the conventional unit of pressure, millimeters of mercury, is itself an arbitrary one, it should be replaced by a more logical unit.

Another recommendation of the committee concerns temperature units. Although the working unit will probably remain the Celsius degree, °C, the SI suggested unit is the kelvin (K), with the zero point on the customary scale, 0°C, being defined as 273.15 K.

Many of the recommendations have been adopted in some European countries, including the expression of the results in millimoles per liter. This latter unit has not yet been adopted in England [25] or in the United States. For the reasons mentioned, the SI units seem a logical development, although it may be some time before they will be universally adopted. There are difficulties in the way of making these changes. When the numerical values of a result are quite different when expressed in SI units, there may be psychological barriers to the change. Thus it may seem strange for uric acid in serum to be given a normal range of 150 to 475 μmol/liter rather than 2.5 to 8.0 mg/100 ml. On the other hand, a physician noting a reported uric acid level of 510 μmol/liter should immediately be aware that this is not a result expressed in milligrams per deciliter as this would be impossible. Another difficult problem in the conversion to new units might be in the use of automated equipment which record the results on calibrated charts, but since the standards could be readily calibrated in the new units, new charts could be easily devised. In those instruments using a computer-controlled printout, the computer could be programmed to calculate the results in any desired units. Substances which do not have a definite molecular weight such as total serum proteins would still be reported in mass units such as grams per liter.

There are some advantages in expressing the results in moles per liter rather than in mass concentration units. Inorganic sulfate in serum or urine is often expressed in terms of milligrams of sulfate (as sulfur) per deciliter, although the methods actually determine sulfate. If the results are expressed in moles, one need not be concerned whether the standard used and the result of the determination were expressed as sulfur or sulfate, since 1 mole (gram atom) of sulfur is equivalent to 1 mole of sulfate ion. Creatine is usually determined by converting to creatinine and measuring the total creatinine before and after the conversion. If the results are expressed in micromoles per liter, the subtraction can be made directly. Many of the methods for triglycerides actually determine the glycerol moiety. In order to convert milligrams of glycerol to milligrams of triglycerides, one must assume an average molecular weight for triglycerides. If, however, the result is expressed in terms of moles of glycerol, then 1 mole of glycerol = 1 mole of triglycerides directly. Similarly, phospholipids are usually determined as lipid phosphorus with a conversion factor of: lipid P $\times$ 25 = phospholipids. If it is assumed that each molecule of phospholipid contains one atom of phosphorus (as is usually the case), then moles of lipid phosphorus = moles of phospholipid. Blood ammonia may be expressed as micrograms of ammonia or of ammonia N. Although the difference in this case is not great, the uncertainty as to which value was actually reported would be eliminated if the results were reported in micromoles.

It is not expected that the conversion to reporting in moles will be very readily adopted in this country, but some of the recommendations of the Committee certainly should be.

It is preferable to use liter rather than an abbreviation to avoid confusion between the letter and the numerical 1. Note that the abbreviations for the various units are to be written without a period after them and are not capitalized unless derived from a proper name (e.g., pascal, Pa, and curie, Ci). Since in this country the mass concentrations will probably be used for some time, these should be abbreviated as g/dl or g/liter using multiples or submultiples of the numerator if needed.

Table 4-5 shows a comparison of conventional and SI units for some common clinical tests.

Table 4-5. Comparison of Conventional Units and New SI Units for Most Common Clinical Tests

Constituent	Conventional Unit		Factor		SI Unit	Normal Range[a] (SI Units)
Glucose	mg/dl	×	0.0555	=	mmol/liter	3.6–6.1
BUN	mg/dl	×	0.357	=	mmol/liter	3.6–7.2
Calcium	mg/dl	×	0.250	=	mmol/liter	2.1–2.6
Inorganic PO_4	mg/dl	×	0.323	=	mmol/liter	0.8–1.5
Creatinine	mg/dl	×	73.6	=	μmol/liter	60–100
Bilirubin	mg/dl	×	17.1	=	μmol/liter	5–25
Cholesterol	mg/dl	×	0.0259	=	mmol/liter	4.0–7.5
Uric acid	mg/dl	×	59.5	=	μmol/liter	150–475
Triglycerides	mg/dl	×	0.0114	=	mmol/liter	0.5–1.5
Magnesium	mg/dl	×	0.411	=	mmol/liter	0.8–1.3
Phospholipids	mg/dl	×	0.0129	=	mmol/liter	1.6–3.9[b]
Iron	μg/dl	×	0.179	=	μmol/liter	11.5–26
PBI	μg/dl	×	78.8	=	nmol/liter	300–620
Thyroxine	μg/dl	×	12.9	=	nmol/liter	70–140
Hemoglobin	g/dl	×	0.620	=	mmol/liter (Fe)	7.5–10.0[c]
Pco_2	mm Hg	×	0.133	=	kPa	2.6–4.0 (venous blood)
Po_2	mm Hg	×	0.133	=	kPa	9.5–11.5 (arterial blood)

BUN, blood urea nitrogen; PBI, protein-bound iodine.
[a] Normal range will vary somewhat with method used and other factors; approximate ranges are given here.
[b] Calculated from lipid phosphorus assuming 1 atom of P per molecule.
[c] The complete hemoglobin molecule containing 4 atoms of iron has a molecular weight of 64.600; often the quarter molecular weight equivalent to 1 atom of iron is used; the absorptivity of cyanmethemoglobin is usually given in quarter mM. The Fe after the unit indicates the basis per atom of iron.

Drug Interference in Clinical Laboratory Tests [26–29]

Although it has been known for many years that the administration of some drugs will influence the results of some laboratory tests, the magnitude of the problem has been more generally recognized only during the last decade or so. This has been in part due to the proliferation of new drugs and to some extent to the development of many new methods. Although the interpretation of the results of a test is not primarily the responsibility of the clinical chemist, he may often be able to aid the physician in this area. He should be able to inform the physician whether the abnormal results of a test could be caused by the administration of certain drugs. Frequently he is asked whether a given drug being administered to a patient would influence the results of a contemplated test. Thus the clinical chemist should have some knowledge of the interferences caused by the more commonly used drugs, and those which are more likely to interfere with chemical testing. He should also have available some source of information concerning the possible interference of the many other drugs in use. The references cited are among the later, more complete sources of this data. The work of Hansten contains an excellent compilation of the effects of many drugs on laboratory tests. The comprehensive treatment includes information on the effects of some drugs in potentiating or antagonizing the effects of other drugs (without necessarily changing the results of a chemical test, and thus of more interest to the physician). The other references are more simply compilations of reported interferences by different drugs with the action of other drugs or with laboratory tests. The one by Young et al. is actually a computer printout of the literature survey. Although the latter references have extensive coverage, they can give little information as to whether the reported interferences were common or were found in only a few cases.

Only a very general survey can be given here. The more common interferences for a given test will be discussed in subsequent chapters under the normal range for that test. The drug interferences with which the clinical chemist would be particularly concerned are those in which the results of a chemical test are different when the drug is administered and when it is not. Some drug interactions do not fall in this category but are the result of potentiation or inhibition of one drug by another. For example, the effect of barbiturates will be enhanced by the presence of alcohol in the body but this would not ordinarily change the level of barbiturate in the blood such as might be determined as a test for the poisoning by this drug. Although this particular interaction is well known, the clinical chemist might still suggest this possibility when asked about the apparent discrepancy between the measured barbiturate level and the clinical symptoms.

Although the chemist might suggest such possibilities, the responsibility for checking them lies with the physician.

In another type of interference, the administration of a drug is known to influence the level of a given substance in the blood. The influence of tolbutamide on the blood glucose level could hardly be called an interference since this is usually the reason for the administration of the drug. Some drugs given for other purposes may also influence the level of blood glucose. It is well established by now that the prolonged administration of oral contraceptives can cause some increase in the blood glucose level over that normally found during a glucose tolerance. The effect usually disappears on withdrawal of the drug. Lack of knowledge of this effect could cause some women to be classified as incipient or mild diabetics when actually they are not. The oral contraceptives also affect some other constituents of blood, particularly the serum proteins.

The administration of thiazide diuretics has been found to increase the level of uric acid in the blood. This is believed to be due to inhibition of the tubular excretion of uric acid. Although the hyperuricemia in itself might not be alarming, the high level in a patient on diuretics in whom the uric acid level had not previously been determined, might lead to an erroneous diagnosis. These undesirable interferences which complicate the diagnosis may shade off into results that indicate actual toxic effects by the drug. A large number of drugs have been found to have hepatoxic effects when administered over a period of time. With such drugs an increase in the blood level of bilirubin or the transaminases might be due not to the progress of the disease itself, but to actual toxic effects on the liver caused by the drug, and the results of such tests cannot be disregarded as mere drug interferences.

The presence of a given drug in the serum (or other body fluid) may interfere with the chemical reactions used in the determination of some constituents. This is the type of interference with which the clinical chemist can be most concerned and of which the physician is less likely to be aware. Thus the presence of abnormal amounts of reducing substances such as ascorbic acid or nalidixic acid (possibly due to the glucuronide formed) will cause a positive error in the determination of glucose by copper reduction methods. The administration of dextran may result in turbidity in the serum which can interfere with a number of tests. Bromide will be measured as chloride in most methods for the latter. The effect of mercury (from mercurial diuretics) on the colorimetric determination of protein-bound iodine is well known. These examples merely illustrate some of the ways in which drugs can interfere with chemical tests. When new drugs are introduced, it may be some time before the reports of various types of interference appear in the literature. The clinical chemist must be aware that interferences may exist. There is no sure way of predicting the physiological type of interference, but from a knowledge of the chemical nature of the drug and of the reactions involved in the determination of the blood constituent, the possibilities can often be suggested.

References

1. Barnett, R. N.   *Clinical Laboratory Statistics*. Boston: Little, Brown, 1971.
2. Mainland, D.   *Clin. Chem.* 17:267, 1971.
3. Elveback, L.   *Hum. Pathol.* 4:9, 1973.
4. Reed, A. H., et al.   *Clin. Chem.* 18:57, 1972.
5. Gindler, E. M.   *Clin. Chem.* 16:124, 1970.
6. Naumann, J. G.   *Clin. Chem.* 14:979, 1968.
7. Roberts, L. B.   *Clin. Chem.* 18:1407, 1972.
8. Reed, A. H., Henry, R. J., and Mason, W. B.   *Clin. Chem.* 17:275, 1971.
9. Harris, E. K., and DeMets, D. L.   *Clin. Chem.* 18:605, 1972.
10. Wilks, S. S.   *Ann. Math. Stat.* 13:400, 1941.
11. Leggate, J., and Crooks, A. A.   *J. Clin. Pathol.* 25:905, 1972.
12. Smith, A. F., and Fogg, B. A.   *Clin. Chem.* 18:1518, 1972.
13. Heggen, D. W., Newman, H. A. I., and Keller, M. D.   *Am. J. Clin. Pathol.* 58:37, 1972.
14. Hoffman, R. G., and Waid, E.   *Am. J. Clin. Pathol.* 43:134, 1965.
15. Reed, A. H.   *Clin. Chem.* 16:129, 1970.
16. Tonks, D.   *Clin. Chem.* 9:217, 1963.
17. Youden, W. J.   *Anal. Chem.* 32:23A, 1960.
18. Barnett, R. N.   *Clinical Laboratory Statistics*. Boston: Little, Brown, 1971.
19. Dybkaer, R., and Jorgensen, K.   *Quantities and Units in Clinical Chemistry*. Baltimore: Williams & Wilkins, 1967.
20. Dybkaer, R.   *Stand. Methods Clin. Chem.* 6:245, 1970.
21. Information for Authors.   *Clin. Chem.* 19:1, 1973.
22. Dybkaer, R.   *Clin. Chem.* 19:135, 1973.
23. Roberts, L. B.   *Clin. Chem.* 18:1407, 1972.
24. Buergi, W.   *Chem. Rundschau* 25:504, 1972.
25. Report of Committee of Royal College of Pathologists.   *J. Clin. Pathol.* 23:818, 1970.
26. Hansten, P. D.   *Drug Interactions*. Philadelphia: Lea & Febiger, 1971.
27. Garb, S.   *Clinical Guide to Undesirable Drug Interactions and Interferences*. New York: Springer, 1971.
28. Young, D. S., Thomas, D. W., Friedman, R. B., and Pestaner, L. C.   *Clin. Chem.* 18:1041, 1972.
29. Constantino, N. V., and Kabat, H. F.   *Am. J. Hosp. Pharm.* 30:24, 1973.

5. Carbohydrates

Protein-Free Filtrates

In the determination of many substances in biological fluids, particularly in whole blood, serum, or plasma, it is often necessary to separate the constituents to be analyzed from the relatively large amounts of protein which would interfere with the reaction. Although more manual methods are being introduced that use serum or plasma directly, it is still often necessary to make a separation, particularly when whole blood is to be analyzed. The usual method is precipitation of the proteins by the addition of a suitable reagent and the separation of the precipitate by filtration or centrifugation. The resulting clear solution is commonly spoken of as a protein-free filtrate, Filtration is relatively simple and usually results in a clear solution if a retentive paper is used. Some precipitates filter very slowly, and if the desired constituent is present in a very small amount there may be the possibility that some of it may be absorbed by the filter paper. On the other hand, reducing substances or salts may be leached from the paper if acid-washed paper is not used. When only small amounts of solutions are available as in micromethods, centrifugation is usually necessary. Sometimes the protein may be precipitated in a very finely divided form so that rapid centrifugation is necessary. Occasionally a few small particles may remain floating at the top of the solution. Often the desired solution can be removed with a pipet without disturbing the particles. In other instances they may be removed by filtering through a small filter paper or pledget of glass wool.

The most commonly used precipitants are tungstic acid [1, 2], trichloracetic acid [3, 4], and zinc hydroxide [5] alone or with barium sulfate [6]. Perchloric acid [7], picric acid, metaphosphoric acid [8], and ethyl alcohol have been used for special tests. Methods have also been devised for extracting the desired constituents from a heat-coagulated serum [9, 10]. In the preparation of the filtrate from blood, serum, or other fluid, the sample is diluted in a definite ratio and this factor must be taken into account in the calculations. In many instances this may be simplified by diluting the standard similarly. One may add, for example, 1 ml of serum to 9 ml of precipitating reagent plus water. The filtrate is then commonly taken to be a 1:10 dilution of the serum. This is not strictly true. Due to the volume occupied by the precipitate, the constituents from the 1 ml of serum are in a liquid volume slightly less than 10 ml. The error will vary with the type of precipitate and dilution, but can be as much as 3 to 4 percent. This error is usually neglected.

Folin-Wu Filtrate [1]

This is probably the oldest method now in use for the preparation of a protein-free filtrate. It was originally used with the Folin-Wu method for glucose and for the determination of nonprotein nitrogen. These particular methods are now rarely used, but modified Folin-Wu filtrates are used for some other determinations.

1. Sodium tungstate, $0.3M$.   Dissolve 50 g of reagent grade sodium tungstate $(Na_2WO_4 \cdot 2H_2O)$ in distilled water to make 500 ml.

2. Sulfuric acid, $0.33M$.   Mix 2 volumes of $.5M$ sulfuric acid with 1 volume of water. (See Chapter 20 for the preparation of standard solutions.)

To prepare a 1:10 dilution from serum or plasma, add in order to 8 volumes of water, 1 volume of serum, 0.5 volume of sodium tungstate, and 0.5 ml of sulfuric acid, mixing after each addition. Allow to stand for about 10 min; then filter or centrifuge. Different dilutions can be made by varying the amount of water; for example, 3 ml of water with the same amount of the reagents and sample will give a 1:5 dilution. For some determinations such as creatinine, it is preferable to have a more acid filtrate; use 1 volume of serum, 0.5 volume of sodium tungstate, and 1 volume of sulfuric acid, adjusting the amount of water to give the desired total volume.

For cerebrospinal fluid, urine, or other fluids containing only small amounts of protein, use less of the precipitating reagents. For a 1:10 dilution take 1 volume of fluid, 8.5 volumes of water, and 0.25 volumes of sodium tungstate and sulfuric acid.

Somogyi Filtrate [6]

This may be used for the Somogyi-Nelson method for glucose as well as for some other glucose methods and other tests.

1. Zinc sulfate solution, $0.175M$ (5%).   Dissolve 50 g of reagent grade uneffloresced zinc sulfate $(ZnSo_4 \cdot 7H_2O)$ in distilled water to make 1 liter.

2. Barium hydroxide, $0.15M$. Dissolve 95 g of barium hydroxide $[Ba(OH)_2 \cdot 8H_2O]$ in recently boiled and cooled distilled water and dilute to 2 liters. Since the salt and the solution tend to absorb carbon dioxide with the formation of insoluble barium carbonate, fresh crystals should be used and the solution protected from air. If the solution remains cloudy, allow to stand undisturbed for several days and then decant the clear supernatant.

The actual concentrations of the two solutions are not so important but they must exactly neutralize each other. This is checked by adding exactly 10 ml of the zinc sulfate solution to a flask containing about 50 ml of water, adding a few drops of phenolphthalein indicator (0.5 g/dl in ethanol), and titrating with the barium hydroxide. The titration must be carried out slowly with constant agitation. Rapid titration may result in a false endpoint. The titration is carried out until one drop of the barium hydroxide solution turns the titrated solution a faint pink which lasts for 1 min. The 10 ml of zinc sulfate should require 10.0 ± 0.05 ml of barium hydroxide. If the solutions are not of equal strength, dilute the stronger solution with the calculated quantity of water to give equal strengths, and repeat the titration. The barium hydroxide solution is fairly stable if kept tightly stoppered out of contact with air.

To prepare a filtrate from whole blood, serum, or plasma, add 1 volume of sample (whole blood should be measured with a "to contain" pipet which is rinsed out) to 5 volumes of water. Add 2 volumes of barium hydroxide solution and mix; then add 2 volumes of zinc sulfate solution and mix. Allow to stand for a short time, then filter or centrifuge. The above proportions produce a 1:10 dilution. Other dilutions can be readily made. For example, 0.2 ml of sample, 3.0 ml of water, and 0.4 ml of each of the two reagents will give a 1:20 dilution for a microprocedure.

Trichloracetic Acid Filtrate [4]

This is used with the analysis of substances requiring a strong acid filtrate for good recovery. For a 1:10 dilution, one could use 1 volume of sample (blood, serum, or plasma) and 9 ml of $0.30M$ (5 g/dl) trichloracetic acid. The acid may be added with agitation to the sample in a small flask. After further mixing, the flask is allowed to stand for about 10 min and the solution then transferred to filter or tube for centrifugation. For different tests, different proportions of acid may be used, usually in concentrations of 0.3 or $0.6M$.

REAGENT

Dilute solutions of trichloracetic acid are somewhat unstable even when kept in the refrigerator. Preferably a more concentrated solution should be made up which can be diluted as required. The acid can be purchased as a 30 or 40% (w/v) solution. The acid is hydroscopic and very corrosive, making it difficult to weight out properly. A convenient method of preparation of a concentrated solution is to dissolve the entire contents of a previously unopened ¼ lb bottle of the reagent grade acid in exactly 115 ml of water. This will give a $1.8M$ solution [30% (w/v)].

Glucose

Many different methods have been used for the determination of glucose in biological fluids; we will not attempt to give a complete survey of them. The methods may be classified as (1) those based on the reducing power of glucose (usually with heating in alkaline solution), (2) those based on the condensation of glucose with aromatic amines to yield a colored complex, (3) those using enzymatic reactions involving glucose, and (4) miscellaneous methods including condensation with phenols and fluorometric and gas chromatographic methods.

Reduction Methods

In the redox methods one oxidant which has been used for many years and is still widely used is the cupric ion which is reduced to the cuprous state by heating with glucose in alkaline solution. The various methods differ in the exact conditions of alkalinity and heating and in the method by which the amount of copper reduced and hence the amount of glucose present is estimated. The reduction reaction is not stoichiometric and thus the various conditions must be carefully controlled. In the well-known Benedict method for urinary glucose [11], the amount of cuprous oxide formed (red to yellow in color) was estimated visually as a semiquantitative determination of the amount of glucose present. In an older method now rarely used, the alkaline copper solution also contains iodide and iodate. After reduction the solution is acidified, liberating free iodine which reoxidizes any cuprous oxide formed. The excess iodine is then titrated with thiosulfate. The difference in titrations between the sample and a blank containing no glucose is a measure of the amount of glucose present [12–14]. In the common colorimetric methods based on copper reduction, the cuprous oxide formed reduces phosphomolybdate [15–17] or arsenomolybdate [18–20] to give the characteristic blue color. In another variation the cuprous ions formed are determined by the yellow color produced with the reagent neocuproine (2,9-dimethyl-1,10-phenanthroline) [21–24]. This reagent is used in some popular automated methods. The copper reduction methods are not entirely specific even for reducing sugars. Ordinarily glucose is the only reducing sugar present in appreciable quantities in blood. In some tests other sugars may be determined in blood by first destroying the glucose, either by incubation with yeast or by the use of glucose oxidase. The copper reduction methods all require either a protein-free filtrate or dialysis. Even then some nonglucose reducing substances will still remain in the filtrate. A Somogyi filtrate eliminates more of these reducing substances than does the older Folin-Wu filtrate [25]. Substances not removed by the preparation of a filtrate or dialysis include ascorbic acid, uric acid, and creatinine. It has been found that in uremia with high levels of these last two substances, the copper reduction methods give glucose values from 15 to 50 mg/dl higher than the enzymatic methods which are more specific for glucose [26].

Another oxidant still commonly used is ferricyanide. This is reduced to ferrocyanide by heating with glucose in alkaline solution. One of the earlier methods involved an iodine titration similar to that given above for the copper reduction [27–30]. This has been largely replaced by colorimetric methods. Since ferricyanide solutions have a marked yellow color whereas those of ferrocyanide are nearly colorless, the simplest colorimetric method is to measure the decrease in color as the ferricyanide is reduced [31–33]. This has been used extensively in automated methods [34–37]. This has the disadvantage that the range of determinations is limited. If the glucose present is sufficient to reduce all the ferricyanide present, additional glucose will cause no further color change. If the initial concentration of ferricyanide is increased to increase the dynamic range, the sensitivity for lower levels of glucose is decreased too much. In another method the ferricyanide formed reduces phosphomolybdate to give the blue color similar to the reaction with copper [38, 39]. Finally by the addition of ferric chloride the ferrous ions may be released from the ferrocyanide and reacted with one of the newer reagents for ferrous iron such as s-tripyridyl triazine [40] or 5-pyridyl-benzodiazepin-2-one [41] which were originally introduced as reagents for the determination of serum iron. The ferricyanide methods have about the same specificity as the copper reduction methods and are subject to the same interferences.

Another reduction method that may be mentioned very briefly is the reduction of picric acid [42] or dinitrosalicylic acid [43, 44] by heating with glucose in alkaline solution. The reduction of the nitro compounds produces an orange-red color. The use of picric acid was introduced many years ago and was soon abandoned as it was found to be relatively nonspecific. Dinitrosalicylic acid is somewhat more specific but is not as good as the more recent methods and now is used very little. The reduction of some tetrazolium salts to colored compounds by heating with glucose has also been used [45, 46].

Condensation with Amines

A number of amines will give a color reaction when heated with simple sugars in glacial acetic acid or other acid solutions. The reaction apparently involves the condensation of the sugar with the amine to form one of a class of compounds known as Schiff bases. The amine most generally used is o-toluidine, but others that have been used are aniline [47, 48], p-bromaniline [49], benzidine [50], and o-aminobiphenyl [51, 52]. This last-named reagent has been discontinued because of its alleged carcinogenic properties. Benzidine is also said to have carcinogenic properties, and some question has been raised concerning o-toluidine. However, a recent investigation indicates that the hazard in the use of this reagent is very small [53].

The use of o-toluidine was first introduced by Hultman [54], by Hyvarinen and Nikkila [55] who suggested the addition of 0.15% thiourea to stabilize the reagent, and by Dubowski [56]. It is now quite commonly used in both manual and automated methods. Although the original method used a solution of o-toluidine in glacial acetic acid, a number of reagents using lower concentrations of acid or replacing the acetic acid by aqueous solutions of other organic acids [57, 58] have been used. Borate has also been added to some of the formulations to increase the sensitivity [59]. The reaction is not specific for glucose in that many aldohexoses and ketohexoses will react, as will some pentoses. Usually glucose is the only one of the reacting sugars that is present normally in the blood in appreciable quantities. In glacial acetic acid solution, glucose produces a green color with a maximum absorbance at about 630 nm. When read at this wavelength with this reagent, galactose and mannose give about one-third the color and fructose does not react. Some pentoses will give about one-fifth the color of glucose. The colored complexes produced by the different sugars do not have the same absorption spectra. By varying the reagent it is possible to determine glucose and xylose simultaneously by reading at two different wavelengths [60]. Many manual methods use glacial acetic acid as solvent [61, 62] but for automated methods by continuous-flow techniques, other reagent formulations containing less acid may be preferred to avoid too rapid deterioration of the pump tubings [58, 63, 64].

Enzymatic Methods

Two different methods are used for the enzymatic determination of glucose. One is based on the use of glucose oxidase (E.C.1.1.3.4). Glucose oxidase was first studied by Keilin and Hartree in 1948 [65]. The first application of it to the determination of blood glucose was given by Teller [66]. Since then the method has been used extensively and several variations have been published. In the presence of oxygen (air), glucose is oxidized by the enzyme to glucuronic acid with the formation of hydrogen peroxide. In the presence of the enzyme peroxidase (E.C.1.11.1.17), the peroxide will liberate oxygen to oxidize a suitable oxygen-acceptor chromogen to give a colored compound suitable for photometric estimation. A number of chromogens have been used; probably the most common is o-dianisidine [67–70]. This is oxidized to an orange compound which on the addition of strong acid is changed to a stable red color. Another chromogen that has been used particularly for automated methods is tolidine [71–74]. This gives a blue color. Other chromogens that have been suggested though not widely used include reduced 2,6-dichlorophenolindophenol [75], epinephrine [76] which is oxidized to a stable colored substance, starch and iodide in the presence of molybdic acid [77], and xylenol orange plus tetravalent titanium which form a stable red chelate with hydrogen peroxide [78].

`Glucose oxidase is a relatively specific enzyme. The only other substance which reacts at an appreciable rate is 2-deoxy-D-glucose which is not present in appreciable amounts in most biological samples. High concentrations of reducing substances, particularly ascorbic acid but also glutathione, uric acid, and glucuronides, can interfere by competing with the chromogen for the hydrogen peroxide and thus cause low results. Hemoglobin may interfere by causing premature decomposition of the hydrogen peroxide. The test can be run directly on unhemolyzed serum. With gross hemolysis or on whole blood, a Somogyi filtrate must be used. Most test strips for urinary glucose use this reaction with tolidine as the chromogen. Strips have also been made for the determination of glucose in whole blood, which is applied to the strip for a designated time; the strip is then rinsed off, and the color reaction observed. By visual inspection one can estimate whether the blood glucose level is low, normal, high, or very high [79–81]. Using a small reflectance meter* which measures the color of the strip photometrically, one can obtain a better estimate of the glucose level [82–84]. If properly calibrated against known standards, the meter can readily give the glucose level to within 10 mg/dl and thus is quite suitable for emergency use.

The other enzymatic method for glucose involves as the primary reaction

$$\text{glucose} + \text{ATP} \xrightleftharpoons{\text{HK}} \text{glucose 6-phosphate} + \text{ADP}$$

where HK represents the enzyme hexokinase (E.C.2.7.1.1), ATP is adenosine triphosphate, and ADP is adenosine diphosphate. The extent of the reaction may be determined in two different ways. One uses the coupled reaction

$$\text{glucose 6-phosphate} + \text{NAD}^+ \xrightleftharpoons{\text{G-6-PDH}} \text{6-phosphogluconic acid} + \text{NADH} + \text{H}^+$$

where G-6-PD is the enzyme glucose 6-phosphate dehydrogenase (E.C.1.1.1.49) and NAD and NADH are the oxidized and reduced forms of niacin adenine dinucleotide [85–87]. The reaction is usually followed by measuring the change in absorbance at 340 nm. At this wavelength NADH has a very marked ab-

* Ames Co., Div. of Miles Laboratories, Elkhart, Ind. 46514.

sorbance and NAD very slight absorbance. The method involves the use of two enzymes together with NAD, and the mixture is not very stable in solution. It is most commonly used with test kits which furnish the reagents in lyophilized or tablet form for reconstitution just before use. It has been used for the determination of small amounts of glucose in normal urine where it is less subject to interference from other urinary constituents than is the glucose oxidase method [88, 89]. The method is very specific for glucose.

The other method for measuring the reaction involves the following steps:

$$ADP + PEP \xrightleftharpoons{PK} ATP + pyruvate$$

where PK is the enzyme pyruvate kinase (E.C.2.7.1.40) and PEP is phosphoenolpyruvate. The next step is:

$$pyruvate + NADH + H^+ \xrightleftharpoons{LDH} lactate + NAD^+$$

where the reaction is catalyzed by the enzyme LDH, lactate dehydrogenase (E.C.1.1.1.27). Again the reaction is followed by the change in absorbance at 340 nm.

In a recent method the above reactions have been used with the NADH formed coupled to reduce a tetrazolium salt with the production of a colored compound [90, 91]. This should make the advantages of the hexokinase method available to laboratories not having an ultraviolet spectrophotometer.

Miscellaneous Glucose Determinations

Glucose has also been determined by heating with anthrone [92, 93] or thymol [94] in strong sulfuric acid, but these methods are now rarely used. Glucose has also been determined fluorometrically by reaction with 5-hydroxy-1-tetralone in sulfuric acid solution [95]. The determination can be readily performed on 10 μl of serum. Glucose and other sugars have been determined by chromatography [96]. Although probably not useful as a routine method, it may be advantageous in the investigation of inborn errors of carbohydrate metabolism since any different sugars present may be accurately quantitated in one run.

Glucose may also be determined by a coulometric method. Strictly speaking, this method should be listed under the glucose oxidase methods since it uses this enzyme. The method uses a coulometric oxygen electrode similar to that used for the measurement of blood Po_2. The electrode is used to measure the changes in oxygen tension in a closed system. As the glucose is oxidized by the action of the enzyme, the amount of dissolved oxygen in the solution decreases. The amount of change in Po_2 under carefully controlled conditions is a measure of the amount of glucose present in the sample [97–99]. A commercial instrument based on this principle is available [100].†

We present in detail three manual methods for the determination of blood glucose. The preferred method is that using the o-toluidine reagent. This can be run directly on small amounts of serum, is rapid, and gives results close to those obtained by the glucose oxidase methods. It now appears to be the most widely used manual method.

If an enzymatic method is desired, the glucose oxidase method is simplest for routine use since the enzyme reagents can be prepared in a fairly stable form. It can be run directly on serum. The only disadvantage is that hemolysis interferes more than with the o-toluidine method. When noticeable hemolysis is present, a Somogyi filtrate must be used.

We also present a copper reduction method, that of Nelson and Somogyi. Although this is rather lengthy for routine use, it is included since it can be used for other reducing sugars as well as glucose.

Glucose by o-*Toluidine Method Using Acetic Acid* [61, 62]

REAGENTS

1. o-Toluidine reagent. Dissolve 1.5 g of thiourea in 940 ml of glacial acetic acid. Add 60 ml of pure o-toluidine and mix well. Store in a brown bottle at room temperature. The reagent should be aged overnight before using. It is stable for at least several months. It should be replaced if it gives a marked increase in the blank reading or becomes much darker in color.

2. Trichloracetic acid, $0.18M$ (3% w/v). Prepare fresh monthly by dilution of a $1.8M$ solution (see Protein-free Filtrates earlier in this chapter). Store in the refrigerator.

3. Glucose standards.

   A. Stock standard, 10 mg/ml. Dissolve exactly 1.00 g of pure glucose in benzoic acid solution (2 g/liter) and dilute to 100 ml in a volumetric flask. This solution is stable at room temperature. In the refrigerator, some of the benzoic acid tends to crystallize.

   B. Working standards. Dilute 5, 10, and 15 ml of the stock standard to 50 ml with the benzoic acid solution. These solutions will contain, respectively, 100, 200, and 250 mg of glucose/dl. For the indirect method using a trichloracetic

† Beckman Glucose Analyzer, Beckman Instruments, Fullerton, Calif. 92634.

acid filtrate, the above standards are further diluted 1:10 with the 0.18M trichloracetic acid.

PROCEDURE—DIRECT METHOD

The reaction is preferably carried out in screw-capped tubes with Teflon-lined caps. To a series of tubes add exactly 3 ml of the toluidine reagent. Using accurate micropipets add 0.05 ml of standards and samples to separate tubes. Reserve one tube for a blank (0.05 ml of water is added). Mix the contents of each tube, cap securely, and heat at 100°C for 12 min. A boiling water bath can be used, but a heating block at this temperature is much more convenient. After heating, cool in running tap water. Read standards and samples against blank at 630 nm.

CALCULATION

$$\frac{\text{absorbance of sample}}{\text{absorbance of standard}} \times \text{conc. of standard} = \text{conc. of sample}$$

COMMENTS

1. If the spectrophotometer has been checked in the procedure and found to be linear up to 300 mg/dl of glucose (as should be the case), only one standard need be run routinely. If the sample reading is too high for convenient reading, the solution may be diluted 1:1 with glacial acetic acid and the reading obtained multiplied by 2. For accurate work the determination should be repeated using a smaller aliquot of serum or a dilution of the serum.

2. Depending upon the photometer used and the size of cuvettes available, slightly different proportions of reagent and sample may be used to give a better range of readings. The absorbance of the 100 mg/dl standard should preferably be between 0.20 and 0.25.

3. For ordinary serum the direct method is satisfactory; for grossly hemolyzed, very hyperlipemic, or icteric serums, one may prefer to use a protein-free filtrate to eliminate possible color interference. The filtrate can also be made from whole blood although this is rarely done.

PROCEDURE WITH TRICHLORACETIC ACID FILTRATE

To 4.5 ml of the 0.18M trichloracetic acid solution add exactly 0.5 ml of serum and mix well. Allow to stand for a few minutes and then filter or centrifuge. To 3.5 ml of the toluidine reagent, add 0.5 ml of filtrate. Also add 0.5 ml of diluted standards (in trichloracetic acid, see above) to separate tubes containing 3.5 ml of reagent and use 0.5 ml of the trichloracetic acid solution as a blank. Cap the tubes, mix well, and heat as in the direct method. The reading and calculations are exactly the same since the serum and working standards are diluted similarly.

COMMENTS

Here, also, slightly different proportions may be used if desired, except that water content of the final mixture should not be more than about 10 percent since a greater amount will reduce the amount of color produced.

Glucose by o-Toluidine Method Using Ethylene Glycol

The acetic acid solution used in the previous procedure is corrosive and unpleasant to work with. We present a method using a less corrosive solution of citric acid and o-toluidine in an ethylene glycol solution [63]. It is based on an automated procedure which can be used as a manual method.

REAGENTS

1. o-toluidine solution. Dissolve 1.5 g of thiourea in about 700 ml of reagent grade ethylene glycol in a 1 liter volumetric flask. Add 50 g of citric acid and dissolve. Add 60 ml of o-toluidine and dilute to 1 liter with ethylene glycol. This solution is stable for at least 6 months when kept in a brown bottle at room temperature.

2. Glucose standards. Same as for the previous method.

PROCEDURE

To separate tubes containing 5 ml of the reagent, add 0.02 ml of serum samples or standards and mix well, reserving one tube as blank (it is not necessary to add 0.02 ml of water). Heat all tubes for 10 min in a boiling water bath, cool, and read standards and samples against blank at 630 nm. The calculations are the same as for the previous method.

COMMENTS

This reagent gives considerably more color for a given amount of glucose than does the acetic acid reagent. As with the previous method it may be necessary to vary the proportion of reagent and sample somewhat to give a suitable absorbance (about 0.2 absorbance for a 100 mg/dl standard). For very hyperlipemic serums this reagent will give results about 10 mg/dl higher than the acetic acid reagent. High levels of bilirubin will also result in some interference, each milligram per deciliter of bilirubin giving additional color equivalent to about 1 mg/dl of glucose. Marked hemolysis will also interfere. In such cases it is best to use a trichloracetic acid filtrate as in the acetic acid procedure, adding

0.5 ml of a 1:10 filtrate to 6 ml of reagent, 0.5 ml of standards diluted 1:10 with the trichloracetic acid, and 0.5 ml of trichloracetic acid solution to blank (see use of trichloracetic acid filtrate with the acetic acid reagent given earlier). The addition of the trichloracetic acid solution decreases the amount of color formed by about 30 percent. Again one may require slightly different proportions of sample filtrate and reagent, but the amount of filtrate added should not be greater than 10 percent of the volume of reagent used.

There are minor disadvantages of this method due to the higher viscosity of the reagent. Difficulty may be encountered in the use of rinse-out micropipets for samples and standards and in the use of automatic dispensors for the reagents. Also extra care must be taken to insure complete mixing of added sample, particularly if the trichloracetic acid filtrate is used. For the filtrate, better mixing may be done by adding the filtrate to the tubes before the reagent. With the filtrate, bubbles may form in the solution on heating and the transfer to cuvettes must be done carefully to avoid entrapment of bubbles. Because of the high viscosity of the solution, these bubbles tend to rise very slowly and cling to the sides of the cuvettes if they are not perfectly clean.

Glucose by Glucose Oxidase Method [67–70]

REAGENTS

1. Phosphate buffer, $0.1M$, pH 7.0. Dissolve 7.80 g of Na_2HPO_4 and 5.30 g of KH_2PO_4 in about 950 ml of water. Measure the pH and adjust to 7.0 if necessary using small amounts of acid or base; then dilute to 1 liter. This solution should be stored in the refrigerator and discarded if mold growth appears.

2. Buffered peroxidase solution. Mix together 125 ml of the phosphate buffer, 175 ml of water, and 200 ml of glycerin. Dissolve 10 mg of peroxidase (Sigma, Type II)‡ in this solution. Dissolve 100 mg of o-dianisidine in 10 ml of methyl alcohol and add to the above solution; mix well. Due to the added glycerol this solution may be stored at $-20°C$, for added stability. Some brands of reagent grade glycerol were found to give a considerable red color on the addition of the dianisidine, and thus a high blank. Merck's reagent grade was found to be satisfactory.

3. Glucose oxidase solution. Mix 20 ml of glycerol and 30 ml of water. Dissolve 500 mg of glucose oxidase (Sigma, Type II) in this solvent. This solution may also be stored at $-20°C$ for added stability.

4. Sulfuric acid, $5M$. Cautiously add 280 ml of concentrated sulfuric acid to 700 ml of water. Mix, cool to room temperature, and dilute to 1 liter.

‡ Sigma Chemical Co., St. Louis, Mo. 63118.

5. Zinc sulfate and barium hydroxide if required for the preparation of a protein-free filtrate (see section on Protein-free Filtrates earlier in this chapter).

6. Glucose standards. Same stock and working standards as used in the preceding *o*-toluidine method.

PROCEDURE—DIRECT METHOD

Add 4.5 ml of the buffered peroxidase solution to a series of test tubes. Warm the tubes to 37°C. Add 0.02 ml of serum, plasma, or other fluid, or standards. Reserve one tube as blank (add 0.02 ml of water). To each tube add 0.5 ml of the glucose oxidase solution, mix, and incubate all tubes at 37°C for 30 min.

Add 3 ml of the 5*M* sulfuric acid to each tube and mix. Read standards and samples against blank at 530 nm.

CALCULATIONS

The standards and samples are treated similarly, hence

$$\frac{\text{absorbance of sample}}{\text{absorbance of standard}} \times \text{conc. of standard} = \text{conc. of sample}$$

COMMENTS

If the method has been checked and found to be linear to over 300 mg/dl, only one standard need be run with each series of samples. If the curve is not linear, it is preferable to run several standards and for the calculation of each sample use the standard having an absorbance closest to the sample.

PROCEDURE WITH PROTEIN-FREE FILTRATE

Since hemoglobin interferes, for serum or plasma with noticeable hemolysis or if it is desired to use whole blood, a Somogyi protein-free filtrate must be prepared (see earlier in this chapter). Make a 1:20 filtrate and use 0.5 ml of the filtrate (or supernatant) in place of the serum or other fluid as given above. It is best to dilute the standards with the precipitating reagents similarly to the dilution of serum or other samples. In this case the calculations are exactly the same as for the direct method. For a micromethod the preparation of the filtrate may be scaled down. With the use of 0.1 ml of serum, 1.5 ml of water, and 0.2 ml of each of the two precipitating reagents, one can readily obtain 0.5 ml of supernatant.

Glucose by Method of Nelson and Somogyi

Although this method is somewhat longer than the others given here and is not used as much as formerly, it is presented because it can also be used for other reducing sugars besides glucose [18, 19]. Since the reactions involved in the copper reduction are not exactly stoichiometric, the conditions of heating etc., must be closely adhered to for consistent results.

REAGENTS

1. Alkaline copper tartrate.   Dissolve 24 g of anhydrous sodium carbonate (Na_2CO_3) and 12 g of sodium potassium tartrate ($NaKC_4H_4O_6 \cdot 4H_2O$) in about 200 ml of water. Dissolve 4 g of copper sulfate ($CuSO_4 \cdot 5H_2O$) in 40 ml of water and add with stirring to the other solution. When completely dissolved, add 16 g of sodium bicarbonate ($NaHCO_3$) and stir until dissolved. Dissolve 180 g of anhydrous sodium sulfate (Na_2SO_4) in 600 ml of water. Heat to boiling to drive out dissolved air, cool to room temperature, and add to the copper solution. Mix well and dilute to 1 liter.

2. Arsenomolybdate solution.   Dissolve 50 g of ammonium molybdate [$(NH_4)_6Mo_7O_{24} \cdot 4H_2O$] in about 900 ml of distilled water. Add slowly with swirling 42 ml of concentrated sulfuric acid. Dissolve 6 g of disodium ortho-arsenate ($Na_2HAsO_4 \cdot 7H_2O$) in 50 ml of water and add to the above solution. Mix well and place in an incubator at 37°C for 2 days. This solution is stable if kept in a brown bottle with a glass stopper and protected from contact with organic matter.

3. Standards.

A. Stock standards.   Same as for previous methods, 10 mg/ml glucose.

B. Working standards.   Dilute 1, 2, and 3 ml of the stock to 100 ml with 0.1% benzoic acid. These contain 10, 20, and 30 mg/dl glucose. When compared with 1:10 dilutions of blood or serum, the standards correspond to 100, 200, and 300 mg/dl glucose in samples. These standards are not very stable and should be made up fresh weekly.

PROCEDURE

Prepare a 1:10 Somogyi filtrate as outlined earlier. Add 1 ml of filtrate to a Folin-Wu sugar tube. To other tubes add 1-ml aliquots of diluted standards or 1 ml of water as blank. To each tube add 1 ml of alkaline copper tartrate and mix well. Heat all tubes in a vigorously boiling water bath for 15 min, remove, and cool in tap water. To each tube add 1 ml of the arsenomolybdate solution, mix well, and dilute to 25 ml with distilled water. Mix each tube by inversion. Read standards and samples against blank at 520 nm. (The maximum absorbance is at much longer wavelengths, but it has been found that more consistent results are obtained by reading at the shorter wavelength.)

$$\frac{\text{absorbance of sample}}{\text{absorbance of standard}} \times 10 \times \text{conc. of standard} = \text{conc. of sample}$$

where the concentration of the standard is taken as that of the final dilution (i.e., 10, 20, or 30 mg/dl).

COMMENTS

1. The procedure may be scaled down as follows. Add 0.2 ml of sample to 3 ml of water. Add 0.4 ml of barium hydroxide solution, mix, then add 0.4 ml of zinc sulfate. Mix well and allow to stand for a few minutes. Centrifuge strongly to obtain a clear supernatant. Using tubes graduated at 12.5 ml, add 1 ml of supernatant to tube. To other tubes add 1 ml of water as blank or 0.5 ml of diluted standard used above plus 0.5 ml of water. To each tube add 1 ml of copper solution, mix well, and heat for 15 min as above. Cool, add 1 ml of arsenomolybdate, mix, and dilute to 12.5 ml with water. Read at 520 nm as previously. Since 1 ml of a 1:20 filtrate is compared with 0.5 ml of standard, the calculations are the same as given above.

2. If it is desired to use the method for a different reducing sugar, the standard would be made up using the sugar to be determined and not glucose. Otherwise the procedure would be the same, though a longer heating time may be advisable.

Normal Values and Interpretation of Results

The normal fasting levels for glucose in venous serum or plasma by the methods given here are from 65 to 110 mg/dl. Many values found in the older literature are for whole blood by the older Folin-Wu method with a range of 80 to 125 mg/dl [101–104]. Capillary blood is similar to arterial blood and contains only a few more milligrams of glucose per deciliter than venous blood in the fasting state. Most methods now use serum or plasma but for comparison the following approximate equation may be used: $P = 1.15 W + 6.6$, where P is the glucose concentration in plasma (or serum) and W the glucose concentration in whole blood [105].

High blood sugar levels (it is still customary to refer to blood sugar levels although the determination is actually made in serum) are found chiefly in diabetes mellitus (up to 500 mg/dl or more, depending upon the severity of the condition). Except for diabetes, the fasting level rarely exceeds 120 mg/dl. Small increases may be found in hyperactivity of the pituitary, thyroid, and adrenal glands. In pancreatitis and pancreatic carcinoma there may be a small increase in blood sugar but the level does not exceed 150 mg/dl except in advanced cases. Moderate increases may also be found in infectious diseases, and in some intracranial diseases such as meningitis, encephalitis, tumors, and hemorrhage. In severe cases the final level may be as high as 200 mg/dl.

Decreased blood sugar level (hypoglycemia) occurs most frequently as the result of insulin overdosage in the treatment of diabetes. The fasting level may also be reduced below normal in hypothyroidism, hypopituitarism, and hypo-adrenalism, with values as low as 20 mg/dl in severe cases of the last two conditions. Low levels may also be found in glycogen storage disease.

Two-Hour Postprandial Glucose

This test is very commonly used by physicians for routine screening for diabetes [106]. Sindoni and other investigators have found the 2-hr postprandial glucose to be a reliable method for the detection of diabetes. Using this method, they demonstrated that the blood glucose concentration was rarely more than slightly elevated in normal individuals, whereas it was significantly increased in diabetics. The test gives results that have correlated well with other standard tests and has the advantage of requiring only one blood specimen.

If possible, though not necessary, the patient should be placed on a high carbohydrate diet for a few days prior to the test. Although a variety of test meals have been used, it is recommended that 100 g of glucose in solution be used routinely. This will make the carbohydrate meal more uniform and the timing more accurate. Two hours after the ingestion of carbohydrate, a blood specimen is drawn and analyzed for glucose by one of the methods described previously in this section.

Interpretation of the results will depend on the method used for analysis and whether serum or whole blood is used. In general, a normal response will have a serum (or plasma) level of less than 125 mg/dl. An abnormal result will have higher values. This figure applies only to adults under age 50; for older individuals it is higher. Factors such as previous diet, drugs, and endocrine disease, discussed in the following sections on glucose tolerance, must be considered in interpreting the results.

Glucose Tolerance Test

In the diagnosis of diabetes mellitus and other disorders of carbohydrate metabolism, a glucose tolerance test is often done. This usually consists in obtaining a fasting blood specimen (after an overnight fast) and then giving orally a definite amount of glucose dissolved in water. Blood samples are taken at definite intervals after ingestion of the glucose. The usual times are 30, 60, 120, and 180 min after ingestion, although some physicians prefer further samples. There is also some variation in the amount of glucose given. For an average adult a set dose of 50 or 75 g is often given, though a dose of 1 g/kg of body weight is recommended. The many details and precautions for the preferred method of carrying out the test may be found elsewhere [106a, 106b]. The samples may be analyzed by any of the methods given here or by automated methods. Previously most tolerance tests were carried out with the analysis of whole blood samples but now most analyses are on serum with methods giving values close to true glucose. A normal glucose tolerance test would give approximately the following levels of serum glucose: fasting, 80 mg/dl; ½ hr, 155 mg/dl; 1 hr, 165 mg/dl; 2 hr, 175 mg/dl; and 3 hr, 80 mg/dl. The main use of the glucose tolerance test is in the diagnosis of diabetes mellitus although some other conditions also give abnormal tolerance curves.

Intravenous Glucose Tolerance Test

This method [107] is used in cases of increased glucose tolerance caused by hypoactivity of the endocrine glands (thyroid, adrenals, and pituitary) or by impaired absorption of carbohydrates (idiopathic steatorrhea, celiac disease, and sprue). All these conditions usually produce flat oral glucose tolerance curves.

A fasting blood specimen is drawn and 0.5 g of glucose/kg of body weight is given intravenously. A 50% sterile glucose solution is used, taking 5 min for its administration. Blood and urine specimens are obtained immediately after administration of glucose and thereafter at the end of 30, 60, 90, and 120 min.

Blood and urine samples are analyzed for glucose. Normally the blood glucose will be increased immediately after infusion is completed, reaching a maximum of about 250 mg/dl. The urine may be positive for glucose. The 30-min specimen shows a marked drop in glucose and the 60-min specimen will show a return to the fasting level (±10 mg/dl).

In diabetes mellitus, even in mild cases, the glucose level remains elevated for 2 hr or longer; in no specimen does the blood glucose level fall below the fasting level. In liver insufficiency, the return to the fasting level may be somewhat delayed.

Insulin Tolerance Test

This test [108] is mainly used in evaluating patients with endocrine disorders or who show insulin resistance.

When insulin is administered to a normal individual in the postabsorptive state, it causes a prompt decrease of the blood glucose and then a gradual return to the original level. This response may be used to determine an individual's sensitivity to insulin (ability to store glycogen) and ability to recover from the induced hypoglycemia.

Before performing the test, it is preferable that the patient be put on a diet containing about 300 g of carbohydrate daily for 2 or 3 days.

A fasting blood specimen is obtained, then 0.1 unit of insulin/kg of body weight is injected intravenously. Further blood specimens are obtained at 30, 60, and 120 min. A physician should be available during the performance of this test to administer glucose promptly if early clinical signs of hypoglycemia develop (hunger, sweating, nervousness, tremulousness). At the end of the test, carbohydrates (glucose solution or fruit juice) should be given to the patient.

A normal response will show a fall to about 50 percent of the fasting level in 30 min, followed by a steady increase and a return to the fasting level in 90 to 120 min. The duration of the hypoglycemia is more significant than the degree.

Two types of abnormal response have been recognized: the insulin-resistant type, in which there may be only a slight or delayed fall in blood glucose, and the hypoglycemia unresponsiveness type, in which the blood glucose falls as in the normal response but shows a delayed rise following the hypoglycemia.

Insulin resistance may occur in hyperfunction of the adrenal cortex or the anterior pituitary, in some cases of diabetes mellitus, and sometimes in the early stages of rheumatoid conditions.

Hypoglycemia unresponsiveness is seen particularly in hyperinsulinism, Addison's disease, hypofunction of the anterior pituitary (Simmonds' disease, pituitary myxedema, pituitary dwarfism), in some cases of hypothyroidism, and in some cases of glycogen storage disease (Von Gierke's disease).

Epinephrine Tolerance Test

Following the administration of epinephrine, there is an increase in blood glucose due to an acceleration in glycogenolysis. This response is an index of the quantity and availability of liver glycogen for the maintenance of normal blood glucose level and is used to evaluate glycogen storage disease [109].

A fasting blood glucose specimen is obtained and then 1 ml of a 1:1000 solution of epinephrine hydrochloride is injected intramuscularly. Obtain blood specimens at 30, 60, and 120 min and analyze for glucose.

In normal individuals the blood glucose will increase 30 mg/dl in 30 min, 35 to 45 mg/dl in 60 min, and return to fasting level in 2 hr.

An abnormal response occurs when the stored glycogen is depleted (hepato-cellular damage, including fatty liver and cirrhosis), in glycogen storage disease (Von Gierke's) (in which the glycogen is not readily available), and in hypoglycemia that does not respond normally to the insulin tolerance test (Addison's disease, pituitary cachexia, hyperinsulinism).

Cortisone-stressed Glucose Tolerance Test

This test [110, 111] is based on the fact that although a large dose of ACTH or other suitable corticosteroid will produce a high glucose tolerance curve and glycosuria in normal individuals, smaller doses will have this effect only in prediabetics. Fajans and Conn have utilized this fact for detecting latent diabetics. They used two oral doses of 50 mg of cortisone (62.5 mg if patient's weight exceeded 160 lb), the first dose 8½ hr before an oral glucose tolerance test and the second dose 2 hr before the test.

A modification of this test has been used in which glucose is measured in urine collected from 10 PM to 6 AM after administration on the previous day (12 M, 4 PM, and 8 PM) of three doses of 25 mg of prednisone.

Tolbutamide Diagnostic Test

This test [112, 113] is based on the response of an individual to the intra-venous administration of tolbutamide (Orinase), a compound that stimulates the pancreas to produce insulin. The same precautions described for the insulin tolerance test should be observed in the performance of this test as the response is similar.

A fasting blood glucose specimen is obtained and 20 ml of Orinase solution injected intravenously at a constant rate over 2 to 3 min. Blood specimens are drawn exactly 20 and 30 min later, timing from the midpoint of the injection. After drawing the 30-min specimen, the test is terminated by feeding a high carbohydrate meal. Blood specimens are analyzed for glucose.

If the blood glucose level at 20 min is 90 percent or more of the fasting level, definite diagnosis of diabetes is established. If it is in the range of 85 to 89 percent of the fasting level, diabetes is probable. The range of 75 to 84 percent represents borderline cases and under 75 percent is considered a normal response. The 30-min specimen is of value in confirming the diagnosis: If it is more than 76 percent of the fasting level, diabetes is almost certain. Lower values have less negative significance.

This test has proved of value in the differential diagnosis of spontaneous hypoglycemia caused by insulomas. In this condition the fall in blood glucose is greater than for normal individuals and the hypoglycemia persists for up to 3 hr.

Effect of Drugs on Blood Glucose Levels

Apart from drugs normally given for their effect on glucose metabolism, a number of other drugs can influence the fasting level of glucose and the glucose tolerance [114–116]. Normal doses of salicylates (aspirin) have little effect on glucose level, but salicylate intoxication may result in hyperglycemia and glucosuria. In infants and young children with carbohydrate depletion, salicylate overdosage can cause the opposite effect with marked hypoglycemia.

Estrogen therapy or the administration of oral contraceptives can result in impaired glucose tolerance. The fasting level may be close to normal, but the levels after glucose administration may be higher than normal. The administration of various thiazide diuretics may result in hyperglycemia, particularly in elderly patients with a tendency toward diabetes. Overdoses of diphenylhydantoin can result in hyperglycemia and glucosuria. A number of other drugs, including adrenogenic agents, corticostereoids, acetazolamide, diazoxide, and reserpine, have been reported to cause hyperglycemia. The effect of drugs must be kept in mind when evaluating abnormal glucose levels in blood. The compilations of drug effects that are cited above do not always distinguish between an effect that has been found in only a few cases and those that are much more common.

When glucose is determined by one of the copper reduction methods such as the Somogyi-Nelson or SMA§ 12/30 or 12/60, high levels of ascorbic acid or dextrans may result in erroneously high values due to the reducing effect of these substances on the copper reagent. A similar effect has been noted in uremia due to the reducing effect of the abnormally high levels of uric acid, creatinine, and possibly other substances.

Nonglucose Sugars

Several methods are available for the determination of nonglucose sugars [117, 118]. When the determination is made in blood or serum with a nonspecific method (such as copper reduction), the effect of the glucose present must be eliminated. In normal urine, which contains only small amounts of glucose, simpler methods may be used. The first method presented here uses glucose oxidase to eliminate any glucose present, and then the remaining carbohydrate can be determined by a copper reduction method.

§ Technicon Instruments Corp., Tarrytown, N.Y. 10591.

Nonglucose Reducing Sugars by Glucose Oxidase Method

REAGENTS

1. Potassium dihydrogen phosphate, $0.26M$. Dissolve 3.5 g of anhydrous potassium dihydrogen phosphate (KH_2PO_4) in water to make 100 ml.

2. Glucose oxidase. Weigh out several 10-mg portions of glucose oxidase (Sigma, Type II†) into separate 10 ml graduated cylinders. Stopper tightly and store in a refrigerator. On the day of use dissolve the contents of one cylinder in 10 ml of the phosphate buffer.

3. Standards. Prepare stock and working standards exactly the same as given for glucose, using a pure sample of the desired sugar instead of glucose. Usually one working standard, 100 mg/dl, is sufficient.

4. Precipitating and color reagents. Same as given earlier for the Somogyi-Nelson method for glucose.

PROCEDURE

Add 0.1 ml of blood or other sample to 0.8 ml of water in a plastic centrifuge tube. Add 0.2 ml of the glucose oxidase solution, mix, stopper lightly, and incubate at 37°C for 90 min. Remove stopper and agitate briefly several different times during the incubation period. Also set up similar tubes containing 0.1 ml of water as blank and 0.1 ml of standard (100 mg/dl). Add 0.2 ml of glucose oxidase and incubate similarly to the sample except that agitation is not necessary during incubation. After incubation, add 0.5 ml of barium hydroxide solution and mix. Add 0.4 ml of the zinc sulfate solution, and mix thoroughly. Centrifuge at high speed to obtain a clear supernatant. If supernatant is not clear, decant and recentrifuge. Add 1.0 ml of the supernatant from each tube to separate Folin-Wu sugar tubes. Add 2 ml of alkaline copper tartrate, mix, and heat in boiling water bath for 20 min. Cool, add 2 ml of arsenomolybdate, and mix. Dilute each tube to 25 ml with water and read standard and samples against blank at 530 nm.

CALCULATION

The standard and samples are treated similarly, hence

$$\frac{\text{absorbance of sample}}{\text{absorbance of standard}} \times \text{conc. of standard} = \text{conc. of sample}$$

† Sigma Chemical Co., St. Louis, Mo. 63118.

Galactose by Galactose Oxidase Method

The principles and procedure are very similar to the previously described determination of glucose by glucose oxidase [119–121]. The galactose oxidase available may not be entirely free of other oxidases; in a galactose tolerance test this is
compensated for by the use of a blank specimen taken before the sugar is
administered.

REAGENTS

1. Phosphate buffer, pH 7.0, 0.02M.   Dissolve 1.74 g of Na_2HPO_4 and 1.06 g
of KH_2PO_4 in about 950 ml of water. Check the pH with a glass electrode, adjust
pH to 7.0 if necessary with small amounts of acid or base, and dilute to 1 liter.

2. Galactose oxidase.   Triturate about 3 mg of dry galactose oxidase (Sigma,
Type I,‡ equivalent to 30 units of enzyme) in a mortar with 50 ml of the phosphate buffer. Filter through Whatman No. 1 paper. This solution is stable for
only a few days in the refrigerator.

3. Color reagent.   Dissolve 10 mg of peroxidase (Sigma, Type II‡) in 500
ml of the phosphate buffer. Dissolve 50 mg of *o*-dianisidine in 5 ml of methyl
alcohol and add to the peroxidase solution. Store in the refrigerator. Discard if
it turns dark or gives a high blank.

4. Sulfuric acid, 5M.   Cautiously add 280 ml of concentrated sulfuric acid to
700 ml of distilled water.

5. Zinc sulfate and barium hydroxide for preparation of Somogyi filtrate.
Same as given earlier.

6. Galactose standards.

A. Stock standard.   Dissolve 1.00 g of pure D(+)-galactose in water to
make 100 ml. This contains 10 mg/ml of the sugar.

B. Working standard.   Dilute 1 and 2 ml of the stock solution to 100 ml
with water. These standards will then contain 10 and 20 mg/dl which are
equivalent to 100 and 200 mg/dl when treated similarly to a 1:10 filtrate.

PROCEDURE

Prepare a 1:10 Somogyi filtrate as outlined earlier. To each of several tubes
add 1 ml of the color reagent and 1 ml of the galactose oxidase solution. To
separate tubes containing the mixed reagent, add 1 ml of the sample filtrate, 1
ml of the diluted standard, or 1 ml of water (as blank), and mix. Incubate for
30 min at 37°C. After incubation, add 2.5 ml of the sulfuric acid solution to each
tube and mix well. Read standards and samples against blank at 530 nm.

‡ Sigma Chemical Co., St. Louis, Mo. 63118.

CALCULATION

$$\frac{\text{absorbance of sample}}{\text{absorbance of standard}} \times \text{conc. of standard} = \text{conc. of sample}$$

COMMENTS

1. Ascorbic acid in high concentrations may interfere, giving high values. This is not of importance in a galactose tolerance test since a blood sample is taken before the galactose is ingested and detects any apparent fasting galactose. The basis for the test is the increase in galactose level.

2. A simple, though not too accurate, estimate of the rise in galactose level after ingestion of the sugar may be made by determining the reducing sugars in the blood samples by the regular Somogyi method for glucose (except using a galactose standard and 20 min heating time). The glucose present will give a fasting level of reducing sugars, but this is assumed to be relatively constant so that any increase in level is due to galactose. The increases in galactose levels may be accurate enough for many clinical purposes.

Pentoses by o-Toluidine Method

Pentoses may also be determined with the *o*-toluidine reagent [122, 123]. These sugars give a color having a maximum absorbance at about 480 nm (glucose gives a color with absorbance maximum at about 630 nm). In the absence of glucose, xylose can be determined with the same reagent as given earlier for glucose; reading is done at 480 nm. In the determination of the xylose in urine for the xylose absorption test, the concentration of xylose will generally be so much greater than that of any glucose present that the interference by the latter is negligible. For this determination one can use 0.1 ml of diluted urine and 6 ml of the toluidine reagent, proceeding as in the above direct method for glucose but using a corresponding standard of 100 mg/dl of xylose (prepared exactly the same way as the glucose standard). If the 5-hr collection of urine is diluted to 1 liter and 0.1 ml of this dilution compared with an equal volume of 100 mg/dl xylose standard, the standard will correspond to a total excretion of 1 g of xylose or 20 percent of a 5-g dose. If an oral dose of 25 g of the sugar is given, a further 1:5 dilution of the urine would be made before analysis to yield the same comparison.

Glucose and xylose can be determined in the presence of each other by reading at the two wavelengths of 630 and 480 nm. The procedure is relatively simple. Separate standards of glucose and xylose are used (usually at concentrations of 100 mg/dl for glucose and 50 mg/dl for xylose, depending upon the blood levels expected).

To a number of tubes add 5 ml of the o-toluidine reagent. To separate tubes containing the reagent, add 0.1 ml of sample, glucose standard, xylose standard, and water (blank). Cap tubes, mix, and heat in bath at 100°C for 12 min as in the glucose method given earlier. After cooling, read all samples and standards against blank at both wavelengths of 630 and 480 nm.

CALCULATION

First compute the K values as follows:

K_1 = absorbance of glucose standard at 630 nm/conc. of glucose standard in mg/dl

K_2 = absorbance of glucose standard at 480 nm/conc. of glucose standard in mg/dl

K_3 = absorbance of xylose standard at 630 nm/conc. of xylose standard in mg/dl

K_4 = absorbance of xylose standard at 480 nm/conc. of xylose standard in mg/dl

Then, if A_1 = absorbance of sample at 630 nm and A_2 = absorbance of sample at 480 nm:

$$\text{mg/dl glucose in sample} = \frac{K_4 A_1 - K_3 A_2}{K_1 K_4 - K_2 K_3}$$

$$\text{mg/dl xylose in sample} = \frac{K_1 A_2 - K_2 A_1}{K_1 K_4 - K_2 K_3}$$

It is important that the same volumes of samples and of both standards be used for the calculations to give the correct results.

Urinary Sugars by Thin Layer Chromatography

Sometimes it may be necessary to identify the exact sugar found in urine. Although the sugar most commonly found in urine is, of course, glucose, in some metabolic disorders other sugars may be present. The simplest way of identification is by thin layer chromatography. A number of different thin layer plates are commercially available. Gelman type SA impregnated glass fiber sheets§ have given good results. The sheets are first thoroughly sprayed with a potassium phosphate solution (13.6 g of KH_2PO_4/liter) until well moistened and then dried at 110°C for 1 hr. These sheets are rather fragile and should be handled carefully.

§ Gelman Instrument Co., Ann Arbor, Mich. 48106.

REAGENTS

1. Developing solution. Just before use mix 30 volumes of chloroform, 35 volumes of glacial acetic acid, and 5 volumes of water.

2. Color reagent. Dissolve 1 g of diphenylamine and 1 ml of aniline in 100 ml of acetone. Just prior to use add 15 ml of 85% phosphoric acid. A precipitate is formed at first but this should dissolve on mixing.

3. Standards. Standards containing 1 mg/ml of the various sugars are made up. It is convenient to make up two standards, one containing 1 mg/ml each of lactose, galactose, and fructose, and a second containing the same quantities of sucrose, glucose, and rhamnose (or other pentose). These should be made up in 0.1 g/dl benzoic acid solution.

PROCEDURE

Ordinarily the urine can be used directly without treatment. A morning specimen is preferable unless it is suspected that the sugar comes from a dietary source; in that case a specimen taken several hours after a meal is better. If the urine is cloudy or contains precipitated material, it should be centrifuged or filtered and the clear supernatant used.

Three lightly penciled dots are made at equal distances from each other across a 5-cm strip about 1.5 cm from one end. The urine sample is applied to the middle spot. About 5 to 25 μl of urine is used, depending upon the concentration. The urine should be applied in small portions to keep the spot as small as possible. The two separate standards are applied to the outer positions, about 5 μl of each standard being used. Traces of moisture are removed from the spots by drying for a few minutes in a stream of warm air. The sheets are developed in the solvent for a distance of about 10 cm, then removed from the developing chamber and dried in air to remove most of the solvent, and then dried in an oven for a few minutes at 110°C to remove the solvent completely. The sheets are then dipped briefly in the color reagent (the two ends are held with forceps), dried in air, and then heated in an oven to develop the color. If a large beaker containing hot water is placed in the oven to increase the humidity, the colors will develop better.

The sugars are identified by means of their Rf values (relative distances traveled by the different spots and the solvent front) and by the colors developed (Table 5-1).

Some impurities in the urine may result in a diffuse background color, but this should be distinguished from a definite spot. The Rf values given above are slightly greater than those given by Haer, [124] but the relative values are the same. Other pentoses will give an Rf value similar to rhamnose. Glucose and

Table 5-1. Identification of Urinary Sugars by Thin Layer Chromatography

Sugar	Rf Value	Color of Spot
Rhamnose (pentose)	0.68	Gray-green
Fructose	0.56	Reddish brown
Glucose	0.50	Gray-blue
Galactose	0.47	Gray-blue
Sucrose	0.36	Gray-blue
Lactose	0.26	Gray-blue

galactose are difficult to distinguish since their Rf values are close together. Having these two sugars in different standards as suggested above may aid in distinguishing them in the sample. Also, glucose will give a positive reaction with the usual urine test strips based on glucose oxidase, whereas galactose will not.

If the apparatus is available, sugars may also be separated and identified by high-voltage electrophoresis [125]. Although sugars ordinarily do not ionize and thus would not move in an electric field, in the presence of borate ions they form complexes that can be separated by electrophoresis.

References

1. Folin, O., and Wu, H.   *J. Biol. Chem.* 38:81, 1919.
2. Haden, R. L.   *J. Biol. Chem.* 56:469, 1923.
3. Greenwald, I.   *J. Biol. Chem.* 21:61, 1915.
4. Greenwald, I.   *J. Biol. Chem.* 34:97, 1918.
5. Somogyi, M.   *J. Biol. Chem.* 86:655, 1930.
6. Somogyi, M.   *J. Biol. Chem.* 160:69, 1945.
7. Neuberg, C., Straus, E., and Lipkin, L. E.   *Arch. Biochem.* 4:101, 1944.
8. Horvath, A. A.   *Anal. Chem.* 18:229, 1946.
9. London, M., and Marymount, J. H.   *Clin. Chem.* 10:298, 1964.
10. London, M., and Marymount, J. H.   *Clin. Chem.* 10:417, 1964.
11. Benedict, S. R.   *J. Biol. Chem.* 5:485, 1909.
12. Schaffer, P. A., and Hartman, A. F.   *J. Biol. Chem.* 45:349, 1921.
13. Schaffer, P. A., and Somogyi, M.   *J. Biol. Chem.* 100:695, 1933.
14. Somogyi, M.   *J. Biol. Chem.* 195:19, 1952.
15. Folin, O., and Wu, H.   *J. Biol. Chem.* 41:367, 1920.
16. Folin, O.   *J. Biol. Chem.* 82:83, 1929.
17. Benedict, S. R.   *J. Biol. Chem.* 76:457, 1928.
18. Nelson, N.   *J. Biol. Chem.* 153:375, 1944.
19. Somogyi, M.   *J. Biol. Chem.* 160:69, 1945.
20. Pope, J. L.   *Am. J. Clin. Pathol.* 20:801, 1950.
21. Brown, M. W.   *Diabetes* 10:60, 1961.
22. Drygert, S., Li, L. H., and Thoma, J. A.   *Anal. Biochem.* 13:367, 1965.

23. Moore, G. R., Barnes, I. C., and Pennock, C. A. *Clin. Chim. Acta* 41:439, 1972.
24. Campbell, D. M., and King, E. J. *J. Clin. Pathol.* 16:173, 1963.
25. Geeting, D. G., Suther, C. A., and Sylbert, P. *Clin. Chem.* 18:976, 1972.
26. Fingerhut, B. *Am. J. Clin. Pathol.* 51:157, 1969.
27. Hagedorn, H. C., and Jensen, B. N. *Biochem. Z.* 135:46, 1923.
28. Hagedorn, H. C., and Jensen, B. N. *Biochem. Z.* 137:92, 1923.
29. Hiller, A., Linder, G. C., and Van Slyke, D. D. *J. Biol. Chem.* 64:626, 1925.
30. Holtz, A. H., Van Dreumel, H. J., and Van Kampen, E. J. *Clin. Chim. Acta* 6:467, 1961.
31. Ceriotti, G. *Clin. Chim. Acta* 8:157, 1963.
32. Davis, D. K. *Tech. Bull. Registry Med. Technol.* 36:121, 1966.
33. Friedmann, T. E., Weber, C. W., and Witt, N. F. *Anal. Biochem.* 4:358, 1962.
34. Edwards, H., and Freier, E. *Am. J. Med. Technol.* 37:9, 1971.
35. Winsten, S. *Ann. N.Y. Acad. Sci.* 102:127, 1962.
36. Rentzsch, K. *Aertzl. Lab.* 11:294, 1967.
37. Fingerhut, B., Ferzola, R., Marsh, W. H., and Miller, A. B., Jr. *Clin. Chem.* 12:570, 1966.
38. Klein, B., Morganstern, S., and Kaufman, J. H. *Clin. Chem.* 12:816, 1966.
39. Fingerhut, B., Ferzola, R., and Marsh, W. H. *Clin. Chim. Acta* 8:953, 1963.
40. Avigad, G. *Carbohydr. Res.* 7:94, 1968.
41. Klein, B., and Lucas, L. B. *Clin. Chem.* 17:1196, 1971.
42. Benedict, S. R. *J. Biol. Chem.* 34:203, 1918.
43. Mohun, A. F., and Cook, I. J. Y. *J. Clin. Pathol.* 15:169, 1962.
44. Schouten, H., and Giterson, A. *Clin. Chim. Acta* 8:802, 1963.
45. Lorentz, K. *Clin. Chim. Acta* 13:66, 1966.
46. Cheronis, N. D., and Zymaris, M. C. *Mikrochim. Acta* 6:769, 1957.
47. Gros, M., and Smrekar, M. *Clin. Chim. Acta* 17:518, 1967.
48. Walborg, E. F., Jr., and Christensson, L. C. *Anal. Biochem.* 13:186, 1965.
49. Deckert, T. *Scand. J. Clin. Lab. Invest.* 20:217, 1957.
50. Huggett, A. St. G. *J. Physiol.* 132:3P, 1956.
51. Forsell, O. M., and Flave, I. P. *Scand. J. Clin. Lab. Invest.* 11:409, 1959.
52. Anthanail, G., and Cabaud, F. C. *J. Lab. Clin. Med.* 51:321, 1958.
53. Thomitzek, W. D., and Bemm, H. *Z. Klin. Chem. Klin. Biochem.* 7:361, 1969.
54. Hultman, E. *Nature* (Lond.) 183:108, 1959.
55. Hyvarinen, A., and Nikkila, E. A. *Clin. Chim. Acta* 7:140, 1962.
56. Dubowski, K. M. *Clin. Chem.* 8:215, 1962.
57. Ceriotti, G. *Clin. Chem.* 17:440, 1971.
58. Winckers, P. L. M., and Jacobs, Ph. *Clin. Chim. Acta* 34:401, 1971.
59. Webster, W. W., Stinson, S. F., and Wong, W. A. *Clin. Chem.* 17:1050, 1971.
60. Goodwin, J. F. *Clin. Chem.* 16:85, 1970.
61. Cooper, G. R., and McDaniel, V. *Stand. Methods Clin. Chem.* 6:159, 1970.
62. Frings, C. S., Ratliff, C. R., and Dunn, R. T. *Clin. Chem.* 16:282, 1970.
63. Snegoski, Sr. M. G., and Freier, E. F. *Am. J. Med. Technol.* 39:140, 1973.
64. Moorehead, W. R., and Sasse, E. A. *Clin. Chem.* 16:285, 1970.
65. Keilin, D., and Hartree, E. F. *Biochem. J.* 42:221, 1948.
66. Teller, J. D. *Abst. 130th Meeting Am. Chem. Soc.*, 1956.

67. Kingsley, G. R., and Getchell, G.　*Clin. Chem.* 6:466, 1960.
68. Huggett, A. St. G., and Nixon, D. A.　*Lancet* 2:368, 1957.
69. Saifer, A., and Gerstenfeld, S.　*J. Lab. Clin. Med.* 51:448, 1958.
70. Washko, M. E., and Rice, E. W.　*Clin. Chem.* 7:542, 1961.
71. Salmon, L. L., and Johnson, J. E.　*Anal. Chem.* 31:453, 1959.
72. Marks, V.　*Clin. Chim. Acta* 4:495, 1959.
73. Cramp, D. G.　*J. Clin. Pathol.* 20:910, 1967.
74. Dick, M.　*J. Clin. Pathol.* 18:249, 1965.
75. Clark, A., and Timms, B. G.　*Clin. Chim. Acta* 20:352, 1968.
76. Trinder, P.　*J. Clin. Pathol.* 22:158, 1969.
77. Aw, S. E.　*Clin. Chim. Acta* 26:475, 1969.
78. Tammes, A., and Nordschow, C. D.　*Am. J. Clin. Pathol.* 49:622, 1968.
79. Parrott, L. H.　*Am. J. Clin. Pathol.* 49:877, 1968.
80. Hollister, L. B., Helmke, E., and Wright, A.　*Diabetes* 15:691, 1966.
81. Rock, J. A., and Gerende, L. J.　*J.A.M.A.* 198:231, 1966.
82. Haworth, J. C., Dilling, L. A., and Van Woert, M.　*Am. J. Dis. Child.* 123:469, 1972.
83. Percy-Robb, I. W., McMaster, R. S., Harrower, A. D. B., and Duncan, L. J. P.　*Ann. Clin. Biochem.* 9:91, 1972.
84. Forman, D. T., Grayson, S. H., and Slonicki, A.　*Lab. Med.* 3:26, 1972.
85. Neeley, W. E.,　*Clin. Chem.* 18:509, 1972.
86. Peterson, J. I., and Young, D. S.　*Anal. Biochem.* 23:301, 1968.
87. Stork, H., and Schmidt, F. H.　*Klin. Wochenschr.* 46:789, 1968.
88. Peterson, J. I.　*Clin. Chem.* 14:513, 1968.
89. Keller, D. M.　*Clin. Chem.* 11:471, 1965.
90. Carroll, J. J., Smith, N., and Babson, A. L.　*Biochem. Med.* 4:171, 1970.
91. Wright, W. R., Rainwater, J. C., and Tolle, L. D.　*Clin. Chem.* 17:1010, 1971.
92. Ree, J. H.　*J. Biol. Chem.* 212:335, 1955.
93. Devor, A. W., Baker, W. L., and Devor, K. A.　*Clin. Chem.* 10:597, 1964.
94. Groger, W. K. L.　*Clin. Chim. Acta* 6:866, 1961.
95. Ohkura, Y., Watanabe, V., and Momose, T.　*Biochem. Med.* 6:97, 1972.
96. Murphy, D., and Pennock, C. A.　*Clin. Chim. Acta* 42:67, 1972.
97. Guilbault, G. G., and Lubrano, G. J.　*Anal. Chim. Acta* 60:254, 1972.
98. Skerry, D. W.　*Clin. Biochem.* 3:319, 1970.
99. Kadish, A. H., and Sternberg, J. C.　*Diabetes* 18:467, 1969.
100. Morrison, B., Scotland, L. J., and Fleck, A.　*Clin. Chim. Acta* 39:301, 1972.
101. Marais, J. P., deWit, J. L., and Quicke, G. V.　*Anal. Biochem.* 15:373, 1966.
102. Seltzer, H. S.　In M. Ellenberg [Ed.], *Diabetes Mellitus: Theory and Practice.* New York: McGraw-Hill, 1970.
103. Forsham, P. H., Steinke, J., and Thorne, G. W.　In T. R. Harrison, et al. [Eds.], *Principles of Internal Medicine,* 5th ed. New York: McGraw-Hill, 1966. Chap. 86.
104. Henry, R. J. [Ed.].　*Clinical Chemistry, Principles and Techniques,* 2nd ed. New York: Hoeber–Harper & Row, 1973.
105. McDonald, G. W., Fisher, S. F., and Burnham, C. D.　*Public Health Rep.* 79:515, 1964.
106. Sindoni, A., Jr.　*Am. J. Dig. Dis.* 13:178, 1946.
106a. *Diabetes* 18:29, 1969.

106b. Winkerson, H. L. C. *Diagnosis and Glucose Tolerance Tests in Diabetes Mellitus Diagnosis and Treatment.* New York: American Diabetes Association, 1964.

107. Langer, P. H., and Fies, H. L. *Am. J. Clin. Pathol.* 11:41, 1941.

108. Thorn, G. W., et al. *J. Clin. Invest.* 19:813, 1940.

109. Soskin, S. *J. Clin. Endocrinol.* 4:75, 1944.

110. Fajans, S. S., and Conn, J. W. *Ann. N.Y. Acad. Sci.* 82:208, 1959.

111. Joplin, G. F., Frazer, R., and Keeley, K. J. *Lancet* 2:67, 1961.

112. Unger, R. H., and Madison, L. L. *Diabetes* 7:455, 1958.

113. Unger, R. H., and Madison, L. L. *J. Clin. Invest.* 19:637, 1958.

114. Hansten, P. D. *Drug Interaction.* Philadelphia: Lea & Febiger, 1971.

115. Garb, S. *Clinical Guide to Undesirable Drug Interactions and Interferences.* New York: Springer, 1970.

116. Young, D. S., Thomas, D. W., Friedman, R. B., and Pestaner, L. C. *Clin. Chem.* 18:1041, 1972.

117. Sonderg, G. *Scand. J. Clin. Lab. Invest.* 10:203, 1958.

118. Waldstein, S. D., Dubin, A., Newcomer, A., and McKenna, C. G. *Clin. Chem.* 10:381, 1964.

119. Frings, C. S., and Pardu, H. L. *Anal. Chem.* 36:2477, 1964.

120. Ford, J. D., and Haworth, J. C. *Clin. Chem.* 10:1002, 1964.

121. Hjelm, M. *Clin. Chim. Acta.* 15:87, 1967.

122. Harriss, A. L. *Clin. Chem.* 15:65, 1969.

123. Goodwin, J. F. *Clin. Chem.* 16:85, 1970.

124. Haer, F. C. *An Introduction to Chromatography on Impregnated Glass Fiber.* Ann Arbor, Mich.: Ann Arbor–Humphrey Science Publishers, 1969.

125. Mabry, C. C., Roekel, I. E., Geuedon, R. E., and Koepke, J. A. *Recent Advances in Pediatric Clinical Pathology.* New York: Grune & Stratton, 1968.

6. Nonprotein Nitrogen Compounds

Alpha Amino Acid Nitrogen

This determination is used as a measure of the total amount of amino acids in blood or urine. All biologically significant amino acids contain an amino group on the carbon atom next to the carboxyl group. Some contain other amino groups as well, but these other amino groups do not react markedly with the reagents used for the alpha amino grouping. One of the earliest methods used for the determination of total amino acids in urine was the formol titration [1]. In this test, neutral formaldehyde is added to react with the alpha amino groups which then no longer act as strongly basic groupings, and the carboxyl groups can be readily titrated with alkali. This method was soon found to be relatively nonspecific. The nitrous acid method introduced by Van Slyke [2–4] involves release of nitrogen gas by the action of nitrous acid on the alpha amino group and measurement of the gas in the Van Slyke apparatus. This method was also found to lack specificity.

The compound ninhydrin (triketohydrindene hydrate) reacts with most amino acids to form a blue color [5, 6]. The reaction is not entirely specific in that other nitrogenous compounds also give a color. In biological fluids the main interference comes from urea and ammonia. In most cases the ammonia can be removed by aeration in a slightly alkaline solution. The interference by urea is very slight if the reaction is carried out in a nonaqueous medium [7, 8]. A satisfactory determination of plasma or urinary alpha amino nitrogen can be made on this basis. The reaction between ninhydrin and the alpha amino group also liberates carbon dioxide under the proper conditions. This is the basis of the method of Van Slyke in which the deproteinized sample is reacted with ninhydrin in a special tube. The carbon dioxide formed is absorbed by alkali in the Van Slyke apparatus and subsequently released for manometric determination [9–11]. Although this may be the most specific method for alpha amino nitrogen, the procedure is too long and tedious for routine use.

In another colorimetric method used for many years, the amino group reacts with sodium β-naphthoquinone-4-sulfonate with the production of a colored compound having a maximum absorbance at around 470 nm [12, 13]. For plasma, a tungstic acid filtrate must be used and for urine the amino acids are best first separated by an ion exchange resin [14, 15].

Another method for the determination of total amino acids uses copper phosphate. The neutral solution containing the amino acids is shaken with a freshly prepared suspension of cupric phosphate. The amino acids react with some of the copper and form soluble copper salts. After removal of the excess unreacted solid cupric phosphate, the solution is analyzed for copper as a measure of the amount of amino acids present. The copper analysis has been carried out in a number of different ways. By the addition of organic amines, sufficient copper-complex blue color may be formed for colorimetric determination directly [16]. Alternatively the copper may be determined by any of a number of reagents used for the colorimetric assay including diethyldithiocarbamate [17], cuprazone I [18], neocuproine [19], acetaldehyde, and oxalyldihydrazone. The copper may also be determined by atomic absorption spectroscopy [20] or by chelometric titration with EDTA (ethylenediaminetetraacetic acid) [21].

More recently the amino groups have been determined by reaction with 2,4,6-trinitrobenzenesulfonate [22] or fluorodinitrobenzene [23–26]. The use of the latter reagent appears to give very satisfactory results. We have chosen this method for presentation here. It is relatively simple and specific. For analysis in blood, plasma (either heparin, oxalate, or EDTA may be used as anticoagulant) is used since some amino acids are liberated in the clotting process. For analysis of urine the ammonia must be removed; this is easily done by simple boiling in a slightly alkaline solution.

Analysis in Plasma

This procedure uses the reaction of dinitrofluorobenzene with amino groups. It has been used for the routine determination of many amino acids. The reagent reacts with amino groups to give a colored product having an absorbance maximum at 420 nm. The colored complex is not very soluble in water, and acidified dioxane is added to dissolve it for the colorimetric reading at 420 nm.

REAGENTS

1. Sodium tetraborate solution, $0.132M$.   Dissolve 50 g of sodium tetraborate decahydrate in water and dilute to 1 liter.

2. Stock dinitrofluorobenzene (DNFB).   Dissolve 0.65 ml of 2,4-dinitrofluorobenzene* in 50 ml of acetone. Store in the refrigerator. This solution should be replaced after 2 to 3 months. Care must be taken in pipetting DNFB as it is very poisonous.

3. Working DNFB solution.   Just before use, mix 1 volume of the stock DNFB with 9 volumes of the borate buffer. This solution is not stable and should be used within a few hours.

4. Acidified dioxane.   Add 2 ml of concentrated hydrochloric acid to 100 ml of dioxane.

5. Plasma precipitating solution.

---

* Sigma Chemical Co., St. Louis, Mo. 63118.

A. Hydrochloric acid, approx. $0.11M$. Add 9.5 ml of concentrated hydrochloric acid to about 500 ml of water in a 1 liter volumetric flask, cool, and dilute to 1 liter.

B. Sodium tungstate, approx. $0.4M$. Dissolve 133 g of sodium tungstate dihydrate in water and dilute to 1 liter.

C. Just before use, mix together 1 volume of the sodium tungstate solution and 9 volumes of the HCl solution.

The plasma precipitating solution is used instead of the usual Folin-Wu method since the presence of sulfate sometimes results in a turbidity in the final colored solution.

6. Standard solutions.

A. Stock standard, 20 mg amino nitrogen/dl. Dissolve 1.050 g of glutamic acid and 0.536 g of glycine in about 150 ml of water in a 1 liter volumetric flask. Add 2.0 g of sodium benzoate and dissolve. Add 700 ml of $1M$ hydrochloric acid, mix, and dilute to 1 liter.

B. Working standards. On day of use, dilute 1, 2, and 4 ml of the stock standard with water to 10 ml in volumetric flasks. These will correspond to 2, 4, and 8 mg/dl of alpha amino nitrogen.

PROCEDURE

In separate tubes place 0.5 ml of plasma samples, 0.5 ml of water (blank), and 0.5 ml of standards. To each tube add 0.5 ml of water and 4.0 ml of the precipitating solution. Mix and allow to stand for a few minutes. Centrifuge tubes containing precipitated protein to obtain a clear supernatant.

Transfer 1 ml of supernatant from each tube to another set of labeled tubes and add to each tube 1 ml of the working DNFB solution. Heat all tubes for 15 min at 70°C. Allow to cool to room temperature and then add 5.0 ml of the acid dioxane to each tube. Mix well.

Read standards and samples against blank at 420 nm.

CALCULATION

$$\frac{\text{absorbance of sample}}{\text{absorbance of standard}} \times \text{conc. of standard} = \text{conc. of sample}$$

For each sample use the standard having an absorbance nearest that of the sample.

Analysis in Urine

REAGENTS

As for analysis in plasma.

PROCEDURE

The ammonia must be removed from the urine by heating in a slightly alkaline solution.

To 5 ml of urine in a 100 ml beaker add about 40 ml of water and a few glass beads. Add a few drops of phenolphthalein indicator solution (0.1 g in 100 ml ethanol); then add dropwise NaOH solution (2 g/dl) until a faint pink color persists. Boil the mixture gently for 15 min, adding more NaOH if necessary to maintain alkalinity, and more water if volume is reduced below 20 ml.

Allow the solution to cool to room temperature and transfer quantitatively to a 100 ml volumetric flask and dilute to the mark. Mix well and filter through a retentive paper.

Prepare standards and blank by diluting the same as for the plasma analysis, but use 1 ml of the urine filtrate without further dilution. To 1-ml aliquots of standards, blank, and urine samples, add 1 ml of the working DNFB and heat at 70°C as for the plasma determination. Complete the determination in the same manner.

CALCULATION

The calculation is similar to the one for plasma except that since the urine is diluted 1:20 instead of 1:10, the standards correspond to 4, 8, and 16 mg/dl amino nitrogen for urine.

COMMENTS

Due to the boiling, the final urine filtrate may have an appreciable yellow color. In this case it may be advisable to run a urine blank. To 1-ml aliquots of water and the urine filtrate, add 1 ml of the stock borate buffer (without DNFB) and heat for 15 min at 70°C. Cool, add 5 ml of the acid dioxane, and read urine sample against water blank. Subtract any absorbance so obtained from the absorbance of the urine sample in the actual determination, for use in the calculations.

NORMAL VALUES AND INTERPRETATION OF RESULTS

The normal level for alpha amino nitrogen in plasma of adults is 3 to 6 mg/dl. The values for children are about the same, with slightly higher values in infants. The urinary excretion of alpha amino nitrogen is 50 to 200 mg/24 hr in adults. For children it has been given as 0.3 to 1.3 mg/kg body weight/24 hr. It will be increased in conditions resulting in increased amino acid excretion, including the various aminoacidurias (see Chapter 7) and in Wilson's disease where levels of up to 1,000 mg/day have been reported.

Urea

Urea is the chief nitrogenous end product of protein catabolism in man and other mammals. The proteins are hydrolyzed to the constituent amino acids. The amino acids are then deaminated (in several different ways for different amino acids) and the ammonia released is converted into urea. This latter step takes place principally in the liver. The over-all reaction is simply:

$$2\,NH_3 + CO_2 \rightarrow NH_2-CO-NH_2 + H_2O.$$

This reaction as written does not take place spontaneously in the body. Instead a series of cyclic reactions are involved. The amino acid arginine is hydrolyzed by the enzyme arginase with the production of ornithine and urea. The ornithine reacts with the addition of CO_2 and H_2 to form citrulline, which then adds another molecule of ammonia and loses a molecule of water to form arginine, thus completing the cycle. The addition of the two molecules of ammonia to form citrulline and arginine are not simple processes but involve several intermediate steps with glutamic and aspartic acids. The essential point is that urea is the end product of protein metabolism and as such is excreted by the kidneys. The urea is filtered from the plasma by the glomeruli and some may be reabsorbed in the tubules. Since urea is continually produced in the body, if the kidneys have a decreased ability to excrete urea, the level in the blood will rise. Thus the urea level in the blood is an indication of the functional ability of the kidneys. Since the urea level in the blood will not rise appreciably above normal until the actual functional tissue has been reduced to about one-half of that originally present, its determination is not a good indicator of beginning renal disease. When the kidney function is impaired, not only the urea level in the blood but also the levels of other nitrogenous end products—uric acid and creatinine—increase. Since the major portion of the nonprotein nitrogen in blood is in the urea, the determination of this compound is now usually used for a simple test of kidney function.

Since the determination of urea was originally made in whole blood and the result conventionally reported as urea nitrogen (for comparison with the total nonprotein nitrogen), the result was usually reported as BUN (blood urea nitrogen). Most automated and many common manual methods now determine the urea nitrogen in serum, but the result is still reported as BUN. The concentration of urea in serum and in whole blood are sufficiently close to each other so that any difference is of no clinical significance. Although the results of these newer methods should be called serum urea nitrogen (SUN), they are good approximations to the actual BUN.

One general method for the determination of urea is by the use of the enzyme urease which hydrolyzes urea to ammonia and carbon dioxide. This enzyme (E.C.3.5.1.5; urea aminohydrolase) is very specific for urea [27]; hence the methods based on its use are also specific. There are a number of methods based on the use of the enzyme; they differ in the manner in which the ammonia formed is determined. The different methods for the determination of the ammonia formed from urea are also used for the determination of ammonia from other sources. The older methods used Nessler's reagent, an alkaline solution of K_2HgI_4 [28], which gives an orange-red color with ammonia. The preparation of a protein-free filtrate was necessary to remove interfering material. This reagent possesses several disadvantages: It is not very stable, the color formed does not follow Beer's law very closely, and the formation of turbidity is often a problem. Even with a protein-free filtrate it was found that the reagent slowly reacted with other substances present so that the color gradually increased [29].

A much more sensitive method for the determination of ammonia involves the Berthelot reaction [30]. In an alkaline solution, ammonia, in the presence of hypochlorite, will condense with phenol to give an intense blue indophenol. The reaction is usually aided by the presence of a catalyst which can be manganous ions or more commonly sodium nitroferricyanide. The reaction is so sensitive that the slightest trace of ammonia will interfere and ammonia-free water must be used throughout. Phenol and hypochlorite are usually used for the reaction although salicylate and dichloroisocyanurate have been suggested [31]. Many slight variations on the method have been published; all of these cannot be mentioned here, but we list a few on which the method to be presented is based [32–34]. The method has also been automated [35–37]. The same reaction has also been used for the determination of ammonia. The reaction may be carried out directly on a microsample of serum. Any preformed ammonia present is also determined in the reaction. The amount of this in blood is so small that the error made by neglecting it is very slight. In urine the amount of preformed ammonia may be appreciable; it can either be removed by treatment with permutit or measured directly by running a sample without urease.

A third method for the measurement of the ammonia formed by the action of urease is by the use of this reaction:

$$NH_4^+ + 2\text{-oxoglutarate} + NADH \rightarrow glutamate + NAD^+ + H_2O$$

which is catalyzed by glutamate dehydrogenase (E.C.1.4.1.2). The reaction is followed by noting the change in absorbance at 340 nm on the conversion of NADH to NAD [38–40]. The sample is incubated with a mixture containing urease, oxoglutarate, NADH, and glutamate dehydrogenase in the proper buffer, and the change in absorbance is determined. This is a satisfactory method, except that like many of the NAD–NADH coupled reactions, the reagents (enzymes and NADH) are not stable in solution and must be kept in lyophilized or frozen form.

A brief mention may be made of a novel method for the determination of urea, which may develop into a method of practical importance. A urea electrode is made by enclosing a glass electrode sensitive to ammonium ions with a layer of polyacrylamide containing urease. When this is immersed in a solution containing urea, the concentration of ammonium ions near the electrode, and hence the electrode potential reading, will depend upon the amount of urea in the solution [41, 42].

Another common method for the determination of urea is based on the reaction with diacetylmonoxime. In the presence of strong acid and an oxidizing agent this reagent will react with urea to produce a color suitable for spectrophotometric determination. The earlier methods used strong acid with arsenic acid as oxidizer [43–45]. Later methods used ferric ions as oxidizing agent [46, 47]. These methods produced a yellow color which did not obey Beer's law even at relatively low concentrations of urea and were somewhat light sensitive. It was later found that the addition of a small amount of thiosemicarbazide produced a red color with the use of weaker acid and gave much better adherence to Beer's law [48–50]. Some variation of this method is now used in most automated and many manual methods. It can be run directly on 50 μl or less of serum [51]. The reaction with monoxime also produces a fluorescent compound and has been used as a basis for a fluorometric method [52].

Among the other reagents which have been used for colorimetric determination of urea are p-dimethylaminobenzaldehyde [53–56], dimethylglyoxime [57, 58], cyclohexanol-1,2-dione dioxime [57], and isonitrosopropiophenone [58], but these are not widely used. There is also a rapid semiquantitative method based on the urease reaction. A strip of impregnated paper is dipped at one end into a small amount of serum in a test tube. As the serum rises in the strip by capillary action, it meets a band of reagent containing urease which hydrolyzes the urea. The ammonium salt produced then comes in contact with a band containing potassium carbonate which liberates free ammonia. A plastic barrier prevents further rise of liquid in the strip but the ammonia diffuses upward to a portion containing an indicator (bromcresol green) and tartaric acid. As the acid is neutralized, the color of the strip changes. Under carefully controlled conditions the height of the changed color of the strip is proportional to the amount of urea in the serum. These commercial strips (Urograph [59, 60], Urostrat [61, 62]) are not very accurate but may be sufficient for screening purposes in a physician's office or an emergency room. Because they depend in part on the diffusion of the ammonia formed, the conditions must be carefully controlled (size of tube, exactly vertical position of the strip) and they should be checked with standard urea solutions.

We present two methods for the determination of urea nitrogen (BUN). The preferred method is the one using the reaction with diacetylmonoxime and thio-semicarbazide. This can be carried out directly on a small sample of serum or, if available, on a Folin-Wu or Somogyi filtrate. Depending upon the photometer used, the method may not follow Beer's law over a very wide range and it is usually best to run at least two standards such as 30 and 60 mg/dl with each set of samples. For an occasional serum giving a result above 80 mg/dl, the test is best performed on a dilution of the serum.

The other method presented is that involving the use of urease and the Berthelot reaction. This method is very specific for urea, but because of the sensitivity of the reaction requires ammonia-free water. It also may be run directly on a small aliquot of serum. As mentioned earlier, the Berthelot reaction may be used to determine ammonia derived from other sources. For the deter-mination of urea in urine, a determination of the preformed ammonia may be required.

Urea by Diacetyl Monoxime Method

REAGENTS

1. Color reagent.   Dissolve 1 g of diacetyl monoxime (2,3-butanedione mon-oxime), 0.2 g of thiosemicarbazide, and 9 g of sodium chloride in water and dilute to 1 liter.

2. Acid solution.   Carefully add 60 ml of concentrated sulfuric acid and 10 ml of 85% phosphoric acid to about 800 ml of water. Dissolve 0.1 g of ferric chloride $(FeCl_3 \cdot 6H_2O)$ in the solution, cool, and dilute to 1 liter. If relatively large amounts of the solution are used, it is convenient to dissolve 1 g of ferric chloride in 100 ml of phosphoric acid and use 10 ml of this mixture for every liter of reagent.

3. Urea standards.

A. Benzoic acid solution for stock standard.   Dissolve 1.5 g of sodium benzoate and 0.7 ml of concentrated sulfuric acid in 1 liter of water.

B. Stock standard. Dissolve 0.644 g of pure urea in the benzoic acid solution to make 100 ml. This contains 3 mg/ml of urea nitrogen.

C. Working standard. Dilute the stock standard solution 1:5 and 1:10 to give solutions containing 60 and 30 mg/dl of urea nitrogen.

PROCEDURE FOR SERUM, DIRECT METHOD

Mix equal volumes of the color reagent and acid solution. The mixed reagent is not very stable and should be used on the day prepared. Pipet 5 ml of the mixed reagent into separate tubes. To the tubes add 0.05 ml of samples and standards (both standards should ordinarily be run), and reserve one tube as blank. Heat all tubes in boiling water bath or heating block at 100°C for 15 min. Cool and read standards and samples against blank at 520 nm.

CALCULATION

Since the standards and samples are treated similarly:

$$\frac{\text{absorbance of sample}}{\text{absorbance of standard}} \times \text{conc. of standard} = \text{conc. of sample}$$

COMMENTS

1. It is preferable to run two standards since the reaction does not always follow Beer's law at higher concentrations. If the result reads over about 75 mg/dl, the test should be repeated on a dilution of the serum. Depending upon the photometer used, the above proportions may not give sufficient color for good readings (absorbance of 30 mg/dl standard about 0.2 to 0.3) or the color intensity may be too great. In the former instance the amount of serum added may be increased somewhat (always adding the same volumes of sample and standards). If the readings are too high, the solutions (including blank) can all be diluted somewhat before reading or a longer wavelength can be used.

2. Although the direct method is most convenient, if a protein-free filtrate is available from another test, it can be used. Add 0.5 ml of filtrate and 0.5 ml of standards diluted an additional 1:10 to actual concentrations of 3 and 6 mg/dl. Since the samples are also diluted, the calculations will be the same. Since such a small amount of serum is needed, there would ordinarily be no need for the use of a filtrate except for extremely hemolyzed or very hyperlipemic serums.

3. If direct methods are used for both urea nitrogen and glucose, it is often convenient to make up one standard containing both constituents. If 1.00 g of glucose is added to the 0.644 g of urea before making to volume, the resulting

solution will contain 10 mg of glucose and 3 mg of urea/ml. The resulting 1:10 and 1:5 dilutions will then contain 100 mg/dl of glucose and 30 mg/dl of urea nitrogen, and 200 mg/dl of glucose and 60 mg/dl of urea nitrogen, respectively. These standards can then be used for both determinations.

PROCEDURE FOR URINE

The procedure can also be carried out directly on urine. The urine should be filtered or centrifuged if not clear, and then diluted. Concentrated urines may be diluted 1:10 with water and those of a lower specific gravity diluted 1:5. The procedure is then carried out exactly as for serum. Because of the wide variation in the concentration of urea in urine, it may be necessary to make additional or different dilutions. The total dilution of the urine must be taken into account in the calculations.

Urea by Urease Method [32–34]

The reaction utilized in this method is very sensitive to traces of ammonia, so a good grade of ammonia-free water must be used throughout. In many water systems, after chlorination a small amount of ammonia is introduced to neutralize excess chlorine. Since ammonia is volatile, traces may not be eliminated by distillation alone. Passage through a bed of mixed ion exchange resin may be necessary to remove the ammonia. Good ion exchange purification systems usually remove all ammonia.

REAGENTS

1. Alkaline hypochlorite.   Dissolve 12.5 g of sodium hydroxide in about 450 ml of water. Cool and add 20 ml of sodium hypochlorite solution (any commercial bleach containing 5.25% available chlorine may be used), dilute to 500 ml, and mix well. Store in a plastic bottle in the refrigerator.

2. Phenol reagent.   Dissolve 0.13 g of sodium nitroprusside [sodium nitroferricyanide, $Na_2NOFe(CN)_5 \cdot 2H_2O$] and 25 g of phenol in water to make 500 ml. (Instead of weighing the phenol crystals, 26 ml of a good grade of liquified phenol can be used.) This solution should be stored in a brown bottle in the refrigerator and should be stable for several months.

3. Buffer solution.   Dissolve 5.0 g of disodium salt of ethylenediaminetetraacetic acid (ethylenedinitrilotetraacetic acid disodium salt) in 250 ml of water and add 200 ml of glycerin. Adjust the pH of the solution to 6.5 using a sodium hydroxide solution (4 g/dl). About 10 ml of the base should be required. Dilute to 500 ml and mix well.

4. Buffered urease. Add 30 mg of urease (Sigma, Type III)* to 100 ml of buffer. This solution is stable for several weeks when kept in the refrigerator.

5. Standards. Same as for the direct method using diacetyl monoxime.

PROCEDURE FOR SERUM

Pipet 0.5 ml of the urease solution into a series of test tubes. Add 0.02 ml of serum or standard to separate tubes. Also reserve one tube of urease solution alone as blank. Incubate all tubes at 37°C for 15 min. After incubation add to each tube 1 ml of the phenol reagent and mix well. Then add 1 ml of the hypo-chlorite solution and mix again. It is important to add the two reagents in this order, mixing well after each addition. Incubate the tubes again for 15 min at 37°C. Add 10 ml of water to each tube, mix, and read standards and samples against blank at 620 nm.

CALCULATION

Since the standards and samples are treated similarly,

$$\frac{\text{absorbance of sample}}{\text{absorbance of standard}} \times \text{conc. of standard} = \text{conc. of sample}$$

If the reaction is linear in that the 60 mg/dl standard gives close to twice the absorbance of the 30 mg/dl standard, only one standard need be run routinely. If necessary, the amount of water added in the final step may be changed so that the 30-mg standard will give an absorbance of between 0.25 and 0.3 units.

PROCEDURE FOR URINE

Although the procedure may be used without change for urine, the latter contains some preformed ammonia which also reacts with the reagents. Thus a correction must be made for this. The total urea plus ammonia nitrogen is determined exactly as the procedure for serum using a 1:15 or 1:10 dilution of urine (filter or centrifuge first if it is not clear). Then the preformed ammonia is determined by running a second determination using water instead of the urease buffer solution. The first incubation prior to the addition of the color reagents is omitted and the final color read against a corresponding blank using water instead of urease solution. The amounts of total nitrogen and ammonia nitrogen are calculated in the usual way for urine samples. Total nitrogen minus ammonia nitrogen equals urea nitrogen.

* Sigma Chemical Co., St. Louis, Mo. 63118.

Normal Values and Interpretation of Results

The normal range of urea nitrogen in serum or whole blood is 10 to 18 mg/dl. If expressed as urea, this is 22 to 40 mg/dl (urea nitrogen × 2.14 = urea). Urea is formed as the end product of protein metabolism, and the usual cause of increased levels of urea in the blood is inadequate excretion, usually due to kidney disease or urinary obstruction. Increased blood levels occur with extensive parenchymatous destruction of kidney tissue, as in advanced nephrosclerosis, pyelonephritis, renal cortical necrosis, renal tuberculosis, malignancy, and chronic gout. Levels below 10 mg/dl may occasionally be found, sometimes in malnutrition.

The excretion of urea nitrogen in the urine is usually in the range of 10 to 15 g/day. It depends greatly on the dietary protein intake. Urinary urea is usually determined only for urea clearance (see discussion of clearance tests, Chapter 11).

In the determinations by the urease method, streptomycin and chloramphenicol may interfere with the enzyme action, thus leading to low results. Mercury compounds, such as mercurial diuretics, will interfere with the enzyme action and the color reaction, also leading to low results. All drugs having a marked diuretic action may lead to lower blood levels of urea nitrogen than expected. Conversely any drug that has a toxic effect on the kidneys will cause an increase in the urea level in the blood.

Nonprotein Nitrogen (NPN)

Before simple methods for urea were available, the degree of kidney function was often assessed by the determination of the NPN. This is a determination of all the nitrogen in a protein-free filtrate by a Kjeldahl digestion. It included the nitrogen from urea, uric acid, creatinine, and amino acids, as well as other substances. It might be expected to vary somewhat with the type of precipitant used. The determination was usually carried out on a Folin-Wu filtrate.

In a normal NPN, about one-third of the nitrogen is from urea. With increased urea levels, the proportion of total nitrogen from urea increases also. At high levels of urea, the proportion is about two-thirds. The proportion of nitrogen from substances such as uric acid and creatinine does not vary greatly, although the levels of these substances will rise in kidney disease. Although some physicians have felt that a rise in NPN was of more significance than a proportionate rise in BUN, there does not appear to be any good reason for using the NPN value rather than the BUN, especially as the BUN can be supplemented by determinations of creatinine and uric acid when necessary. Actually the request for NPN has been very rare in the experience of the authors and consequently the determination is omitted in this section.

Blood Ammonia

Since ammonia is present in blood or plasma in relatively small quantities (in the order of 100 $\mu g/dl$), the separation from interfering substances is an important consideration. In the earlier diffusion methods the sample was treated with alkali and the ammonia separated by diffusion, either in the Conway microdiffusion dish [63] or in the apparatus devised by Seligson [64–66]. The ammonia was trapped in acid and then determined either by microtitration with Nessler's reagent or by the Berthelot reaction. Sodium hydroxide was first used as alkali but it was found that this tended to liberate small amounts of ammonia from other substances in the sample. Potassium carbonate solution was found to be sufficiently alkaline to drive off the ammonia readily and gave less ammonia from interfering materials [65].

Protein-free filtrate (trichloracetic acid [67–69] or Folin-Wu [70, 71]) has been used, in a direct reaction with the color reagent, but it was stated that the precipitating reagents did not remove all interfering substances that might also react with the reagent [72]. The methods now in more general use separate the ammonia by means of an ion exchange resin [73, 74]. Plasma is preferable since it is difficult to remove all interfering hemoglobin, though some resin methods do use whole blood [75]. The diluted plasma is treated with an ion exchange resin which absorbs the ammonia. After washing the resin free of interfering material, the ammonia is eluted with alkali. In batch methods the alkaline reagent (Nessler's or Berthelot) may be added directly to the resin. A number of different resins have been used [76, 77] as well as the Folin Permutite [78]. The enzyme reactions for determining ammonia as given in the determination of urea with urease have been used for the determination of blood ammonia as well [79–81]. The ammonia has also been determined fluorometrically [82], and automated continuous-flow methods have been derived in which the ammonia is separated by dialysis [83, 84].

We present here a method using an ion exchange resin. Because of the small quantity of ammonia present, a good grade of ammonia-free water should be used. That from a high-quality deionizing system is usually satisfactory. It may be convenient to treat the water with some of the same resin as used in the procedure. Heparinized plasma is used. Make certain that the ammonium salt of heparin is not used. The amount of heparin used for a sample plus the volume of water used in the analysis should give a very low blank, preferably an absorbance when read against water of less than 0.05. The ammonia level in blood or plasma increases on standing. If the analysis cannot be made soon after separation of the plasma, the sample should be kept frozen [85–87].

PROCEDURE

1. Phenol reagent and alkaline hypochlorite.   The same reagents as used in the urease method for urea.

2. Ion exchange resin.   Dowex 50W-X4 100–200 mesh† is used. Add 1 volume of resin to 2 volumes of sodium hydroxide solution (10 g/dl) and mix. Allow to stand for about 15 min, decant liquid, and wash resin a number of times with 2 volumes of distilled water by gentle mixing, allowing to settle and decanting supernatant. Continue washings until supernatant is neutral (pH 7) with indicator paper. Usually at least six washes are required. Add 2 volumes of dilute acetic acid (2 ml/dl) and shake. Let stand for 15 min, allow to settle, and decant the supernatant. Wash several times with 2 volumes of water until supernatant is neutral. Store resin under water.

3. Ammonia standards.

A.   Stock standard.   Dissolve 70.8 mg of reagent grade ammonium sulfate in water containing a few drops of sulfuric acid and dilute to 100 ml. The solution is fairly stable if kept in the refrigerator. It contains 15 mg/dl of ammonia nitrogen or 18.2 mg/dl of ammonia.

B. Working standard.   Prepare fresh as required, by diluting 0.1 ml of stock standard and 0.3 ml of sodium chloride solution (24 g/dl) and dilute to 100 ml. This contains 150 μg/dl of ammonia nitrogen or 182 μg/dl of ammonia.

PROCEDURE

To separate tubes add 2 ml of serum sample, 2 ml of working standard, and 2 ml of water as blank. To each tube add about 0.5 g of wet resin. This is most conveniently done by using a wide-tipped pipet (old Mohr pipet with broken tip) to take about 1 ml of wet resin from the bottom of the vessel in which it is stored under water. The exact amount is not critical. Mix well with gentle agitation for several minutes and then allow to settle. Decant upper layer without loss of resin. Wash resin three times in this way with 10-ml portions of water. After final decantation add 1 ml of phenol reagent and agitate gently. Allow to stand for 3 min, add 1 ml of the alkaline hypochlorite, and mix. Incubate at 37°C for 15 min. Dilute with an additional 3 ml of water and mix. Decant into cuvettes and read samples and standards against blank at 630 nm. A small amount of resin in the cuvettes will do no harm as it will settle to the bottom out of the light path. As with the urease method, if the color is too intense, the solutions may all be diluted equally with additional water or a shorter wavelength (580 nm) may be used for reading.

† Dowex resins may be obtained from a number of laboratory suppliers.

CALCULATION

$$\frac{\text{absorbance of sample}}{\text{absorbance of standard}} \times \text{conc. of standard} = \text{conc. of sample}$$

The results may be reported either as ammonia nitrogen or ammonia.

ammonia nitrogen $\times$ 1.22 = ammonia

COMMENTS

If only an occasional determination is done, it will probably be more convenient to use one of the ammonia test kits available. These contain all the reagents and standards for the test, together with some additional resin for treating the water.

NORMAL VALUES AND INTERPRETATION OF RESULTS

The older values for ammonia given in the literature are probably too high and there is still some uncertainty about the normal range. With this method, the normal range may be taken as 35 to 120 μg/dl, as ammonia nitrogen. Increases in blood ammonia occur chiefly in liver disease, and levels of over 150 μg/dl are considered indicative of liver damage.

Creatinine and Creatine

Available Creatinine Methods

Modifications of the Jaffé reaction with alkaline picrate first introduced into clinical chemistry many years ago by Folin [88] are still the most common methods used for the determination of creatinine. It has been shown that with a Folin-Wu or similar filtrate, about three-fifths of the color produced when whole blood is used comes from noncreatinine chromogens [89] and that when serum or plasma is used only about one-fifth of the color is due to noncreatinine chromogens [90, 91]. With urine the excess chromogens amount to only about five percent of the total [90, 91]. For this reason whole blood is, or should be, no longer used for creatinine determinations. Even with serum, several modifications have been proposed to reduce the nonchromogen color. The one most generally used is treatment with Lloyd's reagent [91, 92] (a purified Fuller's earth, which is an aluminum silicate clay). All the creatinine is absorbed on the reagent but most of the nonspecific chromogens are not. After separation of the reagent, the creatinine may be eluted by acid and determined. Whether Lloyd's reagent needs to be used depends upon the purpose for which the creatinine is determined. Merely to judge the level of nonprotein nitrogenous substances in the blood (in some kidney disorders the creatinine may increase proportionately more than the BUN), the simple method without the use of Lloyd's reagent is usually satisfactory. If the creatinine analysis is to be used for the calculation of a creatinine clearance, the determination with Lloyd's reagent should be used. It is usually not necessary to use the reagent when determining the creatinine in urine.

The amount of color developed by the reagent with creatinine will depend upon the final concentration of alkali, and this in turn may be influenced by the original acidity of the solution [90, 94]. It has been found that in the preparation of a protein-free filtrate for the determination of creatinine, the recovery of creatinine is incomplete unless the final filtrate is quite acid (pH near 1) [90]. For this reason in the actual procedure given below, it will be seen that 2 volumes of $0.34M$ sulfuric acid are used for each volume of $0.3M$ tungstate instead of the usual 1:1 ratio. This leads to the problem that the initial acidity of the filtrate will be much greater than that of the standards and blank, hence the amount of color development will be slightly different. Theoretically one could compensate somewhat for this by making the dilutions of the standard and using as a blank not water but the $0.34M$ sulfuric acid diluted 1:5. However, this is hardly necessary as the error involved in the difference in pH is less than that due to the nonspecific chromogens present. When Lloyd's reagent is used, this problem does not arise since the excess acid is eliminated in decanting the supernatant from the precipitated Lloyd's reagent containing the absorbed creatinine. This may be another reason for preferring the procedure with Lloyd's reagent. In the automated systems using dialysis for separation of the proteins, the problem of excess acid does not arise, but it has not been demonstrated that the nonspecific chromogens are eliminated by dialysis.

Other methods have also been used to decrease the amount of nonspecific chromogens for the Jaffé reaction. Taussky [93] uses ether extraction for urine and a combination of treatment with iodine and chloroform extraction for the protein-free filtrate. This makes the procedure considerably more complicated than the one with Lloyd's reagent. The absorption of the creatinine on an ion exchange resin has also been used [95, 96].

A number of other chemical reactions have been reported for creatinine but none is used to any large extent. One is a reaction with 1,4-naphthoquinone sulfonate which was claimed to be quite specific [97]; another uses the conversion of the creatinine to methyl guanidine followed by the colorimetric estimation of this latter compound by treatment with sodium hydroxide, α-naphthol, and hypochlorite (Sakaguchi reaction) to give a red color [98]. Other workers claim that although the former reaction is not satisfactory, the latter gives accurate results but is too long for routine use [99].

Other methods that have been tried are separation by ion exchange resin [100] or Sephadex [101] followed by direct measurement at 235 nm, as well as paper chromatography [102], paper electrophoresis [103], and thin layer chromatography [104].

Some substances that are known to give a color reaction with the alkaline picric acid are the ketone bodies (acetone, acetoacetic acid [105]) often found in uncontrolled diabetes in the blood and urine. If such compounds are suspected, they can be tested for (at least in the urine). If they are found in the urine in appreciable amounts, it would be preferable to use the procedure with Lloyd's reagent for the urine as well.

Although the serum proteins will react with the Jaffé reagent, they react more slowly than creatinine or react differently with changes in pH. Several direct methods for determination of serum creatinine have been suggested based on the above facts. The method based on the difference in reaction rates is not as simple as a manual method since the colorimetric measurements must be made at exact short time intervals, but it should prove satisfactory for an automated method. The method involving the reaction at two different pH values [106–109] requires two tubes and two blanks for each determination, thus requiring more manipulations than the other methods. Also a slight correction must be applied to give accurate results.

Creatinine with Folin-Wu Filtrate Without Use of Lloyd's Reagent

REAGENTS

1. Picric acid, $0.04M$. Dissolve 10.5 g of reagent grade picric acid in water to make 1 liter. The reagent as purchased usually contains about 10 to 12% water (the anhydrous material is much more readily exploded by mechanical shock). The amount of water is taken into account when using the above amount. A saturated solution of picric acid at room temperature is often used.

2. Sodium hydroxide, $0.75M$. Dissolve 30 g of NaOH in water to make 1 liter. Store in a polyethylene bottle.

3. Sulfuric acid and sodium tungstate solutions. See section on Folin-Wu Filtrates in Chapter 5.

4. Creatinine standards.

    A. Stock standard. Dissolve exactly 150 mg of pure creatinine in water containing 0.5 ml of concentrated hydrochloric acid and make up to 100 ml. This solution contains 1.5 mg/ml of creatinine.

B. Working standard.   Dilute 4 ml of the stock to 100 ml with water to give a solution containing 6 mg/dl of creatinine. This solution is not stable and should be made up fresh as needed.

PROCEDURE FOR SERUM

To 3 ml of water in a test tube add 2 ml of serum and 1 ml of the Folin-Wu sodium tungstate solution. Mix well and add 2 ml of the sulfuric acid solution. Stopper and mix well. Allow to stand for a few minutes, then centrifuge or filter. Treat 2 ml of the working standard similarly, diluting with the water and precipitating reagents. This solution need not be filtered.

To one test tube add 3 ml of the sample filtrate. To other tubes add 1, 2, and 3 ml of the diluted working standard and make up to 3 ml with water. Add 3 ml of water to one tube as blank. To each tube add 1 ml of picric acid solution, mix, and add 1 ml of the sodium hydroxide solution. Mix and allow to stand at room temperature for 20 min. Read standards and samples against blank at 520 nm.

CALCULATION

$$\frac{\text{absorbance of sample}}{\text{absorbance of standard}} \times \text{conc. of standard} = \text{conc. of sample}$$

Since the working standard containing 6 mg/dl is treated similarly to the serum, the standard tube containing 3 ml of diluted standard is equivalent to 6 mg/dl for the sample. The other two standard tubes containing 2 and 1 ml of the standard are equivalent to 4 and 2 mg/dl, respectively.

Since Beer's law is not followed very closely, it is preferable to run several standards as noted and use the standard having the absorbance nearest that of a given sample for calculation for that sample. If the result is above 8 mg/dl, the determination should be repeated using 1 ml of the sample filtrate plus 2 ml of water, then multiplying the calculated value by 3.

ALTERNATIVE PROCEDURE FOR SERUM

The following modification of a published micromethod [110] is useful when treatment with Lloyd's reagent is not required and if the Folin-Wu precipitating reagents are not routinely available in the laboratory. In this method picric acid is used to precipitate the proteins with the aid of heat, and the color is developed directly in the filtrate.

1. Picric acid.   Same as for previous method.

2. Sodium hydroxide, $2.5M$.   Dissolve 100 g of sodium hydroxide in water, cool, and dilute to 1 liter. Store in plastic bottle.

3. Creatinine standard.   Same stock solution as for previous method. Dilute 0.5 and 1.0 ml of this to 100 ml to give standards containing 3 and 6 mg/dl.

PROCEDURE

To separate tubes add 0.5 ml of serum samples or standards. To each tube add 1.5 ml of water. To a separate tube add 2.0 ml of water as blank. To each tube add 6.0 ml of the picric acid solution and mix well. Immerse all tubes in a vigorously boiling water bath for 40 sec, then remove to running tap water to cool. Filter (Whatman No. 30, 9 cm) the solutions containing protein precipitate. Transfer 4-ml aliquots from the filtrates from the sample tubes and 4-ml aliquots from the blank and standard mixtures (which need not be filtered) to separate tubes. To each tube add 0.20 ml of $2.5M$ sodium hydroxide solution and mix. Read standards and samples against blank at 520 nm exactly 20 min after the addition of the NaOH.

CALCULATION

$$\frac{\text{absorbance of sample}}{\text{absorbance of standard}} \times \text{conc. of standard} = \text{conc. of sample}$$

Serum Creatinine Using Lloyd's Reagent

REAGENTS

The same reagents are used as in the first method given above, with the addition of the following two reagents.

1. Lloyd's reagent.   This is a purified Fuller's earth obtainable from a number of laboratory supply houses.

2. Oxalic acid, saturated solution.   Add about 18 g of oxalic acid to 100 ml of water and shake until the solution is saturated.

PROCEDURE

Carry out the precipitation of proteins and dilution of standards just as in the first method given above. Add 3 ml of the filtrates to separate tubes as given earlier. Also add 1, 2, and 3 ml of the diluted standards to separate tubes. Add 2 ml of water to each tube containing sample filtrate, and dilute standard tubes to 5 ml rather than 3 ml as in the previous procedure. Also set up a tube containing 5 ml of water as blank. To each tube add 0.5 ml of saturated oxalic acid and approximately 100 mg of Lloyd's reagent. Stopper and shake at frequent intervals during a period of 15 min. Centrifuge all tubes strongly and decant the supernatant. Invert tubes and allow to drain for about 10 min. To each tube add 3 ml of water, 1 ml of picric acid, and 1 ml of the $0.75M$ sodium hydroxide solution. Stopper and shake intermittently during 15 min. Centrifuge strongly, pour supernatant into cuvettes, and read samples and standards against blank at 520 nm. The calculations are the same as in the previous method.

Urinary Creatinine

Since it is usually not necessary to prepare a filtrate for urine nor use Lloyd's reagent as the greater dilution of urine reduces the effect of interfering substances, either of the two procedures given above can be used. When the first method is followed, the standards are prepared from 6 ml of water plus 2 ml of working standard, and the urine is diluted 1:100. The diluted urine and diluted standards are treated exactly like the serum filtrate and diluted standards in this method; then the calculation is the same with an additional factor of 100/4, or 25. Depending upon the concentration of creatinine in the urine, it may be necessary to use a different dilution, with the consequent change in the calculation. In the second method given above, the urine is diluted 1:25 and this factor is used in the calculations for urine since the standards used are actually 3 and 6 mg/dl.

Normal Values and Interpretation of Results

The normal values for serum creatinine are 0.6 to 1.2 mg/dl. As mentioned in the earlier section on urea, all the nitrogenous products in the serum will increase in kidney dysfunction. In urinary tract obstruction and chronic nephritis, the creatinine may be relatively higher than the urea and values of 10 to 15 mg/dl may be found. Low values are found in muscular dystrophy. One reason for the determination of serum and urinary creatinine is for the calculation of the creatinine clearance (see Chapter 11).

The determination of urinary creatinine has often been used for an estimation of the completeness of urinary collection. Folin [111] first suggested that the urinary creatinine excretion in man was relatively constant for a given individual and that it could thus be used to determine the completeness of urinary collection. This has since been studied by a number of other investigators [112–121, 126, 127]. The consensus is that the urinary excretion is not sufficiently constant for use as an accurate test for completeness. It was found [113], for example, that although some subjects would have a day-to-day variation of only 5 percent, others under the same conditions would have variations of up to 20 percent in the urinary excretion of creatinine. Further, in order to use the creatinine excretion as a measure of urine collection in a given individual, one would need to know what the normal excretion was for that individual. The *creatinine coefficient*, milligrams of creatinine excreted per kilogram of body weight per day, is given by various investigators as varying from 21 to 32 for adult males and from 8 to 25 for adult females [122–125], and it is probably influenced by the dietary (meat) intake. This wide range of values makes it difficult to calculate an assumed normal excretion for the individual under test. One would have to measure the actual excretion under controlled conditions for several days to obtain a true baseline. For only a few urine collections it may be simpler to use extra care in collection. The authors' experience in the collection of complete urine specimens from a number of individuals over many months for balance studies is that variations of less than $\pm$ 20% are of little value in assessing the completeness of urine collection.

Available Creatine Methods

Creatine is chemically related to creatinine in that the latter is the inner anhydride of the former.

$$
\underset{\text{creatine}}{
\begin{array}{l}
H_2C\!-\!N\!-\!CH_3 \\
\quad\ \ \ \ |\ \ \ \ \ | \\
\quad\ \ \ \ |\ \ C=NH \\
\quad\ \ \ \ |\ \ \ \ \ | \\
O=C\ \ \ NH_2 \\
\ \ \ O \\
\ \ \ H
\end{array}}
\ \ \rightleftharpoons \ \
\underset{\text{creatinine}}{
\begin{array}{l}
H_2C\!-\!N\!-\!CH_3 \\
\quad\ \ \ \ |\ \ \ \ \ | \\
\quad\ \ \ \ |\ \ C=NH \\
\quad\ \ \ \ |\ \ \ \ \ | \\
O=C\!-\!NH
\end{array}}
\ \ +H_2O
$$

Creatine is found in serum in smaller amounts than is creatinine and its level is less affected by kidney disease. The level may be increased in diseases associated with extensive muscular wasting (muscular dystrophies); such conditions have been the only clinical indications for the determination of creatine.

The determination of creatine was first introduced by Folin [88] who heated the creatine with acid to transform it to the anhydride, creatinine. The difference between the creatinine content, as measured by the Jaffé reaction, before and after heating with acid was taken as a measure of the amount of creatine present. This method is still commonly used [128–131] although it has some disadvantages. The heating can result in the formation of some interfering chromogens and the conversion of creatine may not be complete. Furthermore, since the amount of creatine in the red blood cells is much greater than that in serum, any hemolysis will introduce an error. In normal individuals the creatine in serum is usually at a level about one-half that of the creatinine. Under these circumstances it can be shown that if precision of the determination of creatinine is ± 5% (a very good precision for this method), error in the determination of the creatine by difference can be as much as 25%. For urine the error will be greater. Since the amount of creatine is generally less than 10% of that of creatinine, the error in the determination of creatine by difference can be as much as 100%. Moreover, in the collection of urine for a 24-hr specimen, there may be some intraconversion of creatine and creatinine.

Three other methods have been used for the determination of creatine. In one the creatine reacts with α-naphthol and diacetyl in alkaline solution to form a red color [132–135]. This is fairly specific for creatine, although small amounts of color may be produced by other substances. Another method is an enzymic one using the enzyme creatine phosphokinase [136–138]. The reactions are:

$$creatine + ATP \xrightarrow{CPK} creatine\ phosphate + ADP$$

$$ADP + PEP \xrightarrow{PK} ATP + pyruvate$$

$$pyruvate + NADH \xrightarrow{LDH} lactate + NAD$$

where ATP = adenosine triphosphate, ADP = adenosine diphosphate, PEP = phosphoenolpyruvate, CPK = creatine phosphokinase, PK = pyruvate kinase, LDH = lactate dehydrogenase, NADH = reduced nicotinamide adenine dinucleotide, and NAD = the oxidized form. The reaction is followed by measuring the change in absorbance of NADH at 340 nm. Creatine has also been determined by a fluorometric method by reaction with ninhydrin [139].

Creatine is now rarely determined in the routine laboratory. For that reason we have chosen to present the old Folin method in spite of its shortcomings. The reagents for this are always available. The enzymic method, although probably much more accurate, requires special reagents that are not very stable. If only a few determinations are made during a year, there is a distinct disadvantage in having to make up all new reagents each time.

Serum Creatine by Folin Method

REAGENTS

Same as given for determination of creatinine with Folin-Wu filtrate earlier in this section. Since the same filtrate and diluted standard are used for the total creatinine as for the preformed, it may be advisable to prepare larger quantities of each.

The preformed creatinine is first determined by the method using the Folin-Wu filtrate as presented earlier in this section. Using the same filtrate and diluted standard, add to separate 15 ml graduated centrifuge tubes, 6 ml of water as blank, 3 ml of filtrate plus 3 ml of water, and 1 ml of the diluted standard plus 5 ml of water. To each tube add 1 ml of picric acid solution and heat in a boiling water bath for about 2 hr until the volume is reduced below 4 ml. Remove the tubes from the bath, cool, and dilute each to 4 ml. To each tube add 1 ml of the sodium hydroxide solution $(0.75M)$, mix, and read after 20 min at 520 nm just as for the preformed creatinine. The calculations are the same, the standard used corresponding to 2 mg/dl creatinine. This gives the total creatinine in the sample.

CALCULATION

(total creatinine − preformed creatinine) × 1.16 = creatine

The factor of 1.16 converts creatinine concentration to creatine.

Urinary Creatine

PROCEDURE

The urine is diluted 1:100 and the procedure for preformed creatinine carried out as outlined earlier. The total creatinine is carried out exactly as for serum, using 3 ml of the diluted urine instead of the serum filtrate. The calculations are the same with an additional factor of 25 for the extra dilution of the urine.

Normal Values and Interpretation of Results

The normal values for serum creatine may be taken as 0.2 to 0.5 mg/dl for men and 0.3 to 0.9 mg/dl for women. In urine, men will excrete up to 40 mg/24 hr and women up to 80 mg/24 hr. The creatine values are of little clinical significance except in some of the diseases of muscle. In progressive muscular dystrophy and myasthenia gravis the excretion of creatine in the urine may rise to over 300 mg/day, but this rise is not seen in all patients and the urinary excretion is not a good indicator of the severity of the disease.

Any bromsulphalein (BSP) or phenolsulfonphthalein (PSP) remaining in the serum from other tests will be found in the filtrate and may interfere with the final color reaction as the solution is made distinctly alkaline. Similar dyes or indicators in the urine may be in high enough concentration to interfere with the color reaction. This can be checked by making alkaline another aliquot of the urine and noting if any color develops. High levels of ascorbic acid in the serum or urine may interfere, yielding high results. Ketone bodies also react with the alkaline picrate and in severe ketosis they may be present in sufficient quantities to give high results. The urine can be tested for ketone with one of the dip tests.

Uric Acid

The main methods for the determination of uric acid are based either on the reducing power of urate on phosphotungstate or cupric ions, or on the oxidation of uric acid to allantoin as catalyzed by the enzyme uricase. In the earlier colorimetric methods the uric acid was first isolated from serum or urine by precipitation as the salt of silver, mercury, or other metal [140–143]. The precipitate was later dissolved and caused to react with phosphotungstate to give the characteristic blue color. This procedure was found to be rather lengthy so other methods were developed in which the color reaction was carried out directly on a protein-free filtrate [144, 145]. A Folin-Wu filtrate has been used, but better recovery of added uric acid was obtained using an acid filtrate, such as that prepared with trichloracetic acid. Some methods used phosphotungstic acid as precipitating reagent and later developed the color by the addition of alkali and more phosphotungstic acid [146, 147]. Although the color produced is intensified by the presence of cyanide, the use of this poisonous reagent has been all but abandoned. Sufficient color intensity can be obtained without cyanide by proper adjustment of the alkalinity [148–151].

The color reaction with phosphotungstate is not very specific since other reducing substances present will also produce some color. It was found that by making the solution distinctly alkaline and allowing it to stand for a short time, a number of interfering substances, such as ascorbic acid, are oxidized by the air to substances that cause much less interference [152]. Although the use of phosphotungstic acid is the most common colorimetric method for uric acid, the reduction of cupric ions has also been suggested. The cupric ions are reduced to cuprous ions which are then determined by the yellow color produced with the reagent neocuproine (2,9-dimethyl-1,10-phenanthroline). This reagent is at least as sensitive as the phosphotungstate procedure [153–155] and may be slightly less affected by other reducing substances. The reduction of ferricyanide has also been used for the determination of uric acid [156, 157]. Recently this method has been reintroduced using a new sensitive reagent for the ferrous iron produced in the reduction [158].

The other method for the determination of uric acid uses the enzyme uricase which catalyzes the reaction:

uric acid + $O_2 \rightarrow$ allantoin + H_2O_2

The reaction is usually measured by noting the decrease in absorption at 293 nm. At this wavelength uric acid absorbs strongly whereas allantoin does not. In the method developed by Praetorius the measurements were made directly on a buffered dilution of the serum before and after addition of the enzyme [159, 160]. The presence of protein and other ultraviolet-absorbing substances often gave very high initial readings. This increased the error, particularly in the older instruments which could not be accurately read at absorbances above 1.0 unit. If smaller amounts of serum are used, the change in absorbance is also small. The method was later adapted to the use of a protein-free filtrate [161, 162]. Although the method is undoubtedly very specific for uric acid, it does require an ultraviolet spectrophotometer.

To avoid reading in the ultraviolet and yet retain the specificity of the enzymatic method, the hydrogen peroxide produced has been used to oxidize a chromogen to produce a visible color, usually with the aid of peroxidase or catalase [163–166]. This is similar to the colorimetric methods for glucose using glucose oxidase. These methods have not been widely used. Another variation is to use a phosphotungstic or other reduction method on two aliquots of the same sample, one with and one without treatment with uricase to destroy the uric acid. The difference in absorbance for the two tests is taken as a measure of the uric acid present. The method is thus specific for uric acid but requires additional colorimeter readings [154, 155, 167]. Catalase or peroxidase may also be included to destroy the hydrogen peroxide produced before the reduction step.

Other methods include a fluorometric procedure [168] and a coulometric determination in which the reducing substances in the sample are oxidized by coulometrically produced iodine before and after treatment with uricase [169].

Serum Uric Acid by Uricase Method

The following method has been recommended as a good reference method [155]. It is based on the colorimetric determination of reducing substances before and after the treatment with uricase. The reagent neocuproine is used for the color produced by copper reduction, in such a manner that precipitation of proteins is not required. This makes it simpler than the uricase methods involving the use of phosphotungstate. Catalase is added with the uricase to decompose the hydrogen peroxide formed which would otherwise interfere with the color development (see paragraph 1 in Comments section below).

REAGENTS

1. Borate buffer, $0.5M$, pH approx. 9.5, stock solution. Dissolve 31 g of anhydrous boric acid, 10 g of sodium hydroxide, and 1 g of lithium carbonate in about 700 ml of water in a 1 liter volumetric flask; dilute to the mark.

2. Borate buffer, working solution. Dilute the stock solution 1:10 with distilled water.

3. Copper-neocuproine, stock solution.   Dissolve 0.3 g of cupric sulfate penta-hydrate and 0.6 g of neocuproine hydrochloride* in water, add one drop of $1N$ hydrochloric acid, and dilute to 100 ml. Store in brown bottle.

4. Copper-neocuproine, working solution.   Dilute the stock solution 1:10 with distilled water. This diluted solution should be made fresh each day as required.

5. Uricase,† 2 mg/ml in 50% glycerol.

6. Catalase,† in glycerol.

7. Uricase-catalase mixture.   Mix together 4 volumes of uricase solution, 1 volume of catalase solution, and 3 volumes of water. Store in the refrigerator (see paragraph 2 in Comments section below).

8. Uric acid standards.

A. Stock standard, 50 mg/dl.   Dissolve 1 g of lithium carbonate in about 500 ml of water in a 1 liter volumetric flask. Add 0.500 g of uric acid and after it has dissolved, add 5 ml of 40% formaldehyde and about 400 ml of water. Adjust to a pH of 5.5 by the dropwise addition of $4.35M$ acetic acid (this first adjust-ment may be made with short-range indicator paper). Mix well and allow to stand for 2 days in the dark. Check the pH, preferably with a dipping glass electrode, adjust if necessary, and dilute to the mark. This solution is stable at room temperature in a dark brown bottle; refrigeration is not necessary.

B. Working standards.   Dilute 5 and 10 ml of the stock standard to 50 ml with water to give solutions containing 5 and 10 mg/dl.

PROCEDURE

For each serum set up two tubes labeled A and B. Add 3 ml of diluted borate buffer and 0.1 ml of serum to each tube. Also set up a reagent blank containing 0.1 ml of water and 3 ml of the buffer. Only one reagent blank is needed for each set of serum determinations. Add 0.02 ml of the uricase-catalase mixture to tube B of each set only. Mix and allow to stand for 10 min at room temperature. Add 1.0 ml of color reagent to every tube, mix, and allow to stand at room temper-ature for 10 min. Read all serum tubes against reagent blank at 454 nm. For each serum, subtract absorbance of B tube from that of the corresponding A tube to give net absorbance of sample.

For the standards, only the A tubes are needed since it is assumed that after uricase treatment the color will be the same as for the reagent blank. (See

---

* G. Fredrick Smith Chemical Co., Columbus, Ohio 43223.
† Enzymes obtained from Boehringer-Mannheim Corp. New York, N.Y. 10017; Uricase Cat. No. 15074 EUAC; Catalase Cat. No. 15675 EKAB.

paragraph 3 in Comments section below.) Add 0.1 ml of the two standards to separate tubes containing 3 ml of the diluted buffer, add 1.0 ml of the color reagent, incubate at room temperature for 10 min, and read against reagent blank at 454 nm.

CALCULATION

$$\frac{\text{net absorbance of sample}}{\text{absorbance of standard}} \times \text{conc. of standard} = \text{conc. of sample}$$

COMMENTS

1. Catalase is used, rather than the usual peroxidase, to destroy the hydrogen peroxide because the latter enzyme usually contains impurities that may interfere with the color reaction and give a high blank.

2. The enzyme blank is usually negligible but this should be checked for each batch of enzyme as follows. To each of two tubes add 3.1 ml of the diluted borate buffer. To one tube add 0.02 ml of the mixed enzyme solution and mix. Allow to stand for a few minutes and then add 1.0 ml of diluted color reagent to each tube. Mix and allow to stand for 10 min. Then read the tube containing the enzyme against the one without the enzyme at 454 nm. The absorbance difference should ordinarily be less than 0.005 absorbance units. The reading for this enzyme blank, which may be assumed to be constant for a given batch of enzyme solution, is then *added* to the net readings for all samples (but not standards) before calculation of the final result.

3. It is not necessary to add the enzyme to the standards since they do not contain any other reducing substances. In addition, the small amount of formaldehyde present may interfere with the enzyme. The standards may be used for recovery experiments by adding 0.1 ml of standards to 0.1 ml of serum and 2.9 ml of diluted buffer since the protein present will combine with the formaldehyde present in the standards and prevent its interference with the enzyme action.

Urinary Uric Acid by Uricase Method

Collect 24-hr urine specimen in a bottle containing about 5 ml of saturated lithium carbonate to insure solubility of urates. Mix well and make a 1:5 dilution with water. Treat diluted urine the same as the serum (see preceding section). Multiply the result of the calculation by 5 to correct for the extra dilution. Then: conc. of sample (in mg/dl) $\times$ 10 $\times$ urine volume (in liters) = total 24-hr excretion.

Uric Acid by Phosphotungstate Method [170]

REAGENTS

1. Phosphotungstic acid (Folin and Dennis) [140]. Add 50 g of sodium tungstate ($Na_2WO_4 \cdot 2H_2O$) to about 400 ml of water. After it has dissolved, add 40 ml of 85% phosphoric acid. Reflux in an all-glass apparatus for 2 hr. Cool and dilute to 500 ml. This solution is stable when stored in a brown bottle and protected from contact with organic matter. A portion of the solution is diluted 1:10 for use. The phosphotungstic acid is available commercially. A number of different phosphotungstic acid solutions are sold; be sure to obtain the one that is specified as made according to Folin and Dennis.

2. Sodium carbonate, $0.95M$.   Dissolve 100 g of anhydrous sodium carbonate in water. Filter if not perfectly clear and store in a plastic bottle.

3. Tungstic acid precipitating solution.   To 800 ml of water and 50 ml each of the sodium tungstate and sulfuric acid solutions used for the Folin-Wu precipitation (see section on Folin-Wu filtrate in Chapter 5), add 0.05 ml of 85% phosphoric acid and dilute to 1 liter.

4. Uric acid standards.

   A. Stock standard, 50 mg/dl.   Same as for uricase method above.

   B. Working standard.   As needed each day, dilute 1 ml of the stock to 100 ml with water. This gives a solution containing 0.5 mg/dl. When compared with an equal volume of a 1:10 serum filtrate, it is equivalent to 5 mg/dl.

PROCEDURE

Add 1 ml of serum to 9 ml of the tungstic acid precipitating solution. Mix well and allow to stand for 10 min. Centrifuge or filter through a retentive paper; the former procedure may be more rapid. To separate tubes add 5 ml of the filtrate or supernatant, 5 ml of working standard, and 5 ml of water as blank. To each add 1 ml of the sodium carbonate solution, mix, and allow to stand for 10 min. Add 1 ml of the diluted phosphotungstic acid to each tube, mix, and allow to stand for 10 min. Read standard and samples against blank at any convenient wavelength between 660 and 720 nm (660 nm is usually satisfactory and is available with most photometers).

PROCEDURE FOR URINE

If the urine is cloudy, warm to 60°C to dissolve any precipitated urates. Dilute 1 ml urine to 100 ml with water. Treat 5 ml of this dilution similarly to the serum filtrate. Note that the urine has been diluted 10 times more than blood (1:100 instead of 1:10) so that in the calculation given the figure 5 is replaced by 50. Depending upon the urine concentration, other dilutions may be necessary to bring the readings to a suitable range.

CALCULATION

Since 5 ml of standard containing 0.5 mg/dl is compared with 5 ml of a 1:10 dilution of the serum:

$$\frac{\text{absorbance of sample}}{\text{absorbance of standard}} \times 5 = \text{conc. of sample (in mg/dl)}$$

COMMENTS

The determination of uric acid by the phosphotungstic acid reduction methods may be influenced by the presence of other reducing agents. Large amounts of ascorbic acid may give high results as may therapeutic doses of L-dopa and methyldopa. The uricase method is not affected by these drugs. A number of drugs such as the thiazide diuretics, ethacrynic acid, furosemide, and chlorthalidone act by decreasing the urinary excretion of uric acid and thus raising the level in the blood. Other drugs reported to increase the uric acid levels in blood are quinethazone, acetazolamide, diazoxide, 6-mercaptopurine, and nitrogen mustard. Usual doses of salicylates have been reported to increase the uric acid level when given along with phenylbutazone or probenecid, but large continued doses of the drug have been stated to decrease the serum level. A number of other drugs have been reported to decrease uric acid levels in serum but the decrease was found in only a fraction of all cases. The drugs mentioned in this connection include phenylbutazone, azathioprine, chlorprothixene, acetohexamide, and clofibrate.

Normal Values and Interpretation of Results

For the phosphotungstate reduction methods the normal range of uric acid in plasma or serum is 3.5 to 8.0 mg/dl for males and 2.5 to 7.0 mg/dl for females. Slightly lower values may be obtained using the uricase method. Some of the older reported values are for whole blood. Since most of the uric acid is in the plasma, the values for whole blood would vary with the proportion of cells in the blood.

Increased levels of uric acid are associated with nitrogen retention and the increase in urea, creatinine, and the other nonprotein nitrogenous constituents of the blood. This is often interpreted as another indication of decreased kidney function. Uric acid is formed from the breakdown of the cell nucleic acids and is often increased in the blood in conditions in which excessive cell breakdown and catabolism of nucleic acids occur. Increases in the blood level have been reported in the acute stages of infectious diseases, excessive exposure to roentgen rays, multiple myeloma, and leukemia. Increased levels of blood uric acid have been found in gout, but the increase may be slight in the early stages of the disease and the degree of increase is not directly related to the severity of the disease.

References

1. Henriques, V., and Sorensen, S. P. L.   *Z. Physiol. Chem.* 64:120, 1909.
2. Van Slyke, D. D., MacFayden, D. A., and Hamilton, P. B.   *J. Biol. Chem.* 150:251, 1943.
3. Van Slyke, D. D.   *J. Biol. Chem.* 9:181, 1911.
4. Van Slyke, D. D., and Kirk, E.   *J. Biol. Chem.* 102:651, 1933.
5. Saifer, A., Gerstenfeld, S., and Harris, A. F.   *Clin. Chim. Acta* 5:131, 1960.
6. Mathews, D. M., Muir, G. G., and Baron, D. M.   *J. Clin. Pathol.* 17:150, 1960.
7. Khachadurian, A., Knox, W. E., and Cullen, A. M.   *J. Lab. Clin. Med.* 56:321, 1960.
8. Sienst, D., Sienst, G., and Besson, S.   *Bull. Soc. Pharm. Nancy* 51:19, 1961.
9. Van Slyke, D. D., Dillon, R. T., MacFayden, D. A., and Hamilton, P. B.   *J. Biol. Chem.* 141:627, 1941.
10. MacFayden, D. A.   *J. Biol. Chem.* 145:387, 1942.
11. Frame, E. G.   *Stand. Methods Clin. Chem.* 4:1, 1963.
12. Folin, O.   *J. Biol. Chem.* 51:377, 1922.
13. Folin, O.   *J. Biol. Chem.* 51:393, 1922.
14. Frame, E. G., Russell, J. A., and Wilhelmi, A. E.   *J. Biol. Chem.* 149:225, 1943.
15. Russell, J. A.   *J. Biol. Chem.* 156:467, 1944.
16. Clayton, C. C., and Steele, B. F.   *Clin. Chem.* 13:49, 1967.
17. Vincent, W. A.   *Nature* (Lond.) 185:530, 1960.
18. Kekki, M.   *Scand. J. Clin. Lab. Invest.* 11:311, 1959.
19. Wells, M. G.   *Clin. Chim. Acta* 25:27, 1969.
20. Hall, F. F., Payton, G. A., and Wilson, S. D.   *Tech. Bull. Registry Med. Technol.* 39:89, 1969.
21. Nydr, J., and Stahlavska, A.   *Clin. Chim. Acta* 38:457, 1972.
22. Palmer, D. W., and Peters, T., Jr.   *Clin. Chem.* 15:891, 1969.
23. Dubin, D. T.   *J. Biol. Chem.* 235:783, 1960.
24. Goodwin, J. F.   *Clin. Chim. Acta* 21:231, 1968.
25. Goodwin, J. F.   *Clin. Chem.* 14:1080, 1968.
26. Goodwin, J. F.   *Stand. Methods Clin. Chem.* 6:89, 1970.
27. Wang. J. H., and Tarr, D. A.   *J. Am. Chem. Soc.* 77:6205, 1955.
28. Koch, F. C., and McMeekin, T. L.   *J. Am. Chem. Soc.* 46:2066, 1924.
29. Henry, R. J., and Chamori, H.   *Am. J. Clin. Pathol.* 29:277, 1958.
30. Noble. E. D.   *Anal. Chem.* 27:1413, 1955.
31. Searcy, R. L., Reardon, J. E., and Foreman, J. A.   *Am. J. Med. Technol.* 33:15, 1967.
32. Fawcett, J. K., and Scott, J. E.   *J. Clin. Pathol.* 13:149, 1960.

33. Mathies, J. C. *Clin. Chem.* 10:366, 1964.
34. Naftalin, L., Whitaker, J. F., and Stephens, S. *Clin. Chim. Acta* 14:771, 1966.
35. Wilson, B. W. *Clin. Chem.* 12:360, 1966.
36. Searcy, R. L., Foreman, J. A., Ketz, A., and Readon, J. *Tech. Bull. Registry Med. Technol.* 37:107, 1967.
37. Creno, R. J., Wenk, R. E., and Bolig, P. *Am. J. Clin. Pathol.* 54:828, 1970.
38. Kirsten, E., Gerez, G., and Kirsten, C. *Biochem. Z.* 337:312, 1963.
39. Kaldwasser, H., and Schegel, H. G. *Anal. Biochem.* 16:138, 1966.
40. Hallett, C. J., and Cook, J. G. H. *Clin. Chim. Acta* 35:33, 1971.
41. Guilbault, G. G., and Montalvo, J. G., Jr. *J. Am. Chem. Soc.* 92:2533, 1970.
42. Guilbault, G. G., and Hrabankovna, E. *Anal. Chim. Acta* 52:287, 1970.
43. Natelson, S., Scott, M. L., and Belfa, C. *Am. J. Clin. Pathol.* 21:277, 1951.
44. Friedman, H. S. *Anal. Chem.* 25:660, 1953.
45. Caraway, W. T., and Fanger, H. *Am. J. Clin. Pathol.* 26:1475, 1956.
46. Marsh, W. H., Fingerhut, B., Dirsch, E. *Am. J. Clin. Pathol.* 28:681, 1957.
47. Richter, H., and Lapointe, Y. *Clin. Chem.* 8:335, 1962.
48. Coulombe, J. J., and Favaeau. L. *Clin. Chem.* 9:102, 1963.
49. Marsh, W. H., Fingerhut, B., and Miller, H. *Clin. Chem.* 11:624, 1965.
50. Annino, J. S. *Am. J. Clin. Pathol.* 48:147, 1967.
51. Crocker, C. L. *Am. J. Med. Technol.* 33:361, 1967.
52. McCleskey, J. E. *Anal. Chem.* 36:1646, 1964.
53. Geiger, P. J. *Microchem. J.* 13:481, 1968.
54. Bailly, M., Fonty, P., and Leger, N. *Ann. Biol. Clin.* (Paris) 25:1221, 1967.
55. Yatzidis, H., Garidi, M., Vassilikos, C., Mayopoulou, D., and Akilas, A. *J. Clin. Pathol.* 17:163, 1964.
56. Nakagawa, H., Heirwegh, K., and De Grotte, J. *Int. Congr. Clin. Chem.* (*Proc.*) *7th* 1:35, 1969.
57. Siest, G. *Clin. Chim. Acta* 18:155, 1967.
58. Timmermans, C. J. *Clin. Chim. Acta* 7:887, 1962.
59. Logan, J. E. *Can. Med. Assoc. J.* 98:341, 1963.
60. Tarnoky, A. L. *Clin. Chim. Acta* 10:253, 1964.
61. Napier, J. A. F., and Lines, J. G. *Ann. Clin. Biochem.* 6:59, 1969.
62. McNair, R. D. *Clin. Chem.* 11:74, 1965.
63. Conway, E. J. *Microdiffusion and Volumetric Error*, 4th ed. London: Coxby & Lockwood, 1957.
64. Seligson, D., and Seligson, H. *J. Lab. Clin. Med.* 38:324, 1951.
65. Seligson, D., and Hirahara, K. *J. Lab. Clin. Med.* 49:962, 1957.
66. Stone, W. E. *Proc. Soc. Exp. Biol. Med.* 93:589, 1956.
67. Hoffman, M. *Z. Med. Labortech.* 10:86, 1969.
68. Gips, C. H., and Wibbens, A. M. *Clin. Chim. Acta* 22:183, 1968.
69. Wueller-Beissenhirtz, W., and Keller, H. *Klin. Wochenschr.* 43:43, 1965.
70. Leffler, H. H. *Am. J. Clin. Pathol.* 48:233, 1967.
71. McCullough, H. *Clin. Chim. Acta* 17:297, 1967.
72. Lorentz, K., and Ossenberg, W. *Med. Lab.* (Stuttg.) 20:77, 1967.
73. Kingsley, G. R., and Tager, H. S. *Stand. Methods Clin. Chem.* 6:115, 1970.
74. Rahiala, E. I., and Kekomaki, M. P. *Clin. Chim. Acta* 30:761, 1970.

75. Szabo, E., Wein, I., and Szam, I.   *Z. Gesamte Inn. Med.* 25:27, 1970.
76. Dienst, G., and Morris, B.   *J. Lab. Clin. Med.* 64:495, 1964.
77. Foreman, D. T.   *Clin. Chem.* 10:497, 1964.
78. Wenzel, E., Oppolzer, R., and Schuster, D.   *Microchim. Acta* 197:680, 1971.
79. Oreskes, I., Hirsch, C., and Kupfer, S.   *Clin. Chim. Acta* 26:185, 1969.
80. Reichelt, K. L., Kvamme, E., and Tveit, B.   *Scand. J. Clin. Lab. Invest.* 16:433, 1964.
81. Kirsten, E., Gerez, C., and Kirsten, R.   *Biochem. Z.* 337:312, 1963.
82. Sardesai, V. M., and Provido, H. S.   *Microchem. J.* 14:550, 1969.
83. Imler, M., Frick, A., Stahl, A., Peter, B., and Stahl, J.   *Clin. Chim. Acta* 37:246, 1972.
84. Manston, R.   *Biochem. Med.* 4:486, 1970.
85. Lowe, W. C.   *Clin. Chem.* 14:1074, 1968.
86. Conn, H. O., and Kuljian, A.   *J. Clin. Lab. Med.* 63:1033, 1964.
87. Orloff, M. J., and Stevens, C. O.   *Clin. Chem.* 10:991, 1964.
88. Folin, O.   *Z. Physiol. Chem.* 41:223, 1904.
89. Miller, B. F., and DuBois, R.   *J. Biol. Chem.* 121:447, 1937.
90. Owen, J. A., Iggo, B., Scandrett, F. G., and Stewart, C. P.   *Biochem. J.* 58:426, 1954.
91. Hare, R. S.   *Proc. Soc. Exp. Biol. Med.* 74:148, 1950.
92. Ralston, M.   *J. Clin. Pathol.* 8:160, 1955.
93. Taussky, H. H.   *Stand. Methods Clin. Chem.* 3:99, 1961.
94. Paget, M., Gontier, M., and Liefooghe, J.   *Ann. Biol. Clin.* (Paris) 13:535, 1955.
95. Polar, E., and Metcoff, J.   *Clin. Chem.* 11:763, 1965.
96. Stotten, A.   *J. Med. Lab. Technol.* (Lond.) 25:240, 1968.
97. Sullivan, M. X., and Irreverre, F.   *J. Biol. Chem.* 233:530, 1958.
98. Van Pilsum, J. F., Martin, R. P., Kito, E., and Hess, J.   *J. Biol. Chem.* 222:225, 1956.
99. Cooper, J. M., and Biggs, H. G.   *Clin. Chem.* 7:665, 1961.
100. Adams, W. S., Davis, F. W., and Hansen, L. E.   *Anal. Chem.* 34:854, 1962.
101. McEvoy-Bowe, E.   *Anal. Biochem.* 16:153, 1966.
102. Paumgartner, G., Kramp, O., and Fischer, F. X.   *Clin. Chim. Acta* 8:960, 1963.
103. Fischl, J., Segal, S., and Yulzari, Y.   *Clin. Chim. Acta* 10:73, 1964.
104. Rink, M., and Krebber, D.   *J. Chromatogr.* 25:80, 1966.
105. Husdan, H., and Rapoport, A.   *Clin. Chem.* 14:222, 1968.
106. Bartels, H., Boehmer, M., and Heierli, C.   *Clin. Chim. Acta* 37:193, 1972.
107. Zender, R., and Jacot, P.   *Anal. Lett.* 5:143, 1972.
108. Raabo, E., and Wello-Hansen, P.   *Scand. J. Clin. Lab. Invest.* 29:297, 1972.
109. Larsen, K.   *Clin. Chim. Acta* 41:209, 1972.
110. Natochin, Y. V., and Shakhmatova, E. L.   *Lab. Delo* 1:17, 1973.
111. Folin, O.   *Am. J. Physiol.* 13:45, 1905.
112. Anker, R. M.   *J. Lab. Clin. Med.* 43:798, 1954.
113. Vestergaard, P., and Leverett, R.   *J. Lab. Clin. Med.* 51:211, 1958.
114. Smith, R. L., Loewenthal, H., Lehmann, H., and Ryan, E.   *Clin. Chim. Acta* 4:384, 1959.
115. Kennedy, W. P.   *Arch. Dis. Child.* 36:325, 1961.
116. Bleiler, R. E., and Schedl, H. P.   *J. Lab. Clin. Med.* 59:945, 1962.
117. Albanese, A. A., and Wangerin, D. M.   *Science* 100:58, 1944.
118. Wray, P. M., and Russell, C. S.   *J. Obstet. Gynaecol. Br. Emp.* 67:623, 1960.
119. Cramer, W., Cramer, H., and Selander, S.   *Clin. Chim. Acta* 15:331, 1967.

120. Scott, P. J., and Hurley, P. J.   *Clin. Chim. Acta* 21:411, 1968.
121. Chattaway, F. W., Hullin, R. P., and Odds, F. C.   *Clin. Chim. Acta* 26:567, 1969.
122. Wang, E.   *Acta Med. Scand. (Suppl.)* 105:1, 1939.
123. Clarke, J. T.   *Clin. Chem.* 7:271, 1961.
124. Ryan, R. J., Williams, J. D., Ansell, B. M., and Bernstein, L. M.   *Metabolism* 6:365, 1957.
125. Clark, L. C., Jr., Thompson, H. L., Beck, E. I., and Jacobson, W.   *Am. J. Dis. Child.* 81:774, 1951.
126. Petersen, N.   *Clin. Chim. Acta* 18:57, 1967.
127. Edwards, O. M., Bayless, R. I. S., and Millen, S.   *Lancet* 2:1165, 1969.
128. Martinek, R. G.   *Am. J. Med. Technol.* 28:323, 1966.
129. Dunicz, B. L.   *Clin. Chim. Acta* 9:203, 1964.
130. Biggs, H. G., and Cooper, J. M.   *Clin. Chem.* 7:655, 1961.
131. Brinnerink, P. C.   *Clin. Chim. Acta* 6:531, 1961.
132. Wong, T.   *Anal. Biochem.* 40:18, 1971.
133. Gundlach, G., Hoppe-Seyler, G., and Johann, H.   *Z. Klin. Chem. Klin. Biochem.* 6:415, 1968.
134. Lauber, K.   *Z. Klin. Chem.* 4:119, 1966.
135. Griffiths, W. J.   *Clin. Chim. Acta* 9:210, 1964.
136. Barlet, H. A.   *Clin. Chim. Acta* 20:149, 1968.
137. Marymount, J. H., Jr., Smith, J. N., and Klotsch, S.   *Am. J. Clin. Pathol.* 49:289, 1968.
138. Kibrick, A. C., and Milhorat, A. T.   *Clin. Chim. Acta* 14:201, 1966.
139. Conn, R. B.   *Clin. Chem.* 6:537, 1960.
140. Folin, O., and Dennis, L.   *J. Biol. Chem.* 13:469, 1912–13.
141. Folin, O., and Merenzi, A. D.   *J. Biol. Chem.* 83:109, 1929.
142. Brechner-Mortensen, K.   *Acta Med. Scand. (Suppl.)* 84:1, 1937.
143. Bergman, F., and Dikstein, S.   *J. Biol. Chem.* 211:149, 1954.
144. Benedict, S. R.   *J. Biol. Chem.* 51:187, 1922.
145. Brown, H.   *J. Biol. Chem.* 158:610, 1945.
146. Pileggi, W. J., DiGiorgio, J., and Wybenga, D. R.   *Clin. Chim. Acta* 36:141, 1972.
147. Carroll, J. J., Coburn, H., Douglass, R., and Babson, A. L.   *Clin. Chem.* 17:158, 1971.
148. Archibald, R. H.   *Clin. Chem.* 3:102, 1957.
149. Kern, A., and Stransky, E.   *Biochem. Z.* 290:419, 1937.
150. Caraway, W. T.   *Am. J. Clin. Pathol.* 25:840, 1955.
151. Kantner, S. L.   *Clin. Chem.* 13:406, 1967.
152. Caraway, W. T.   *Clin. Chem.* 15:720, 1969.
153. Bittner, D. L., Hall, S., and McCleary, M.   *Am. J. Clin. Pathol.* 40:423, 1963.
154. Morgenstern, S., Flor, R. V., Kaufman, J. H., and Klein, B.   *Clin. Chem.* 12:748, 1966.
155. Bittner, D. L., and Gambino, S. R.   *Uric Acid Assays*. Chicago: Commission on Continuing Education, American College of Pathologists, 1970.
156. Silverman, H., and Gubernick, I.   *J. Biol. Chem.* 167:363, 1947.
157. Bulger, H. A., and John, H. E.   *J. Biol. Chem.* 140:427, 1941.
158. Klein, B., and Lucas, L. B.   *Clin. Chem.* 19:67, 1973.
159. Praetorius, E.   *Scand. J. Clin. Lab. Invest.* 1:222, 1949.

160. Praetorius, E., and Poulsen, H. *Scand. J. Clin. Lab. Invest.* 5:273, 1955.
161. Feichtmeier, T. V., and Wrenn, H. T. *Am. J. Clin. Pathol.* 25:833, 1955.
162. Remp, D. C. *Stand. Methods Clin. Chem.* 6:1, 1970.
163. Domagk, G. E., and Schlicke, H. H. *Anal. Biochem.* 22:219, 1965.
164. Lorentz, K., and Berndt, W. *Anal. Biochem.* 18:58, 1967.
165. Gochman, N., and Schmitz, J. M. *Clin. Chem.* 17:1154, 1971.
166. Kegeyams, N. *Clin. Chim. Acta* 31:421, 1971.
167. Caraway, W. T., and Marable, H. *Clin. Chem.* 12:18, 1966.
168. Block, P. L., and Lata, G. F. *Anal. Biochem.* 3:1, 1970.
169. Troy, R. T., and Purdy, W. C. *Clin. Chim. Acta* 27:410, 1970.
170. Caraway, W. T. *Stand. Methods Clin. Chem.* 4:239, 1963.

7. Proteins and Amino Acids

Serum Total Proteins

The reference method still used for the determination of serum total proteins is the Kjeldahl method introduced over 90 years ago. It is based on the conversion of all nitrogenous compounds to ammonia by heating with sulfuric acid (usually with the aid of catalysts). The solution is then made alkaline and the ammonia distilled over and titrated in the distillate with standard acid. In the digestion procedure, potassium sulfate may be added to give a somewhat higher digestion temperature. Phosphate may be added for its catalytic action. Another catalyst is usually used, either copper, mercury, or selenium. Copper has been used for many years. Mercury is also a good catalyst but has the disadvantage that in alkaline solution the mercuric ions will tend to combine with the ammonia and prevent its distillation. This may be overcome by adding some sodium thiosulfate along with the alkali; the mercury is precipitated as the sulfide so that it does not interfere.

The method for serum proteins has now been fairly well standardized [1]. The result is obtained as nitrogen, and the figure must be converted into equivalent protein by multiplying by an appropriate factor. For many years the factor was taken as 6.25, i.e., N × 6.25 = protein. It has been shown that the factor is different for each type of protein such as albumin, globulin, etc. [2–4]. The problem is, of course, to be certain that the different proteins have been isolated in a pure form. One investigation gave a factor of 6.54 for the total protein in serum [5]. Another worker judged the factor to be close to 6.4 [6, 7]. The difference between the two factors is about 4.5 percent. Many of the literature values for the normal ranges for total protein were probably obtained with methods standardized by the Kjeldahl method using the older factor of 6.25.

The discussion in this section will deal mainly with the determination of protein in serum; the next section will present the methods that are more applicable to the determination of the much smaller amounts of protein in urine and cerebrospinal fluid. Some of the methods mentioned here, however, are sensitive enough to be used for urine. One theoretically simple method is the measurement of the absorbance of a dilute protein solution in the ultraviolet. This has been used by a number of workers [8–10]. The measurement may be made at 280 nm or by using the difference in absorbance at 215 and 225 nm [11, 12]. Measurements at 210 nm and the difference measurements at 215 and 225 nm are somewhat more accurate, but the measurements at 280 nm show greater precision and less interference from nonprotein substances. Some workers claim that the variation in absorbance among the different proteins is too great to allow the method to be used for a variable mixture of proteins such as serum

[13]; as a result the method has not been used very widely for serum, but because of its sensitivity it is suitable for urine and spinal fluid.

A physical method that has been used for serum protein utilizes the measurement of the index of refraction. All dissolved material in a solution such as serum will contribute to the increase in refractive index over that of water. The great amount of protein makes the largest contribution whereas that due to the other substances is more constant [14–16]. The method is subject to errors due to lipemia and icterus as well as increased amounts of other dissolved material in the serum [17, 18]. Since the refractive index method was standardized on serums whose protein content was determined usually by the Kjeldahl method, it has been suggested that the formula used for the calculation of serum proteins by refractive index be changed to agree with the new factor for conversion of nitrogen to protein [19]. In one method that seeks to overcome some of the interference, the refractive index of the solution is measured before and after the precipitation of the proteins by heat [20]. This should give greater accuracy; however, it reduces the great advantages of the refractometric method—its rapidity and the use of only a small amount of serum [21].

Another physical method is determination of the specific gravity. As with the refractive index, all the dissolved substances in the serum will contribute to the specific gravity, but here also the major contribution to the specific gravity is made by the protein. Because of the presence of the other substances, an empirically determined equation must be used relating the serum protein and the specific gravity. Such an equation was derived by Moore and Van Slyke using the specific gravity as determined with a pycnometer [22]. The most popular method using this principle is the copper sulfate method introduced by Van Slyke and associates [22–24] but even this method is seldom used, even as a screening test. It is not very accurate with abnormal serums or with hyperlipemic or highly icteric serums.

The most widely used method for the determination of serum total protein is the colorimetric method based on the biuret reaction. The reaction occurs between cupric ions in alkaline solution and a number of compounds containing at least two peptide linkages (which would include all proteins), as well as a number of simpler compounds containing two $-CONH_2$, $-CH_2NH_2$, or similar groups joined together directly or through a C or N atom [25]. A number of different biuret reagents have been used, the most common ones containing either tartrate [26, 27] or citrate [28] to keep the cupric ions in solution in the alkaline medium. One other method uses a stabilized suspension of cupric hydroxide, the excess being centrifuged down before making the readings [29].

The difference in absorbance between the blank and the blank plus sample has a maximum near 550 nm and the readings are usually made at wavelengths between 545 and 555 nm. The biuret complex has a high absorbance in the near ultraviolet and measurements have been made at 300 nm [30] or 263 nm [31]. Even the weaker reagents, reading at these short wavelengths, give over tenfold increase in sensitivity. Such methods are often used on a micro scale or for the determination of protein in cerebrospinal fluid or urine. There is often interference from turbidity, hyperlipemia, or icteric serums. The interference is greater when the readings are made in the ultraviolet. The interference can be minimized by reading a blank prepared from an aliquot of the serum in a solution that contains all the components of the biuret reagent except the copper. Sometimes the interference from hyperlipemia can be reduced by first extracting the serum with ether. Another method of correcting for these interferences is to make readings before and after the addition of cyanide which destroys the copper-biuret complex [32].

The methods mentioned above have been used chiefly for serum proteins and other solutions containing appreciable amounts of protein. The ultraviolet method is sensitive enough to be used with cerebrospinal fluid [37]. Although the biuret reaction as used for serum proteins is hardly sensitive enough for use with spinal fluid, it has been used for this purpose, either by first precipitating the proteins and then adding the biuret reagent to dissolve the precipitate and form the color [38], or by measuring the biuret color in the ultraviolet [39, 42]. A sensitive method uses the Folin-Ciocalteu phenol reagent with the addition of copper which increases the sensitivity [33, 34]. A disadvantage of this method is that a number of drugs such as salicylates, sulfonamides, and phenacetin among others also react with the phenol reagent to give appreciable color [35, 36]. Suggested methods for eliminating this interference include first precipitating the protein [40] or making readings with two solutions, one containing the Folin reagent and the other containing the Folin reagent plus copper [41]. The protein in spinal fluid may also be determined by first separating the protein by means of gel filtration with Sephadex G-50 and then either measuring the absorbance in the ultraviolet at 220 nm [43] or by use of the Folin reagent with copper [44].

The small amounts of protein in urine and cerebrospinal fluid may also be determined by turbidimetric methods. One of the earliest reagents used for this purpose was sulfosalicylic acid [45]. It was found that different proteins gave different amounts of turbidity with this reagent [46]. The addition of $0.49M$ anhydrous sodium sulfate [47, 48] or 20% of a saturated urea solution [49] improved the reagent. The solution containing sodium sulfate in addition to the sulfosalicylic acid is often used. The other reagent used to produce the turbidity is trichloracetic acid added to a final concentration of about 2.5% [46, 50, 51].

This reagent is simple to prepare and is preferred by many workers. Usually the turbidity is read in an ordinary colorimeter using a wavelength (which is not critical) in the short end of the visible spectrum. A direct nephelometric method has been suggested in which a fluorometer was used as a nephelometer [52].

More recently the reaction between proteins and trinitrobenzenesulfonic acid has been shown to give a stable color suitable for the determination of proteins [53]. This method has not been extensively used. Although more sensitive than the biuret reaction, it is not as simple a procedure.

In the electrophoresis of proteins the amount of proteins in the separate fractions is estimated by the color of the stained proteins. This has been applied as a micromethod for protein. A small aliquot of the protein solution is placed on paper or cellulose acetate strips such as used for electrophoresis. After fixing and staining, the color is compared with that produced by spots from standard amounts of protein [54, 55].

Total Proteins by the Kjeldahl Method

The Kjeldahl method [56–58] is still the reference method used for the determination of protein in spite of the uncertainty already mentioned about the conversion factor from nitrogen to protein. The method is much too long for routine use. The nitrogen in the protein is converted entirely into ammonium sulfate by digestion with sulfuric acid in the presence of a catalyst. A number of catalysts have been used. A mixed one containing mercury and selenium [59] has been found very satisfactory. When mercury is used, it must be removed after digestion as it would interfere with the distillation of the ammonia. Either zinc, which reduces and amalgamates with the mercury, or sodium thiosulfate, which forms an insoluble mercuric sulfide, can be used. Potassium sulfate and potassium phosphate are added to the digestion mixture to increase the temperature and aid in digestion. After digestion, the solution is made alkaline and the ammonia distilled over into a boric acid solution. The boric acid aids in trapping the ammonia without interfering with the subsequent titration. The distilled ammonia is then titrated with standard acid. A blank should be run on all the reagents and the procedure checked by analyzing a standard solution of ammonium sulfate. Deionized water should be used throughout; ordinary distilled water may contain traces of ammonia.

REAGENTS

1. Digestion mixture. Mix together 90 g of anhydrous potassium sulfate (K_2SO_4), 45 g of dibasic potassium phosphate (K_2HPO_4), 9 g of mercuric oxide,

and 1.5 g of selenium powder. The potassium salts are first powdered in a large mortar if necessary and then all the powders are well mixed.

2. Concentrated sulfuric acid, reagent grade.

3. Sodium hydroxide, 50 g/dl, plus sodium thiosulfate, 10 g/dl. Dissolve 100 g of sodium hydroxide in about an equal amount of water, cool, add 20 g of sodium thiosulfate, dissolve, and dilute to 200 ml.

4. Methyl red indicator, 0.1 g/dl. Dissolve 0.1 g of methyl red indicator in 100 ml of 95% ethanol.

5. Bromcresol green indicator, 0.1 g/dl. Dissolve 0.1 g of bromcresol green (acid form, not water-soluble sodium salt) in 100 ml of 95% ethanol.

6. Boric acid solution with indicator [60]. Dissolve 20 g of reagent grade boric acid in 1 liter of water with the aid of gentle heating. To 400 ml of the cool boric acid solution add 1 ml of the methyl red solution and 5 ml of the bromcresol green solution, and mix. It is advisable to check the indicator endpoint color change as it may vary with the exact amount of indicators used, and some individuals can detect the endpoint better than others. Add 20 ml of the boric acid solution with indicator to a flask. Add 20 ml of water and a few drops of ammonia solution. Titrate with the standard acid and note the endpoint. The change should be from green to pink, with the solution being almost colorless at the endpoint. Different proportions of the indicators can be tried to give the sharpest endpoint. Once the optimum proportion has been found, it can be used for subsequent preparation of the boric acid solution with indicator.

7. Sulfuric acid, $0.1M$. Dilute 20 ml of accurately standardized $0.5M$ acid to 1 liter with water.

8. Ammonium sulfate standard. Dissolve exactly 3.3035 g of reagent grade ammonium sulfate, which has been previously dried at 105°C for several hours, in water containing about 0.5 ml of concentrated sulfuric acid and dilute to 500 ml. This solution contains 1.401 mg of nitrogen per ml. One milliliter of this solution when carried through the entire procedure and distillation should give a titration (corrected for the blank) of exactly 5.0 ml of the $0.02N$ acid. This solution is used to check the procedure and the $0.1M$ sulfuric acid.

SPECIAL EQUIPMENT REQUIRED

1. Kjeldahl digestion flasks, 30 ml.

2. Digestion apparatus for flasks with fume hood for removing acid fumes. For occasional use, a holder for the flask may be improvised from a ringstand and clamps. Most of the fumes may be removed by positioning a small funnel over the mouth of the flask and connecting the stem of the funnel to a glass or plastic (not metal) water aspirator by means of tubing (preferably Tygon or

other plastic). A micro-Kjeldahl digestion rack* is more convenient. These racks have a glass fume manifold which may be connected to a water aspirator to remove the fumes and condensate.

3. Micro-Kjeldahl distillation apparatus. The Kemmerer-Hallett type† is a convenient and compact unit; however, it is all glass with one ground joint and thus more subject to breakage. The Pregl-Parnas-Wagner design [61] is made with a number of rubber tubing connections and thus the separate parts may be easily replaced; connecting tubings must be changed occasionally and may be more subject to leakage than glass. If the latter type of apparatus is used, a borosilicate glass, rather than silver, condenser tube is preferred.

PROCEDURE

To a dry 30 ml Kjeldahl flask add about 1 g of the digestion mixture (this need not be accurately weighed). After the volume occupied by 1 g has been noted, it may be conveniently measured by volume in a small test tube or plastic spoon. Add exactly 0.2 ml of serum, taking care not to get any on the upper neck of the flask. Add 3 ml of concentrated sulfuric acid, rinsing down the walls of the flask. Digest by heating the flask, cautiously at first as there may be foaming, then more strongly. Heat until the solution is clear and the sulfuric acid boils vigorously. A few small glass beads added at the beginning may help prevent bumping. Digest for 1 hr after clear. Remove from the flame and cool somewhat. Cautiously add 10 ml of water while still warm (to prevent condensation of salts), then cool to room temperature. Transfer the solution quantitatively to the distillation apparatus with the aid of the minimum amount of water. Place a 125 ml flask containing 20 ml of the boric acid indicator solution under the condenser tip. The tip should just dip below the surface of the boric acid solution. Add 15 ml of the sodium hydroxide solution and steam distil until about 20 ml of distillate has been collected. Remove the flask from below the condenser, rinsing off the tip of the condenser with a stream of water from a wash bottle before stopping the distillation.

When the boiling in the steam generation flask stops and the water cools, a negative pressure develops which draws out the spent alkaline digestate from the flask. About 20 ml of water should be immediately added to the distillation flask before suction ceases, to wash out the flask. The time required to completely distil over the ammonia may vary somewhat with the type of apparatus. This should be checked when the apparatus is first set up by adding 2 ml of the

* A. H. Thomas Co., Philadelphia, Pa. 19105; Labconco, Kansas City, Mo. 64132.
† Sargent-Welch Scientific Co., Skokie, Ill. 60076.

nitrogen standard and 3 ml of sulfuric acid (diluted with about 10 ml of water and cooled) to the distillation flask. The sodium hydroxide is added and the solution distilled into 20 ml of boric acid indicator solution in the usual manner, until 25 ml of distillate has been collected. Then without stopping the distillation a second flask containing fresh boric acid indicator solution is placed under the condenser and an additional 10 ml of distillate collected. The titration of the second flask should be less than 0.2 ml of acid. If it is more, the first period was not long enough to distil all the ammonia.

The distillate from each sample is titrated with the 0.02N acid to the regular endpoint.

CALCULATION

Since 1 ml of the acid is equivalent to 0.2801 mg nitrogen,

$$(\text{titration of sample} - \text{titration of blank}) \times 0.2801 \times \frac{100}{0.2} = \text{mg/dl protein nitrogen}$$

or

net titration $\times$ 140 = mg/dl protein nitrogen in sample

When a regular serum is analyzed, a correction must be made for the small amount of nonprotein nitrogen (NPN) present. If this is not determined, it may be assumed to be 30 mg/dl, or 2 $\times$ BUN value. The correction is: mg/dl NPN $\times$ 0.00642, which is subtracted from the total protein as previously determined. If a solution of pure protein such as the albumin solutions mentioned as standards is used, the correction for NPN is neglected. Then:

corrected protein nitrogen in mg/dl $\times$ 0.00642 = g protein/dl

COMMENTS

This procedure could also be used for checking the protein content of albumin or other material used to prepare the protein standard, though it is better to check the final solution in order to include any errors made in preparing the solution. For protein determination by this micro method, only about 10 mg of protein is needed. This requires the use of a semimicro balance capable of weighing to 0.01 mg. To transfer these small samples quantitatively to the Kjeldahl digestion flask, the sample is weighed out on a small square of nitrogen-free paper. The paper is then carefully rolled and folded to enclose the sample, and the paper containing the sample is dropped into the digestion flask. It is then digested in the usual manner. Originally cigarette papers (i.e., for making roll-your-own cigarettes) were used. Any other fine paper such as lens paper or tissue paper may also be used, provided the sample does not sift through, and a blank is run on five squares to determine any blank nitrogen content.

Total Proteins by the Biuret Reaction

This is the most common method used for the determination of total protein in serum or other biological fluids containing at least 1% protein [26, 62]. The reagent gives a purple color with the —CONH— linkage found in all proteins. It gives practically the same amount of color on a weight basis with all the proteins found in serum.

REAGENTS

1. Sodium hydroxide, $0.2M$.   Dissolve 16 g of reagent grade NaOH in water and dilute to 2 liters.

2. Stock biuret solution.   Dissolve 45 g of sodium potassium tartrate in about 400 ml of the sodium hydroxide solution in a 1 liter volumetric flask. When dissolved, add slowly with stirring 15 g of copper sulfate pentahydrate as fine crystals and continue to stir until all the copper sulfate has dissolved. Add 5 g of potassium iodide and dilute to 1 liter with the $0.2M$ sodium hydroxide solution.

3. Dilute biuret working solution.   Dilute 200 ml of the stock biuret solution to 1 liter with the $0.2M$ sodium hydroxide solution and add 4 g of potassium iodide.

4. Tartrate–iodide solution for blank.   Dissolve 9 g of sodium potassium tartrate and 5 g of potassium iodide in the $0.2M$ sodium hydroxide solution and dilute to 1 liter.

5. Standard.   The simplest standard to use is one of the commercially available lyophilized control serums. The serum blank should be carefully checked (see procedure below). The blank result should be small and constant for a given lot of the serum and thus need not be determined for every run. Other standards can also be used (see Comments below).

PROCEDURE

To a series of appropriately labeled tubes add 5 ml of the biuret reagent. To the separate tubes add exactly 0.1 ml of standard, samples, or water (reagent blank). Mix and allow to stand at 30°C for 10 min or at room temperature for 20 min. Read standard and samples against blank at 550 nm.

Also set up serum blanks using the same procedure except that the tartrate-iodide blank solution is used instead of the biuret solution. Read these against a water blank (using water instead of serum) at 550 nm. Subtract the absorbance obtained for the blank from that obtained with the respective serums to obtain corrected readings. The serum blank will be very small for clear, nonicteric serums and may usually be omitted for them, but it should be used for lipemic or icteric serums or those with any visible degree of hemolysis.

CALCULATION

$$\frac{\text{corr. absorbance of sample}}{\text{corr. absorbance of standard}} \times \text{conc. of standard} = \text{conc. of sample (in g/dl)}$$

COMMENTS

Other standards that may be used are those prepared from purified bovine albumin. The standard solutions available for automated analysis may be used. These stock solutions are usually made up to 10 g/dl; they should be carefully diluted to 6 g/dl with sodium chloride solution (0.85 g/dl). Alternatively 0.05 ml of the 10 g/dl standard can be compared with 0.1 ml of serum samples, in which case the equivalent concentration of the standard would be 5.0 g/dl. It is difficult to prepare a solution from the powdered albumin (usually that labeled Cohen Fraction V). The solution is made with a solvent containing 8.5 g of sodium chloride and 1.0 g of sodium azide/liter of water. Correction must be made for the moisture content of the albumin powder. This is usually around 5% but can vary from 3 to 9% with different lots of the material. It has been recommended [63] that the moisture content be determined by heating at 110°C in a vacuum oven for several days. This is not a practical procedure for most laboratories. Theoretically the protein content can be checked by a Kjeldahl nitrogen determination. Most clinical laboratories are not equipped for accurate Kjeldahl analysis, and one would still be faced with the problem, discussed earlier, of what factor to use to convert the figure for nitrogen to protein. For serums with high protein content (over 10 g/dl as often found in multiple myeloma), it is advisable to use 0.05 ml of sample rather than 0.1 ml to bring the sample and standard closer in absorbance. An appropriate correction is then made in the calculations. For exudates and some other samples it may be necessary to use somewhat larger samples such as 0.2 ml; when read against a standard of 0.1 ml of the regular standard and 0.1 ml of water, the concentration of the standard is taken as one-half the labeled amount.

The biuret method is not sensitive enough for the small amounts of protein usually found in cerebrospinal fluid and urine, and a turbidimetric method should be used.

Total Proteins by Refractometry

The Hand Refractometer† is a simple instrument for the rapid determination of total serum protein. It requires only a few drops of serum and the measurement can be made in less than a minute. It is based on the experimentally determined relationship between the refractive index and total serum protein [64–66]. Gross hemolysis or marked lipemia may make readings difficult or impossible but these conditions may also cause errors in other methods for protein determination. Since the relation is based on measurement of refractive index and total protein in a large number of relatively normal serums, the method may be slightly in error in grossly abnormal serums such as found in myeloma. The curve indicating the relation between the two quantities does not pass through the origin, i.e., other substances in serum contribute to the refractive index. Therefore it is inadmissible, for example, to attempt to obtain a reading on a lipemic serum by diluting with an equal volume of saline and then multiplying by two the total protein obtained. Reconstituted lyophilized serum may differ in some respects from ordinary serum and the refractometer may give results differing slightly from those obtained by the biuret method. Also, the refractometer will not necessarily give accurate results if used to check a protein standard composed principally of a solution of bovine albumin. For example, one such solution containing the equivalent of 10 g/dl of protein by the biuret reaction gave a reading equivalent to only 8.7 g/dl with the refractometer.

If not subjected to thermal or mechanical shock, the instrument is stable and should rarely require recalibration. Provision is made for adjusting the zero point with pure water. Calculations from the table furnished with the instrument [67] and other published data [68] indicate that a solution containing exactly 7.50 g/dl of pure NaCl should give a refractive index equivalent to 5.15 g/dl of protein, and a solution containing 10.00 g/dl of NaCl, a reading corresponding to 7.20 g/dl of protein. These two points can be used for checking the instrument.

Normal Values and Interpretation of Results

This discussion is included in the section on Albumin and Globulins later in this chapter.

Total Proteins in Urine and Cerebrospinal Fluid

Turbidimetric methods are generally used for total protein determination in urine and CSF; they are sufficiently accurate for most clinical purposes. Either

† T-S Meter, American Optical Co., Buffalo, N.Y. 14215.

trichloracetic acid or a modified Exton's reagent (sulfosalicylic acid plus sodium sulfate) may be used. The method presented here uses the former reagent [46, 69].

REAGENTS

1. Trichloracetic acid, $0.75M$. Dissolve 12.5 g of trichloracetic acid in water and dilute to 100 ml.
2. Sodium chloride, $0.15M$. Dissolve 0.88 g of sodium chloride in water and dilute to 100 ml.
3. Protein standard. Any clear serum of known protein content may be used, or a commercially available lyophilized serum. This is diluted with the sodium chloride solution to 25 mg/dl or any other known value near that. For example, a standard serum containing 6.4 g/dl may be conveniently diluted 1:200 to give a solution containing 32 mg/dl.

PROCEDURE FOR URINE

Set up three tubes for standard, sample, and blank. To the standard tube add 4 ml of the diluted protein standard; to the sample tube add 4 ml of clear urine (which has been previously centrifuged if originally cloudy). To the blank tube add 4 ml of clear urine plus 1 ml of water. Add 1 ml of the trichloracetic acid solution to sample and standard tubes and mix at once. Let stand for 10 min, then read in a photometer at 420 nm. The tubes should be mixed by inversion several times just before reading. Read the standard tube against water and the sample tube against the urine blank.

CALCULATION

$$\frac{\text{absorbance of sample}}{\text{absorbance of standard}} \times \text{conc. of standard} = \text{conc. of sample}$$

COMMENTS

A number of drugs and some foods will give a dark color to the urine. This color may change on the addition of acid or may be so great as to render the method inaccurate. If the urine appears to change color on the addition of the acid, a dilute solution of hydrochloric acid rather than water should be used in the blank.

PROCEDURE FOR CEREBROSPINAL FLUID (CSF)

For a standard add 4 ml of the diluted protein standard to one test tube. To another tube add 1 ml of CSF and 3 ml of water and mix. To each tube add 1 ml of the trichloracetic acid solution and mix at once. Allow to stand for 10 min and read sample and standard against a water blank at 420 nm. The tubes should be mixed by inversion several times just before reading.

CALCULATION

Since the 4 ml of standard is compared with 1 ml of sample:

$$\frac{\text{absorbance of sample}}{\text{absorbance of standard}} \times \text{conc. of standard} \times 4 = \text{conc. of sample}$$

COMMENTS

For samples giving values over about 300 mg/dl, the determination should be repeated using a smaller aliquot of sample. Occasionally one may encounter a xanthochromic spinal fluid. This may interfere with the determination, and a blank, similar to that used for urine, should be run.

The modified Exton's reagent (5 g of sulfosalicylic acid and 7 g of anhydrous sodium sulfate made up to 100 ml with water) may be used similarly to the trichloracetic acid solution [48].

Normal Values and Interpretation of Results [70]

The normal amount of protein in urine is small, and ranges from 25 to 75 mg/24 hr. Normal infants may have a significant proteinuria during the first few days of life. Large increases in urinary protein are found in nephrosis, where the excretion may be as much as several grams per day. Variable amounts of protein are found in the urine in other destructive lesions of the kidneys, but usually less than in nephrosis.

The protein content of normal lumbar spinal fluid is usually 15 to 45 mg/dl. Cisternal fluid has slightly lower levels and ventricular fluids still lower levels, averaging around 10 mg/dl. An increase in protein is the most commonly found abnormality in spinal fluid. This increase is small in many conditions, rarely exceeding 100 mg/dl. In polyneuritis, tumors, and different types of meningitis, increases up to 400 mg/dl may be found.

Albumin and Globulins

The earlier methods for the separation of serum proteins into albumin and one or more globulin fractions involved the precipitation of the globulins by salt or other means and the measurement of the protein in the filtrate by the biuret reaction or other means. Howe in 1921 suggested the use of sodium sulfite for the precipitation [71]. This method or methods using sodium sulfite-sodium sulfate mixture for precipitation [72, 73] were used extensively until the advent of simple electrophoretic methods. Another method used different concentrations of potassium dihydrogen phosphate to precipitate different protein fractions [74]. Other precipitation methods less used involved the precipitation with a solution of hydrochloric acid [75] or trichloracetic acid in ethanol [76, 77]. Cohn and associates developed a method for the separation of the different blood proteins on a large scale by precipitation with buffered ethanol at refrigerator temperature [78]. This was developed into a laboratory procedure using methanol at refrigerator temperature, but the low temperature requirement made the method inconvenient [79].

For an exact quantitation of albumin and the different globulin fractions, the electrophoretic methods are preferred, but often it is sufficient to determine merely the ratio of albumin to total globulins. There are a number of methods based on dye binding for the determination of albumin. When the total protein is known, the globulins may then be determined by difference. There are also some methods for the direct determination of globulins but these are less widely used and are not as simple as the dye-binding methods for albumin.

A number of dyes will bind albumin with a change in color in a buffered solution. Most of the dyes used are actually pH indicators, and the color change is related to the fact, known for many years, that indicator papers may not give the same color in two solutions of the same actual pH, if one contains an appreciable amount of protein and the other does not (protein error). One of the first dyes used for the determination of albumin was the well-known indicator methyl orange [80–84]. This proved fairly satisfactory, though subject to some interference by bilirubin and hemolysis. The introduction of 2-(4'-hydroxy-benzeneazo) benzoic acid (HABA), for the continuous-flow automated methods in particular, made this dye the most widely used [85–89]. High levels of bilirubin will cause low results by this method also. In general, the most satisfactory dye seems to be bromcresol green which is becoming more widely used [90–97]. It is much less subject to interference by bilirubin than the other dyes and is very sensitive. It is the method that will be presented here.

Other dyes that have been suggested for the determination of albumin include a dye closely related to bromcresol green [98], bromcresol purple [99], as well as other sulfonphthalein dyes [100]. Also used has been a dye known as Spectru AB-2 [101, 102].

Although not strictly a dye method, mention can be made of some very sensitive fluorometric methods for the determination of albumin. Albumin combines with the reagent to give an intensely fluorescent compound. Earlier methods used vasoflavine [103] or 8-anilinonaphthalene-1-sulfonic acid [104] but a new, extremely sensitive reagent, *bis*(1-anilino-8-naphthalene sulfonate), has been more recently introduced. It is claimed that as little as 10 nanograms (ng) of albumin can be determined with it [105].

The fact that the globulins may be precipitated by sodium sulfate or sodium sulfite has been used as a direct turbidimetric method for globulin, the turbidity produced by the precipitated globulins being measured. These methods require careful attention to detail in order to obtain the same conditions of precipitation for every sample. They have been used chiefly in the automated continuous-flow methods where more constant conditions are obtained [106, 107].

Another method used is the precipitation of the globulins by isopropyl alcohol [108], rivinol [109], or alcoholic trichloracetic acid [110], and determination of the protein in the precipitate by the biuret reaction or by the glyoxal reaction with tryptophan. Its use in the determination of globulins is based on the fact that the globulins contain about 2.5% tryptophan [111]. Since albumin contains a much smaller percentage of tryptophan, contamination of the precipitate with a small amount of albumin will not cause a serious error. In fact, methods have been developed for the direct determination of globulin without separation from the albumin [112–114]. Since the amount of tryptophan in the globulins is about 15 times that in albumin, the usual changes in the amount of albumin in the serum will not significantly affect the determination of the globulins. Thus, when the method is standardized with serums of known globulin content, an accurate measurement of globulin can be made.

We present here a method based on the albumin binding of the dye bromcresol green. The binding by the albumin changes the color of a buffered solution of the dye. A surface-active agent is added to give a better color change. We also give a method for globulin based on the determination of tryptophan in a direct reaction.

Serum Albumin by Bromcresol Green Binding

The binding of the indicator by albumin in a buffered solution changes the color of the solution [97]. A surface-active agent is added which potentiates the change. Globulins do not react. Bilirubin does not interfere in concentrations up to 20 mg/dl. Because of the high dilution of the serum used, small amounts of hemolysis or moderate lipemia do not interfere markedly. A serum concentration of hemoglobin of 100 mg/dl will result in an increase in the apparent albumin content by only 0.1 g/dl.

REAGENTS

1. Stock succinate buffer, $0.1M$, pH 4.0. Dissolve 11.8 g of succinic acid and 100 mg of sodium azide in about 800 ml of water. Adjust the pH to 4.0 by the addition of sodium hydroxide solution (40 g/liter). About 40 to 50 ml may be required. Dilute to 1 liter and store in a polyethylene bottle in the refrigerator.

2. Stock bromcresol green, 0.6 mM. Dissolve 419 mg of the acid form of the indicator in 10 ml of $0.1M$ sodium hydroxide and dilute to 1 liter. Alternatively use 432 mg of the sodium salt which will dissolve without the use of alkali. Store in the refrigerator.

3. Working dye reagent. Mix 1 volume of stock bromcresol green with 3 volumes of the stock buffer. Add 4 ml of 30% Brij-35‡ per liter. Adjust to a pH of 4.20 ± 0.05. Store in the refrigerator. A concentrated solution of this reagent is available commercially§ which can be used after proper dilution with water.

4. Albumin standard. For occasional use one of the commercially available lyophilized serums may be used. The albumin value given for this material as determined by electrophoresis should be used unless the package insert specifically designates a bromcresol green value. Pure albumin standards may be used. Human serum albumin (Cohen fraction V) is preferable, but bovine albumin (Cohen fraction V) may be used; it gives an absorbance only about 3 percent higher than human albumin. Determine the moisture content of the standard material by weighing about 0.1 g into a tared weighing bottle and drying for 24 hr at 105°C, then cooling in a desiccator over phosphorus pentoxide. Reweigh and calculate the percentage of moisture. Keep the remainder of the sample in a tightly stoppered bottle. For a 10 g/dl standard, weigh out the equivalent of exactly 10 g of albumin corrected for the moisture content and any globulin present. For example, if the determined moisture content was 4.5% and the label stated that the sample contained 97% albumin and 3% globulins, one would weigh out $10.00/(0.955 \times 0.97) = 10.91$ g of the undried material. Dissolve in a solution containing 8.5 g of sodium chloride and 0.5 g of sodium azide/liter of water. The albumin may go into solution very slowly. Probably the best method of preparation is to transfer carefully all the albumin to a dry 100 ml volumetric flask, carefully add about 80 ml of the saline solution, and mix by gentle swirling. Allow to stand until all the albumin has dissolved, then dilute to the mark. Mix well, avoiding excessive foaming. The solution is stable for at least 6 months in the refrigerator. Standards of 2, 4, and 6 g/dl of albumin can be prepared by

‡ Technicon Instruments Corp., Tarrytown, N.Y. 10591.
§ Albustrate, General Diagnostics Division, Warner-Lambert, Morris Plains, N.J. 07950.

careful dilution of 2, 4, and 6 ml of the stock standard to 10 ml with saline. This solution could also be used for a total protein standard when account is taken of the amount of globulin present.

PROCEDURE

To 5 ml of the working dye reagent, add 25 μl of the serum and mix. Treat a standard similarly. Mix and allow to stand for 10 min. Read standard and samples against a reagent blank at 628 nm.

CALCULATION

$$\frac{\text{absorbance of sample}}{\text{absorbance of standard}} \times \text{conc. of standard} = \text{conc. of sample (in g/dl)}$$

COMMENTS

Depending upon the instrument used, the proportion of serum and reagent may be varied slightly to give readings in the desired absorbance range, the best range being an absorbance of about 0.5 for a 4 g/dl standard.

Globulin by Direct Determination [112–114]

REAGENTS

1. Color reagent.   Dissolve 1 g of cupric sulfate ($CuSO_4 \cdot 5H_2O$) in 90 ml of water in a 1 liter volumetric flask. Add 400 ml of glacial acetic acid and 1 g of glyoxalic acid monohydrate and mix well. Add without delay 60 ml of concentrated sulfuric acid slowly with swirling. Cool to room temperature and dilute to the mark with glacial acetic acid. This reagent is stable for 1 year under refrigeration.

2. Globulin standard.   Theoretically any of the commercially available lyophilized serums of known globulin content could be used as a standard. It has been stated that acetyltryptophan has been used as a stabilizer in some of these serums. Since this compound would react with the reagent, such serums are not suitable. The one recommended by the original authors is the protein standard obtained from Dow Diagnostics.|| This contains 3.0 g/dl of globulin and 4.5 g/dl of albumin and so could also be used for an albumin and total protein standard. One can also use a synthetic standard. Dissolve 175 mg of N-acetyltryptophan, C.P. grade,# in 0.8 ml of $1M$ sodium hydroxide and dilute to 100 ml

|| Dow Diagnostics, Indianapolis, Ind. 46206.
Schwartz/Mann Div., Becton, Dickinson & Co., Orangeburg, N.J. 10962

with water. When used like a serum, this is said to correspond to 3.0 g/dl of globulin. The usual control serums could be checked against this standard, and if the results agree with the labeled value, the control serums could be used as secondary standards.

PROCEDURE

Use screw-capped tubes with Teflon-lined caps. To a series of tubes containing 4.0 ml of the color reagent, add 0.020 ml of serum or other sample or standard. Reserve one tube as blank. Cap tubes, mix by vigorous shaking, and place in a boiling water bath for 4 min. A heating block set at 100°C can be used with 5 min heating. Remove tubes and cool in tap water. Read standards and samples against blank at 555 nm (or any convenient setting between 540 and 570 nm). The color is stable for at least 1 hr after heating.

CALCULATION

$$\frac{\text{absorbance of sample}}{\text{absorbance of standard}} \times \text{conc. of standard} = \text{conc. of sample (in g/dl)}$$

Normal Values and Interpretation of Results

The normal range for serum total proteins is 6 to 8 g/dl; for albumin, 4.0 to 5.5 g/dl; and for globulins, 1.5 to 3.0 g/dl. The albumin–globulin (A/G) ratio is usually between 1.5 and 2.5.

Increases in total protein are found in hemoconcentration due to dehydration from extreme loss of fluid (such as with vomiting or diarrhea). These cases will also show increases in albumin and globulins concentration, with the A/G ratio remaining unchanged. This is the only instance in which an increase in albumin is found.

In severe liver disease there is a significant increase in the globulins. Increases are also found in some infectious diseases and in multiple myeloma where the total globulin level may reach 6 g/dl or more. The globulin increases found in some infectious diseases may be due to increases in γ (gamma)-globulins which are the proteins concerned with antibody formation.

A decrease in the total protein level is usually associated with low albumin levels, which are usually accompanied by a lesser change in globulins and a low A/G ratio.

Low albumin levels may be due to significant loss of albumin in the urine, decreased formation in the liver, or insufficient protein level. In edema of nephritic origin, the albumin is decreased significantly, as is the A/G ratio.

A rare genetic disease with an almost complete absence of albumin has been reported. Another rare condition due to genetic metabolic defect is agamma-globulinemia in which the γ-globulins are very low or nearly absent; other blood proteins are relatively normal.

Fibrinogen

Fibrinogen is the plasma protein involved in the clotting of blood. The complex mechanism of blood clotting cannot be given here in detail. One step is the conversion of fibrinogen to fibrin clot by the action of calcium ions and other factors found in plasma or formed during the clotting process. Fibrinogen has been determined in a number of ways. The simplest methods are the turbidi-metric ones which measure the turbidity formed in the diluted plasma by the action of some substances that produce a turbidity with fibrinogen but not with the other proteins. Protamine sulfate [115] or thrombin [116, 117] has been used to convert the fibrinogen into a suspension of fibrin. An antibody to fibrinogen has also been used to form a finely divided precipitate with the fibrinogen [118]. The most common turbidimetric methods use concentrations of salts that pre-cipitate the fibrinogen but not the other proteins. Salts that have been used are sodium sulfate [119], sodium sulfite [120, 121], and particularly ammonium sulfate [122–124]. A method using heat to form the fibrinogen precipitate is also used [125].

In another type of procedure, the fibrin clot is formed in the diluted plasma by the addition of calcium ions and thrombin. The clot is then separated from the solution and washed free of other proteins. The amount of protein in the clot may then be determined gravimetrically after drying [126–128], or the clot may be dissolved in alkali or urea solution and the protein determined with the biuret reaction [129–131] or the Folin-Ciocalteu reagent [132]. The clot has also been digested with sulfuric acid and the resulting ammonia determined with Nessler's reagent [133]. The clotting times of diluted plasma have also been used to estimate the fibrinogen content [134–137]. The fibrinogen may also be deter-mined by radial immunodiffusion [138].

We present here two methods, the first a simple turbidimetric method using ammonium sulfate that is suitable for routine use, and the second one in which the fibrin clot is formed, separated, and analyzed by the biuret reaction. This is considerably longer but can serve as a reference method for checking the first method.

Fibrinogen by Turbidimetric Method

REAGENTS

1. Parfentjev's reagent.   Dissolve 133.33 g of ammonium sulfate and 10.11 g of sodium chloride in sufficient water to make 1 liter. Add 0.025 g of thimerosal (Merthiolate) as preservative.

2. Normal saline.   8.5 g of sodium chloride/liter in water.

3. Fibrinogen standard.   Small lyophilized vials containing 300 or 600 mg of pure fibrinogen may be obtained commercially.* These are diluted with water, as directed, by gentle mixing, to make a standard containing 300 mg/dl of fibrinogen. Standards containing 150 and 75 mg/dl are prepared by dilution of the high standard 1:2 and 1:4.

PROCEDURE

To collect the specimen, 4.5 ml of blood is anticoagulated with 0.5 ml of $0.14M$ sodium citrate. These are the same anticoagulant and proportions regularly used for prothrombin time. The supernatant is obtained by centrifuging for 10 min at 2,500 rpm.

The blank and sample are set up by adding 0.5 ml of plasma to 4.5 ml of saline for blank, and 0.5 ml of plasma to 4.5 ml of the reagent for sample tube. The tubes are mixed by inversion and the sample tube read against blank in the photometer at 510 nm exactly 3 min after addition of reagent. If sample has clotted, give it a vigorous shake prior to reading. The blank compensates for the color of the serum. A standard curve is prepared by treating standards containing 75, 150, and 300 mg/dl similarly. No blank is needed for the standards as they are colorless; the tubes may be read against saline. The calibration curve should be checked from time to time, and a sample from a normal individual should be run frequently. The curve can also be checked by analyzing several serums by the turbidimetric method and the longer clot analysis method given below.

Fibrinogen by Formation of Fibrin Clot [129]

REAGENTS

1. Sodium chloride, $0.15M$.   Dissolve 0.88 g of sodium chloride in water to make 1 liter.

2. Sodium hydroxide, $0.25M$.   Dissolve 10 g of sodium hydroxide in water and dilute to 1 liter.

* General Diagnostic Div., Warner-Lambert, Morris Plains, N.J. 07950.

3. Biuret reagent.   This is the stock biuret as used for total protein determination.

4. Thrombin.   Available as topical thrombin. It is usually marketed in vials containing 5,000 NIH units. The material is diluted to 100 units/ml. It may be convenient to dilute the entire contents of the vial with saline to this concentration and place 0.2-ml aliquots in small, tightly stoppered tubes, which are kept frozen until used.

5. Protein standard.   Any standard used for total protein analysis can be used here. It is diluted to give a concentration of about 2 mg/ml (200 mg/dl) with the $0.25M$ sodium hydroxide solution. Thus 1 ml of the standard containing 6.5 g/dl added to 30 ml of sodium hydroxide solution would give a concentration of 210 mg/dl.

PROCEDURE

The blood sample is obtained using double oxalate as anticoagulant. Vacutainers are available with this anticoagulant. If it is desired to prepare this anticoagulant, dissolve 1.2 g of ammonium oxalate monohydrate and 0.8 g of potassium oxalate monohydrate in water to make 100 ml. Pipet 0.5 ml of this mixture to collection tubes and evaporate the water with a stream of air and gentle heating. Collect 5 ml of blood in a tube; mix well with anticoagulant to avoid formation of clots. Centrifuge to separate the plasma and allow the plasma to stand at room temperature for 2 hr to minimize fibrinolysis, but if necessary standing can be omitted.

Add 0.2 ml of the thrombin solution to 5 ml of the sodium chloride solution in a test tube. Add 1 ml of plasma and mix well. Clotting should begin within a short time. Allow the tube to stand for 30 min.

The clot is then loosened from the walls of the tube by gentle shaking. The clot may be rimmed with a fine-pointed glass rod if necessary. Usually the clot can be removed in one piece. Transfer the clot to the middle of a pile of several sheets of hardened filter paper resting upon several sheets of absorbent paper. Cover the clot with several sheets of the hardened paper and then add several sheets of absorbent paper on top. Place a weight of about 250 g on top (a flask containing about 250 ml of water). Allow to stand for about 15 min. Remove the top sheets. The fibrin clot should be seen as a glistening membrane in the middle of the sheet. Gently loosen the clot with a needle or fine-pointed glass rod. Rinse the clot with saline and drop it into a graduated centrifuge tube. Be sure that all the clot is removed. Add 5 ml of the $0.25M$ sodium hydroxide solution and place in a boiling water bath for about 15 min until the clot dissolves. This may be hastened by breaking up the clot with a small glass stirring rod. When the clot has completely dissolved, cool, remove the rod, and add more water if necessary to bring the volume back to 5.0 ml. Set up a blank by adding 5 ml of the sodium hydroxide solution to a test tube. Prepare two standards by adding 1 and 2 ml of the diluted standards to two other tubes, bringing the volume up to 5 ml with the sodium hydroxide solution. To each tube including dissolved clot, add 1 ml of the stock biuret (this is the concentrated stock, not the diluted one used for total protein determination), mix, and allow to stand at room temperature for 30 min. Read standards and samples against blank at 555 nm.

CALCULATION

$$\frac{\text{absorbance of sample}}{\text{absorbance of standard}} \times \text{conc. of standard} \times 100 = \text{conc. of sample (in mg/dl)}$$

where conc. of standard represents the actual milligrams of protein in the 1 ml (or 2 ml) of standard used.

Normal Values and Interpretation of Results

The normal range of plasma fibrinogen is 200 to 400 mg/dl. Fibrinogen is the plasma component that affects the sedimentation rate the most and thus is usually increased in rheumatic fever and other diseases in which there is an increased sedimentation rate. In patients in whom bleeding may be caused or increased by low levels of fibrinogen, serial determination of fibrinogen may be of value.

Electrophoresis of Proteins in Serum

The general principles for the separation of proteins by electrophoresis on a supporting medium have been discussed in Chapter 3. Electrophoresis is used not only for serum proteins (the proteins in other body fluids such as urine and cerebrospinal fluid are generally derived from the serum), but also for other substances such as hemoglobins, isoenzymes, and lipoproteins. We are here concerned with the serum proteins. In the classic method devised by Tiselius [139], electrophoresis was carried out in solution and various protein fractions were quantitated by refined optical methods based on refractive index gradients. Although this method may still be used for some research purposes, much more rapid and simple methods have been developed for use in the clinical laboratory. These are carried out on a supporting medium. One of the earlier methods used an absorbent paper similar to filter paper [140–143]. This was generally used for some years. One of its disadvantages was the relatively long time required for a determination (up to 16 hr). The relative amounts of proteins were usually determined by scanning of a stained strip and the different proteins did not all stain exactly the same. Other media that have been used are agar or agarose gel [144–147], starch gel [148–151], and polyacrylamide gel [152–156]. These are mainly used for some special purpose such as hemoglobin or isoenzyme electrophoresis. For serum proteins the cellulose acetate strips are now used to a large extent [157–161]. The cellulose acetate or similar strips made by different manufacturers are not exactly the same in regard to their staining and clearing properties and one cannot always use them interchangeably. A general discussion of the procedures for protein electrophoresis on cellulose acetate is given in the books by Cawley and Nerenberg mentioned in the earlier discussion of electrophoresis in Chapter 3. Some general information is also given in the two publications by the Commission on Continuing Education of the American Society of Clinical Pathologists [162, 163].

A number of manufacturers supply complete systems of protein electrophoresis including power supplies, electrophoresis chamber, and densitometer. Of the systems for cellulose acetate electrophoresis, the authors are most familiar with those by Helena,† Gelman,‡ and Beckman§ which have proved satisfactory. No doubt many others are equally satisfactory, but a complete listing cannot be given here.

Some general directions will be given here which are applicable to most electrophoretic systems. A number of different buffers have been used for protein electrophoresis; the most common one is a barbital buffer of pH 8.6 and ionic strength 0.05. The buffers are often specified in terms of ionic strength rather than molarity since the number of ions has a greater effect than the actual salt concentration. The ionic strength may be calculated according to its definition, one-half the sum of the concentration of each ion multiplied by the square of the charge (see discussion of buffers in Chapter 20). The barbital buffer may be prepared by dissolving 1.84 g of diethylbarbituric acid (barbital) and 10.26 g of the sodium salt of the barbituric acid in water and diluting to 1 liter. The barbital may dissolve slowly; this can be increased by gentle heating. A few crystals of thymol may be added as preservative. The buffer should be stored in a refrigerator. Most suppliers of electrophoretic apparatus furnish packets of mixed salts for preparation of the buffers. This is convenient and insures that the proper buffer is used for the particular technique and apparatus involved. The cellulose strips are usually moistened with the buffer before use. This is pref-

† Helena Laboratories, Beaumont, Tex. 77704.
‡ Gelman Instrument Co., Ann Arbor, Mich. 48106.
§ Beckman Instruments, Inc., Fullerton, Calif. 92634.

erably done by placing the strips carefully on the surface of some of the buffer in a flat dish and allowing the buffer to soak in from the underside of the strip, before wetting completely. The buffer is then blotted somewhat and the serum applied. The instruments are usually furnished with an applicator for the application of the serum; only very small amounts are usually required. In some types of apparatus the moistened strip is placed in the electrophoresis chamber before the serum is applied, whereas in others this is not required. In the application of the electric potential, the point of application of the serum is placed nearest the negative electrode. A current of 10 to 30 milliamperes (ma) at 200 to 300 v for 30 to 45 min is used. These specifics will vary with the apparatus used and the size of the electrophoresis strips.

After the electrophoresis, the strip is removed and stained, usually with Ponceau Red S. The staining solution usually contains a fixative to prevent the dye from washing off. A typical solution may be made by dissolving 1 g of Ponceau Red S, 37.5 g of trichloracetic acid, and 37.5 g of sulfosalicylic acid in water to make 500 ml. Various manufacturers may recommend somewhat different dye solutions for their cellulose strips. After staining, the strip is washed free of excess dye with a solution such as $0.09M$ acetic acid. After washing, the strips are usually cleared and made almost transparent by use of a clearing solution. Because of differences in the manufacture of the cellulose strips, it is important to use the clearing solution recommended by the manufacturer, as the wrong solution may tend to dissolve the strips.

The resulting strip will appear similar to Figure 7-1, showing the stained bands for the different protein fractions. It is generally assumed that the amount of staining in each band is proportional to the amount of protein present in the fraction. In routine determinations this may not actually be true. In addition, the different fractions will not absorb the dye to the same extent, so that the same amounts of albumin and globulin will not give exactly the same amount of color [164–166]. Since, however, almost all the normal ranges are obtained without making corrections for nonlinearity or differences in binding of the different proteins, the results from a patient can also be used without corrections. It also appears that the dye Ponceau Red S is not always uniform in quality and purity, also resulting in some differences in staining [167]. Most general instructions neglect these factors.

The simplest way of estimating the amount of protein in the fractions is to wash the strip free of excess dye and dry. The strip is then cut into separate portions (at vertical lines in the figure), each containing one of the protein fractions, plus an additional strip as a control. The separate portions are then dissolved in a solvent such as 1 part absolute ethanol plus 9 parts chloroform, or

Figure 7-1. Serum protein electrophoresis scan showing normal distribution of protein fractions.

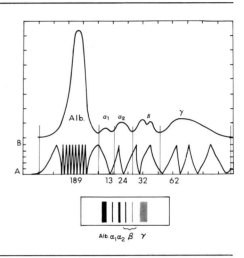

1 part formic acid plus **9** parts dimethylsulfoxide [**168**]. The absorbance of the separate solutions is then determined in a photometer, reading against the control strip at **520** nm. If equal volumes of the solvent are added to each tube, the absorbances will be proportional to the amount of protein in each fraction. This method requires a relatively large cellulose acetate strip in which the fractions are clearly separated, as in the illustration. Often they are not and it is difficult to decide where to make the separation. The fractional percentage of the individual proteins multiplied by the total protein as determined by the refractometer or the biuret reaction gives the absolute amount of each fraction present.

The protein fractions are usually quantitated by some type of densitometer. The strip is moved mechanically between a filtered light source and a photocell. The resulting absorbance is then automatically plotted on a moving chart paper, giving a result similar to that shown in Figure 7-1. The height of the curve above the baseline represents the absorbance of the strip at that point. The curve can generally be resolved into 5 major fractions: albumin and alpha-1 (α_1), alpha-2 (α_2), beta (β), and gamma (γ) globulins. In abnormal serums the separation between the fractions is often not as complete as shown in the illustration. The area under each peak is proportional to the amount of protein in that fraction. The relative areas of the different peaks are usually determined by some type of integrator. The older apparatus uses a mechanical (disc) integrator but in

Table 7-1. Calculation of Protein Fractions

Fraction	Relative Area	Percentage	Amount (g/dl)
Albumin	189	59.06	4.16
Globulins			
α_1	13	4.06	0.29
α_2	24	7.50	0.53
β	32	10.00	0.70
γ	62	19.38	1.37
Total	320	100.00	7.05

the newer models the integration is done electronically. A brief discussion of these is given in the introductory section on electrophoresis (Chapter 3). From the integrator section, one determines the relative area of the successive peaks. The calculation of the protein fractions from Figure 7-1 is shown in Table 7-1, based on total protein of 7.05 g/dl.

NORMAL VALUES AND INTERPRETATION OF RESULTS

The normal values for the serum proteins by cellulose acetate electrophoresis are given in Table 7-2. (The use of different supporting media may lead to slightly different values.) A complete discussion of the diagnostic significance of the plasma protein abnormalities cannot be given here, only a few general remarks. The occurrence of increased amounts of albumin in the serum is rare except in instances of severe dehydration. In some types of kidney disease, albumin may

Table 7-2. Normal Ranges of Serum Proteins by Cellulose Acetate Electrophoresis

Fraction	Amount (g/dl)	Percentage
Albumin	2.7–5.7	54–74
Globulins		
α_1	0.1–0.3	1.1–4.2
α_2	0.4–1.0	4.6–13.0
β	0.5–1.0	7.3–13.5
γ	0.5–1.5	8.1–19.9
Total	6.5–8.2	100

be excreted in relatively large amounts in the urine, leading to a decreased serum level. A rare genetic disease leading to an almost complete absence of albumin in the serum has been observed.

As the globulins are synthesized to some extent in the liver, changes in their level in the serum are often associated with liver disease and the total amount of globulins may be greater than the amount of albumin. The various immune antibodies are associated with the globulins, particularly the gamma globulins, and these may be increased in infectious diseases, especially those of a chronic nature. The globulins may be markedly elevated in multiple myeloma and related conditions.

Electrophoresis of Proteins in Urine and Cerebrospinal Fluid

The proteins in urine and CSF are derived almost exclusively from the serum but are usually present in much smaller amounts and generally in different proportions. Ordinarily most of the protein in these fluids is albumin.

Because of the small amount of protein present, urine and CSF must usually be concentrated before electrophoresis. A method has been given for the electrophoresis of undiluted cerebrospinal fluid [169], but it is preferable to concentrate it; this is always necessary for urine in order to remove the excess salts. The concentration of the fluid may be carried out in a number of different ways which may be classified as dialysis, ultrafiltration, and gel absorption. In dialysis the fluid is placed in a cellophane bag and the bag placed in a concentrated solution of polyvinylpyrrolidone (molecular weight at least 40,000). The water and salts will diffuse outward into the concentrated solution and the proteins will remain behind with a small amount of liquid. Although this method will produce a relatively high concentration of protein in the residue, it is lengthy and it is sometimes difficult to recover the small amount of protein solution from the interior of the bag. In place of the polyvinylpyrrolidone it has been suggested that the cellophane bag containing the fluid be placed in contact with powdered sucrose which will readily remove the water [170]. Two other procedures using cellophane bags only (purchased as cellophane dialysis tubing which is cut into lengths and the ends tied securely to form the bag) have been used. One method suggests placing a filled bag in a stream of warm air—such as from a hair dryer—at a sufficient distance from the air source so that the evaporation of the water keeps the bag cool. As the water evaporates, the bag is retied to give a smaller volume and keep the enclosed liquid under some tension. After reducing to a small volume, the fluid is dialyzed against the electrophoresis buffer for a few hours [171]. Another suggestion is to tie the bag securely and place it between two horizontal glass plates. On the top plate is placed a weight of 10 to 15 kg (taking care not to break the bag). As the volume of liquid is reduced, the volume of the bag is reduced by retying [172]. These simple methods may be satisfactory for occasional use when no other apparatus is available.

One method of ultrafiltration uses a special collodion bag assembly.|| The solution to be concentrated is placed in the bag [173], which is then immersed in a glass holder containing distilled water. When a vacuum is applied to the outer container, water and salts are filtered through the bag but the protein and high-molecular-weight substances remain inside. The filtration rate is slow but concentration factors of up to 25 times or more can be attained. The collodion bags are fairly substantial and can be washed and reused several times. A somewhat simpler procedure is the use of the Minicon sample concentrators# [174]. These consist of a plastic cell containing the sample. One side of the cell is a semipermeable membrane, the other is a thick layer of absorbent material. The water and salts diffuse through the membrane and are absorbed on the other side, while the protein remains in the cell which is tapered at the bottom so that the small amount of concentrated fluid remaining may be easily removed. The apparatus comes in several sizes; the larger (B-15) will hold 5 ml of urine and concentrate it to a volume of about 0.1 ml in approximately 2 hr (the rate will vary with the solute content of the urine). This appears to be a very satisfactory way of concentration.

A somewhat different apparatus (with which the authors have had no experience) is the single hollow-fiber concentrator.* This would appear to be more adapted to concentrating larger volumes of solution than needed for electrophoresis or immunodiffusion. A gel absorption method is exemplified by the use of Lyphogel,† which is a polyacrylamide hydrogel. This is added as relatively large granules to the solution to be concentrated. The water and accompanying inorganic salts are absorbed into the fine pores of the gel but the larger protein molecules cannot enter and remain in a more concentrated solution outside. Sometimes there is difficulty in separating the small amount of concentrated solution from the gel granules.

After concentration, the electrophoresis of urine or cerebrospinal fluid is carried out just as for serum, although it may occasionally be necessary to add more sample to the cellulose strip. In normal urine and cerebrospinal fluid the chief protein found will be albumin which may constitute as much as 90 percent of the entire protein. In myelomas, in which abnormal amounts of globulins are

|| Schleicher and Schuell, Inc., Keene, N.H. 03431.
Scientific Systems Division, Amicon Corp., Lexington, Mass. 02173.
* Biomed Instruments, Chicago, Ill. 60602.
† Gelman Instrument Co., Ann Arbor, Mich. 48106.

found in the serum, these may also be found in the urine in relatively large amounts.

Other Proteins by Immunodiffusion or Immunoelectrophoresis

There are a large number of other proteins or protein-like substances in the serum that are usually detected or quantitated by immunodiffusion or immunoelectrophoresis [162, 163, 175]. These proteins are generally present in smaller amounts (usually under 300 mg/dl) and are found associated with one of the other five fractions found in the usual electrophoresis, mostly with the globulins. Neglecting the enzymes and enzyme inhibitors (which are protein in nature) and the factors concerned with coagulation and fibrinolysis, the following is a list of some of these proteins that have been studied.

Associated with α_1-globulin:
α_1-lipoprotein
α_1-thyroxine-binding globulin
α_1-glycoprotein
transcortin

Associated with α_2-globulin:
haptoglobin
α_2-macroglobulin
Gc-globulin

Associated with β-globulin:
transferrin
β-lipoprotein
hemopexin
C′3 (β_1C globulin)
C′4 (β_1E globulin)
C′5 (β_1F globulin)

Associated with gamma globulin:
γM globulin (or IgM)
γA globulin (or IgA)
γG globulin (or IgG)

This is by no means a complete listing, but only those more frequently determined. Antibodies to most of these proteins are available from several sources.‡ The lipoproteins will be discussed with the lipids in Chapter 12, and the thyroxine-binding globulin with the thyroid function tests in Chapter 13.

Transferrin (siderophilin) is the serum protein concerned with the transport of iron in the plasma. Although it may be estimated by means of immunodiffusion or immunoelectrophoresis, the simpler determination of the iron binding capacity (see Serum Iron in Chapter 8) is usually used to estimate the amount of this protein present. Haptoglobin is a glycoprotein which combines with free hemoglobin. The hemoglobin in serum is much more stable when combined with haptoglobin. For instance, methemoglobin will be formed only when there is present in the serum, hemoglobin in excess of that required to saturate the haptoglobin present. This protein may be determined by immunomethods but is most commonly determined by other methods to be mentioned later. The proteins most commonly determined by immunodiffusion are the immune globulins, which are associated with γ-globulin. Those listed, IgM, IgA, and IgG, are readily determined by radial immunodiffusion. There is also an IgD globulin but it is less well characterized.

The procedure is relatively simple [176, 177]. A thin layer of 2% agar dissolved in barbital buffer (pH 8.6) is placed in a Petri dish. The agar also contains the antibody to the particular globulin to be determined. Although the plates can be prepared in the laboratory, this is not recommended. Prepared plates can be obtained from the suppliers listed earlier for antibodies. Standards containing known amounts of the protein being determined are also furnished. A number of small wells, 1 mm in diameter, are cut into the agar. In these wells are placed 2 μl of the various sample serums or standards. It is not always emphasized in the directions accompanying the plates that for best results the amounts of serum added to each well should be very closely the same and that the wells should be just filled to the top. The plates are allowed to stand at room temperature in a moist chamber for 24 to 48 hr, or until diffusion appears complete. When examined, each well will be surrounded by a precipitate ring (Fig. 7-2). The diameters of the rings are measured with a micrometer microscope (low power) or with a calibrated hand lens. A number of different methods have been suggested for preparing the standard curves. The square of the ring diameters may be plotted against the concentration, or the ring diameters may be

‡ Behring Diagnostics, Somerville, N.J. 08876; Hyland Laboratories, Costa Mesa, Calif. 92626; Meloy Laboratories, Biological Products Div., Springfield, Va. 22151; Kallestad, Minneapolis, Minn. 55416; Diagnostic Reagents, Div. Hoechst Pharmaceuticals, Inc., Cincinnati, Ohio 45229.

Figure 7-2. Radial immunodiffusion. Diameter of the ring formed by the precipitate is a measure of the concentration of antigen in the well.

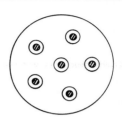

plotted against the logarithm of the concentration or against the square root of the concentration. One can try the different methods to see which gives the best curve with the experimental data. The concentrations of the various samples are then read from the standard curve.

Another immunological method has been introduced for these immunoglobulins. The appropriate antibody is mixed with the diluted serum and the turbidity produced is read in a fluorometer adapted to function as a nephelometer using a small, square flow-through cuvette [179, 180]. The procedure may be performed manually or may be automated in a continuous-flow system. The latter might prove very useful as a screening procedure. For an occasional test the radial immunodiffusion is simpler.

The normal levels for these immunoglobulins in adults are:

IgA   100 to 220 mg/dl

IgG   900 to 1,400 mg/dl

IgM   70 to 130 mg/dl

The levels of IgA and IgM are very low in the newborn and increase gradually to adult levels at puberty. The level of IgG just after birth is about equal to that in cord blood. It gradually decreases over the next few months to about 400 mg/dl, and then increases with age, reaching adult level near puberty.

A large number of proteins can be identified and semiquantitated by immunoelectrophoresis; other sources provide details of the procedures [162, 163, 181].

Haptoglobin

Haptoglobin may be determined by radial immunodiffusion [182] or by immunoelectrophoretic techniques using the antibody to this protein, and also by an antibody turbidimetric method similar to that mentioned for the immunoglobulins (see Albumin and Globulins earlier in this chapter). Other types of assay are based on the fact that haptoglobin combines strongly with hemoglobin, and the haptoglobin-hemoglobin complex has somewhat different properties from free hemoglobin. In the electrophoretic methods, advantage is taken of the fact that the haptoglobin complex migrates at a different rate from free hemoglobin. If a known excess of hemoglobin is added to serum, the haptoglobin-hemoglobin complex will be separated from the free hemoglobin on electrophoresis. The separation has been made using paper [183, 184], plaster of paris [185], agar gel [186, 187], starch gel [188, 189], polyacrylamide gel [190, 191], and cellulose acetate [192–195] as the supporting medium. By scanning with a densitometer, the relative amounts of the complex and of free hemoglobin may be estimated. The scanning may be made directly at 415 nm [192] where hemoglobin has a strong absorption band, or the strip may be stained with a peroxidase stain [195]. Knowing the total amount of hemoglobin added, the amount of the complex and hence of the haptoglobin may be calculated. By these methods the haptoglobin is frequently reported as the equivalent amount of hemoglobin that combines with the haptoglobin in 100 ml of serum. The separation has also been carried out using chromatography on a dextran gel [196].

In another method advantage is taken of the fact that the haptoglobin-hemoglobin complex has a greater peroxidase activity than free hemoglobin. The activity is usually measured by the addition of hydrogen peroxide and a chromogen (such as guaiacol or o-dianisidine). The peroxidase liberates free oxygen from the peroxide. The oxygen reacts with the chromogen to produce a colored product which can be measured photometrically. One can incubate the samples for a definite period of time and measure the color produced [197], or measure the time required to produce a specified absorbance [198]. This procedure has been automated [199].

As the amount of hemoglobin added to a definite amount of serum is increased, the amount of peroxidase activity is increased at a proportionate rate. When sufficient hemoglobin has been added to saturate the haptoglobin completely, the further addition of hemoglobin causes a much smaller rate of increase in the peroxidase activity, since the free hemoglobin has much less activity than the complex with haptoglobin. The break in the curve obtained by plotting the amount of hemoglobin added against the peroxidase activity represents the amount of haptoglobin present. In practice, methemoglobin is usually used instead of hemoglobin but the principle is the same. The production of such a curve for each determination is a lengthy process, though the time required has been reduced by automation [199]. This procedure, as devised by Owen [197], is lengthy and will not be given here; the details are also given by Bauer [200].

We present a relatively simple method for determining haptoglobin using hemoglobin electrophoresis. Cellulose acetate is the most widely used supporting medium. Two suggested scanning methods are given, either direct scanning at 415 nm [192] or peroxidase staining [195] with dianisidine.

Haptoglobin by Electrophoretic Method

REAGENTS

1.   Phosphate buffer, pH 7.0, 0.05M.   Dissolve 2.65 g of KH_2PO_4 and 4.34 g of anhydrous Na_2HPO_4 in about 900 ml of water. Check the pH and adjust to 7.0 if necessary with small amount of 1M NaOH or HCl. This is the recommended buffer, but if the manufacturer of the particular system recommends a different buffer for hemoglobin electrophoresis, it could be used.

2. Modified Drabkin's solution.   Dissolve 200 mg of potassium ferricyanide [$K_3Fe(CN)_6$], 140 mg of KH_2PO_4, and 50 mg of potassium cyanide in about 900 ml of water, add 0.5 ml of Sterox SE (or similar nonionic detergent), and dilute to 1 liter. This solution is used for the determination of the total hemoglobin. If the hemoglobin can be accurately determined in the hematology laboratory, this solution is not needed.

3. Staining solution (if needed).   Dissolve 100 mg of o-dianisidine in 70 ml of ethyl alcohol. Add 10 ml of acetate buffer (made by dissolving 11.2 g of sodium acetate and 4.7 ml of glacial acetic acid in water and diluting to 100 ml) and 18 ml of water. Just before use, add 2 ml of 3% hydrogen peroxide.

4. Hemoglobin solution.   About 5 ml of heparinized blood from a healthy subject is transferred to a tube and centrifuged. The supernatant serum is removed and the cells washed several times with 2 volumes of isotonic sodium chloride solution (0.85 g/dl). After the final washing, the saline is removed as completely as possible. Add to the packed cells 2 volumes of water and 0.5 volume of carbon tetrachloride and mix well by inversion. Allow to stand in the refrigerator for 15 min and then centrifuge strongly. Remove the clear hemolysate to a separate tube. Determine the hemoglobin content of the hemolysate by adding exactly 20 μl of the hemolysate to 5.0 ml of the Drabkin's solution and mix well. Allow to stand for about 5 min and read in exactly 1 cm light path cells at 540 nm. Then: absorbance × 36.8 = g/dl of hemoglobin. Alternatively any other calibrated method can be used for the determination of hemoglobin. Dilute the hemolysate to 5 g/dl. The concentration need not be exactly 5 g/dl but it should

be accurately determined. The solution can be kept for some time when frozen at $-15°C$.

PROCEDURE

Add 10 μl of the hemoglobin solution to 100 μl of the serum sample (or other convenient volumes in this proportion). Mix and allow to stand for 5 min. Apply an aliquot to the cellulose acetate strip which has been moistened with the buffer. Following the general directions given for the apparatus, carry out the electrophoresis as for hemoglobin. After the electrophoresis the strip may be treated in several different ways. It may be made translucent by treating with liquid petrolatum and scanned at 410 nm, or it may be cleared by the method suggested by the manufacturer of the strips and scanned. The strip may also be stained by immersing in the staining solution for several minutes, washing, drying, and scanning.

CALCULATION

The percentages in the two bands are calculated in the usual manner. Then, if A is the fraction in the haptoglobin-hemoglobin complex and B the concentration of the added hemoglobin in the serum in mg/dl,

$$A \times B \times 100 = mg/dl\ haptoglobin$$

Example: The hemoglobin solution is added to the serum in the ratio of 1:10 so that if a hemoglobin solution containing exactly 5 g/dl is used, the concentration in the serum would be 500 mg/dl. If half of this were bound by the haptoglobin, the concentration of the latter would be reported as 250 mg/dl.

NORMAL VALUES AND INTERPRETATION OF RESULTS

The normal haptoglobin values by the electrophoretic methods have been found to be 60 to 150 mg/dl. Higher levels have been found in leukemias, Hodgkin's disease, and a number of inflammatory conditions. Some workers have reported higher values in myocardial infarction.

Cryoglobulins [201]

These abnormal globulins are occasionally found in serum and will give abnormal electrophoretic patterns. They can precipitate or form a gel when the serum is cooled to refrigerator temperature (4°C) and redissolve when rewarmed to 37°C. In vivo, cryoglobulins can impair circulation in the toes and fingers by precipitation or gelling out in the blood. Even death may result from cryoglobulin blockage of key blood vessels in a vital organ.

Qualitative Test

Serum or plasma (oxalated, citrated, or heparinized) may be used. It is important that blood samples are kept at 37°C after being drawn. Specimens should be allowed to clot at this temperature for an hour, then they are centrifuged. The serum or plasma should be stored at 37°C until tested.

Transfer 2-ml aliquots of serum or plasma to two test tubes. Incubate one in the refrigerator (4°C) for 4 hr (or preferably overnight) and the other at 37°C. If cryoglobulins are present, the sample incubated at 4°C may show the following changes: The serum or plasma may appear to be completely clotted or to be divided in two layers, the upper layer containing normal serum or plasma and the lower one containing the cryoglobulins. The sample incubated at 37°C will show no changes.

To confirm the presence of cryoglobulins and to differentiate from cryofibrinogen, remix the tube containing the gelled specimen and warm to 37°C. The clot should dissolve and the serum or plasma should become clear.

Normal serum or plasma does not show clotting, precipitation, or cloudiness when incubated at 4°C. The temperature at which cryoglobulins precipitate is influenced by factors such as pH, the concentration of other proteins, and the concentration and type of cryoglobulins. Precipitation has been observed at temperatures as high as 32°C in severe cases of cryoglobulinemia.

Quantitative Test

Cryoglobulins can be quantitated in the following manner. Separate the clot or precipitate formed in the qualitative test and centrifuge in a refrigerated centrifuge. Decant the supernatant and wash the precipitate with cold saline and recentrifuge. Pour off the supernatant and dissolve the precipitate in a known volume of saline at 37°C. Analyze for protein using any of the methods described in this section.

Cryoglobulins can also be quantitated with the biuret method by determining the protein in the serum before and after low-temperature precipitation. The difference between the two analyses represents the cryoglobins.

Cryoglobulins may be found associated with multiple myeloma, lupus erythematosus, lymphosarcoma, and rheumatoid arthritis. In these conditions the cryoglobulin concentration may range from 0.04 to 4.0 g/dl.

Macroglobulins [202]

These globulins are very large molecules and constitute a very small percentage (less than 5 percent) of the total serum proteins. In macroglobulinemia the electrophoretic pattern will show abnormally high globulins, resembling multiple myeloma. Macroglobulinemia is best studied by ultracentrifugation of the serum. It has also been determined by gel filtration on Sephadex-200 by using either column or thin layer chromatography.

The following qualitative test is based on the fact that in macroglobulinemia, serum or plasma will form a white precipitate (have poor solubility) when diluted with water containing low salt concentration (Sia test).

PROCEDURE

Place 5 ml of $0.01M$ phosphate buffer, pH 7.1, in a test tube. Using a pipet brought close to the meniscus of the solution, add slowly down the sides of the tube 0.10 ml of serum or plasma (oxalated, citrated, or heparinized). If macroglobulin levels in excess of 0.7 g/dl are present, the entire solution will turn a whitish color due to the condensation of transparent, slimy masses forming a precipitate that quickly sinks to the bottom. This precipitate is packed by centrifugation and the supernatant decanted. Upon the addition of normal saline solution, the precipitate will dissolve.

Normal serum, when treated by this method, shows no cloudiness or precipitate. Serum from patients with malaria or kala-azar will give a positive test.

Amino Acids

In the fasting state, plasma contains over 20 different amino acids, ranging in concentration from about 10 to over 400 μmol/liter depending upon the particular amino acid and the nutritional state of the individual. These amino acids are also found in urine and are excreted in the amounts of 10 to 3,000 μmol/24 hr. In general, the higher the level in the plasma, the larger the amount excreted in the urine, but the proportion will be different for the different amino acids.

In a number of inborn metabolic disorders, the concentrations of some amino acids may be high in the plasma or urine or both. These are often caused by genetically controlled deficiencies in certain enzymes concerned with the metabolism of the particular amino acids. This is exemplified by phenylketonuria, or PKU. In this disease the concentration of phenylalanine is increased in the plasma due to the deficiency of the enzyme phenylalanine hydroxylase (E.C.1.14.16.1) which oxidizes phenylalanine to tyrosine [203]. In the body's effort to eliminate the excess phenylalanine, some is metabolized to phenylpyruvic acid and other phenylketones which are excreted in the urine. For this condition a screening test may be based on the color produced by ferric chloride with the phenylketones in the urine. Usually the diagnosis of PKU is established by the estimation of the serum level of phenylalanine.

In many metabolic disturbances involving the amino acids, the amounts of the particular amino acids found in the urine will be 10 to 20 times normal [204], and the conditions can thus be detected by relatively simple means without need of accurate quantitation. For accurate studies the amino acids in the urine or plasma can be determined by the automated liquid chromatographic method using ion exchange resin originally introduced by Stein and Moore [205, 206]. Commercial instruments are now available which will automatically determine some 20 or more amino acids in a sample in a few hours. These instruments are rather expensive and are found in clinical laboratories usually only in connection with research projects. The complete analysis can also be made by means of gas-liquid chromatography but this is less widely used [207, 208]. For most purposes, certainly for screening, simpler methods are sufficient for estimating abnormal concentrations of one or more amino acids in the urine.

High-voltage electrophoresis can be used for a relatively good separation of the urinary amino acids, at least into a number of major groups, in a short time [209–211]. After electrophoresis, the strips can be stained with ninhydrin and scanned in a densitometer to give a better estimate of the amounts of the different amino acids present. Although the electrophoretic apparatus is not very expensive, it is not found in many clinical laboratories. It might be very useful in a laboratory having a large pediatric work load since most of the aminoacidurias are first seen in infants.

The most common methods for preliminary screening use one- or two-dimensional chromatography on paper or thin layers. Formerly paper chromatography was used [212–215], but the thin layer methods are much more rapid [216–219]. Some metabolic disturbances involve only one or a few particular amino acids. For these instances the plasma or urine may be analyzed for the particular amino acids involved by more specific means. These analyses may be used not only for the detection of the abnormality but also for prognosis of the course of the disease or for following the results of therapy. A few of the amino acids commonly determined separately will be mentioned here but detailed methods will not be given for all.

The amino acid most commonly determined in blood is *phenylalanine*. A microbiological screening test for this amino acid developed by Guthrie and Susi [220] has been extensively used [221]. In this method agar gel containing spores of a special strain of *Bacillus subtilis* is used. This organism requires phenylalanine for growth. A medium containing a minimal amount of the amino acid is added to Petri dishes. The medium also contains a definite quantity of β-2-thienylalanine which is a biological antagonist to phenylalanine. It counteracts the effect of the traces of phenylalanine present and ordinarily no growth will occur on incubation.

Small discs containing blood from the patient are placed on the surface of the agar plate. If the concentration of phenylalanine in the blood is sufficiently high, the amino acid diffusing out from the disc will be sufficient to overcome the antagonism of the thienylalanine, and a ring of bacterial growth will surround the disc. The concentrations of the substances in the medium are adjusted so that with normal blood levels of phenylalanine not enough of the amino acid will diffuse out to counteract the inhibitor and no growth will occur.

For accurate diagnosis and to follow treatment with a low phenylalanine diet, the concentration of phenylalanine in the blood is accurately determined by chemical means. In one method the phenylalanine is oxidized to pyruvic acid by an L-amino oxidase and the product determined by absorption measurements in the ultraviolet [222, 223] or by a colorimetric method [224]. Phenylalanine has also been determined specifically by gas-liquid chromatography [225, 226]. The most common method is that described by McCaman and Robins [227], in which a fluorescent compound is produced by the reaction between phenylalanine and ninhydrin in the presence of cupric ions and a peptide (a number of different ones have been used). This method is widely used [228–230] and has been automated [231], particularly for use with the small amounts of blood collected on filter paper discs [232, 233]. These discs are then autoclaved and the phenylalanine eluted in the sampler cups. The method is suitable for screening procedures since the dried blood on the discs is stable. Later in this section we will present a manual micromethod for the analysis of phenylalanine in blood.

Another amino acid determined in blood or urine is *tyrosine*. (The ion exchange chromatographic methods can usually be modified so that particular amino acids of interest are eluted near the beginning of the test and it may not be necessary to determine the entire series of amino acids [234, 235], but we are concerned here with specific methods for individual amino acids.) The metabolism of tyrosine is connected with that of phenylalanine, and tyrosine is frequently determined in PKU [236]. Also it appears that children with cystic fibrosis may have abnormal levels of tyrosine in their blood or urine [237]. The determination of the blood level of tyrosine was suggested as a method of investigating thyroid function [238, 239], but this has not met with wide acceptance. The method of LaDu [222] mentioned earlier for phenylalanine can also be used for tyrosine, but the usual method is based on condensation with nitrosonaphthol to yield a fluorescent product. Details of the method [240–243] will not be given here.

Hydroxyproline has also been studied, particularly its excretion in the urine. This amino acid is derived almost exclusively from collagen [244]; other tissues contain a very small proportion of the acid. Thus the urinary excretion has been studied in diseases involving connective tissue and bone [245]. The urinary excretion is increased during active bone growth in children [246] and will be below normal in failure to develop properly. It is increased in adults with bone disease [247, 248]. It has been reported to be increased in a number of endocrine conditions but it is not always certain whether this is part of a more generalized aminoaciduria or a specific increase in hydroxyproline [249, 250]. If the urinary excretion is measured in adults, substances containing gelatine must be eliminated from the diet for several days since this substance contains a relatively large amount of hydroxyproline. A number of methods have been investigated for the determination of the amino acid; most are based on the treatment of the amino acid with an oxidizing agent such as hydrogen peroxide or chloramine-T to yield a condensation product which gives a red color with Ehrlich's reagent (p-dimethyl-aminobenzaldehyde in a strong acid solution) [251–253]. The methods vary in the way in which the proline is first separated from interfering substances.

Tryptophan is the precursor of serotonin in the body. The serotonin is excreted as the metabolic product 5-hydroxyindoleacetic acid. The excretion of this compound is greatly increased in the carcinoid syndrome. The method of LaDu [222] is applicable to tryptophan as well as tyrosine and phenylalanine. The method requires the determination of the absorption spectra in the ultraviolet. Fluorometric [254–256] and colorimetric methods [257, 258] are also available.

Histidine can also be determined in blood or urine. A modification of the method of LaDu and Michael [259, 260] can be used or a fluorometric method based on the condensation of the amino acid with o-phthaldialdehyde [261, 262].

Cystine and *cysteine* are present in large quantities in the urine in one type of genetic aminoaciduria. These are usually detected by high-voltage electrophoresis or chromatography. Occasionally a quantitative determination may be needed and two colorimetric methods are available, one using 1,2-naphthoquinone-4-sulfonate [263] and the other using 2,6-dichloro-p-benzoquinone [264]. In the first method the amino acid is isolated by use of an ion exchange resin and in the second by precipitation as cuprous mercaptide. On this basis the first method would seem preferable. There is also a semiquantitative screening method based on the reaction of aliphatic thiols with 5-5'-dithiobis(2-nitrobenzoic acid) [265]. This will detect both cystinuria and homocystinuria.

Two other amino acids will be mentioned very briefly. These are not concerned with the usual aminoacidurias. One of these is δ-aminolevulinic acid, a precursor of porphobilinogen. This is often measured in the urine in cases of suspected lead poisoning and the porphyrias. Most of the methods are modifications [266, 267] of the original one of Mauzerall and Granick [268] in which the amino acid is isolated by absorption on an ion exchange column, eluted, and heated with acetyl acetone. The condensation product gives a red color with Ehrlich's reagent.

The other amino acid is *glutathione* which is usually determined in red blood cells. Its determination is helpful in the study of certain types of hereditary red cell fragility. Several methods have been used: reaction with nitroprusside [269, 270], use of the reagent 5,5′-dithiobis(2-nitrobenzoic acid) [271, 272] which is specific for thiol groups, and a sensitive fluorometric method with *o*-phthaldehyde [273].

Phenylalanine in Blood

The procedure used is the modification by Ambrose of the original method of McCaman and Robins [228, 230]. Ninhydrin condenses with phenylalanine in the presence of a peptide to give a fluorescent product. The method is fairly specific for phenylalanine and it is not necessary to make any preliminary separation of the amino acid; there is usually some nonspecific fluorescence when blood filtrates are used. In attempting to eliminate some of the extra fluorescence, Ambrose filtered all reagents through Millipore filters and made other changes in the original method.

REAGENTS

The water and all reagents should preferably be filtered through 0.45 micrometer Millipore filters.

1. Phthalate buffer, $0.5M$, pH 5.0. Place 51.05 g of potassium acid phthalate in a 500 ml volumetric flask along with 420 ml of water and 25 ml of $5M$ NaOH solution (200 g/liter). The phthalate is dissolved by agitation and the pH adjusted to 5.0 by the addition of more NaOH solution (about 20 to 25 ml more may be required). The solution is then brought to volume and mixed well. (Stable at room temperature for 3 months.)

2. Copper reagent, 2.4 mM. Dissolve 600 mg of $CuSO_4 \cdot 5H_2O$ and 677 mg of sodium potassium tartrate ($NaKC_4H_4O_6 \cdot 4H_2O$) separately in about 300 ml portions of water. Add the two solutions to a volumetric flask, mix, and dilute to 1 liter with water. (Stable for 6 months in refrigerator.)

3. Pyrophosphate reagent, $0.125M$. Dissolve 27.88 g of sodium pyrophosphate ($Na_4P_2O_7 \cdot 10H_2O$) in about 250 ml of water in a 500 ml volumetric flask. Add 50 ml of the copper reagent, mix, and adjust the pH to 7.2 by the cautious addition of hydrochloric acid (approx. $6M$; concentrated acid diluted with equal volume of water); then dilute to the mark and mix well. The solution is stable for about 1 month in the refrigerator.

4. Ninhydrin reagent, $0.03M$. Dissolve 54 mg of ninhydrin in 10 ml of the phthalate buffer just before use.

5. Dipeptide reagent, 5mM. Just before use, dissolve 10 mg of L-leucyl-L-alanine in 10 ml of the phthalate buffer. (When determinations are made frequently,

it may be convenient to weigh out the proper amounts of the materials for reagents 4 and 5 in a number of separate stoppered vials for future use.)

6. Ninhydrin-peptide mixture.   Mix 5 ml of phthalate buffer, 10 ml of the ninhydrin solution (reagent 4), and 10 ml of the peptide solution (reagent 5). This should be prepared just before use.

7. Peptide control mixture.   Mix 6 ml of phthalate buffer and 4 ml of the ninhydrin solution. Prepare just before use.

8A. Trichloracetic acid, $0.6M$.   Dissolve 50 g of trichloracetic acid in water to make 500 ml. This solution, in particular, should be filtered through a Millipore filter. Store in refrigerator.

8B. Trichloracetic acid, $0.06M$.   Dilute the $0.6M$ acid 1:10 with filtered water. Store in refrigerator.

9. Phenylalanine standards.

A. Stock standard.   Dissolve 50 mg of pure phenylalanine in $0.06M$ trichloracetic acid to make 100 ml. Store in refrigerator. Stable for about 1 month.

B. Working standard.   Dilute 2 ml of the stock solution to 100 ml with $0.06M$ trichloracetic acid. Preferably this solution should be made up fresh at least once a week. Store in refrigerator.

PROCEDURE

To 0.5 ml of serum or heparinized plasma add 0.5 ml of $0.6M$ trichloracetic acid. Mix well on a vortex mixer. Allow to stand 10 min and centrifuge. Add 4.0 ml of water to supernatant, mix, and centrifuge again. For microsamples, the amount of serum can be reduced, keeping the other reagents in the same proportion.

It is recommended that the reactions be carried out in small 13×100 mm tubes with Teflon-lined screw caps as these can be used as fluorometer cuvettes in the measurement, but other somewhat larger tubes with similar caps can be used and the liquid transferred to fluorometer cuvettes for final reading. If enough filtrate is available, it is preferred that the serum samples and serum blanks be run in duplicate, and also the standards and reagent blank. In the following procedure the use of duplicate samples is not explicitly mentioned but its use is preferred.

The different tubes are set up in the following way. After the first treatment they are all treated identically.

Reagent Blank (RB). Add 0.5 ml of $0.06M$ trichloracetic acid to tube, then add 1 ml of the ninhydrin-peptide mixture, cap tubes, and mix well on a vortex mixer.

Serum Blank (SB). For each serum sample set up a serum blank by adding 0.5 ml of the serum supernatant and 1.0 ml of the peptide control mixture, cap, and mix well with vortex mixer.

Serum Samples (SS). For each serum use 0.5 ml of serum supernatant and 1.0 ml of ninhydrin-peptide reagent and treat as above.

Standards. To a series of tubes add 0.0, 0.1, 0.2, 0.3, 0.4, and 0.5 ml of the working standard and make up to 0.5 ml with the 0.06M trichloracetic acid. Add 1.0 ml of ninhydrin-peptide mixture, cap, and mix well.

All tubes are heated at 85°C for exactly 6 min (or at 60°C for 15 min). The tubes are then placed in a water bath at 30°C for 10 min. Add 5.0 ml of pyrophosphate solution to each tube, mix well by inversion, and replace in 30°C bath for 30 min. The tubes are then heated at 85°C for 6 min (or at 60°C for 15 min) and then replaced in the 30°C bath. After about 15 min the tubes are read in the fluorometer. Since the fluorescence is influenced by the temperature, it is recommended that a constant temperature door be used. Fairly satisfactory results can be obtained by keeping all the tubes in the water bath until just before reading and making the final reading at a definite time (say, 30 seconds) after the tube is inserted in the fluorometer. For use with the Turner No. 110 or 111 fluorometer§ with high-intensity kit, a 1-X aperture is used with a primary filter No. 7-60 (360 nm) and a secondary filter No. 2A plus 65A (495 nm). A neutral density filter is used if required so that the highest standard gives a reading between 50 and 75 on the dial.

CALCULATION

For each serum sample (average if run in duplicate), subtract readings of serum blank and reagent blank. For readings of standard, subtract only the reagent blank. The standards as made up correspond to 2, 4, 6, 8, and 10 mg phenylalanine/dl serum. A calibration curve is drawn and the samples read from the curve.

NORMAL VALUES AND INTERPRETATION OF RESULTS

The normal values given for this modification are:

Premature infants	1.0–3.0 mg/dl
Full-term newborns	0.7–2.1 mg/dl
Adults	0.6–1.9 mg/dl

§ G. K. Turner Associates, Palo Alto, Calif. 94303.

For screening purposes the upper limit of normal for newborn may be taken as 4.0 mg/dl. Infants who are heterozygous for the genetic trait may have levels between 4 and 8 mg/dl, but not be phenylketonuric. The typical infant with phenylketonuria will have a level between 5 and 8 mg/dl at birth and rise to levels greater than 30 mg/dl on milk feedings.

Urinary Amino Acids by Thin Layer Chromatography

The chromatography of urinary amino acids follows the same general procedure as given in the section on Chromatography in Chapter 3. Here also it is generally simpler to purchase the prepared plates. One can use plates coated with silica gel G or with cellulose powder. Although the cellulose may give results more similar to those obtained by paper chromatography, silica gel seems to be more generally used. Glass fiber sheets impregnated with silica gel can also be used. For cellulose-coated plates a good developing solvent for one-dimensional chromatography is isopropanol-formic acid-water in the ratio 40:2:10 (by volume). For two-dimensional chromatography, one can use for the first solvent, n-butanol-acetone-diethylamine-water in the ratio 10:10:2:5, and for the second solvent, phenol-water in the ratio 75:25 (by weight), adding a small amount of potassium cyanide to the mixture and saturating the chamber with the vapor from 3% ammonia solution. Generally for cellulose plates the chromatograms are run without prior saturation of the chamber space.

For silica gel plates, a good solvent for one-dimensional work is butanol-acetic acid-water in the ratio 8:2:2. For two-dimensional chromatography one can use for the first solvent, chloroform-methanol-17% ammonium hydroxide in the ratio 2:2:1, and for the second solvent, phenol-water in the ratio 75:25 (by weight). Generally for silica gel plates the chamber is first saturated with the solvent vapor. Similar solvents can be used for the glass fiber sheets; for these a good solvent is chloroform-methanol-ammonium hydroxide in the ratio 80:80:20. For two-dimensional chromatography a good first solvent is methyl ethyl ketone-pyridine-water-glacial acetic acid in the ratio 70:15:15:2. The disadvantage of this solvent is the disagreeable smell from pyridine; a good fume hood is required. For a second solvent one can use phenol-water as given above. Theoretically the excess salts should be removed from the urine before chromatography, but if only about 10 μl of urine is used, the desalting can be omitted. Also it is advisable to oxidize the cystine and cysteine to the more stable cysteic acid. Without oxidation the two amino acids tend to give several diffuse spots on the chromatogram. The oxidation is easily carried out by mixing an aliquot of the urine with an equal volume of freshly prepared performic acid (9 volumes of 97% formic acid and 1 volume of 30% hydrogen peroxide). The mixture is allowed to stand at room temperature for about 1 hr and is then applied to the chromatogram, using twice the usual volume of liquid.

For one-dimensional chromatography, a number of samples can be applied to one sheet. These may be unknown urines, control (normal) urines, and standards. It is difficult to suggest standards that will meet all requirements. If one is interested in a particular type of aminoaciduria, standards containing this amino acid should certainly be used. The following standards can be used (figures in parentheses are approximate molecular weights): tryptophan (204), leucine (131), methionine (149), serine (105), proline (115), lysine (146), cystine (204), and glycine (75). The designated amounts in milligrams are weighed out for each acid. Each is dissolved in 20 ml of $0.1M$ hydrochloric acid, 20 ml of isopropanol, and water to make 100 ml. In some cases gentle warming may be necessary to aid solution. These solutions contain 10 μmol/ml. As a standard, for example, 1 ml of each of the above could be added to a test tube, and aliquots of the mixture applied to the sheet. Usually it is necessary to try several different mixtures to locate the positions of the spots for the different amino acids since they vary with exact conditions used. The references cited will give some helpful information, but one must usually establish his own conditions [275–277].

For visualization of amino acids the usual spray is ninhydrin. One can purchase the spray cans ready for use‖ or make up a solution for spraying by dissolving 0.3 g of ninhydrin in 100 ml of butanol and adding 3 ml of glacial acetic acid. If the developing solvent contains ammonia, one must be certain that this is completely removed from the sheet before spraying since ammonia will give some color with ninhydrin. After the spraying, the sheets are heated in an oven at 110°C for a short time until the color develops. Most of the amino acids give a red, blue, or purple color, but proline and hydroxyproline will give a yellow color. The ninhydrin spots tend to fade on standing. They may be stabilized to some extent by spraying with an acid copper solution. Mix together 1 ml of a saturated solution of cupric nitrate, 0.2 ml of nitric acid (1 ml of concentrated acid diluted with 5 volumes of water), and 99 ml of absolute ethanol. Spray with this reagent and allow to dry at room temperature.

Another spray that is sometimes used is isatin which is also available commercially in ready-to-use spray cans. This solution can be prepared by dissolving 200 mg of isatin in 98 ml of acetone and 2 ml of pyridine. The sheet is sprayed with this and then heated in an oven in a fume hood at 60°C for a short time.

Most of the amino acids give some shade of blue. Those which are most distinct are proline, phenylalanine, and tyrosine.

‖ Sigma Chemical Co., St. Louis, Mo. 63118.

Table 7-3. Some Conditions That May Exhibit Markedly Increased Amino Acid Excretion in Urine

Condition	Amino Acids Increased
Maple syrup disease	Valine, leucine, isoleucine
Phenylketonuria	Phenylalanine
Histidinemia	Histidine
Hyperglycinemia	Glycine
Cystinuria	Cystine
Homocystinuria	Homocystine
Hartnup disease	General aminoaciduria: histidine, serine, threonine, phenylalanine, tryptophan, tyrosine
Lowe's syndrome	General aminoaciduria, particularly glutamine
Fanconi syndrome	General aminoaciduria
Wilson's disease	General aminoaciduria

A number of conditions result in an excess excretion of one or more amino acids in the urine (Table 7-3) [204]. Many of these, when present from birth and not corrected, result in mental retardation and other serious defects.

Glycoproteins and Mucoproteins (Seromucoids)

Seromucoids are proteins containing carbohydrates (usually galactose or mannose), hexose amines, and sialic acid as constitutents of the molecule. Those containing less than 4 percent hexose amines and less than 15 percent total carbohydrate are *glycoproteins* and the ones with larger amounts of these substances are *mucoproteins*. This class of proteins includes a number of different specific proteins including transferrin, ceruloplasmin, and haptoglobulin, which may be determined by specific methods given elsewhere in this book. Many of the other proteins have been determined by electrophoretic [278, 279] or immunodiffusion methods [280, 281], but the total quantity of seromucoids is generally determined by less specific chemical methods [282]. The seromucoids may be separated from other proteins by the fact that they are soluble in dilute perchloric acid but precipitated by phosphotungstic acid. The other proteins are first precipitated with perchloric acid and then the seromucoids are precipitated from the perchloric acid filtrate with phosphotungstic acid [283]. The precipitate is washed and the amount of seromucoid determined colorimetrically. One may determine either the carbohydrate moiety by reaction with orcinol [284], or the protein portion by the biuret reagent [285, 286], ultraviolet absorption at 280 nm [279], or use of the Folin-Ciocalteu reagent [282, 285, 286]. These methods cannot be exact since the amount of carbohydrate in the seromucoid may vary with the patient, particularly in those with high levels, and the proportion of protein may also be variable. When the protein is determined with the Folin-Ciocalteu reagent, the standard used is tyrosine with the assumption that the protein contains 4.2 percent of this amino acid, although this may not always be the case. We present two methods, one in which the carbohydrate is determined and the other in which the protein is determined. The initial separation of the seromucoids is the same for both methods.

Determination as Carbohydrate [287]

REAGENTS

1. Perchloric acid, $1.8M$. Dilute 15.6 ml of 70% perchloric acid to 100 ml with water.

2. Phosphotungstic acid, approx. $0.009M$. Dissolve 5 g of phosphotungstic acid ($P_2O_5 \cdot 24WO_3 \cdot nH_2O$) in dilute hydrochloric acid ($2M$, made by diluting 84 ml of concentrated acid to 500 ml) and make up to 100 ml with the dilute acid. This compound as obtained contains variable amounts of water and is deliquescent. It should be kept in a tightly stoppered bottle. The solution is fairly stable.

3. Sodium chloride, $0.15M$. Dissolve 8.5 g of sodium chloride in water to make 1 liter.

4. Ethanol, 95%.

5. Sulfuric acid, approx. $11M$. Cautiously add 60 ml of concentrated sulfuric acid to 40 ml of water. Cool and store in glass-stoppered bottle.

6. Sodium hydroxide, $0.1M$. Dissolve 4 g of sodium hydroxide in water to make 1 liter, or appropriately dilute a stronger solution if available.

7. Carbohydrate standards.

A. Stock standard, 200 mg/dl. Dissolve 100 mg of anhydrous galactose and 100 mg of mannose in $0.01M$ benzoic acid (1.2 g dissolved in 1 liter of water) and dilute to 100 ml with the benzoic acid solution. This standard is stable for several months in the refrigerator.

B. Working standard, 20 mg/dl. Just before use, dilute 1 ml of the stock to 10 ml with water.

8. Orcinol reagent, $0.16M$. Dissolve 2 g of orcinol (5-methylresorcinol, anhydrous form) in about 30 ml of water in a 100 ml volumetric flask. Add 50 ml of the $11M$ sulfuric acid and dilute to the mark with water. Some lots of the chemical will give high blanks. The material should be off-white or slightly pink in color. Lots that are dirty brown in color are usually not satisfactory. The material furnished by Eastman (Cat. No. 2112)# and Matheson, Coleman and Bell (Cat.

Eastman Kodak Co., Rochester, N.Y. 14650.

No. Mx1411)* have generally proved satisfactory. The orcinol may be purified by recrystallization if necessary (see paragraph 1 under Comments).

PROCEDURE

1. Separation of seromucoids.   Add 0.5 ml of serum sample and 4.5 ml of the sodium chloride solution to a small Erlenmeyer flask. Add dropwise with constant swirling, 2.5 ml of the 1.8M perchloric acid solution. Mix well and allow to stand for 10 min. Filter through a 7 cm retentive paper (Whatman No. 50 or 42). Transfer a 5-ml aliquot to a 12 ml centrifuge tube and add 1 ml of the phosphotungstic acid solution. Mix well by inversion and allow to stand for 10 min. Centrifuge strongly for 15 min. Pour off the supernatant fluid and wash the precipitate by adding 5 ml of 95% ethanol by blowing the alcohol in strongly from a pipet. Centrifuge again and pour off the supernatant wash alcohol. Invert the tubes and allow to drain for about 5 min.

2. Color development.   Dissolve the precipitate in 0.5 ml of the sodium hydroxide solution. Set up a blank tube containing 0.5 ml of water and a standard tube containing 0.5 ml of the working standard. To each tube add 1 ml of the orcinol reagent and 7.5 ml of the 11M sulfuric acid solution. Mix well on a vortex mixer. Cap the tubes with a glass marble or an inverted 10 ml beaker and heat in a water bath at 80°C for exactly 20 min. Remove tubes from water bath and cool in running water for 10 min. Read standards, samples, and blank against water at 520 nm. (See paragraphs 2 and 3 under Comments.)

CALCULATION

Since 0.5 ml of sample is diluted to 7.5 ml in the first precipitation, with 5 ml of this taken for analysis, and compared with 0.5 ml of standard,

$$\frac{\text{absorbance of sample} - \text{absorbance of blank}}{\text{absorbance of standard} - \text{absorbance of blank}} \times \frac{7.5}{5.0} \times 20 = \text{hexose in sample (in mg/dl)}$$

when the standard containing 20 mg/dl is used.

COMMENTS

1. The orcinol may be recrystallized as follows. Add 20 g of orcinol and 30 ml of reagent grade benzene to a flask. Warm to 75°C. (*Caution!* Use hot plate [low heat] or steam bath in fume hood; do NOT use open flame.) The solution will separate into a slightly yellow upper layer and a brown lower layer. Carefully decant the upper layer into a warmed 20 ml sintered glass filter (medium porosity)

* Matheson, Coleman and Bell, Norwood, Ohio 45212.

and filter at once by suction. Cool solution to about 5°C in refrigerator for at least 1 hr. Filter the precipitated crystals on sintered glass filter, wash crystals with about 20 ml of ice-cold benzene, and dry by passing air through filter. Store crystals in tightly stoppered bottle in refrigerator.

2. Samples and blank are read against water to check on the absorbance of the blank. When this increases more than about 50 percent, a fresh orcinol solution should be prepared.

3. Beer's law may not be followed when absorbance of the sample is much greater than that of the standard. If the concentration of the sample is more than about 35 mg/dl, the test should be repeated using a smaller aliquot of sample. Repeat precipitation and dissolve the final precipitate in 1.0 ml of NaOH solution instead of 0.5 ml. Take 0.5 ml of this for the test, multiplying the result obtained by 2.

Determination as Protein (Tyrosine) [282]

REAGENTS

1–3. The same first three reagents as used in the previous procedure for precipitation of seromucoids.

4. Sodium carbonate solution, $1.9M$. Dissolve 20 g of anhydrous sodium carbonate (Na_2CO_3) in water to make 100 ml.

5. Folin-Ciocalteu reagent. This reagent can be prepared in the laboratory but requires 10 hr refluxing in an all-glass apparatus. It is simpler to purchase the reagent, unless large quantities are used. It is usually listed as Phenol Reagent, Folin-Ciocalteu, $2M$. If it is desired to prepare the reagent, the directions are as follows. In a 1 liter round-bottom flask add 50 g of sodium tungstate ($Na_2WO_4 \cdot 2H_2O$), 12.5 g of sodium molybdate ($Na_2MoO_4 \cdot 2H_2O$), and about 350 ml of water. Swirl until all the salts are dissolved. Then add 25 ml of 85% phosphoric acid and 50 ml of concentrated hydrochloric acid. Attach a ground glass reflux condenser and reflux gently for 10 hr. Cool somewhat and add 75 g of lithium sulfate ($LiSO_4$), 25 ml of water, and 3 to 4 drops of 30% hydrogen peroxide. Reheat and boil for about 15 min, then cool. Dilute to 1 liter and mix well. The solution should be free of any greenish tint. If it is turbid, it may be filtered. The working reagent is prepared by diluting 1 part of the reagent as prepared or purchased, with 2 parts of water.

6. Tyrosine standards.

A. Stock standard, 20 mg/dl. Dissolve 50 mg of pure tyrosine in $0.1M$ hydrochloric acid (8.4 ml of concentrated acid diluted to 1 liter) to make 250 ml. Store in the refrigerator.

B. Working standard, 5 mg/dl.   On the day of use, dilute 2 ml of the stock with 6 ml of the 0.1M hydrochloric acid.

PROCEDURE

The precipitation and separation of the seromucoids are done exactly as in the previous method except that the final wash of the precipitate is carried out with 2 ml of the phosphotungstic acid solution. A small glass rod is used to break up the precipitate. The washed precipitate is then dissolved in 1 ml of the sodium carbonate solution by gentle agitation and 3.5 ml of water added. The blank is made up from 3.5 ml of water and 1 ml of the sodium carbonate solution. A standard is prepared by mixing 0.5 ml of the diluted standard, 3.0 ml of water, and 1 ml of sodium carbonate solution. Add 0.5 ml of the Folin reagent to each tube and mix. Place all tubes in a water bath at 37°C for 15 min. Remove tubes, cool in running water, and add to each 2.0 ml of water with mixing. Read samples and standard against blank at 680 nm.

CALCULATION

There is 0.5 ml of sample diluted to 7.5 ml, and 5 ml of this is compared with 0.5 ml of standard. Thus if the 5 mg/dl standard is used:

$$\frac{\text{absorbance of sample}}{\text{absorbance of standard}} \times \frac{7.5}{5} \times 5 = \text{tyrosine in sample (in mg/dl)}$$

If the seromucoids are assumed to contain 4.2 percent tyrosine, the factors above would be multiplied by $1/0.042 = 23.8$; accordingly

$$\frac{\text{absorbance of sample}}{\text{absorbance of standard}} \times 178.5 = \text{mucoprotein (in mg/dl)}$$

NORMAL VALUES AND INTERPRETATION OF RESULTS [288–290]

When determined as hexose, the normal values for seromucoids are 9 to 15 mg/dl. When determined as mucoprotein, the normal values are 75 to 135 mg/dl. Increases in serum levels are associated with many inflammatory and proliferative conditions such as rheumatoid arthritis, tuberculosis, acute infections, carcinoma, and lymphosarcoma. In rheumatoid arthritis the levels (as mucoprotein) may be from 100 to 400 mg/dl and in acute infections up to 450 mg/dl.

References

1. Archibald, R. M.   *Stand. Methods Clin. Chem.* 2:91, 1958.
2. Armstrong, S. J., Jr., Budka, M. J. E., and Morrison, K. C.   *J. Am. Chem. Soc.* 69:416, 1947.

219

3. Armstrong, S. H., Jr., Budka, M. J. E., Morrison, K. C., and Hasson, M. *J. Am. Chem. Soc.* 69:1747, 1947.
4. Strickland, R. D., Mack, P. A., Gurule, F. T., Podelski, T. R., Salome, O., and Chiles, W. A. *Anal. Chem.* 31:199, 1959.
5. Sunderman, F. W., Jr., Sunderman, F. W., Falvo, E. A., and Kallick, C. L. *Am. J. Clin. Pathol.* 30:112, 1958.
6. Watson, D. *Clin. Chim. Acta* 16:322, 1967.
7. Sunderman, F. W. *Am. J. Clin. Pathol.* 46:679, 1966.
8. Tombs, M. P., Souter, F., and MacLagan, N. F. *Biochem. J.* 73:167, 1959.
9. Dubowski, K. M. In F. W. Sunderman and F. W. Sunderman, Jr. [Eds.], *Serum Proteins and Dysproteinemias.* Philadelphia: Lippincott, 1964. P. 81.
10. Gandolfi, E., Gossani, M. R., and Fabrini, G. *Quad. Sclavo Diagn.* 1:437, 1965.
11. Waddell, W. J. *J. Lab. Clin. Med.* 48:311, 1956.
12. Webster, G. C. *Biochim. Biophys. Acta* 207:371, 1970.
13. Strickland, R. D., Mack, P. A., Podelski, T. R., and Chiles, W. A. *Anal. Chem.* 32:199, 1960.
14. Drickman, A., and McKeon, F. A., Jr. *Am. J. Clin. Pathol.* 38:393, 1962.
15. Naumann, H. N. In F. W. Sunderman and F. W. Sunderman, Jr. [Eds.], *Serum Proteins and Dysproteinemias.* Philadelphia: Lippincott, 1964. P. 86.
16. Martinek, R. G. *Proc. Assoc. Clin. Biochem.* 3:264, 1965.
17. Guillien, C. O., Wahl, R., and Laurencin, M. L. *Boll. Soc. Chim. Biol.* 11:387, 1929.
18. Linder, G. C., Lundsgaard, C., and Van Slyke, D. D. *J. Exp. Med.* 39:887, 1924.
19. Sunderman, F. W. *Am. J. Clin. Pathol.* 46:679, 1966.
20. Wolf, A. V., Fuller, J. E., Goldman, E. J., and Mahoney, T. D. *Clin. Chem.* 8:158, 1962.
21. Barry, K. C., McLaurin, A. W., and Parnell, B. L. *J. Lab. Clin. Med.* 55:803, 1960.
22. Moore, N. S., and Van Slyke, D. D. *J. Clin. Invest.* 8:337, 1930.
23. Van Slyke, D. D., Hiller, A., Phillips, R. A., Hamilton, P. B., Dole, V. P., Archibald, R. M., and Eder, H. A. *Biol. Chem.* 183:331, 1950.
24. Van Slyke, D. D., Phillips, R. A., Dole, V. P., Hamilton, P. B., Archibald, R.M., and Plazin, J. *J. Biol. Chem.* 183:349, 1950.
25. Strickland, R. D., Freeman, M. L., and Gurule, F. T. *Anal. Chem.* 33:545, 1961.
26. Weichselbaum, T. E. *Am. J. Clin. Pathol.* 16: (Tech. Sec.)40, 1946.
27. Gornall, A. G., Bardawell, C. J., and David, M. M. *J. Biol. Chem.* 177:751, 1949.
28. Henry, R. J., Sobel, C., and Berkman, S. *Anal. Chem.* 29:1491, 1957.
29. Siltanen, P., and Kekki, M. *Scand. J. Clin. Lab. Invest.* 12:228, 1960.
30. Mehl, J. W. *J. Biol. Chem.* 157:173, 1945.
31. Elman, G. L. *Anal. Biochem.* 3:40, 1962.
32. Bode, C., Goebell, H., and Straehler, E. *Z. Klin. Chem. Klin. Biochem.* 6:418, 1968.
33. Daughaday, W. H., Lowry, O. H., Rosebrough, N. J., and Fields, W. S. *J. Lab. Clin. Med.* 39:663, 1952.
34. Dowry, O. H., Rosebrough, N. J., Farr, A. L., and Randall, R. J. *J. Biol. Chem.* 218:641, 1956.
35. Rieder, H. P. *Clin. Chim. Acta* 6:188, 1961.
36. Svensmark, O. *Scand. J. Clin. Lab. Invest.* 10:50, 1958.
37. Watson, D. *Clin. Chem.* 10:412, 1964.

38. Savory, J., Pu, P. H., and Sunderman, F. W., Jr.   *Clin. Chem.* 14:1160, 1968.
39. Burgi, W., Richterich, R., and Briner, M.   *Clin. Chim. Acta* 15:181, 1967.
40. Rieder, H. P.   *Klin. Wochenschr.* 42:803, 1964.
41. Rieder, H. P.   *Klin. Wochenschr.* 44:1036, 1966.
42. Goa, J.   *Scand. J. Clin. Lab. Invest.* 5:218, 1953.
43. Patrick, R. L., and Thiers, R. E.   *Clin. Chem.* 9:283, 1963.
44. Joergensen, M. B.   *Acta Med. Scand.* 181:153, 1967.
45. Denis, W., and Ayer, J. B.   *Arch. Intern. Med.* 26:436, 1920.
46. Henry, R. J., Sobel, C., and Segalov, M.   *Proc. Soc. Exp. Biol. Med.* 92:748, 1956.
47. Gernand, K., and Hajek, E.   *Dtsch. Gesundheitsw.* 21:510, 1966.
48. Muelemans, O.   *Clin. Chim. Acta* 5:757, 1956.
49. Krause, H., and Seidler, E.   *Z. Med. Labortech.* 10:105, 1969.
50. Schriever, H., and Gambino, S. R.   *Am. J. Clin. Pathol.* 44:667, 1965.
51. Rice, E. W., and Loftis, J. W.   *Clin. Chem.* 8:56, 1962.
52. Shin, Y. S.   *Clin. Chem.* 15:1234, 1969.
53. Goodwin, J. F., and Choi, S. Y.   *Clin. Chem.* 16:24, 1970.
54. Searcy, R. L., Gough, G. S., and Bergquist, L. M.   *Am. J. Med. Technol.* 29:241, 1963.
55. Bramhall, S., Noack, N., Wu, M., and Loewenberg, J. R.   *Anal. Biochem.* 31:146, 1969.
56. Hiller, A., Plazin, J., and Van Slyke, D. D.   *J. Biol. Chem.* 176:1401, 1949.
57. Peters, J. P., and Van Slyke, D. D.   Quantitative Clinical Chemistry. Vol. 2, *Methods.* Baltimore: Williams & Wilkins, 1972.
58. Bradstreet, R. B.   *Anal. Chem.* 26:235, 1952.
59. Kirk, P. L.   *Adv. Protein Chem.* 3:139, 1947.
60. Wagner, E. C.   *Ind. Eng. Chem. Anal. Ed.* 12:771, 1940.
61. Parnas, J. K., and Wagner, R.   *Biochem. Z.* 125:253, 1921.
62. Reinhold, J. G.   *Stand. Methods Clin. Chem.* 1:88, 1953.
63. Doumas, B. T., Watson, W. A., and Biggs, H. G.   *Clin. Chim. Acta* 31:87, 1971.
64. Rubini, M. E., and Wolf, A. V.   *J. Biol. Chem.* 225:860, 1957.
65. Remp, D. G., and Schelling, V.   *Clin. Chem.* 6:400, 1960.
66. Barry, K. G., McLaurin, A. W., and Parnell, B. L.   *J. Lab. Clin. Med.* 55:803, 1960.
67. *Instructions for Use of the T. S. Meter.*   Buffalo: American Optical Co., 1968.
68. Weast, R. C. [Ed.].   *Handbook of Chemistry and Physics*, 46th ed. Cleveland: Chemical Rubber Co., 1965. P. D-158.
69. Muelemans, O.   *Clin. Chim. Acta* 5:757, 1960.
70. Ellis, A. W.   *Lancet*, 1:1, 1942.
71. Howe, P. E.   *J. Biol. Chem.* 49:93, 1921.
72. Campbell, W. R., and Hanna, M. I.   *J. Biol. Chem.* 119:9, 1937.
73. Cohn, C., and Wolfson, W. Q.   *J. Lab. Clin. Med.* 33:367, 1948.
74. Aull, J. C., and McCord, W. M.   *J. Lab. Clin. Med.* 46:476, 1955.
75. Fernandez, A., Sobel, C., and Goldenberg, A.   *Clin. Chem.* 12:194, 1966.
76. Debro, J. R., Tarver, H., and Korner, A.   *J. Lab. Clin. Med.* 50:728, 1957.
77. Kachani. Z. F. C.   *Z. Klin. Chem.* 2:91, 1964.
78. Cohn, E. J., et al.   *J. Am. Chem. Soc.* 72:465, 1950.
79. Pillemer, L., and Hutchinson, M. C.   *J. Biol. Chem.* 158:299, 1959.

80. Wrenn, H. T., and Fetchmeyer, T. V. *Am. J. Clin. Pathol.* 26:960, 1956.
81. Keyser, J. W. *Clin. Chim. Acta* 6:445, 1961.
82. McDonald, C., and Gerarde, H. W. *Microchem. J.* 7:57, 1963.
83. Crowley, L. V. *Clin. Chem.* 10:1131, 1964.
84. Watson, D., and Nankiville, D. D. *Clin. Chim. Acta* 9:359, 1964.
85. Rutstein, D. D., Ingenito, E. F., and Reynold, W. E. J. *Clin. Invest.* 33:211, 1954.
86. Niall, M. M., and Owen, J. A. *Clin. Chim. Acta* 7:155, 1962.
87. Goodwin, J. F. *Clin. Chem.* 10:309, 1964.
88. Ness, A. T., Dickerson, H. C., and Pastewka, J. V. *Clin. Chim. Acta* 12:532, 1965.
89. Martinek, R. G. *Clin. Chem.* 11:441, 1965.
90. Rodkey, F. L. *Clin. Chem.* 11:478, 1965.
91. Dow, D., and Pinto, P. V. C. *Clin. Chem.* 15:1006, 1969.
92. Doumas, B. T., Watson, W. A., and Biggs, H. G. *Clin. Chim. Acta* 31:87, 1971.
93. Leonard, P. J., Persaud, J., and Motwani, R. *Clin. Chim. Acta* 35:409, 1971.
94. McPherson, I. G., and Everard, D. W. *Clin. Chim. Acta* 37:117, 1972.
95. Westgard, J. O., and Poquette, M. O. *Clin. Chem.* 18:647, 1972.
96. Bigat, T. K., and Saifer, A. *Clin. Chem.* 18:630, 1972.
97. Doumas, B. T., and Biggs, H. G. *Stand. Methods Clin. Chem.* 7:175, 1972.
98. Carter, P. *Microchem. J.* 15:531, 1970.
99. Carter, P. *Int. Congr. Clin. Chem. (Proc.)* 7th. Vol. 1, p. 1, 1961.
100. Bartholomew, R. J., and Delaney, A. M. *Proc. Aust. Assoc. Clin. Biochem.* 1:214, 1966.
101. Gindler, E. M., and Hanson, K. L. *Adv. Automat. Anal. Technicon Int. Congr.* Vol. 1, p. 43, 1969.
102. Pemberton, J. R., and DeJong, J. *Anal. Biochem.* 43:575, 1971.
103. Betheil, J. J. *Anal. Chem.* 32:560, 1960.
104. Rees, V. H., Fieldes, J. E., and Laurence, D. J. R. *J. Clin. Pathol.* 7:336, 1954.
105. Rosen, C. G., Pesce, A. J., and Gaizutis, M. *Microchem. J.* 16:218, 1971.
106. Glenn, J. H. *J. Clin. Pathol.* 18:131, 1965.
107. Jones, D. D. *J. Lab. Med. Technol.* 23:25, 1966.
108. Read, E. A. N. *Clin. Chem.* 15:1186, 1969.
109. Saifer, A., Gerstenfeld, S., and Vecsler, F. *Clin. Chem.* 7:626, 1961.
110. Saifer, A., and Gerstenfeld, S. *Clin. Chim. Acta* 7:149, 1962.
111. Saifer, A., and Marven, T. *Clin. Chem.* 12:414, 1966.
112. Rodkey, F. L. *Clin. Chem.* 11:486, 1965.
113. Savory, J., Heintges, M. G., and Sobel, R. E. *Clin. Chem.* 17:301, 1971.
114. Goldenberg, H., and Drewes, P. A. *Clin. Chem.* 17:358, 1971.
115. Rieche, K. *Z. Med. Labortech.* 8:277, 1967.
116. Burmester, H. B. C., and Aulton, H. G. *J. Clin. Pathol.* 23:43, 1970.
117. Ellis, B. C., and Stransky, A. *J. Lab. Clin. Med.* 58:477, 1961.
118. Farrell, G. W., and Wolf, P. *J. Immunol. Methods* 1:217, 1972.
119. Goodwin, J. F. *Clin. Chem.* 13:1057, 1967.
120. Phillips, L. L., Jenkins, E. B., and Hardaway, R. M. *Tech. Bull. Regist. Med. Technol.* 39:130, 1969.
121. Goodwin, J. F. *Am. J. Clin. Pathol.* 35:227, 1961.

122. Parfentjev, I. A., Johnson, M. L., and Cliffton, E. E.   *Arch. Biochem.* 46:470, 1953.
123. Fowell, A. H.   *Am. J. Clin. Pathol.* 25:340, 1955.
124. Grannis, G. F.   *Clin. Chem.* 16:486, 1970.
125. Hunter, D. T., Jr., and Allenworth, J. L.   *Tech. Bull. Regist. Med. Technol.* 35:145, 1965.
126. Rice, E. W., and Muesse, D. E. R.   *Clin. Chem.* 18:73, 1972.
127. Farrell, G. W., and Wolf, P.   *Med. Lab. Technol.* 28:310, 1971.
128. Emar, A.   *Feuill. Biol.* 9:47, 1968.
129. Reiner, M., and Cheung, H. L.   *Stand. Methods Clin. Chem.* 3:114, 1961.
130. Kloske, W.   *Aerztl. Lab.* 9:316, 1963.
131. Emar, A.   *Feuill. Biol.* 9:51, 1968.
132. Astrup, T., Brakman, P., and Nissen, U.   *Scand. J. Clin. Lab. Invest.* 17:57, 1965.
133. Wycoff, H. D.   *J. Lab. Clin. Med.* 47:645, 1956.
134. Donati, M. B., and Vancoetsem, T.   *Pathol. Biol.* 18:697, 1970.
135. Rosenberg, A. A., and Peabody, R. A.   *Clin. Chem.* 18:666, 1972.
136. Okuno, T., and Selenko, V.   *Am. J. Med. Technol.* 38:196, 1972.
137. Davey, F. B., Carrington, C. E., and Nelson, D. A.   *Clin. Chem.* 18:1360, 1972.
138. Brittin, G. M., Rafinia, H., Raval, D., Werner, M., and Brown, B.   *Am. J. Clin. Pathol.* 57:89, 1972.
139. Tiselius, A.   *Trans. Faraday Soc.* 33:524, 1937.
140. Block, R. J., Durrum, E. L., and Zweig, G.   *A Manual of Paper Chromatography and Paper Electrophoresis.* New York: Academic, 1955.
141. McCoy, K. L.   *Am. J. Med. Technol.* 29:83, 1963.
142. Cremer, H., and Tiselius, A.   *Biochem. Z.* 320:273, 1950.
143. Sunderman, F. W., Jr.   In Sunderman, F. W., and Sunderman, F. W., Jr. [Eds.], *Serum Proteins and the Dysproteinemias.* Philadelphia: Lippincott, 1964.
144. Glazer, L. G., and Wagener, G. N.   *Am. J. Clin. Pathol.* 54:766, 1970.
145. Elevitch, F. R., Aronson, R. B., Feichtmeir, T. V., and Enterline, M. L.   *Am. J. Clin. Pathol.* 46:692, 1966.
146. van der Helm, H. G., and Holster, M. G.   *Clin. Chim. Acta* 10:483, 1964.
147. Matson, C. F.   *Am. J. Clin. Pathol.* 37:143, 1962.
148. Roesler, B.   *Acta Biol. Med. German* 16:106, 1966.
149. Lange, V.   *Z. Klin. Chem.* 3:168, 1965.
150. Cheng, F. H. F., and Steinberg, B.   *J. Lab. Clin. Med.* 63:694, 1964.
151. Daams, J. H.   *J. Chromatogr.* 10:450, 1963.
152. Nandi, M., and Lewis, G. P.   *J. Clin. Pathol.* 23:727, 1970.
153. Man, E. B., and Whithead, R. J., Jr.   *Clin. Chem.* 14:1002, 1968.
154. Pastewka, J. V., Ness, A. T., and Peacock, A. C.   *Clin. Chim. Acta* 14:219, 1966.
155. Clarke, J. T.   *Ann. N.Y. Acad. Sci.* 121:428, 1965.
156. Klein, U. E.   *Clin. Chim. Acta* 16:163, 1967.
157. Keyser, J. W., and Watkins, G. L.   *Clin. Chem.* 18:1541, 1972.
158. Friedman, H. S.   *Clin. Chim. Acta* 6:775, 1961.
159. Alonso, E.   *Clin. Chim. Acta* 6:883, 1961.
160. Webster, D.   *Clin. Chim. Acta* 11:101, 1965.
161. Durstein, F. H., MacCallum, D. B., Anson, J. H., and Mohammed, A.   *Clin. Chem.* 10:853, 1964.

162. Cawley, L. P., Penn, G. M., Itano, M., Bell, H. E., and Minard, B. *Basic Electrophoresis, Immunoelectrophoresis and Immunochemistry.* Chicago: American Society of Clinical Pathologists, 1972.
163. Fuller, J. B. *Selected Topics in Clinical Chemistry, Serum Protein Analysis.* Chicago: American Society of Clinical Pathologists, 1972.
164. Keyser, J. W., and Watkins, G. L. *Clin. Chem.* 18:1541, 1972.
165. Busse, V., and Dulce, H. J. *Z. Klin. Chem. Klin. Biochem.* 7:486, 1969.
166. Busse, V. *Z. Klin. Chem. Klin. Biochem.* 6:273, 1968.
167. Kohn, J., and Mepham, B. L. *Clin. Chim. Acta* 41:151, 1973.
168. Rice, E. W., and Hammer, P. A. *Tech. Bull. Regist. Med. Technol.* 39:17, 1969.
169. Sherwin, R. M., and Moore, G. J. *Am. J. Clin. Pathol.* 55:705, 1971.
170. McFarlane, H. *Clin. Chim. Acta* 9:378, 1964.
171. Burgi, W. *Z. Klin. Chem. Klin. Biochem.* 5:277, 1967.
172. Peetoom, F., and Gerald, P. S. *Clin. Chim. Acta* 10:375, 1964.
173. Kaplan, A., and Johnstone, M. *Clin. Chem.* 12:717, 1966.
174. Windisch, R. M., and Bracken, M. M. *Clin. Chem.* 16:416, 1970.
175. Weiner, L. N., and Zak, B. *Stand. Methods Clin. Chem.* 7:305, 1972.
176. Behm, E., Friemel, H., Hofmann, R., and Doerfling, P. *Z. Med. Labortech.* 12:177, 1971.
177. Uffelman, J. A., Engelhard, W. E., and Jolliff, C. R. *Clin. Chim. Acta* 28:185, 1970.
178. Sander, G., and Urosevic, A. *C.R. Acad. Sci.* [D] (Paris) 276:2753, 1973.
179. Killingsworth, K. M., and Savory, J. *Clin. Chem.* 18:335, 1972.
180. Savory, J., Heintges, M. G., Killingsworth, L. M., and Potter, J. D. *Clin. Chem.* 18:37, 1972.
181. Iammarino, R. M. *Stand. Methods Clin. Chem.* 7:163, 1972.
182. Braun, H. J., and Aly, F. W. *Z. Klin. Chem. Klin. Biochem.* 9:508, 1971.
183. Javid, J., and Horowitz, H. I. *Am. J. Clin. Pathol.* 34:35, 1960.
184. Herman, E. C. *J. Lab. Clin. Med.* 57:825, 1961.
185. Affanso, A. *Clin. Chim. Acta* 22:466, 1968.
186. Bernier, G. M. *Clin. Chim. Acta* 18:309, 1967.
187. van Ros, G., and van Sande, M. *Clin. Chim. Acta* 10:62, 1964.
188. Baur, E. W. *Clin. Chim. Acta* 9:252, 1964.
189. Bundschuh, G. *Dtsch. Gesundheitsw.* 15:2103, 1960.
190. Dietz, A. A., Lubrano, T., and Rubinstein, H. M. *Clin. Biochem.* 4:59, 1967.
191. Ferris, T. G., Easterlig, R. E., Nelson, K. J., and Budd, R. E. *Am. J. Clin. Pathol.* 46:385, 1966.
192. Pantlitschko, M., and Weippl, G. *Clin. Chim. Acta* 19:439, 1968.
193. Hall. R. *J. Med. Lab. Technol.* (London) 21:64, 1964.
194. Valeri, C. R., Bond, J. C., Fowler, K., and Sobucki, J. *Clin. Chem.* 11:581, 1965.
195. Colis, B., and Verheyden, J. *Clin. Chim. Acta* 12:470, 1965.
196. Lionetti, F. J., Valeri, C. R., Bond, J. C., and Fortier, N. F. *J. Lab. Clin. Med.* 64:519, 1964.
197. Owen, J. A., Better, F. C., and Hoban, J. *J. Clin. Pathol.* 13:163, 1960.
198. Veneziale, C. M., and McGuckin, W. F. *Mayo Clin. Proc.* 40:751, 1965.
199. Lupovitch, A., and Katase, R. Y. *Clin. Chim. Acta* 11:566, 1965.

200. Bauer, J. D.   In S. Frankel, S. Reitman, and A. C. Sonnenwirth [Eds.], *Gradwohl's Clinical Laboratory Methods and Diagnosis,* 7th ed. St. Louis: Mosby, 1970. P. 463.
201. Stefanini, M., and Dameshak, W.   *The Hemorrhagic Disorders.* New York: Grune & Stratton, 1955.
202. Thompson, R. H. S., and King, E. J.   *Biochemical Disorders in Human Disease.* New York: Academic, 1964.
203. Hsai, D. Y-Y.   *Inborn Errors of Metabolism,* 2nd ed. Chicago: Year Book, 1966. P. 133.
204. Mabry, C. C., Roeckel, I. E., Gevedon, R. E., and Koepke, J. A.   *Recent Advances in Pediatric Clinical Pathology.* New York: Grune & Stratton, 1968.
205. Moore, S., Spackman, D. H., and Stein, W. H.   *Anal. Chem.* 30:1185, 1958.
206. Spackman, D. H., Stein, W. H., and Moore, S.   *Anal. Chem.* 30:1190, 1958.
207. Gehrke, C. W., Lamkin, W. M., Stalling, D. L., and Shahrokhi, F.   *Biochem. Biophys. Res. Commun.* 19:328, 1965.
208. Roach, D., and Gehrke, C. W.   *J. Chromatogr.* 43:303, 1969.
209. Mabry, C. C.   *Crit. Rev. Clin. Lab. Sci.* 1:135, 1970.
210. Holmgren, G., Jeppson, J. O., and Samuelson, G.   *Scand. J. Clin. Lab. Invest.* 26:313, 1970.
211. Samuels, S., and Ward, S. S.   *J. Lab. Clin. Med.* 67:669, 1966.
212. Porter, C. A., Margolis, D., and Sharp, P.   *Contrib. Boyce Thompson Inst.* 18:465, 1957.
213. Efron, M. L., Young, D., Moser, H. W., and MacReady, R. A.   *N. Engl. J. Med.* 270:1378, 1964.
214. Levy, H. L., Shih, V. E., Madigan, P. M., Karolewicz, V., and MacReady, R. A.   *Clin. Biochem.* 1:200, 1968.
215. Spinella, C. J.   *Clin. Chem.* 15:1011, 1969.
216. Munier, R., and Sarrazin, G.   *Bull. Soc. Chim. France* 10:2959, 1965.
217. Kutter, D., and Humbel, R. E.   *Pharm. Acta Helv.* 45:553, 1970.
218. Kraffeczyk, F., Helger, R., Lang, H., and Bremer, H. J.   *Clin. Chim. Acta* 35:345, 1971.
219. Ersse, R. S.   *J. Med. Lab. Technol.* 27:142, 1970.
220. Guthrie, R., and Susi, A.   *Pediatrics* 32:338, 1963.
221. Newman, R. L., Stern, J., Starr, D. J., Lindsay, G., Kennedy, R., Lilly, P., Obank, W. B., and Watt, C. R.   *J. Clin. Pathol.* 24:576, 1971.
222. LaDu, B. N., and Michael, P. J.   *J. Lab. Clin. Med.* 55:491, 1960.
223. Woolf, L. I., and Goodwin, R. L.   *Clin. Chem.* 10:146, 1964.
224. Jones, D. D.   *J. Med. Lab. Technol.* (London) 24:301, 1967.
225. Steed, E., Pereira, W. D., Halpern, B., Solomon, M. D., and Duffield, A. M.   *Clin. Biochem.* 5:166, 1972.
226. Halpern, B., Pereira, W., Solomon, M. D., and Steed, E. C.   *Anal. Biochem.* 39:156, 1971.
227. McCaman, M. W., and Robins, E.   *J. Lab. Clin. Med.* 59:885, 1962.
228. Ambrose, J. A.   *Clin. Chem.* 15:15, 1969.
229. Terlingen, J. B. A., and Van Dreumel, H. J.   *Clin. Chim. Acta* 22:643, 1968.
230. Ambrose, J. A., Ingerson, A., Garrettson, L. G., and Chung, C. W.   *Clin. Chim. Acta* 15:497, 1967.

231. Hill, J. B., Summer, G. K., Pender, M. W., and Roszel, N. O. *Clin. Chem.* 11:541, 1965.
232. Clark, P. T., and Rice, J. D. *Am. J. Clin. Pathol.* 46:486, 1966.
233. Dourdillon, J., and Vanderlinde, R. W. *Public Health Reports* (U.S.) 81:991, 1966.
234. Mondino, A., and Bongiovanni, G. *J. Chromatogr.* 67:49, 1972.
235. Levy, H. L., Baullinger, P. C., and Madigan, P. M. *Clin. Chim. Acta* 31:447, 1971.
236. Berman, J. L., Cunningham, G. C., Day, R. W., Ford, R., and Hsia, D. Y-Y. *Am. J. Dis. Child.* 117:54, 1969.
237. Robinson, R. *Clin. Chim. Acta* 14:166, 1966.
238. Melmon, K. L., Rivlin, R., Oates, J. A., and Bjoersdama, A. *J. Clin. Endocrinol. Metab.* 24:691, 1964.
239. Siersback-Nielson, K. *Acta Med. Scand.* 179:417, 1966.
240. Robins, E. *Methods Biochem. Anal.* 17:287, 1969.
241. Scott, H. P. *Clin. Chim. Acta* 35:17, 1971.
242. Ambrose, J. A., Sullivan, P., Ingerson, A., and Brown, R. L. *Clin. Chem.* 15:611, 1969.
243. Waalkes, T. P., and Udenfriend, S. *J. Lab. Clin. Med.* 50:733, 1957.
244. Dull, T. A., and Henneman, P. H. *N. Engl. J. Med.* 268:132, 1963.
245. Raab, W. *Z. Klin. Chem. Klin. Biochem.* 10:195, 1972.
246. Jones, C. R., Bergman, M. W., Kittner, P. J., and Pigman, W. W. *Proc. Soc. Exp. Biol. Med.* 115:85, 1964.
247. Goidanich, I. F., Lenzi, L., and Silva, E. *Clin. Chim. Acta* 11:35, 1965.
248. Leroy, E. C. *Adv. Clin. Chem.* 10:213, 1967.
249. Benoid, F. L., Theil, G. B., and Watten, R. H. *Metabolism* 12:1072, 1963.
250. Emmerich, R., Haentzschel, H-J., and Haentzschel, H. *Z. Gesamte Inn. Med.* 22:193, 1967.
251. Klein, L. *Stand. Methods Clin. Chem.* 6:41, 1970.
252. Parekh, A. C., and Jung, D. H. *Biochem. Med.* 4:446, 1970.
253. Goverde, B. C., and Veenkamp, F. J. N. *Clin. Chim. Acta* 41:29, 1972.
254. Wapnir, R. A., and Stevenson, J. H. *Clin. Chim. Acta* 26:203, 1969.
255. Lehmann, J. *Scand. J. Clin. Lab. Invest.* 23:249, 1969.
256. Denchla, W. D., and Dewey, H. K. *J. Lab. Clin. Med.* 69:160, 1967.
257. Saifer, A., and Gerstenfeld, S. *Clin. Chem.* 10:970, 1964.
258. Search, R. L. *Arch. Biochem. Biophys.* 81:275, 1959.
259. Baldridge, R. C., and Greenberg, N. *J. Lab. Clin. Med.* 61:700, 1963.
260. Hill, H. D., Jr., Summer, G. K., and Newton, D. A. *Clin. Chim. Acta* 36:361, 1972.
261. Ambrose, J. A., Crim, A., Burton, J., Paullin, K., and Ross, C. *Clin. Chem.* 15:361, 1969.
262. Gerber, D. A. *Anal. Biochem.* 34:500, 1970.
263. Haux, P., and Natelson, S. *Clin. Chem.* 16:366, 1970.
264. Fernandez, A. A., and Henry, R. J. *Anal. Biochem.* 11:190, 1965.
265. Moench, E., and Siemes, H. *Z. Klin. Chem. Klin. Biochem.* 8:516, 1970.
266. Sun, M-W., Stein, E., and Gruen, F. W. *Clin. Chem.* 15:183, 1969.
267. Davis, J. R., and Andelman, S. L. *Arch. Environ. Health* 15:53, 1967.
268. Mauzeral, D., and Granick, S. *J. Biol. Chem.* 219:435, 1956.
269. Mortensen, E. *Scand. J. Clin. Lab. Invest.* 16:87, 1964.

270. Watson, D. *Scand. J. Clin. Lab. Invest.* 16:587, 1964.
271. Beutler, E., Duron, O., and Mikus, B. *J. Lab. Clin. Med.* 61:862, 1963.
272. Owens, C. W. L., and Belcher, R. J. *Biochem. J.* 94:705, 1965.
273. Cohn, V. N., and Lyle, J. *Anal. Biochem.* 14:434, 1966.
274. Elevitch, F. R. *Fluorometric Techniques in Clinical Chemistry.* Boston: Little, Brown, 1973. P. 111.
275. Haer, F. C. *An Introduction to Chromatography on Impregnated Glass Fiber.* Ann Arbor: Ann Arbor-Humphrey, 1969.
276. Kirchner, J. G. *Thin-Layer Chromatography.* New York: Wiley-Interscience, 1967.
277. Randerath, K. *Duennschicht-Chromatographie.* Weinheim: Verlag Chemie, 1965.
278. Schultze, H. E., Heide, K., and Haupt, H. *Clin. Chim. Acta* 7:854, 1962.
279. Keyser, J. W. *Anal. Biochem.* 12:395, 1965.
280. deVaux St. Cyr, C. *Exp. Biol. Med.* 52:571, 1963.
281. Glenn, W. G., Lanchantin, G. F., Mitchell, R. B., and Marable. I. W. *Tex. Rep. Biol. Med.* 1:320, 1958.
282. Winzler, R. J. In F. W. Putman [Ed.], *The Plasma Proteins,* vol. 1. New York: Academic, 1960. Pp. 309–347.
283. Winzler, R. J. *Methods Biochem. Anal.* 2:279, 1955.
284. Bruckner, J. *Biochem. J.* 55:126, 1953.
285. Weimer, H. E., Redlick-Moshin, J., Salkin, D., and Boak, R. A. *Proc. Soc. Exp. Biol. Med.* 87:102, 1954.
286. Asher, T. M., and Cooper, C. R. *Clin. Chem.* 6:189, 1960.
287. Lauchantin, G. F. *Stand. Methods Clin. Chem.* 6:137, 1970.
288. Kushner, D. S., Honig, K., Dubin, A., Dyniewicz, H. A., Bronsky, D., de la Huerga, J., and Popper, H. *J. Lab. Clin. Med.* 47:409, 1956.
289. Bottiger, L. E., and Carlson, L. A. *Clin. Chim. Acta* 5:664, 1958.
290. Gilmore, H. R., and Schwartz, C. J. *Aust. J. Exp. Biol. Med. Sci.* 36:575, 1958.

8. Inorganic Ions

Calcium

One of the oldest methods still in use for the determination of calcium in serum is the precipitation of the calcium as the oxalate, followed by determination of oxalate in the precipitate by oxidimetric titration [1, 2]. In spite of the limitations of the method, it has been considered, at least until recently, as one of the reference methods. In the usual procedure the calcium is precipitated directly from diluted serum by the addition of ammonium oxalate and adjustment of the pH to between 4 and 7. Below pH 3.5 the calcium is incompletely precipitated and above 7.5 magnesium ammonium phosphate may be precipitated [3]. The original method called for only 30 min standing for precipitation, but others have claimed that at least 3 hr are required at room temperature [4] and that overnight standing may be desirable. The precipitation is complete in $\frac{1}{2}$ hr at 56°C [5]. After precipitation the tube is centrifuged strongly and the packed precipitate washed to remove excess reagent. The washing may cause slight losses of calcium oxalate due to solubility or mechanical removal of traces of the precipitate [6]. On the other hand, small amounts of potassium or sodium oxalate, as well as other reducing substances such as protein or uric acid, may be included in the precipitate causing positive errors. The positive and negative errors appear to balance out fairly well so that the method gives accurate results.

In the original methods the precipitate was dissolved in sulfuric acid and titrated with standard potassium permanganate at about 70°C. The potassium permanganate serves as its own indicator, the slight excess of reagent at the endpoint imparting a pink color to the solution. The first trace of pink may be difficult to detect and the color will fade on standing. The usual directions state that the endpoint is reached when the pink color persists for 15 sec. Standards must be run with each set of samples as the permanganate solution is not stable. More recently the use of ceric ions as oxidant has been introduced. In $2M$ perchloric acid, perchloratocerate has a sufficiently high oxidizing potential to react with oxalate at room temperature. An indicator must be used, usually the o-phenanthroline-ferrous complex which changes in color from orange to blue at the endpoint [7].

A number of complex colored organic compounds—which for the sake of brevity we will refer to as dyes—at the proper pH give one color in the presence of calcium ions and another color in their absence. Theoretically only about half the calcium present in serum is found in the free ionic form, the other half being combined mostly with protein. The dyes have a stronger affinity for calcium ions than for the proteins so that all the calcium in the serum will react with the dye.

These dyes can be used to determine calcium in two ways, one titrimetric and

227

the other colorimetric. They may be used as a titration indicator. The solution is titrated with a compound that forms a very stable complex with calcium ions. As long as there are any calcium ions in the solution, the added dye will form the colored dye-calcium complex. When all the calcium has been complexed by the added titrant, the dye will give only the color it has in the absence of calcium ions. The difficulty with most of these titration procedures is that the endpoint is not very sharp, particularly in the presence of other ions.

The substance most commonly used as a titrant is EDTA [ethylenediamine-tetraacetic acid], or, more correctly, [(ethylenedinitrilo)tetraacetic acid] [8]. Another substance that has been used is EGTA [ethylene glycol bis-(aminoethyl ether) tetraacetic acid], or, more correctly, [ethylene bis-(oxyethylenenitrilo) tetraacetic acid] [9]. These substances will form complexes with a large number of different metallic ions and have been used for their determination for some time [10].

Many different dye indicators have been used. They are usually rather complex organic molecules and many are known simply by their trade names. Some of these compounds are: murexide [11], Cal-Red [12], hydroxynaphthol blue [13], Calcein [14, 15], Plasma Corinth B [16], Eriochrome Blue SE [17], and glyoxal bis-(2-hydroxyanil) [18]. The fact that so many different indicators have been tried implies that a completely satisfactory one has not been found [19].

An accurate but somewhat cumbersome method is to use a photometric titration. The color change of the dye is noted by measuring the absorbance at a given wavelength as successive increments of the titrant are added. The endpoint is obtained graphically by plotting the change in absorbance against the amount of titrant added [20]. Some of the dyes, such as Eriochrome Black T, will titrate both calcium and magnesium, and one subtracts from the titrated value the assumed value of 1.64 mEq/liter for magnesium to obtain the corrected value for calcium. This could lead to an error of as much as 0.4 mEq/liter for calcium [21]. Some dyes may give a sharper endpoint for magnesium than for calcium, and advantage can be taken of this by adding an excess of EDTA and then back-titrating with a standard magnesium solution [22].

Some of the indicators are not very stable in solution and are used in a diluted solid form made by grinding about one part of the dye with 100 parts of an inert salt such as potassium chloride. In this way the small amount of indicator can be conveniently measured by taking a small scoop of the diluted dye. The color of hemolyzed or highly icteric serums may interfere with the detection of the titration endpoint. Two methods have been used to eliminate this: The calcium may be separated from the serum by absorption on an ion exchange resin and later eluted with strong base, or the serum may be treated with hydrogen peroxide or sodium hypochlorite to oxidize the interfering chromogens.

The other method of determination of calcium using the dye complex involves addition of an excess of dye, more than sufficient to combine with all the calcium present. The amount of the color due to the calcium-dye complex is proportional to the amount of calcium present and is determined in a colorimeter against a blank of the dye solution alone. With some of these methods the interference of magnesium is minimized by the addition of 8-hydroxyquinoline or its sulfonic acid derivative which form stable complexes with the magnesium. In the colorimetric methods a large number of different dyes have also been used. These include cresolphthalein complexone [23, 24], Plasma Corinth B [25], sodium alizarine-sulfonate [26], nuclear fast red [27], murexide [28], Eriochrome Blue SE [29], glyoxal *bis*-(2-hydroxyanil) [30], and methyl-thymol blue [30a].

If the serum is added directly to the dye solution (without the use of a tri-chloracetic acid filtrate), hemolysis, icterus, or lipemia may cause some interference with the colorimeter readings. One way of compensating for this is to measure the absorbance of the diluted serum in a blank solution not containing the dye (read against a similar solution without the serum). The other method measures the absorbance of the calcium-dye complex along with that of any interfering substances. Then EDTA is added to both sample and blank to destroy all color due to calcium. The remaining difference in absorbance is due to the interfering substances [26, 31].

In some methods the calcium is separated from the magnesium and other interfering substances by precipitation as the oxalate. The precipitate is then dissolved and determined by one of the dye methods. This is only a little shorter than the oxidimetric titration except that in this case the precipitate need not be washed, only the supernatant decanted off [32, 33].

Calcium may be determined by precipitation as the chloranilate. After washing of the precipitate to remove excess reagent, the precipitate is dissolved in EDTA solution or other solvent and the color of the chloranilate determined [34, 35]. Calcium has also been precipitated as the naphthohydroxamate. After washing, the precipitate is redissolved and the hydroxamate determined colorimetrically with ferric ions [36].

The dye calcein forms a fluorescent complex with calcium ions. It has been used as a titration indicator, viewed under ultraviolet light, or as the basis for a direct fluorescent method for calcium [37, 38]. The fluorescence is not linear with concentration over a very wide range. A careful adjustment must be made of the dye concentration and alkalinity to obtain a fairly linear curve over the range of 3 to 7 mEq/liter of calcium in serum. Only 20 μl of serum is required, an advantage in pediatric work. Due to the low concentration of calcium, absorption of the calcium on glass becomes a problem and plastic tubes must be used. Not all workers have been able to obtain good results with this method.

Calcium has also been determined by flame photometry. In general, more difficulties are experienced with this metal than with sodium and potassium [39–41]. If a simple dilution of the serum is used, the aspiration rate of the diluted serum may be different from that of the aqueous standards. One may use either serum standards or a trichloracetic acid filtrate [42]. Phosphate interferes by forming refractory phosphates in the flame, and sodium causes an enhancement of the calcium emission which depends upon the amount of sodium present. These interferences may be overcome by first precipitating the calcium as the oxalate, dissolving the precipitate, and using this in the flame photometer [43, 44]. This adds considerably to the length of the procedure; one of the advantages of the flame photometric method is that it is very rapid and simple. Sodium can be corrected for by using standards containing about the same amount of sodium as the sample, making corrections based on the determined amount of sodium present, or by adding a large excess of sodium to both standards and samples [45].

The addition of EDTA or preferably lanthanum chloride prevents interference by phosphate [46]. In spite of the apparent simplicity of the flame photometric method, it has not been very successful for calcium determinations. The best results require the use of a hot flame (acetylene-oxygen), the addition of lanthanum, and the use of a surfactant such as Sterox.

Calcium can probably be most accurately determined by atomic absorption spectrophotometry [47–49]. Phosphate will cause some interference but this is eliminated by the addition of lanthanum chloride to a final concentration of about 0.5%. Usually a simple dilution of the serum is aspirated, although a trichloracetic acid filtrate can be used [50]. A recent modification of the method has been the use of a double-beam instrument with an internal standard of strontium. This is said to increase greatly the accuracy of the determination (and also the complexity and cost of the instrument) [51]. The use of the internal standard appears to be the best reference method available [52].

We present two methods for the determination of calcium. The first is a colorimetric method using cresolphthalein complexone, which requires a maximum of 0.05 ml of serum. The other method is the chloranilate precipitation procedure, which is longer and requires more serum. Some general directions are also given for the determination of calcium by atomic absorption; specific directions depend upon the particular instrument.

Calcium by Cresolphthalein Complexone Micromethod

In this method [53, 54] the serum reacts directly with a cresolphthalein complexone reagent containing dimethyl sulfoxide. Hemolysis and bilirubin are corrected for by addition of a chelating agent to destroy the calcium color and compensate for color interference by these substances.

REAGENTS

1. Color reagent. Dissolve 40 mg of cresolphthalein complexone in 1 ml of concentrated hydrochloric acid in a small beaker. Rinse into a 1 liter volumetric flask with 100 ml of dimethyl sulfoxide. Add 2.5 g of 8-hydroxyquinoline, dissolve, and dilute to 1 liter with water.

2. Diethylamine solution, $0.38M$. Dissolve 0.5 g of potassium cyanide and 40 ml of diethylamine in water to make 1 liter.

3. EGTA reagent. Dissolve 0.5 g of ethylene glycol *bis*(2-aminoethyl ether) tetraacetic acid [EGTA, also known as ethylene *bis*(oxyethylenenitrilo)tetraacetic acid] in water to make 100 ml. If the material appears to dissolve very slowly, add a few milliliters of diethylamine.

4. Calcium standards.

A. Stock standard. Dissolve 3.122 g of reagent grade calcium carbonate, which has been previously dried at 110°C for several hours, in about 10 ml of water and 5 ml of concentrated hydrochloric acid. Transfer to a 500 ml volumetric flask and dilute to the mark with water. This solution contains 2.5 mg/ml of calcium $(0.0625M)$.

B. Working standards. Dissolve 3, 4, and 5 ml of the stock standards to 100 ml with water. These standards will contain 7.5, 10.0, and 12.5 mg Ca/dl $(1.75, 2.50,$ and $3.25mM)$. Atomic absorption standards can be purchased for a number of metals. These usually contain 1,000 ppm, which is equivalent to 1 mg/ml. To use these, dilute 7, 10, and 13 ml to 100 ml to give working standards containing 7, 10, and 13 mg/dl.

PROCEDURE

In a number of plastic test tubes add 2.5 ml of the color reagent. (The use of plastic tubes is advised; because of the small amount of calcium present, absorption on glass may cause considerable error.) Add 0.05 ml of sample or standard, reserving one tube as blank (it is not necessary to add 0.05 ml of water to the blank). To each tube add 2.5 ml of the diethylamine reagent and mix. Read standards and samples against blank at 525 nm. Add 0.1 ml of the EGTA solution to each tube (including blank) and mix. Read all tubes again against blank. Subtract any absorbance obtained in the second reading from the absorbance obtained in the first reading for the corresponding tubes.

CALCULATION

$$\frac{\text{corrected absorbance of sample}}{\text{corrected absorbance of standard}} \times \text{conc. of standard} = \text{conc. of sample}$$

Beer's law should be followed closely but it may be advisable to use the absorbance of the standard nearest that of the sample for calculations. The second reading after addition of the EGTA is not necessary for aqueous standards or clear, non-icteric serum without hemolysis, but it may be used routinely since it is so simple. The method may be scaled down by using 1 ml of each reagent and 0.02 ml of sample if cuvettes are available that accept 2 ml of volume.

Calcium by Chloranilate Method

In this method [34, 55] the calcium is precipitated as the insoluble chloranilate salt. The precipitate is centrifuged down, washed, and dissolved with the aid of EDTA to give a colored solution that is read in the photometer.

REAGENTS

1. Chloranilate reagent. Dissolve 1.5 g of chloranilic acid and 3.0 g of tris-(hydroxymethyl)aminomethane in water and dilute to 250 ml. Filter before using.

2. Wash solution. Mix equal volumes of ethylene glycol monomethyl ether and water.

3. EDTA solution. Dissolve 25 g of the tetrasodium salt of ethylenediaminetetraacetic acid in water to make 500 ml. If only the disodium salt is available, use 25 g of this plus 5 g of NaOH.

4. Calcium standard. Same as for previous method.

PROCEDURE

Pipet 2 ml of samples and standards to 15 ml conical centrifuge tubes. These tubes should be scrupulously clean. Add to each tube 1 ml of the chloranilate reagent. Add reagent rapidly by blowing pipet out. Mix well and allow to stand in water bath for at least 2 hr at 37°C. Remove from water bath and centrifuge strongly to pack precipitate. Carefully decant supernatant and allow tubes to drain in an inverted position for 5 min. Carefully wipe off any liquid adhering to the lip of the tube. Add about 0.15 ml of the wash solution to each tube. Tap the tube vigorously until all the precipitate is suspended in the solution; there should be no precipitate adhering to the bottom of the tube. Add 6 ml of the wash solution to each tube, mix gently, and then centrifuge strongly. Decant wash solution and drain tubes for 5 min. Wipe off any liquid adhering to the mouth of the tube. Add 0.2 ml of EDTA solution to each tube. Again tap tubes until all precipitate is resuspended in solution. Add an additional 6.0 ml of EDTA solution to each tube, mix, and read standards and samples against water blank at 525 nm.

CALCULATION

$$\frac{\text{absorbance of sample}}{\text{absorbance of standard}} \times \text{conc. of standard} = \text{conc. of sample}$$

The standard concentration is usually 10 mg/dl, equivalent to 2.5 mmol/liter.

The above method requires 2 ml of serum. The following variation based on the work of Spandrio [56] requires less serum. He found that at a pH of 1.5 to 1.6 the chloranilic acid has several times the absorbance it has at pH of near 7. This method requires only one additional reagent.

5. Glycine buffer, $0.05M$, pH 1.5.   Dissolve 3.8 g of glycine in about 700 ml of water in a 1 liter volumetric flask. Add hydrochloric acid ($1M$, 84 ml concentrated acid diluted to 1 liter) to bring the pH to 1.5. About 75 ml of acid will be required, so about 60 ml can be added before checking the pH. Then dilute to 1 liter.

PROCEDURE

Using 0.5 ml of sample or standard and 0.25 ml of chloranilate reagent, precipitate, centrifuge, and wash just as in the procedure given above. After the washing, again suspend the precipitate in 0.2 ml of EDTA solution. Then add 5 ml of the acid buffer and mix. Read at 525 nm.

CALCULATION

Same as in the previous procedure.

Calcium by Atomic Absorption Spectrophotometry

This is in general a very satisfactory method for the determination of calcium in serum and urine. Interference from phosphate is greatly reduced by the addition of lanthanum. A good diluent is made by dissolving 11.7 g of pure lanthanum oxide in 30 ml of concentrated hydrochloric acid and diluting to 1 liter with water. This contains 1 g/dl of lanthanum ($0.072M$). The serum or urine is diluted 1:25 or 1:50 with this solution (depending upon the instrument used). The emission line at 422.7 nm and the air-acetylene flame are generally used. More specific details will depend upon the particular instrument. Generally the instrument is zeroed with the diluent alone and several standards are run, bracketing the expected values. The standards given earlier for the chemical methods are satisfactory when diluted similarly to the samples.

Calcium Determination in Urine

Any of the above methods can be used for most urines. On becoming alkaline, the urine may have a precipitate of calcium phosphate, which must be dissolved before sampling. Acidify the entire urine specimen to a pH of about 1 with acid (HCl), warm for a few minutes to 60°C, then cool and mix well. Measure the total volume and take a proper aliquot. If the urine for calcium determination is collected with about 10 ml of concentrated hydrochloric acid, any further acidification is unnecessary.

In the determination of calcium in urine it may be necessary to use a dilution of the urine or a slightly larger sample. If it is assumed that the total 24-hr urine specimen contains 200 mg of calcium, the proper dilution can be readily calculated to give a final sample containing 0.1 mg/ml. With urine the interference in the atomic absorption method by large amounts of phosphate may be appreciable in spite of the use of lanthanum. In the colorimetric methods the presence of EDTA in the urine may interfere. These interferences can be eliminated by first precipitating the calcium as oxalate, then dissolving the precipitate for the determination. To 2 ml of urine in a conical centrifuge tube add 0.5 ml of ammonium oxalate solution (10 g/dl in water), a few drops of bromcresol green indicator solution (0.1 g in 100 ml of ethanol), and a few drops of concentrated ammonium hydroxide. Add dilute acetic acid (10 ml concentrated acid diluted to 100 ml with water) until the color just turns yellow. Place in a bath at 37°C for half an hour, then centrifuge and decant the supernatant. Drain the tube in an inverted position for a few minutes. For atomic absorption the precipitate can be dissolved in 2 ml of $1M$ hydrochloric acid and 2 ml of sodium citrate solution (1.5 g/100 ml in water). It is then diluted the required amount with lanthanum solution. For the colorimetric methods, add a few drops of concentrated nitric acid and heat in a boiling water bath for 10 min. Evaporate all the acid with the aid of a stream of air, and dissolve the residue in 2 ml of diluted hydrochloric acid (5 ml concentrated HCl diluted to 100 ml with water). Use this in place of the urine, since the calcium concentration in this digested sample is the same as in the original urine.

Normal Values and Interpretation of Results

The normal range of calcium in serum is 8.5 to 10.5 mg/dl (2.1 to 2.6 mmol/liter). Since about half the calcium is bound to the serum proteins, the serum level of calcium may be lower in hypoproteinemia. Low values may also be found in hypoparathyroidism, in which the level may drop to 6 mg/dl. The highest levels are found in hyperparathyroidism when values up to 20 mg/dl may be found. Excessive administration of vitamin D will also increase the serum level, though usually not above 15 mg/dl. The urinary excretion of calcium will depend to a considerable extent on the dietary intake but is usually between 50 and 400 mg/day. It may be increased somewhat in osteoporosis.

Ionized Calcium

Approximately half the calcium in serum is bound to the serum proteins in a nondiffusable form. A small portion is complexed with citrate and phosphate and the balance is present in the ionized form (Ca^{2+}). The ionized calcium is the form that is physiologically active, and theoretically it should be more valid to measure this fraction, which may be referred to as ionized calcium, diffusible calcium, or ultrafilterable calcium. This fraction has been determined in a number of ways. In one of the older methods the color reagent murexide is used. By reading at two different wavelengths, the proportion of ionized calcium can be calculated [57–60]. The ionized and bound calcium have also been separated by the use of gel filtration [61, 62], ion exchange resin [63], or coated charcoal [64]. The fraction may also be determined by ultrafiltration through a membrane that will pass the calcium ions but not allow the calcium bound to protein to pass through [65–68], or by means of the calcium ion specific electrode [69–73]. With the further development of these electrodes, this will be the preferred method and one that will be available to any laboratory with an expanded-scale pH meter, which can also be used for the measurement of other ion potentials.

The formula of McLean and Hastings is often used for approximating the ionized calcium in serum. This may be written:

$$\% \text{ of Ca in ionized form} = 100 \times \frac{[6 - (P/3C)]}{P + 6}$$

where P is the total protein content of the serum in g/dl and C is the calcium content in mg/dl. This formula gives fairly accurate results for normal serums but may be somewhat in error in those with abnormally high or low protein content. Normally about 50 to 58% of the calcium in serum is in the unbound form. Since cerebrospinal fluid is similar to an ultrafiltrate of plasma, the level of calcium found in it is generally only about 0.2 mg/dl greater than the ionized calcium in serum [74].

Phosphorus

Phosphorus is present in serum in a number of different chemical forms or combinations—as phosphate ions, as phosphate esters, and as phospholipids. The only determinations now usually made in serum are inorganic phosphorus and phospholipids (lipid phosphorus). Although the inorganic phosphorus is present as the phosphate ion, it is conventionally reported as phosphorus. The lipid phosphorus is found in combination with glycerol fatty acid esters such as lecithin. The phospholipids may be extracted from the serum with an organic solvent and the solvent evaporated. The residue is then treated with an oxidizing agent to eliminate organic matter and convert the phosphorus to phosphate. Since the phospholipids are generally combined with proteins, they may be precipitated by trichloracetic acid and the phosphorus determined in the precipitate after oxidation. Phospholipid phosphorus determination is discussed in Chapter 12.

The majority of the methods for the determination of phosphate are based on its reaction with molybdate to form a complex phosphomolybdate. This is then reduced with the production of a blue complex commonly known as molybdenum blue. Although this method has been used for many years, the exact chemical nature of the blue complex is not completely known. A large number of different reducing agents have been employed. These include stannous chloride [75–77] which gives maximum color production but which is rather unstable, hydrazine alone [78, 79] or in combination with stannous chloride [80], ferrocyanide [81], ferrous sulfate [82–84] which produces a very stable color, ascorbic acid [85–87], and a number of other organic reducing substances, including hydroquinone [88, 89], p-methylaminophenol [90–93], aminonaphtholsulfonic acid [94], n-phenyl-p-phenylenediamine [95], methampyrone [96], and p-phenylenediamine [97, 98].

The most common methods used in the past have been those based on the procedures of Gomori [90] or Fiske and Subbarow [94], at least for manual methods. These require a trichloracetic acid precipitation of the proteins before analysis. Some other methods use an acid molybdate solution to precipitate the proteins, the reducing agent being added later to a portion of the filtrate, or the reducing agent may be added to the trichloracetic acid used for precipitation [82] and the molybdate added to the filtrate. Methods have also been developed that do not require precipitation of the proteins before analysis (besides the automated methods which use dialysis [87, 93]). In these methods after the development of the blue color the solution is made alkaline to solubilize the proteins.

Some methods have been developed that do not use the reduction step. Phosphomolybdate has an absorbance maximum in the ultraviolet at about 325 nm [99]. Satisfactory results can be obtained by reading at 340 nm, a wavelength available in many instruments [100]. If the molybdate is replaced by vanadate, the absorption maximum is then about 410 nm [101–103], a wavelength more convenient for use with many simple photometers.

Another method used for the determination of phosphate is based on the fact that in the presence of phosphomolybdate the color produced by certain indicator dyes is changed. The dyes that have been used include malachite green [104–107] and methyl green [108], although other dyes could probably be used [104]. An indirect method using atomic absorption spectroscopy has also been used [109]. The phosphomolybdate complex is extracted from the aqueous solution with isobutyl acetate and the molybdenum determined in the organic phase by atomic absorption. The unreacted ammonium molybdate is not extracted by the organic solvent.

We present three methods for the determination of phosphorus. The first is an older method based on the work of Gomori [90], but one that is still used frequently; it utilizes a trichloracetic acid filtrate. The second method uses a strongly acid molybdate solution to precipitate the proteins, and the reducing agent is added to an aliquot of the filtrate [98]. In the third method the proteins are not precipitated but in the final step are solubilized with the aid of an alkaline amine [93].

Phosphorus—Method I

REAGENTS

1. Trichloracetic acid, $0.3M$.   Dilute 1 volume of the $1.8M$ solution (prepared according to the directions in the section on protein-free filtrates, Chapter 5) with 5 volumes of water. Store in refrigerator.

2. Acid molybdate solution.   Dissolve 5.0 g of sodium molybdate ($Na_2MoO_4\cdot 2H_2O$) in about 700 ml of water in a 1 liter volumetric flask. Add 14 ml of concentrated sulfuric acid, cool, and dilute to 1 liter.

3. Reducing solution.   Dissolve 1 g of p-methylaminophenol sulfate (Elon, Metol) and 3 g of sodium bisulfite ($NaHSO_3$) in water to make 100 ml. Filter before use and make up fresh monthly.

4. Phosphate standards.

   A. Stock standard.   Dissolve 0.4394 g of potassium dihydrogen phosphate (KH_2PO_4) in water and dilute to 1 liter. This solution contains 10 mg/dl of phosphorus (P).

   B. Working standard.   Dilute 4 ml of the stock to 50 ml with the $0.3M$ trichloracetic acid. This solution contains 0.8 mg/dl. When 1 ml of this solution is treated similarly to 2 ml of a 1:10 filtrate, the standard is equivalent to 4 mg/dl P.

PROCEDURE

Add to 1 ml of serum in a test tube, 9 ml of the trichloracetic acid solution. Mix thoroughly, allow to stand for 5 min, and then filter or centrifuge strongly. To one test tube add 2 ml of the filtrate, to a second tube add 2 ml of the trichloracetic acid solution for a blank, and to a third tube add 1 ml of the standard and 1 ml of the trichloracetic acid solution. To each tube add 5 ml of the molybdate solution and mix; then add 0.25 ml of the reducing solution and mix. Allow to stand for 45 min and then read samples and standard against blank at 660 nm (any convenient wavelength in the range 650 to 720 nm can be used).

CALCULATION

$$\frac{\text{absorbance of sample}}{\text{absorbance of standard}} \times \text{conc. of standard} = \text{conc. of sample}$$

As mentioned, when the volumes given above are used, the standard may be taken as equivalent to 4 mg/dl P.

Phosphorus—Method II

REAGENTS

1. Acid molybdate.   Cautiously add 30 ml of concentrated sulfuric acid to about 50 ml of water; add 5 g of ammonium molybdate [$(NH_4)_6Mo_7O_{24} \cdot 4H_2O$; 82% MoO_2], cool, and dilute to 100 ml.

2. Trichloracetic acid, 1.80M.   For preparation see section on protein-free filtrates, Chapter 5.

3. Combined reagent.   Mix together 4 volumes of water, 3 volumes of the acid molybdate, and 2 volumes of the 1.8M trichloracetic acid.

4. Reducing agent.   Dissolve 0.5 g of p-phenylenediamine hydrochloride in water containing 5 g of sodium bisulfite ($NaHSO_3$) and dilute to 100 ml.

5. Phosphorus standard.   The stock standard given in the previous method (10 mg/dl) is diluted 1:2 to give a standard containing 5 mg/dl.

PROCEDURE

Add 0.2 ml of sample, standard, and water (blank) to separate centrifuge tubes. Add 1.8 ml of the combined reagent to each tube and mix well. Allow to stand for about 5 min, then centrifuge strongly. Pipet 1-ml aliquots of the supernatants to separate tubes (there will be no precipitate in the standard or blank). To each tube add 4 ml of the reducing solution and mix well. Allow to stand for 25 min and then read standard and samples against blank at 700 nm (any wavelength between 650 and 720 nm is satisfactory).

CALCULATION

$$\frac{\text{absorbance of sample}}{\text{absorbance of standard}} \times \text{conc. of standard} = \text{conc. of sample}$$

In the proportions as given, the standard corresponds to 5 mg/dl P.

Phosphorus—Method III (No Precipitation)

REAGENTS

1. Acid molybdate.   Dissolve 44 g of ammonium molybdate (same as used in previous method) and 90 ml of concentrated sulfuric acid in about 700 ml of water, cool, and dilute to 1 liter. This solution is stable several months if kept in a poly-ethylene bottle.

2. Reducing reagent.   Dissolve 5 g of *p*-methylaminophenol sulfate and 15 g of sodium bisulfite in water to make 1 liter. This solution is stable for several months.

3. Monoethanolamine.   Any reagent grade of the chemical may be used.

4. Phosphorus standard.   Dilute the stock phosphorus standard of the first method 1:2 to give a standard containing 5 mg/dl.

PROCEDURE

To a series of tubes add 2 ml of the acid molybdate, then add 0.1 ml of standard or serum sample to separate tubes keeping one tube as blank. Mix well, add 1 ml of monoethanolamine, and mix well. Immediately after mixing, pour into clean colorimeter cuvettes. Allow the tubes to stand *undisturbed* at room temperature for 10 min after addition of the amine; agitation may cause some oxidation of the molybdate complex, causing low results. For this reason the solution is poured into the cuvettes just after mixing. Read samples and standards against blank at 660 nm.

CALCULATION

$$\frac{\text{absorbance of sample}}{\text{absorbance of standard}} \times \text{conc. of standard} = \text{conc. of sample}$$

The standard corresponds to 5 mg/dl P.

If a volume larger than 3 ml is required for the cuvettes, all the reagents and sample volumes may be doubled.

Phosphorus Determination in Urine

Any of the above methods can be used for urine, a diluted urine being treated in all ways similarly to a serum sample. Since the concentration of phosphates in the urine may vary over a wide range, several different dilutions may be necessary. For concentrated urines, 1:10 and 1:20 dilutions are used; for more dilute urine, 1:5 and 1:10. Methods II and III are preferable. Since in alkaline urines there may be some precipitation of phosphates, the urine sample should be acidified to a pH

of about 1 and mixed well before sampling. The urine phosphate is usually calculated as grams of phosphorus excreted per day.

Normal Values and Interpretation of Results

The normal level of serum inorganic phosphorus is 2.5 to 5.0 mg/dl; it is 1 to 2
mg/dl higher in actively growing children. The level may be greatly elevated in
severe nephritis and decreased to 2 mg/dl or less in rickets. There is a moderate increase in hyperparathyroidism. The use of purgatives or enemas containing large
amounts of phosphates usually causes an increase in serum inorganic phosphorus
which lasts for several hours.

The urinary excretion of phosphate depends upon the dietary intake as well as
on the kidney function. The usual excretion is 0.9 to 1.3 g/day calculated as P.

Sulfate

Sulfur in serum is present as inorganic sulfate (SO_4^{2-}), as conjugated sulfate, and
as a constituent of such amino acids as glutathione and cysteine. The total sulfur
has been determined by oxidizing a protein-free filtrate with nitric acid and peroxide and determining the resulting sulfate. There is very little clinical indication
for this determination and it is now rarely done. There is no good clinical reason
for the determination of the organic sulfate in urine or serum and this procedure
has also fallen into disuse. The most common method for sulfate determination
was precipitation as benzidine sulfate. The benzidine in the precipitate was then
determined by a variety of methods including color produced after diazotization
with thymol [110] or phenol [111], color produced with naphthoquinone sulfonate
[112, 113], or photometric determination of the color produced with ferric chloride
and hydrogen peroxide [114] or with iodine and ammonia [115]. The presence of
phosphate often interferes and the serum may be deproteinized and the phosphate
removed by precipitation with uranyl acetate [112]. Sulfate has also been determined, both manually [116] and in an automated procedure [117, 118], by the
turbidity produced with barium chloride, but these turbidimetric methods are not
very satisfactory. In another procedure insoluble barium chloranilate is added,
with the production of barium sulfate and an equivalent amount of chloranilic
acid. After centrifugation to remove the barium sulfate and excess reagent, the
liberated chloranilic acid is determined photometrically [119, 120]. This method
is not very sensitive.

Sulfate has also been indirectly determined by flame photometry. A known excess of barium chloride is added and, after removal of the precipitated barium
sulfate, the remaining barium in the supernatant is determined by flame photometry [121, 122]. The sulfate has also been determined by precipitation as the

barium salt and the separated barium sulfate dissolved in a known excess of standard EDTA solution. The excess EDTA is then titrated with standard magnesium solution [123].

Since its determination is rarely requested, a method for inorganic sulfate is not presented here. If one is needed, the method of Kleeman et al. [113] as modified by Henry [124] appears to be as satisfactory as any available.

Magnesium

Since references to many of the earlier methods are given in two reviews [125, 126], only a few of these will be given here for background, with more emphasis on later studies. One of the oldest methods for the determination of magnesium is precipitation as the magnesium ammonium phosphate followed by the determination of the phosphate in the washed precipitate [127]. This method is lengthy and requires the prior precipitation of the calcium as oxalate followed by the precipitation of the phosphate in the filtrate. It is still recommended by some recent workers as a very accurate method [128, 129]. The methods suggested below are much more rapid.

A colorimetric method for magnesium is based on the formation of a colored lake with magnesium hydroxide and a dye in alkaline solution. The most commonly used dye is probably Titan yellow (also known as Clayton Yellow or Thiazole yellow) which forms a red lake with magnesium hydroxide [130–132, 136]. The colored lake does not form a true solution and some additional substances are needed to stabilize the color. Gum Ghatti has been used, but the most common additive is polyvinyl alcohol which has the added advantage of increasing the color nearly twofold [133]. The analysis is usually performed on a trichloracetic acid filtrate although a Folin-Wu filtrate has also been used [134]. Some workers feel that calcium will slightly enhance the color and add calcium equivalent to a normal serum level to the standard [131]. Several other dyes have also been used. These include Congo red [135], Xylidyl Blue II [137], and another dye that does not have a simple name but is often referred to as Mann and Yoe reagent [138, 139].

Magnesium may be titrated with EDTA using Eriochrome Black T as indicator. This dye can also be used as a colorimetric reagent for magnesium in the presence of the strontium salt of EGTA to prevent interference by calcium [140]. Magnesium can also be determined by a simple fluorometric method. The metal forms a fluorescent complex with 8-hydroxyquinoline or 8-hydroxyquinoline-5-sulfonic acid [141, 142]. Since there may be some other fluorescent material present in serum, the procedure measures the fluorescence at a pH of 6.5 where the magnesium complex is strongly fluorescent and at a pH of 3.5 where it is not fluorescent. If the fluorescence of the interfering material is not influenced by the change in pH, the difference between the readings at the two pH values will be a measure of the magnesium. The reaction may be carried out in an alcoholic solution; this precipitates the small amount of protein present which is then separated by centrifugation. The method has been automated [143].

Magnesium has been determined by flame photometry, but in general this requires a very hot flame (oxy-acetylene), or the use of an enriched flame produced by aspirating solutions containing 80% acetone [144]. Strontium and EDTA may be added to reduce the interference by other elements [145] and sodium and potassium may be added to the standards. To reduce interference, the magnesium may be first precipitated as the phosphate [146], and the precipitate separated and dissolved for aspiration into the flame. Magnesium can be determined very satisfactorily by atomic absorption spectrophotometry. Strontium or lanthanum is sometimes added to reduce interference by phosphate [147] but generally a simple dilution with water is all that is required [148–150].

We present here a colorimetric method using Titan yellow and a simple microfluorometric method. Some observations on the determination by atomic absorption are included. Either serum or plasma may be used for magnesium determinations.

Magnesium by Titan Yellow Colorimetric Method [131, 132]

REAGENTS

1. Sodium hydroxide, 2.5M.   Dissolve 50 g of NaOH in water, cool, and dilute to 500 ml. The concentration should be checked by titration against standard acid and adjusted to $2.50 \pm 0.05M$.

2. Polyvinyl alcohol, 0.1 g/dl.   Suspend 1 g of polyvinyl alcohol* in about 40 ml of ethyl alcohol; then pour into about 500 ml of warm water while swirling. Warm further if necessary to dissolve, cool, and dilute to 1 liter.

3. Titan yellow, stock solution.   Dissolve 75 mg of Titan yellow in 100 ml of the polyvinyl alcohol solution.

4. Titan yellow, working solution.   Dilute the stock solution 1:10 with the polyvinyl alcohol as needed.

5. Trichloracetic acid, 0.3M.   Dissolve 50 g of trichloracetic acid in water to make 1 liter, or dilute the 1.8M solution (as prepared according to the section on protein-free filtrates, Chapter 5) 1:6 with water. Store in the refrigerator.

6. Magnesium standards.
    A. Stock standard.   Dissolve 2.145 g of magnesium acetate [reagent grade

---

* Elvanol 70-05, E.I. DuPont de Nemours and Co. Inc., Niagara Falls, N.Y. 14310.

$Mg(C_2H_3O_2)_2 \cdot 4H_2O$] in water to make 1 liter. This solution contains 10 mmol/liter.

   B. Working standard.   Dilute the stock standard 1:10 as required to give a solution containing 1 mmol/liter.

PROCEDURE

Add 1 ml of serum to 5 ml of the trichloracetic acid solution in a test tube. Mix well, allow to stand for about 10 min, then centrifuge strongly. Pipet 3 ml of the supernatant to a second tube. To other tubes add 0.5 ml of water and 2.5 ml of the trichloracetic acid solution (blank), or 0.5 ml of the diluted standard plus 2.5 ml of trichloracetic acid. To each tube add 2 ml of the Titan yellow working solution. Mix, then add 1 ml of the sodium hydroxide solution. Mix and read standards and samples against blank at 540 nm 10 min after the addition of the alkali.

CALCULATION

Since the 3 ml of supernatant is equivalent to 0.5 ml of serum and this is in the same final volume as 0.5 ml of standard:

$$\frac{\text{absorbance of sample}}{\text{absorbance of standard}} \times 1.0 = \text{conc. in sample (in mmol/liter)}$$

Magnesium is often reported in mg/dl; mmol/liter $\times$ 2.4 = mg/dl.

COMMENTS

The final solution is highly colored even in the absence of any magnesium. If too much dye is present, it may be difficult to zero the instrument with the blank. If too little dye is present, the color may not develop completely with the magnesium. The dye will vary somewhat with different lots. It may be necessary to experiment to find the best dilution to use in making the working Titan yellow solution. Since the magnesium content of erythrocytes is about three times that of plasma, marked hemolysis will cause some error.

Magnesium by Fluorometric Method [141, 142]

REAGENTS

   1. Oxine stock solution.   Dissolve 1 g of reagent grade 8-hydroxyquinoline (also called oxine) in 20 ml of absolute ethanol. Store in a brown bottle in refrigerator. The solution is not very stable and if only an occasional test is done, it is best to prepare a small amount of the solution fresh each time.

   2. Acetate buffer, 2M, pH 3.5.   Add 11.5 ml of glacial acetic acid to 70 ml of

water. Adjust to a pH of 3.5 ± 0.2 by the addition of sodium hydroxide solution (8 g/dl in water) ; about 10 ml of the alkali should be required. Then dilute to 100 ml with water.

3. Acetate buffer, $2M$, pH 6.5.   Dissolve 27.2 g of reagent grade sodium acetate [Na($C_2H_3O_2$) $\cdot$ 3H$_2$O] in about 70 ml of water. Adjust the PH to 6.5 ± 0.1 by the dropwise addition of glacial acetic acid (less than 1 ml is usually required) and dilute to 100 ml.

4. Reagent A.   Prepare fresh as required by mixing together 2 volumes of the pH 3.5 buffer, 2 volumes of the oxine solution, 30 volumes of absolute ethanol, and 5 volumes of water.

5. Reagent B.   This is freshly prepared similarly to reagent A except that the pH 6.5 buffer is used instead of the pH 3.5 buffer.

6. Magnesium standard.   Dilute the same stock standard as given in the previous method 1:10 to prepare a standard containing 1 mmol/liter.

PROCEDURE

To each of two 15 ml glass-stoppered centrifuge tubes add 0.1 ml of serum labeling the tubes A and B. Also set up two tubes containing 0.1 ml of standard (standard A and standard B). To each of the A tubes add 3.9 ml of reagent A. To each of the B tubes add 3.9 ml of reagent B. Stopper the tubes, shake vigorously for 2 min, and centrifuge. Decant the supernatant into fluorometer tubes (the standards need not be centrifuged). Read the tubes in a fluorometer with an exciting wavelength of 470 nm and a secondary filter for emitted light peaking at 580 nm. (For the Turner Model† 110 or 111 fluorometer, the primary filter is 47B and the secondary filter 2A-12.) Since the difference between readings is used, the fluorometer may be zeroed with water or with the A tubes.

CALCULATION

$$\frac{R \text{ serum}}{R \text{ standard}} \times 1.0 = Mg \text{ (in mmol/liter)}$$

where R is the reading of the B tubes when the respective A tubes are set to zero, or the difference in scale readings between the B tube and the A tube for each sample or standard when another zero setting is used.

Magnesium Determination in Urine

The fluorometric method is also adapted to urine. Since there is a possibility of the precipitation of magnesium salts in alkaline urine, the urine should be acidified

† G.K. Turner Associates, Palo Alto, Calif. 94303.

somewhat before sampling. Reagents A and B are made up slightly differently for the urine determination and an additional reagent is needed.

7. Urine reagent A. Mix together just before use 5 volumes of pH 3.5 buffer, 2 volumes of oxine solution, and 31 volumes of ethanol.

8. Urine reagent B. Prepare similarly to reagent A using the pH 6.5 buffer instead of the pH 3.5 buffer.

9. EDTA solution, $0.01M$. Dissolve 3.7 g of reagent grade disodium salt of ethylenediaminetetraacetic acid in water to make 1 liter.

PROCEDURE

The urine is first diluted 1:5 with water. Aliquots of 0.2 ml of the diluted urine are treated with 3.8 ml of reagents A and B as in the serum procedure. Standards are similarly treated. If, after the addition of the reagent, the urine tubes appear perfectly clear, they need not be centrifuged; if they are not clear, they should be centrifuged. To compensate for additional fluorescent material in urine, dilute another aliquot of urine 1:5 with the EDTA solution (this effectively prevents any fluorescence by the magnesium complex) and carry this through the same procedure as for the urine sample diluted with water.

CALCULATION

If R_1 is the difference between the scale readings for the B and A tubes for the urine diluted with water, R_2 the difference for the urine diluted with EDTA, and R_3 the reading for the standard,

$$\frac{R_1 - R_2}{R_3} \times 5 = \text{Mg in urine (in mmol/liter)}$$

(The additional factor of 5 is due to dilution of the urine.)

Magnesium by Atomic Absorption Spectroscopy

Magnesium may be readily determined by atomic absorption. Only an air-acetylene flame is required. The line at 285.2 nm is used. Serum samples may be diluted 1:25 or 1:50 (depending upon the instrument used) with water. However, if the sample is diluted with lanthanum chloride solution (see calcium above), the same dilution may be used for both calcium and magnesium determinations.

Normal Values and Interpretation of Results

The normal level for magnesium in serum may be taken as 0.7 to 1.1 mmol/liter (1.8 to 2.6 mg/dl). Decreased levels have been found in a number of diseases including acute pancreatitis, malabsorption syndrome, and chronic alcoholism.

Increased levels have been found in Addison's disease, dehydration, and decreased kidney function. Large doses of magnesium-containing antacids may cause increased levels in the blood. The administration of calcium gluconate may interfere with the colorimetric method, as will citrate when used as anticoagulant. Little is known concerning the significance of urinary magnesium. The normal range of excretion is from 1 to 12 mmol (20 to 300 mg)/day.

Iron and Iron Binding Capacity

Serum Iron—Available Methods

Two general methods are used for the determination of serum iron [151]. In one the iron is split off from the binding protein with hydrochloric acid, and the protein precipitated with trichloracetic acid. The iron is determined colorimetrically in the supernatant after centrifugation. The automated continuous-flow methods use this scheme except that dialysis is used to separate the iron from the protein. In the other general method the color reagent is added directly to the serum and the color read directly without precipitation of the proteins. The direct methods have the advantages of requiring less manipulation and pipetting as well as fewer reagents. Iron is a ubiquitous contaminant of most reagents, and the trichloracetic acid used for protein precipitation is often a source of iron contamination. The direct methods have the disadvantage of the accuracy being poor for turbid, icteric, or lipemic serums. Since the serum itself contributes significantly to the absorption of the solution, a blank must be run for each serum. Either of two procedures is used: (1) Two tubes are set up, each containing the serum and buffer, but only one having the color reagent. The difference in absorption between the two constitutes the absorbance due to the iron complex. (2) Only one tube is set up containing serum and buffer only. The absorption is read and then a very small volume of color reagent is added. After color development, the increase in absorbance due to the iron complex is read. A correction may be applied for the small change in volume due to the addition of the color reagent. With icteric, lipemic, or turbid serums, the absorbance without the color reagent may constitute a significant proportion of the total absorbance after the addition of the color reagent. One cannot always be certain that the absorbance due to the serum is exactly the same in the two readings. This difficulty may be experienced with lyophyilized serums often used in quality control. With extremely lipemic serums accurate results are not possible. The precipitation methods do not suffer from these disadvantages, though the reagents may often contain enough iron to give relatively high blanks. Although the direct methods are more rapid, with good reagents the precipitation methods are preferable and more accurate. The International Committee on Standardization in Hematology has recommended a precipitation method as a reference procedure [154].

Although a number of colorimetric reagents have been proposed for the determination of iron, the two most commonly used at present are sulfonated bathophenanthroline [152–155] and 2,4,6-tripyridyl-s-triazine (TPTZ) [156–159]. The former gives a red color with an absorbance maximum at about 535 nm, and the latter a blue color with an absorbance maximum at about 595 nm. Both have approximately the same molar absorptivities for the iron complexes (around 23,000) and both give the color only with ferrous iron. Bathophenanthroline may give a satisfactory color over a slightly wider pH range than TPTZ but both are useful over a relatively wide pH range. Since these reagents give a color only with ferrous iron, a reducing agent must be added. The most commonly used reducing agents are ascorbic acid [157–160], thioglycollic acid [154, 161, 162], and hydroxylamine [152, 163]. If precipitation is used, it is preferable to add the reducing agent before precipitation. Both color reagents have been used in the automated continuous-flow methods. In some of the older methods, bathophenanthroline is used as the color reagent and the colored iron complex extracted into an immiscible solvent such as amyl alcohol for colorimetric determination [164, 165].

More recently several other complex organic chelating agents have been introduced for the colorimetric determination of iron. These include Ferrozine [166–168] [disodium salt of 3-(2-pyridyl)-5,6-bis(4-sulfonic acid-1,2,4-triazine)], terosite sulfonate [169] [the sulfonate of 2,6-bis(4-phenyl-2-pyridyl)-4-phenyl-pyridine], and the compounds 2,6-di(2-pyridyl)-4-(p-methoxyphenyl)pyridine [170] and 7-bromo-1,3-dihydro-1-(3-dimethylamin-opropyl)-5-(2-pyridyl)-2H-1,4-benzodiazepin-2-one dihydrochloride [171, 172]. These compounds have somewhat higher molar absorptivities than TPTZ and bathophenanthroline sulfonate, and have absorbances maximums in the range of 560 to 580 nm. They may prove to be superior to the more commonly used color reagents but the higher absorptivity usually increases the sensitivity by no more than about 20 percent.

Other color reagents that have been used for the colorimetric determination of iron include hematoxylin [173], nitroso-R salt [174–176], 2,2'-dipyridyl [177, 178], 2,2',2''-tripyridyl [179], and thiocyanate [180]. These are generally less sensitive than the newer reagents and are now rarely used for serum iron, although they may be used for other determinations where iron is present in larger amounts.

Atomic absorption spectroscopy has also been used for the determination of serum iron [181–184]. The difficulty is that the amount of iron in serum is often near the lower limit of sensitivity of many atomic absorption instruments, particularly as the serum must be diluted somewhat before aspiration. With a dilution of only 1:2 or 1:3, the relatively large amount of organic matter in the aspirate may tend to clog the burner and will usually necessitate that the standards be also aspirated in a similar matrix. Theoretically the proteins could first be precipitated and the iron determined in the supernatant, but this usually requires a greater final dilution. With some colorimetric methods, traces of hemoglobin may not interfere since the iron may not be split from the hemoglobin by the mild conditions used, but in atomic absorption all the iron would be determined. It is possible that the newer methods such as ultrasonic atomization [185, 186] or the carbon rod analyzer [187, 188] will permit the analysis of serum iron on small samples by atomic absorption.

Another method for the determination of serum iron and iron binding capacity uses radioactive iron ^{59}Fe. This is discussed in the next section.

Serum Iron Binding Capacity—Available Methods

The iron in serum is combined with a particular glycoprotein called siderophilin or transferrin. The protein serves to transport the iron from the absorption sites in the intestines to the bone marrow or other erythropoietic sites where it is used for the synthesis of hemoglobin. Ordinarily the transferrin is only about one-third saturated with iron; it could combine with more iron than is actually present in the serum. The total amount of iron with which the protein would be combined when saturated is termed the *total iron binding capacity* (TIBC). The additional iron with which the protein could combine over that actually bound is termed the *latent or unsaturated iron binding capacity* (LIBC) or (UIBC) [161]. These are usually determined by adding excess iron to the serum to saturate the transferrin completely. Then either the excess unbound iron is removed from the solution and the iron present in the saturated serum is determined as TIBC, or the conditions may be adjusted so that the excess iron will react with the color reagent but that bound to the serum will not [189, 190]. From this, one can calculate the amount of added iron that has combined with the serum, or the UIBC. In the determination of TIBC the excess iron may be removed from the solution either by precipitation with magnesium carbonate [177, 190, 191] or by the use of an ion exchange resin [159, 161, 192, 193]. The precipitation method is widely used, but care must be taken to adjust the pH of the solution carefully for accurate results [194–196]. In the removal of iron by an ion exchange resin, the methods usually use a cation exchange resin which removes the iron as a complex with added citrate. The bound and unbound iron have also been separated by Sephadex gel filtration [197] though this method is not widely used.

Another method for the determination of serum iron and iron binding capacity uses radioactive iron, ^{59}Fe [198–200]. For the determination of UIBC an excess of iron labeled with ^{59}Fe is added to the serum. After separation of the serum bound iron from that remaining free in the solution by any of the methods given above or by the use of hemoglobin-coated charcoal [201], the radioactivity of one or the other of the fractions is measured. Knowing the initial added radioactivity and total iron, the amount of iron absorbed by the serum protein can be easily calculated. The radioactive method can also be used for the determination of serum iron although the procedure is somewhat more complicated. In one method, the original iron combined with the serum is split off by hydrochloric acid and removed from the solution with an ion exchange resin. The iron binding capacity of this serum is then determined as usual. The difference between the TIBC determined by this procedure and that determined without first splitting off the iron represents the original serum iron. This requires careful technique for low serum iron levels (often those of greatest clinical interest), since the final result is a relatively small difference between two larger numbers. In the other method a slightly different procedure is used. The pH of the solution is decreased until the iron is split off from the protein. Radioactive iron is added and the pH increased until the iron is reabsorbed by the protein. From the amount of radioactivity in the serum compared with that in a serum similarly treated but without prior splitting off of the iron, one can calculate the amount of serum iron originally present [202].

Urinary Iron—Available Methods

In some instances it may be desirable to determine the iron excreted in the urine. For the total iron, particularly in cases of iron poisoning, the iron may be readily determined by atomic absorption spectroscopy, using direct aspiration of the urine into the flame [203, 204]. Another indication for the determination of urinary iron is the measurement of the amount excreted after the administration of desferrioxamine. This is used to estimate the iron stored in the body [205, 206]. The iron can be measured directly in the urine after splitting off from the chelating agent and reduction by any of the color reagents mentioned earlier [207–209]. Also the urine can be digested with acid and the iron measured in the digest [208].

Serum Iron by Precipitation Method [161]

Iron is a very common contaminant in most reagents used in the clinical laboratory. Minute amounts of iron are determined in this procedure; this contamination must be avoided as much as possible. Deionized or double-distilled water should be used, and all glassware should be washed with dilute nitric acid and thoroughly rinsed with deionized water. It is preferable to use disposable plastic or glass tubes and pipets whenever possible. In the collection of the samples, hemolysis must be avoided since the erythrocytes have a very high content of iron. Although the hemoglobin iron may not always be split off as easily as the serum iron, some error can be caused by hemolysis. Serum is usually used for the determination; oxalate and citrate may interfere with some of the methods.

REAGENTS

1. Trichloracetic acid, $1.80M$.   See section on protein-free filtrates, Chapter 5.

2. Thioglycollic acid, reagent grade.   Use as purchased. Store in the refrigerator.

3. Hydrochloric acid, $0.2M$.   Dilute 17 ml of concentrated hydrochloric acid to 1 liter with deionized water.

4. Sodium acetate, $4.3M$.   Add 150 g of sodium acetate trihydrate ($NaC_2H_3O_2 \cdot 3H_2O$) to water to make 250 ml. This will make a solution that may be near saturation; one may use a saturated solution as such. The trihydrate generally gives a clearer solution than the anhydrous salt.

5. Color reagent.   Dissolve 50 mg of the sodium salt of sulfonated bathophenanthroline‡ in water to make 100 ml. This may dissolve slowly.

6. Iron standards.

A. Stock standard, 1 mg/ml.   Accurately weigh about 1 g of reagent grade iron wire, dissolve it in about 100 ml of $0.2M$ HCl with the aid of gentle heat, then cool and dilute to 1 liter. One may also use 3.512 g of ferrous ammonium sulfate (Mohr's salt, [Fe $(NH_4)_2(SO_4)_2 \cdot 6H_2O$]) dissolved in water containing 1 ml of concentrated HCl and made up to 1 liter. The ferrous sulfate solution will contain exactly 1 mg/ml. The solution made up from the iron wire will contain an accurately known amount of iron which may not be exactly 1 mg/ml (if exactly 1 g of iron was not weighed out). In our discussion we will assume that the stock standard contains exactly 1 mg/ml and the appropriate change in calculations must be made if this is not the case.

B. Working standard.   Dilute 1 ml of the stock standard to 500 ml with water. This solution contains 2 μg/ml or 200 μg/dl.

PROCEDURE

To 2 ml of serum add 3 ml of the hydrochloric acid solution and 0.1 ml of thioglycollic acid, mix well, and allow to stand at room temperature for 30 min. Treat 2 ml of water (blank) and 2 ml of working standard similarly. To each tube add 1 ml of the trichloracetic acid solution and mix well by covering the tubes with Parafilm and using a vortex mixer. Allow to stand at room temperature for about 20 min and then centrifuge strongly. To separate tubes, carefully pipet

‡ G. Frederick Smith Chemical Co., Columbus, Ohio 43223.

4 ml of supernatant from each tube—standard, blank, and samples. To each tube add 0.5 ml of the sodium acetate solution and 1 ml of the bathophenanthroline solution. Mix and allow to stand for 10 min; then read standard and samples against blank at 535 nm.

CALCULATION

Since the standard containing 200 μg/dl is treated similarly to the samples,

$$\frac{\text{absorbance of sample}}{\text{absorbance of standard}} \times 200 = \text{conc. of iron (in } \mu\text{g/dl)}$$

(μg/dl $\times$ 0.18 = μmol/liter)

Serum Iron Binding Capacity by Precipitation Method [161]

REAGENTS

1. Ferric ammonium sulfate, 50 μg/ml of Fe. Dissolve 216 mg of ferric ammonium sulfate [Fe(NH$_4$)(SO$_4$)$_2 \cdot$12H$_2$O] in water containing 25 ml of 0.2M hydrochloric acid and dilute to 500 ml. (*Note:* Most methods use ferric ammonium citrate as the source of iron since the iron is removed by the resin as the ferric citrate complex. The citrate solution is rather unstable. In the variation presented here, the citrate is combined with the ion exchange resin [49].)

2. Barbital buffer, pH 7.5. Dissolve 6.4 g of sodium chloride, 2.3 g of sodium diethylbarbiturate, and 6.0 g of diethylbarbituric acid in about 950 ml of water, warming if necessary to aid in solution. Cool, check pH, adjusting if necessary with small amounts of NaOH or HCl (0.5M), and dilute to 1 liter. This may be stored at room temperature.

3. Ion exchange resin (Amberlite IRA-410§). Suspend about 500 g of the analytical grade resin in 1 liter of citric acid solution (190 g of citric acid monohydrate in water to make 1 liter). Allow to stand overnight, then wash several times by decantation with distilled water, at the same time removing any "fines" present. Suspend the washed resin in 2 volumes of the barbital buffer, and adjust the pH of the suspension to 7.5 by the addition of small amounts of sodium hydroxide solution (4 g/dl). Then filter by suction and dry at 95°C. Store in dessicator or tightly stoppered container.

4. Working standard. Dilute 2 ml of the stock iron solution (see preceding section) to 500 ml with water to give a solution containing 400 μg/dl. The working

§ Rohn and Hass, Philadelphia, Pa. 19105.

solution used for serum iron determination could be used, but the stronger solution is preferable.

PROCEDURE

Add 1 ml of serum (heparinized plasma can be used, but oxalate or citrate will interfere) to a centrifuge tube. Add 0.1 ml of the ferric ammonium sulfate solution and mix well. Allow to stand at room temperature for 10 min. Add 0.4 ml of dry resin (measured by volume with a prepared scoop). Mix and allow to stand for 10 min with occasional swirling. Also set up standard and blank using 1 ml of the working standard or 1 ml of water. To each of these tubes add 0.1 ml of water (in place of ferric sulfate). Do not add resin to standard or blank, otherwise treat these tubes the same as the sample tubes in the rest of the procedure (except they need not be centrifuged). Add 5 ml of buffer to each tube and mix well. Allow to stand for 10 min, smacking the sample tubes occasionally during this time. Centrifuge sample tubes briefly to pack resin. Carefully pipet 5 ml of supernatant from each sample tube and place in labeled tube. Also pipet 5 ml from standard and blank tubes. To each tube add 0.1 ml of concentrated hydrochloric acid and 0.1 ml of thioglycollic acid and mix. Allow to stand for 30 min, then add 1 ml of the trichloracetic acid solution. Cover with Parafilm and mix well on a vortex mixer. Allow to stand for 20 min, then centrifuge strongly. Pipet 4 ml of supernatant from each tube. To this add 0.5 ml of sodium acetate solution and 1 ml of the color reagent. Mix and allow to stand for 10 min. Read samples and standard against blank at 535 nm.

CALCULATION

Since the standard containing 400 µg/dl is treated exactly the same as a serum sample,

$$\frac{\text{absorbance of sample}}{\text{absorbance of standard}} \times 400 = \text{total iron binding capacity (in } \mu g/dl)$$

Serum Iron and Iron Binding Capacity by Direct Method

The following method [157] will determine both the iron and the iron binding capacity on the same serum sample without precipitation of the proteins. As mentioned, the direct methods may give inaccurate results with turbid, lipemic, or icteric serums. The pH is lowered so that the iron is split off from the serum and the iron determined colorimetrically. Then additional iron is added and the pH increased so that the serum is now saturated with iron, and the iron not bound to the protein is determined.

REAGENTS

1. Acid buffer, pH 2.2. To about 400 ml of water add 7.0 g of potassium chloride (KCl) and 4 ml of hydrochloric acid solution (1.0M, 84 ml of concentrated acid diluted to 1 liter). When dissolved, dilute to 500 ml.

2. Color reagent. Dissolve 1.16 g of tripyridyl-s-triazine|| (TPTZ) in 400 ml of water containing 1.5 ml of concentrated hydrochloric acid and dilute to 500 ml.

3. Alkaline buffer, pH 9.0. Dissolve 182 g of tris(hydroxymethyl)amino-methane in 250 ml of water and 170 ml of 1.0M hydrochloric acid (see above). Check the pH and add more acid if required. Dilute to 500 ml.

4. Working standards.

A. Dilute 1 ml of the stock standard used in the previous method to 100 ml with water. This gives a standard containing 10 μg/ml.

B. Dilute standard A 1:5 to give a solution containing 2 μg/ml. Note that in the procedure both different standards are used.

PROCEDURE

Two tubes are set up for each sample, one labeled serum test and the other, serum blank. Colorimetric cuvettes should be used as the entire procedure is carried out in these tubes. Tubes labeled reagent blank and standard are also set up. Just before use, dissolve 0.3 g of ascorbic acid in 10 ml of the acid buffer. Use this solution when the directions below call for acid buffer. Add sample, standard, and reagents as follows:

	Serum Test	Serum Blank	Reagent Blank	Standard
Serum	1.0 ml	1.0 ml	—	—
Water	—	0.25 ml	1.0 ml	—
Standard B (2 μg/ml)	—	—	—	1.0
Acid buffer	2.5 ml	2.5 ml	2.5 ml	2.5 ml

Mix and allow to stand for 5 min. Then add:

| Color reagent | 0.25 ml | — | 0.25 ml | 0.25 ml |

Mix, allow to stand for 5 min, then read in photometer at 595 nm. Read test serum and standard against the reagent blank, and read serum blank against water.

|| G. Frederick Smith Chemical Co., Columbus, Ohio 43223.

Subtract absorbance of the serum blank from that of the serum test. After reading, add reagents as follows:

	Serum Test	Serum Blank	Reagent Blank	Standard
Alkaline buffer	2.0 ml	2.0 ml	2.0 ml	2.0 ml
Standard A (10 μg/ml)	0.5 ml	—	—	—
Standard B (2 μg/ml)	—	—	—	0.5
Water	—	0.5	0.5	—

Cap the tubes with Parafilm and incubate in a water bath at 37°C for 1 hr. Carefully remove tubes from bath and read in photometer without undue agitation. If the cuvettes are of such a size that the readings for serum iron (total volume 3.75 ml) can be readily made but the tubes will not conveniently hold the final volumes of 6.25 ml, proceed as follows. After the first readings, carefully decant the solution from each cuvette into larger clean tubes. Add the second set of reagents and mix well. Then decant some of the solutions back into the original cuvettes, cap these, and incubate. As with the previous readings, read the serum test and standard against the reagent blank, and the serum blank against water. Subtract any absorbance obtained with the serum blank from that obtained for the serum test.

CALCULATION

Serum iron (first readings).   Since the standard containing 200 μg/dl is treated the same as the serum sample,

$$\frac{\text{corrected absorbance of sample}}{\text{absorbance of standard}} \times 200 = \text{serum iron (in } \mu\text{g/dl)}$$

Iron binding capacity (second readings).   The actual measurement is that of the iron not absorbed by the serum. The standard for this reading is equivalent to 300 μg/dl [(1.0 + 0.5) ml of the 200 μg/dl standard compared with 1 ml of serum], then

$$\frac{\text{corrected absorbance of sample}}{\text{absorbance of standard}} \times 300 = \text{nonabsorbed iron (in } \mu\text{g/dl)}$$

The added iron was equivalent to 500 μg/dl (0.5 ml of 10 μg/ml standard compared with 1 ml serum) ; thus:

500 μg/dl — nonabsorbed iron (μg/dl) = absorbed iron, or unsaturated iron binding
capacity (in μg/dl)

Then, as defined, serum iron plus unsaturated iron binding capacity = total iron binding capacity.

Normal Values and Interpretation of Results

The normal range of serum iron is 70 to 150 μg/dl for males and 60 to 135 μg/dl for females. Low values are found in hypochromic anemias and frequently in patients with various infectious diseases. Increased levels have been found in anemias characterized by decreased hemoglobin formation not due to iron deficiency, such as pernicious anemia. The serum iron at birth is usually in the range of 150 to 200 μg/dl but falls rapidly to around 100 μg/dl. There appears to be a diurnal variation in serum iron which can be as much as 20 percent. The level is highest in the morning and falls during the day. There are also other random variations in the serum iron level, so that a single determination is not satisfactory for an accurate diagnosis. If several determinations are to be made, the samples should all be drawn at the same time of day, preferably in the early morning such as just before breakfast.

The total iron binding capacity in serum averages around 300 μg/dl with about 40 percent saturation in men and 35 percent in women. Various investigators have given different ranges so that a range of 200 to 400 μg/dl could be taken as normal. Steroid administration is said to decrease the serum iron level. It is increased by the administration of estrogens or the use of oral contraceptives. In iron deficiency anemia the total iron binding capacity is normal or elevated and the serum iron may be significantly decreased.

Copper

Copper in serum and urine has been determined by a number of different reagents. These have included diethyldithiocarbamate [210, 211], bathocuproine [212], dithizon [213], 1,5-diphenylcarbohydrazide [214], oxalyl dihydrazide [215–217], and more recently two new reagents, morpholinium-4-carbodithionate [218] and dicyclopentamethylene thiuram disulfide [219]. In the earlier methods for serum, wet digestion was used, but in the more recent procedures the copper is split off from the protein with acid and the proteins precipitated with trichloracetic acid. The copper is then determined colorimetrically. Because of interfering substances in urine, wet digestion is usually necessary, although extraction methods have been used. Many of the more recent methods use atomic absorption spectrophotometry. This is a very convenient method as it can usually be done on diluted serum or urine without further treatment [220–224]. The flameless carbon rod atomizer can also be used for the determination of copper by atomic absorption [225]. We present here a simple colorimetric method for copper [219], and some notes on the use of atomic absorption [210].

Serum Copper by Colorimetric Method

The copper is split off from the proteins with hydrochloric acid and the proteins then precipitated with trichloracetic acid. The precipitate is centrifuged and the copper determined colorimetrically in the supernatant. Disposable plastic test tubes and pipets are recommended, to avoid contamination by traces of copper. A good grade of copper-free water is also necessary. This is best prepared by passing ordinary distilled water through an ion exchange column. The blood samples should be collected using stainless steel needles.

REAGENTS

1. Hydrochloric acid, $2M$.   Dilute 168 ml of concentrated hydrochloric acid to 1 liter with water.

2. Trichloracetic acid, $1.38M$.   Quantitatively transfer the contents of a previously unopened $\frac{1}{4}$ lb bottle of reagent grade trichloracetic acid into a 500 ml volumetric flask and dilute to the mark with water.

3. Color reagent.   Dissolve 50 mg of oxalyl dihydrazide in 50 ml of the $2M$ hydrochloric acid. Larger quantities can be made up but since the reagent is stable for only 2 months, it is simpler to make up as required.

4. Concentrated ammonium hydroxide, reagent grade.

5. Disodium ethylenediaminetetraacetic acid (disodium EDTA).

6. Citric acid crystals, reagent grade.

7. Acetaldehyde solution.   Mix 1 volume of cold water with 1 volume of reagent grade acetaldehyde. Store in refrigerator.

8. Copper standards.

A. Stock standard.   Dissolve 1.964 g of fresh uneffloresced crystals of copper sulfate ($CuSO_4 \cdot 5H_2O$) in water containing a few drops of concentrated sulfuric acid and dilute to 500 ml. This solution contains 1 mg/ml of copper.

B. Working standard.   Dilute 1 ml of the stock standard to 500 ml with water, giving a standard containing 200 μg/dl of copper. This solution is not stable and should be made up fresh as needed.

PROCEDURE

Pipet 1 ml of sample (serum or heparinized plasma) into a plastic tube. Also set up a standard of 1 ml of the diluted standard and two blank tubes with 1 ml of water. Add 1 ml of the color reagent to each tube and mix well. Allow to stand for about 10 min, then add 1 ml of the trichloracetic acid. Mix well on vortex mixer or with a small footed glass rod. Cover with Parafilm and allow to stand for 15 min. Centrifuge strongly. To separate tubes add 2 ml of the supernatant, the standard, and the two blanks. Add a few crystals of citric acid to each tube and mix. To only one of the blanks add a pinch of the EDTA. To each tube add 0.5 ml of the acetaldehyde solution. Mix and allow to stand for 30 min at room temperature. Read all the tubes, including the blank without EDTA, against the blank tube containing EDTA at 542 nm. Subtract the absorbance obtained for blank without EDTA from the absorbance of samples and standard.

CALCULATION

Since the 200 μg/dl standard is treated the same as the samples,

$$\frac{\text{corrected absorbance of sample}}{\text{corrected absorbance of standard}} \times 200 = \text{copper in sample (in } \mu\text{g/dl)}$$

Copper Determination in Urine [226]

The above method cannot be applied directly to urine because of the presence of interfering substances. The urine can be wet-ashed, but this is a tedious process and all traces of oxidizing material must be removed after digestion to avoid interference with the color development. The following simple extraction method will give a rather low absorbance for normal copper levels, but can readily detect abnormal conditions giving increased copper excretion.

The urine is heated with hydrochloric acid to release any copper bound to protein. The copper is extracted into carbon tetrachloride with dibenzyldithiocarbamate and determined colorimetrically. A urine blank is extracted with carbon tetrachloride to compensate for any color extracted from the urine by the solvent.

REAGENTS

1. Zinc salt of dibenzyldithiocarbamate (DBDC).# Dissolve 75 mg of the reagent in 500 ml of reagent grade carbon tetrachloride.

2. Hydrochloric acid, concentrated, reagent grade.

3. Standard. Same working standard as used in the previous method for serum (200 μg/dl).

PROCEDURE

Pipet 20-ml aliquots of urine into two separate tubes, add 2 ml of concentrated hydrochloric acid to each, mix, and heat at 95°C for 15 min; cool. Transfer urine

Cat. No. 7308, Eastman Kodak Co., Rochester, N.Y. 14605.

samples to separate 60 ml Squibb-type separatory funnels. To another funnel add 19 ml of water, 1 ml of working standard, and 2 ml of hydrochloric acid (standard); to a fourth funnel add 20 ml of water and 2 ml of hydrochloric acid (blank). To one of the urine funnels, the standard, and the blank, add 5 ml of the DBDC reagent. To the other urine funnel, add 5 ml of carbon tetrachloride (urine blank). Shake each funnel vigorously for 1 min and then allow the layers to separate completely. Carefully draw off the lower carbon tetrachloride layers, filtering into separate tubes through a 7 cm Whatman No. 31 filter paper. The filtrate should be perfectly clear. If it is not, centrifuge. Read standard against blank and urine against urine blank at 436 nm.

CALCULATION

Since the standard (2 μg/ml) is compared with 20 ml of urine,

$$\frac{\text{absorbance of sample}}{\text{absorbance of standard}} \times \frac{2}{20} \times 1000 = \text{copper in urine (in } \mu\text{g/liter)}$$

The following alternative procedure may be preferable for urines containing small amounts of copper; it uses an internal standard. Set up three tubes, each containing 20 ml of urine. To two tubes labeled U (urine) and UB (urine blank), add 1 ml of water and 2 ml of hydrochloric acid. To a third tube labeled US (urine plus standard), add 1 ml of working standard and 2 ml of hydrochloric acid. Heat all tubes for 15 min at 95°C, cool, and transfer to separatory funnels. Extract U and US with 5 ml of the DBDC reagent and extract UB with 5 ml of carbon tetrachloride. Filter into tubes as previously and read U and US against UB at 436 nm.

CALCULATION

$$\frac{\text{absorbance of U}}{\text{absorbance of US} - \text{absorbance of U}} \times 100 = \text{copper in urine (in } \mu\text{g/liter)}$$

Normal Values and Interpretation of Results

The normal level of serum copper may be taken as 75 to 160 μg/dl in adults. It is somewhat lower in children, being about 15 to 65 μg/dl in newborn infants and 30 to 150 μg/dl in young children. Pregnancy and the administration of estrogens (oral contraceptives) will increase the copper level in serum to as high as 300 μg/dl. Hypercupremia has also been reported in cirrhosis, rheumatoid arthritis, and myocardial infarction. Decreased copper levels have been reported in a number of conditions such as nutritional disorders, nephrosis, and particularly

Wilson's disease (hepatolenticular degeneration) in which levels of 40 to 60 μg/dl have been reported.

The urinary excretion of copper is somewhat variable, but the usual levels are around 40 to 50 μg/day. The excretion will be higher in conditions resulting in proteinuria and aminoaciduria since these conditions tend to carry copper along with them into the urine. In Wilson's disease the excretion of copper is greatly increased and may be as high as 500 to 1,000 μg/day.

Copper and Zinc by Atomic Absorption Spectroscopy

Since the determinations of copper and zinc are so similar, they are discussed together here. Giving the complete details of a procedure using a particular instrument would be little help to individuals having different instruments. We present the general details of two slightly different approaches that should be applicable to most instruments. The zinc is usually determined using the line at 213.9 nm and copper with the line at 324.7 nm. The stock standards are most simply purchased as solutions containing 1 mg/ml of the metal (usually labeled as 1,000 ppm). (Those obtained from Fisher Scientific have proved satisfactory.) The colorimetric methods given here include the preparation of stock standards of 1 mg/ml.

Method I

The first method uses a 1:10 dilution of serum for both metals. This may be too great a dilution for some instruments but it can be tried.

REAGENTS

1. Diluting fluid. Dilute 60 ml of reagent grade n-butyl alcohol to 1 liter with water.

2. Sodium chloride solution, 1.5M. Dissolve 8.77 g of sodium chloride in water and dilute to 100 ml.

3. Copper and zinc standards.

A. Stock standards, 1 mg/ml (1,000 ppm). Purchased as such or prepared as given under the colorimetric methods for copper (above) and zinc (below).

B. Intermediate standard. Add 10 ml of the copper stock standard and 10 ml of the zinc stock standard to the same 100 ml volumetric flask and dilute to the mark with water. This solution contains 100 μg/ml of Cu and Zn.

C. Working standard. Dilute 1 ml (or 2 ml) of the intermediate standard plus 10 ml of the sodium chloride solution to 100 ml with water. This gives a solution containing 100 (or 200) μg/dl of copper and zinc and 150 mmol/liter of sodium chloride.

PROCEDURE

The working standards and the fluids to be analyzed (serum, urine, or cerebro-spinal fluid) are separately diluted 1:10 with the diluting fluid and the dilution aspirated directly into the flame using the diluting fluid alone for zero setting. The levels in urine may be much lower than in serum (depending upon the urine volume excreted) and it may be preferable to dilute the urine 1:5 instead of 1:10. If the normal dilution of the standard is used, divide the result obtained by two.

Method II

The dilutions are made with hydrochloric acid $(0.1M)$ with the addition not only of sodium chloride but also of a number of other inorganic ions [227, 228].

REAGENTS

1. Salt solution.   Dissolve the following quantities of reagent grade reagents in water and dilute to 1 liter: sodium chloride, 5.08 g; potassium chloride, 2.86 g; calcium carbonate, 0.312 g; magnesium chloride hexahydrate, 0.418 g; concentrated sulfuric acid, 0.67 ml; concentrated hydrochloric acid, 8.7 ml; and ammonium dihydrogen phosphate, 3.09 g.

2. Hydrochloric acid, $0.1M$.   Dilute 8.4 ml of concentrated acid to 1 liter.

3. Standards.   The stock and intermediate standards are made up just as in the previous procedure. For serum determinations the working standard is made by diluting the intermediate standard 1:100 with water, giving a standard containing 100 μg/ml.

PROCEDURE

The standard and serum samples are diluted 1:5 or 1:10 with the hydrochloric acid solution and aspirated into the flame using the hydrochloric acid solution alone for zero setting. For urine the working standard is made by diluting 1 ml of the intermediate standard to 100 ml with the salt solution. This standard and the urine samples are then diluted 1:5 with the hydrochloric acid and aspirated into the flame. For a zero setting a corresponding 1:5 dilution of the salt solution in hydrochloric acid is used.

Depending upon the sensitivity of the instrument, the above procedures may give very low readings so that a scale expansion is necessary. In this case better results are obtained by using a recorder so that the sequence of blank, samples, and standards can be repeated several times for each set of samples.

Zinc

The more common methods for the colorimetric determination of zinc use either dithizon (diphenylthiocarbazone) [229–231] or zincon (2-carboxy-2'-hydroxy-5'-

sulfoformazylbenzene) [232–234]. Neither is entirely specific for zinc, but by adjusting the pH and using masking reagents they can be used to determine zinc in the presence of fair amounts of interfering metals. Other compounds that have been used for colorimetric methods are rhodamine B [235] and di-β-naphthyl-thiocarbazone [236]. A fluorometric method has also been developed using 8-hydroxyquininol [237]. Many of the colorimetric methods require wet digestion with strong acids [231]. This not only lengthens the procedure greatly but also increases the possibility of contamination from the reagents. In some methods for serum the zinc is split off with hydrochloric acid and the proteins then precipitated by trichloracetic acid similar to methods for iron and copper [230, 234].

Most modern methods for the determination of zinc in serum and urine use atomic absorption [228, 238–240]. This is relatively simple, usually requiring only a simple dilution of the serum or urine with water, with sometimes other substances added to reduce matrix effects. The procedures for copper and zinc by atomic absorption are described in the preceding section.

The general precautions against contamination mentioned under the determination of copper should also be observed for zinc determinations.

Serum Zinc by Colorimetric Method [234]

After precipitation of the proteins by trichloracetic acid, zinc is determined in the filtrate by use of zincon compound (2-carboxy-2'-hydroxy-5'-sulfoformazyl-benzene). Some other metals will also react with the reagent but these may be eliminated by the use of proper masking agents. Cyanide will prevent the reaction of most metals with the reagent, and the addition of chloral hydrate will destroy the effect of cyanide on zinc, but not on the other elements. Only zinc will give a color with the reagent.

REAGENTS

1. Trichloracetic acid, 0.6M.   Dilute 1 volume of 1.8M trichloracetic acid (prepared as directed in the section on protein-free filtrates, Chapter 5) with 2 volumes of water, or dissolve 100 g of trichloracetic acid in water to make 1 liter.

2. Hydrochloric acid, 1M.   Dilute 84 ml of concentrated acid to 1 liter with water.

3. Sodium hydroxide, 4M.   Dissolve 160 g of sodium hydroxide in water, cool, and dilute to 1 liter.

These three solutions need not be made up exactly but they should be checked by the following procedure. Mix together 3 ml water, 1.5 ml of the hydrochloric acid, and 1.5 ml of the trichloracetic acid solution. Transfer a 4-ml aliquot of this mix to a small flask and titrate to the phenolphthalein endpoint with an accurately made 1:10 dilution of the sodium hydroxide solution. About 4.0 ml should be required, equivalent to 0.4 ml of the undiluted sodium hydroxide solution. Record this titration. In the procedure as given later the directions will call for the addition of 0.4 ml of the sodium hydroxide solution. If the titration as performed above is not equivalent to exactly 0.4 ml of the $4M$ base, use the appropriate value. For example, if the titration required the equivalent of 0.45 ml of the sodium hydroxide solution, use this amount instead of 0.4 ml.

4. Borate buffer, $0.5M$, pH 9.0.   Dissolve 31 g of boric acid in about 600 ml of water, add 53 ml of $4M$ sodium hydroxide solution, and dilute to 1 liter. Check the pH and adjust if necessary.

5. Potassium cyanide, $0.12M$.   Dissolve 0.78 g of potassium cyanide in water to make 100 ml.

6. Color reagent.   Dissolve 65 mg of zincon* in 1 ml of Acationox† and dilute to 50 ml with water.

7. Zinc standards.

A. Stock standard, 1 mg/ml.   It is difficult to find a good standard for zinc. Most zinc salts are hydrated and easily gain or lose water on standing. The best substance is probably zinc oxide. Dissolve 1.245 g of reagent grade zinc oxide, which has been previously dried at 200°C, in about 20 ml of water and 2 ml of concentrated hydrochloric acid. When dissolved, transfer to volumetric flask and dilute to 1 liter. Atomic absorption standards for several metals are available from a number of suppliers. These are usually labeled as 1,000 ppm (1,000 parts per million = 1 mg/ml). These standards are very convenient to use and are stable.

B. Working standard, 100 μg/dl.   Dilute 0.5 ml of the stock standard to 500 ml with water. This standard must be made up on the day of use. If preferred, an intermediate stock solution can be prepared by diluting 10 ml of the stock solution and 0.5 ml of concentrated hydrochloric acid to 100 ml with water. This solution contains 100 μg/ml. It is stable for a few weeks in the refrigerator if kept in a plastic bottle. The working standard is made by diluting this intermediate stock 1:100 with water.

8. Chloral hydrate solution, $3.6M$.   Dissolve 6 g of chloral hydrate in water and dilute to 10 ml.

PROCEDURE

To 3 ml of serum in a centrifuge tube add 1.5 ml of the hydrochloric acid solution, mix well, and heat in a boiling water bath for 5 min. Cool, add 1.5 ml of the

* Obtainable from most laboratory supply houses.
† Scientific Products Division, American Hospital Supply Corp., McGaw Park, Ill. 60085.

trichloracetic acid solution, and mix well. Allow to stand for 5 min, then centrifuge strongly to obtain a clear supernatant. Carefully pipet 4 ml of supernatant to a tube. In other tubes set up a blank of 2 ml of water plus 1 ml of the hydrochloric acid solution and 1 ml of the trichloracetic acid solution, and a standard containing 2 ml of working standard with 1 ml of HCl solution and 1 ml of the trichloracetic acid solution. To each tube add 1 ml of the borate buffer and 0.4 ml of the sodium hydroxide solution. Then add to each tube 0.2 ml of the potassium cyanide solution (*caution*) and mix. Add 0.6 ml of color reagent to each tube.

Read the tubes as follows. Set wavelength to 630 nm. Place cuvette with solution in photometer and zero instrument with the solution in place. Then add 0.1 ml of the chloral hydrate solution, mix, and read again within 30 sec. If the instrument is set to zero before addition of the chloral hydrate, the increase in absorbance is read directly. If it is not possible to zero, set to a definite value of absorbance, as low as possible. Then after the addition of the chloral hydrate, note the difference in absorption. Read samples, standards, and blank separately in this way.

CALCULATION

Three milliliters of serum are diluted to 6 ml in the precipitation step, and 4 ml of this taken for analysis. This is equivalent to 2 ml of serum and is compared with 2 ml of standard containing 100 μg/dl; thus:

$$\frac{\Delta \text{absorbance of sample} - \Delta \text{absorbance of blank}}{\Delta \text{absorbance of standard} - \Delta \text{absorbance of blank}} \times 100 = \text{zinc in serum (in } \mu\text{g/dl)}$$

The procedures have not been tried with urine except following wet-ashing which is a very tedious procedure. For urine as well as serum, the preferred method is atomic absorption, as discussed in the previous section.

Normal Values and Interpretation of Results

The normal level of zinc in healthy adults may be taken as 60 to 120 μg/dl. It may be slightly higher in children. Although zinc is an essential nutritional element and a variety of pathological conditions have been found in animals fed a zinc-deficient diet, the correlation between zinc serum levels and definite diseases is not yet clearly established.

References

1. Kramer, B., and Tisdall, F. F.   *J. Biol. Chem.* 47:574, 1921.
2. Clark, E. P., and Collip, J. B.   *J. Biol. Chem.* 63:461, 1925.
3. Sendroy, J., Jr.   *J. Biol. Chem.* 152:539, 1944.

4. Clark, G. W. *J. Biol. Chem.* 49:487, 1921.
5. Elert, B. T. *Am. J. Med. Technol.* 20:263, 1954.
6. Van Slyke, D. D., and Sendroy, J., Jr. *J. Biol. Chem.* 84:216, 1929.
7. Ellis, G. H. *Anal. Chem.* 10:112, 1938.
8. Schwarzenbach. G. *Anal. Chim. Acta* 7:141, 1952.
9. Dunsbach, F. *Clin. Chim. Acta* 8:481, 1963.
10. Schwarzenbach, G., and Ackermann, H. *Helv. Chim. Acta* 30:1978, 1949.
11. Greenblatt, I. J., and Hartman, S. *Anal. Chem.* 23:1708, 1963.
12. Pappenhagen, A. R., and Jackson, H. D. *Clin. Chem.* 6:582, 1960.
13. Catledge, G., and Biggs, H. G. *Clin. Chem.* 11:521, 1965.
14. Appleton, H. D., West, M., Mandel, M., and Sala, A. M. *Clin. Chem.* 5:36, 1959.
15. Klass, C. S. *Tech. Bull. Regist. Med. Technol.* 32:77, 1952.
16. Sourdais, C. M. *Rev. Fr. Etud. Clin. Biol.* 12:391, 1967.
17. Flaschka, H., Abd, E., Raheem, A. A., and Sadek, F. *Z. Physiol. Chem.* 310:97, 1958.
18. Burr, R. G. *Clin. Chem.* 15:1191, 1969.
19. Sadek, F. S., and Reilly, C. N. *J. Lab. Clin. Med.* 54:621, 1959.
20. Copp, D. H. *J. Lab. Clin. Med.* 61:1029, 1963.
21. Leifheit, H. C. *J. Lab. Clin. Med.* 47:623, 1956.
22. Rehell, B. *Scand. J. Clin. Lab. Invest.* 6:335, 1954.
23. Arvan, D. A. *Am. J. Clin. Pathol.* 45:357, 1966.
24. Kessler, G., and Wolfman, M. *Clin. Chem.* 10:686, 1964.
25. Kingsley, G. R., and Robnett, O. *Am. J. Clin. Pathol.* 27:223, 1957.
26. Connerty, H. V., and Briggs, A. R. *Clin. Chem.* 11:716, 1965.
27. Kingsley, G. R., and Robnett, O. *Anal. Chem.* 33:522, 1961.
28. Chilcote, M. E., and Wasson, R. D. *Clin. Chem.* 4:200, 1958.
29. Smith, P., Jr., Kurtzman, C. H., and Ambrose, M. E. *Clin. Chem.* 12:418, 1966.
30. Singer, L., Armstrong, W. D., and Coleman, L. M. *Anal. Biochem.* 9:21, 1964.
30a. Gindler, E. M., and King, J. D. *Am. J. Clin. Pathol.* 58:376, 1972.
31. Bellinger, J. F., and Campbell, R. A. *Clin. Chem.* 12:90, 1966.
32. Harrison, H. E., and Harrison, H. C. *J. Lab. Clin. Med.* 46:662, 1955.
33. Toribara, T. Y., and Koval, L. *Talanta* 7:248, 1961.
34. Ferro, P. V., and Han, A. B. *Am. J. Clin. Pathol.* 28:689, 1957.
35. Webster, W. W., Jr. *Am. J. Clin. Pathol.* 37:330, 1963.
36. Trinder, P. *Analyst* 85:889, 1960.
37. Kepner, B. L., and Hercules, D. M. *Anal. Chem.* 35:1238, 1963.
38. Lewin, M. R., Wills, M. R., and Baron, D. N. *J. Clin. Pathol.* 22:222, 1969.
39. Mosher, R. E., Itano, M., Boyle, A. J., Myers, G. B., and Isen, L. T. *Am. J. Clin. Pathol.* 21:75, 1951.
40. Chen, P. S., and Toribara, T. Y. *Anal. Chem.* 25:1642, 1953.
41. Denson, J. R. *J. Biol. Chem.* 209:233, 1954.
42. Humoller, F. L., and Walsh, J. R. *J. Lab. Clin. Med.* 48:127, 1956.
43. Poulos, P. P., and Pitts, R. F. *J. Lab. Clin. Med.* 49:300, 1957.
44. Woollen, J. W., and Walker, P. G. *J. Clin. Pathol.* 12:149, 1959.
45. Brandstein, M., Castellano, A., and Mezzacappa, C. *Am. J. Clin. Pathol.* 40:583, 1963.
46. Rick, W., and Herrmann, R. *Z. Gesamte Exp. Med.* 136:221, 1962.

265

47. Willis, J. B. *Anal. Chem.* 33:556, 1961.
48. Sideman, L., Murphy, J. J., Jr., and David, T.   *Clin. Chem.* 16:597, 1970.
49. Trudeau, D. F., and Frier, E. F.   *Clin. Chem.* 13:101, 1967.
50. Johnson, J. R. K., and Riechmann, G. O.   *Clin. Chem.* 14:1218, 1968.
51. Pybus, J. N., Feldman, F. J., and Bowers, G. N., Jr.   *Clin. Chem.* 16:998, 1970.
52. Cali, P. S., Mandel, J., Moore, L., and Young, D. S.   *Standard Reference Materials; A Referee Method for the Determination of Calcium in Serum.* Washington, D.C.: National Bureau of Standards Special Publication 260–36, 1972.
53. Baginski, E. S., Marie, S. S., Clark, W. L., Salancy, J. A., and Zak, B.   *Microchem. J.* 17:293, 1972.
54. Baginski, E. S., Marie, S. S., Clark, W. L., and Zak, B.   *Clin. Chim. Acta* 46:49, 1973.
55. Ham, A. B.   *Am. J. Med. Technol.* 35:807, 1969.
56. Spandrio, L.   *Clin. Chim. Acta* 10:376, 1964.
57. Ettori, J., and Scoggan, S. M.   *Clin. Chim. Acta* 6:861, 1961.
58. Harnach, F., and Coolidge, T. B.   *Anal. Biochem.* 6:477, 1963.
59. Pedersen, K. O.   *Scand. J. Clin. Lab. Invest.* 25:199, 1970.
60. Farese, G., Mager, M., and Blatt, W. F.   *Clin. Chem.* 16:226, 1970.
61. Usher, D. J., and Deegan, T.   *Clin. Chim. Acta* 29:361, 1970.
62. Lomax, G. D.   *J. Med. Lab. Technol.* (London) 24:103, 1967.
63. Frizel, D. E., Malleson, A. G., and Marks, V.   *Clin. Chim. Acta* 16:45, 1967.
64. Briscoe, A. M., and Ragan, C.   *J. Lab. Clin. Med.* 69:351, 1967.
65. Robertson, W. G., and Peacock, M.   *Clin. Chim. Acta* 20:315, 1968.
66. Farese, G., Mager, M., and Blatt, W. F.   *Clin. Chem.* 16:226, 1970.
67. Halver, B.   *Clin. Chem.* 18:1488, 1972.
68. Putnan, J. M.   *Clin. Chim. Acta* 37:33, 1972.
69. Hattner, R. S., Johnson, J. W., Bernstein, D. S., Wachman, A., and Brackman, J.   *Clin. Chim. Acta* 28:67, 1970.
70. Li, T. K., and Piechocki, J. T.   *Clin. Chem.* 17:411, 1971.
71. Subryan, V. L., Popovtzer, M. M., Parks, S. D., and Reeve, E. D.   *Clin. Chem.* 18:1459, 1972.
72. Burr, G. G.   *Clin. Chim. Acta* 43:311, 1973.
73. Moore, E. W.   In R. A. Durst [Ed.], *Ion-Selective Electrodes.* Washington, D.C.: National Bureau of Standards Special Publication 314, 1969.
74. McLean, F. C., and Hastings, A. B.   *J. Biol. Chem.* 108:285, 1935.
75. Kuttner, T., and Cohen, H. R.   *J. Biol. Chem.* 75:517, 1927.
76. Kuttner, T., and Lichtenstein, L.   *J. Biol. Chem.* 86:671, 1930.
77. Polley, J. R.   *Can. J. Res.* 27:265, 1969.
78. Harrop, G. A., Jr.   *Proc. Soc. Exp. Biol. Med.* 17:162, 1919.
79. Johnston, J. F.   *J. Med. Lab. Technol.* (London) 17:25, 1960.
80. Hunter, D. T., Jr., and McGuire, L.   *Am. J. Med. Technol.* 36:374, 1970.
81. Tisdal, F. F.   *J. Biol. Chem.* 50:329, 1922.
82. Goldenberg, H., and Fernandez, A.   *Clin. Chem.* 12:871, 1966.
83. Yee, H. Y., and Blackwell, L.   *Clin. Chem.* 14:898, 1968.
84. Harbin, M. T., Mooers, C. D., and Thomas, W. C.   *Am. J. Med. Technol.* 36:425, 1970.
85. Chen, P. S., Toribara, T. Y., and Warner, H.   *Anal. Chem.* 28:1756, 1956.

86. Baginski, E. S., and Zak, B.   *Clin. Chim. Acta* 5:834, 1960.
87. Goodwin, J. F.   *Clin. Chem.* 16:776, 1970.
88. Briggs, A. P.   *J. Biol. Chem.* 53:13, 1922.
89. Canellakis, E. S., and Tomlinson, M. C.   *Am. J. Med. Technol.* 28:195, 1962.
90. Gomori, G. J.   *Lab. Clin. Med.* 27:955, 1941.
91. Young, D. S.   *J. Clin. Pathol.* 19:397, 1966.
92. Delsal, J. L., and Manhouri, M.   *Bull. Soc. Chim. Biol.* 40:1623, 1968.
93. Drewes, P. A.   *Clin. Chim. Acta* 39:81, 1972.
94. Fiske, C. H., and Subbarow, Y. J.   *Biol. Chem.* 66:375, 1925.
95. Simon, K. H.   *Med. Monatsschr.* 21:235, 1967.
96. Guirgis, F. K., and Habib, Y. A.   *Clin. Chem.* 17:78, 1971.
97. Parekh, A. C., and Jung, D. H.   *Clin. Chim. Acta* 27:373, 1970.
98. Jung, D. H., and Parekh, A. C.   *J. Clin. Pathol.* 25:263, 1972.
99. Delsal, J. L., and Manhouri, H.   *Bull. Soc. Chim. Biol.* 40:1169, 1958.
100. Amador, E., and Urban, J.   *Clin. Chem.* 18:601, 1972.
101. Pulss, G.   *Z. Anal. Chem.* 176:412, 1960.
102. Robinson, R., Roughan, M. E., and Wagstaff, D. F.   *Ann. Clin. Biochem.* 8:168, 1971.
103. Davies, J. L., Andrews, G. S., Miller, R., and Owen, H. G.   *Clin. Chem.* 19:411, 1973.
104. Itaya, K., and Ui, M.   *Clin. Chim. Acta* 14:361, 1966.
105. Bastiaanse, A. J., and Meijers, C. A. M.   *Z. Klin. Chem. Klin. Biochem.* 6:48, 1968.
106. Stepanova, I.   *Clin. Chim. Acta* 16:330, 1967.
107. Hohenwalner, W., and Wimmer, E.   *Clin. Chim. Acta* 45:169, 1973.
108. Van Belle, H.   *Anal. Biochem.* 33:132, 1970.
109. DeVoto, G.   *Boll. Soc. Ital. Biol. Sper.* 44:424, 1968.
110. Cuthbertson, D. P., and Tompsett, S. L.   *Biochem. J.* 25:1237, 1931.
111. Joergensen, M. B.   *Scand. J. Clin. Lab. Invest.* 6.303, 1954.
112. Leetonoff, T. V., and Reinhold, J. G.   *J. Biol. Chem.* 114:147, 1936.
113. Kleeman, C. R., Taborsky, E., and Epstein, F. H.   *Proc. Soc. Exp. Biol. Med.* 91:480, 1956.
114. Hubbard, R. S.   *J. Biol. Chem.* 74:v, 1927.
115. Yoshimatsu, S.   *Tohoku J. Exp. Med.* 14:29, 1920.
116. Berglund, F., and Sorbo, B.   *Scand. J. Clin. Lab. Invest.* 12:147, 1960.
117. Haff, A. C.   *Adv. Automat. Anal. Technicon Int. Congr.* 1:81, 1969.
118. Dieu, J. P.   *Clin. Chem.* 17:1183, 1971.
119. Haekkinen, I., and Joergensen, L. M.   *Scand. J. Clin. Lab. Invest.* 11:294, 1959.
120. Wainer, A., and Koch, A. L.   *Anal. Biochem.* 3:457, 1962.
121. Alt, D.   *Landsirtsch. Forsch.* 16:278, 1964.
122. Strickland, R. D., and Maloney, W. M.   *Am. J. Clin. Pathol.* 24:1100, 1954.
123. Lewis, D. A.   *Analyst* 87:566, 1962.
124. Henry, R. J.   *Clinical Chemistry, Principles and Techniques.* New York: Harper & Row, 1964. P. 417.
125. Sunderman, F. W., Jr., and Sunderman, F. W.   In F. W. Sunderman, Jr., and F. W. Sunderman [Eds.], *Clinical Pathology of Serum Electrolytes.* Springfield, Ill.: Thomas, 1966. P. 56.
126. Alcock, N. W., and MacIntyre, I.   *Methods Biochem. Anal.* 14:1, 1966.
127. Simonson, D. G., Westover, L. M., and Wertman, I. W.   *J. Biol. Chem.* 169:39, 1947.

128. Aikawa, J. K., and Rhoades, E. L.   *Am. J. Clin. Pathol.* 31:314, 1959.
129. Heaton, F. W.   *J. Clin. Pathol.* 13:358, 1960.
130. Orange, M., and Rheim, H. C.   *J. Biol. Chem.* 189:379, 1951.
131. Neill, D. W., and Neely, R. A.   *J. Clin. Pathol.* 9:162, 1956.
132. Andreasen, E.   *Scand. J. Clin. Lab. Invest.* 9:138, 1957.
133. Heagy, F. C.   *Can. J. Res.* 26:295, 1948.
134. Anast, C. S.   *Clin. Chem.* 9:544, 1963.
135. Ellis, N. J., and Bishop, D. M.   *Can. J. Biochem.* 42:1225, 1965.
136. Sky-Peck, H. H.   *Clin. Chem.* 10:391, 1964.
137. Chromy, V., Svoboda, V., and Stepanova, I.   *Biochem. Med.* 7:208, 1973.
138. Leskovar, R., and Leskovar, C.   *Z. Klin. Chem. Klin. Biochem.* 8:477, 1970.
139. Straumfjord, J. V., Jr., and Doumas, B.   In F. W. Sunderman, Jr., and F. W. Sunderman [Eds.], *Clinical Pathology of Serum Electrolytes.* Springfield, Ill.: Thomas, 1966. P. 62.
140. Gitelman, H. J., Hurt, C., and Lutwak, L.   *Anal. Biochem.* 14:106, 1966.
141. Schachter, D.   *J. Lab. Clin. Med.* 54:763, 1959.
142. Schachter, D.   *J. Lab. Clin. Med.* 58:495, 1961.
143. Klein, B., and Oklander, M.   *Clin. Chem.* 13:26, 1967.
144. Van Fossen, D. D., Baird, E. E., and Tekell, G. S.   *Am. J. Clin. Pathol.* 31:368, 1969.
145. Montgomery, R. D.   *J. Clin. Pathol.* 14:400, 1961.
146. Andersen, C. J., Jensen, J. N., and Rud, N.   *Scand. J. Clin. Lab. Invest.* 14:560, 1962.
147. McDonald, M. A., and Watson, L.   *Clin. Chim. Acta* 14:233, 1966.
148. Hunt, B. J.   *Clin. Chem.* 15:979, 1969.
149. Hansen, J. L., and Frier, E. F.   *Am. J. Med. Technol.* 33:158, 1967.
150. Iida, C., Fuwa, F., and Wacker, W. E. C.   *Anal. Biochem.* 18:18, 1967.
151. Ramsay, W. N. M.   *Adv. Clin. Chem.* 1:1, 1958.
152. Jung, D. H., and Parekh, A. C.   *Am. J. Clin. Pathol.* 54:813, 1970.
153. Bouda, J.   *Clin. Chim. Acta* 21:159, 1968.
154. Lewis, S. M.   *Am. J. Clin. Pathol.* 56:543, 1971.
155. Zak, B., and Epstein, E.   *Clin. Chem.* 11:641, 1965.
156. Picardi, G., Nyssen, M., and Dorche, J.   *Clin. Chim. Acta* 40:219, 1972.
157. O'Malley, J. A., Hassen, A., Shiley, J., and Traynor, R.   *Clin. Chem.* 16:92, 1970.
158. Young, D. S., and Hicks, J. M.   *Clin. Pathol.* 18:98, 1965.
159. Lehmann, H. P., and Kaplan, A.   *Clin. Chem.* 17:941, 1971.
160. Friedman, H. S., and Cheek, C. S.   *Clin. Chim. Acta* 31:315, 1971.
161. Giovanniello, T. J., and Pecci, J.   *Stand. Methods Clin. Chem.* 7:127, 1970.
162. Williams, H. L., and Conrad, M. E.   *Clin. Chim. Acta* 37:131, 1972.
163. Fischer, D. S., and Price, D. C.   *Clin. Chem.* 10:21, 1964.
164. Forman, D. T.   *Tech. Bull. Regist. Med. Technol.* 34:93, 1964.
165. Kingsley, G. R., and Getchell, G.   *Clin. Chem.* 2:175, 1956.
166. Carter, P.   *Anal. Biochem.* 40:450, 1971.
167. Yee, H. Y., and Zin, A.   *Clin. Chem.* 17:950, 1971.
168. Persijn, J. P., Van der Slik, W., and Riethorst, A.   *Clin. Chim. Acta* 35:91, 1971.
169. Zak, B., Baginski, E. S., Epstein, E., and Weiner, L. M.   *Clin. Chim. Acta* 29:77, 1970.

170. Schmidt, R., Weis, W., Kingmueller, V., and Staudinger, H. *Z. Klin. Chem. Klin. Biochem.* 5:304, 1967.
171. Klein, B., Lucas, L., and Searcy, R. L. *Clin. Chim. Acta* 26:517, 1969.
172. Klein, B., Kleinman, N., and Searcy, R. L. *Clin. Chem.* 16:495, 1970.
173. Mikac-Devic, D. *Clin. Chim. Acta* 24:293, 1969.
174. Ness, A. T., and Dickerson, H. C. *Clin. Chim. Acta* 12:579, 1965.
175. Pre, J., Giraudet, P., and Cornillot, P. *Clin. Chim. Acta* 22:429, 1968.
176. Martinek, R. G. *Clin. Chim. Acta* 43:73, 1973.
177. Ramsey, W. N. M. *Clin. Chim. Acta* 2:221, 1957.
178. Bothwell, T. H., and Mallett, B. *Biochem. J.* 59:599, 1955.
179. Levy, A. L., and Vitacca, P. *Clin. Chem.* 7:241, 1961.
180. Fischl, J., and Cohen, S. *Clin. Chim. Acta* 7:121, 1962.
181. Dreux, C., Bouchet, R., and Girard, M. L. *Ann. Biol. Clin.* (Paris) 29:251, 1971.
182. Olson, A. D. *Clin. Chem.* 15:438, 1969.
183. Tavenier, P., and Hellendoorn, H. B. A. *Clin. Chim. Acta* 23:47, 1969.
184. Rogerson, D. O., and Helfer, R. E. *Clin. Chem.* 12:338, 1966.
185. Uny, G., Brule, M., and Spitz, J. *Ann. Biol. Clin.* (Paris) 27:387, 1969.
186. Korte, N. E., Moyers, J. L., and Denton, M. B. *Anal. Chem.* 45:530, 1973.
187. Matousek, J. P., and Stevens, B. J. *Clin. Chem.* 17:363, 1971.
188. Amos, M. D., Benett, P. A., Brodie, K. G., Lung, P. W. Y., and Matousek, J. P. *Anal. Chem.* 43:211, 1971.
189. Schade, A. L., Oyama, J., Reinhardt, R. W., and Miller, J. R. *Proc. Soc. Exp. Biol. Med.* 87:443, 1954.
190. Ressler, N., and Zak, B. *Am. J. Clin. Pathol.* 30:87, 1958.
191. Caraway, W. T. *Clin. Chem.* 9:188, 1963.
192. Birdsall, N. J. M., Kok, D'A., and Wild, F. *J. Clin. Pathol.* 18:453, 1965.
193. Brownstein, H. *Am. J. Clin. Pathol.* 47:714, 1967.
194. Legatte, J., and Crooks, A. E. *J. Clin. Pathol.* 25:905, 1972.
195. Cook, J. D. *J. Lab. Clin. Med.* 76:497, 1970.
196. Williams, H. L., and Conrad, M. E. *Clin. Chim. Acta* 37:131, 1972.
197. Nielsen, I. *Z. Klin. Chem. Klin. Biochem.* 6:103, 1968.
198. Hara, M., and Kayamori, R. *Radioisotopes* 21:19, 1972.
199. Van de Wal, G. *Clin. Chim. Acta* 36:570, 1972.
200. Lehmann, H. P. *Clin. Chem.* 17:941, 1971.
201. Herbert, V., Gottlieb, C. W., Lau, K-S., Fisher, M., Gevirtz, N. R., and Wasserman, L. R. *J. Lab. Clin. Med.* 67:855, 1966.
202. Herbert, V., Gottleib, C. W., Lau, K-S., Gevirtz, N. R., Sharney, L., and Wasserman, L. R. *J. Nucl. Med.* 8:529, 1967.
203. Zettner, A., and Mansbach, L. *Am. J. Clin. Pathol.* 44:517, 1965.
204. Arroyo, M., Coca, M. C., and Diaz Rubio, C. *Rev. Clin. Esp.* 125:43, 1972.
205. Losowsky, M. S. *J. Clin. Pathol.* 19:165, 1966.
206. Fielding, J., O'Shaughnessy, M. C., and Brumstroem, G. M. *J. Clin. Pathol.* 19:159, 1966.
207. Lundvall, O., and Weinfeld, A. *J. Clin. Pathol.* 20:611, 1967.
208. Barry, M. *J. Clin. Pathol.* 21:166, 1968.

269

209. Werkman, H. P. T., Trijbels, J. M. F., Van Munster, P. J. J., Schretien, E. D. M. A., and Moerkerk, C. *Clin. Chim. Acta* 31:395, 1971.
210. Deszo, I., and Fulop, T. *Mikrochim. Acta* 592, 1959.
211. Risse, E. *Aertztl. Lab.* 5:221, 1959.
212. Zak, B. *Clin. Chim. Acta* 3:328, 1958.
213. Butler, E. J., and Newman, G. E. *Clin. Chim. Acta* 11:452, 1965.
214. Mikac-Devic, D. *Clin. Chim. Acta* 26:127, 1969.
215. Wilson, J. F., and Klassen, W. H. *Clin. Chim. Acta* 13:766, 1966.
216. Rice, E. W. *Stand. Methods Clin. Chem.* 4:57, 1964.
217. Beale, R. N., and Croft, D. *J. Clin. Pathol.* 17:260, 1964.
218. Bayer, W. *Clin. Chim. Acta* 38:119, 1972.
219. Carter, P. *Clin. Chim. Acta* 39:497, 1972.
220. Meret, S., and Henkin, R. I. *Clin. Chem.* 17:369, 1971.
221. Spector, H., Glusman, S., Jatlow, P., and Seligson, D. *Clin. Chim. Acta* 31:5, 1971.
222. Griffiths, W. C., Ullucci, P. A., and Martin, H. F. *Clin. Biochem.* 3:189, 1970.
223. Pybus, J. N., Pfau, P., and Begansky, T. A., Jr. *Hartford Hosp. Bull.* 25:138, 1970.
224. Ichida, T., and Nobuka, M. *Clin. Chim. Acta* 24:299, 1969.
225. Matousek, J. P., and Stevens, V. J. *Clin. Chem.* 17:363, 1971.
226. Giorgio, A. J., Cartwright, G. E., and Wintrobe, M. M. *Am. J. Clin. Pathol.* 41:22, 1964.
227. Dawson, J. B., Ellis, D. J., and Newton-John, H. *Clin. Chim. Acta* 21:33, 1968.
228. Dawson, J. B., and Walker, B. E. *Clin. Chim. Acta* 26:465, 1969.
229. Kagi, J. H. R., and Vallee, B. L. *Anal. Chem.* 30:1951, 1958.
230. Hellwege, H. H., Schmalfuss, H., and Goschenhofer, D. *Z. Klin. Chem. Klin. Biochem.* 7:56, 1969.
231. Helwig, H. L., Hoffer, E. M., Thielen, W. C., Alcocer, A. E., Hoteling, D. R., and Rogers, W. H. *Am. J. Clin. Pathol.* 45:160, 1966.
232. Stojanovski-Bubanj, A., and Keler-Bacoka, M. *Clin. Chim. Acta* 25:478, 1969.
233. Platte, J. A., and March, V. M. *Anal. Chem.* 31:1226, 1959.
234. Williams, L. A., Cohen, J. S., and Zak, B. *Clin. Chem.* 8:502, 1962.
235. Tvoroha, B., and Mala, O. *Mikrochim. Acta* 634, 1962.
236. Mikac-Devic, D. *Clin. Chim. Acta* 23:499, 1969.
237. Mahanand, D., and Houck, J. C. *Clin. Chem.* 14:6, 1968.
238. Pekarek, R. S., Beisel, W. R., Bartelloni, P. J., and Bostian, K. A. *Am. J. Clin. Pathol.* 57:506, 1972.
239. Hackley, B. M., Smith, J. C., and Halsted, J. A. *Clin. Chem.* 14:1, 1968.
240. Meret, S., and Henkin, R. I. *Clin. Chem.* 17:369, 1971.
241. Prasad, A. S., Overleas, D., Wolf, P., Horwitz, R. C., and Vazquez, J. M. *J. Clin. Invest.* 46:549, 1967.
242. Sanstead, H. H., Prasad, A. S., Schulert, A. R., Farid, Z., Maiale, A., Jr., Bassilly, S., and Darby, W. J. *Am. J. Clin. Nutr.* 20:422, 1967.

9. Electrolytes

Sodium and Potassium

Before the introduction of flame photometers, sodium and potassium were determined by chemical methods which were all rather lengthy. Since flame photometers are now generally available, the use of the chemical methods has declined markedly. A brief survey of these methods will be given, however. Sodium (Na) forms a series of relatively insoluble salts of the type $NaX (UO_2)_2 (C_2H_3O_2)_9 \cdot 6H_2O$, where X is a divalent metallic ion such as Mg, Zn, Co, Cu, or Ni. The Mg, Zn, or Co complexes are most generally used in the chemical methods. The sodium is precipitated (as the sodium zinc uranyl acetate, for instance), and the precipitate is then separated, washed, and determined by colorimetric or other reaction. In the simplest method the precipitate is dissolved in water and the yellow color due to the uranyl ion measured at about 430 nm [1, 2, 3]. Alternatively the reaction between the uranyl ion and ferricyanide [4] or salicylate [5] is used to produce a color for measurement. In another variation the precipitate may be dissolved in water and the acetic acid formed titrated with alkali [6, 7]. Finally the zinc may be titrated with EDTA (ethylenediaminetetraacetic acid) using murexide as indicator [8].

The most commonly used chemical method for potassium (K) is the precipitation of the metal as the potassium sodium cobaltinitrite. After the precipitate is separated and washed, it is determined by colorimetric or other methods. It may be determined by the reaction between the cobalt and the Folin-Ciocalteu reagent in the presence of a small amount of an amino acid such as glycine to give a molybdenum blue color [9, 10], or by the color produced with choline and ferrocyanide [11, 12]. The cobalt may also be determined by titration with EDTA [13, 14]. A number of other methods have also been occasionally used for the determination of the cobalt. Potassium has also been precipitated as the silver potassium cobaltinitrite.

A method that has been used for many years in inorganic analysis is the precipitation of potassium as the chloroplatinate [15]. Its use in biological samples required a preliminary ashing which made it lengthy. A simple method involves the measurement of the turbidity produced in an acid filtrate by the addition of tetraphenyl boron which forms a relatively insoluble potassium salt [16, 17].

By far the most widely used method for the determination of sodium and potassium is with the flame photometer. Relatively inexpensive flame photometers (which are nearly as accurate as the more sophisticated models) are available for almost every clinical laboratory. The operation of the different flame photometers, although generally the same in principle, will vary in detail with the instrument. We will give here only a few references to general reviews [18, 19] plus a few citations of selected methods [20–23]. There is some interference

271

between sodium and potassium when they are determined in serum since both are present. Since the sodium is present in a much higher concentration, its interference on the potassium determination is greater than the reverse effect. In some of the older methods this was minimized by adding a large excess of sodium to all potassium samples and standards, so that they all have essentially the same sodium concentration. By this technique sodium and potassium cannot be determined on the same dilution of serum, which would be advantageous. The effect is minimized by a greater dilution of the serum and by the use of an internal lithium standard. Both of these are used in the newer instruments. Usually the ratio of lithium to sodium in the diluted serum is of the order of 10:1 (on an equivalents basis) or greater. This large excess of lithium reduces interference since it is the same in all samples.

In theory sodium and potassium can also be determined by atomic absorption spectrophotometry, but this offers little advantage over flame photometry, which is satisfactory for these two elements. The flame photometers are less expensive and are preferable for these two elements, leaving the atomic absorption instrument, if available, for the determination of those elements not readily determined by flame photometry.

Another method recently developed for the determination of sodium and potassium is the use of the specific ion electrodes (see Chapter 2). Although glass electrodes sensitive to sodium and potassium ions have been used for some time in determining these ions in biological fluids [24–27], the technique was rather exacting. With the development of newer electrodes including the valinomycin electrode [28, 29] for potassium, the method is coming into routine use. An automated procedure is available for the simultaneous determination of sodium, potassium, and chloride in serum using three different electrodes [30].*

Related to sodium and potassium is the determination of total serum base by conductivity (this must not be confused with the buffer base as the term is used by Astrup [31, 32]). The total serum base is equivalent to the sum of all the inorganic cations in the serum. It was formerly estimated by measuring the conductivity of the serum. Since the greatest proportion of the cations is the sodium ion (usually over 90 percent on a molar basis), the determination of total base was used, before the introduction of flame photometers, to follow the electrolyte balance, since it is a very rapid method. Although not entirely specific, it was much simpler than a chemical determination of sodium. It also has the advantage that it is nondestructive—after the conductivity has been measured, the serum can be used for other tests [33–35].

---

* Stat/Ion, Technicon Instruments, Tarrytown, N.Y. 10591.

Serum Sodium and Potassium by Flame Photometry

The serum and standards are first diluted with the proper diluent containing lithium (Li). Some instruments have an attached diluting mechanism so that the serum is drawn in and automatically diluted before aspiration into the flame. In other instruments the serum must be diluted manually before aspiration. For these an automatic dilutor is very convenient. The dilution is usually 1:100 though this may vary with the particular instrument. The dilution may not be exactly 1:100 (or other nominal value) but this is not important as long as all the samples and standards are diluted in exactly the same way.

The standard used usually contains 140 (or 150) mmol/liter for sodium and 5 mmol/liter for potassium. (Note that we are expressing the concentrations in millimoles per liter rather than milliequivalents. Since the two expressions are numerically the same for sodium and potassium, no confusion should result.) A zero standard containing the lithium diluent is also used. Some instruments may require additional standards. For most purposes it is more convenient to use the commercial standards available from the instrument manufacturer or supplier. If it is desired to prepare the standards, the following directions may be used.

Sodium and potassium standards. Fine crystals of reagent grade sodium chloride and potassium chloride are dried for several hours at 110°C, then cooled in a dessicator. If the crystals are not very small, they should be crushed in a mortar before drying. For accurate primary standards, the two salts are available from the National Bureau of Standards† (KCl, SRM No. 918; NaCl, SRM No. 919). To prepare a solution containing exactly 140 mmol/liter of Na and 5 mmol/liter of K, 8.182 g of NaCl and 0.3728 g of KCl are accurately weighed out and together dissolved in deionized water to make 1 liter. If a standard containing 150 mmol/liter of Na is desired, use 8.766 g of NaCl instead. Some instruments may require more than one standard, in which case it may be more convenient to make up stock solutions for dilution as required. A stock standard containing 1,000 mmol/liter of NaCl may be made by dissolving exactly 58.44 g of the salt in water to make 1 liter. A stock solution of 100 mmol/liter of KCl is prepared by dissolving 3.728 g of KCl in water to make 500 ml. Fifteen milliliters of the NaCl stock and 5 ml of the KCl stock are added to a 100 ml volumetric flask and diluted to the mark to give a solution containing 150 mmol/liter of Na and 5 mmol/liter of K. Other dilutions may be made as required.

The various instruments use slightly different concentrations of lithium in the diluent and it is preferable to use the stock solutions furnished with the instrument since each is designed for use with a particular lithium concentration. Usually

† National Bureau of Standards, Washington, D.C. 20025.

this is around 15 mmol/liter in the diluent. A stock solution may be made by weighing out 37.0 g of reagent grade lithium carbonate and transferring to a 1 liter volumetric flask. Add about 200 ml of water. Dilute 85 ml of concentrated HCl with about 200 ml of water and add in divided portions to the flask containing the lithium carbonate. (*Caution:* Avoid excess foaming.) Swirl until dissolved. A few drops of additional HCl may be added to dissolve completely. When all the carbonate has dissolved, cool to room temperature and dilute to the mark with water. This solution contains 1,000 mmol/liter. If 15 ml of this stock is diluted to 1 liter, the resulting diluting fluid will contain 15 mmol/liter of Li. Even if the flame photometer does not use an internal standard, it may be advisable to use a diluent containing lithium as this reduces the interference of the sodium with the potassium as mentioned earlier.

Urinary Sodium and Potassium by Flame Photometry

Occasionally it is desired to measure the daily output of sodium and potassium in the urine. The concentration of these ions will vary much more in urine than in serum, depending upon dietary intake, urine volume, and endocrine factors. Ordinarily the daily excretion of sodium will be in the range of 40 to 220 mmol and that of potassium 25 to 125 mmol. Because of the different ratio of Na to K in urine, it may not be possible to determine the concentrations of both ions on a single dilution as is done with serum. Some flame photometers have expanded scales in which a wide range of concentrations may be determined with appropriate standards. With instruments having a narrower range some preliminary dilutions of the urine may be required. With these instruments one can first try using an aliquot of the filtered urine and treating it exactly as a serum sample for the determination of sodium, and a 1:10 dilution of the urine (with water) treated exactly like a serum sample for potassium. If both these results fall on scale, the potassium reading is multiplied by 10 to obtain the true concentration of potassium. The concentrations in millimoles per liter are then multiplied by the 24-hr urine volume (in liters) to obtain the 24-hr excretion in millimoles. If the potassium reading is too high, off scale, another dilution must be tried; if the reading is very low, a smaller dilution should be run for better accuracy. It may be convenient initially to make 1:5, 1:10, and 1:20 dilutions of the urine and run these for potassium similar to a serum, multiplying the reading nearest midscale by the appropriate dilution factor. If the sodium reading is too high, off scale, a 1:2 or 1:4 dilution may be run similar to a serum, and the reading multiplied by the appropriate dilution factor. Some instruments may not give readings below 100 mmol/liter of sodium. In this instance, a smaller dilution of the urine must be made. This is most conveniently done manually. If a 1:100 dilution is ordinarily used, 1 ml of standard is diluted to 100 ml with the lithium solution in a volumetric flask and 2 ml of the urine similarly diluted. These solutions are then aspirated into the flame without further dilution. If the instrument has an attached automatic dilutor, this must be bypassed for direct aspiration. The reading obtained is then divided by 2 to give the actual sodium concentration.

Chloride

Reviews of the methods for the determination of chloride in biological materials have been published by Cotlove [36] and de la Huerga and associates [37]. The following discussion will not be a complete review but merely give the principles of some of the more common methods, together with some pertinent references. In general analytical chemistry the titration of chloride with silver nitrate has been used for many years. This was adapted to a protein-free filtrate by White-horn [38]. An excess of silver nitrate was added and the silver remaining after precipitation of all the chloride was titrated with thiocyanate solution using ferric ions as indicator. The endpoint for this titration is not very sharp and is subject to fading [39]. An alternative procedure has been suggested in which the excess silver in the solution is determined by atomic absorption [40]. The chloride has also been titrated directly with silver nitrate using dichlorofluorescein as an absorption indicator [41]. Electrometric methods for detecting the endpoint have also been used [42, 43]. The potential developed between a silver electrode (silver ions) or a silver-silver chloride electrode (chloride ions) and an indifferent electrode is measured as increments of the titrant are added and the endpoint determined by graphical means or by noting a sudden change in potential.

Cotlove has developed a coulometric procedure in which the titration is done automatically [44–46]. In the instrument (chloridimeter) the silver ions are generated at a constant rate by passage of a constant current between a pair of electrodes, one of pure silver. Another pair of electrodes measures the silver ion concentration. As long as any chloride ions are present, the concentration of free silver ions remains low. When all the chloride ions have been precipitated, the silver ion concentration begins to rise. This change is detected by the other pair of electrodes which activate a relay to stop the titration. Since the silver ions are generated at a constant rate, the time for the titration as measured by an associated clock will be proportional to the amount of chloride present. The instrument is accurate, simple to operate, and now very widely used. A number of different instruments based on this principle are on the market. Some give a direct digital readout in milliequivalents per liter, use smaller samples than the original instrument, and require only infrequent standardization.

Another commonly used procedure is the mercurimetric titration developed by Schales and Schales [39, 47]. The chloride is titrated with mercuric nitrate in a slightly acid solution. The mercuric chloride formed is only very slightly ionized so that as long as any chloride ions are present, the concentration of mercuric ions remains very low. When a slight excess of mercuric ions is present, they react with the indicator (s-diphenylcarbazone) to give an intense blue to purple color. The titration may be carried out on a protein-free filtrate (Folin-Wu) or on urine (adjusted to the proper pH). The titration can be carried out directly on serum, but this procedure results in a slight positive error, presumably due to the effect of the proteins. It has been claimed that the addition of ether to the solution to be titrated eliminated this error [48]. It has also been stated that titrations on urine samples that have stood for several days in the refrigerator often give erratic results, but Henry [49] has not been able to confirm this. The color of highly icteric serums may interfere in the direct titration; this has been eliminated by a preliminary treatment with hydrogen peroxide [50].

Another method involves the use of silver iodate [51, 52]. When a solution of chloride ions is shaken with solid silver iodate, the more insoluble silver chloride is formed and an equivalent amount of iodate goes into solution. After separation of the excess silver iodate, potassium iodide is added and the liberated iodine either titrated with thiosulfate or determined colorimetrically [53, 54]. Usually the silver iodate is added along with the reagents for the preparation of the protein-free filtrate, so that only one filtration is needed. If the amount of chloride present is very low, a correction may be made for the slight solubility of silver iodate.

There are a number of strictly colorimetric methods for chloride. In one the chloride solution is shaken with the relatively insoluble mercuric chloranilate. Mercuric chloride is formed and an equivalent amount of chloranilic acid dissolves. After removal of the excess reagent, the chloranilic acid is determined colorimetrically. Usually some solvent such as isopropyl alcohol is added to aid in the solubility of the chloranilic acid [55, 56].

In another colorimetric method the solution containing the chloride is added to a reagent containing mercuric thiocyanate and ferric chloride in an acid solution. The mercuric thiocyanate is only very slightly ionized, giving only a few thiocyanate ions to react with the ferric nitrate to produce the characteristic orange-red color of ferric thiocyanate [57, 58]. When chloride ions are present, the mercuric thiocyanate reacts to form mercuric chloride and free thiocyanate ions.

The latter react with the ferric ions to produce the color. The amount of color produced is proportional to the amount of chloride present. This method is used in some continuous-flow automated methods. In the automated procedure some mercuric ions are added to the reagent so that when the sample is added, no color is produced until the chloride concentration exceeds about 70 mmol/liter. This serves to expand the photometric scale to a useful range since serum chloride levels are rarely below 70 mmol/liter. The same procedure can be used for a manual method [59]. The procedure uses a mercuric salt and, particularly in the automated method, considerable amounts of waste solution containing appreciable amounts of mercury may be produced. This may constitute a disposal problem. A different reagent has been suggested for the automated procedure [60]. Ferric perchlorate in dilute perchloric acid solution is practically colorless. On the addition of chloride ions the intensely yellow ferric chloride is produced.

In the methods for chloride mentioned so far, bromide and iodide will react on a molar basis exactly the same as chloride. The amount of iodide in serum is not likely to be more than a few percent of the amount of chloride. In acute bromide intoxication, an appreciable fraction of the measured chloride may actually be bromide. Ordinarily bromide replaces chloride in the serum on a molar basis so that the total measured halide may not differ greatly from the normal value. It has been found that when using the automated mercuric thiocyanate method, the bromide present gave proportionately more color than an equivalent amount of chloride [61]. Thus in cases of bromide intoxication the apparent chloride concentration was very high in comparison with the other electrolyte values.

A much more specific method for chloride is based on the chloride specific ion electrode, which will respond almost exclusively to chloride ions. (For a general discussion of these types of electrodes see Durst [62].) These electrodes can measure the chloride ion concentration in a solution on a logarithmic scale similar to that used for pH (the hydrogen ion electrode is the most commonly used specific ion electrode). They have been used for several years for the measurement of sweat chloride in testing for cystic fibrosis [63, 64]. After sweating has been induced by iontophoresis or the application of heat, the electrode combination is merely applied to the moist skin and the reading taken. More recently, smaller electrodes have been developed for the determination of chloride in serum.‡

We present the mercury titration method for chloride and some comments on the coulometric titration method (the exact details for this method will depend upon the particular instrument used).

‡ Stat/Ion, Technicon Instruments, Tarrytown, N.Y. 10591.

Serum Chloride by Mercury Titration (Schales and Schales)

REAGENTS

1. Folin-Wu reagents for preparing protein-free filtrate.   See Chapter 5.

2. Mercuric nitrate solution, approx. $0.006M$.   Dissolve 2 g of reagent grade mercuric nitrate and 3 ml of concentrated nitric acid in about 200 ml of water and dilute to 1 liter. Since the reagent is standardized against the sodium chloride solution, the salt need not be weighed out accurately. In fact, some suppliers of the chemical furnish the monohydrate and others the anhydrous salt. The crystals absorb water and liquefy readily. For this reason the chemical should be purchased in small quantities. An alternative procedure is to dissolve 1.2 g of reagent grade red mercuric oxide in 5 ml of nitric acid and 20 ml of water and dilute to 1 liter.

3. Indicator solution.   Dissolve 100 mg of s-diphenylcarbazone in 100 ml of ethyl or methyl alcohol. This solution is not stable even when kept in the refrigerator; it should be replaced when it becomes appreciably darker in color. Some bottles of the reagent chemical may be labeled s-diphenylcarbazide; the material as supplied is actually a mixture of the two compounds in either case.

4. Chloride standard, $0.1M$.   Dry reagent grade sodium chloride (or potassium chloride) at 120°C for several hours. Weigh out exactly 5.845 g of NaCl (or 7.455 g of KCl), dissolve in water, and dilute to 1 liter.

PROCEDURE

Prepare a Folin-Wu filtrate from 1 volume of serum, 7 volumes of water, 1 volume of sodium tungstate solution, and 1 volume of sulfuric acid solution. This gives a slightly acid filtrate which is preferable. Transfer 2 ml of filtrate to a 25 ml Erlenmeyer flask, add 2 to 3 drops of indicator, and titrate with mercuric nitrate using a microburet calibrated in 0.01 ml intervals. The buret should have a fine glass or platinum tip that will deliver about 100 drops/ml. (Some of the older microburets had small hypodermic needles as tips; these are not satisfactory as the reagent reacts with the metal of the tip.) The clear solution will become intensely violet on the addition of the first drop of excess titrant. Also titrate a standard similarly using 2 ml of a 1:10 dilution, or 0.2 ml of the standard added to 1.8 ml of water in the titrating vessel.

CALCULATION

With the volumes used above:

$$\text{ml titration of sample} \times \frac{100}{\text{ml titration of standard}} = \text{conc. of sample (in mmol/liter)}$$

Serum may be titrated directly but with some possible loss in accuracy. Add 0.2 ml of serum to 1.8 ml of water, and titrate. The solution may appear purple after the addition of the first few drops of mercuric nitrate; this color will disappear with further titration and reappear at the true endpoint. This technique may give results several millimoles per liter higher than the use of a filtrate but it may be suitable for some purposes. In the direct titration the color of highly icteric serum will interfere with the detection of the endpoint. The following method [50] has been suggested to eliminate most of the color. To 0.2 ml of serum in the titrating flask, add 0.2 ml of 3% hydrogen peroxide and place in a boiling water bath for approximately 30 sec or on the top of a water bath for a slightly longer time. Cool, add 1.6 ml of water and 2 to 3 drops of indicator, and titrate.

Urinary Chloride

Two milliliters of a diluted urine (1:10 is usually satisfactory) may be titrated as described for the filtrate; 2 ml of a 1:10 dilution of the standard is also titrated. If the chloride concentration is very low, the titration may be repeated using a different dilution. The urine sample must not be too alkaline at the start of the titration (test with wide-range pH paper). If the diluted urine is alkaline, add a few drops of diluted nitric acid (1:20) to bring the pH to about 4. If the acidity is too high, the indicator loses sensitivity. It may be convenient to add the acid dropwise to an aliquot of the diluted urine, counting the drops required to bring the pH to 4, and then add the same number of drops to the aliquot to be titrated. Another way of adjusting the acidity is to make a dilution exactly as in preparing the Folin-Wu filtrate.

This method sometimes gives slightly high results with urines that have been standing for several days, even if refrigerated. Thus only fresh specimens should be used.

CALCULATION

If 1:10 dilutions of the urine and standard are titrated, the calculations are the same as for serum. If a different dilution of urine is used, the results are multiplied by 10 B/A, where B ml of urine is diluted up to volume A.

If the total amount of chloride excreted is to be reported, the result in millimoles per liter is multiplied by the urine volume in liters. If the results are to be expressed as grams of sodium chloride: mmol $\times$ 0.0585 = g NaCl.

Sweat Chloride

The chloride in sweat can be determined by the titration method given earlier in this section. This usually requires several tenths of a milliliter of sweat, which is sometimes difficult to obtain by iontophoresis. The sweat is collected on an absorbent filter paper pad which is carefully weighed in a closed weighting bottle before and after collection. The sweat is then eluted with a measured quantity of deionized water and the solution, or an aliquot, analyzed for chloride. Unless a microtitration is made, the equivalent of at least 150 mg of sweat should be analyzed. If, for example, A g (or ml) of sweat was eluted with B ml of water and the titration of 2 ml of this eluate is compared with the titration of 2 ml of the diluted standard (10 mmol/liter), then:

$$\frac{\text{titration of sample}}{\text{titration of standard}} \times 10 \times \frac{A + B}{A} = \text{chloride in sweat (in mmol/liter)}$$

Chloride by Coulometric Determination

There are a number of different coulometric chloride titrators on the market.§ The operation may differ somewhat with the various models, but they are all basically similar. All the newer models have digital readout to give the result directly in millimoles (milliequivalents) per liter. The sample is introduced into the diluting fluid and the titration started. The titration is automatically stopped at the endpoint and the result displayed. With many of the models the equivalent of 10 to 20 μl of serum can be accurately titrated, and a number of titrations can be run on the same cup of diluting fluid merely by adding another sample after each titration is finished. These instruments are very satisfactory for the determination of sweat chlorides as they accurately determine a few millimoles of chloride.

Normal Values and Interpretation of Results

The normal values for serum chlorides are 98 to 109 mmol/liter and for spinal fluid 122 to 132 mmol/liter. The urinary excretion of chloride will depend upon the dietary intake and may be subject to wide variation. For individuals on a normal diet the normal range is 170 to 250 mmol/day. Vomiting may lead to a loss of chloride from the body with a lowering of the serum level.

In normal infants the chloride level in the sweat is usually in the range of 10 to 40 mmol/liter, and levels above 60 mmol/liter are taken as indicative of cystic fibrosis.

§ The London Co., Cleveland, Ohio 44145; Buchler Instruments, Fort Lee, N.J. 07024; Fiske Associates, Uxbridge, Mass. 01569; American Instrument Co., Silver Springs, Md. 20910; Instrumentation Laboratory, Inc., Lexington, Mass. 02173.

Carbon Dioxide and pH

The carbon dioxide in blood is present in two forms, as dissolved CO_2 (carbonic acid, $H_2CO_3 \rightleftharpoons H_2O + CO_2$) and as bicarbonate ($HCO_3^-$). The relation between these two forms and the pH is given by the Henderson-Hasselbalch equation:

$$pH = pK_a' + \log \frac{[HCO_3^-]}{[H_2CO_3]}$$

where the square brackets denote concentrations (in moles per liter) and pK_a' is usually taken as 6.10. For accurate estimation of the acid-base balance in an individual, two of the three quantities (pH, $[HCO_3^-]$, and $[H_2CO_3]$) must be measured. This point is discussed more fully later in the interpretation of results. Depending upon the methods used for measurement, either $[HCO_3^-]$ may be replaced in the equation by $T - [H_2CO_3]$, where T is the total carbon dioxide content ($[HCO_3^-] + [H_2CO_3]$) as determined by measuring the total amount of CO_2 liberated by treatment with acid, or $[H_2CO_3]$ may be replaced by $\alpha \cdot Pco_2$, where Pco_2 is the partial pressure of CO_2 in the sample (as measured by the Pco_2 electrode) and α is a proportionality factor usually taken as 0.0308 at 37°C when the Pco_2 is reported in millimeters of mercury (for Pco_2 in kiloPascals, the factor would be 0.230).

The pH is usually measured with a glass electrode, preferably one of the microelectrodes [65]. Depending upon the method, either heparinized whole blood or plasma obtained from this may be used. The relation between the results obtained with the two types of samples and various methods of collection of the samples will be discussed later. We present here a general survey of the various methods that have been used.

The total CO_2 (T) is generally measured on plasma or serum, particularly when using the manometric Van Slyke [66, 67] or Natelson apparatus [68, 69]. The sample is treated with lactic acid in an evacuated chamber and the pressure exerted by the liberated CO_2 measured, usually before and after absorption of the CO_2 by alkali. The Natelson apparatus is simpler to use than the Van Slyke and requires smaller samples. The liberated CO_2 has also been measured by infrared spectrophotometry [70] and gas chromatography [71, 72], but these methods are not widely used. A simple kit is available for the determination of total CO_2 in which the volume of gas liberated under specified conditions is measured in a special syringe attached to the apparatus [73].‖ This is essentially a volumetric

‖ CO_2 Apparatus; Harleco, Philadelphia, Pa. 19143.

apparatus and to compensate for day-to-day changes in barometric pressure, a standard is run each time for comparison. This apparatus is admittedly linear only up to about 40 mmol/liter and thus may not give accurate results in the elevated range.

In automated continuous-flow analysis [74–76] the CO_2 is liberated from serum by treatment with acid in a closed system and the liberated gas passed into a buffered solution of an indicator such as phenol red. The addition of the CO_2 changes the pH of the solution and hence the color indicator. The change in absorbance of the solution is proportional to the amount of CO_2 present. In another device known as the Cediometer [77], the CO_2 liberated in a closed system is circulated past a P_{CO_2} electrode (for a discussion of this electrode, see below) and the amount of CO_2 liberated estimated from the electrode reading.

The bicarbonate concentration $[HCO_3^-]$ in serum may be estimated by titration as first introduced by Van Slyke [65, 78]. An excess of standard acid is added to liberate all the CO_2 and the remaining acid backtitrated with standard alkali. The endpoint in the titration is rather unstable and careful technique is required. A kit is available which uses this method.#

Another procedure uses the determination of $[H_2CO_3]$ by measurement of P_{CO_2}, the partial pressure of carbon dioxide in the sample. This is done by means of the P_{CO_2} electrode [79–81]. In essence this is a small pH electrode assembly separated from the sample by means of a membrane permeable to gaseous CO_2 but not to ionic species. The carbon dioxide diffusing to the electrode changes the pH of the buffer surrounding the electrode. The resulting pH change is usually calibrated to read out directly in P_{CO_2} (usually as millimeters of mercury). The electrode is usually calibrated by equilibration of gases of known CO_2 content. The procedure is similar to that for the P_{O_2} electrode given elsewhere, and the same precautions and comments in regard to the gas samples apply here as well. There are on the market a number of different blood gas instruments with electrodes for measuring pH, P_{CO_2}, and P_{O_2}.* The instruments are relatively expensive but allow one to make measurements of the three quantities in a few minutes. Some of the instruments may have readouts giving other derived quantities obtained from the Henderson-Hasselbalch equation or other formulas. The exact details of operation will vary somewhat from instrument to instrument, but some general precautions will be given later.

There is also an indirect method for measuring P_{CO_2} and calculation of the various parameters involved in the evaluation of acid-base balance. This is the

---

Bicarbonate Kit, Oxford Laboratories, San Mateo, Calif. 94401.
* Instrumentation Laboratories, Lexington, Mass. 02173; Corning Glass Works, Corning, N.Y. 14830; Radiometer, Inc., The London Co., Cleveland, Ohio 44145.

method of Astrup [82, 83]. The pH of the original blood sample is measured. Then aliquots of the same sample are equilibrated in tonometers at $37°C$ with two gases having two different known P_{CO_2} values. The pH of the equilibrated samples is then measured. These pH values are plotted against the logarithm of the known P_{CO_2} values. A straight line is drawn between the two points and extended as necessary. The pH of the original sample is then used to find the corresponding P_{CO_2} from the curve. This corresponds to the P_{CO_2} of the original sample. With the nomogram available for this method [84] the total CO_2 content and other derived values such as base excess, standard bicarbonate, and buffer base can be obtained. These latter are calculated values that are said to be helpful in assessing the acid-base balance. This method does not use a P_{CO_2} electrode or other means of directly measuring the CO_2 content, but does require the use of two different concentrations of calibrated gases and a good tonometer for equilibrating the sample at $37°C$.

As mentioned, some methods use plasma or serum and others use whole blood. With the proper precautions in obtaining and preparing the sample, satisfactory results can be obtained with either type of sample. This is discussed in detail by Gambino [85]. Since it is desired to measure the actual CO_2 concentration and pH of the blood sample as existing in vivo, contact of the sample with air must be avoided as far as possible to prevent loss or absorption of CO_2 and changes in pH. If serum or plasma is used, one must be certain that the equilibrium between cells and plasma existing in vitro at the time of separation is the same as that existing in vivo.

The manometric methods (Van Slyke and Natelson) as well as the continuous-flow automated methods use serum or plasma for the determination of total CO_2 concentration as does the titration method for bicarbonate. Conventionally these methods are used only for the determination of CO_2 and not in combination with pH determinations for complete acid-base assessment. As discussed later under interpretation of results, it will be seen that the determination of total CO_2 or of bicarbonate alone does not give an accurate picture of the acid-base status but they may be satisfactory for detecting gross abnormalities. Even for these methods the sample should theoretically be collected and separated out of contact with air to prevent loss of CO_2. Originally it was suggested that the sample be collected and centrifuged under a layer of mineral oil to prevent contact with air. The sample is collected in an oiled glass syringe and centrifuged in a tube with a thin layer of oil on top. This is not very satisfactory and it is preferable to collect the sample in evacuated collection tubes (such as Vacutainers), filling the tube as completely as possible and centrifuging with the stopper in place. The stopper is removed just before the sample is taken for analysis. This is fairly satisfactory for the manometric methods. With the continuous-flow automated methods there may be a loss of CO_2 as the sample stands in the plastic cups before analysis. Gambino [86] has suggested that the loss of CO_2 may be eliminated by adding a drop of 1% ammonia to all the cups (standards as well as samples) but this may cause some interference with other tests in the multiple-channel analyzers. It has been claimed that the latter are unsatisfactory for detecting any but gross abnormalities as a screening procedure [87].

Because of the difficulty in the anaerobic collection of samples, the concept of carbon dioxide combining power was introduced. This is theoretically the carbon dioxide content after the sample has been equilibrated with a gas having a partial pressure of CO_2 of 40 mm of mercury. The gas originally suggested was alveolar air from the technician's lungs [66], which can actually vary considerably from the stated partial pressure. A tank of gas containing the appropriate amount of CO_2 (about 5.5% by volume) can be used instead. However, since the carbon dioxide combining power is of no greater value than the carbon dioxide content as ordinarily determined, it is now rarely done.

When the Pco_2 and pH are done on heparinized plasma or whole blood, further precautions are necessary. When whole blood is allowed to stand at room temperature, the pH will decrease due to glycolysis. The ratio of $[HCO_3^-]$ to $[H_2CO_3]$ will also change, but in a closed container the total CO_2 content should not change. Thus the sample should be placed in ice immediately after collection and analyzed as soon as possible. The change in pH is about 0.030 units/hr at 25°C and roughly one-fourth as much at 4°C [88]. If plasma is used for the pH determination, the sample must be warmed at least to room temperature before centrifugation (it will usually warm further during centrifugation). The pH of whole blood and plasma change by different amounts for a given change in temperature. If ice-cold plasma is centrifuged, the resulting pH when measured at 37°C may be over 0.1 pH unit higher than that of the whole blood also measured at this temperature [89]. Thus for measurement it may be more convenient to use whole blood. The same precautions for collection must be used as given earlier for Po_2 except that the pH and Pco_2 of capillary samples are not significantly different from that collected in syringes (in contrast to the differences sometimes found for Po_2).

Total CO_2 in Serum or Plasma with the Natelson Microgasometer

The general principles of manometric analysis have been discussed earlier. In the Natelson instrument, all the samples and reagents are introduced through the tip which is calibrated at 0.01, 0.02, 0.03, and 0.10 ml. The bulb in which the extraction is carried out has a volume of 3 ml and the extracted gas is compressed to a volume of 0.12 ml. The mercury is moved by a plunger attached to a handwheel.

The shaking of the sample for extraction of the gas may be done manually or by means of a mechanical shaker. Newer models have a motorized handwheel for raising and lowering the mercury level and a small magnetic stirrer for gas extraction, but the basic principles remain the same. The general precautions for the collection of the sample have been given in the previous discussion.

REAGENTS

1. Lactic acid, $1.0M$. Dilute 90 ml of 85% lactic acid to 1 liter with water.
2. Sodium hydroxide, $3.0M$. Dissolve 60 g of reagent grade sodium hydroxide in water and dilute to 500 ml.
3. 1-Octanol (caprylic alcohol).

GENERAL DIRECTIONS

It is convenient to keep portions of the reagents for immediate use in small screw-capped vials. One vial is filled about one-third full of mercury, and on top of this is placed about an equal volume of the lactic acid with a layer of caprylic alcohol on top. The second vial will contain mercury and water and a third vial, which should be kept tightly stoppered, will have mercury and the sodium hydroxide solution. One may also have additional vials with water and lactic acid for washing the apparatus. One should also have some sort of vessel below the measuring tip to collect any mercury which may fall out.

The procedure for transferring a reagent from a vial to the chamber of the apparatus is as follows. The sampling tip is dipped below the surface of the liquid in the vial, and with the upper stopcock open, the plunger is advanced slightly so that a small droplet of mercury is forced from the tip. The appropriate amount of liquid is then drawn into the tip as measured by the graduations on the tip. The tip is then lowered below the surface of the mercury in the bottom of the vial and about 0.03 ml of mercury drawn up into the tip, if another reagent is to be added next.

If the material in the capillary is to be transferred to the chamber, the tip is kept in the mercury and the plunger retracted so that the liquids are carried completely into the extraction chamber by the flow of mercury. The sampling tip should not be held forcibly against the bottom of the vial so as to impede the flow of mercury. Also the bore of the stopcock should be precisely aligned with the other capillaries; otherwise there is a tendency for the mercury to break up into drops on entering the chamber.

The same procedure can be used for adding the serum sample if the serum overlays some mercury in a small tube. The mercury must be added first to the tube, then the serum. If one adds mercury to the serum already in the tube, the mercury will be broken up into drops and difficulty will be experienced in adding mercury to the chamber. Another method for the addition of serum is to add it to somewhat above the 0.03 mark, then place the tip in mercury, expel the excess serum exactly to the mark, and draw up some mercury.

PROCEDURE

Add successively, measuring with the tip and sealing with mercury after each addition, 0.03 ml of sample, 0.03 ml of lactic acid, 0.01 ml of caprylic alcohol, and 0.1 ml of water. After adding the water, draw mercury in to carry all the solutions into the extraction chamber. The mercury should be brought to the top of the extraction chamber.

Close the stopcock and lower the mercury to near the bottom of the extraction chamber. Shake for 2 min. If a small bubble of liquid remains in the tube above the extraction chamber, warm above the bubble by holding the glass stem with the fingers. This will cause the gas to expand and expel the bubble. Raise the level in the chamber to the 0.12 ml mark and read the pressure (P_1).

Advance the mercury to release the pressure and add 0.03 ml of $3M$ sodium hydroxide (be sure that the sampling tip is in the NaOH solution before opening the stopcock) followed by mercury to bring the NaOH solution into the chamber. Lower the liquid briefly to the bottom of the chamber, then bring up to the 0.12 mark. Read the pressure (P_2). The amount of CO_2 is then calculated with the use of appropriate factor (F) (see Table 9-1).

As mentioned in the general discussion, the various corrections due to water vapor, reabsorption of carbon dioxide, deviations from the gas laws, and others, which would be used in the calculations, are all included in the factors given in the table. The factor varies with the temperature and the one corresponding to the

Table 9-1. Factors for Calculating Total CO_2 with the Natelson Microgasometer

Temp. (°C)	Factor	Temp. (°C)	Factor
17	0.242	25	0.232
18	0.240	26	0.231
19	0.238	27	0.230
20	0.237	28	0.229
21	0.236	29	0.228
22	0.235	30	0.227
23	0.234	31	0.225
24	0.233	32	0.224

$(P_1 - P_2) \times F =$ amount of CO_2 (in mmol/liter)

ambient temperature should always be used. The factor is valid only for a sample size of 0.03 ml. If a different amount is used, the appropriate correction should be made. If the total volume of sample and all reagents including water are more than slightly different from the directed amounts, the factor may be somewhat in error. For example, in the variation of the method used by Martinek [69], the reagent volumes are slightly different so a different set of factors is required.

The procedure may be standardized by using a sodium carbonate solution containing 25 mmol/liter. This is prepared by dissolving 0.265 g of reagent grade sodium carbonate (dried at $110°C$ for several hours) in recently boiled and cooled water and diluting to 100 ml. This solution will change in concentration with time unless tightly stoppered and is preferably made up fresh as required. When the solution is run in the apparatus as a check, the same procedure is followed as for a serum except that caprylic alcohol is omitted. A discussion of the normal values and interpretation will be deferred until after a consideration of the determination of Pco_2 and pH.

If the pH of the sample has also been determined, the Pco_2 and bicarbonate concentration may be calculated by means of the factors and formulas given in Table 9-2. A number of different tables or nomograms have been used to indicate

Table 9-2. Factors (F) Obtained from pH to Calculate Pco_2 from Total CO_2 Content

pH	0.00	0.01	0.02	0.03	0.04	0.05	0.06	0.07	0.08	0.09
6.80	6.30	6.15	5.99	5.87	5.73	5.59	5.46	5.36	5.24	5.12
6.90	5.01	4.89	4.77	4.66	4.55	4.44	4.35	4.26	4.17	4.08
7.00	3.99	3.90	3.82	3.73	3.65	3.56	3.49	3.42	3.34	3.27
7.10	3.20	3.44	3.07	3.00	2.93	2.87	2.81	2.75	2.69	2.62
7.20	2.56	2.51	2.45	2.39	2.33	2.27	2.22	2.16	2.12	2.08
7.30	2.03	1.98	1.94	1.89	1.85	1.81	1.76	1.72	1.69	1.65
7.40	1.61	1.58	1.54	1.51	1.46	1.43	1.40	1.37	1.34	1.31
7.50	1.28	1.25	1.22	1.19	1.16	1.14	1.11	1.08	1.06	1.04
7.60	1.02	1.00	0.97	0.95	0.92	0.90	0.88	0.86	0.84	0.83
7.70	0.81	0.79	0.77	0.75	0.73	0.72	0.70	0.69	0.67	0.66

pH = left column + top line

Pco_2 = F × CO_2 content

$[H_2CO_3]$ = Pco_2 × 0.03

$[HCO_3^-]$ = CO_2 content − $[H_2CO_3]$

the relationships among the three quantities of the Henderson-Hasselbalch equation; the different factors do not all agree. One reason is that pK′ of the equation is not exactly constant but must be determined experimentally. We have chosen to use a set of factors intermediate between those given by Gambino [85] (cited as modified from Weissberg [90]) and the values obtained from the use of the nomogram of Siggaard-Andersen [91]. The values obtained by the different authors agree well at pH above about 7.2. At pH of 6.8 the values in Table 9-2 are about 2 percent lower than those given by Gambino and about 2 percent higher than those derived from the nomogram.

Blood P_{CO_2}

The determination of P_{CO_2} requires the use of a special electrode which is usually a part of a blood gas apparatus [85]. The principle involved in the determination has already been mentioned (see Chapter 2). The electrode chamber usually contains a P_{O_2} electrode as well, and both are kept at a constant temperature (usually 37°C). The expendable part of the electrode is the membrane which must be replaced periodically. Most instruments have a device for indicating when the membrane has deteriorated. The electrode is calibrated with two gases of different known CO_2 content. The exact procedure for the calibration and measurement of samples will vary somewhat with the particular instrument. In some models the sample must be injected with a syringe, in others the sample is drawn through a small plastic tip. Most instruments are equipped to handle capillary whole blood. For P_{CO_2} the difference between capillary blood and that collected in a syringe is not great enough to be of clinical significance [92]. The P_{CO_2} is usually determined on heparinized whole blood. Often the P_{O_2} is also requested; this is done on the same sample and, of course, requires whole blood. From the P_{CO_2} and the pH, the total carbon dioxide content and bicarbonate concentration may be calculated by means of the factors given.

Blood pH [85, 93]

The general determination of pH with the glass electrode has been discussed earlier (Chapter 2). We give here some special precautions necessary for blood pH. For most other pH determinations in the clinical laboratory, such as checking of buffers, an accuracy of 0.01 pH unit is more than adequate. For blood pH, since the normal variation is only about 0.1 unit, a greater accuracy is desired. The pH meter should be capable of giving readings to 0.001 unit with a reproducibility of 0.002 unit. An instrument with a digital readout to three decimals is convenient but not necessary.

The buffers used for standardization of the instrument must be very accurate.

Most manufacturers of blood pH instruments supply special buffers for standardization. Often these are supplied in bottles holding 500 ml. On continued exposure to air with reopening of the bottle, the pH of the buffer may change slightly. It would be preferable if the buffers were supplied in ampules or small well-sealed bottles so that a fresh ampule or bottle could be used for each standardization. This would, however, add considerably to the cost. As an alternative, one could transfer the solution from the large bottle to a number of smaller bottles of borosilicate glass which are kept tightly stoppered, only one being opened as needed. It has been suggested [94] that a few crystals of thymol be added to each large bottle as preservative and that it be kept in the refrigerator.

The standardization uses two buffers of different pH values, one of which should be close to 7.40 pH. The stability of this latter buffer is the more important. One commonly used pair of phosphate buffers has pH values of 6.84 and 7.38 at 37°C (the temperature at which the pH of blood is preferably measured). Some instrument manuals may suggest a nitrophenol buffer with a pH of 6.98 and an acetate buffer of 4.64. These buffers are generally more stable and are often suggested with instruments of European manufacture. Some blood gas instruments have only a limited pH range (e.g., 5.8 to 8.8) and the nitrophenol and acetate could not be used with these instruments.

For checking of the pH buffers, two accurate solutions may be made up from potassium dihydrogen phosphate and disodium hydrogen phosphate obtained from the National Bureau of Standards.† These are dried at 110°C for a few hours and then cooled in a dessicator. If exactly 1.179 g of KH_2PO_4 and 4.302 g of Na_2HPO_4 are dissolved in CO_2 and ammonia-free water and diluted to exactly 1 liter at 25°C, the solution should have a pH at 37°C of 7.385. If 3.40 g of KH_2PO_4 and 3.53 g of $NaHPO_4$ are similarly dissolved and diluted to 1 liter, the resulting solution should have a pH of 6.84 at 37°C.

The pH of blood or serum is usually measured at 37° (or 38°)C, using a micro-electrode which is kept at the desired temperature by means of a circulating water jacket or electrical heating. It has been suggested that the KCl junction be also kept at the elevated temperature [95], but not all instruments have provision for this and the buffers used may not always be standardized under these conditions. If the pH of the blood or plasma (or serum) cannot be measured immediately after obtaining the sample or, when plasma is used, centrifuged at once, it is recommended that the blood be stored in ice until ready for analysis. If whole blood is thus cooled and later warmed to 37°C (in the instrument), this will not

† Samples No. 186 Ic (KH_2PO_4) and No. 186 IIc (Na_2HPO_4). National Bureau of Standards, Washington, D.C. 20025.

influence the final pH. If, however, blood is cooled and then centrifuged while cold and the resulting serum or plasma warmed to 37°C for pH measurement, the result will not be the same as if the blood has been centrifuged at 37°C. This is because the change in pH with temperature is not the same for whole blood as it is for serum or plasma [96, 97]. The pH of the plasma when separated in the cold will be 0.1 pH unit (or more) higher than when separated at 37°C. There is no difference in the pH between serum and heparinized plasma when both are treated the same. Thus if blood has been kept in ice, it should be warmed at least to room temperature before centrifuging (it will usually warm up somewhat more in the centrifuge).

If it is necessary to measure a sample at a temperature other than 37°C, an approximate correction can be made for this. The pH of whose blood increases about 0.015 unit for every degree below 37°C (i.e., at 27°C the pH would be 0.15 unit higher than at 37°C). For plasma the change is about 0.0215 pH unit/degree (change in the same direction as for whole blood). The corrections can only be approximate since the change with temperature is not the same for all samples [96, 98].

The actual pH determination is similar for most instruments. The blood or standard is drawn up into the capillary electrode and the tip inserted in the connecting KCl solution. One must be certain that the small plastic tip is completely filled with blood and that no air bubbles are drawn up into the capillary electrode. After blood or plasma is drawn up in the capillary, it should always be rinsed with saline or buffer (old buffer that is no longer suitable for standardization may be used for this). If the capillary is rinsed with water, a film of protein may be deposited on the surface of the electrode and this will cause erratic results. The electrode should be rinsed with a mild cleaning solution occasionally. Some instrument makers furnish a cleaning solution that contains pepsin in dilute hydrochloric acid. This is said to dissolve readily any protein film [99]. Gambino [85] states that good results may be obtained with a mild detergent such as 1% solution in normal saline of a material such as Dreft or Lux Liquid. The electrode should always be rinsed immediately after a measurement. The inner capillary electrode should never be allowed to dry out. It is best kept filled with some old pH 6.84 buffer. If it does dry out, it will give very erratic results. It can usually be regenerated by allowing it to stand filled with very dilute HCl for some time. Other precautions may be given in the instrument manuals.

NORMAL VALUES AND INTERPRETATION OF RESULTS

Several different ranges for the normal values have been given in the literature [100–103]. We present some average values.

	Arterial	Venous
pH, whole blood	7.35–7.45	7.32–7.42
Pco$_2$, whole blood	41–51 mm Hg	38–48 mm Hg
Total CO$_2$, plasma	26–31 mmol/liter	25–30 mmol/liter

The values for men may average a few percent higher, and for women a few percent lower than those given here. The pH and Pco$_2$ are more variable in children than in adults [104, 105]. The average pH of capillary blood from premature infants was found to be only 7.31 and that from full-term newborns 7.34; it rose to about 7.40 after a few hours.

Acid-Base Balance

Since the pH, [HCO$_3^-$], and [H$_2$CO$_3$] are related by the Henderson-Hasselbalch equation, the determination of only one of these parameters will not give a complete picture of the acid-base balance. There are a number of graphical methods for illustrating this relationship. A convenient one is that devised by Weissberg [90]. Figure 9-1 is based on this. The Pco$_2$ isopleths have been recalculated and redrawn to correspond to the factors given in Table 9-2. In the figure, pH is plotted along the horizontal axis and the total carbon dioxide content along the vertical axis. From the table, one can calculate the Pco$_2$ for any given pH and total CO$_2$. The lines drawn in the figure are lines of constant Pco$_2$ (isopleths), each line having the Pco$_2$ value given at the top and left. The circle in the center represents the approximate normal range. This has a center at a pH of 7.40 and a total CO$_2$ of 28 mmol/liter corresponding to a Pco$_2$ of 45 mm of Hg. Since the actual normal range differs for arterial and venous blood (with some possible differences for capillary blood as well) and may differ for men and women, the circle represents only an average value.

When the pH and total CO$_2$ have been measured (or the latter calculated from Pco$_2$), these two values determine a unique point in Figure 9-1 representing the acid-base status of the patient. Points A and C, for example, represent patients with a pH lower than normal—acidosis. At point A the Pco$_2$ is increased proportionately more than the total CO$_2$. This is usually due to the retention of CO$_2$ caused by the inability of the lungs to eliminate the gas (hypoventilation). The condition is known as respiratory acidosis. At point C the total CO$_2$ is low but the Pco$_2$ has not changed markedly. This condition is found in metabolic acidosis. The production of excess acidic substances in the body, such as the keto acids in diabetes, converts some of the bicarbonate to carbonic acid which may be eliminated by the lungs, thus decreasing the total CO$_2$.

Points B and D represent patients with a pH higher than normal—alkalosis.

Figure 9-1. Weisberg's nomogram illustrating relationship between pH, CO_2, and Pco_2. See text for further explanation.

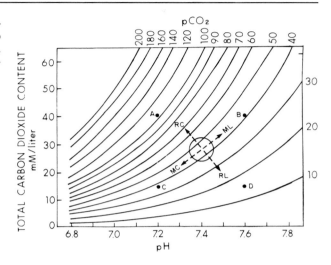

Point D, the opposite of A, represents respiratory alkalosis; the Pco_2 is decreased proportionately more than the total CO_2, due to hyperventilation of the lungs eliminating more than normal amounts of CO_2. Point B, the opposite of C, represents metabolic alkalosis with the total CO_2 increased proportionately more than the Pco_2. The increased total CO_2 (actually an increase in bicarbonate) is due to the presence in the blood of basic ions derived from the dietary intake which convert carbonic acid into bicarbonate.

The four arrows pointing roughly in the direction of the points A, B, C, and D indicate the general direction of change for respiratory acidosis (RC), metabolic alkalosis (ML), respiratory alkalosis (RL), and metabolic acidosis (MC), respectively. Note that points A and C have the same pH value yet represent quite different conditions. Points A and B have the same total CO_2 content, and points B and C have only slightly different Pco_2 values. This illustration indicates that the determination of one value alone, pH or Pco_2, cannot give a complete picture of the acid-base status. There may also exist intermediate conditions with both respiratory and metabolic acidosis, for example. These would be illustrated by intermediate points.

Plotting the results in this way is a convenient way of visualizing the acid-base balance. The use of paper containing the Pco_2 isopleths is convenient but not

Figure 9-2. Plotted hypothetical results to visualize acid-base balance.

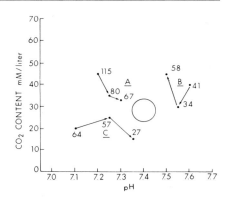

absolutely necessary. This is illustrated in Figure 9-2 where the various points are located by plotting the pH against the total CO_2, with the calculated Pco_2 written opposite each point. The portions A, B, and C represent successive hypothetical determinations on the same patient as indicated by the arrows. Thus in A note that the successive points are approaching the normal range indicating good results of the therapy. At B the second determination indicates an improvement; the third point indicates a deterioration in the condition of the patient. At C, although the pH is approaching normal, the total CO_2 content remains low. This might be considered compensated metabolic acidosis—the underlying condition remains but the body has compensated to bring the pH back to near normal by increased elimination of CO_2 (Pco_2 decreased).

Oxygen Saturation

Blood Sample Collection Methods

The oxygen content of arterial blood is a measure of the ability of the lungs to oxygenate the blood adequately. Thus in collection, care must be taken to insure that the sample does not come in contact with air which would change the oxygen content. For oxygen saturation studies, true arterial blood must be obtained. The blood sample is usually drawn from the brachial or femoral artery though other arteries may be used. Arterial blood is preferably collected in heparinized glass syringes. A small amount of heparin solution is introduced into the syringe, the walls of the syringe coated with the solution, and the excess expelled. After collection, any bubbles present are expelled through the tip and the syringe sealed either by inserting the tip of the needle into a rubber stopper or preferably by means of a metal cap. The blood is then mixed with the anticoagulant by rolling the syringe between the palms of the hands. All samples, no matter how collected, should be analyzed as soon as possible and kept in ice until the analysis is performed. Plastic syringes may be used, but it has been shown that these syringes are somewhat permeable to air [106] and thus the oxygen tension may change with time. Evacuated, heparinized blood collection tubes have also been used. The ordinary tubes contain a small amount of residual air which may cause some change in oxygen tension in the sample, and special nitrogen tubes are available in which the air in the tubes is replaced by nitrogen before evacuation. There is some controversy in the literature as to whether these tubes give the same results as the use of glass syringes [107–109]. The use of the latter is thus preferable. Some workers find the arterial puncture easier to perform with syringes.

When micro methods of analysis are available for blood from infants or other patients in which arterial puncture is difficult or inadvisable, arterialized capillary blood may be used. For this the hand (or heel, in infants) is first immersed in water at 45°C for about 10 min and wiped dry. Then blood is obtained by puncture and drawn into a heparinized capillary tube. A small iron stirring "flea" is then inserted and the tube sealed with plastic. The blood is mixed with the anticoagulant by moving the flea back and forth by means of an external magnet. For somewhat larger samples the Natelson collection tubes may be used. Arterialized blood may also be obtained from the earlobe but here the hyperthermia necessary for arterialization is more difficult. An electric heater has been devised for this purpose [110], and there are a number of pastes on the market which when rubbed on the earlobe (or finger) will induce increased blood flow. A small but significant difference has been found between arterialized fingertip blood and that obtained directly from an artery, the former showing a higher oxygen tension [111]. This may be due to some extent to exposure of the blood to air during collection. Theoretically a large drop of blood should be allowed to collect on the finger and the capillary inserted into the interior of the drop. On the other hand, a negligible difference was found using blood from the earlobe. With high oxygen tension (such as found in patients under oxygen therapy), the differences between arterial and fingertip blood were much larger and more variable.

Oxygen Saturation—Methods of Determination

The oxygen content and oxygen saturation of whole blood may be determined by three different general methods: gasometric, spectrophotometric, and coulometric.

GASOMETRIC METHODS

The gasometric methods are exemplified by those using the Van Slyke [112] or Natelson [113] manometric gas apparatus. In these the oxygen content is determined by treating the blood in a closed chamber with ferricyanide to liberate the oxygen. After absorption of the carbon dioxide which is also liberated, the pressure exerted by the oxygen is measured. From this the amount of oxygen present in the blood may be calculated directly or the oxygen may be absorbed by cuprous chloride or hyposulfite and the difference in pressure used to calculate the oxygen content. Usually a correction must be made for the small amount of dissolved nitrogen in the direct method and for the small amount of physically dissolved oxygen in both variations. In the Van Slyke apparatus the reagents may be de-aerated within the apparatus, whereas with the Natelson apparatus the reagents must be de-aerated externally by the application of a vacuum. For finding oxygen saturation, two determinations must usually be made, one on the original sample of blood and one after the blood has been saturated with oxygen in a tonometer. A difficulty with this procedure is that when the blood is equilibrated in the tonometer, the plasma tends to adhere more to the sides of the tonometer than do the red cells so that it is difficult to obtain saturated blood with exactly the same percentage of red cells as in the original sample. For a simplified procedure, the oxygen content at 100 percent saturation may be calculated (as volume percent) by multiplying the hemoglobin concentration (in grams per deciliter) by the factor 1.34, provided that the blood does not contain appreciable amounts of methemoglobin or other hemoglobin derivatives not readily converted into oxyhemoglobin or that the method used for hemoglobin does not include these derivatives.

The use of gas chromatography [114–117] should be included among the gasometric measurements. By the use of special columns, the gases extracted from blood, O_2, CO_2, CO, and N_2O, can be readily determined. Only a small sample of blood is needed and the procedure is fairly rapid. Special instrumentation is required. Small gas chromatographs are made particularly for such measurements, but they are limited in their application to other analyses.

SPECTROPHOTOMETRIC METHODS

The spectrophotometric method uses the measurements at two different wavelengths for determinations in a two-component system. The general principles involved in this method have been given in Chapter 1. A discussion of the application to the determination of hemoglobin derivatives appears in the monograph by Van Assendelft [118]. The details of one procedure for the determination of oxygen saturation by the spectrophotometric method will be given later. Only a general discussion will be presented here. Contact with air must be avoided lest the oxygen content be changed. On the other hand, the red cells must be lysed to form a clear solution for spectrophotometric measurement. If the lysis is produced by the addition of an agent such as Triton X-100, the solution added must be in as small a volume as possible to avoid contact with excess dissolved oxygen. This means that the blood will be only slightly diluted and a very short light path is necessary. In the visible range light paths of 0.1 mm may be used and in the near infrared, paths of 1.0 mm. A number of different pairs of wavelengths have been used, one usually being an isobestic point. The short light paths are usually obtained by inserting glass spacers into a regular spectrophotometric cell [108]. As explained in the earlier description of the two-component method, the exact light path is not important as long as it is constant.

In a slightly different method the blood is collected in heparinized micro-hematocrit tubes which are then centrifuged. The packed cells are lysed by alternate freezing and thawing several times. The lysed blood is then introduced into a special cuvette which is essentially an unruled hemocytometer chamber with the light beam passing vertically through it. Thus it is possible to obtain a very short light path [119].‡ In another commercial instrument the blood and lysing agent are mixed out of contact with air by automated means, the measurements made at different wavelengths, and the percent oxygen saturation computed and read out on a digital scale.§

Measurement at two wavelengths assumes that there are no other hemoglobin derivatives present besides oxyhemoglobin and reduced hemoglobin. These are indeed the main derivatives present, but small amounts of carboxyhemoglobin, sulfhemoglobin, or methemoglobin may also be present. These other derivatives would have different absorption characteristics and would thus cause a small error if present. Theoretically one could compensate for the presence of carboxyhemoglobin by reading at three different wavelengths to calculate the percentage of all three components. This is exactly what the Instrumentation Laboratory instrument does. For a detailed discussion of the principles involved in this, see the discussion by de la Huerga and Sherrick [120].

Ordinary spectrophotometric measurements use only two wavelengths and do not take into account the small amounts of other derivatives present. Since the absorption curve for carboxyhemoglobin, for example, is not markedly different from that of oxyhemoglobin, the error caused by the presence of the former

‡ Radiometer Type OSM-1, London Company, Cleveland, Ohio 44145.
§ I.L. CO-oximeter, Instrumentation Laboratory, Lexington, Mass. 02173.

substance may not be as great as might be expected. That is, the presence of 5% carboxyhemoglobin will probably not result in a 5% error in the oxygen saturation, depending upon which particular wavelengths are used for the measurements.

In a slightly different procedure the light diffusely reflected from a blood sample is measured at two wavelengths. At 805 nm oxyhemoglobin and reduced hemoglobin reflect the same amount of light, whereas at 650 nm the former reflects much more light than the latter. An instrument has been designed for making measurements in this way [121].|| It has the advantage that the blood is not lysed. In the micro modification, only 0.2 ml of blood is required. The method appears to give good correlation with other spectrophotometric methods.

COULOMETRIC METHODS

An entirely different method for the determination of the oxygen content of blood is that based on the oxygen electrode, a polarographic technique. The electrode consists of a platinum cathode and a silver chloride electrode in a suitable buffer. The solution containing the electrodes is separated from the blood by a membrane permeable to gaseous oxygen but not to dissolved substances. If a potential of about 0.65 is applied between the two electrodes, a reaction takes place which may be written as follows:

$$O_2 + 4e^- \rightleftharpoons 2\,O^{2-}$$

although the actual reaction is probably more complicated. The essential point is that under specified conditions the current passing between the electrodes is proportional to the partial pressure of the oxygen in the solution. Thus with proper standardization the amplified current can be read as a measure of the oxygen tension [122, 123]. The temperature of the electrode and other conditions must be kept constant for consistent results. A number of different instruments are available which use this principle for the measurement of oxygen tension. The exact procedure may vary slightly from instrument to instrument. An excellent discussion of the principles and details of operation has been given by Gambino [124]. The result of measurement is given as partial pressure of oxygen, Po_2 (conventionally given in millimeters of mercury). For some clinical purposes this value may be sufficient once a normal range has been established. However, the actual percentage of oxygen saturation may be desired. This can be obtained from the Po_2 but the exact relationship depends upon the temperature (most

|| AO Reflectance Oximeter, American Optical Co., Buffalo, N.Y. 14215.

Figure 9-3. Nomogram for the calculation of oxygen saturation when blood Po$_2$ and pH are known.

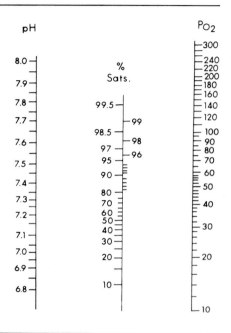

measurements are now made at 37°C) and on the pH of the blood sample so this latter must also be measured. A number of nomograms are available for the calculation; one type is illustrated in Figure 9-3.

Oxygen Saturation—Standardization

The *gasometric* methods have the theoretical advantage that they yield absolute values with reference to any external standards (provided, of course, that the chamber volumes and manometer scales are accurately calibrated).

The *spectrophotometric* methods when used with a regular spectrometer require calibration using samples containing zero and 100% saturation. The 100% saturation sample is obtained by equilibrating the blood in a tonometer with 100% oxygen gas. The blood should be preferably obtained from a nonsmoker (low carboxyhemoglobin content) and one who has not been on medication that could result in any methemoglobinemia or sulfhemoglobinemia. The zero percent saturation is usually obtained by the addition of sodium dithionite (sodium hyposulfite, $Na_2S_2O_4$). When one of the wavelengths chosen is an isobestic point, the oxygen saturation will be a linear function of the ratio of the absorbances at the two wavelengths. It is advisable to check the calibration curve from time to time [109]. The commercial instruments for spectrophotometric determination of oxygen saturation usually give calibration factors for use with the particular instrument or may have direct readout in percent saturation. The calibration for these should be checked from time to time by the use of the zero and 100% samples. Theoretically, intermediate values can be obtained by mixing different proportions of the two standards, but accurate pipetting of whole blood samples is difficult.

The *polarographic* electrodes are usually calibrated by the use of gases of known oxygen content, which can be purchased from a number of supply houses and are usually accurately standardized. Since these gases are also used for standardization of the P_{CO_2} electrode, the usual gas mixtures are either 5% CO_2, 12% O_2, and 83% N_2, or 10% CO_2 and 90% N_2 (zero percent O_2). It has been suggested [125] that one or more tanks of gas whose contents have been accurately determined be obtained for use as reference standards for each new tank of gas actually used for calibration of the instruments. These reference tanks may also be accurately analyzed by the use of a modified Hemple apparatus [126]. This is satisfactory for the CO_2 concentrations, for which the analysis is relatively simple, but is less satisfactory for oxygen analysis and the latter is usually not attempted. Since small variations in oxygen saturation are usually not of great clinical importance, the commercial gas mixtures are satisfactory for the calibration of P_{O_2}. The tanks of accurately analyzed gas are relatively expensive; another solution has been the use of proportioning pumps [127].# These pumps will take gases from less expensive tanks of 100% O_2 and 100% N_2, for example, and deliver a mixture of the two gases in any desired proportion. This method may be more economical if large quantities of the calibrating gas mixtures are used. In another method the high oxygen concentration is obtained by equilibrating pure water with air at 37°C [128]. This water is then introduced into the apparatus similarly to a blood sample and is considered to have the same oxygen tension as air which contains 20.95% oxygen by volume. For the zero P_{O_2} standard, a small amount of sodium sulfite (about 0.1 g/dl) is dissolved in $0.01M$ sodium borate solution just before use.

In all these methods the actual partial pressure of oxygen in the mixture will depend upon the total external (barometric) pressure. Thus if the gas is stated to contain a volume fraction A of, say, oxygen (A = percent by volume/100), P is the external barometric pressure, and p is the vapor pressure of water at the

Godart Gas Mixing Pumps, Instrumentation Associated, New York, N.Y. 10023.

temperature of measurement, then $Po_2 = A \times (P - p)$. The pressures are usually stated in millimeters of mercury, although the use of kiloPascals (kPa) is recommended. The vapor pressure of water is taken as 47.1 mm of Hg (6.28 kPa) at 37°C and 47.9 mm of Hg (6.62 kPa) at 38°C. For routine work the day-to-day variations in the barometric pressure may be ignored and an average value used for the calculation of Po_2. If a good mercurial barometer is not available, the barometric pressure may be obtained from the local office of the National Weather Service. The barometric pressure usually given by the office is that reduced to sea level (for easier comparison between different stations) and not the actual barometric pressure. If the laboratory is not close to sea level, the actual pressure should be requested. This is often cited in inches of mercury (1 inch of Hg = 25.4 mm of Hg = 3.39 kPa). As a good approximation it may be taken that an increase in height above sea level of 100 meters will cause a decrease in barometric pressure of 8 mm of Hg or 1.07 kPa.

Oxygen Saturation by Spectrophotometry [118, 129, 130]

The precautions for the collection and preservation of the specimens have been noted earlier in this section. Although it is claimed that better accuracy is obtained by using cuvettes with a 0.1 mm light path, these are not always available. We present methods that can use either 0.1 or 1.0 mm cuvettes. The procedures are essentially similar except for the cuvettes and the wavelengths used.

SPECIAL APPARATUS

Glass syringes, Luer-Lok, 2 and 5 ml sizes.   Preferably the specimens should be collected in Luer-Lok syringes.

Three-way stopcock to fit syringes (B-D Type MS10-T).*

Spectrophotometer cells with 1.0 mm light path.   These may be obtained from a number of manufacturers.† A holder may be required to enable these cuvettes to fit in the regular 1 cm light path cell holder. A plastic holder with four 1 mm cells is available.‡ This is satisfactory but care must be taken not to scratch the plastic. Cells with a 0.1 mm light path are also available. These are of the demountable type and are more difficult to fill without exposure of the sample to air. Van Assendelft uses 1 mm cells with a glass insert to reduce the light path to approximately 0.1 mm. These inserts are fragile and not very readily available.

---

* Becton-Dickenson Co., Rutherford, N.J. 07070.
† Helma Cells, Inc., Forest Hills, N.Y. 11375; Precision Cells, Inc., Hicksville, N.Y. 11801; Markus Science Supply Co., Del Mar, Calif. 92104.
‡ Waters Instruments, Rochester, Minn. 55901; Rohn and Haas, Philadelphia, Pa. 19105.

REAGENTS

1. Triton X-100 solution. Mix 15 ml of Triton X-100§ and 85 ml of water containing 0.1 g of sodium carbonate. Either the solution should be made up with recently boiled and cooled distilled water and the material dissolved in a filled closed vessel without undue agitation, or the solution should be de-aerated after being made. This may be done by placing an open flask containing the solution in a vacuum dessicator and evacuating (reducing the pressure slowly to avoid foaming). This solution is used for the procedure with 1.0 mm light path cells. For the 0.1 mm cells the Sterox solution given below is preferred.

2. Sterox solution. Mix 2 ml of Sterox SE and 98 ml of water. Like the previous solution this should be made with recently boiled and cooled distilled water with the minimum of agitation, or de-aerated before use. This solution is preferred when the 0.1 mm light path cuvettes are used.

3. Reducing solution. Add 1 g of sodium hydrosulfite (sodium dithionite, $Na_2S_2O_4$) to 10 ml of either of the above solutions as required, contained in a test tube of such a size that it is nearly filled by the liquid. Stopper tightly and dissolve the salt by gentle agitation. If the solution is not clear, centrifuge with the stopper in place. Prepare just before use from fresh crystals of the salt. Keep tightly stoppered until used. This solution is used to prepare the zero oxyhemoglobin standard.

PROCEDURE

The procedure will be first described for use with the 1.0 mm cuvettes; any modifications necessary for use with the 0.1 mm cuvettes will be mentioned later. The three-way stopcock has two female joints for syringes and one male joint for a needle. One syringe when connected will be in line with the needle. This will be called syringe A or the syringe on joint A. The position of the stopcock lever when this syringe is connected to the needle will be called position A of the lever. The other syringe will be connected at right angles to the needle and will be noted as syringe B or the syringe at joint B. Also the position of the stopcock lever when this syringe is connected to the needle will be noted as position B. At the other position of the stopcock lever, the connection for liquid is between the two syringes but not to the needle. This will be noted as position C of the lever.

To facilitate mixing of blood in the sample syringes, a small amount (0.5 ml) of mercury is injected into the syringe from a small syringe and needle. Any air bubbles introduced are expelled and the sample syringe is recapped. The mercury drop aids in mixing the blood just before sampling. A 2 ml Luer-Lok syringe con-

§ Rohn and Haas, Philadelphia, Pa. 19105.

taining about 0.5 ml of mercury is firmly connected to joint A. A bent needle is attached to the needle joint to aid in collecting excess mercury. With the syringe in a vertical position with the tip upwards, the excess mercury is expelled through the tip leaving about 0.2 ml in the syringe. The syringe is left attached and the lever turned to position B. The sample in the sampling syringe is well mixed and the syringe attached to joint B, after expelling any air bubbles in the syringe. Some blood is now expelled through the needle and the lever turned to position C so that about 1 ml of blood can be transferred to syringe A by pressure on the plunger of syringe B. When 1 ml of blood has been transferred to A, the lever is moved back to sampling and syringe B removed. Attach a 2 ml syringe completely filled with the hemolyzing (Triton X-100 or Sterox) solution and from which any air bubbles have been expelled. Some of this solution is expelled through the needle; then the lever is moved to position C and some of the solution added to the blood in syringe A (about 0.2 ml of Triton X-100 or 0.8 ml of Sterox). The lever is now moved back to position B and the syringe containing the hemolyzing solution removed. The blood in syringe A is mixed and hemolyzed by inversion a number of times. The needle is replaced and the stopcock lever is moved to position A, some of the blood is expelled, and a part of the remainder is used to fill the cuvette. Since the syringe also contains some mercury, care must be taken not to get any mercury into the cuvette. The best way is to use a very short needle and attach to this a 10 cm length of fine polyethylene tubing.|| In this way the syringe can be held nearly vertical with the mercury at the bottom and the tubing inserted to the bottom of the cuvette. The tubing should have an outside diameter of less than 1 mm so that it will enter the cuvette easily. It is important to insert the tubing to the bottom of the cuvette to prevent trapping air bubbles. The absorbance of the blood in the cuvette is now measured at 650 and 805 nm against a blank of the Triton X-100 solution. The ratio of the absorbance at 650 nm to that at 805 nm is calculated and the percent saturation read from a calibration curve.

CALIBRATION

To obtain the point corresponding to 100% saturation, blood from a normal individual is equilibrated with pure oxygen. Almost any kind of gentle mechanical rotator can be used. About 4 ml of blood is added to a test tube having a capacity of about 15 ml. The air space above the blood is flushed out with oxygen and the tube is tightly stoppered and gently rotated. About every 10 min the rotation is stopped, and the tube flushed again with oxygen. Three or four changes of oxygen

---

|| Intramedic polyethylene tubing, No. PE50, Clay Adams Co., Parsippany, N.J. 07054.

should be sufficient. This blood is then carried through the procedure as given above. To obtain the point corresponding to zero percent saturation, any non-oxygenated blood sample is carried through the procedure except that instead of the simple Triton X-100 solution, the reducing solution containing the sodium dithionite is used. Since one of the points used is an isobestic point, the curve should be linear between the two calibration points of zero and 100%. Using the method as given, the ratio for 100% saturation should be close to $A_{650}/A_{805} = 0.50$, and for zero saturation (all reduced hemoglobin) the ratio should be close to $A_{650}/A_{805} = 4.0$. Slightly different values may be obtained with different spectrophotometers, but duplicate determinations should agree fairly well.

The method using the 0.1 mm cuvettes is essentially the same. One uses the solutions containing Sterox SE rather than Triton X-100, but otherwise the procedure is identical. Difficulty may be encountered in filling the demountable cuvettes. Those with an open end are preferred. If the two sections are placed so that the space to be occupied by the blood is open at both ends, the blood should enter almost by capillary action similar to the filling of a hemocytometer chamber. For these cuvettes, the readings are made at 560, 522, and 506 nm against a blank of the hemolyzing solution. The two ratios, A_{560}/A_{522} and A_{560}/A_{506}, are calculated and the percent saturation obtained from a calibration chart. The values for zero and 100% oxygen saturation are obtained exactly as given above. For these wavelengths the ratio A_{560}/A_{522} should be close to 1.9 for zero percent and 1.3 for 100%. For the ratio A_{560}/A_{506} the values should be about 2.55 for zero and 1.75 for 100% saturation.

Note that the ratios using the 1 mm cuvettes are numerically farther apart than those with the 0.1 mm cuvettes so that the former method on this basis should be more precise. The spectrophotometric methods give good results on low and moderate levels of saturation, but exact values near 100% saturation will depend so much on the 100% calibration point that one may at times obtain figures of over 100% saturation. For high saturation levels, the Po_2 electrode yields better results.

The normal level of oxygen saturation in arterial blood is around 95% (94 to 97%); for venous blood the range is much larger, usually 60 to 70% saturation. The arterial saturation will be low in any condition that interferes with the gaseous exchange in the lungs.

Osmolality

The measurement of the osmolality of serum and urine [131, 132] has been used as an additional test in the study of electrolyte and water balance. The electrolytes present contribute the most to the osmolality, and the actual osmolality of body fluids influences the passage of water and electrolytes across cell membranes. The physical and instrumental bases for the determination of osmolality have been given in Chapter 2. We will discuss briefly here some of the clinical applications. The osmolality is measured in either serum or urine. Spinal, pleural, or peritoneal fluids have also been used. The osmolality of these ordinarily does not differ greatly from that of serum. The serum is collected by venipuncture with a minimum of stasis. Use perfectly dry needles and syringes and separate the serum as soon as possible after clotting. Small amounts of hemolysis do not interfere but it should be avoided as far as possible. After centrifugation, the serum is carefully removed. It is preferable to recentrifuge the serum to remove any particulate matter that might interfere by causing premature freezing during the determination. If the determination cannot be done at once, stopper the serum container tightly and store in the refrigerator or freeze. If the sample is frozen, complete mixing must be done after thawing. The results are generally somewhat lower for samples that have stood for considerable lengths of time. Plasma anticoagulated with a minimum amount of heparin can be used, but inorganic anticoagulants should not be. The urine should be collected in a clean dry bottle without preservative. Cap the collection bottle tightly to prevent any evaporation of water. If a 24-hr specimen is desired, store in the refrigerator; otherwise run the determination as soon as possible after collection. Clear centrifuged specimens are used. If the sample has been refrigerated, warm to room temperature and mix before centrifugation.

The osmolality of serum in normal individuals ranges from 280 to 300 mosmols/kg. The urine osmolality varies over a considerable range, depending in part on the water and salt intake. Values of 350 to 1,100 mosmol/kg have been found in random specimens with a 24-hr excretion of 750 mosmol/kg for males and 450 to 1,150 mosmol/kg for females. During maximal urine concentration the range is 850 to 1,350 mosmol/kg. The osmolality of urine in bed patients has been given as about 20 percent lower than the values given above; part of this may be due to differences in dietary intake. The ratio of the osmolality of the serum and urine is also important. A high urine and low serum osmolality or vice versa indicate a disturbed electrolyte balance.

Various formulas have been devised for calculating the osmolality of serum from the concentration of some constituents. One such formula is:

$$\text{mosmol/kg} = 1.86\,\text{Na} + (G/18) + (N/2.8) + 5$$

where Na is the sodium content in millimoles per liter, G is the glucose concentration in milligrams per deciliter, and N is the BUN in milligrams per deciliter.

These calculated osmolalities are generally 5 to 10 mosmol/kg lower than the measured value. A larger difference usually indicates electrolyte or water imbalance. Several investigators [133, 134] have found that in a variety of conditions the persistence of a measured osmolality more than 30 mosmol/kg greater than the value calculated from the above or similar formula to be a grave prognostic sign. The serum osmolality is a measure of the amount of osmotically active substances dissolved in the serum water. Ordinarily the water content of serum is fairly constant, but the amount of water per milliliter of serum may be decreased in extreme hyperlipemia due to the volume occupied by the chylomicrons which are osmotically relatively inactive. Thus in such hyperlipemic serums the concentration of electrolytes would appear to be diminished. This may be illustrated by a numerical example: A serum having a sodium level of 135 mmol/liter of serum would have an equivalent concentration of 147 mmol/liter of serum water (serum containing 92% water). But if the water content of the serum were reduced to 85% by the presence of hyperlipemia, a level of 125 mmol/liter of sodium in the lipemic serum would also represent a sodium concentration of 147 mmol/liter of serum water. Thus the apparent low sodium level is actually due to the presence of the large amounts of lipids. There is no simple way to correct for this but it should be kept in mind in interpreting the electrolyte values in such hyperlipemic serums.

The ratio of urinary to serum osmolality has been used in the study of kidney function by calculation of an osmolal clearance. This will be treated in the section on kidney function (Chapter 11).

References

1. Albanese, A. A., and Lein, M. D.   *J. Lab. Clin. Med.* 33:246, 1948.
2. Maruna, R. F. L.   *Clin. Chim. Acta* 2:581, 1957.
3. Katchman, B. J., and Zipf, R. E.   In F. W. Sunderman and F. W. Sunderman, Jr. [Eds.], *Clinical Pathology of the Serum Electrolytes.* Springfield, Ill.: Thomas, 1966. P. 16.
4. Salit, P. W.   *J. Biol. Chem.* 96:659, 1932.
5. Butterworth, C. E.   *J. Clin. Pathol.* 4:99, 1951.
6. Weinbach, A. P.   *J. Biol. Chem.* 110:95, 1935.
7. Clark, W. G., Levitan, N. I., Gleason, D. F., and Greenberg, G.   *J. Biol. Chem.* 145:85, 1942.
8. Dugandzic, M.   *Clin. Chim. Acta* 4:819, 1959.
9. Lochhead, H. B., and Purcell, M. K.   *Am. J. Clin. Pathol.* 21:877, 1951.
10. Gadsden, R. H., and Cannon, A.   In F. W. Sunderman and F. W. Sunderman, Jr. [Eds.], *Clinical Pathology of the Serum Electrolytes.* Springfield, Ill.: Thomas, 1966. P. 27.
11. Hoffman, W. S.   *J. Biol. Chem.* 120:57, 1937.

12. Simon, K. H.   *Med. Monatsschr.* 20:571, 1966.
13. Flaschka, H., Holasek, A., and Mosenthal, M.   *Z. Physiol. Chem.* 308:183, 1957.
14. Holasek, A., and Pecar, M.   *Clin. Chim. Acta* 6:125, 1961.
15. Shohl, A. T., and Bennett, H. B.   *J. Biol. Chem.* 78:643, 1928.
16. Teeri, A. E., and Sesin, P. G.   *Am. J. Clin. Pathol.* 29:86, 1958.
17. Hillman, G., and Beyer, G.   *Z. Klin. Chem.* 5:93, 1967.
18. Margoshes, M., and Vallee, B. L.   *Methods Biochem. Anal.* 3:353, 1956.
19. Telph, H. A.   In F. W. Sunderman and F. W. Sunderman, Jr. [Eds.], *Clinical Pathology of the Serum Electrolytes.* Springfield, Ill.: Thomas, 1966. P. 189.
20. Hald, P. M., and Mason, W. B.   *Stand. Methods Clin. Chem.* 2:165, 1958.
21. Zak, B., Mosher, R. E., and Boyle, A. J.   *Am. J. Clin. Pathol.* 23:60, 1953.
22. Kingsley, G. R., and Schaffert, R. R.   *J. Biol. Chem.* 206:807, 1954.
23. Dryer, R. L.   *Clin. Chem.* 2:112, 1956.
24. Fridman, S. M., Wong, S. L., and Walton, J. H.   *J. Appl. Physiol.* 18:950, 1963.
25. Portnoy, H. D., and Gurdjian, E. S.   *Clin. Chim. Acta* 12:429, 1964.
26. Jacobson, H.   *Anal. Chem.* 38:1951, 1966.
27. Annino, J. S.   *Clin. Chem.* 13:227, 1967.
28. Wise, W. M., Kurey, M. J., and Baum, G.   *Clin. Chem.* 16:103, 1970.
29. Pioda, L. A. R., Somon, W., Bosshard, H. R., and Curtius, H. C.   *Clin. Chim. Acta* 29:289, 1970.
30. Dahms, H.   *Clin. Chem.* 13:437, 1967.
31. Astrup, P.   *Scand. J. Clin. Lab. Invest.* 8:32, 1956.
32. Astrup, P.   *Clin. Chem.* 7:1, 1961.
33. Sunderman, F. W.   *Am. J. Clin. Pathol.* 46:679, 1966.
34. Sunderman, F. W.   In F. W. Sunderman and F. W. Sunderman, Jr. [Eds.], *Clinical Pathology of the Serum Electrolytes.* Springfield, Ill.: Thomas, 1966. P. 166.
35. Stevenson, G. F..   In F. W. Sunderman and F. W. Sunderman, Jr. [Eds.], *Clinical Pathology of the Serum Electrolytes.* Springfield, Ill.: Thomas, 1966. P. 172.
36. Cotlove, E.   *Methods Biochem. Anal.* 12:277, 1964.
37. de la Huerga, J., Smetters, G. W., aand Sherrick, J. C.   In F. W. Sunderman and F. W. Sunderman, Jr. [Eds.], *Clinical Pathology of the Serum Electrolytes.* Springfield, Ill.: Thomas, 1966. P. 74.
38. Whitehorns, J. C.   *J. Biol. Chem.* 45:449, 1921.
39. Schales, O., and Schales, S. S.   *J. Biol. Chem.* 140:879, 1941.
40. Bartels, H.   *Atomic Absorption Newsletter* 6:132, 1967.
41. Bauman, M. L., and Hatmaker, A. L.   *Clin. Chem.* 7:356, 1961.
42. Seligson, D., McCormack, G. J., and Sleeman, S.   *Clin. Chem.* 4:159, 1958.
43. Kirman, D., Morgenstern, S. W., and Feldman, D.   *Am. J. Clin. Pathol.* 30:564, 1958.
44. Cotlove, E., Trantham, H. V., and Bowman, R. L.   *J. Lab. Clin. Med.* 51:461, 1958.
45. Cotlove, E., and Nishi, H. H.   *Clin. Chem.* 7:285, 1961.
46. Cotlove, E.   *Stand. Methods Clin. Chem.* 3:81, 1961.
47. Schales, O.   *Stand. Methods Clin. Chem.* 1:37, 1953.
48. Rice, E. W.   *Am. J. Clin. Pathol.* 28:654, 1957.
49. Henry, R. J.   *Clinical Chemistry, Principles and Techniques.* New York: Harper & Row, 1964. P. 406.
50. Fingerhut, B., and Marsh, W. H.   *Clin. Chem.* 9:204, 1963.

51. Sendroy, J., Jr. *J. Biol. Chem.* 120:405, 1937.
52. Sendroy, J., Jr. *J. Biol. Chem.* 120:335, 1937.
53. Kingsley, G. R., and Dowdell, L. A. *J. Lab. Clin. Med.* 35:637, 1950.
54. Rodkey, F. L., and Sendroy, J., Jr. *Clin. Chem.* 9:668, 1963.
55. Itano, M., Williams, L. A., and Zak, B. *Am. J. Clin. Pathol.* 32:213, 1959.
56. Baer, S. *Clin. Chim. Acta* 7:642, 1962.
57. Sobel, S., and Fernandez, A. *Proc. Soc. Exp. Biol. Med.* 113:187, 1963.
58. Schoeldfield, R. G., and Lewellen, C. J. *Clin. Chem.* 10:533, 1964.
59. Hamilton, R. H. *Clin. Chem.* 12:1, 1966.
60. Fingerhut, B. *Clin. Chim. Acta* 41:247, 1972.
61. Driscoll, J. L., and Martin, H. F. *Clin. Chem.* 12:314, 1966.
62. Durst, R. A. [Ed.]. *Ion-Selective Electrodes.* Washington, D.C.: National Bureau of Standards Special Publication 314, 1969.
63. Kopito, L., and Schwachman, H. *Pediatrics* 43:798, 1968.
64. Hansen, L., Bueschele, M., Koroshec, J., and Warwick, W. J. *Am. J. Clin. Pathol.* 49:834, 1968.
65. Straumfjord, J. V., Jr. *Stand. Methods Clin. Chem.* 2:107, 1958.
66. Peters, J. P., and Van Slyke, D. D. Quantitative Clinical Chemistry. Vol. 2, *Method.* Baltimore: Williams & Wilkins, 1932.
67. Turpin, F. H., and Roberts, B. L. *Am. J. Med. Technol.* 28:217, 1962.
68. Natelson, S., and Manning, C. M. *Clin. Chem.* 1:655, 1955.
69. Martinek, R. G. *Clin. Chem.* 10:153, 1964.
70. Peterson, J. I. *Clin. Chem.* 16:144, 1970.
71. Buggs, H., and Jackson, D. C. *Am. J. Med. Technol.* 37:355, 1971.
72. Yee, H. Y. *Anal. Chem.* 37:924, 1965.
73. Strever, B. C., Johnson, C. A., and Gladsden, R. H. *Clin. Chem.* 19:1075, 1973.
74. Kenny, M. A., and Cyen, M. H. *Clin. Chem.* 18:352, 1972.
75. Bonnie, L. *Am. J. Med. Technol.* 37:361, 1971.
76. Skeggs, L. T. *Am. J. Clin. Pathol.* 33:181, 1960.
77. Rispens, P., Brunsting, J. R., Zijilstra, W. G., and Van Kampen, E. J. *Clin. Chim. Acta* 22:261, 1968.
78. Eschenbach, C., and Rausch-Strooman, J. D. *Klin. Wocherschr.* 39:693, 1961.
79. Battistini, A., and Waring, W. W. *Am. Rev. Resp. Dis.* 100:237, 1969.
80. Gaudebout, C., Blayo, M. C., and Pocidalo, J. J. *Ann. N.Y. Acad. Sci.* 133:66, 1966.
81. Stow, R. W., Baer, R. F., and Randall, B. F. *Arch. Phys. Med. Rehabil.* 38:646, 1957.
82. Astrup, P., and Siggaard-Andersen, O. *Adv. Clin. Chem.* 6:1, 1963.
83. Astrup, P. *Scand. J. Clin. Lab. Invest.* 8:33, 1956.
84. Siggaard-Andersen, O. *Scand. J. Clin. Lab. Invest.* 14:568, 1962.
85. Gambino, S. R. *Stand. Methods Clin. Chem.* 5:169, 1965.
86. Gambino, S. R., and Schrieber, H. *Am. J. Clin. Pathol.* 45:406, 1966.
87. Went, J., and Whitehead, T. P. *Clin. Chim. Acta* 26:559, 1959.
88. Siggaard-Andersen, O. *Scand. J. Clin. Lab. Invest.* 13:196, 1961.
89. Rosenthal, T. B. *J. Biol. Chem.* 52:501, 1948.
90. Weisberg, H. F. *Water, Electrolyte and Acid-Base Balance,* 2nd ed. Baltimore: Williams & Wilkins, 1962.
91. Siggaard-Andersen, O. *Scand. J. Clin. Lab. Invest.* 15:151, 1963.

92. Sofford, J. M., Dowling, A. S., and Pell, S.   *J.A.M.A.* 224:1297, 1973.
93. Straumfjord, J. V., Jr.   *Stand. Methods Clin. Chem.* 2:107, 1958.
94. Fleischer, W. R.   *Stand. Methods Clin. Chem.* 2:175, 1958.
95. Austin, W. H., and Littlefield, S. C.   *J. Lab. Clin. Med.* 67:516, 1966.
96. Rosenthal, T. P.   *J. Biol. Chem.* 173:25, 1948.
97. Cullen, G. E.   *J. Biol. Chem.* 52:501, 1922.
98. Austin, W. H., Lacombe, E. H., and Rand, P. W.   *J. Appl. Physiol.* 19:893, 1964.
99. Sanz, M. C.   *Clin. Chem.* 3:406, 1957.
100. Gambino, S. R.   *Am. J. Clin. Pathol.* 32:294, 1959.
101. Shock, N. W., and Hastings, A. B.   *J. Biol. Chem.* 104:585, 1948.
102. Baldwin, E. de F., Blackwood, F. C., Palmer, L. E., and Sloman, K. G.   *Medicine* 27:243, 1948.
103. Albritton, E. C.   *Standard Values in Blood.* Philadelphia: Saunders, 1952. P. 120.
104. Graham, B. D., Wilson, J. L., Tsao, M. U., Bauman, M. L., and Brown, S.   *Pediatrics* 8:68, 1951.
105. Cassels, D., and Morse, M.   *J. Clin. Invest.* 32:824, 1953.
106. Scott, P. V., Horton, J. N., and Maoleson, W. W.   *Br. Med. J.* 3:512, 1971.
107. Fleisher, M., and Schwartz, M. K.   *Clin. Chem.* 17:610, 1971.
108. Lang, G. E., Mueller, R. G., and Hunt, P. K.   *Clin. Chem.* 19:559, 1973.
109. Gambino, S. R.   *Am. J. Clin. Pathol.* 32:285, 1959.
110. Laughlin, D. E., McDonald, J. S., and Bedell, G. N.   *J. Lab. Clin. Med.* 64:330, 1964.
111. Koch, G.   *Scand. J. Clin. Lab. Invest.* 21:10, 1968.
112. Peters, J. P., and Van Slyke, D. D.   Quantitative Clinical Chemistry. Vol. 2, *Methods*. Baltimore: Williams & Wilkins, 1932.
113. Natelson, S., and Manning, C. M.   *Clin. Chem.* 1:165, 1955.
114. Davis, D. D.   *Br. J. Anaesth.* 42:19, 1970.
115. Malmlund, H. O.   *Scand. J. Clin. Lab. Invest.* 28:471, 1971.
116. Chambless, K. W., and Nouse, D. C.   *Clin. Chem.* 8:654, 1962.
117. Galla, J. J., and Ortenstein, D. M.   *Ann. N.Y. Acad. Sci.* 102:4, 1962.
118. Van Assendelft, W. O.   *Spectrophotometry of Hemoglobin Derivatives.* Springfield, Ill.: Thomas, 1970.
119. Siggaard-Andersen, O., Joergensen, K., and Naerra, N.   *Scand. J. Clin. Lab. Invest.* 14:298, 1962.
120. de la Huerga, J., and Sherrick, J. C.   *Ann. Clin. Lab. Sci.* 1:261, 1971.
121. Gambino, S. R., Goldenberg, L. S., and Polanyi, M. L.   *Am. J. Clin. Pathol.* 42:364, 1964.
122. Clark, C. C., Wolf, R., Granger, D., and Taylor, Z.   *J. Appl. Physiol.* 6:189, 1953.
123. Polger, G., and Forster, R. E.   *J. Appl. Physiol.* 15:706, 1960.
124. Gambino, S. R.   *Stand. Methods Clin. Chem.* 6:171, 1970.
125. Gambino, S. R.   *Stand. Methods Clin. Chem.* 5:169, 1965.
126. Siggaard-Andersen, O., and Joergensen, K.   *Scand. J. Clin. Lab. Invest.* 13:149, 1961.
127. Astrup, P.   *Scand. J. Clin. Lab. Invest.* 19:203, 1966.
128. Johnstone, J. H.   *J. Clin. Pathol.* 19:357, 1966.
129. Johnstone, G. W.   *Stand. Methods Clin. Chem.* 4:183, 1963.
130. Johnstone, G. W., Holtkamp, F., and Eve, J. R.   *Clin. Chem.* 5:421, 1959.
131. Johnson, R. B., Jr., and Hoch, H.   *Stand. Methods Clin. Chem.* 5:159, 1965.

132. Weisberg, H. F.   *Osmolality*. #CC-71. Chicago: American Society of Clinical Pathologists, 1972.
133. Boyd, D. R., and Mansberger, A. R., Jr.   *Am. Surg.* 34:744, 1968.
134. Mansberger, A. R., Jr., Boyd, D. R., Cowley, R. A., and Buxton, R. W.   *Ann. Surg.* 169:672, 1969.

10. Liver Function Tests

The liver is intimately connected with a number of different metabolic processes in the body and many tests have been developed to measure its function. The liver is the main route in the excretion of the substances resulting from the breakdown of the heme portion of the hemoglobin molecule, and the determination of bilirubin—the main bile pigment—is widely used as a liver function test. The liver plays an important role in the synthesis of many serum proteins, particularly the globulins, and measurement of variations in the amount and proportion of the plasma proteins constitutes an important test of liver function. In relation to protein metabolism, the amounts of amino acids and ammonia in plasma are also related to liver function. The liver is also connected with the formation of some of the factors necessary in the coagulation of blood, including fibrinogen. The liver also plays an important part in carbohydrate metabolism and storage. Other functions of the liver are the detoxification of deleterious substances and the formation of glucuronate or other conjugates with many steroid hormones. The various functions of the liver involve a large number of different enzymes and the determination of a number of these in plasma or serum is also of value in assessing liver disease. This brief summary indicates that a large number of different types of tests have been used for study of liver function. Those mentioned here will be grouped according to the particular function of the liver being studied—a classification that cannot be exclusive and may in some cases be somewhat arbitrary. For many of the tests, such as protein electrophoresis and determination of specific carbohydrates or enzymes, the procedures have already been given in earlier sections of this book and their application as liver function tests will be noted in this chapter.

Tests Based on Bile Pigments

Bilirubin

Bilirubin is derived from the breakdown of hemoglobin from aged erythrocytes. It is then conjugated in the liver and excreted in the bile. The measurement of bilirubin in the serum may be used to detect either increased breakdown of hemoglobin (erythrocytes) or decreased excretion by the liver. The use of bilirubin determinations as an index of hemolysis is confined mainly to the detection of erythroblastosis and the chief use of bilirubin determinations is as a measure of liver function. Bilirubin is a pigment wth a marked yellow color and its increase in the serum to the point at which a yellow color is noted in the skin is known as *jaundice*. As will be discussed more fully in the interpretation of results, it may be noted here that the failure of the liver to excrete bilirubin properly may be due either to the inability of the liver cells to excrete the compound into the biliary tract or to an obstruction somewhere in the tract which prevents flow of bile into the intestines.

It has long been known that the bilirubin in serum exists in two forms. One, relatively water soluble, reacts readily in aqueous solution with a reagent such as diazotized sulfanilic acid. The other form reacts very slowly unless a solubilizing agent is added. The readily reacting form was formerly called the direct reacting bilirubin and the other the indirect. Conventionally the direct bilirubin was determined by reaction without the solubilizing agent, and the total bilirubin by reaction with the presence of the solubilizing agent. The difference was labeled the indirect bilirubin. At one time, other forms such as the "one minute" or "five minute" bilirubin, the amount determined by reaction for the specified length of time, were reported, but these are now rarely reported.

It is now generally accepted that the more soluble, direct reacting bilirubin is conjugated with glucuronic acid, most of it in the form of a diglucuronide and some as a monoglucuronide. The proportion of the two may be changed by abnormal conditions [1]. The conjugation renders the molecules more soluble. The indirect, less soluble bilirubin, is unconjugated and usually exists in the serum combined with albumin. As will be seen later, the ratio of conjugated to unconjugated bilirubin in the serum may vary in different types of liver disease.

Since bilirubin has an absorbance maximum at about 450 nm (the exact wavelength depends upon the solvent), one method of determination is to measure its absorbance directly in the serum or other body fluid or after extraction with a solvent. The direct method has been used for the determination of total bilirubin in infants, particularly those with suspected erythroblastosis [2–5]. Hemoglobin interferes with the direct methods and a correction is made by reading at several different wavelengths. The direct method is not so well adapted to determination in adults since the serum of the latter contains other pigments such as carotene which also absorbs near 450 nm. The interference by these pigments is said to be equal usually to not more than 0.5 mg/dl of bilirubin. The bilirubin in amniotic fluid has also been estimated by direct means. The absorbance of the fluid is measured at wavelengths from about 350 to 600 nm. After plotting the results, the difference in absorbance at 450 nm between the sample curve and a normal curve is estimated. This is usually reported as Δ 450, and is a measure of the amount of bilirubin present [6–10]. The extraction methods for direct determination may be adjusted so as to separate the free and conjugated bilirubin so that both can be estimated [11–14].

Most laboratories use some modification of the diazo reaction for bilirubin determinations. The most common reagent is diazotized sulfanilic acid which reacts with bilirubin to produce a compound that is red-violet in acid solution and blue in alkaline solution. In the older method of Malloy and Evelyn [15–17] the reaction was carried out in acid solution with methyl alcohol added to a final concentration of about 50% as a solubilizing agent for total bilirubin. The product formed is actually an indicator and the color produced will vary with the pH. As the solution is not very well buffered, small variations in the procedure or amount of protein present can result in changes in color and thus an error in the determination. The high concentrations of alcohol may tend to precipitate some of the proteins. Although a blank is usually run, this often does not exactly compensate for the turbidity.

Other solubilizing agents have also been used, such as solutions of urea and sodium benzoate [18], sodium acetate and antipyrine [19], acetamide [20, 21], diphylline [22, 23], and caffeine plus sodium benzoate buffered with sodium acetate [24–26]. These solubilizers are often used in methods in which the final color is read in an alkaline solution such as the method of Jendrassik and Grof [24]. The addition of the alkali changes the color from red to blue. The pH of the solution is now highly buffered so that the color of the solution is more constant. Also the alkali will dissolve any protein that tended to precipitate out, so that the turbidity of the final solution is greatly reduced. In this method the interference of small amounts of hemoglobin can be eliminated or at least greatly reduced by the addition of ascorbic acid just before the final alkalinization. This is not possible with the Malloy and Evelyn method.

Although diazotized sulfanilic acid is the most common color reagent used, a number of other diazotized amines have also been tried. These include 2-chloro-4-nitroaniline [27], 2,4-dichloroaniline [28], 5-nitroanisidine [29], and Fast Red RC [30]. These are said to be more stable or to give a greater amount of color but have not been widely adopted.

In a different type of procedure the bilirubin is oxidized to biliverdin or other compounds by means of oxidizing agents. Ferro and Ham [31, 32] used ferric chloride, a relatively mild oxidizer, to produce a blue color. Zaroda [33] used mercuric nitrate and nitric acid to produce a stable green color. These methods are not very specific and are somewhat lacking in sensitivity. A micromethod for elevated total bilirubin levels in infants has been suggested which is rapid and simple [34]. The serum is diluted with saline and read at 460 nm. The bilirubin color is then destroyed by the addition of a small amount of sodium hypochlorite solution (commercial bleach) and a second reading taken. The absorbance due to any hemoglobin present is not changed significantly by the hypochlorite, so that the difference in absorbance is proportional to the amount of bilirubin present.

Total bilirubin has also been determined fluorometrically [35]. The method is very simple; the only reagent used is 85% phosphoric acid. It is said to measure the degradation products produced by light, so that protection of the sample from light is not as necessary as in most bilirubin methods.

For general laboratory use our choice is some modification of the method of Jendrassik and Grof [24]. Although requiring more reagents, it has several advantages over the popular method of Malloy and Evelyn [15]. The former method is less influenced by the presence of hemoglobin and by variations in the amount of protein present. Since the reaction is carried out in a strongly buffered solution, the color is more constant and the turbidity is greatly reduced. A discussion of the advantages of the Jendrassik and Grof method has been given by Gambino and Di Re [36].

BILIRUBIN STANDARDIZATION

One problem in the determination of bilirubin is the availability of satisfactory standards. For many purposes the use of lyophilized control serums is adequate if these have been standardized against pure bilirubin. The values furnished by the manufacturer are fairly reliable but they must be checked occasionally. In the past it has often been difficult to obtain a pure bilirubin standard or to know if one that was obtained was pure. A joint committee [37] from a number of interested organizations has agreed upon some specifications. A bilirubin standard should have a molar absorptivity of 60,700 ± 800 in chloroform at 25°C. (This means that the calculated absorbance for a solution containing 584.6 g of bilirubin in 1 liter of solution in chloroform with a 1 cm light path would be the given figure.) The committee recommended that any bilirubin having an absorptivity between 59,100 and 62,300 would be acceptable. Currently a number of manufacturers offer a bilirubin standard stated to meet this requirement. The determination of the molar absorptivity requires a good spectrometer with an accurate wavelength calibration and cuvettes of accurately known light path. The determination is not very simple; it is given later in this section.

A very pure sample of bilirubin may be obtained from the National Bureau of Standards.* The lot is stated to contain 99.0% bilirubin (with 0.8% chloroform, the solvent from which the material was recrystallized). The calculated absorptivity for this sample is 61,100. This material is much too expensive for even routine checking and can be used only for checking a secondary standard. This latter would be a good grade of commercial bilirubin which after checking should be divided into a number of small portions and kept in tightly sealed vials under nitrogen in a refrigerator.

A chloroform solution of bilirubin has been used as a standard with the Malloy

* National Bureau of Standards, Washington, D.C. 20025; Sample no. 916.

and Evelyn method. The addition of the methyl alcohol makes the chloroform solution miscible with the aqueous reagents. It has been claimed that the absorption peak for this solution is not exactly the same as that obtained when the source of bilirubin is an aqueous dilution of serum. If the determinations are made with a filter photometer or a wide-band simple spectrophotometer, the apparent differences may not be great, but with a narrow-band instrument the error may be considerable. The preferred method for standardization which can be used with any method is to add small quantities of a concentrated aqueous bilirubin solution to a fresh serum of low bilirubin content.

The procedure is given below. Since bilirubin solutions, particularly those with a high pH, are sensitive to light, the entire procedure from the time of beginning to dissolve the bilirubin to the time of the final spectrophotometric measurements must be carried out under very subdued light, preferably in a darkroom with only the usual darkroom safelight. The concentration of bilirubin in the standard solution and the amount added to the serum can be varied somewhat to suit the particular method used. The amounts given will illustrate the procedure.

PROCEDURE

Obtain a pool of about 10 ml of fresh, nonhemolyzed serum of low bilirubin content (less than 1.0 mg/dl). The serum should preferably be not more than a few hours old and not have been exposed to heat or excess light. If it is not perfectly clear, it should be centrifuged. Weigh out, say, exactly 40 mg of pure bilirubin and dissolve in exactly 40 ml of sodium carbonate solution ($0.1M$, 10.6 g anhydrous salt/liter). As mentioned above, all these steps should be carried out in the darkroom. The pure bilirubin should be readily soluble in the carbonate solution. A sodium hydroxide solution, $0.05M$, has also been used but the bilirubin is less stable in this solution. Clarke [38] found that in the carbonate solution the decrease in azobilirubin color formed from the solution on standing was about 2%/hr for the carbonate solution and about 10%/hr for the NaOH solution even when they were kept in the dark. Thus the solution should be used as soon as possible. A solvent of formamide containing 0.65 g of potassium cyanide/dl can also be used. The bilirubin is said to be stable in this solution for up to 4 hr, but it is less convenient to use.

One should have ready sets of four tubes, the tubes of each set containing a definite amount of the clear serum plus diluent as required by the method. To one pair of the set is added a definite amount of the bilirubin solution, such as 0.02 ml, from an accurate micropipet. One set of each pair is treated as a blank and the other as a sample and processed by the regular method being standardized. Thus the tubes might be set up as follows:

Tube no.	Contents	Treatment
1	serum	blank
2	serum	sample
3	serum + bilirubin	blank
4	serum + bilirubin	sample

The absorbances of the tubes are then measured after color development against a suitable reagent blank. Then if A_1 to A_4 are absorbances of the solutions as given above,

$$(A_4 - A_3) - (A_2 - A_1) = \text{net absorbance due to added bilirubin}$$

One can then calculate the corresponding concentration, if all the added bilirubin was in the sample. Suppose, for example, the tubes each contained 0.2 ml of serum and the bilirubin standard contained exactly 1 mg/ml so that the added 0.02 ml would be equal to 0.02 mg and the corresponding point on the calibration curve would be:

$$\frac{0.2}{0.02} \times 100 = 10 \text{ mg/dl bilirubin when 0.2 ml of serum is analyzed as directed}$$

As a check, if the measurements can be made in cuvettes of exactly known light path, the molar absorptivity may be calculated. If the Jendrassik and Grof method is used as given here, the calculated absorptivity is given by the formula:

$$\frac{584 \times A \times V}{B}$$

where A is the measured absorbance in a 1 cm light path, V is the final volume of the solution (5.1 ml in the method as given), and B is the actual amount of bilirubin added (0.02 mg in the above example). The absorptivity is calculated on the basis of a molecular weight of bilirubin of 584. If the net absorbance as given above were 0.470, the absorptivity would calculate to 70,000; the expected value for the above method is 73,000 [36]. Even when cuvettes with accurate light paths are used, the absorbance of the purified bilirubin should be carefully recorded for future comparison. A value of 0.445 has been given for round Coleman 12 mm cuvettes, compared with 0.490 for square cuvettes [36]. Using similar cuvettes one should obtain values near these figures. For a calibration curve one must also use lower concentrations of bilirubin which can be obtained by carefully diluting the stronger bilirubin solution with the sodium carbonate solution. Note again that the solution is not stable for extended periods of time so that all the determinations must be made rather rapidly and the serum and diluents should all be ready before the bilirubin standard is made up.

EFFECT OF LIGHT

Bilirubin in serum samples is slowly oxidized in the light. The degree of decomposition varies with the sample and with the intensity of the light. In some experiments reported by Gambino and Di Re [36], serum exposed to indirect daylight (samples placed near a window on a clear day with no direct sunlight entering the window) gave an azobilirubin color of about 90 percent of the original after 1 hr exposure to the light. When the serum was placed in direct sunlight, the color developed dropped to only 30 percent after ½ hr exposure. The effect of fluorescent lighting was variable. With ordinary lighting, very little effect could be noted in 1 hr. When the samples were close to fluorescent lights, some decrease was noted. The glucuronides are much more sensitive to light than the free bilirubin [40]. The proportions of the two forms in the light experiments were not mentioned. Most commercial control serums contain chiefly the unconjugated bilirubin. All samples for bilirubin determination should be protected from undue exposure to light and should never be placed in direct sunlight.

The oxidation methods and the fluorometric method are said to be less sensitive to the effect of light as they measure the oxidation products of bilirubin as the unchanged bilirubin; the diazo methods react only with the unchanged bilirubin.

BILIRUBIN IN AMNIOTIC FLUID

As mentioned in the introductory section on bilirubin, the determination of bilirubin pigments in amniotic fluid is used in the diagnosis of erythroblastosis. One method measures the absorbance of the fluid over a range of from 350 nm to about 550 nm. The logarithm of the absorbance is then plotted against the wavelength and a smooth curve drawn. A baseline curve is drawn and the excess absorbance at 450 nm measured. This method has been used by a number of investigators [6–8, 41–43]. This is a rather cumbersome method unless one has a recording spectrophotometer in which the logarithm of the absorbance is plotted directly. Plotting the absorbance itself against the wavelength can lead to errors in determining the true baseline [9]. Appreciable amounts of hemoglobin will interfere by causing increased absorbance near 410 nm. In some instances the presence of other pigments may distort the curve so as to give a negative value for the 450 nm difference [44]. The preferred method is to determine the total bilirubin using the Jendrassik and Grof method [24, 36] with an undiluted sample.

FASTING HYPERBILIRUBINEMIA

A cause of increased bilirubin levels in serum—and one that has been somewhat neglected—is that of several days' fasting. Barrett [45] found that in a number of normal and abnormal subjects a fast of about 48 hr increased the serum bilirubin level by about 240 percent in normal individuals and only slightly less in abnormal subjects. Although this phenomenon had been observed earlier [46, 47], it apparently was not widely known. The cause of the change is apparently due to a decrease in the ability of the liver to remove unconjugated bilirubin from the plasma [48].

REAGENTS AND STANDARDIZATION

We present a regular macromethod, a micromethod, and one for amniotic fluid, all using the same reagents, and a simple direct method for the determination of bilirubin in newborn infants.

REAGENTS

1. Caffeine mixture. Dissolve 50 g of pure caffeine alkaloid, 75 g of sodium benzoate, 125 g of sodium acetate trihydrate ($NaC_2H_3O_2 \cdot 3H_2O$), and 1 g of the disodium salt of ethylenediaminetetraacetic acid in water with stirring. When the substances are dissolved, dilute to 1 liter with water. (The anhydrous sodium acetate [82 g] could be used instead of the hydrated crystals, but the latter usually gives a clearer solution.) If the solution is turbid, filter through filter paper. This solution is stable for at least 6 months at room temperature.

2. Diazo solution I. Dissolve 5.0 g of sulfanilic acid and 15 ml of concentrated hydrochloric acid in water to make 1 liter.

3. Diazo solution II. Dissolve 500 mg of sodium nitrite in 100 ml of water. This solution is stable for at least 1 week in the refrigerator. For convenience, 500-mg portions of the salt may be weighed out beforehand and kept in stoppered vials for later use.

4. Diazo reagent. Mix 20 ml of diazo solution I and 0.5 ml of diazo solution II. Although this solution is stated to be stable for several hours, it is preferable to make it up just before use, as the preparation is so simple.

5. Alkaline tartrate. Dissolve 75 g of sodium hydroxide and 265 g of sodium tartrate ($Na_2C_4H_4O_6 \cdot 2H_2O$) in water and dilute to 1 liter. (Sodium tartrate is used instead of the original Rochelle salt, as the latter sometimes causes turbidity.) This solution should be stored in an alkali-resistant bottle.

6. Ascorbic acid, 4 g/dl. Dissolve 200 mg of ascorbic acid in 5 ml of water. This solution should be made up just before use. Small sterile vials of ascorbic

acid are convenient; the solution may be easily diluted to the required concentration.

7. Hydrochloric acid, $0.05M$. Dilute 4.5 ml of concentrated hydrochloric acid to 1 liter with water.

STANDARDS

The problem of standardization has been discussed earlier in this section. Usually one uses a calibration curve prepared as given, or a commercial control serum, as standard.

BILIRUBIN BY MACROMETHOD

PROCEDURE FOR TOTAL BILIRUBIN

In a series of tubes place 2.1 ml of the caffeine mixture. For each sample and serum standard, if run, set up two tubes labeled test and blank. To each pair of tubes add 1.0 ml of serum or diluted serum and mix. (For serums of low bilirubin content, one can use a 1:2 dilution of the serum with saline; for higher bilirubin levels, as judged by the color of the serum, a 1:4 or 1:6 dilution may be used.) Add to the test 0.5 ml of diazo reagent, and to the blank add 0.5 ml of diazo solution I (sulfanilic acid). Mix and allow to stand 10 min. Add 1.5 ml of alkaline tartrate to each tube, mix, allow to stand 10 min, and read sample against appropriate blank at 600 nm. If a number of clear nonhemolyzed serums of relatively low bilirubin content are run together at the same dilution, one blank tube will usually suffice for the group. If the serum contains visible hemolysis, modify the procedure as follows. Use 2.0 ml of the caffeine mixture instead of 2.1 ml. Add the diazo reagent or diazo I as above. After 10 min add 0.1 ml of ascorbic acid to each tube and immediately after mixing add 1.5 ml of the alkaline tartrate. It is important that the alkaline tartrate be added immediately after the ascorbic acid.

CALCULATION

If comparison is made with a standard serum run at the same time,

$$\frac{\text{absorbance of sample}}{\text{absorbance of standard}} \times \text{conc. of standard} = \text{conc. of sample}$$

provided the standard serum and sample are both diluted the same. Otherwise a factor must be added. If S is the actual amount of serum used (e.g., 1 ml of a 1:4 dilution would represent 0.25 ml of serum used) and T is the amount of standard serum used, the left side of the equation would be multiplied by the factor T/S. The same principle would apply if a calibration curve is used. Here it would be best to make the units on the curve in terms of the absorbance for an undiluted serum, and the results read off the curve would be multiplied by the dilution factor.

PROCEDURE FOR DIRECT BILIRUBIN

The procedure is carried out exactly the same as for total bilirubin except that $0.05M$ hydrochloric acid is substituted for the caffeine mixture.

BILIRUBIN BY MICROMETHOD

PROCEDURE FOR TOTAL BILIRUBIN

The same reagents are used as in the macromethod with the same order of addition; the only difference is in the quantities used: caffeine mixture, 1.00 ml; sample, 50 μl plus 200 μl of water, or for low bilirubin values, 100 or 250 μl with water if necessary to total 250 μl. Add 250 μl of diazo reagent or diazo I and mix. Let stand 10 min, add 0.50 ml of alkaline tartrate, and mix. Read at 600 nm after 10 min. This will require microcuvettes.

PROCEDURE FOR DIRECT BILIRUBIN

This is carried out the same way as total bilirubin, except that $0.05M$ HCl is used in place of the caffeine mixture.

TOTAL BILIRUBIN IN AMNIOTIC FLUID

PROCEDURE

This is essentially the same procedure described in the macromethod above. To 2 ml of caffeine mixture add 1 ml of amniotic fluid (centrifuged), setting up test and blank tubes. Also set up a standard. Since most bilirubin levels in the fluid will be below 1 mg/dl, 1 ml of a serum standard diluted to near this concentration with saline may be used. As above, add 0.5 ml of diazo reagent to the test tubes and 0.5 ml of diazo I to the blank tubes. Mix and after 10 min add 1.5 ml of alkaline tartrate to each tube. Mix, let stand 10 min, and read at 600 nm against respective blanks.

CALCULATION

Since the comparison is made with a suitable standard,

$$\frac{\text{absorbance of sample}}{\text{absorbance of standard}} \times \text{conc. of standard} = \text{conc. of sample}$$

MICROBILIRUBIN IN INFANTS

This rapid procedure [49] is satisfactory for the determination of total bilirubin in neonatal jaundice. It is based on the fact that a dilute solution of hypochlorite will destroy the absorbance of bilirubin at 460 nm but not immediately affect other substances.

REAGENTS

1. Sodium chloride, $0.15M$. Dissolve 8.8 g of sodium chloride in water and dilute to 1 liter.
2. Sodium hypochlorite solution. Any commercial bleach containing 5.25% available chlorine is satisfactory.
3. Standard. The procedure is best standardized against the Versatol† Pediatric Control serum which contains about 20 mg/dl of bilirubin.

PROCEDURE

To 1.5 ml of the saline solution, add 50 μl of serum. Mix and read at once in colorimeter at 460 nm. Add 20 μl of the hypochlorite solution, mix, and read at once at 460 nm.

CALCULATION

The difference in absorbance will be the measure of the bilirubin present, when compared with a standard similarly treated.

$$\frac{\text{absorbance difference for sample}}{\text{absorbance difference for standard}} \times \text{conc. of standard} = \text{conc. of sample}$$

NORMAL VALUES AND INTERPRETATION OF RESULTS

The normal values for total bilirubin are 0.3 to 1.3 mg/dl and for direct bilirubin 0.1 to 0.4 mg/dl. The difference between the two is sometimes reported as indirect bilirubin. In hemolytic jaundice the increase is greatest in the indirect fraction whereas in obstructive jaundice the increase is more significant in the direct fraction. The explanation for this is that in obstructive jaundice the direct reacting bilirubin, which is in the conjugated form in the liver, cannot be properly excreted and diffuses back into the serum. In hemolytic jaundice the bilirubin formed by the breakdown of the red cells reaches a high concentration in the serum before

† General Diagnostics Div., Warner Lambert, Morris Plains, N.J. 07950.

reaching the liver. The ratio of direct to indirect bilirubin is thus helpful in distinguishing between the two types of jaundice.

Hemolysis will tend to interfere with the determination of bilirubin. The use of ascorbic acid in the Jendrassik and Grof method reduces the interference of small amounts of hemolysis, but strongly hemolyzed serum may give erroneous results. The prior administration of dextran may lead to a turbidity which will give high results. Radiopaque media or other substances such as bromsulfophthalein (BSP) may give a color in the final alkaline solution, leading to erroneous results. As noted in the micromethod above, hypochlorite will readily oxidize bilirubin, and traces of this substance in glassware can lead to erroneous results.

A large number of drugs are reported to have hepatoxic effects, particularly after prolonged administration. This may lead not only to increased bilirubin levels but also to positive results with the other liver function tests given later.

At birth the bilirubin level in the cord blood of normal infants is around 2 mg/dl. The level in the blood rises to about 7 mg/dl at 2 days, then decreases toward the normal adult levels. In erythroblastosis the cord level may be over 5 mg/dl and the level in the blood rapidly rises to over 20 mg/dl. A high or rapidly increasing level indicates the need for an exchange transfusion.

Urobilinogen

The determinations of the urinary bilirubin and the urinary and fecal urobilinogen have also been used as tests for liver function. Bilirubin may be found in the urine in obstructive jaundice but not in hemolytic jaundice since the conjugated form is more readily excreted in the urine. Urinary bilirubin usually is tested only qualitatively, and we present semiquantitative methods for urobilinogen in urine and stool [50, 51]. Bilirubin excreted as bile into the intestines is reduced by bacterial action to urobilinogen. Part of this is reabsorbed and excreted in the urine. The amount excreted in urine and stool is a rough measure of the amount of bilirubin excreted into the intestines from the liver.

The urobilinogen found in stool and urine will give a red color on reaction with Ehrlich's reagent (dimethylaminobenzaldehyde). For a quantitative determination any urobilin present is reduced to urobilinogen by ferrous sulfate. The pigments are extracted from urine with petroleum ether to concentrate them. Pure urobilinogen is not available as a standard. The present methods report the amount of color produced in comparison with a standard amount of phenolsulfonphthalein in alkaline solution; 0.20 mg/dl of the dye is assumed to give the same color as 0.35 mg/dl of urobilinogen when treated with the reagent.

REAGENTS

1. Ehrlich's reagent (modified, Watson). Dissolve 0.7 g of p-dimethylamino-benzaldehyde in 150 ml of concentrated hydrochloric acid and add 100 ml of water.

2. Sodium hydroxide, 2.5M. Dissolve 100 g of sodium hydroxide in water and dilute to 1 liter.

3. Ferrous sulfate, approx. 0.7M. Dissolve 5 g of ferrous sulfate ($FeSO_4 \cdot 7H_2O$) in 23 ml of water. Prepare fresh as needed.

4. Sodium acetate, saturated solution. Add 100 g of sodium acetate to 60 ml of warm water. Warm further to dissolve, then cool. Some of the salt should crystallize out.

5. Glacial acetic acid, AR.

6. Petroleum ether, AR.

7. Urobilinogen standards.

A. Stock standard, 20 mg/dl. An artificial standard is used. Add 20 mg of phenolsulfonphthalein to a 100 ml volumetric flask. Add about 50 ml of water and 0.5 ml of the 2.5M sodium hydroxide solution. Swirl to dissolve the dye, then dilute to 100 ml with water and mix well. This solution is stable for several months in the refrigerator.

B. Working standard. As required, dilute 1 ml of the stock standard and 0.5 ml of the sodium hydroxide solution to 100 ml with water. This gives a solution containing 0.20 mg/dl of the dye, which is equivalent to 0.35 mg/dl of urobilinogen.

Urobilinogen in Urine

PROCEDURE

Usually a 24-hr specimen is used. Collect in a brown bottle containing about 50 ml of petroleum ether and 5 g of sodium carbonate. Measure the volume of the total urine and transfer a 50-ml aliquot to a flask of about 250 ml capacity. Add 25 ml of the ferrous sulfate solution and mix. Add 25 ml of the sodium hydroxide solution and mix again. Stopper the flask and allow to stand for about 1 hr in the dark at room temperature. Filter the solution through filter paper. Transfer a 50-ml aliquot of the filtrate to a separatory funnel. Acidify with about 5 ml of the glacial acetic acid (test for acidity with wide-range paper; the pH should be below 3). Extract the solution with three 50-ml portions of petroleum ether, shaking well. Combine the ether extracts and wash once with about 30 ml of water, discarding the lower wash layer.

Add 2 ml of the Ehrlich's reagent to the combined petroleum ether extracts in the separatory funnel. Shake well, add 4 ml of the saturated sodium acetate solution, and shake well again. Allow the layers to separate completely and drain the lower aqueous layer into a test tube. Repeat the extraction with the reagent and sodium acetate solution exactly as above, and add this extract to the same test tube. Mix the contents and read in a photometer at 540 nm against a blank of 2 ml of Ehrlich's reagent and 4 ml of the saturated sodium acetate solution. If the extract appears cloudy, centrifuge briefly before reading. Also read an aliquot of the working standard against a water blank at the same wavelength.

CALCULATION

Since 50 ml of urine was taken for analysis, diluted to 100 ml, and 50 ml of the dilution taken for the colorimetric reaction, this was the equivalent of 25 ml of the original urine; the final volume of the colored solution was 12 ml.

$$\frac{\text{absorbance of sample}}{\text{absorbance of standard}} \times \frac{12}{25} \times 0.35 = \text{urobilinogen (in mg/dl)}$$

$$\frac{\text{urobilinogen (in mg/dl)}}{100} \times \text{total urine volume (in ml)} = \text{urobilinogen (in mg) excreted/24 hr}$$

Normal values and interpretation of results follow the procedure for feces.

UROBILINOGEN IN FECES

PROCEDURE

Either an aliquot of a single stool specimen may be used, or more accurately an aliquot of a 24- or 48-hr specimen. The total stool specimen is mixed in a blender with the minimum amount of water. The weight of the total homogenized specimen is noted for use in the calculation. Weigh out about 2 to 2.5 g of stool to the nearest milligram. Transfer to a small flask with 8 ml of water. The sample may be weighed in the tared flask with sufficient accuracy. Add 10 ml of the ferrous sulfate solution and mix. Add 10 ml of the sodium hydroxide solution, mix well, stopper tightly, and allow to stand in the dark for 1 hr. Filter the mixture and pipet a 2 ml-aliquot of the filtrate to a test tube. Add 2 ml of Ehrlich's reagent, mix, and allow to stand for 10 min. Add 6 ml of saturated sodium acetate solution and mix. Read in the photometer at 540 nm, using as a blank a solution containing 2 ml of water, 2 ml of Ehrlich's reagent, and 6 ml of saturated sodium acetate solution. Read the diluted working standard against water as in the urine determination.

CALCULATION

Since the stool sample used for analysis was made up to 10 ml, and 2 ml of the filtrate from this dilution made up to 10 ml final volume,

$$\frac{\text{absorbance of sample}}{\text{absorbance of standard}} \times \frac{30 \times 10}{W \times 2} \times 0.35 = \text{urobilinogen (in mg/100 g)}$$

where W is the weight of sample in grams. The total amount of urobilinogen in the entire stool sample is then calculated:

$$\text{urobilinogen (in mg/100 g)} \times \frac{\text{total weight of stool (in g)}}{100} = \text{total urobilinogen (in g)}$$

NORMAL VALUES AND INTERPRETATION OF RESULTS

Normally 0.4 to 1.0 mg of urobilinogen will be excreted in the urine per day. The values will be increased in hemolytic jaundice (up to 10 mg/day) but normal or low in obstructive jaundice.

The normal range for feces is 40 to 280 mg urobilinogen/day. The amount to be expected in a random stool sample is more variable, but may be taken as 25 to 250 mg/100 g stool. The amount in stool is increased in conditions resulting from increased breakdown of hemoglobin such as hemolytic jaundice; up to 400 to 1,400 mg/day may be excreted. In conditions such as obstructive jaundice in which the amount of bile reaching the intestines is decreased, the fecal level of urobilinogen may be low.

Bromsulfophthalein (BSP) Excretion

The excretion of BSP, or rather its removal from the blood stream by the liver, has been used as a test of liver function. The procedure has been to inject a certain amount of the dye intravenously and determine the amount remaining in the blood stream after a definite period of time. Some other dyes have also been used, such as indocyanine green [52, 53] and rose bengal [54–56], but BSP is by far the most widely used. Originally a dose of 2 mg/kg body weight was used [57], but later workers found better results were obtained with a dose of 5 mg/kg [58–60]. This dose is most often used now although some advocate higher dosages such as 7.5 mg/kg [61] or even 20 mg/kg [62]. The dosage is most often calculated on the basis of body weight, but it is claimed that more consistent results are obtained when the calculation is made on the basis of lean body mass [63] or surface area [64, 65].

If the concentration of the dye in the blood is measured, say, 30 min after the injection, one would need to know the initial concentration in the blood in order to calculate the percentage excreted. Theoretically one could measure the concentration at frequent intervals after injection and extrapolate back to zero time. This would require at least four or five blood samples and is too tedious for routine use. The assumption is usually made that the effective plasma volume is 50 ml/kg body weight, so that when a dose of 5 mg/kg is injected the initial concentration is 5 mg/50 ml or 10 mg/dl. Thus the measured concentration of 5 mg/dl after 30 min would correspond to 50% excretion in that time. This can be only an approximation. The data cited by Jablonski and Owen [66] give average values of 40 to 48 ml/kg as obtained by different investigators for the volume of distribution of the dye. Moreover, the volume of distribution could vary more in diseased subjects than with the normal individuals studied. It would appear that the assumption of a plasma volume of 50 ml/kg would tend to overestimate the actual excretion. However, since most normal values are obtained on this assumption, using it will usually give satisfactory clinical results. It would probably still be best to make some sort of correction in markedly obese or emaciated individuals whose plasma volume would vary more from the average. Such corrections are given later.

The determination of BSP is usually made colorimetrically by measuring the purple color produced in alkaline solution. This has an absorption peak at 580 to 585 nm but the peak is shifted somewhat by the presence of proteins. The correction for hemolysis or the presence of other colored substances is usually made by reading the color of the alkaline solution, then making acid, which destroys the color of the BSP, and reading again. If the solution is made too acid, any hemoglobin present will be changed to acid hematin which will have a different absorbance from the hemoglobin in alkaline solution. In the method of Seligson [16, 17] presented here, the pH adjustment is made by means of buffers to give a pH of 10.7, which is sufficiently alkaline to develop the full color of the BSP, and a pH of 7.0, which is acid enough to eliminate the BSP color but not affect the hemoglobin. Toluenesulfonic acid is added to combine with the protein to eliminate its effect on the BSP color. A standard is made by diluting an aliquot of the same lot of dye as was injected, to a concentration of 5 mg/dl which will correspond to a retention of 50%. This method may be unsatisfactory for grossly hemolyzed or hyperlipemic serums. In this case the method of Henry et al. [69] may be used in which the proteins are precipitated by acetone. The standard must also be diluted with the acetone solution since the latter affects the BSP color.

BSP EXCRETION BY SELIGSON METHOD [67, 69]

REAGENTS

1. Alkaline buffer. Dissolve 0.76 g of anhydrous trisodium phosphate (Na_3PO_4), 6.46 g of anhydrous disodium phosphate (Na_2HPO_4), and 3.2 g of sodium p-toluenesulfonate in water to make 500 ml. This should have a pH between 10.6 and 10.7.

2. Acid buffer. Dissolve 25.8 g of anhydrous monosodium phosphate (NaH_2PO_4) or 27.6 g of the hydrated salt ($NaH_2PO_4 \cdot H_2O$) in water to make 100 ml.

3. Bromsulfophthalein standards.

A. Stock standard. An aliquot of the same lot of the dye as injected is used for the standard. The dye is usually furnished in ampules containing 50 mg/ml. Since the different lots may vary slightly in concentration, it is important to use the same lot of dye for the standard as was injected. Dilute an aliquot of dye 1:10 to give a solution containing 5 mg/ml. This solution is stable for several months if kept in a dark bottle.

B. Working standard. As needed, dilute 1 ml of the stock standard to 100 ml with water. This contains 5 mg/dl of the dye.

PROCEDURE

The equivalent of 5 mg/kg body weight is injected (0.1 ml/kg). The dye should be injected slowly over a period of at least 1 min. Care should be taken so that all the solution enters the vein as the dye is irritating to tissue. Reactions to the dye, possibly allergic, have been reported which may be severe and even fatal. The injection should be made only under the supervision of a physician. Exactly 30 and 45 min after the injection of the dye, blood samples are taken without anticoagulant from a vein in the other arm. (Different times of collection may be desired by the physician, but 30 and 45 min are the most commonly used.) Care must be taken to avoid hemolysis as much as possible.

To 1 ml of serum add 7 ml of the alkaline buffer and mix. Treat a 1-ml aliquot of the working standard similarly. Read sample and standard against water blank at 580 nm (or any convenient wavelength between 560 and 600 nm). Add 0.2 ml of the acid buffer, mix, and read again.

CALCULATION

Subtract the absorbance of the acid solution (second reading) from that of the alkaline solution (first reading) to obtain corrected absorbance. Since the standard is assumed to represent 50% retention:

$$\frac{\text{corr. absorbance of sample}}{\text{corr. absorbance of standard}} \times 50 = \% \text{ retention}$$

BSP EXCRETION BY METHOD OF HENRY ET AL. [69]

This method is used for grossly hemolyzed or hyperlipemic serums.

REAGENTS

The first three reagents are the same as for the previous method.

4. Acid acetone solution.   Add 176 ml of acetone and 0.1 ml of glacial acetic acid to a 200 ml volumetric flask and dilute to the mark with water.

5. Sodium hydroxide, 2.5M.   Dissolve 10 g of sodium hydroxide in water to make 100 ml.

PROCEDURE

Injection of the dye is done according to the procedure for the previous method.

Add 2 ml of serum to 8 ml of the acetone reagent, stopper, and mix well. Treat 2 ml of working standard similarly. Centrifuge sample tube strongly and read supernatant and standard mixture (which contains no precipitate) against a blank of the reagent at 580 nm. Add 0.1 ml of the sodium hydroxide solution, mix, and read again. Note that here the acid solution is read first so that one subtracts the first reading from the second. The calculations are the same as in the previous method. For both methods the reading of the standard in acid solution is essentially zero and can usually be omitted.

Corrections have been suggested for the fact that the volume of dilution is not always equal to 50 ml/kg. One of these is as follows: For weights between 50 and 67 kg add 1% to the measured retention; for weights between 68 and 73 kg no correction is applied. For weights between 74 and 86 kg subtract 1%; for weights between 87 and 105 kg subtract 2%; for weights above 105 kg subtract 3%.

NORMAL VALUES AND INTERPRETATION OF RESULTS

The normal values are often taken as not more than 15% retention after 30 min and not more than 5% after 45 min. Quittner [70] gives slightly different values: not more than 35% after 15 min, 16% after 20 min, 12% after 30 min, 7.5% after 40 min, 6% after 45 min, and 5% after 50 min. BSP excretion is considered one of the most sensitive tests for liver function. In increased retention it does not distinguish between disease due to hepatocellular damage and that due to biliary obstruction.

Tests Based on Detoxicating Functions

An important function of the liver is its role in the detoxification of deleterious substances. One method for detoxification is the formation of conjugated substances that are more readily excreted by the kidneys than the original ones and that may also be less toxic. The test based on this function involves the administration (oral or intravenous) of sodium benzoate, which is conjugated with glycine by the liver and excreted as hippuric acid. The amount of hippuric acid excreted after a definite dose of benzoate is a measure of the ability of the liver to detoxify. In the earlier methods for the determination of the hippuric acid, the urine was acidified and the hippuric acid allowed to crystallize out or the acid extracted from the urine with ether. The separated hippuric acid was titrated with standard sodium hydroxide. This is a rather lengthy method. We present also a very simple colorimetric method that uses pyridine.

Hippuric Acid Excretion Test [71–73]

In the oral test the patient is given 6 g of sodium benzoate dissolved in about 250 ml of water and the bladder is emptied and the urine discarded. All the urine voided during the next 4 hr is saved and pooled together with that obtained by emptying the bladder at the end of the period. The sodium benzoate is sometimes given intravenously; it is available in sterile ampules containing the equivalent of 1.5 g of benzoic acid. The patient empties the bladder and the urine is discarded. One ampule of the benzoate is injected slowly over a 5-min period. The patient drinks at least 250 ml of water and all the urine for the next hour is collected and used for analysis.

Hippuric acid by crystallization method

REAGENTS

1. Sodium chloride, $5.2M$. Dissolve 30 g of sodium chloride in water to make 100 ml. Store in the refrigerator because this solution should be cold when used. The crystallization of a small amount of the salt is of no importance.

2. Sodium hydroxide, $0.1M$. Accurately standardized or made by careful dilution of a stronger standard solution; see Chapter 20.

3. Concentrated sulfuric acid, AR.

4. Sodium chloride crystals, AR.

PROCEDURE

Measure the total urine volume and mix well. Take an accurately measured aliquot of the urine equivalent to about one-tenth of the total volume for the oral test and about one-fifth of the total volume for the intravenous test. Place the sample aliquot in a suitable-size centrifuge tube and saturate with sodium chloride (add 3 g of NaCl for each 10 ml volume). Dissolve the NaCl completely, warming slightly if necessary. Add 0.1 ml of concentrated sulfuric acid for each 10 ml of aliquot used, mix well, and place in the refrigerator for at least 1 hr. If a precipitation of hippuric acid has not occurred after this time, scratch the inside of the tube below the surface of the liquid with a glass rod and place in refrigerator for another ½ hr or longer. It may be advisable to leave in the refrigerator overnight. Centrifuge at high speed and carefully decant and discard the supernatant. Wash the precipitate by adding 10 ml of cold 5.2M sodium chloride solution, rinsing down the sides of the tube. Mix and centrifuge again. Decant the supernatant and repeat the washing. After the second washing, dissolve the residue in about 10 ml of boiling water. Titrate with 0.1M sodium hydroxide using phenolphthalein as indicator. Some of the hippuric acid may dissolve slowly and one must make certain that it has all been titrated at the endpoint.

CALCULATION

Each milliliter of 0.1M sodium hydroxide is equal to 0.0179 g of hippuric acid.

$$0.0179 \times \text{ml titration} \times \frac{\text{total urine volume}}{\text{volume of aliquot taken}} = \text{hippuric acid (in g)}$$

If the sodium hydroxide used is not exactly 0.1M, the left side of the equation must be multiplied by the factor: actual molarity/0.1. To the calculated value, add 0.12 g of hippuric acid for each 100 ml of total urine volume. This is to compensate for the slight solubility of the hippuric acid in the sodium chloride solution.

HIPPURIC ACID BY COLORIMETRIC METHOD [74]

REAGENTS

1. Pyridine, reagent grade.   Store in tightly stoppered bottle.
2. Benzenesulfonyl chloride, reagent grade.   This material has a melting point around 15°C. It will solidify if kept in a cold place but can easily be liquefied by warming.
3. Ethanol, 95%.
4. Hippuric acid standards.
    A. Stock standard.   Dissolve 100 mg of hippuric acid in water and dilute to 100 ml. This contains 1 mg/ml.
    B. Working standard.   On the day of use, dilute 2 and 4 ml of the stock to 10 ml to give standards containing 0.2 and 0.4 mg/ml.

PROCEDURE

Assuming the excretion of 4 g of hippuric acid in the oral test and 2 g in the intravenous test, dilute an aliquot of the well-mixed urine sample to a concentration of between 0.3 and 0.4 mg/ml. For example, if the total volume in an oral test were 350 ml of urine, the assumed concentration would be 4,000 mg/350 ml = 11.4 mg/ml. A 1:30 dilution of the urine would give a concentration of 0.38 mg/ml.

Pipet 0.5 ml of diluted urine samples, standards, and water as a blank to separate test tubes. To each tube add 0.5 ml of pyridine and mix. Then add 0.2 ml of benzenesulfonyl chloride and mix well with a vortex mixer. (*Caution:* Take great care in pipetting these two reagents.) Allow the mixtures to stand for 30 min at room temperature. Add 4 ml of ethanol and mix. The solutions may be slightly turbid, particularly the samples. Centrifuge the tubes strongly to clarify, then decant into photometer cuvettes and read samples and standards against the blank at 410 nm.

CALCULATION

Since 0.5 ml of the working standard and 0.5 ml of the diluted urine are treated similarly,

$$\frac{\text{absorbance of sample}}{\text{absorbance of standard}} \times \text{conc. of standard} \times \frac{D \times V}{1,000} = \text{hippuric acid excreted (in g)}$$

where D is the dilution factor (30 in the example given), V is the total volume of urine collected, and the concentration of the standard is 0.2 or 0.4 mg/ml, taking the standard whose absorbance is closest to that of the sample. The factor of 1,000 converts from milligrams to grams.

NORMAL VALUES AND INTERPRETATION OF RESULTS

For the oral test 3 to 3.5 g of hippuric acid should be excreted; for the intravenous test, 0.7 to 1.6 g. Higher values have no significance. The excretion is essentially normal in hemolytic jaundice, uncomplicated gallbladder disease, and obstructive jaundice. It is decreased in hepatitis, liver tumors, and other conditions causing liver impairment.

Tests Based on Plasma Proteins

Since the advent of the rapid cellulose acetate methods for protein electrophoresis, these have generally been used for the determination of the amounts of the different plasma proteins, with the addition in selected cases of immunodiffusion or immunoelectrophoretic methods for specific proteins. The techniques, together with some information on the patterns obtained in liver disease, have been discussed in Chapter 3. In general, the amount of albumin may decrease somewhat in liver disease and the amount of globulins increase, so that the A/G ratio is inverted. Fibrinogen, the plasma protein concerned with the coagulation process, is produced almost exclusively in the liver. Except in severe forms of liver disease, its concentration in the plasma is not greatly changed from normal. Methods for the determination of this protein have been given in Chapter 7.

The amino acid and ammonia concentrations in the serum may also be related to protein metabolism. Any condition leading to severe liver damage will result in marked increases in the amino acids in the plasma and correspondingly in the urinary excretion. The most significant increases in amino acids are found in Wilson's disease (hepatolenticular degeneration). The level of blood ammonia is also increased in severe liver disease, but it does not appear to be characteristic of any particular type of liver disease. The procedures for the determination of amino acids have been given in Chapter 7 and for ammonia nitrogen in Chapter 6.

Flocculation Tests

Before simple electrophoretic methods became available, flocculation or turbidity tests were commonly used for the study of liver disease. These tests are based on the fact that changes in the proportion or concentration of the serum albumin and globulins will result in a precipitation or flocculation with the reagents that does not occur with normal serum. These changes in proteins usually occur in liver disease, and the tests are used for the detection of such conditions. These tests are not specific for liver disease; any condition resulting in changes in protein ratio will give a positive test. Not all liver conditions develop the protein changes necessary for a positive test. Because of their simplicity, the flocculation tests are still used, particularly for following the course of a disease. For an initial diagnosis the electrophoretic pattern is preferred, but serial determinations of one of the flocculation tests may be helpful in following the course. A number of such tests have been devised, including the Takata-Ara [75] and Weltman [76, 77] tests, colloidal gold serum test [78], zinc sulfate [79, 80] and cadmium sulfate turbidity [81, 82] tests, thymol turbidity [83–85] and cephalin cholesterol flocculation test [86, 87]. Methods for only the last two named will be given here.

THYMOL TURBIDITY AND FLOCCULATION TEST [83–85]

The original procedure for this test used a barbital buffer which was not very stable. This has been replaced by the more stable tris buffer. The buffer originally used has a pH of 7.8, but it was later found that a pH of 7.55 gave greater sensitivity. If a commercially prepared buffer is used, it should be of the latter pH.

1. Thymol buffer. Place 1,000 ml of deionized water in a 2 liter flask and heat to boiling. Allow to cool to about 85°C, then pour the hot water into another flask containing 6 g of thymol and 1.21 g of tris buffer [tris-(hydroxymethyl) aminomethane] and mix. When the solution has cooled to about 25°C, add a few crystals of thymol and shake vigorously. When the excess thymol has crystallized out, filter. To the filtrate add 7 ml of $1M$ hydrochloric acid (84 ml of concentrated acid to 1 liter with water). Check the pH and adjust to 7.55 $\pm$ 0.03 by the addition of small amounts of HCl or tris buffer. If only a few tests are run, it may be more convenient to use a commercially prepared buffer—either a completely prepared buffer or a vial containing all the reagents for the preparation of a definite quantity of buffer.

PROCEDURE

Place 6 ml of the buffer in a test tube and add 0.1 ml of serum and mix. Allow to stand at room temperature for 30 min. Then read at 650 nm against a blank of 0.1 ml of water and 6 ml of the buffer. Allow to stand at room temperature in the dark for 18 hr and then examine for visual flocculation. Record on the basis of zero for no flocculation to 4+ for almost complete flocculation and precipitation.

The original method introduced by Maclagan called for a visual comparison against the Kingsbury turbidity standards [88] used for urinary protein. The protein standard of 100 mg/dl was chosen as corresponding to 10 thymol turbidity units. Albumin and thymol turbidity standards are available commercially in permanent form in sealed tubes. If the photometer used can be adapted to take the standard tubes and the unknowns are run in similar size tubes, then:

$$\frac{\text{absorbance of sample}}{\text{absorbance of standard}} \times \text{units of standard} = \text{units in sample}$$

Standards can also be prepared in the laboratory according to the method of Shank and Hoagland [89]. There was an error in the original publication so that the standard described by these authors as being 20 units actually corresponded to 10 Maclagan units. Thus 1 Maclagan unit is the same as 2 Shank-Hoagland units. This should be kept in mind when using commercial standards. The following procedure for the preparation of standards gives the results as Shank-Hoagland units as they are more commonly used in this country.

Dissolve 1.17 g of barium chloride ($BaCl_2 \cdot 2H_2O$) in water to make exactly 100 ml. Prepare a sulfuric acid solution exactly $0.21M$. This may be prepared by accurately diluting a $1M$ solution which has been accurately prepared as outlined in Chapter 20. Pipet 0.7 ml and 1.35 ml of the barium chloride solution to two 100 ml volumetric flasks. Cool a sufficient quantity of the $0.21M$ sulfuric acid to 10°C and add about 95 ml to each flask. Warm the flasks to room temperature and dilute accurately to the mark. Mix well. Add portions to cuvettes and mix again just before reading. The absorbances of the two solutions correspond to 5 and 10 Shank-Hoagland units. The preparation of the standards is lengthy and it may be preferable to purchase a stabilized latex particle‡ suspension prepared according to the suggestion of Ferro and Ham [90]. This constitutes a fairly stable turbidity standard.

Serum for this test is usually stable for a few days in the refrigerator. The room temperature should not vary greatly from 25°C when performing the tests; if there are large fluctuations it is preferable to place the samples in a pan of water at 25°C during the incubation period. Lipemia will cause marked errors in increased turbidity and such serums cannot be used.

The normal range is given as up to 5 units with only slight flocculation after 18 hr. The interpretation of results is similar to that of the cephalin cholesterol flocculation test and will be discussed later.

Cephalin cholesterol flocculation test [86, 87]

REAGENTS

1. Cephalin cholesterol reagent.   The mixture for the preparation of the reagent is available commercially.§ Dissolve the material from 1 unit in 5 ml of ether. If a clear solution is not obtained, add a small drop of water and mix. The solution is stable for some time when kept in a tightly stoppered container in the refrigerator. To prepare the reagent add 35 ml of water to a 50 ml beaker with a graduation mark at 30 ml. Warm to about 65 to 70°C and add 1 ml of the ether slowly with stirring. Heat slowly to boiling and let simmer until the volume is reduced to 30 ml. Cool to room temperature and store in the refrigerator. This solution is stable for only about 1 week. When only a few tests are run, it may be more convenient to obtain a commercially lyophilized preparation.‖ The reagent is prepared from this merely by reconstitution with water.

‡ Dade Latex Particle Standard, Dade Division, American Hospital Supply Corp., Miami, Fla. 33152.
§ Difco Laboratories, Detroit, Mich. 48232.
‖ Ceph-Flox, Hopper Laboratories, Inc., San Antonio, Tex. 78205.

2. Sodium chloride, $0.15M$.   Dissolve 8.7 g of sodium chloride in water to make 1 liter.

PROCEDURE

To a test tube add 4 ml of the sodium chloride solution and 0.2 ml of serum sample. Mix and add 1 ml of the cephalin cholesterol reagent. Mix well, stopper with a pledget of cotton, and allow to stand at room temperature in the dark. The reaction is light and temperature sensitive so the tubes must be kept in the dark at a relatively constant temperature close to 25°C. Observe the tubes after 24 and 48 hr, noting the amount of precipitation or flocculation. Normal serum will show little change in the state of flocculation and would be graded as negative. The positive reactions are graded from 1+ to 4+. The 4+ indicates complete precipitation with a clear supernatant. The intermediate values of 1+, 2+, and 3+ indicate intermediate degrees of flocculation. Positive and negative controls should also be run. A serum giving a 4+ reaction may be used for about 1 month if kept frozen. Traces of acids or heavy metals on the tubes may cause false positive reactions; the tubes should be scrupulously cleaned and free of acid.

A 1-hr test [91] has been suggested to replace the above test with a considerable saving in time for completion. The same reagents are used as in the longer test, and the serum and reagents are mixed in the same way. A saline control is also set up with 4.2 ml of the sodium chloride solution and 1 ml of the reagent. After preparation, place all tubes in a refrigerator (4 to 8°C) for exactly 1 hr. Remove all tubes from the refrigerator and centrifuge all except the saline control. This tube is not disturbed until making the final photometric readings. When the other tubes have been centrifuged (about 2,000 rpm for 10 min), carefully remove as much as possible of the supernatant fluid without disturbing any precipitate in the bottom. To the sample tubes but not the saline control, add 5.0 ml of the sodium chloride solution and mix by inversion to suspend any precipitate. Mix the saline control by inversion. Read all tubes in a photometer at 415 nm against a water blank. The suspensions are stable for about ½ hr.

CALCULATION

$$\frac{\text{absorbance of sample}}{\text{absorbance of saline control}} \times 100 = \% \text{ flocculation}$$

The results are compared with the previous method as follows: $0–20\% =$ negative; $21–40\% = 1+$; $41–60\% = 2+$; $61–80\% = 3+$; $81–100\% = 4+$. One important precaution is that false positive results may be obtained if the serum is refrigerated prior to testing; only fresh serum should be used.

Normal serum will show no flocculation or at the most 1+. Only results greater than 2+ should be considered indications of significant liver damage. Both the thymol turbidity and the cephalin cholesterol flocculation tests are used to follow the course of liver disease. The cephalin cholesterol test becomes positive earlier in hepatitis than the thymol turbidity test but the latter may persist for a longer time. The tests may be negative in uncomplicated obstructive jaundice, as they are indications of hepatic parenchymal damage.

Tests Based on Carbohydrate Metabolism

Although the liver plays an important role in glucose metabolism, tests with this sugar cannot be used for liver function studies because of the other factors greatly influencing glucose metabolism. However, other sugars can be used. Galactose is rapidly converted by the normal liver to glucose and glycogen and only small amounts of galactose will be found in the blood after an oral dose of the sugar. With impaired liver function the galactose is converted to glucose less rapidly and the amount of galactose will—assuming adequate absorption—increase in the blood. A similar test based on fructose (levulose) has also been used.

Galactose Tolerance Test [92, 93]

The patient should be in the fasting state. Take a blood sample for a blank determination, then give the patient 40 g of galactose dissolved in about 250 ml of water. Take blood samples at 30, 60, 90, and 120 min after the ingestion of the galactose, and determine the galactose concentration in each sample. Galactose may be determined by the use of galactose oxidase (similar to the method for glucose by glucose oxidase, see Chapter 5), or by first incubating the serum with baker's yeast to ferment and eliminate the glucose and then determining the galactose using the Somogyi-Nelson copper reagent. The former method is somewhat simpler, but the latter may be more convenient if only an occasional test is done and the copper reagent is available.

REAGENTS

1. Sulfuric acid, $0.33M$, and sodium tungstate, $0.30M$. These are the solutions used to prepare a Folin-Wu filtrate (Chapter 5).

2. Alkaline copper tartrate and arsenomolybdate. These are the reagents for determination of glucose by the Somogyi-Nelson method (Chapter 5).

3. Yeast suspension. Suspend about 50 g of baker's yeast in about 100 ml of water, mixing well. Centrifuge and decant the supernatant and discard. Repeat the washing of the yeast several times until the final supernatant is perfectly clear. Decant the final washing and prepare a 20% (by volume) suspension of the yeast.

4. Galactose standard, 100 mg/dl. Dissolve 100 mg of galactose in water to make 100 ml. This solution is not stable for more than a week in the refrigerator. For a stable standard, prepare a stock solution in benzoic acid solution as directed for the glucose standards (Chapter 5) and dilute a working standard as required.

PROCEDURE

To a number of tubes add 6 ml of the well-mixed yeast suspension. To separate tubes containing the yeast add 1 ml of the various blood samples including the sample taken before ingestion of the galactose (blank). To a final tube add 1 ml of this blank blood sample and 1 ml of the galactose standard. Mix well and incubate all the tubes at 37°C for 20 min. Cool and add to each tube 1 ml of the sulfuric acid solution and mix. Then add 1 ml of the sodium tungstate solution. Mix well and filter or centrifuge. Analyze the filtrates by the Somogyi-Nelson method (Chapter 5), increasing the heating time to 20 min. In the procedure as outlined, both the blank and standard contain the blood obtained before ingestion of the galactose. This compensates for other reducing substances in the blood or from the yeast. Read all samples and standards against the blank at 535 nm.

CALCULATION

Since 1 ml of standard and 1 ml of blood samples are treated similarly, for a standard of 100 mg/dl:

$$\frac{\text{absorbance of sample}}{\text{absorbance of standard}} \times 100 = \text{galactose in blood (in mg/dl)}$$

The galactose may also be determined by the use of galactose oxidase as given in Chapter 5.

NORMAL VALUES AND INTERPRETATION OF RESULTS

Various values have been given for the normal galactose level after ingestion of the sugar. The maximum level is usually in the range of 40 to 80 mg/dl. Another criterion that has been used is that the sum of the galactose levels of the four samples should not exceed 160 mg/dl. Increased levels may be found in liver disease. In infectious and toxic hepatitis, levels two or three times those given here may be found. Generally the amount of galactose found in repeated tests decreases with clinical improvement of the patient.

Levulose Tolerance Test

The procedure for the glucose tolerance test is followed (Chapter 5) but the patient receives 50 g of levulose (fructose) instead of glucose. Blood samples are drawn as in the glucose tolerance test. The samples are analyzed by the Somogyi-Nelson copper reduction method (Chapter 5). This gives both glucose and fructose levels, but the glucose may be considered to be relatively constant during the test and any increase in reducing sugars as due to fructose. In normal individuals the rise in apparent blood sugar is usually not more than 25 to 30 mg/dl above the fasting level. The rise is greater in liver disease, up to 60 to 90 mg/dl, but the differences are rarely striking.

Enzymes in Liver Disease

A large number of enzymes have been suggested at one time or another for the diagnosis of liver disease. Not all of these have been found to be of value. Listed below are the enzymes more commonly used for the detection of liver disease. The enzymes are discussed in detail in Chapter 14; we merely list them here. Further information may be found in Zimmerman et al. [94].

1. Enzymes markedly elevated in patients with obstructive jaundice, intra-hepatic cholestasis, and metastatic carcinoma, but only slightly elevated in patients with hepatitis and other hepatocellular diseases.

   Alkaline phosphatase*   Leucine aminopeptidase
   5′-Nucleotidase   γ-Glutamyl transpeptidase

2. Enzymes only slightly elevated in obstructive jaundice but markedly elevated in acute hepatitis and somewhat elevated in infectious mononucleosis and metastatic carcinoma; also found in other tissues and may be elevated in other diseases.

   Aspartate aminotransferase (GOT)*   Fructose diphosphate (aldolase)
   Malate dehydrogenase   Isocitrate dehydrogenase
   Glucosephosphate isomerase

3. Enzymes slightly elevated in obstructive jaundice but markedly elevated in infectious hepatitis; found chiefly in the liver.

Alanine aminotransferase (GPT)* Glutamate dehydrogenase
Ornithine carbamoyltransferase Guanine deaminase (guanase)
Iditol dehydrogenase (sorbitol dehydrogenase)

4. Enzymes not markedly elevated in liver disease but which may be used to differentiate liver disease from other conditions.

Lactate dehydrogenase* 2-Hydroxybutyrate dehydrogenase

5. Enzyme decreased in liver disease.

Cholinesterase (pseudocholinesterase)

6. Enzyme decreased in Wilson's disease.

Ceruloplasmin

Note: The four enzymes marked * are still the most commonly used though not necessarily the best ones. A discussion of the merits of the various enzymes is given by Zimmerman et al. [94]. These authors found that these enzymes are still useful, particularly if a profile of the four enzymes is used.

Australia Antigen: Hepatitis Associated Antigen (HAA)

This antigen has been found in individuals during the early phases of viral hepatitis [95–98]. When multiple samples are taken at regular intervals during the acute phase of the disease, positive results have been reported in about 98 percent of the cases. The first positive test occurs prior to appearance of symptoms. Positive tests may be found for several years in certain cases of chronic hepatitis.

The antigen can be present in many of the common liver diseases such as chronic hepatitis, primary liver cirrhosis, and hepatocellular carcinoma. In early hepatitis the antigen usually becomes positive before there is an elevation of serum glutamic pyruvate transaminase (SGPT).

The antigen has not been found in nonviral forms of liver disease, a fact that makes the viral antigen useful for the differential diagnosis of hepatitis.

HAA has been determined by several methods which include immunoelectrophoresis, immunodiffusion, complement fixation, and radioimmunoassay (RIA). RIA is the most specific and sensitive of these methods. It is capable of detecting twice as many carriers as the other methods. Details of the method will be described in Chapter 15.

References

1. Harper, H. A.   *Review of Physiological Chemistry* (10th ed.) Los Altos, Calif.: Lange, 1965. P. 321.
2. Michaelsson, M.   *Scand. J. Clin. Lab. Invest.* 30:387, 1972.
3. Jackson, S. H., and Hernandez, A. H.   *Clin. Chem.* 16:462, 1970.
4. Levkoff, A. H., Westphal, M. C., and Finklea, J. F.   *Am. J. Clin. Pathol.* 54:562, 1970.
5. Walters, M. I., and Gerarde, H. W.   *Microchem. J.* 13:253, 1968.
6. Liley, A. W.   *Am. J. Obstet. Gynecol.* 82:1359, 1961.
7. Freda, V. J.   *Am. J. Obstet. Gynecol.* 92:341, 1965.
8. Lewi, S.   *Ann. Biol. Clin.* (Paris) 22:797, 1964.
9. Burnett, R. W.   *Clin. Chem.* 18:150, 1972.
10. Bartsch, F. K.   *Ann. Ostet. Ginecol.* 92:482, 1970.
11. Brodersen, R., and Vind, I.   *Scand. J. Clin. Lab. Invest.* 15:225, 1963.
12. Stevenson, C. W., Jacobs, S. L., and Henry, R. J.   *Clin. Chem.* 10:95, 1964.
13. Bratlid, D., and Winsnes, A.   *Scand. J. Clin. Lab. Invest.* 28:41, 1971.
14. Ham, A. B.   *Clin. Chem.* 18:1547, 1972.
15. Malloy, H. T., and Evelyn, K. A.   *J. Biol. Chem.* 119:481, 1937.
16. Ducci, H., and Watson, C. J.   *J. Lab. Clin. Med.* 30:293, 1945.
17. MacDonald, R. P.   *Stand. Methods Clin. Chem.* 5:65, 1965.
18. O'Hagan, J. E., Hamilton, T., LeBreton, E. G., and Shaw, A. E.   *Clin. Chem.* 3:609, 1957.
19. Bruckner, J.   *Clin. Chim. Acta* 6:370, 1961.
20. Boutwell, J. H.   *Clin. Chem.* 10:197, 1964.
21. van den Bossche, H.   *Clin. Chim. Acta* 11:379, 1965.
22. Michaelsson, M., Nosslin, B., and Sjolin, S.   *Pediatrics* 35:925, 1965.
23. Thompson, R. P. H.   *J. Clin. Pathol.* 22:439, 1969.
24. Jendrassik, L., and Grof, P.   *Biochem. Z.* 297:81, 1936.
25. Fog, J.   *Scand. J. Clin. Lab. Invest.* 10:241, 1958.
26. Gambino, S. R.   *Stand. Methods Clin. Chem.* 5:55, 1965.
27. Bartels, H., and Boehmer, M.   *Z. Klin. Chem. Klin. Biochem.* 9:133, 1971.
28. Hillman, G., and Beyer, G.   *Z. Klin. Chem.* 5:92, 1967.
29. Keyser, J. N., and Spillane, M. T.   *Clin. Chem.* 10:375, 1964.
30. Kulhanek, V., and Appelt, J.   *Clin. Chim. Acta* 20:29, 1968.
31. Ferro, P. V., and Ham, A. B.   *Am. J. Clin. Pathol.* 47:472, 1967.
32. Ferro, P. V., and Ham, A. B.   *Tech. Bull. Regist. Med. Technol.* 33:111, 1963.
33. Zaroda, R. A.   *Am. J. Clin. Pathol.* 45:70, 1966.
34. Lipsitz, P. J., and London, M.   *J. Lab. Clin. Med.* 81:625, 1973.
35. Roth, M.   *Clin. Chim. Acta* 17:487, 1967.
36. Gambino, S. R., and Di Re, J.   *Bilirubin Assay* (Revised). Chicago: American Society of Clinical Pathologists, 1968.
37. Recommendations on a uniform bilirubin standard.   *Clin. Chem.* 8:405, 1962.
38. Clarke, J. T.   *Clin. Chem.* 11:681, 1965.
39. Dybaeker, R., and Hertz, H.   *Scand. J. Clin. Lab. Invest.* 25:151, 1970.
40. Cremer, R. J., Perryman, P. W., Richards, D. H., and Holbrook, B.   *Biochem. J.* 66:60p, 1957.

341

41. Walker, A. H. C. *Br. Med. J.* 2:376, 1957.
42. Crosby, W. M., and Merrill, J. A. *Am. J. Obstet. Gynecol.* 92:341, 1965.
43. Fleming, A. F., and Woolf, A. J. *Clin. Chim. Acta* 12:67, 1965.
44. Kapitulnik, J., Kaufmann, N. A., and Blondheim, S. H. *Clin. Chem.* 16:756, 1970.
45. Barrette, P. V. D. *J.A.M.A.* 217:1349, 1971.
46. Stengle, J. M., and Schade, A. L. *Br. J. Haematol.* 3:117, 1957.
47. Bruschke, G., and Volkheimer, G. *Z. Gesamte Inn. Med.* 11:804, 1956.
48. Bloomer, J. R., Barrett, P. V. D., Rodkey, F. L., and Berlin, N. I. *Gastroenterology* 61:479, 1971.
49. Lipsitz, F. J., and London, M. *J. Lab. Clin. Med.* 81:625, 1973.
50. Watson, C. J. *Am. J. Clin. Pathol.* 18:84, 1936.
51. Maclagan, N. F. *Br. J. Exp. Pathol.* 125:234, 1944.
52. Cherrick, G. R., Stein, S. W., Leevy, C. M., and Davidson, C. S. *J. Clin. Invest.* 39:592, 1960.
53. Hunton, D. B., Bollman, J. L., and Hoffman, H. N. *J. Clin. Invest.* 40:1648, 1961.
54. Delprat, G. D. *Arch. Intern. Med.* 34:537, 1924.
55. Kerr, W. J., Delprat, G. D., Epstein, N. N., and Dunievitz, M. *J.A.M.A.* 85:942, 1925.
56. Snell, A. J., and Magath, T. G. *J.A.M.A.* 110:167, 1938.
57. Rosenthal, S. M., and White, E. C. *J.A.M.A.* 84:1112, 1925.
58. Helm, J. D., and Machella, T. E. *Am. J. Dig. Dis.* 9:141, 1942.
59. Mateer, J. G., Galtz, J. I., Comanduras, P. D. M., Steelf, P. D., and Brouwer, S. *Gastroenterology* 8:52, 1947.
60. O'Leary, P. A., Greene, C. H., and Rowntree, L. G. *Arch. Intern. Med.* 44:155, 1929.
61. Tovey, J. E. *Clin. Chim. Acta* 15:149, 1967.
62. Castenfors, J., and Hultman, E. *Scand. J. Clin. Lab. Invest.* 14(Supp. 64):45, 1962.
63. Broholt, J. *Scand. J. Clin. Lab. Invest.* 19:67, 1967.
64. Allen, T. H. *Metabolism* 5:328, 1956.
65. Alliot, M., and Paraf, A. *Presse Med.* 59:501, 1951.
66. Jablonski, P., and Owen, J. A. *Adv. Clin. Chem.* 12:309, 1969.
67. Seligson, D., Marino, J., and Dodson, E. *Clin. Chem.* 3:638, 1959.
68. Seligson, D., and Marino, J. *Stand. Methods Clin. Chem.* 2:186, 1957.
69. Henry, R. J., Chiamori, M., and Ware, A. G. *Am. J. Clin. Pathol.* 32:201, 1959.
70. Quittner, H. In F. W. Sunderman and F. W. Sunderman, Jr. [Eds.], *Laboratory Diagnosis of Liver Disease.* St. Louis: Green, 1968.
71. Weichselbaum, T. E., and Probstein, J. G. *J. Lab. Clin. Med.* 24:636, 1939.
72. Gaebler, O. H. *Am. J. Clin. Pathol.* 15:452, 1945.
73. Lavers, G. D., et al. *J. Lab. Clin. Med.* 34:965, 1949.
74. Tomokumi, K., and Ogata, M. *Clin. Chem.* 18:349, 1972.
75. Ragins, A. B. *J. Lab. Clin. Med.* 20:902, 1935.
76. Weltman, O. *Med. Klin.* 26:240, 1930.
77. Weltman, O., and Medvei, C. V. *Z. Klin. Med.* 118:670, 1931.
78. Maclagan, N. F. *Br. J. Med.* 2:863, 1944.
79. Kunkel, H. G. *Proc. Soc. Exp. Biol. Med.* 66:217, 1947.
80. Maclagan, N.F. *Br. J. Exp. Pathol.* 25:334, 1944.
81. Wunderley, C., and Wuhrmann, F. *Schweiz. Med. Wochenschr.* 74:185, 1944.

82. Wunderley, C., and Wuhrmann, F.   *Klin. Wochenschr.* 22:587, 1943.
83. Reinhold, J. G.   *Adv. Clin. Chem.* 3:84, 1960.
84. Reinhold, J. G., and Yanan, V. L.   *Am. J. Clin. Pathol.* 26:669, 1956.
85. Reinhold, J. G.   *Clin. Chem.* 8:475, 1962.
86. Hanger, F. M.   *J. Clin. Invest.* 18:261, 1939.
87. Hanger, F. M., and Patek, A. J.   *Am. J. Med. Sci.* 202:48, 1941.
88. Kingsbury, F. P., Clark, C. P., Williams, G., and Post, A. L.   *J. Lab. Clin. Med.* 11:981, 1926.
89. Shank, R. E., and Hoagland, C. L.   *J. Biol. Chem.* 162:133, 1946.
90. Ferro, P. V., and Ham, A. V.   *Am. J. Clin. Pathol.* 45:166, 1966.
91. Ratliff, C. R., Warren, P. E., and Casey, A. E.   *Am. J. Gastroenterol.* 55:589, 1971.
92. Maclagan, N. F.   *Q. J. Med.* 9:151, 1940.
93. Maclagan, N. F., and Rundel, F. F.   *Q. J. Med.* 9:215, 1940.
94. Zimmerman, H. J., and Seeff, L. B.   In E. L. Coodley [Ed.], *Diagnostic Enzymology.* Philadelphia: Lea & Febiger, 1970. P. 1.
95. Blumberg, B. S., Sutnick, A. I., and London, W. T.   *J.A.M.A.* 207:1895, 1969.
96. Soloway, R. D., and Sommerkill, W. H. J.   *Postgrad. Med.* 53:77, 1973.
97. Feinman, S. V.   *Mod. Med. Can.* 28:21, 1973.
98. Blumberg, B. S.   *Med. Times* 101:50, 1973.

11. Kidney Function Tests

Since the main function of the kidney is to excrete metabolic waste products and excess water, the usual tests for kidney function are based on the organ's ability to excrete either endogenous substances such as urea or creatinine or administered exogenous substances such as phenolsulfonphthalein or inulin. The kidney is made up of a large number of individual units (nephrons). In the simplest aspect each nephron is composed of two distinct parts—the glomerulus, in which essentially a protein-free ultrafiltrate is produced from the plasma, and the tubule, in which those substances not excreted are reabsorbed. Those substances not reabsorbed are excreted in the urine. Some substances are also secreted by the tubules. The formation of the ultrafiltrate is essentially a passive process but the secretion of substances by the tubules is an active process involving the expenditure of energy as is the reabsorption of substances from the glomerular filtrate.

In the passage of a given amount of blood through the kidney, only a fraction of the amount of a substance in the blood that will be excreted is actually removed. The amount removed is usually measured in terms of a clearance. If in 1 min the amount of a substance in 100 ml of blood was actually excreted by the kidneys, then the clearance for that substance would be 100 ml/min. Actually, 500 ml of blood might have passed through the kidneys in this time and one-fifth of the amount in each milliliter of blood have been removed by the kidneys, with the net result that the amount of substance in 100 ml of blood was excreted per minute, or the equivalent of 100 ml of blood was cleared of the substance in question per minute.

Clearance is calculated by the formula:

$$\text{clearance} = \frac{U}{P} \times V$$

where U is the concentration of the substance in the urine, P the concentration in the plasma, and V the urine volume, usually expressed as milliliters per minute (ml/min). For a more detailed description of the action of the kidneys, reference is made to any textbook on physiology such as the one by Ganong [1].

A number of substances have been measured in the plasma and urine for the determination of clearance. Two that have been widely used are urea and creatinine. Since these are endogenous substances normally in the blood, the administration of any foreign substance is not required. Exogenous substances that have been used for clearance are inulin, p-aminohippuric acid (PAHA), and Diodrast. These substances have the disadvantage that they must be given by continuous infusion to keep the plasma level fairly constant during the test period, and thus are not often used for routine testing. The amount of blood passing through and filtered by the kidney will depend upon the size of the kidney and thus on the size of the individual. For precise work the measured clearance should be corrected to the standard size, commonly used for the determination of basal metabolic rate, 1.73 m² surface area. For the urea clearance and often for the creatinine clearance in adults, this correction is not made. Obviously it should be made in individuals differing greatly from the average adult size and certainly should be used for children. Since the basal metabolic rate is not determined as frequently as formerly, tables relating the height and weight to the surface area may not be readily available. A nomogram can be found in a book, available to most clinical chemists, by Peters and Van Slyke [2].

The osmolality of the urine or an osmolal clearance has also been used as a measure of kidney function since these quantities are related to the ability of the kidneys to form a concentrated urine. Finally, some dilution and concentration tests have been used in which the maximum specific gravity of the excreted urine under standard conditions of restricted water intake or the decrease in specific gravity under water loading is noted. These tests may be more reliable if the osmolality is measured rather than the specific gravity.

Clearance Tests

The clearance tests are performed by collecting the urine for timed periods and obtaining corresponding blood samples. The concentration of the substance in urine and blood (usually in plasma or serum rather than whole blood) is determined and the clearance is calculated according to the formula given. With the endogenous substances urea and creatinine, the blood concentration is relatively constant and only one sample is usually taken at some time during the urine collection period. With creatinine the 24-hr clearance is often measured, using a complete 24-hr urine collection and a blood sample taken in the fasting state. The urine may be collected for shorter periods such as 1 or 2 hr, but since the term entering into the equation is milliliters per minute, the timing of the urine collection must be accurate. Usually with these shorter periods two or more collection periods are used and the average clearance for the periods reported. With the clearance of exogenous substances, blood samples must be taken at more frequent intervals since the level of the substance in the blood will not be constant. Usually for these substances, two 2-hr collection periods are used with blood samples taken at the beginning and end of each period. Usually some water is given orally at the beginning of the test to produce a normal urine flow.

The methods for the determination of creatinine and urea in blood and urine have been given in Chapter 6. The calculation of the actual clearance is given above. For urea several different clearance tests have been used.

Urea Clearance

The urea clearance has often been used as a measure of kidney function though it is known that urea is to some extent reabsorbed by the kidney tubules and thus the clearance may vary with the urine volume. This was recognized by Moeller and associates [3] who introduced the concepts of maximal clearance and standard clearance. The former was calculated in the usual way: $C_m = (U/B)V$, where U is the concentration of urea or urea nitrogen in the urine, B the corresponding concentration in the blood, and V the urine volume in milliliters per minute. The maximal clearance was used when the urine flow was 2 ml/min or more. When the urine flow was less than this, the standard clearance was used; this was calculated according to the formula: $C_s = (U/B) \cdot \sqrt{V}$, where the letters have the same significance as in the earlier equation. Technically speaking, this latter quantity is not a clearance. It does not possess the proper dimensions of milliliters per minute, and hence cannot really be the measure of any volume of blood cleared by the kidneys. The usual normal values for the two clearances are given as 75 ml/min and 54 ml/min for the maximal and standard clearance, respectively. Since $\sqrt{2} \times 54 = 76$, the two formulas give the same result at 2 ml/min. The earlier work on clearances was done using the concentrations measured in whole blood. Currently serum is often used, particularly in automated analysis. However, the difference in concentrations of urea in blood and serum is very small, particularly in the fasting state, so there is no significant difference in the clearance results.

Van Slyke and associates [4] found that in patients with an increased amount of preformed ammonia in the urine, the clearance was decreased unless one used the total of ammonia + urea nitrogen instead of urea nitrogen alone as the concentration in the urine. They suggested that for best results this calculation be used at all times for all clearances. Methods such as the urease methods, which can readily determine the total of urea + ammonia, are recommended for this purpose. The diacetyl monoxime methods determine only urea. It is stated that when ammonia plus urea is measured, the normal values for the maximal and standard clearances should be increased about 5 percent to 79 and 57 ml/min, respectively [5]. Chesley [6] found that when the urine volume was below about 0.35 ml/min, the ratio U/B became practically constant. He introduced the minimal clearance, calculated as $C_{mi} = (U/B) \times 0.35$. V was taken as 0.35 as long as the volume per minute was less than this. The normal value for this clearance was given as 32 ml/min. It is customary to report the urea clearances as a percentage of the normal values. Thus one would have for the clearances in terms of percentage of normal:

$$\text{maximal clearance} = 1.33 \times \frac{UV}{B}$$

$$\text{standard clearance} = 1.85 \times \frac{U\sqrt{V}}{B}$$

$$\text{minimal clearance} = 1.09 \times \frac{U}{B}$$

If the ammonia is included with the urea for the urine analysis, the factors given above will be changed to 1.26, 1.76, and 1.04, respectively.

The work of Brulles and associates [7] indicates that the urea clearance calculated in the usual way, as $C = (U/B)V$, varies directly with the urine volume over a wide range. Thus the above formulas are only approximations for different ranges of urine volumes, and even above 2 ml/min the clearance is not constant. Although there was considerable variation in the results on different subjects, their results suggest that if a urea clearance is to be used, one must recognize the dependence on urine volume. These authors suggest the use of a chart similar to Table 11-1. Although further studies may indicate that the actual values given in the table should be changed somewhat, the fact remains that the urea clearance is greatly dependent upon the urinary volume. This makes urea clearance tests less satisfactory than the creatinine clearances.

Table 11-1. Relationship Between Urea Clearance and Urine Volume

Urine Vol. (ml/min)	Normal Clearance (ml/min)	Urine Vol. (ml/min)	Normal Clearance (ml/min)
1	29	5	94
2	52	6	100
3	72	8	109
4	85	10	115
		14	122

Creatinine Clearance

The measurement of the endogenous creatinine clearance has been used for many years to estimate the glomerular filtration rate (GFR). This would be exactly equal to the clearance of a substance that was filtered through the glomeruli but not reabsorbed or excreted by the tubules; that is, all of the substance that is filtered through the glomeruli is found in the excreted urine. Inulin, a naturally occurring polysaccharide, is generally considered to be such a substance (see below). It was thought that the creatinine clearance was a good indicator of the GFR. It was found that in normal individuals the ratio of creatinine to inulin clearance was 1.02 but when the serum concentration was increased to around 4 to 5 mg/dl (either by the injection of creatinine into normal individuals or as found in patients with renal insufficiency), the ratio increased to approximately 1.4 [8]. Others have also felt that the creatinine clearance was not as reliable as the inulin clearance as it showed much greater variation [9]. Still other investigators claim that the creatinine clearance is sufficiently reliable as an estimate of the GFR [10]. In view of the relative simplicity of the creatinine clearance as compared with the inulin clearance, it will continue to be used for clinical purposes.

Inulin Clearance

Inulin is a polysaccharide (composed of fructose molecules) having a molecular weight of about 5,000. It is obtained from the tubers of certain members of the *Compositae* family. The substance is completely filtered from the plasma by the glomeruli and is not secreted or reabsorbed by the tubules. It is therefore a good substance for the measurement of GFR. The method usually requires continuous intravenous infusion of the inulin (to maintain a fairly constant level in the blood) and generally requires catheterization for accurate urine collection. It is therefore not a routine test and the details of the procedure will not be given here. Procedures for the analytical determination of both inulin and PAHA (see below) are given by Winston and Dalal [11], and by Bauer et al. [12].

p-*Aminohippuric Acid Clearance*

p-Aminohippuric acid (PAHA) is excreted both by the glomeruli and the tubules and can be used to measure the total blood flow in the kidney. At low plasma concentrations the kidneys will remove up to 90 percent of the PAHA from the blood in one passage. Under these conditions the PAHA clearance is considered as the effective renal plasma flow (ERPF), a measure of the actual plasma flow through the kidneys. This procedure, like inulin clearance, requires an intravenous infusion of the substance and thus is not a routine procedure. It has been used, with the aid of a urethral catheter, to measure the blood flow through the separate kidneys.

Normal Values and Interpretation of Results

The various normal values for the urea clearances have been given earlier in this section. These are usually expressed as a percentage of a standard value and thus the normal range would be 90 to 110%. The normal range for the creatinine clearance is 100 to 120 ml/min corrected to 1.73 m² surface area. This range is usually assumed for adults without correction for surface area.

The clearances are decreased in most types of kidney disease. Generally the amount of decrease in clearance is proportional to the severity of the disease.

Urine Osmolality [13]

The glomerular filtrate has an osmolality similar to the plasma from which it is derived. During the passage through the tubules, water and other substances may be added to or removed from the filtrate. These changes involve changes in the osmolality of the filtrate and thus the final osmolality may be a measure of kidney function. One can, in fact, calculate an osmolal clearance using the usual clearance formula:

$$C_o = \frac{U}{P} \times V$$

where U is the urine osmolality, P the plasma osmolality, and V the urine volume in milliliters per minute. However, it is usually sufficient to consider only the U/P ratio, which is a measure of the amount of filtered water that has been removed to form the final urine. If the urine has a higher osmolality than the plasma, some water must have been removed from the filtrate. If the urine has a lower osmolality than the plasma, additional water must have been added by the tubules. The U/P ratio may vary from 0.2 to 4.7. Normally the ratio is above 1, whereas in intrinsic renal disease the ratio is about unity. In circulatory shock, renal function is generally markedly decreased. As an indicator of the status of renal function, it was found that if the U/P ratio was less than 1.5 this was evidence of progressive renal failure, whereas if the ratio was over 1.5, renal failure was not likely [14].

Phenolsulfonphthalein (PSP) Excretion Test

This test is used as a measurement of the secretory activity of the proximal renal tubules [15–17]. About 60 to 70 percent of the dye is removed from the plasma by one passage through the kidney. The removal is mainly by tubular excretion, only a small fraction being removed by glomerular filtration. About 20 percent of the dye is removed by the liver. A fixed amount of the dye is injected intravenously and the amount excreted in the urine at various time intervals after injection is measured. Usually 1 ml of a sterile solution of the dye containing 6 mg/ml is injected.

REAGENTS

1. Sodium hydroxide, $2.5M$. Dissolve 100 g of sodium hydroxide in water to make 1 liter.

2. PSP standard. Pipet exactly 1 ml of the dye solution (of the same lot number as used for injection) into a 1 liter volumetric flask. Add 10 ml of sodium hydroxide solution and dilute to the mark with water. This represents a 100% standard. Prepare 30, 50, and 70% standards by diluting 3, 5, and 7 ml of the stock to 10 ml with water.

PROCEDURE

About 20 min before the start of the test, give the patient about 500 ml of water to drink to assure good urine flow. At the beginning of the test the bladder is emptied and the urine discarded. Exactly 1 ml of the dye (6 mg) is injected intravenously. The urine is collected at 15, 30, 60, and 120 min after the injection of the dye. The collection periods should be accurately timed and each specimen should have a volume of at least 40 ml. Preferably an indwelling catheter should be used but this may not always be practicable. Each urine sample is filtered into a 1 liter volumetric flask, about 500 ml of water and 10 ml of sodium hydroxide solution are added, and the flask is filled to the mark with water. The 60- and 120-min specimens may be diluted to 500 ml and the readings obtained divided by 2, as these specimens may contain only small amounts of the dye. Mix all samples thoroughly and measure absorbance of samples and standards against a water blank at 545 nm.

CALCULATION

$$\frac{\text{absorbance of sample}}{\text{absorbance of standard}} \times \text{conc. of standard} = \text{conc. of sample}$$

The concentrations are expressed as percent of the total dose excreted; the standards given above are 30, 50, and 70%. Routinely the 50% standard may be used.

NORMAL VALUES AND INTERPRETATION OF RESULTS

The normal amounts excreted are: in the 15-min specimen, 25 to 45%; in the 30-min specimen, 15 to 25%; in the 60-min specimen, 10 to 15%, and in the 120-min specimen, about 5%. Note that the first two specimens are the most important in regard to the total amount excreted. The excretion is lower in uncompensated chronic nephritis and parallels the nitrogen retention in the blood. Carinamide inhibits the tubular excretion of the dye.

References

1. Ganong, W. F.   *Review of Medical Physiology*, 2nd ed. Los Altos, Calif.: Lange, 1965.
2. Peters, J. P., and Van Slyke, D. D.   Quantitative Clinical Chemistry. Vol. 2, *Methods*. Baltimore: Williams & Wilkins, 1932. P. 208.
3. Moeller, E., McIntosh, J. F., and Van Slyke, D. D.   *J. Clin. Invest.* 6:427, 1928.
4. Van Slyke, D. D., Page, I. H., Hiller, A., and Kirk, E.   *J. Clin. Invest.* 14:901, 1935.
5. Bodansky, O.   *The Biochemistry of Disease*, 2nd ed. New York: Macmillan, 1952. P. 216.
6. Chesley, L. C.   *J. Clin. Invest.* 16:653, 1937.
7. Brulles, A., Gras, J., Magriñá, M., Torres, N., and Caralps, A.   *Clin. Chim. Acta* 24:261, 1969.
8. Gayer, J.   *Med. Welt* 1:38, 1964.
9. Dodge, W. F., Travis, L. B., and Deutschner, C. W.   *Am. J. Dis. Child.* 113:683, 1967.
10. Heierli, C., Thoelen, H., and Bertels, H.   *Dtsch. Med. Wochenschr.* 97:67, 1972.
11. Winston, S., and Dalal, F.   *Manual of Clinical Laboratory Procedures for Non-Routine Problems*. Cleveland: Chemical Rubber Co., 1972.
12. Bauer, J. D., Ackermann, P. G., and Toro, G.   *Clinical Laboratory Methods*, 8th ed. St. Louis: Mosby, 1974.
13. Smith, H. W.   *The Kidney, Structure and Function in Health and Disease*. New York: Oxford University Press, 1951.
14. Jones, L. W., and Weil, M. H.   *Am. J. Med.* 51:314, 1971.
15. Rowntree, L. G., and Geraghty, T. J.   *J. Pharmacol. Exp. Ther.* 1:579, 1910.
16. Goldring, W., Clarke, R. W., and Smith, H. W.   *J. Clin. Invest.* 15:221, 1936.
17. Pierce, J. M., Jr., Rusumma, R., and Segar, R.   *J.A.M.A.* 175:711, 1961.

12. Lipids

The lipids are a rather heterogeneous group of substances that resemble fats in their solubility characteristics and are mostly involved in the metabolism of the fatty acids. A complete discussion of lipid analyses would include a number of substances that are not found to any appreciable extent in the human blood stream and these will not be covered here. There are also lipids or lipid-like substances in the erythrocytes but these lipids are not concerned with fat metabolism and are rarely determined. The main serum lipids are: cholesterol, cholesterol esters, phospholipids and triglycerides (neutral fats), and small amounts of monoglycerides and diglycerides, nonesterified or free fatty acids, bile acids, other steroids, and the fat-soluble vitamins (A and D). The lipids are not soluble in aqueous solutions and are present in the serum either in combination with proteins (lipoproteins) or as emulsified fats (chylomicrons).

Requests for the determination of one or more of the blood lipid components may have one of several objectives. One may be to pinpoint the malfunctioning of the digestion and absorption of fats that occurs in pancreatic disease or steatorrhea, in which the fasting levels of the blood lipids may be low, but blood lipid analyses are of little value in assessing the malabsorption syndromes. A fat tolerance test, the measurement of the changes in the blood total fatty acids (or esterified fatty acids), or simply the measurement of the changes in turbidity after a fat meal, may be helpful. A fat balance test, which involves the determination of the fat in the stool, is often more satisfactory.

Perhaps the main reasons for the study of the serum lipids is their apparent close relationship with cardiovascular disease. High serum lipid levels are correlated with a higher incidence of atherosclerotic heart disease. The studies may be carried out by two general methods. In one, the specific individual components such as cholesterol (free and esterified), phospholipids, triglycerides, and nonesterified fatty acids are determined separately along with the total lipids. In the other approach, the emphasis is on the lipid-protein complexes, the lipoproteins. These may be separated into a number of classes by ultracentrifugation, immunological methods, or more generally by electrophoresis. The electrophoresis may be carried out in cellulose acetate or polyacrylamide or other gels, with either subsequent fat staining to locate the separated fractions or the determination of the amount of a given constituent such as cholesterol in the various fractions. Even when the electrophoresis is carried out, determination is also made of some of the major constituents such as cholesterol and triglycerides.

Normal blood serum or plasma in the postabsorptive state contains cholesterol and phospholipids in approximately equal amounts, 150 to 300 mg/dl. The triglycerides are present in concentration of 40 to 150 mg/dl and the other lipid constituents are present in even smaller amounts. The amount of total lipid is 450 to 900 mg/dl. We will first present methods for the determination of the total and individual lipid components—total lipid, cholesterol, phospholipids, triglycerides, and nonesterified fatty acids—and then the procedure and interpretation of lipoprotein electrophoresis.

Total Lipids

The total lipids have been determined by turbidimetric, gravimetric, and colorimetric methods. In the *turbidimetric* methods [1–3] the lipids are first extracted from the serum, the extract redissolved, and a reagent added to form a turbid suspension of the lipids. The *gravimetric* methods [4–10], in which the lipids are extracted from the serum, purified, and finally determined by weighing the amount of extract, are conceded to be the most accurate, but they are also time consuming. Although not suitable for routine use, gravimetric methods are generally used as reference methods and for the determination of the lipids in a serum pool to be used as a standard or control. The gravimetric method, being an absolute one, is not dependent upon other standards. The *colorimetric* method based on the reaction introduced by Chabrol and Chardonnet [11] is now widely used. It is relatively simple though not absolutely specific for lipids [12]. A small amount of serum is heated with concentrated sulfuric acid and the mixture then reacted with a phosphoric acid vanillin reagent to give a red to purple color. The reaction is apparently a complex one and is given by a number of substances besides lipids. Good results are obtained when a material such as olive oil or a pooled serum is used as a standard [13–15]. Because of the heterogeneous nature of the total lipids, its determination is not as meaningful as the determination of the individual components. It is useful as a screening procedure and in checking the values of the individual components. The colorimetric method has been completely automated [16–19].

Total Lipids by Gravimetric Method

We will give in detail the method of Sperry [4]. In this method the lipids are extracted with methanol-chloroform and the extract washed with a special washing solution to remove inorganic impurities. An aliquot of the washed extract is then evaporated to dryness and weighed.

REAGENTS

1. Methanol, reagent grade.
2. Chloroform, reagent grade.

3. Wash solution. Dissolve approximately 100 mg of calcium chloride in water to make 500 ml. Mix together 100 ml of chloroform and 50 ml of methanol. In a separatory funnel, pipet 96.3 ml of the chloroform-methanol mixture and 23.7 ml of the calcium chloride solution. Shake gently for a few minutes and then allow the layers to separate completely. Draw off and discard the lower chloroform layer and save the upper aqueous layer for use as the wash solution.

PROCEDURE

Add 2 ml of serum to 16.7 ml of methanol in a 50 ml volumetric flask. Add 16.7 ml of chloroform and heat just to boiling on a steam bath. Cool the flask to room temperature and dilute to the mark with chloroform. Stopper and mix well. Filter through paper into a 125 ml Erlenmeyer flask, covering the top of the funnel with a watchglass to prevent evaporation of the solvent. Pipet 40 ml of the filtrate to a 50 ml glass-stoppered graduated cylinder. Add exactly 8 ml of water and mix vigorously. Allow the cylinder to stand undisturbed until the two layers separate completely. This may require allowing to stand overnight. Remove and discard the upper aqueous layer with a Pasteur pipet, taking care not to remove any of the lower layer. Add 6 ml of the wash solution slowly with a pipet, washing down the walls of the cylinder above the organic layer. Swirl the cylinder gently, then remove wash solution with a pipet. Repeat the washing twice with the wash solution. After removing the last aliquot of wash solution, add 4 ml of methanol to the liquid in the cylinder, mix, and transfer to a 100 ml beaker. Rinse out the cylinder several times with small aliquots of the 2:1 chloroform-methanol mixture and add rinsings to the beaker. Evaporate to a small volume on a steam bath, preferably at a temperature not over 50°C, with the aid of a slow stream of N_2 or CO_2. Quantitatively transfer the solution to a tared beaker or weighing bottle, rinsing the beaker several times with chloroform and adding washings to the tared vessel. Evaporate completely to dryness, store in vacuum dessicator over calcium chloride overnight. Weigh tared vessel and contents to the nearest 0.1 mg. Subtract weight of vessel without lipid to obtain the weight of the lipids.

CALCULATION

In the procedure as given, the aliquot evaporated represents 80% of the original aliquot (2 ml) of the serum taken, or 1.6 ml. This must be taken into account in the calculations. Then:

$$\frac{\text{weight of lipid (in mg)}}{\text{vol. of serum (in ml) (1.6)}} \times 100 = \text{total lipid (in mg/dl)}$$

Total Lipids by Colorimetric Method

Although the reactions involved in this determination are not specific for lipids, the method [13] usually gives results comparable with the gravimetric method. The method is very simple and since the measurement of total lipids cannot be of the highest clinical significance, it is satisfactory for most purposes.

REAGENTS

1. Vanillin reagent, 0.04M.   Dissolve 6.1 g of vanillin in water and dilute to 1 liter. This solution is stable for about 2 months in a brown bottle at room temperature.

2. Phosphovanillin reagent.   Add 350 ml of the vanillin reagent and 50 ml of water to a flask. Add with constant stirring, 600 ml of concentrated (85%) phosphoric acid. This solution is also stable for about 2 months in a brown bottle at room temperature.

3. Sulfuric acid, concentrated, reagent grade.

4. Standard solution.   A good U.S.P. grade of olive oil may be used as a standard. In two tared 100 ml volumetric flasks add approximately 0.5 and 1 ml of the olive oil and weigh again to obtain the exact weight of oil added. (It is time consuming to try to weigh out exactly 500 mg, or any other definite weight, of the oil; the approximate amounts are added and the exact weight determined.) The above standards should be about 500 and 1,000 mg/dl. Dissolve the oil in absolute ethanol and dilute to the mark with the ethanol. This solution is stable for about 1 month in the refrigerator.

PROCEDURE

In separate tubes add 20 μl of water (blank), 20 μl of samples, and 20 μl of standards. To each tube add 0.2 ml of concentrated sulfuric acid. Mix well, preferably on a vortex mixer. Place all tubes in boiling water bath for 10 min, remove, and cool in water to room temperature. To each tube add 10 ml of the phosphovanillin reagent and mix well. Incubate at 37°C in a water bath for 15 min. Cool and read standards and samples against blank at 540 nm.

CALCULATION

$$\frac{\text{absorbance of sample}}{\text{absorbance of standard}} \times \text{conc. of standard} = \text{conc. of sample (total lipid in mg/dl)}$$

The method should be linear up to over 1,000 mg/dl, but in the calculations one may use for each sample the standard having the absorbance nearest that of the sample.

The normal range of total lipids is 400 to 900 mg/dl. The total lipids will obviously be increased when any of the various separate lipid fractions is increased. The interpretation of these increases will be discussed later. In extremely hyperlipemic serums the total lipids may be over 2,000 mg/dl.

Cholesterol and Cholesterol Esters

There are many variations of the different methods for cholesterol and its esters. Reviews of the methodology may be consulted for details [20–23]. We will give here only a very brief survey. Most of the methods for cholesterol may be classified as those using the Lieberman-Burchard reaction (acetic anhydride and sulfuric acid), those using ferric salts in addition to acetic acid and sulfuric acid [24–26], those using p-toluenesulfonic acid [23–29], and some fluorometric methods [30–32]. Some methods hydrolyze the esters before determining total cholesterol, others do not. Some methods make the determination directly on serum, others use an extract of the serum made with an organic solvent such as given in the gravimetric method for total lipid. The direct methods are more subject to interference by bilirubin and hemolysis.

We present three methods for the determination of cholesterol. The first, that of Abell, is a reference method for total cholesterol and may be used for determining accurately the cholesterol content of pooled serums for standards or control serums. The second method is an extraction method for free and esterified cholesterol. The third method given is a direct method without extraction. Although this may not be as accurate as the earlier methods, it is very simple and usually sufficiently accurate for most clinical purposes.

Total Cholesterol by Method of Abell

In this method [33–35] the cholesterol is saponified before extraction with hexane. An aliquot of the extract is evaporated to dryness and treated with the Lieberman-Burchard reagent.

REAGENTS

1. Stock potassium hydroxide, 33% by weight. Dissolve 20 g of potassium hydroxide in 40 ml of water. Keep in tightly stoppered plastic bottle.

2. Working potassium hydroxide solution. Just before use dilute 6 ml of the stock hydroxide solution with 94 ml of absolute ethanol.

3. Color reagent. Prepare just before use. Add 100 ml of acetic anhydride to a glass-stoppered bottle or flask. Cool in ice and add gradually with swirling, 5

ml of concentrated sulfuric acid. Cool again in ice, then add 50 ml of glacial acetic acid. Allow this solution to come to room temperature before use.

4. Petroleum ether (bp, 30°–60°C).

5. Cholesterol standard. Dissolve 250 mg of pure cholesterol in absolute ethanol to make 100 ml. This gives a standard containing 250 mg/dl cholesterol.

PROCEDURE

Set up a series of 25 ml glass-stoppered centrifuge tubes. To each tube add 0.5 ml of the working hydroxide solution. To separate tubes add 0.5 ml of serum samples or standard. Stopper the tubes securely and mix well. Incubate the tubes in a water bath at 37°C for 1 hr. During the incubation period, mix the tubes at several intervals by gentle inversion. Cool the tubes to room temperature and add exactly 10 ml of petroleum ether to each tube, followed by 5 ml of water. Stopper the tubes securely and shake viborously for 1 min; allow to stand until the layers separate completely. Remove most of the lower aqueous layer with a Pasteur pipet and centrifuge the tubes. Pipet off exactly 4-ml aliquots of the supernatant ether layers to separate tubes, taking care not to include any of the aqueous layer. Evaporate the ether by placing the tubes in a warm water bath, hastening the evaporation with a stream of N_2 or CO_2. (*Caution:* The ether vapor is highly flammable and the evaporation should be carried out in a good fume hood.) Place the tubes containing the evaporated extract and an empty tube as blank in a water bath at 25°C (a large pan of water is satisfactory). To each tube, at timed intervals (usually 30 sec), add 6 ml of the color reagent and mix for a few seconds on a vortex mixer. Place the tubes in the water bath in a dark place for 30 min. Read against the blank exactly 30 min after the addition of the color reagent, at 630 nm.

CALCULATION

Since the standard and samples are treated similarly,

$$\frac{\text{absorbance of sample}}{\text{absorbance of standard}} \times 250 = \text{cholesterol in sample (in mg/dl)}$$

Total and Free Cholesterol by Method of Parekh and Jung

Of the many methods that determine the total and free cholesterol in serum we have chosen to present the method of Parekh and Jung [36, 37]. This is a relatively simple method that uses the reaction with an iron salt. The total cholesterol is determined directly on the serum by extracting with one of the reagents and, after centrifugation to remove a small amount of precipitate, developing a color in an aliquot of the extract with a second reagent. The free and esterified cholesterol are separated by precipitating the free with digitonin. The precipitate is treated with the color reagent direcly after washing. A standard is treated similarly to compensate for any color due to the digitonin.

TOTAL CHOLESTEROL

REAGENTS

1. Ferric acetate-uranyl acetate. Dissolve 0.5 g of ferric chloride ($FeCl_3 \cdot 6H_2O$) in about 10 ml of water in a centrifuge tube. Add sufficient concentrated ammonium hydroxide to precipitate all the iron as ferric hydroxide (about 3 ml will be required, a slight excess does no harm). Centrifuge the mixture and decant the supernatant. Wash the precipitate several times with about 10 ml of water containing a few drops of ammonium hydroxide. Dissolve the precipitate in glacial acetic acid, transfer to a 1 liter volumetric flask, and dilute to the mark with acetic acid. Add to this 0.1 g of powdered uranyl acetate [$UO_2 (C_2H_3O_2)_2 \cdot 2H_2O$] and mix well. Allow to stand overnight, then mix well. This solution is stable for about 6 months when stored in a brown bottle at room temperature.

2. Sulfuric acid-ferrous sulfate. In a 1 liter volumetric flask place 0.1 g of anhydrous ferrous sulfate and 100 ml of glacial acetic acid. Swirl to dissolve as completely as possible. Then add, with constant swirling, 100 ml of concentrated sulfuric acid. Cool to room temperature and dilute to the mark with sulfuric acid. *Caution:* This reagent is very corrosive. It should be stored in a tightly closed bottle to prevent absorption of moisture.

3. Cholesterol standard, 250 mg/dl. Dissolve 250 mg of pure cholesterol in chloroform to make 100 ml. This solution is stable but it must be kept in a tightly stoppered bottle to prevent evaporation of chloroform.

PROCEDURE

Using screw-capped tubes with Teflon-lined caps (glass-stoppered tubes may also be used), add 50 μl of serum samples or standard to separate tubes. To each tube add 10 ml of reagent 1. Cap and mix gently by inversion a few times, then mix well on a vortex mixer. Allow the tubes to stand for about 5 min and then centrifuge strongly. Transfer 3-ml aliquots of the clear supernatants to another set of tubes. Also set up a blank tube containing 3 ml of reagent 1. To each tube add 2 ml of reagent 2, cap, and mix well by inversion. The corrosive reagent is preferably delivered by some type of all-glass dispenser. Allow the tubes to stand at room temperature for 20 min; then read standard and samples against blank at 560 nm.

CALCULATION

Since the standard and samples are treated similarly:

$$\frac{\text{absorbance of sample}}{\text{absorbance of standard}} \times 250 = \text{cholesterol in sample (in mg/dl)}$$

FREE AND ESTERIFIED CHOLESTEROL

ADDITIONAL REAGENTS

4. Digitonin solution.   Dissolve 1 g of purified digitonin in 100 ml of 50% ethanol by warming to about 60°C with gentle agitation.

5. Isopropanol, reagent grade.

6. Cholesterol standard for free cholesterol.   Dissolve 75 mg of pure cholesterol in isopropanol to make 100 ml. This contains 75 mg/dl.

PROCEDURE

Add 0.2 ml of serum or standard to separate screw-capped centrifuge tubes. Add 2 ml of isopropanol to each tube and mix well. Allow to stand for about 5 min and then centrifuge strongly. Transfer 1-ml aliquots from the supernatant to another set of tubes. Add to each tube 0.5 ml of the digitonin solution. Mix well and place tubes in refrigerator for 30 min. Centrifuge strongly, then decant the supernatant and discard it. Wash the precipitate remaining in the tube with two 10-ml portions of acetone, centrifuging each time and decanting the acetone. After removal of the second wash, invert the tubes and allow to drain for a few minutes; then add 5 ml of reagent 1 to each tube. Mix thoroughly to dissolve the precipitate. Add 2 ml of reagent 2 to each tube by running it down the inner wall of the tube. Cap the tubes and mix by holding the capped end and swinging in an 180° arc ten times. Allow the tubes to stand at room temperature for 20 min, then read samples and standards at 560 nm against a blank of a mixture of 3 ml of reagent 1 and 2 ml of reagent 2.

CALCULATION

Since the standard and samples are treated similarly,

$$\frac{\text{absorbance of sample}}{\text{absorbance of standard}} \times 75 = \text{free cholesterol (in mg/dl)}$$

The esterified cholesterol is estimated by difference:

total cholesterol − free-cholesterol = esterified cholesterol

Direct Method for Cholesterol [37a]

Although the direct method is more subject to interference by bilirubin and hemolysis than the extraction procedures, it is very simple and rapid. It is used in many automated methods and as a simple manual method. Increased bilirubin levels will cause high results for cholesterol. If the bilirubin level is known, an approximate correction can be applied (see calculations below). Visible hemolysis (more than 100 mg of hemoglobin per deciliter of serum) will render the sample unsuitable for analysis.

REAGENTS

1. Lieberman-Burchard reagent. Cool acetic anhydride and concentrated sulfuric acid to about 0°C in a freezer. In a 1 liter flask, cooled in ice water, place 220 ml of cold acetic anhydride. Add with swirling 200 ml of glacial acetic acid (at room temperature), then add slowly with mixing 30 ml of cold concentrated sulfuric acid. This reagent is stable for at least one month when kept in a brown bottle in a refrigerator.

2. Cholesterol standard, 200 mg/dl. Dissolve 200 mg of pure cholesterol in glacial acetic acid and dilute to exactly 100 ml. This standard is stable at room temperature.

PROCEDURE

To separate test tubes labeled blank, standard, and sample add 6 ml of the reagent. (Because of the strongly corrosive nature of the reagent it is preferable to carry out the color development in the cuvettes to be used, to avoid transfer of the reagent.) To standard tube add 0.1 ml of standard, to sample tube add 0.1 ml of serum, and to blank tube add 0.1 ml of water. Carefully mix each tube well and incubate all tubes for 18 min at 37°C. Read standard and sample against blank at 625 nm.

CALCULATION

$$\frac{\text{absorbance of sample}}{\text{absorbance of standard}} \times \text{conc. of standard (200 mg/dl)} = \text{conc. of sample}$$

Comment. An approximate correction for high bilirubin levels may be made by subtracting 6 mg/dl cholesterol from the result obtained for each mg/dl of bilirubin above 1 mg/dl.

Normal Values and Interpretation of Results

The range of normal total cholesterol values for healthy young adults on the usual American diet may be taken as 160 to 270 mg/dl. The level increases somewhat with age until after the sixth decade when it decreases somewhat. The proportion of the total cholesterol present as esters is between **68** and **74%** in normal individuals.

Cholesterol and other lipids are increased in the serum in uncontrolled diabetes but not in proportion to the severity of the disease. The cholesterol level is variable in liver disease. Increased cholesterol levels are found in hypothyroidism, but the level is not a good indicator of the severity of the disease. Cholesterol levels are influenced by the dietary intake, particularly of saturated (animal) fats. Individuals with high cholesterol levels are much more likely to be subject to cardiovascular disease than those with lower levels. This is discussed further in the section on serum lipoproteins later in this chapter.

Phospholipids

The phospholipids in serum are composed of a glycerol molecule in which two of the hydroxyl groups are esterified with long-chain fatty acids and the third is esterified with phosphoric acid. The phosphate has also attached to it a grouping containing an amino nitrogen (in lecithin the grouping is choline, and in cephalin the grouping is ethanolamine). In brain and nervous tissue a different type of phospholipid is found. These contain no glycerol, and upon hydrolysis yield a fatty acid, phosphoric acid, choline, and a complex amino alcohol. Blood phospholipids are usually determined, after separation from other material, by means of their phosphate content. The phospholipids may be precipitated from the serum along with other proteins by an agent such as trichloracetic acid. After separation, the precipitate is digested with sulfuric and nitric acids and the phosphate determined in the digest by one of the usual methods. The phospholipids may also be extracted from the serum by an organic solvent, an aliquot of the extract evaporated, and the phosphorus in the residue determined, after digestion, by any method for phosphate. In either method it is assumed that all the phosphorus found is derived from the phospholipid present. The figure for phosphorus content is then converted to phospholipid by multiplying by a factor of 25. This is an average value for the ratio in the phospholipids commonly found in blood.

Phospholipid Phosphorus by Method of Baginski et al. [38–40]

The phospholipids are extracted from the serum with alcohol-ether. An aliquot of the extract is evaporated and digested with nitric acid to destroy organic matter. A small amount of calcium is added to prevent loss of phosphate during digestion. The phosphorus is then determined by a modification of the reaction with acid molybdate, using ascorbic acid as a reducing agent.

1. Trichloracetic acid with ascorbic acid, 0.6M. A trichloracetic acid solution (0.6M) is prepared by dissolving 100 g of trichloracetic acid in water to make 1 liter or by dilution of the 1.8M acid prepared as directed in Chapter 5 for protein-free filtrates. As required, 1 g of ascorbic acid is dissolved in 50 ml of the 0.6M trichloracetic acid. This solution is not stable more than a few weeks in the refrigerator.

2. Ammonium molybdate, 1 g/dl. Dissolve 1 g of ammonium molybdate [$(NH_4)_6Mo_7O_{24} \cdot 4H_2O$] in water to make 100 ml.

3. Arsenite-citrate solution. Dissolve 20 g of trisodium citrate ($Na_3C_6H_5O_7 \cdot 2H_2O$) and 20 g of sodium arsenite ($NaAsO_2$) in water, add 20 ml of glacial acetic acid, and dilute to 1 liter.

4. Nitric acid-calcium solution. Dissolve approximately 30 mg of calcium carbonate in 1 liter of concentrated nitric acid.

5. Antibumping granules.* The granules are prewashed by boiling with nitric acid and then rinsed thoroughly with water and dried. A coarse carborundum powder could probably be similarly washed and used.

6. Extraction mixture (alcohol-ether 3:1). Mix 3 volumes of absolute ethanol and 1 volume of ether.

7. Phosphate standards.

A. Stock standard, 1 mg/ml of phosphorus. Dry reagent grade monopotassium phosphate (KH_2PO_4) for several hours at 100°C and cool in a dessicator. Weigh out exactly 439.4 mg and dissolve in water to make 100 ml.

B. Working standard. Dilute 5 ml of the stock standard to 100 ml with water. This contains 0.05 mg/ml of P.

Pipet 2 ml of the extraction mixture into a tube; add 50 μl of serum with continuous mixing. Mix thoroughly on a vortex mixer, allow to stand for 5 min, and centrifuge strongly to obtain a clear supernatant. Pipet 1 ml of the supernatant into a 25 × 150 mm borosilicate test tube and evaporate to dryness on a steam bath. Also set up one tube containing 50 μl of the working standard and a third empty tube as a blank. To each tube (including blank) add 2 ml of the nitric acid solution and a few antibumping granules. Heat the tubes over a microburner, gently at first, until the nitrous oxide fumes have completely disappeared and the acid is entirely evaporated. A slight excess heating will do no harm but prolonged heating after complete evaporation of the nitric acid should be avoided. Cool the

* Gallard-Schlesinger Chemical Manufacturing Corp., Carle Place, N.Y. 11514.

tubes and add 1.0 ml of the ascorbic acid-trichloracetic acid mixture to each tube. Agitate to dissolve any precipitate present. Add 0.5 ml of the molybdate solution and mix: then add 1 ml of the arsenite-citrate solution and mix again. Allow the tubes to stand for 20 min and then read standard and samples against the blank at 700 nm. If the cuvettes used require larger volumes, all the quantities (including extract) can be doubled to obtain a volume of 5 ml.

CALCULATION

Since 0.05 ml of serum is diluted with 2.0 ml of extraction solvent in making the extraction, 1 ml of the extract is equivalent to $(1 \times 0.05)/(2.00 + 0.05) = 0.0244$ ml of serum. This is compared with 0.05 ml of standard containing 0.05 mg/ml P, or 0.0025 mg P in aliquot used.

$$\frac{\text{absorbance of sample}}{\text{absorbance of standard}} \times \frac{0.0025 \times 100}{0.0244}$$

$$= \frac{\text{absorbance of sample}}{\text{absorbance of standard}} \times 10.25 = \text{lipid P (in mg/dl)}$$

The figure for phosphorus is converted to phospholipid by use of the factor 25:

lipid P $\times$ 25 = phospholipid

COMMENTS

Many detergents contain phosphate in large quantities and tubes washed with such cleansers must be thoroughly rinsed before use, preferably a few times with dilute nitric acid before the final water rinses.

NORMAL VALUES AND INTERPRETATION OF RESULTS

The normal serum level of lipid phosphorus is 6 to 14 mg/dl which corresponds to 150 to 350 mg/dl of phospholipids. Generally increases in phospholipid parallel increases in cholesterol in serum.

Triglycerides

The triglycerides have been determined by two general methods, a chemical determination after extraction from the serum and by enzymic methods which may be carried out directly on the serum. In one extraction method the serum is extracted with isopropanol and the extract treated with the silicates zeolite or Florisil to remove interfering material [41, 43]. In the other extraction method, a mixture of nonane, isopropanol, and dilute acid is used [44, 45]. This extracts only the triglycerides and further purification is not needed. After extraction, the triglycerides are hydrolyzed with alkali to the salts of fatty acids and glycerol (glycerin). The glycerol is oxidized by periodate to formaldehyde, which is determined colorimetrically. In the earlier methods the formaldehyde combined with chromotropic acid in strong acid to form a blue color [46–48]. More recent methods use the condensation of formaldehyde with ammonium salts and acetylacetone to form a substituted lutidine having an absorption peak near 410 nm [49, 50]. The yellow compound formed is also fluorescent so that the method can be used with a fluorometer as well as a photometer [51, 52]. The principle of the reaction is as follows:

triglycerides + sodium methoxide + isopropyl alcohol → glycerol + fatty acid esters

glycerol + sodium metaperiodate → formaldehyde

formaldehyde + acetylacetone + ammonium acetate → 3,5-diacetyl-1,4-dihydrolutidine

The triglycerides may also be determined enzymatically. After enzymatic hydrolysis or saponification by alkali, the liberated glycerol is determined. One method uses the coupled series of reactions similar to those of the hexokinase method for glucose: the phosphorylation of the glycerol by glycerol kinase (GK) and ATP, with the formation of ADP. The ADP reacts with phosphoenolpyruvate (PEP) in the presence of pyruvate kinase (PK), with the formation of pyruvate which is then determined with LDH and NADH [53–56]. Since there is a small amount of free glycerol in the serum, this must also be determined and subtracted from the total glycerol determined after hydrolysis. The enzymatic method is simple in that it requires no extraction or other prior treatment of the serum. The enzymes necessary are available in kit form from a number of suppliers. The principle of this reaction is as follows:

$$\text{glycerol} + \text{ATP} \xrightarrow{\text{GK}} \text{GP} + \text{ADP}$$

$$\text{ADP} + \text{PEP} \xrightarrow{\text{PK}} \text{pyruvate} + \text{ATP}$$

$$\text{pyruvate} + \text{NADH} \xrightarrow{\text{LDH}} \text{lactate} + \text{NAD}$$
$$\qquad\text{(colored}\qquad\qquad\text{(colorless}$$
$$\qquad\text{at 340 nm)}\qquad\quad\text{at 340 nm)}$$

We present two slightly different modifications of an extraction method using nonane or heptane.

Triglycerides by Nonane Extraction

REAGENTS

1. *n*-Nonane, reagent grade.
2. Isopropanol, reagent grade.
3. Sulfuric acid, $0.04M$.   Dilute 40 ml of $1M$ sulfuric acid to 1 liter with water.
4. Acetic acid, $1M$.   Dilute 58.5 ml of glacial acetic acid to 1 liter with water.
5. Sodium methylate.   Dissolve 50 mg of sodium methylate (sodium methoxide) in 100 ml of isopropanol as needed. This solution is stable for at least 1 week in the refrigerator.
6. Sodium metaperiodate, stock solution, $0.025M$.   Dissolve 5.347 g of sodium metaperiodate ($NaIO_4$) in acetic acid $(1M)$ to make 1 liter.
7. Sodium metaperiodate, working solution.   As needed, dilute 12 ml of stock sodium metaperiodate and 20 ml of isopropanol with $1M$ acetic acid to make 100 ml. Prepare fresh daily.
8. Ammonium acetate, $2M$. Dissolve 154.2 g of ammonium acetate ($NH_4C_2H_3O_2$) in water to make 1 liter. This solution is stable for several months at room temperature.
9. Acetylacetone.   Dissolve 0.75 ml of acetylacetone (2,4-pentanedione) and 2.5 ml of isopropanol in 100 ml of $2M$ ammonium acetate solution. This reagent is stable for about 1 month when stored in the refrigerator in a brown bottle.
10. Triolein standard.   Dissolve 300 mg of pure triolein in isopropanol to make 100 ml. This solution is stable if kept in a tightly stoppered bottle in the refrigerator. A good (U.S.P.) grade of olive oil can be used as a secondary standard if checked against the triolein standard.

PROCEDURE

Use 16×150 mm screw-top test tubes with caps having Teflon liners. To a series of tubes add 2 ml of nonane, 3.5 ml of isopropanol, and 1.0 ml of the sulfuric acid solution. (These tubes can be made up beforehand and kept tightly stoppered in the refrigerator.) To separate tubes add 0.5 ml of serum (samples) and to one tube add 0.5 ml of water (blank). Also set up a standard tube with 0.5 ml of standard, 0.5 ml of water, 2.0 ml of nonane, 3.0 ml of isopropanol, and 1 ml of sulfuric acid. Mix the tubes well for 30 sec on a vortex mixer, then allow the layers to separate completely. Use aliquots of the upper layer for analysis.

In separate test tubes add 0.2-ml aliquots from upper layer of standard, sample, and blank tubes. Add to each tube 3 ml of the sodium methylate solution, mix, and incubate at 60°C for 15 min. Add 0.2 ml of the working metaperiodate solution and mix. Add 1 ml of the acetylacetone reagent, mix, and incubate at 60°C for

15 min. Cool to room temperature, and read samples and standard against blank at 410 nm. Color is stable for at least 1 hr.

Since the standard and samples are treated similarly,

$$\frac{\text{absorbance of sample}}{\text{absorbance of standard}} \times 300 = \text{triglycerides (in mg/dl)}$$

The triglycerides may also be expressed as millimoles per liter by multiplying the above result by 0.0113 (when triolein is used as standard), or as milliequivalents per liter by multiplying the above result by 0.036.

First the linearity of the photometer and the procedure should be checked. Prepare a supernatant from a number of diluted standards by the procedure given above. To five test tubes add 1, 2, 3, 4, and 5 ml of the triglyceride standard, add isopropanol, when needed, to make a final volume of 5 ml, and mix thoroughly. Treat 0.5 ml of the standards as described in the procedure. These correspond to 60, 120, 180, 240, and 300 mg/dl of triglycerides. Plot the calibration curve. If it is not linear, several different standards should be run with each set of samples and the standard having an absorbance nearest that of a given sample should be used in the calculation for that sample.

Alternative Procedure (Gottfried and Rosenberg)

This procedure [57] uses heptane instead of nonane, and an aqueous solution of potassium hydroxide instead of sodium methylate; solutions of KOH are more stable.

REAGENTS

Only the following numbered reagents are different from the corresponding ones in the previous procedure; the others are the same and are not repeated here.

1A. *n*-Heptane, Spectro grade AR.

3A. Potassium hydroxide solution, 6.25M. Dissolve 35 g of potassium hydroxide in water, cool, and dilute to 100 ml with water.

7A. Working periodate solution (made up directly). Dissolve 3 g of sodium metaperiodate and 25 ml of glacial acetic acid in water to make 500 ml.

PROCEDURE

The extraction procedure is exactly the same as given in the previous method except for the use of heptane in place of nonane. After extraction, pipet into separate tubes 0.4 ml of the upper layer. To each tube add 2 ml of isopropanol and 1 drop of the potassium hydroxide solution. Mix well, stopper, and incubate at 70°C for 10 min. Allow to cool to room temperature and add to each tube 0.2 ml of periodate solution and mix. Add 1 ml of the acetylacetone solution, stopper, and incubate at 70°C for 10 min. Cool to room temperature and read standards and samples against blank at 410 nm. The calculations are the same as for the previous method. Since the reaction may not be exactly linear above 200 mg/dl, it is advisable to run three standards with each group of samples. Dilution aliquots of the 300 mg/dl standard (1:3 and 2:3 to obtain 100 and 200 mg/dl standards) can be used.

If for either modification it is desired to obtain larger final volumes for the colorimetric readings, all the volumes used after the extraction step can be doubled.

Normal Values and Interpretation of Results

The normal values found by this method range from 40 to 145 mg/dl (1.4 to 4.9 mEq/liter). Values below the normal range are of no significance. Elevated levels are often indicative of cardiovascular disease (see later discussion in this chapter of lipoprotein phenotyping).

Nonesterified (Free) Fatty Acids

Although they represent but a small fraction of the total serum lipids, the nonesterified fatty acids represent the major endogenous lipid transport system. Although often called free fatty acids, they are present in the serum bound in albumin-fatty acid complexes. They are free, however, in the sense that they are not esterified. The free fatty acids enter the blood at increased rates under conditions in which carbohydrates are not available in adequate amounts or cannot be utilized because of hormonal abnormalities, and thus they are usually increased in diabetes.

The determination of the free fatty acids in serum or plasma after extraction with an organic solvent may be accomplished in a number of ways. In one of the earlier procedures the free fatty acids were titrated with standard alkali [58–60]. In another type of procedure the organic solvent extract of the fatty acids is shaken with an aqueous solution of a copper or cobalt salt, which forms metallic soaps with the fatty acids. These soaps, being soluble in the organic phase, introduce some of the metal into this phase. After careful separation from the excess aqueous reagent, the amount of copper or cobalt in the organic phase is determined colorimetrically [61–66] or by atomic absorption spectrophotometry [67]. The amount of fatty acids has also been estimated by subjecting the extract to thin layer chromatography and using the size of the resulting spots as a semi-quantitative determination of the fatty acids [68, 69].

We present two slightly different extraction procedures for the colorimetric determination of free fatty acids. The first method requires less manipulation and gives a better separation of the copper solution (a major source of error in the methods is the carryover of traces of the copper reagent to the final colorimetric solution), but requires special apparatus. The second method is slightly longer but requires no special apparatus and is readily adapted to final measurement by atomic absorption.

Free Fatty Acids by Method of Soloni and Sardina [62]

REAGENTS

1. Copper reagent. Dissolve 40 g of cupric nitrate $[Cu(NO_3)_2 \cdot 3H_2O]$ in about 500 ml of water, add 120 ml of reagent grade triethanol amine, and dilute to 1 liter with water.

2. Cuprizone reagent. Dissolve 20 mg of cuprizone† in 5 ml of chloroform and dilute to 500 ml with isopropyl alcohol. This solution is stable for 3 months at room temperature.

3. Ammonia reagent, $1.5M$. Dilute 100 ml of concentrated ammonium hydroxide to 1 liter with water.

4. Fatty acid-free albumin. Shake 5 g of crystallized human albumin with 500 ml of methanol for 15 min at room temperature. Filter the albumin by suction through paper. Break up the cake and shake again for 15 min in a glass-stoppered bottle with 300 ml of acetone. Filter the albumin by suction and dry under vacuum or a stream of nitrogen until free of acetone.

5. Chloroform, AR.

6. Standard fatty acid solution, 50 mg/dl. Dissolve 50 mg of pure palmitic acid in 4 ml of water and 1 ml of $1M$ sodium hydroxide in a 100 ml volumetric flask by warming to 80°C for 15 to 20 min. Add 40 ml of water and warm again for 15 min. Dissolve 4.5 g of the purified albumin and 10 mg of sodium azide in water to make 50 ml. While the palmitic acid solution is still warm (about 50°C), add the albumin solution (reagent 4), mix, and dilute to 100 ml. This solution is stable at room temperature for more than 4 months. With some loss of accuracy and precision, a simpler standard can be made by dissolving 50 mg of palmitic acid in 95% ethanol to make 100 ml. The extraction with this standard is more erratic than with the albumin-containing standard.

7. Filter paper pads. Circles of 12.5 mm diameter are cut from filter paper

† Oxalic acid *bis*(cyclohexylidenehydrazide); Aldrich Chemical Co., Milwaukee, Wis. 53233.

sheets of 3 to 4 mm thickness.‡ After cutting, the pads are washed twice with heptane and then blown dry with a stream of nitrogen under a hood and stored in a stoppered vial. After washing, the pads should not be handled with the fingers but only with a metal forceps to avoid contamination by lipids from the skin.

SPECIAL APPARATUS

Vortex mixer with a 16-tube shaker head.§ Hand shaking tends to give erratic results due to copper reagent adhering to the tube walls. Mixing in a vertical position with a vortex mixer was most satisfactory.

PROCEDURE

Add to a series of 16 × 100 mm disposable test tubes: (1) 100 μl of sample (serum or heparinized or oxalated plasma but not citrated), (2) 100 μl of standard, (3) 50 μl of standard plus 50 μl of water, and (4) 100 μl of water as blank. To all tubes add 0.3 ml of the copper reagent and 2.0 ml of chloroform and stopper with neoprene stoppers (ordinary rubber stoppers cannot be used). Shake for 10 min on vortex mixer. To each tube add one filter paper pad (handle with tweezers) and stopper again. Allow to stand for a few minutes until the pad has completely absorbed the blue aqueous layer. Gently decant the chloroform layer into another tube so that the pad remains at the bottom of the tube. Do not attempt to drain completely. Stopper at once with the same stopper as used previously after wiping it dry. From each tube, pipet exactly 1 ml of the chloroform extract into one of a third set of tubes. To each tube add 0.9 ml of the cuprizone reagent, stopper, and shake gently. Then add 0.1 ml of the ammonia reagent and mix gently. Read the samples and standards against blank at 620 nm. The volume of liquid used above is sufficient for 12 × 75 mm cuvettes. Other cuvettes may require larger volumes and it may be necessary to double the amounts of samples, reagents, and standards to obtain a sufficient volume.

CALCULATION

Since the standards and samples are treated similarly,

$$\frac{\text{absorbance of sample}}{\text{absorbance of standard}} \times \text{conc. of standard} = \text{free fatty acid (in mg/dl)}$$

where the two standards used are equivalent to 25 and 50 mg/dl. The results may be converted to micromoles (μmol) per liter by multiplying by the factor 39.0.

‡ Seitz S-3T, filter sheets, Republic Seitz Filter Corp., Milldale, Conn. 06467.
§ Mixer No. 58223 and 16-tube shaker head No. 58228-2, Scientific Products Div., American Hospital Supply Corp., McGaw Park, Ill. 60085.

Free Fatty Acids by Alternative Method

This method, adapted from the method of Lehane and Werner [67], was designed for the determination of the copper in the final extract by use of atomic absorption, but the determination can also be made colorimetrically.

1. Extraction solvent.   Mix together 100 ml of chloroform, 75 ml of n-heptane, and 3.5 ml of methanol.

2. Silicic acid, 100 mesh powder.|| Activate by heating for 6 hr at 120°C before use.

3. Ethanol, 95%.

4. Copper nitrate, 0.5M.   Dissolve 60.4 g of cupric nitrate [$Cu(NO_3)_2 \cdot 3H_2O$] in water to make 500 ml.

5. Triethanolamine, 1M.   Dissolve 74.5 g (67 ml) of triethanolamine (reagent grade) in water to make 500 ml.

6. Sodium hydroxide, 1M.   Dissolve 20 g of sodium hydroxide in water to make 500 ml.

7. Copper reagent.   Mix together 30 ml of the copper nitrate solution, 30 ml of the triethanolamine solution, and 10.5 ml of the sodium hydroxide solution. Dilute to 300 ml with water. Dissolve 99 g of sodium chloride in this solution (almost saturated) and adjust pH to 7.9. This reagent is stable for only about 1 week.

8. Color reagent (if used).   Same as in the previous procedure.

9. Standard, 50 mg/dl.   Dissolve 50 mg of pure palmitic acid in 95% ethanol to make 100 ml.

To separate tubes add 100 μl of sample or one of two standards, one containing 100 μl of standard solution and the other 50 μl of standard solution plus 50 μl of water. To another tube add 100 μl of water as blank. To all tubes add 4 ml of the extraction solvent and about 250 mg of the silicic acid (this can be measured roughly by volume after determining the volume occupied by 250 mg in a small test tube). Shake the tubes well and allow to stand for 15 min; shake again. Centrifuge for 10 min at 5,000 rpm and transfer the supernatants to a second series of tubes. This may be done by careful decanting so that none of the silicic acid is transferred. To each tube containing a supernatant add 1.5 ml of the copper reagent and shake vigorously for 1 min (neoprene stoppers may be used). Centrifuge for 5 min at 5,000 rpm. Carefully pipet 2 ml of the supernatants to a

|| Mallinckrodt Chemical Works, St. Louis, Mo. 63160.

third set of tubes. Great care must be taken not to include any of the lower aqueous layer in the transfer.

For colorimetric determination add 0.9 ml of the color reagent and 0.1 ml of the ammonia solution as in the previous procedure and read in the photometer at 620 nm. The calculations are the same as in the previous method since the same standards are treated similarly to the samples. Again, depending upon the cuvettes used, it may be necessary to vary the final volume.

For determination of the copper by atomic absorption, evaporate the solvent in each tube to dryness using a water bath and dissolve the residue in 3 ml of ethanol. The exact details of the determination will depend upon the instrument used, but the wavelength of 324.7 nm is used.

Normal Values and Interpretation of Results

The normal range of free fatty acids in the fasting state may be taken as 4 to 18 mg/dl (160 to 700 μmol/liter). The concentration of free fatty acids is related to carbohydrate metabolism and the level is generally much higher in diabetes. In a glucose tolerance test the level of free fatty acids falls as the glucose level rises. The changes in level are much greater with the diabetic type of curve, and measurement of free fatty acids has been used along with the measurement of the glucose level in some tolerance tests as an aid to diagnosis [68, 69].

Lipoproteins

The various serum lipids are present chiefly in combination with the proteins as lipoproteins. For some diagnostic purposes the amounts and proportions of the various lipoprotein fractions are more important than the total amounts of the various lipids present. The different lipoproteins have been separated by means of ultracentrifugation, but this is strictly a research procedure requiring special instrumentation. The different lipoproteins may also be characterized by means of electrophoretic methods. The serum proteins move in the electric field and carry the associated lipids along with them. The different serum proteins have different amounts of lipids associated with them. Thus by use of a lipid-staining dye, the amounts of lipids in the different fractions may be determined, or the actual amounts of a given lipid (e.g., cholesterol) may be determined in the different fractions. The lipoprotein electrophoresis has been carried out in a number of different media including starch gel [70, 71], polyacrylamide gel [72–76], agar or agarose gel [77–81], paper [81–85], Cellogel [86–88], and other types of cellulose acetate [89–93]. The various methods have advantages and disadvantages. Cellulose acetate electrophoresis is relatively simple and widely used and is the method presented in this section.

A number of screening methods have been developed for detecting abnormal amounts of certain lipoproteins, particularly the β-lipoproteins. In some of these methods the serum is mixed in a capillary tube with a precipitating reagent such as amylosulfate or other sulfonated polysaccharide [94]. After centrifugation, the amount of precipitate in the end of the tube is taken as a measure of the amount of lipoprotein present [95, 96]. The precipitating reagent may be an antibody to the lipoprotein [97, 98]. Turbidimetric methods have also been developed using potassium agar [99, 100], dextran sulfate [100], or heparin [101, 102]. These methods may be useful for mass screening to detect subjects with possible lipoprotein abnormalities but they do not yield as much information as the electrophoretic method.

Electrophoresis of Lipoproteins

As with the electrophoresis of serum proteins (Chapter 7), the exact details of the procedure depend somewhat upon the particular apparatus and type of cellulose acetate used. The following procedure [91] typifies those in use and may be varied to suit the conditions.

REAGENTS

1. Barbital buffer, pH 8.6, ionic strength 0.075. Dissolve 15.40 g of sodium diethylbarbiturate and 2.76 g of diethylbarbituric acid in about 900 ml of water with the aid of gentle heating if necessary, cool, and dilute to 1 liter. This is substantially the same as the Beckman B-2 buffer# furnished for protein electrophoresis. Others [90] may prefer to dilute the buffer to 1,400 ml rather than 1,000 ml to obtain an ionic strength of 0.05.

2. Dye solution. Dissolve 0.4 g of Oil Red O (Color Index No. 26125) in a mixture of 700 ml of methanol and 300 ml of water by heating to boiling with continuous stirring. Allow to cool to room temperature, then store at 37°C.

3. Bleaching solution. Mix together 5 ml of commercial sodium hypochlorite bleach containing 5.25% available chlorine, 95 ml of water, and 5 ml of glacial acetic acid. This solution should be made fresh each time and used only once.

PROCEDURE

The cellulose strip is moistened with the buffer. Usually this is done by floating the strip carefully on the surface of some buffer and allowing it to absorb the buffer. It is then blotted slightly and the serum applied with the special applicator, either before or after placing the strip in the electrophoresis chamber, depending upon the type of apparatus used. The electrophoresis is then carried

Beckman Instruments, Inc., Fullerton, Calif. 92634.

out, generally using a voltage of 200 to 300 v for 30 to 45 min. Again these factors will vary somewhat with the apparatus used and may be adjusted somewhat to give the best separation. After electrophoresis, the strip is removed from the chamber and immersed in the staining solution. The staining technique used and the time depend on the particular system. After staining, the strip is removed and either washed in running water or placed in the bleaching solution to remove stain precipitate and background color. The strip may be inspected visually and the density of the various bands noted, or it may be scanned in a densitometer without clearing, or the strip may be cleared before inspection or scanning. The various manufacturers of the cellulose strips recommend somewhat different solutions for clearing the strips and their recommendations should be followed.

Different types of cellulose strips have been used. Beckering and Crowson [93] state that the Beckman* membranes were not as satisfactory as those made by Gelman† or Sartorius,‡ with the last-named giving the better separation. On the other hand, Fletcher [91] has obtained good results with the Beckman membranes. The authors have used the membranes furnished by Helena§ with very good results. With any membrane some experience is necessary to produce a satisfactory pattern.

Figure 12-1 indicates the types of results obtained. This shows the approximate appearance of the stained strip and the results of scanning. In normal serum, three bands are noted; these are labeled as indicated, α, pre-β, and β, with the β band the heaviest as indicated by scanning. The abnormal patterns have been classified into five different types as shown; this will be discussed in detail later. Some workers carry out the classification merely by visual inspection of the strips. This is easily done in the clear-cut examples that are shown in the illustration but, in practice, intermediate curves may often be found which are more difficult to classify by visual inspection and a densitometer scan may be necessary in these instances. From the densitometer scan one can readily calculate the relative areas of the different peaks. If these are taken to represent the relative amounts of lipids in the different fractions, and the total serum lipid is known, one can calculate the amount of lipid in the separate fractions. This calculation is actually of little significance, since the different lipids are not stained to the same extent by the dye and the different lipoprotein fractions do not contain the same proportion of the different lipids.

The hyperlipoproteinemias have been divided into five different classes or types

---

* Beckman Instruments, Inc., Fullerton, Calif. 92634.
† Gelman Instrument Co., Ann Arbor, Mich. 48106.
‡ Sartorius Div., Brinkman Instruments, Inc., Westbury, N.Y. 11590.
§ Helena Laboratories, Beaumont, Tex. 77704.

Figure 12-1. Serum lipoprotein patterns; the stained strip is shown at the bottom of each scan. The dashed line indicates the point of application. A normal pattern and the five Fredrickson's patterns are shown. See text for further explanation.

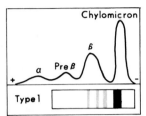

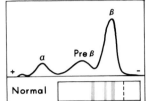

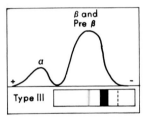

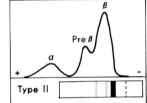

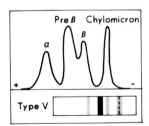

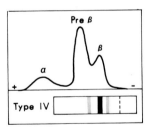

following the classification of Fredrickson and Lees [103, 104]. These will be discussed briefly here; for further information the reader is referred to several reviews [105–108]. Although a preliminary classification is made on the basis of the lipoprotein electrophoresis, the serum levels of cholesterol, triglycerides, and total lipids are also considered. In type I, the concentration of chylomicrons is greatly increased and the serum usually appears creamy even when the patient is in the fasting state. Since the chylomicrons consist mainly of true fats, the triglyceride and total lipid will be significantly increased and may in some cases exceed 2,000 mg/dl. The cholesterol level is usually only slightly increased and may not be over 300 mg/dl. This condition is not common and is apparently due to a rare abnormal allele. The glucose tolerance is normal and these patients apparently do not have the tendency toward atherosclerosis found in other types. The α and pre-β lipoproteins are decreased.

In type II the β-lipoprotein level (and to some extent the pre-β level) is markedly increased. The individuals found in this group have marked increases in the serum cholesterol levels, which are often above 400 mg/dl. The triglyceride levels are normal and the serum appears clear. Individuals with essential familial hypercholesterolemia are found in this type. Xanthomatosis is common and there is an increased tendency toward atherosclerotic heart disease. The carbohydrate metabolism as indicated by the glucose tolerance test is essentially normal.

In the electrophoretic pattern for type III the β and pre-β fractions occur together as a single band. The cholesterol and triglyceride levels are elevated. The levels of cholesterol are often in the range of 350 to 450 mg/dl and the triglyceride levels often reach 350 mg/dl. These patients usually have an abnormal glucose tolerance curve with a tendency toward diabetes and atherosclerosis.

Type IV patterns are characterized by a marked increase in the pre-β fraction. The triglyceride level in the serum is increased but the increase is variable, the levels ranging in different patients from 200 to 400 mg/dl. Unless the triglycerides are very markedly increased, the amount of cholesterol generally remains in normal limits. Depending upon the increase in total lipids, the serum may be clear, slightly cloudy, or even milky. Many individuals of this type have impaired carbohydrate metabolism but this is not as common as for those of type III. There is some evidence indicating that young individuals of this type of lipoprotein pattern show special proneness to coronary artery disease.

In type V the serum is usually creamy due to the presence of very high levels of chylomicrons. It is distinguished from type I by a marked elevation of the pre-β fraction. The levels of cholesterol in the serum are generally increased to about the same extent as in type I. The triglyceride levels are very markedly increased, but may not reach quite as high levels as in type I.

A simple summary of the data on the various types of hyperlipoproteinemia is given in Table 12-1.

Although the use of the lipoprotein phenotyping is increasing, a small laboratory may not receive sufficient samples to warrant the performance of the procedure every day. Thus the question of the stability of the samples becomes important. According to a recent study using lipoprotein electrophoresis on cellulose acetate, samples that had been standing at room temperature for 3 days gave almost the same results as those obtained with fresh samples [109]. With

Table 12-1. Lipid Fractions in Hyperlipoproteinemia

Type	Cholesterol	Triglycerides	Pre-β	β	Chylomicrons
I	+	++++	±	±	++++
II	++	±	+	++	0
III	++	+		+++	0
IV	+	++	+++	+	0
V	++	+++	+++	+	+++

0, not present.
±, normal or only slightly increased.
+, noticeably increased.
++, +++, and ++++, increasing levels of the fraction, ++++ representing a tenfold or greater increase in concentration.

longer standing, poorer correlation was obtained. The samples may be kept for a longer period if refrigerated or frozen. If frozen, only one thawing is permissible. Cholesterol is stable for at least 5 days at room temperature [110] and for longer periods if refrigerated or frozen [111]. Less information is available concerning the stability of the triglycerides in serum. Levy [45] found no difference in serums that had been kept frozen for up to 5 days.

Fat in Feces

The determination of fecal fat may be used to determine the amount of absorption of fat in the intestines. The method we present is that of van de Kamer [112, 113]. This involves the hydrolysis of the fecal fats to the free fatty acids, the extraction of the fatty acids with petroleum ether, and titration of the acids with standard alkali after the evaporation of the solvent. Although this method has been used for a number of years, recent work indicates that with some modifications it is still a satisfactory method. In those instances in which the fecal fats contain appreciable amounts of medium-chain triglycerides (fatty acids with 8 to 14 carbon atoms), these acids will be underestimated [114, 115], but the difference is not great enough to vitiate the use of the test for clinical purposes, particularly since in most instances the fat intake is not accurately known and thus the percentage excreted can only be estimated.

COLLECTION OF SPECIMEN

Specimen collection is a very important part of the procedure. The determination of fat in a random stool specimen is of little value. The total stool excreted over 2 or preferably 3 days should be collected. The patient should have been on a fairly constant diet of approximately known fat content for a few days before the start of the collection period. The specimens can be stored in a refrigerator after collection. Empty 1-gallon paint cans with tight-fitting lids are convenient containers. The cans can be weighed before the collection period and at the end to obtain the total weight of stool. The tight-fitting lids decrease the odor from the specimen and the large opening makes it easy to add the specimens. If a shaker mixer such as used by paint stores for these cans is available, after the collection period and after the cans have been weighed, some water can be added to make a thick slurry and the stool thoroughly homogenized by vigorous shaking. This is much more convenient than the alternative method of using a blender, although for only an occasional analysis the purchase of the can shaker may not be justified. The feces are transferred to the blender with a minimum amount of water and thoroughly blended. The material is then transferred to a tared container of suitable size and the weight of the homogenized sample determined.

REAGENTS

1. Ethyl alcohol with 0.4% amyl alcohol. Mix 0.4 ml of amyl alcohol with 100 ml of ethyl alcohol. The amyl alcohol reduces the foaming that occurs with some specimens.

2. Potassium hydroxide, $12M$. Dissolve 67 g of potassium hydroxide in about 50 ml of water, cool, and dilute to 100 ml.

3. Hydrochloric acid, $12M$. Dilute 250 ml of concentrated hydrochloric acid with 125 ml of water.

4. Thymol blue indicator solution. Dissolve 0.2 g of thymol blue in 100 ml of 95% ethyl alcohol.

5. Sodium hydroxide, $0.1M$. Dilute a standardized $1M$ solution 1:10 to obtain a $0.1M$ solution. The dilution must be made accurately.

6. Petroleum ether, AR (bp, 30°–60°C).

PROCEDURE

Weigh out 7 to 8 g of the feces to the nearest 10 mg. This is conveniently done in a disposable plastic vial with a tight cover and a capacity of about 20 ml. Add 5 ml of the potassium hydroxide to the sample and mix. Transfer to a 250 ml Erlenmeyer flask, using 40 ml of the ethyl alcohol with amyl alcohol to rinse the material completely into the flask. Boil the mixture gently under a reflux condenser for 20 min. If the mixture shows a tendency to bump, add a few boiling granules.‖ Cool, add 17 ml of the $12M$ hydrochloric acid, and cool again,

‖ Boileezers, Fisher Scientific Co., Pittsburgh, Pa. 15219.

preferably to below room temperature in the refrigerator. Add exactly 50 ml of petroleum ether, stopper the flask, and shake vigorously for 2 min. Allow the solution to stand in the refrigerator until the layers have completely separated. Carefully pipet exactly 25 ml of the upper petroleum ether layer and add to a 125 ml flask. Care must be taken not to include any traces of the lower aqueous layer which would produce high results. A procedure that may be used to obtain a complete separation is as follows. Transfer the solution to a 120 ml separatory funnel, decanting from the flask into the funnel. As much as possible of the ether layer must be decanted but all of the aqueous layer containing some precipitated material need not be transferred to the funnel. Allow the layers to separate, then draw off the lower layer and discard. Wipe off the top of the funnel and carefully decant the ether from the top of the funnel through a rapid filter paper (Whatman No. 31) into a flask. A few droplets of the aqueous layer will be retained by the paper but avoid adding any large amount of this to the paper. Pipet exactly 25 ml of the filtrate to a clean flask. Evaporate the petroleum ether aliquot to dryness on a steam bath in a fume hood. (*Caution:* The petroleum ether vapors are very flammable.)

To 100 ml of ethyl alcohol add about 20 drops of the thymol blue indicator solution, and mix. The color will usually be yellow due to traces of acid in the alcohol. Add a small amount of the sodium hydroxide (0.1M) to the alcohol until the solution turns green. Avoid any excess of the base. Add 20 ml of the neutralized alcohol to the flask containing the evaporated extract and warm in hot water for a few minutes to dissolve any residue. Titrate with the 0.1M sodium hydroxide solution, using a 5 ml buret, to a distinct blue endpoint. In the presence of yellow pigments extracted from the feces the endpoint will be greenish blue, but it should be distinct.

CALCULATION

$$284 \times 1.04 \times \frac{50}{25} \times \frac{100\,N}{10,000\,W} = \text{fatty acids (in g/100 g feces)}$$

where 284 is taken as the average molecular weight of the fatty acids, 1.04 is a factor to compensate for a slight increase in volume of the petroleum ether layer on extraction, N is the milliliters of 0.1M sodium hydroxide used in the titration, and W is the weight (in grams) of the stool sample taken.

The equation reduces to

$$\frac{5.91\,N}{W} = \text{fat (in g/100 g feces)}$$

Then

$$g/100\ g \times \frac{\text{total sample weight (in g)}}{100} = \text{fat in whole sample (in g)}$$

This figure is divided by the number of days over which the specimen was collected to give the grams excreted per day.

NORMAL VALUES AND INTERPRETATION OF RESULTS

The normal amount of fat in the stool will depend upon the dietary intake. For normal individuals this will be 1 to 7 g/day, which is usually less than 10 percent of the intake. In steatorrhea the excretion of fat is increased and as much as 40 percent of the ingested fat may be excreted, but at least the approximate fat intake must be known to estimate the percentage excreted.

Amniotic Fluid Phospholipids

A test that also involves the determination of some lipids but different in scope than the preceding ones is the measurement of the amount of phospholipids in amniotic fluid and in particular the ratio of lecithin to sphingomyelin (L/S ratio), as an index of fetal pulmonary maturity. Up to about the 26th week of gestation the amount of lecithin in amniotic fluid is less than that of sphingomyelin (L/S ratio less than unity). The amount of lecithin then gradually increases so that at about the 34th week the amounts of the two lipids are approximately equal. During the later weeks of gestation the amount of lecithin increases greatly and the L/S ratio is greater than 2.

Lecithin is a necessary component of the surface-active layer of the lung alveoli. This surface-active agent stabilizes the alveoli and prevents their collapse on expiration with consequent progressive atelectasis and resulting in the respiratory distress syndrome (RDS). This can terminate in fatal hyaline membrane disease [116]. The L/S ratio in the amniotic fluid reflects the amount in the fetal lung and is thus a good index of fetal lung maturity. If the infant is delivered when the L/S ratio is still low, the respiratory distress syndrome may result. There have been several methods of testing for fetal lung maturity. The most commonly used one extracts the phospholipids from the amniotic fluid with methanol and chloroform. The solvent is evaporated and a thin layer chromatography separation is carried out on the extract. The resulting spots after development are compared with standards of the two phospholipids of interest and the relative amounts estimated by visual inspection or densitometric scanning [117–119].

In some methods the amount of total phospholipid in a purified extract is used as the measure of fetal maturity. The phospholipids are estimated by phosphorus determinations as given earlier in this chapter under Phospholipids [120]. The base material on the chromatographic plate at the positions of the various spots may be scraped from the plate and the lipid phosphorus determined by the usual method. For some procedures the extract may be purified before chromatography by precipitation with cold acetone, but in other methods this has not been found necessary.

We present a method based on that of Coch et al. [121, 122] which uses glass fiber impregnated sheets for the chromatography. This gives a rapid separation and has provision for different methods of visualization of the spots. The method relies solely on the visual estimation of the size of the spots as compared with standards. This is generally satisfactory when two or more different methods of visualization of the spots are used. In all the methods the presence of appreciable amounts of blood interferes with the procedure. Phospholipids may be present from the hemolyzed cells in different proportions from those in the amniotic fluid. The fluid is centrifuged and if 10 ml of fluid in a 15 ml centrifuge tube yields more than about 0.2 ml of packed cells, the specimen is unsuitable for analysis.

REAGENTS

1. Chloroform, reagent grade.

2. Methanol, reagent grade.

3. Developing solvent. Mix together, just before use, 170 ml of chloroform, 20 ml of methanol, and 3 ml of concentrated ammonium hydroxide.

4. Iodine crystals.

5. Acetic acid solution, $3.8M$. Dilute 200 ml of glacial acetic acid to 1 liter with water.

6. Bismuth subnitrate solution. Dissolve 17 g of bismuth subnitrate in $3.8M$ acetic acid and dilute to 1 liter with the acid.

7. Potassium iodide, $2.4M$. Dissolve 40 g of potassium iodide in water and dilute to 100 ml. Store in a brown bottle and discard when solution turns yellow.

8. Spray reagent. Mix together 70 ml of the acetic acid solution (reagent 5), 20 ml of the bismuth subnitrate solution (reagent 6), and 5 ml of the potassium iodide solution (reagent 7). This is stable in a brown bottle for 1 week.

9. Standard 1. Dissolve 20 mg of phosphatidylethanolamine and 20 mg of lysolecithin in chloroform to make 10 ml.#

10. Standard 2. Dissolve 40 mg of dipalmitoyl lecithin and 20 mg of sphingomyelin in chloroform to make 10 ml.#

These phospholipids may be obtained from General Biochemicals, Chagrin Falls, Ohio 44022.

11. Standard 3.   Dissolve 100 mg of dipalmitoyl lecithin and 20 mg of sphingomyelin in chloroform to make 10 ml.#

SPECIAL APPARATUS

1. Thin layer chromatography (TLC) sheets, 20 × 20 cm, silica gel impregnated glass fiber sheets.*

2. TLC chamber for sheets.†

3. Spotting guide.   This is not absolutely necessary but it greatly facilitates the application of the spots at uniform distances apart.‡

4. Chromatographic sprayer.   See section on thin layer chromatography in Chapter 3.

PROCEDURE

The amniotic fluid is centrifuged to remove cells and the supernatant is used. A convenient volume (preferably 4 or 5 ml) is added to a screw-capped tube with Teflon-lined cap. An equal volume of methanol is added and mixed well. Then two volumes of chloroform (twice that of the methanol) are added and the tube capped and shaken vigorously for several minutes. The tube is then centrifuged to break any emulsion formed. The upper aqueous layer is aspirated off and discarded. The chloroform layer is filtered through a small Whatman No. 1 paper into an evaporating dish and the filter washed with a few milliliters of chloroform. The combined filtrate is evaporated to about 0.5 ml on a steam bath. This solution is transferred to a small test tube with the aid of 1 to 2 ml of chloroform and further evaporated to about 50 μl. A slow stream of nitrogen may be used to aid in the final evaporation.

The residue and the standards are applied to the chromatographic sheet. Two samples can be applied to one sheet. Using the 20 × 20 cm sheets, the applications points are marked across the sheet 2.5 cm from one end. Eleven points, 2 cm apart, can be marked across the sheet with the end points 1 cm from the edges of the sheet. Numbering the points 1 to 11 from left to right, the order of application of the standards and samples is as follows: nos. 1 and 11, standard 1; nos. 2 and 8, sample 1; nos. 3 and 7, standard 2; nos. 4 and 10, sample 2; nos. 5 and 9, standard 3; nos. 6, vacant. Five-microliter portions of the standards are applied. The extract is applied in 5-μl portions to keep the spots as small as possible. Half of the extract is applied to each of the two locations. The sheet

---

These phospholipids may be obtained from General Biochemicals, Chagrin Falls, Ohio 44022.

* Gelman Type SG, Cat. No. C4943-28, Gelman Instrument Co., Ann Arbor, Mich. 48106.

† Eastman Chromatographic Chamber Cat. No. 6071, Eastman Kodak Co., Rochester, N.Y. 14650.

‡ Cat. No. 51333, Gelman Instrument Co., Ann Arbor, Mich. 48106.

is then developed in the chromatographic chamber with the developing solvent. The development is allowed to proceed for about 20 min until the solvent front has traveled about 15 cm. The sheets are then removed and air-dried. Care should be taken in handling the sheets as lipids or other material from the fingers may give spurious spots. The sheet is then divided vertically along the line of the empty position (no. 6).

One half of the sheet is sprayed with the bismuth spray reagent until the standards appear as white spots against an orange background. The half-sheet is then air-dried for a few minutes and then destained by immersing for a few minutes in the acetic acid solution. The lecithin and sphingomyelin standards should appear as bright yellow spots against a light background. The position and size of the spots are outlined lightly with pencil, as the color of the background begins to reappear in a short time. The size and color intensity of each spot are noted and an estimate made of the lecithin-sphingomyelin ratio of the sample spots in comparison with the standard spots (no. 2, L/S ratio, 2:1; no. 3, L/S ratio, 5:1).

The other half of the sheet is placed in a desiccator or other air tight container with some iodine crystals. On exposure to iodine vapors, the phospholipids appear as yellow to brown spots. The iodine treatment is less specific than the bismuth spray but is more sensitive. The spots are outlined in pencil as they tend to fade on standing. The Rf values found will depend somewhat on the conditions but should be approximately 0.67 for lecithin and 0.48 for sphingomyelin.

The evaluation of the L/S ratio is based on the spots obtained with the bismuth spray. Values of the ratio above 5:1 indicate a definitely mature fetus; values between 3:1 and 4:1 are transitional (some immaturity with chance of RDS but chance of survival); values below 2:1 indicate immature fetus with good chance of RDS and poor prognosis. It should be noted that the dividing line for the L/S ratio depends upon the method of detection, and procedures using other techniques (such as acid charring) will have different ranges for the pertinent L/S values.

References

1. Canal, J., Lelattre, J., and Girard, M. *Ann. Biol. Clin.* (Paris) 30:324, 1972.
2. Lazaroff, N. *Z. Med. Labortech.* 7:242, 1966.
3. Castaldo, A., Petri, L., and Ragno, I. *Rass. Med. Sper.* 11:201, 1964.
4. Sperry, W. M. *Stand. Methods Clin. Chem.* 4:173, 1963.
5. Folch, J., Lees, M., and Sloan-Stanley, G. H. *J. Biol. Chem.* 226:497, 1957.
6. Sperry, W. M., and Brand, F. C. *J. Biol. Chem.* 213:69, 1955.
7. Jacobs, S. L., and Henry, R. J. *Clin. Chim. Acta* 7:270, 1962.
8. Pernokis, E. W., Freeland, M. R., and Kraus, L. *J. Lab. Clin. Med.* 26:1978, 1941.

9. Friedman, H. S.   *Clin. Chim. Acta* 25:173, 1969.
10. Friedman, H. S.   *Clin. Chim. Acta* 19:291, 1968.
11. Chabrol, E., and Charonnet, R.   *Presse Med.* 45:1713, 1937.
12. Zoellner, M., and Kirsch, K.   *Z. Gesamte Exp. Med.* 135:545, 1962.
13. Frings, C. S., Fendley, T. W., Dunn, R. T., and Queen, C. A.   *Clin. Chem.* 18:673, 1972.
14. Knight, J. A., Anderson, S., and Rawle, J. M.   *Clin. Chem.* 18:199, 1972.
15. Woodman, D., and Price, C. P.   *Clin. Chim. Acta* 38:39, 1972.
16. Coudon, B., and Bouige, D.   *Ann. Biol. Clin.* (Paris) 31:3, 1973.
17. Bolacco, F.   *Quad. Sclavo Diagn.* 7:698, 1971.
18. Genetet, F., Nabet, P., and Paysant, P.   *Ann. Biol. Clin.* (Paris) 26:1129, 1968.
19. Ratliff, C. R., Culp, T. W., and Gevedon, R. E.   *Adv. Automat. Anal. Technicon Int. Congr.* 1:117, 1969.
20. Martinek, R. G.   *J. Am. Med. Technol.* 32:64, 1970.
21. Zak, B., Epstein, E., and Baginski, E. S.   *Ann. Clin. Lab. Sci.* 2:101, 1972.
22. Vanzetti, G.   *Clin. Chim. Acta* 10:389, 1964.
23. Ham. A. B.   *Am. J. Med. Technol.* 28:99, 1962.
24. Fasce, C. F., and Vanderlinde, R. E.,   *Clin. Chem.* 18:901, 1972.
25. Hawthorne, B. E.   *Clin. Chem.* 10:258, 1964.
26. Tonks, D. B.   *Clin. Biochem.* 1:12, 1967.
27. Pearson, S., Stern, S., and McGavack, T. H.   *Anal. Chem.* 25:813, 1953.
28. Turner, T. J., and Eales, L.   *Scand. J. Clin. Lab. Invest.* 9:210, 1957.
29. Jamieson, A.   *Clin. Chim. Acta* 10:530, 1964.
30. Solow, E. B., and Freeman, L. W.   *Clin. Chem.* 16:472, 1970.
31. Carpenter, K. J., Gotsis, A., and Hegsted, D. M.   *Clin. Chem.* 3:233, 1957.
32. McDougal, D. B., Jr., and Farmer, H. S.   *J. Lab. Clin. Med.* 50:485, 1957.
33. Abell, L. L., Levy, B. B., Brodie, B. B., and Kendall, F. E.   *J. Biol. Chem.* 195:357, 1952.
34. Abell, L. L., Levy, B. B., Brodie, B. B., and Kendall, F. E.   *Stand. Methods Clin. Chem.* 2:26, 1958.
35. De la Huerga, J., and Sherrick, J. C.   *Ann. Clin. Lab. Sci.* 2:360, 1972.
36. Parekh, A. C., and Jung, D. H.   *Anal. Chem.* 42:1423, 1970.
37. Jung, D. H., and Parekh, A. C.   *Clin. Chim. Acta* 35:73, 1971.
37a. Kim, E., and Goldberg, M.   *Clin. Chem.* 15:1171, 1969.
38. Baginski, E. S., Epstein, E., and Zak, B.   *Ann. Clin. Lab. Sci.* 2:255, 1972.
39. Baginski, E. S., and Zak, B.   *Clin. Chim. Acta* 5:834, 1960.
40. Baginski, E. S., Foa, P. P., and Zak, B.   *Clin. Chem.* 13:326, 1967,
41. Fletcher, M. J.   *Clin. Chim. Acta* 22:393, 1968.
42. Timms, A. R., Kelly, L. A., Spirito, J. A., and Engstrom, R. G.   *J. Lipid Res.* 9:675, 1968.
43. Galetti, F.   *Clin. Chim. Acta* 15:184, 1967.
44. Soloni, F. G.   *Clin. Chem.* 17:529, 1971.
45. Levy, A. L.   *Ann. Clin. Lab. Sci.* 2:474, 1972.
46. Van Handel, E., and Zilversmit, D. B.   *J. Lab. Clin. Med.* 50:152, 1957.
47. Carlson, L., and Waldstrom, L. B.   *Clin. Chim. Acta* 4:197, 1959.
48. Rice, E. W.   *Stand. Methods Clin. Chem.* 6:215, 1970.

49. Foster, L. B., and Dunn, R. T.   *Clin. Chem.* 19:338, 1973.
50. Sardesai, V. M., and Manning, J. A.   *Clin. Chem.* 14:156, 1968.
51. Kessler, G., and Lederer, H.   *First Technicon International Symposium.* New York: Medical, 1965. P. 341.
52. Goedicke, W., and Gerike, U.   *Clin. Chim. Acta* 30:727, 1970.
53. Bucolo, G., and David, H.   *Clin. Chem.* 19:476, 1973.
54. Seitz, H. J., and Tarnowski, W.   *Z. Klin. Chem. Klin. Biochem.* 6:411, 1968.
55. Altmann, A., Bach, A., and Metais, P.   *Ann. Biol. Clin.* (Paris) 25:439, 1967.
56. Da Fonseca-Wollheim, F.   *Aerztl. Lab.* 19:65, 1973.
57. Gottfried, S. P., and Rosenberg, B.   *Clin. Chem.* 19:1077, 1973.
58. Dole, V. P., and Meintertz, H.   *J. Biol. Chem.* 235:2595, 1960.
59. Novak, M.   *J. Lipid Res.* 6:431, 1965.
60. Thomitzek, W. D.   *Z. Med. Labortech.* 11:227, 1970.
61. Mikac-Devic, D., Stankov, H., and Boskovic, K.   *Clin. Chim. Acta* 45:55, 1973.
62. Soloni, F. G., and Sardina, L. C.   *Clin. Chem.* 19:419, 1973.
63. Pinelli, A.   *Clin. Chim. Acta* 44:385, 1973.
64. Mansterski, A., Watkins, R., and Zak, B.   *Microchem. J.* 17:682, 1972.
65. Itaya, K., and Kadowski, T.   *Clin. Chim. Acta* 26:401, 1969.
66. Duncombe, W. B.   *Clin. Chim. Acta* 9:122, 1964.
67. Lehane, D. P., and Werner, M.   *Am. J. Clin. Pathol.* 59:10, 1973.
68. Schlierf, G., and Wood, P.   *J. Lipid Res.* 6:317, 1965.
69. Hagenfeldt, L.   *Clin. Chim. Acta* 13:266, 1966.
70. Kunkel, H. G., and Slater, R. J.   *J. Clin. Invest.* 31:677, 1952.
71. Ackermann, P. G., Toro, G., and Kountz, W. B.   *J. Lab. Clin. Med.* 44:517, 1954.
72. Naito, H. K., Wade, M., Ehrhart, L. A., and Lewis, L. A.   *Clin. Chem.* 19:228, 1973.
73. Hall, F. F., Ratliff, C. R., Westfall, C. L., and Culp, T. W.   *Biochem. Med.* 6:464, 1972.
74. Moran, R. F., Castelli, W. P., and Moran, M. W.   *Clin. Chem.* 18:217, 1972.
75. Frings, C. S., Foster, L. B., and Cohen, P. S.   *Clin. Chem.* 17:111, 1971.
76. Dangerfield, W. G., and Pratt, J. J.   *Clin. Chim. Acta* 30:273, 1970.
77. Elphick, M. C.   *J. Clin. Pathol.* 24:83, 1971.
78. Papadopoulos, N. M., and Kintzios, J. A.   *Clin. Chem.* 17:427, 1971.
79. Dyerberg, J., and Hjoerne, N.   *Clin. Chim. Acta* 28:203, 1970.
80. Iammarino, R. M., Humphrey, M., and Antolik, P.   *Clin. Chem.* 15:1218, 1969.
81. McGlashan, D. A. K., and Pilkington, T. R.   *Clin. Chim. Acta* 22:646, 1968.
82. Pyrovolaskia, J., and Hatzioannou, J.   *J. Clin. Pathol.* 24:368, 1971.
83. Buckley, G. C., Little, J. A., Csima, A., Koenig, E., Yano, R., and Sullivan, K.   *Can. Med. Assoc. J.* 102:943, 1970.
84. Moinuddin, M., and Taylor, L.   *Lipids* 4:186, 1969.
85. Lees, R. S., and Hatch, F. T.   *J. Lab. Clin. Med.* 61:518, 1963.
86. Berends, G. T., De Jong, J., and Zondag, H. A.   *Clin. Chim. Acta* 41:187, 1972.
87. Messerschmidt, H. J. M., and Sedee, P. D. J. W.   *Clin. Chim. Acta* 36:51, 1972.
88. De Baets, J., and Lezy, W.   *Clin. Chim. Acta* 32:142, 1971.
89. Magnani, H. N., and Howard, A. N.   *J. Clin. Pathol.* 24:837, 1971.
90. Beckering, R. E., Jr., and Ellefson, R. D.   *Am. J. Clin. Pathol.* 53:84, 1970.
91. Fletcher, M. J., and Stylion, M. H.   *Clin. Chem.* 16:362, 1970.

92. Winkelman, J., Wybenga, D. R., and Ibbott, F.   *Clin. Chim. Acta* 27:181, 1970.
93. Beckering, R. E., Jr., and Crowson, M.   *Am. J. Clin. Pathol.* 56:765, 1971.
94. Searcy, R. L., Colaianni, W., Young, E., Drnec, J., and Magoe, T.   *Clin. Chim. Acta* 38:291, 1972.
95. Searcy, R. L., Carrol, V. P., Jr., Carlucci, J. S., and Bergquist, L. M.   *Clin. Chem.* 8:166, 1962.
96. Bergquist, L. M., Carrol, V. P., Jr., and Searcy, R. L.   *Lancet* 1:537, 1961.
97. Burstein, M., and Samaille, J.   *Rev. Fr. Etud. Clin. Biol.* 3:780, 1958.
98. Orvis, H. H., and Burger, D.   *Med. Ann. D.C.* 32:44, 1963.
99. Boyle, E., and Moore, R. V.   *J. Lab. Clin. Med.* 53:272, 1959.
100. Dangerfield, W. G., and Faulkner, G.   *Clin. Chim. Acta* 10:123, 1964.
101. Lopez, A., Vial, R., Gremillion, L., and Bell, L.   *Clin. Chem.* 17:994, 1971.
102. Berenson, G. S., Srinivasan, S. R., Lopez, S. A., Radhakrishnamurthy, B., Pargaonkar, P. S., and Deupree, R. H.   *Clin. Chim. Acta* 36:175, 1972.
103. Fredrickson, D. S., and Lees, R. S.   *Circulation* 31:321, 1965.
104. Fredrickson, D. S., and Lees, R. S.   In J. V. Stanbury, J. B. Wyngaarden, and D. S. Fredrickson [Eds.], *The Metabolic Basis of Inherited Disease*, 2nd ed. New York: McGraw-Hill, 1966. P. 426.
105. Fredrickson, D. S., Levy, R. I., and Lees, R. S.   *N. Engl. J. Med.* 276:94, 148, **215**, 273, 1973.
106. Harlan, W. R., and Shaw, W. A.   *C.R.C. Crit. Rev. Clin. Lab. Sci.* 3:45, 1972.
107. Lipo, J. F., and Preston, J. A.   *C.R.C. Crit. Rev. Clin. Lab. Sci.* 2:461, 1971.
108. Beaumont, J. L., et al.   *Bull. W.H.O.* 43:891, 1970. Reprinted in *Stand. Methods Clin. Chem.* 7:79, 1971.
109. Winkelman, J., Wybenga, D. R., and Ibbott, F. A.   *Clin. Chem.* 16:507, 1970.
110. De Traverse, R. M., Lavergne, G. H., and Depraitère, R.   *Ann. Biol. Clin.* (Paris) 14:236, 1956.
111. Anderson, J. R., and Keys, A.   *Clin. Chem.* 2:145, 1956.
112. van de Kamer, J. H., ten Bokkel Huinink, H., and Weyers, H. A.   *J. Biol. Chem.* 177:347, 1949.
113. van de Kamer, J. H.   *Stand. Methods Clin. Chem.* 2:347, 1958.
114. Braddock, L. I., Fleisher, D. R., and Barbaro, G. J.   *Gastroenterology* 55:165, 1968.
115. Losowski, M. S., Kelleher, J., and Walker, B. E.   *Clin. Chim. Acta* 30:267, 1970.
116. Forman, D. R.   *Ann. Clin. Lab. Sci.* 3:242, 1973.
117. Kneiser, M. R., Hurst, R., and Tuegel, C. R.   *Am. J. Clin. Pathol.* 58:579, 1972.
118. Sarkozi, L., Kovacs, H. N., Fox, H. A., and Kerenyi, T.   *Clin. Chem.* 18:956, 1972.
119. Nelson, G. H.   *Am. J. Obstet. Gynecol.* 112:827, 1972.
120. Nelson, G. H.   *Am. J. Obstet. Gynecol.* 105:1072, 1969.
121. Coch, E. H., and Kessler, G.   *Clin. Chem.* 18:490, 1972.
122. Coch, E. H., Meyer, J. S., Goldman, G., and Kessler, G.   *Clin. Chem* 19:967, 1973.

13. Hormones

Hormones are substances produced by the glands of internal secretion (endocrine glands), secreted directly into the blood stream, and carried to various organs and tissues where they exert their effects [1–3]. The main function of these substances is to influence the rate at which biological reactions proceed. A deficiency or excess in hormone production leads to serious derangement of bodily function.

All hormones appear to have a short physiological half-life; therefore they must be continually synthesized and secreted by the glands. Their effect is rapidly exerted and they in turn are rapidly inactivated by metabolic processes.

Most hormones are present in blood in minute amounts and their measurement is tedious and difficult. Before the introduction of the competitive protein binding and radioimmunoassay procedures, very few of the hormones could be determined directly in blood.

Because some of the hormones or their metabolites are excreted in the urine, analyses are often performed in this fluid. A 24-hr urine specimen is preferred as there are diurnal variations in the hormone excretion. The hormones most often found in measurable amounts in the urine are those with low molecular weight such as the steroid hormones.

Various methods have been used for the analyses of hormones. The older ones are generally colorimetric or fluorometric procedures. In these the hormone is extracted from blood or urine with an immiscible organic solvent, and the extract is purified either by chemical or chromatographic means and then treated with a reagent to produce a color or fluorescence for measurement. Problems are usually encountered in completely removing interfering substances that also react with the reagent. In urine the hormones or metabolites may be present as conjugated glucuronides or sulfates and must be hydrolyzed prior to extraction. This is usually accomplished with the use of acid or enzymatic hydrolysis. The urinary steroid hormones and metabolites have also been determined by gas chromatography [4] (see Chapter 3).

More recently better and simpler methods have been introduced for hormone determinations. These use competitive protein binding (CPB) [5] and radioimmunoassay (RIA) [6] principles and will be described in Chapter 15. These methods have become extremely popular in the clinical laboratory and numerous procedures are available as complete kits from several manufacturers.

The Thyroid Hormones

The hormones secreted by the thyroid gland [7, 8] are thyroxine and triiodothyronine. These compounds contain a relatively large percentage of iodine and they exert their effects on the general metabolic activity of the body. The thyroid

Figure 13-1. Schematic diagram of negative feedback mechanism controlling thyroid function.

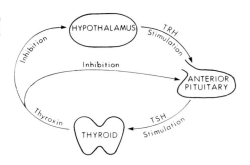

gland also produces a second type of hormone, thyrocalcitonin, a polypeptide, involved in the regulation of calcium metabolism. This hormone contains no iodine and its action is to counteract the effects of hypercalcemia in the blood.

The synthesis and utilization of the thyroid hormones involve several steps. The inorganic iodide ingested with the food is absorbed into the circulation and selectively trapped by the thyroid gland. The iodides are converted into active free iodine by means of enzymatic oxidation. The active iodine combines with tyrosine to form monoiodotyrosine and diiodotyrosine, which combine to form triiodothyronine (T_3) and tetraiodothyronine (T_4). The T_3 and T_4 are stored in the thyroid gland as thyroglobulin. Under the influence of thyroid stimulating hormone (TSH) from the anterior pituitary gland, T_3 and T_4 are released in the free form and secreted into the blood stream. The release of TSH is regulated by a TSH releasing hormone (TRH) produced by the hypothalamus. The TSH releasing hormone and the thyroid stimulating hormone are inversely related to the level of thyroid hormone in the blood. The higher the level of thyroid hormone, the less TRH is released by the hypothalamus and consequently the less TSH is secreted by the pituitary. Similarly, the opposite action will occur when there is decreased thyroid hormone in the blood. This process is known as a negative feedback mechanism (Fig. 13-1) and is responsible for maintaining the proper blood levels of these hormones.

In the blood the thyroid hormones are primarily bound by three plasma proteins: an α-globulin designated thyroxine binding globulin (TBG), thyroxine binding prealbumin, and albumin. The blood levels of these proteins may vary significantly between individuals.

Of the total circulating thyroid hormones, approximately 96 percent is thyroxine (T_4) and 4 percent triiodothyronine, largely in bound form. Only about 0.05 percent exists in a free or unbound state. It has been stated that the concentration of free thyroxine determines the thyroid status of the individual by regulating the amount of TSH released from the pituitary and that the measurement of this fraction would give the most reliable indication of thyroid status [9–11].

Tests for the evaluation of thyroid function may be divided into five main categories: (1) tests based on the ability of the thyroid gland to concentrate iodine, e.g., radioactive iodine uptake (RAI); (2) tests based on the proteins responsible for the transport of the thyroid hormone, e.g., thyroxine binding globulin (TBG), thyroid binding index (TBI), and resin T_3 uptake; (3) tests measuring total circulating thyroid hormone level, e.g., protein bound iodine (PBI) and thyroxine (T_4), or test for the direct measurement of free (unbound) thyroxine; (4) tests based on the measurement of other hormones influencing the thyroid gland, e.g., thyroid stimulating hormone (TSH) and long-acting thyroid stimulator (LATS); and (5) tests that reflect the peripheral effects of thyroid hormone, e.g., basal metabolic rate (BMR), serum cholesterol level, the Achilles tendon reflex time, plasma tyrosine level, and serum enzyme concentrations.

Tests in the third listed category—those measuring total circulating thyroxine —are the most commonly used. Most of the tests mentioned above will be fully discussed in this chapter. They can furnish valuable information for the diagnosis of thyroid disease. Because of the complexity of thyroid physiology, however, no single thyroid test can answer all diagnostic problems and a battery of these tests is usually recommended.

Thyroid Function Test Based on Iodine Concentration

Radioactive Iodine Uptake (RAI)

This test measures the ability of the thyroid gland to absorb and retain iodine [12, 13]. Following the administration of a solution containing radioactive iodine, the isotope is distributed evenly throughout the body's iodine pool. The radioactive iodine behaves similarly to the nonradioactive iodine and is absorbed by the thyroid gland. In general, the rate of absorption as measured by the increased radioactivity in the gland is a measure of the ability of the gland to concentrate iodine.

The patient is given orally, in one or more doses, a tracer dose of radioactive iodine (^{131}I). The radioactivity absorbed by the gland is measured at 2, 6, or 24 hr and expressed as a percentage of the total administered radioactivity. The 24-hr measurement is preferred because at this time the uptake value is at or near its peak. Normal individuals will have an absorption of 1 to 13% after 2 hr, 2 to 25% after 6 hr, and 15 to 45% after 24 hr. The RAI uptake is elevated in hyperthyroidism and endemic goiter and decreased in hypothyroidism, hypopituitarism, and thyroxine treatment.

A number of factors influence the RAI uptake, primarily by interfering with the iodine trapping mechanism of the thyroid gland. Diagnostic x-ray procedures using organic iodides are the most common influences; these will produce a low RAI uptake (hypothyroid level) in a normal individual. This depressing effect on RAI uptake is produced by all iodide-containing medications. Other substances that will falsely depress the RAI uptake are thyroxine, cortisone, ACTH, and antithyroid drugs (thioamides, aniline derivatives, and polyhydric phenols).

Thyroid Function Tests Based on Transport Proteins

Thyroxine Binding Globulin

Thyroxine binding globulin (TBG) is the major binding protein for thyroxine [13, 14]. It has been estimated that it is normally less than one-third saturated with thyroxine and that it has a binding capacity of about 25 μg T_4/dl of serum.

The test measures the total thyroxine binding capacity of the specific thyroxine binding globulin of serum. A large excess of ^{131}I-labeled thyroxine is added to the serum and equilibration is allowed to take place. This is followed by reverse-flow electrophoretic separation. The paper strip is then scanned radiometrically for the presence and distribution of ^{131}I. By relating the proportion of ^{131}I in the TBG area to the total thyroxine content in the supplemented serum, the TBG present can be calculated. This value is expressed in terms of micrograms of thyroxine binding capacity per deciliter of serum. Normal values are 12 to 20 μg/dl.

The test can be affected by a number of factors. The value may be increased in hypothyroidism, pregnancy, oral contraceptive administration, and liver disease. It may be decreased in the nephrotic syndromes, hypoproteinemia, liver disease, and anabolic steroid administration.

Thyroid Binding Index (TBI) and T_3 Uptake

This is an in vitro test that, under specific conditions, indirectly measures the concentration of free thyroxine binding globulin in serum [15–18]. Normally the TBG is not saturated with thyroxine. Except in cases of severe dysproteinemia, the amount of TBG is relatively constant and the number of binding sites available for added hormone will depend upon the amount of endogenous hormone. Thus in hyperthyroidism excess amounts of thyroxine and triiodothyronine are secreted; these excess hormones combine with the TBG, causing less than the normal number of binding sites to be available. Conversely in hypothyroidism, because of the decrease in hormone concentration, there is more than the normal number of binding sites available. These facts are used as the bases for the TBI and T_3 uptake tests.

A known amount of radioactive labeled (^{131}I) T_3, in excess of that required for complete saturation of the TBG, is added to serum under specified conditions. The T_3 will not displace any hormone already bound to the TBG, but will combine with the unsaturated binding sites. After equilibration, the unbound radioactive T_3 is separated from the serum and the radioactivity of the serum or that in the free T_3 is compared with the initial total amount of radioactivity added. The separation is usually made using an ion exchange resin or Sephadex filtration. These absorb the free T_3 and can be separated from the serum by centrifugation.

A number of test kits are available commercially. In some the radioactive ^{131}I T_3 is combined with the resin (beads or strips), in others the labeled T_3 is added separately. All kits use similar methods of separation of the free T_3 from that bound to the TBG.

The results can be expressed as either the thyroid binding index (TBI) or the T_3 uptake. In the TBI the radioactivity of the saturated TBG in the serum is compared with that obtained in a standard control serum similarly treated. This leads to an inverse proportion between the TBI and thyroid function. Serum from hypothyroid individuals having increased amount of free TBG and low thyroxine concentration will show a high index, whereas that from hyperthyroid subjects having decreased amounts of free TBG and high thyroxine concentration will have a low index. Thus the lower the thyroid function, the higher the TBI.

In the T_3 uptake the results are expressed as the percentage of the added radioactive T_3 taken up by the resin. The results are expressed as the percentage of unbound thyroxine. The greater the proportion taken up by the resin, the smaller the proportion taken up by the TBG, indicating a large amount of circulating thyroxine (hyperthyroidism). Conversely, when the proportion of radioactivity taken up by the serum is greater than that taken up by the resin, there is indication of low thyroxine hormone concentration (hypothyroidism). Thus in this test we find a direct relationship, the greater the amount of circulating hormone, the higher the T_3 uptake.

These methods are not subject to interference from x-ray contrast media or other iodine compounds from exogenous sources.

The normal value for serum TBIs are 0.9 to 1.10. With the inverse relationship existing in this test, hyperthyroidism produces values of less than 0.90 and hypothyroidism, values greater than 1.10.

The normal values for T_3 uptake by the usual methods are 25 to 35%. In hyperthyroidism the uptake is greater than 35% and in hypothyroidism, less than 25%.

The TBI is significantly increased (T_3 uptake decreased) in hypothyroidism and pregnancy and by oral contraceptives. It is decreased (T_3 uptake increased) by hyperthyroidism, nephrosis, severe liver disease, metastatic malignancies, pulmonary insufficiency, hyperandrogenic states, and anticoagulant therapy (Dicumarol, heparin).

Thyroid Function Tests Measuring Circulating Hormone Level

Protein-Bound Iodine (PBI)

Protein bound iodine testing is the oldest method used for the determination of total thyroid hormones [19–22]. Since the serum also contains a small amount of inorganic iodine, this must first be separated from the PBI. Earlier methods used the precipitation of the proteins with barium hydroxide and zinc sulfate, trichloracetic acid, or perchloric acid. The PBI was precipitated along with the proteins. The precipitate was then washed free of inorganic iodine and analyzed for organic iodine. More recent and simpler methods remove the inorganic iodine from the serum by treatment with an ion exchange resin.

After removal of the iodide, all the organic matter in the serum or precipitate must be destroyed. This can be accomplished by drying the serum or precipitate with alkali and then heating in a furnace to about 600°C to oxidize all the organic matter. Precautions must be taken to avoid loss of iodine during incineration. Alternatively the organic matter may be destroyed by wet digestion. Manually this may be done by heating the serum or precipitate with chloric acid on a hot plate. Using smaller samples the wet digestion can be performed in large test tubes in a heating block. The wet oxidation may also be done in the AutoAnalyzer automatic digestor which uses a mixture of sulfuric, nitric, and perchloric acids. This is a very convenient method when a large number of samples are run.

After destruction of the organic matter, the iodine present is determined by the reaction between ceric ions (Ce^{4+}) and arsenious acid (As^{3+}). This is a two-step reaction:

$$2Ce^{4+} + 2I^- \rightarrow 2Ce^{3+} + I_2$$

$$AsO_2^- + I_2 + H_2O \rightarrow AsO_3^- + 2I^- + 2H^+$$

The ceric ions give a strong yellow color to the solution whereas the reduced cerous ions impart only a very faint color. Thus the reaction is followed by the decrease in yellow color. Ordinarily this reaction proceeds very slowly, even at slightly elevated temperatures. It is catalyzed by the presence of iodide ions. Very small amounts of iodide will greatly increase the rate of the reaction. The amount of reaction in a given time is proportional to the concentration of iodide ions. Thus by measuring the rate of reaction in the presence of various concentrations of standard iodide in comparison with the samples, the concentration of iodide in the samples can be determined. This can be done manually or in the Technicon AutoAnalyzer* PBI system. After incineration or digestion, the iodide is usually in the form of iodate but it is rapidly reduced to iodide by the excess of arsenious acid which is added before the cerate. The greater the amount of iodide present, the faster the rate of the reaction. The change in absorbance in a given period of time will be a measure of the amount of iodide present in the solution.

Great care must be taken to prevent contamination by extraneous iodine in the reagents or other sources. One milliliter of serum used for a determination may contain less than 0.05 μg of iodine; thus even the smallest amount of extraneous iodine will give erroneous results. Traces of mercury will prevent the reaction from proceeding properly (probably by complexing the iodide) and thus give low results. The determinations should be carried out in a separate room in which neither iodine nor mercury are used in any other procedure.

X-ray contrast media usually contain large amounts of iodine. Some of these substances may be loosely bound to the serum proteins and cause erroneously high results for several weeks or longer after their use.

WET ASHING METHOD USING ION EXCHANGE RESIN [19]

The serum is treated with the anion exchange resin Amberlite IRA-401 to remove the inorganic iodine. The serum is digested with chloric acid and chromate to convert the iodine to iodate. The iodate in the digestion residue is reduced by arsenious acid and determined as iodide by its catalytic action on the reduction of ceric sulfate by arsenious acid.

EQUIPMENT

All glassware should be washed once with nitric acid or sulfuric-chromic acid cleaning solution. Rinse three or four times with tap water, three or four times with distilled water, and once with redistilled water. The redistilled water is obtained from an all-glass still or from distilled water purified by passing through an anion exchange column.

1. Electrolytic beakers (250 ml) for digestion and oxidation.

* Technicon Instruments Corp., Ardsley, N.Y. 10591.

2. Fisher Speedyvap or similar watch glasses for covering the electrolytic beakers.

3. Glass beads (5 mm).

The digestion and oxidation are performed on a hot plate at 150°C under a fume hood.

PROCEDURE: RESIN TREATMENT

If the resin beads are used, treat about 2 ml of serum with 0.5 ml volume of resin (0.3 g) in a 12 × 100 mm test tube, stopper tightly, mix thoroughly, and let stand about 10 min. Centrifuge for about 5 min to settle the resin and pipet 1-ml aliquots of serum for PBI determinations.

When using the disposable columns, about 2 ml of serum is passed through the column and a 1-ml aliquot of the serum is used for the PBI determination.

REAGENTS

1. Amberlite IRA-401, AR (Cl⁻).† This resin is also conveniently available in disposable columns by several manufacturers.‡

2. Concentrated sulfuric acid, reagent grade.

3. Chloric acid reagent with chromate. Dissolve 500 g of potassium chlorate (KClO₃) and 200 mg of sodium dichromate (Na₂CrO₄) in 1 liter of redistilled water with the aid of heat. While the solution is still hot, add 370 ml of perchloric acid (72%), stir, and cool under running water. Store in the freezer compartment overnight and filter while the solution is cold. Store in a dark place below 4°C.

4. Arenious acid with sodium chloride. Dissolve 2.64 g of arsenic trioxide and 1.9 g of sodium hydroxide in about 100 ml of redistilled water. Dilute to approximately 400 ml and neutralize to phenolphthalein with concentrated sulfuric acid. Add 30 ml more of concentrated sulfuric acid and 6.6 g of sodium chloride and dilute to 1 liter.

5. Ceric ammonium sulfate, $0.028M$. Dissolve 17.7 g of ceric ammonium sulfate [$(NH_4)_4Ce(SO_4)_4 \cdot 2H_2O$] and 103 ml of concentrated sulfuric acid in 700 ml of redistilled water. Let stand overnight; then dilute to 1 liter. Filter if cloudy and store in an amber bottle.

6. Iodate standards.

A. Stock standard. Dissolve 168.5 mg of potassium iodate (KIO₃) that has been dried in a vacuum desiccator in redistilled water to make 1 liter. This stock solution contains 100 μg iodine/ml.

B. Working standard. Dilute 1 ml of stock standard to 2 liters with re-

† Mallinckrodt Chemical Co., St. Louis, Mo. 63101.
‡ Whale Scientific, Denver, Colo. 80239; Bio-Rad Laboratories, Richmond, Calif. 94804; Curtis Nuclear, Los Angeles, Calif. 90058.

distilled water. This solution contains 0.05 μg iodine/ml and should be made fresh every 2 weeks.

PROCEDURE: ACID DIGESTION

To the 1-ml aliquots of resin-treated serum in 250 ml electrolytic beakers containing 8 to 10 glass beads, add 10 ml of chloric acid. Cover with a watch glass and heat on the hot plate. After crackling caused by the decomposition of chloric acid ceases and white fumes appear, add 0.5 ml more chloric acid. Add a few drops of chloric acid from time to time to maintain the chromium in its orange hexavalent state. (*Caution:* Reduction of chromium to the green trivalent state for more than a few seconds is associated with loss of iodine.)

When the volume has been reduced to about 5 ml and dense white fumes have been present for about 5 min, the watch glasses may be removed. The rate of digestion should be such that about 45 min is required to reach this point from the beginning of the digestion.

Evaporate to an estimated volume of 0.3 to 0.5 ml. Iodine may be lost if complete dryness is obtained. With practice, the visual estimate of the desired volume can be made with negligible error. The final volume should just wet the glass beads with little excess liquid. Set up several blanks containing 1 ml of water and 10 ml of chloric acid. Also run standards containing 0.5, 1.0, 1.5, 2.0, and 3.0 ml of working standard with 10 ml of chloric acid. Digest and treat the same as the unknowns.

PROCEDURE: COLORIMETRIC DETERMINATIONS

Allow beakers to cool after digestion and add 6 ml of arsenious acid solution to each. Warm the beakers gently on the hot plate to effect solution and allow to cool to room temperature. Pipet 5-ml aliquots from each sample into labeled cuvettes and place in a water bath at 30°C for about 15 to 20 min. Also warm an amount of ceric ammonium sulfate solution in the water bath. Since the temperature markedly affects the rate of reaction, the water bath should have an accurate thermostatic control.

At timed intervals of 30 sec, add 0.5 ml of ceric ammonium sulfate to each of eight or ten cuvettes and mix well by inversion, with a piece of Saran wrap or Parafilm over the top.

At 30-sec intervals read the percent transmission at 420 nm against a water blank. Thus each tube is read every 4 to 5 min. Two people can handle as many as 20 tubes at 15-sec intervals; one technician removes the tubes from the bath at the proper time and the other takes the reading or adds the ceric ammonium sulfate. This can be done easily with a little practice. Keep the tubes in the water bath between readings.

Read the tubes at accurately timed intervals, each tube being read at exactly 5, 10, 15, and 20 min after the addition of ceric ammonium sulfate solution. Take readings until the percent transmission is close to 60%.

CALCULATION AND READING

Calculate the rate of reaction by subtracting the spectrophotometric reading closest to 35% from that closest to 60% transmission (ΔT) and dividing by the time interval in minutes (t). With the use of the blank for zero iodine and the five standards corresponding to 2.5, 5.0, 7.5, 10.0, and 15 μg/dl iodine, plot the $\Delta T/t$ for each standard against the iodine concentration. From the $\Delta T/t$ for the samples, read the PBI values in micrograms per deciliter directly from the curve.

As the curve is nearly straight, it may be extrapolated somewhat. An immediate decolorization of the sample tube indicates contamination, and the result may be reported merely as over 20 μg/dl because in this case the exact value is of no significance.

It is necessary to digest a series of standards each time samples are run because there will be some variation from day to day. Also quality control serums of various known concentrations of PBI should be included with each set of determinations.

NORMAL VALUES AND INTERPRETATION OF RESULTS

Normal serum PBI values range from 4 to 8 μg/dl. In hyperthyroidism, values ranging from 8 to 20 μg/dl are found; in hypothyroidism, levels of 0 to 4 μg/dl. Low values are sometimes found in nephrosis or other conditions with very low serum proteins. In pregnancy there may be a mild elevation of the PBI level. There may be erroneously low results for a few weeks after the administration of mercurial diuretics or other drugs containing mercury. Many x-ray contrast media (such as Priodax, Neo-Iopax, and Skiodan) used in the diagnosis of gallbladder and kidney disorders contain iodine and are apparently absorbed on the serum proteins. They will cause erroneously high values of protein-bound iodine for 6 months or longer after their use. Drugs containing inorganic iodine (e.g., Lugol solution) will cause false high values for a week or so after administration.

PROTEIN-BOUND IODINE BY AUTOMATION [23–25]

Technicon Instruments Corp.§ has developed a fully automated procedure using an automated digestion apparatus and standard automated colorimetric equip-

§ Technicon Instruments Corp., Ardsley, N.Y. 10591.

ment for PBI determinations. The method is a modification of the Zak wet ash method [20]. The serum is treated manually with an ion exchange resin (Amberlite IRA-401) for removal of inorganic iodine before analysis. This system has been thoroughly evaluated by our laboratories as well as by many other workers in the field of clinical chemistry and has been found to be accurate and precise. Good correlation has been found in comparing the results obtained by automation with those using the dry ash method of Barker [19] and the wet ash method of Zak. Although the automated system is designed to analyze samples at the rate of 20/hr with a sampling-wash ratio of 2:1, our laboratories have increased this rate to 30 samples/hr by changing the sample-wash ratio to 1:1. The results obtained at these rates agree very well with those obtained at the slower rate.

Currently the automated procedure is the method of choice when a large number of samples are measured. In addition, it has also proved the reliability of the use of ion exchange resin for removal of inorganic iodine, a method we strongly recommend as a substitute for precipitation and washing of proteins in the manual methods.

Thyroxine (T_4)

Two methods are commonly used for the determination of T_4, a chemical method using column chromatography and a competitive protein binding method [26–31].

In the chemical method the serum is adjusted to a pH of 10 to 11 with ammonium hydroxide and placed in an ion exchange column to remove thyroxine (T_4), triiodothyronine (T_3), iodide, and other iodoamino acids, if present. The column is washed successively with water, acetate buffer (pH 4.0), and acetic acid (pH 1.9) to remove proteins, iodotyrosines, and some organic iodine compounds. The thyroxine and other iodotyronines are quantitatively removed from the column with $8.7M$ acetic acid (pH 1.3). The eluate is collected in two fractions; the first one usually contains 70 to 90 percent of the thyroxine and the second fraction 10 to 30 percent. If the second fraction contains more than 30 percent of the thyroxine, it usually indicates contamination from other organic iodides and renders the results invalid.

The acetic acid containing the thyroxine is then treated with bromine water and used directly (without wet ashing) in the ceric-arsenite reaction.

Exogenous organic iodine compounds used in radiography and therapy interfere with this method less than with PBI determination.

The competitive protein binding method (CPB) was first described by Murphy and Pattee [32] and is based on the specific binding properties of the thyroxine binding globulin (TBG). The principles of CPB are discussed in Chapter 15. The thyroxine is separated from the serum by precipitating the proteins with alcohol, in which thyroxine is soluble. After centrifugation, an aliquot of the supernatant is used for analysis. This aliquot is either evaporated to dryness or treated directly with TBG and an excess of radioactively labeled thyroxine. After equilibration, the free thyroxine is separated from that bound to the protein and the radioactivity of either fraction determined. The amount of thyroxine in the sample is obtained from a standard curve treated similarly. The separation can be made by means of an ion exchange resin, Sephadex filtration, or dextran-coated charcoal. The extraction efficiency of alcohol is about 80 percent. If pure thyroxine is used for the standard curve, a correction must be made for the incomplete extraction of the thyroxine. This method has the advantage that neither inorganic nor organic iodine contaminants interfere with the results.

THYROXINE BY COLUMN CHROMATOGRAPHY

REAGENTS

1. Ion exchange resin.   Dowex AG1-X2 (200 to 400 mesh) in acetate form.||

2. Ammonium hydroxide, 0.8M.   Dilute 53 ml of concentrated reagent grade ammonium hydroxide to 1 liter.

3. Acetic acid, 0.2M.   Dilute 11.6 ml of glacial acetic acid to 1 liter.

4. Sodium acetate, 0.3M.   Dissolve 27.2 g of sodium acetate trihydrate in water to make 1 liter.

5. Acetate buffer, pH 4.0   Mix 4.16 volumes of 0.2M acetic acid with 1 volume of 0.2M sodium acetate. Check pH and adjust if necessary to a pH of 4.0 ± 0.2.

6. Acetic acid, pH 1.9.   Mix 1 volume of glacial acetic acid with 4.55 volumes of water. Adjust pH if necessary to 1.9 ± 0.1.

7. Acetic acid, pH 1.3.   Mix equal volumes of glacial acetic acid and water. Adjust pH if necessary to 1.30 ± 0.05.

8. Bromine solution.   Dissolve 1.67 g of potassium bromate and 7.14 g of potassium bromide in water to make 100 ml.

9. Arsenious acid reagent.   Dissolve 1.44 g of arsenious trioxide in 15 ml of 0.5M sodium hydroxide with the acid of gentle warming. Dilute to about 850 ml in a 1 liter volumetric flask. Add with mixing 28 ml of concentrated hydrochloric acid and then 58 ml of concentrated sulfuric acid. Cool to room temperature and dilute to 1 liter.

10. Ceric ammonium sulfate reagent.   Place in a 1 liter volumetric flask 20.0 g of ceric ammonium sulfate and about 600 ml of water. Add slowly while mixing

|| Bio-Rad Laboratories, Richmond, Calif. 92112.

49 ml of concentrated sulfuric acid. Mix until the ceric ammonium sulfate is dissolved, cool, and dilute to the mark.

11. Thyroxine standards. Dissolve 5.77 mg of the sodium salt of thyroxine[#] in 50 ml of a mixture of 99 parts of methanol and 1 part of ammonium hydroxide. The solution contains the equivalent of 100 μg of thyroxine/ml and is the stock solution. A working stock is prepared by diluting 1 ml of the stock solution to 100 ml with a mixture of equal volumes of glacial acetic acid and water. This solution contains 1 μg/ml and should be made up fresh weekly. Actual working standards are prepared by further dilution (see Procedure).

PROCEDURE

Preparation of columns. The chromatographic columns may be plastic or glass. The columns may be purchased filled with the resin[*] and discarded after use or emptied, washed, and refilled with fresh resin. The latter is less expensive and also insures that freshly made columns are being used. The resin is suspended in the 0.8M ammonium hydroxide and poured into the column, which has a plug of glass wool at the bottom, to a height of 3 to 4 cm. The column is allowed to drain and more resin is added if necessary to obtain the desired height. The columns should be prepared just before use and the resin slurried in the ammonium hydroxide should be discarded after about 1 week.

Add a 1-ml aliquot of serum to 3 ml of 0.8M ammonium hydroxide and mix well. Allow the mixture to stand for 15 to 30 min and then transfer to a chromatographic column. Collect the eluate from the column in the original container and pass it through the column a second time. Discard the second eluate. Rinse the original container with 4 ml of water and add to the column and discard eluate. Now add 4 ml of the acetate buffer to the column and discard the eluate. When the column has drained dry, add 4 ml of acetic acid, pH 1.9, directly to the column and discard the eluate, allowing the column to drain completely. Add exactly 4 ml of acetic acid, pH 1.3, to the column and collect the eluate in a test tube, allowing the column to drain completely. Repeat with a second 4 ml of the pH 1.3 acetic acid, collecting the second eluate in a separate tube. Label these tubes as fractions A and B for the sample.

Preparation of standards. Dilute 4 ml of the working stock to 100 ml with the 8.7M acetic acid (pH 1.3). To a series of tubes add 0.5, 1, 2, 3, and 4 ml of the diluted standard and make up to 4 ml with 8.7M acetic acid where necessary. Also set up a tube containing 4 ml of 8.7M acetic acid as a blank. Pipet 2-ml

[#] L-Thyroxine, sodium salt with 9–11% H_2O; Calbiochem, San Diego, Calif. 49112.
[*] Bio-Rad Laboratories, Richmond, Calif. 92112.

aliquots of the standards, blank, and the tubes containing the column eluates to colorimeter cuvettes. To each tube add 1 drop of the bromide solution and mix. This should impart a distinct yellow tinge to the solution. Allow to stand for 5 min and then add 4 ml of the arsenious acid solution, mix, and transfer to a 37°C water bath. At timed intervals add 1 ml of the ceric sulfate to the tubes and proceed exactly as in the colorimetric determination for PBI. The colorimetric analysis has been automated by Kessler and Pileggi [33], using Technicon's AutoAnalyzer. Complete details are given in their publication.

CALCULATION

The column eluates represent the thyroxine from 1 ml of serum. The working stock solution contains 1 μg of thyroxine/ml and the dilution of this contains 0.040 μg/ml. Thus when 0.5, 1.0, 2.0, 3.0, and 4.0 ml of the dilution are used and made up to 4 ml, and these are treated the same as the 4 ml elutions from the column from 1 ml of serum, the standards are equivalent to 2, 4, 8, 12, and 16 μg/dl of thyroxine. If one wishes to report results in terms of thyroxine iodine (to compare with the PBI), the above figures are multiplied by 0.654, the proportion of iodine in thyroxine. Thus the above standards would correspond to 1.3, 2.6, 5.2, 7.8, and 10.5 μg of thyroxine iodine/dl. Normally 70 to 90 percent of the measured iodine is in the first fraction. The amounts in the two fractions are added to give the total.

NORMAL VALUES AND INTERPRETATION OF RESULTS

The normal range for this procedure is 3.0 to 6.8 μg/dl of thyroxine iodine or 4.6 to 10.5 μg/dl of thyroxine. The changes with thyroid diseases are similar to those given for PBI. Normally in the elution pattern approximately 70 to 90 percent of the total thyroxine is found in the first fraction. In exogenous organic iodine contamination, significant amounts of iodine may be found in the second fraction; the iodine content may be higher than in fraction one. This usually renders the results unacceptable. Contaminants that interfere with the results include Hippuran, Itrumil, Miokon, Orabilix, Telepaque, and Teridax.

THYROXINE BY COMPETITIVE PROTEIN BINDING (CPB)

This method, first described by Murphy and Pattee [32], is based on the specific binding properties of the thyroxine binding globulin (TBG) and does not depend on iodide analysis [34–39]. It is considered the most specific method for the determination of serum thyroxine.

Before assaying for thyroxine, all endogenous protein that might interfere by binding the T_4 must be completely removed or denatured. In most of the methods the thyroxine is extracted from the serum by precipitating the proteins with ethanol.

Some methods use an aliquot of the alcohol extract directly in the analysis,† others evaporate the extract to dryness and use the residue for analysis.‡ Another method uses a surface absorption technique§ to isolate the T_4 from serum; the T_4 is dissociated from the binding proteins in dilute acid and absorbed onto an inorganic matrix (silicate particles). The T_4 is then removed from the absorbent by an alkaline buffer and assayed by standard CPB technique.

Because of the difficulty in preparing and standardizing the reagents used in these procedures, it is best to purchase the necessary material which is available from several manufacturers as complete kits or bulk reagents.

The remainder of the procedure is about the same for all kits. To the T_4 extract is added the binding protein (TBG) and a definite excess of radioactive labeled T_4. After the mixture is allowed to come to equilibrium, the thyroxine in the free state is separated from that bound to the TBG. The radioactivity of either of the fractions is then determined. The amount of thyroxine is calculated from a standard curve obtained by processing standards of known thyroxine content in the same manner as the unknowns. The separation of the unbound T_4 from the bound is made by using either an ion exchange resin (beads, sponge, or strips), Sephadex filtration, or dextran-coated charcoal.

The methods using ethanol extraction require that a correction be made for the incomplete extraction by thyroxine. The extraction efficiency of ethanol is usually between **78** and **82** percent. This should be checked by every laboratory under their conditions of assay by analyzing a serum with a known thyroxine content and comparing the results with the pure T_4 standards.

The results by this method are usually expressed in micrograms of T_4 per deciliter and compare well with those by the column chromatographic method.

The most significant advantage of this method when compared to the other methods for the determination of thyroxine is the fact that it is not subject to interference by other iodine compounds, whether organic or inorganic. Dilantin and large doses of salicylates may cause falsely low T_4 values by competing for TBG binding sites. Choloxin, a cholesterol-reducing drug, produces falsely highly elevated results by displacing the radioactive T_4 from the TBG in the reaction mixture.

Other kits available for the determination of T_4 by this method are: T-T_4

† Mallinckrodt Nuclear, Division Mallinckrodt Chemical Works, St. Louis, Mo. 63160.
‡ Tetrasorb-125 Abbott Laboratories, North Chicago, Ill. 60064.
§ Tetra-Tab-Nuclear Medical Laboratories, Inc., Dallas, Tex. 75247.

(Curtis),|| Thyrox-I-Tale (Dade),# Tetralute (Ames),* and THE-T$_4$ (Bio Nuclear).†

The normal values for T$_4$ by this method are 5.0 to 13.7 μg/dl T$_4$ or 3.3 to 9.0 μg/dl T$_4$ iodine. The changes with thyroid disease are similar to those given for PBI earlier in this chapter.

Free Thyroxine Index [40–44]

The concentration of free thyroxine in the blood is considered the best indicator of the thyrometabolic state of the individual. Of the total circulating thyroxine, approximately 99.95 percent is bound to the plasma proteins and 0.05 percent is found in the free (unbound) state. At present the methods available for determination of the free thyroxine are equilibrium dialysis ultrafiltration and gel filtration. These tests determine the percentage of free thyroxine, which must be multiplied by the total T$_4$ in order to determine the actual concentration of free T$_4$.

The procedures are very difficult and time consuming, requiring reagents of the highest purity. Consequently these techniques have not been generally adapted for routine clinical use.

It appears that the level of free T$_4$ in serum is directly related to the total T$_4$ concentration and the concentration of the TBG, reflected by the T$_3$ uptake test. A number of methods have been devised for the estimation of an index that is proportional to the concentration of free thyroxine in serum. The first was introduced by Clark and Horn [42], based on the determination of PBI and T$_3$ uptake measurement (PBI $\times$ T$_3$) and called the free thyroxine index.

This measurement has been referred to by various names: FT$_4$ index, T$_7$, and T$_{12}$. It is calculated from the analysis of T$_4$ and T$_3$ uptake and has been shown to have a high degree of correlation with the thyrometabolic status of the individual.

Abbott's T$_7$, using Triosorb and Tetrasorb, is obtained by multiplying the T$_3$ uptake (expressed as a decimal) by the T$_4$ value (in μg T$_4$/dl) and gives a normal range of 1.3 to 4.4. Mallinckrodt's FT$_4$, using Res-o-Mat T$_3$ and T$_4$, is obtained by multiplying the T$_3$ uptake (expressed as T$_3$ TBG index) by the T$_4$ (in μg T$_4$/dl). The normal range is given as 4.4 to 15.7. Nuclear Medical's (NML) T$_{12}$, using Tri-Tab and Tetra-Tab, is obtained by multiplying the T$_3$ uptake (expressed as a decimal) by the T$_4$ (in μg T$_4$/dl) and gives a normal range of 1.58 to 5.18. When Ames' Trilute and Tetralute are used, the manufacturer pro-

|| Curtis Nuclear Corp., Los Angeles, Calif. 90058.
Dade Reagents, Division American Hospital Supply Corp., Miami, Fla. 33152.
* Ames Co., Division Miles Laboratories, Elkhart, Ind. 46514.
† Bio Nuclear Laboratories, San Juan Capistrano, Calif. 92675.

vides a slide rule to calculate the free thyroxine index from the two values. This is approximately T_4 (in μg T_4/dl) $\times$ [T_3 uptake (in %)/50] and gives a normal range of 3.0 to 9.1.

There is a significant lack of uniformity in the range of normals given for free thyroxine index by the different manufacturers. This is due primarily to the different range of normals given for T_3 and T_4 by the various kits. More uniformity of these values seems desirable and the various manufacturers should make an effort to accomplish this. It certainly would simplify the interpretation of the results.

A recently introduced kit by Mallinckrodt (Res-O-Mat ETR) combines the T_3 and T_4 determinations into a single test. The results are expressed as the effective thyroxine ratio (ETR), with a normal range of 0.88 to 1.10. Unlike the TBI, this ratio has a direct relationship with thyroid function; it is increased in hyperthyroidism and decreased in hypothyroidism.

The free thyroxine index is not affected by the administration of estrogens, androgens, salicylates, diphenylhydantoin, pregnancy, nephrosis, or factors causing abnormal T_3 uptake results due to variations in the level of TBG or to interference with its normal binding.

Thyroid Function Tests Based on Other Hormones

Thyroid Stimulating Hormone (TSH) [48–52]

The plasma concentration of this pituitary hormone is extremely low (approx. 0.2 μg/dl), yet it is essential for the maintenance of normal thyroid function. The role of TSH in the negative feedback mechanism of the thyroid gland has already been noted in the introductory section on the thyroid hormones earlier in this chapter. The release and possibly the synthesis of this hormone are controlled by the thyrotropine releasing factor (TRF), a hormone produced by the hypothalamus. This hormone has been isolated, characterized, synthesized, and prepared in highly purified form. When injected into man, it produces a significant increase in serum TSH within a few minutes. This action may prove to be useful for the clinical evaluation of pituitary TSH availability.

The TSH is best determined by radioimmunoassay; the methodology is discussed in Chapter 15. TSH analysis is primarily used for the diagnosis of myxedema and for the differentiation of primary and secondary myxedema. High levels are found in untreated cases of primary myxedema; undetectable or very low levels are found in myxedema secondary to pituitary failure. In Hashimoto's thyroiditis, TSH levels are elevated in cases with clinical evidence of hyperthyroidism and in about 40 percent of the cases considered clinically euthyroid.

Untreated cases of hyperthyroidism have undetectable or low-normal TSH levels. Normal TSH levels have been found in Cushing's syndrome, acromegaly, euthyroid ophthalmopathy, and a variety of other endocrine disorders.

A TSH suppression test [52] has also been used for the evaluation of TSH production. The administration of thyroxine to euthyroid individuals inhibits the secretion of endogenous thyroxine hormone and also suppresses the secretion of TSH by the pituitary gland (feedback mechanism). In hyperthyroidism the thyroid activity and TSH secretion cannot be suppressed by the administration of exogenous thyroxine.

Long-acting Thyroid Stimulator (LATS) [53–56]

This is an abnormal thyroid stimulating substance found in the blood of patients with hyperthyroidism. It is distinguished from TSH by its longer duration of action. This substance does not appear to have its origin in the pituitary, and it is classified as a 7-S γ-globulin.

LATS is frequently found in the serum of patients with malignant exophthalmos or Graves' disease. It has also been reported present in infants born of mothers suffering from Graves' disease.

The methods used for the detection of LATS in serum require bioassay techniques and will not be described here. Several methods have been published that can be used for both estimation of TSH and detection of LATS. Normally no LATS is detectable in serum.

Miscellaneous Thyroid Function Tests

Basal Metabolic Rate (BMR)

Basal metabolic rate [57, 58] was the most commonly used thyroid function test before the introduction of PBI and other more accurate and dependable tests. It refers to the metabolism (heat production) during complete mental and physical rest in the postabsorptive state and is a measure of the vital metabolic processes. BMR is essentially an index of thyroid activity and is measured as the percentage variation from the normal found for an individual of the same height, weight, age, and sex. Unfortunately this test, besides being very time consuming, is affected by many extraneous factors. Currently its use is generally restricted for the diagnosis of such conditions as euthyroid hypometabolism and hypermetabolism without hyperthyroidism.

Antithyroid Antibodies

Circulating antithyroid autoantibodies [59–61] have been demonstrated in cases of thyroiditis, Hashimoto's disease, and other thyroid diseases (e.g., myxedema).

They were first detected by immunoprecipitation in agar gels and hemagglutination and complement fixation techniques. The action of the antibodies does not appear to be directly destructive to thyroid tissue but is rather a response of the thyroid tissue to infectious damage, causing it to produce "foreign" antigen, and a subsequent autoimmune response. A commercially available latex thyroglobulin reagent (TA-test‡) gives a reaction with serums containing precipitin antibodies to human thyroglobulins.

Achilles Tendon Reflex

This test involves the known effects of hypofunction and hyperfunction of the thyroid gland on the Achilles tendon reflex. The duration of the reflex response upon tapping of the Achilles tendon with a percussion hammer has been found to be increased in hypothyroidism and decreased in hyperthyroidism as compared with normal subjects. A simple device (Photomotograph§) is available for the photoelectric measurement of the reflex time using an electrocardiograph machine as a recorder. The test is very easy to perform and measurement of the time from the electrocardiograph record is relatively simple. With limits of under 290 msec for hyperthyroid individuals, Squires and Langhorne [62] found the test to be slightly more reliable than the PBI and the radioiodine uptake. It is not subject to the interferences encountered in other methods. Although possibly not strictly a laboratory procedure, it is mentioned as an additional method for testing thyroid function. For details the publications by Chaney [63] and Gilson [64] may be consulted.

The Steroid Hormones [65, 66]

A number of important hormones as well as other biologically active compounds have the basic steroid nucleus:

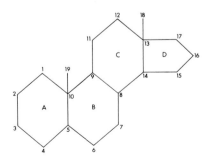

‡ Hyland Laboratories, Costa Mesa, Calif. 92626.
§ Burdick Corp., Milton, Wis. 53563.

In this structural formula each angle in the rings represents a carbon atom. These are usually numbered as indicated (1 to 17). There may also be some additional carbon atoms outside the rings which are numbered as shown (18, 19). Conventionally the hydrogen atoms are not indicated. The remaining valences of the (tetravalent) carbon atoms are assumed to be satisfied by attachment to hydrogen atoms unless other substituents are indicated in the diagram. In general, the estrogens (ovarian hormones) contain only the carbon atoms numbered 1 to 18. Most of the androgens (testicular hormones) contain the additional carbon 19. Many of the hormones of the adrenal cortex (adrenocorticosteroids) and the hormones of the corpus luteum also contain the additional carbon atoms 20 (attached to 17) and 21 (attached to 20). Cholesterol also has this basic structure with an additional six-carbon chain attached to carbon 20 (C-20). This compound is believed to be a precursor in the biosynthesis of many of these hormones. In addition to the basic structure outlined, the actual compounds will have additional —OH or —O— groups as well as one or more double bonds between carbon atoms.

The compounds containing 18 carbon atoms are labeled as derivatives of a parent hydrocarbon, estrane. Thus estradiol, an important estrogen, is named 1,3,5-estratriene-3,17β-diol. The *ene* indicates a double bond; thus es*tra*tri*ene* has three double bonds, in this case from carbon atoms 1 (to 2), 3 (to 4), and 5 (to 10) as indicated by the prefix numbers. The suffix *ol* indicates an —OH group (alcoh*ol*); thus the compound has two —OH groups, one on C-3 and one on C-17. The β (beta) indicates a type of stereoisomerism. If the —OH group is known to be spatially on the same side of the ring structure as C-18, the —OH is labeled beta, and if it is on the opposite side it is labeled alpha (α).

The compounds having 19 carbon atoms are named as derivatives of a parent hydrocarbon, androstane. Thus testosterone, an important testicular hormone, is named 4-androsten-17β-ol-3-one. The ending *one* indicates a C=O group (ket*one*). Testosterone has 19 carbon atoms with a double bond from atom 4 (to 5), an —OH at C-17, and a C=O group at C-3.

The compounds containing 21 carbon atoms are named as derivatives of pregnane. Thus corticosterone, a hormone of the adrenal cortex, is 4-pregnene-11β,21-diol-3, 20-dione. It has a double bond at atom 4 (to 5), —OH groups on carbon atoms 11 and 21, and =O on atoms 3 and 20. Another adrenal hormone has an =O group instead of an —OH group at atom 11 and is often called 11-dehydro-corticosterone (a hydrogen atom has been removed from corticosterone). Only slight differences in chemical structure may give large differences in physiological activity. This makes the separation of the different types of hormones by chemical means difficult.

The concentrations of most of these hormones in the blood are low and difficult to analyze by chemical means. The radioimmunoassay procedures are the most practical methods available for their determination (see Chapter 15). The simpler methods are based on the determination of the amounts excreted in the urine, where the concentrations of the hormones may be higher than in blood and larger samples are readily obtainable.

The evaluation by urinary excretion involves a number of problems. The content of the physiologically active hormones excreted may be low and a number of more or less inactive metabolites of each hormone may actually be excreted. In addition, very similar metabolites may be obtained from hormones actually secreted by the testes, the adrenal cortex, and even the ovaries. Thus it may be difficult to obtain an evaluation of the functioning of a given organ unless very complex chemical separations are made.

Most of the steroids excreted in the urine are in a conjugated form as glucuronates or sulfates. The conjugated steroids must be hydrolyzed before they can be extracted by the organic solvent (ether, benzene, chloroform, heptane). They may be hydrolyzed by heating with acid. Insufficient heating or too weak acid will give incomplete hydrolysis, and prolonged heating or too strong acid will tend to break down some of the steroids. The steroids may also be hydrolyzed by incubation with certain enzymes, but this is a lengthy procedure usually done for use with gas chromatography.

Steroid hormones influence the entire body. They control salt and water metabolism; affect carbohydrate, protein, and fat metabolism; control the development of the primary and secondary sex characteristics; and affect the distribution of hair and development of the muscular and skeletal systems. The physiological differences of the various hormones result from variations in the nature and position of the side chains in the molecules.

Androgenic Steroids

The androgenic steroids have methyl groups at positions 18 and 19 and an —OH or =O group at position 17. Those with the C=O structure at position 17 are the 17-ketosteroids. Testosterone is not a ketosteroid, but a number of its metabolites such as androsterone, dehydroepiandrosterone, and etiocholanolone are ketosteroids found in the urine. Their concentration in urine, in part, represents androgenic activity.

17-Ketosteroids

These steroids have 18 or 19 carbon atoms with a C=O (keto) group at C-17. The principal 17-ketosteroids found in urine are androsterone, dehydroepiandrosterone, etiocholanolone, 11-ketoandrosterone, 11-ketoetiocholanolone, 11-β-hy-droxyandrosterone, and 11-β-hydroxyetiocholanolone. These compounds will react with alkaline dinitrobenzene (Zimmerman reaction) to give a purple-red color. The urine is hydrolyzed with acid and the steroids extracted with ether or ben-zene. The estrogens and some other interfering materials are removed by washing the extract with dilute alkali. The extract is then filtered, an aliquot evaporated, and the residue treated with the reagent to develop the color. There are a number of variations in the reagents used to develop the color but the principle remains the same. Not all the ketosteroids give the same amount of color with the reagent. The results are usually reported in terms of total ketosteroids in comparison with a standard that is usually the particular steroid, dehydroepiandrosterone.

Testosterone

Testosterone [67–70] is the most biologically active of the androgenic steroids and is synthesized mainly by the interstitial cells of Leydig in the testis in the male. In the female it is produced by the adrenals and ovaries.

Testosterone is responsible for the development of secondary sex characteristics in the male. In the female it serves primarily as a precursor for estrogen. Physio-logically, testosterone is anabolic in its action and promotes nitrogen retention for the maintenance of nitrogen balance.

Testosterone has been determined by a variety of methods including bioassay, gas chromatography, double isotope derivative, competitive protein binding radio-assay, and radioimmunoassay. The method of choice is radioimmunoassay (see Chapter 15).

The normal levels in men range from 350 to 1,000 nanograms (ng)/dl and in females, from 15 to 70 ng/dl. Testosterone is decreased in the adult male in Klinefelter's syndrome, orchidectomy, estrogen therapy, hypopituitarism, hypo-gonadism, and cirrhosis of the liver. In the female it is increased during preg-nancy, administration of oral contraceptives, the Stein-Leventhal syndrome, idiopathic hirsutism, and in virilizing tumors.

Ketogenic Steroids

These steroids are 21-carbon molecules which contain an —OH group at C-17. They can be oxidized by treatment with sodium bismuthate or sodium meta-periodate to yield 17-ketosteroids. The ketosteroids are then extracted and determined as above. If only the oxidation is done, the determination will be of total ketosteroids: original ketosteroids plus those produced from the ketogenic steroids. The latter are then determined by difference. Preferably the urine may be first treated with sodium borohydride to reduce the original ketosteroids to compounds that will not be reoxidized to ketosteroids. On subsequent oxidation, only the ketogenic steroids will produce 17-ketosteroids to react with the color reagent. Appreciable amounts of glucose in the urine interfere with the results. This can be overcome by using larger amounts of borohydride and periodate.

Some hydroxycorticoids such as hydroxycortisone may be determined by the Porter-Silber reaction. The serum or hydrolyzed urine is extracted with methylene chloride or other solvent and the extract washed with alkali and filtered. An aliquot of the extract is shaken with a reagent containing phenylhydrazine in a mixture of 2 parts $10.8M$ sulfuric acid and 1 part ethanol. After separation, the sulfuric acid layer is further incubated to develop a yellow color. This is read in a colorimeter against a blank prepared by the same procedure without the phenylhydrazine.

Plasma or serum cortisol may be determined by a fluorometric method. The plasma or serum is extracted with ethylene dichloride. The aqueous layer is discarded and an aliquot of the solvent layer shaken with a mixture of $11.7M$ sulfuric acid and 35% ethanol. The fluorescence of the sulfuric acid layer is then read in a fluorometer in comparison with a standard similarly extracted. This is a very simple method which, although not entirely specific, gives adequate estimate of adrenal function.

Any of these methods can be used for the estimation of adrenal function by ACTH stimulation. The secretion of the steroid hormones by the adrenal cortex is regulated by the pituitary. This gland secretes a hormone (ACTH, adrenocorticotropic hormone) which stimulates the adrenal cortex to produce its hormones. In cases of decreased secretion of the adrenal hormones it is often desirable to know whether this is due to the fact that the pituitary no longer secretes sufficient ACTH to stimulate the adrenal or to lack of response by the adrenal to the endogenous ACTH. The ketosteroids, ketogenic steroids, or other corticosteroids are measured in the blood or urine before and after the administration of exogenous ACTH. A marked increase in the amount of hormones found after stimulation indicates that the adrenal can respond to ACTH but presumably the pituitary is not producing a sufficient amount of the hormone. If the steroid levels increase only slightly, this indicates that the adrenal is not responding to ACTH.

Determination of 17-Ketosteroids and Hydroxysteroids
(Ketogenic Steroids) [71–74]

REAGENTS

1. Methyl alcohol, AR, acetone-free.

2. Potassium hydroxide, $4M$ in methanol. Add 34 g of potassium hydroxide pellets to 120 ml of purified methanol and dissolve, preferably with the aid of a

magnetic stirrer. Filter through hardened paper or centrifuge if necessary and transfer the clear liquid to a polyethylene bottle. Add exactly 2 ml of solution to about 20 ml of water and titrate with standard acid with phenolphthalein indicator. Calculate the normality and dilute to $4.00 \pm 0.15M$. This reagent is stable; a small amount of potassium carbonate may settle out but this does not affect the results.

3. Ethylene glycol monomethyl ether (Methyl Cellosolve). The reagent grade of the solvent may be used.

4. Dinitrobenzene, 1 g/dl. Dissolve 1 g of dinitrobenzene (recrystallized from alcohol or special Sigma grade)[Δ] in 100 ml of Methyl Cellosolve. Store in refrigerator.

5. Sodium hydroxide, $5M$. Dissolve 100 g of sodium hydroxide in water to make 500 ml. This solution need not be standardized.

6. Sodium hydroxide, $1M$ and $0.1M$. Prepare from the $5M$ solution by appropriate dilutions (1:5 and 1:50).

7. Sulfuric acid, $0.5M$. Add 14 ml of concentrated acid to about 400 ml of water. Cool and dilute to 500 ml.

8. Acetic acid, $4.35M$. Mix 100 ml of glacial acetic acid and 300 ml of water.

9. Acetic acid, $1.05M$. Dilute 30 ml of glacial acetic acid to 500 ml with water.

10. Sodium borohydride, 10 g/dl. Just before use, dissolve 1 g of sodium borohydride in 10 ml of $0.1M$ sodium hydroxide solution.

11. Sodium metaperiodate, $0.47M$. Just before use, dissolve 2 g of sodium metaperiodate in 20 ml of water. This solution is near saturation and the salt may dissolve slowly.

12. Steroid standards.

A. Stock standard. Dissolve 250 mg of dehydroepiandrosterone in methyl alcohol and dilute to 100 ml. Store in tightly stoppered bottle in refrigerator. One milliliter contains 2.5 mg of steroid.

B. Working standard. Dilute 2 ml of stock standard to 100 ml with methyl alcohol. This standard contains 25 μg in 0.5 ml. Store in refrigerator.

14. Ether, AR.

15. Sulfuric acid, $9.0M$. Cautiously add 100 ml of concentrated sulfuric acid to 100 ml of water. Mix and cool.

16. Sodium hydroxide, AR, solid pellets.

GENERAL PROCEDURE

Measure total volume of a well-mixed 24-hr urine specimen. Record volume and reserve a few hundred milliliters for analysis. Run determinations in duplicate.

[Δ] Sigma Chemical Co., St. Louis, Mo. 63118.

Perform extractions in 25×150 mm screw-capped tubes with Teflon-lined caps. During the extraction procedure with ether, keep tubes cooled in ice water or cold, running tap water. In warm weather it is helpful to cool the solutions used for washing the ether extract. This is conveniently done by storing bottles of these solutions in the refrigerator. (*Caution:* Care must be taken in the evaporation of ether extracts; even a hot plate can ignite high concentrations of ether vapors.) The lower aqueous layer is removed from the ether extract in the tubes by means of a 1 ml serological pipet attached with rubber tubing to a source of suction and a trap. Suction is controlled by pressure on the rubber tubing at the top of the pipet.

PROCEDURE FOR KETOSTEROIDS

To the extraction tube add 7.5 ml of urine and 1 ml of $9.0M$ H_2SO_4. Heat in a boiling water bath for 15 min, cool well, add exactly 25 ml of ether, cap tightly, and shake for 3 min. Allow the layers to separate while cooling, then remove the lower aqueous layer. Add 5 ml of cold $1M$ sodium hydroxide solution, cap, and shake for 15 sec. Allow layers to separate and remove lower aqueous layer. Add 20 to 30 pellets of solid sodium hydroxide, cap, and.shake at intervals during the next 10 min. Cool well and centrifuge briefly. Filter ether extract through rapid paper, covering top of funnel with watch glass to prevent evaporation. Transfer a 10-ml aliquot of the filtrate (corresponding to 3 ml of urine) to a 25×150 mm tube. (A larger aliquot may be used if the original total urine volume was over 1,500 ml.) Evaporate the ether on a water bath under a fume hood. (*Caution:* No flame!) Dissolve the residue of 0.5 ml of methyl alcohol and again evaporate to dryness on a water bath. Similarly, evaporate a standard of 25 μg of steroid in 0.5 ml of alcohol and a blank of 0.5 ml of methyl alcohol. Allow tubes to cool to room temperature before proceeding with color development. Prepare a sufficient quantity of color reagent just before use by mixing 3 volumes of $4M$ alcoholic potassium hydroxide and 4 volumes of 1% metadinitrobenzene. Add 0.7 ml of the mixture to each tube of sample, standard, and blank, and mix gently to dissolve the residue. Allow to stand for 1 hr at room temperature in the dark. Then add to each tube 5 ml of Methyl Cellosolve and mix. Read standards and samples against blank at 520 and 430 nm.

CALCULATION

Calculate the corrected absorbance (A) for each tube as follows:

$$\text{corr. } A = \frac{A_{520} - 0.6 \times A_{430}}{0.73}$$

The corrected A for the standard should be close to the uncorrected A_{520} (within 5 percent). A lower corrected absorbance usually indicates faulty reagents. Then:

$$\frac{\text{corr. } A \text{ sample}}{\text{corr. } A \text{ standard}} \times \frac{\mu\text{g of standard}}{\text{ml of urine extracted}} \times \frac{\text{volume of ether added}}{\text{volume of ether evap.}} = \mu\text{g steroid/ml urine}$$

Since $\mu\text{g/ml} = \text{mg/liter}$,

$\mu\text{g/ml} \times$ volume in liters $=$ total mg ketosteroids extracted

For the aliquots as given:

$$\frac{\text{corr. } A \text{ sample}}{\text{corr. } A \text{ standard}} \times \frac{25 \times 25 \times \text{urine volume in liters}}{7.5 \times 10} = \text{total mg excreted}$$

ALTERNATIVE PROCEDURE FOR KETOSTEROIDS

As an alternative to reading at two wavelengths, proceed as follows: When the color has been developed after standing 1 hr, add to each tube 2.5 ml of Methyl Cellosolve and mix. Add 2 ml of water and mix. Then add 2.5 ml of ethylene dichloride and mix. The reagents must be added in the order given, mixing after each addition. Agitate strongly and transfer to a smaller test tube and centrifuge at high speed. Remove top aqueous layer completely. If lower layer is not perfectly clear, add a few drops of methyl alcohol. Transfer lower layer to cuvette and read at 520 nm. The calculations are the same as given above except that A at 520 nm is used without any correction.

PROCEDURE FOR HYDROXYSTEROIDS (KETOGENIC STEROIDS)

Test urine for sugar with Clinistix or similar test strip. If no more than a trace of sugar is present, proceed as outlined here. If more than a trace of sugar is present, see the comments at the end of this section for necessary modifications: Pipet 7.5 ml of urine to 25×150 mm tube. Check pH and adjust to 6.5 to 7.5 with short-range paper, using $1M$ sodium hydroxide or $1.05M$ acetic acid as needed. Add a few milliliters of ether to aid in reducing foaming, then add 0.9 ml of freshly prepared sodium borohydride solution. Allow the tubes to stand at room temperature for 1 hr. Cautiously add 0.5 ml of $4.35M$ acetic acid, adding a few milliliters more ether if necessary to prevent foaming. Allow to stand for a few minutes, check the pH, and add a few more drops of acid if necessary to bring the pH to near 7. Allow to stand for 15 min with occasional swirling. Add 3 ml of

$0.47M$ sodium metaperiodate solution and 0.5 ml of 1M sodium hydroxide. Incubate at 37°C for 1 hr. Add 0.5 ml of $5M$ sodium hydroxide and incubate for 15 min more. Cool tubes well, add exactly 25 ml of ether, cap tightly, and shake for 3 min. Remove the lower aqueous layer. It may be necessary to centrifuge briefly at this point to break any emulsion formed on shaking. Gently stir up any white precipitate in the bottom of the tube and remove as much of it as possible along with the aqueous layer. Add 3 ml of $0.5M$ sulfuric acid, cap, and shake vigorously. Allow layers to separate while cooling, then remove aqueous layer. Wash the ether layer with 5 ml of cold $1M$ sodium hydroxide solution, shake again for about 15 sec, allow the layers to separate, and remove the lower layer by suction. Add 20 to 30 pellets of solid sodium hydroxide to the ether. From this point proceed exactly as with the ketosteroids. All final details and calculations are the same.

COMMENTS

If more than a trace of sugar is present in the urine, proceed as follows: After the addition of the borohydride, allow to stand for 45 min and add 0.3 ml of $4.35M$ acetic acid and about 60 to 70 mg more of solid borohydride. Allow to stand for an additional 45 min. Then add 0.5 ml of $4.35M$ acetic acid as in the regular procedure and allow to stand for 10 min. More acetic acid may be necessary to bring the pH to 7. After the addition of the periodate, incubate for at least $1\frac{1}{2}$ hr, add an additional 200 mg of periodate after 45 min, and then agitate to dissolve the salt after addition.

NORMAL VALUES AND INTERPRETATION OF RESULTS

Increases in the ketosteroids are found in testicular or other virilizing tumors and in precocious puberty. Decreased levels are found in gonadal agenesis or dysgenesis and in hypothyroidism. A fraction of the ketosteroids measured in the urine also are metabolites of the various hormones secreted by the adrenal cortex; hence changes in the secretion of these hormones will also cause changes in the urinary excretion of ketosteroids. The normal levels of the ketosteroid excretion per day are as follows: birth to 3 years, less than 1 mg; 3 to 8 years, 0.5 to 2.5 mg; 12 to 16 years, 4 to 9 mg; adult males, 10 to 18 mg; and adult females, 6 to 15 mg. After the age of about 50 years, excretion of the ketosteroids begins to decrease and may be only in the range of 4 to 8 mg/day in either sex after the age of 65. The changes in ketosteroids as reflections of the changes in adrenocortical hormones are included in the discussion of those hormones.

The normal levels of the ketogenic steroids are 4 to 12 mg/day for both males and females. There is a decrease in ketogenic steroid (and ketosteroid) excretion in Addison's disease, hypopituitarism, Simmonds' disease, and cretinism. An increased level is found in Cushing's syndrome, in precocious puberty due to adrenal hyperplasia, and in physiological stress (surgery, burns, infectious diseases).

URINARY 17-HYDROXYSTEROIDS BY MODIFIED PORTER-SILBER METHOD

This method utilizes enzymatic hydrolysis to split the corticosteroids that are excreted as glucuronide conjugates [75–77]. β-Glucuronidase is used for the hydrolysis at the optimum pH of 6.8. The corticosteroids are extracted with chloroform and the extract purified by washing with sodium hydroxide. An aliquot of the extract is treated with phenylhydrazine-alcoholic sulfuric acid reagent to form a yellow chromogen. A urine control is run with each specimen to compensate for the nonsteroid chromogens present.

This method determines only the dihydroxyacetone side chain compounds. These include cortisone (17-hydroxy-11-dehydrocorticosterone), hydrocortisone (17-hydroxycorticosterone), and their tetrahydro derivatives.

REAGENTS

1. Chloroform, reagent grade. Redistil before use and add 1 ml ethanol per liter to prevent formation of phosgene.

2. NaOH, $0.1M$. Dissolve 4 g of NaOH in 1 liter of distilled water.

3. Absolute ethyl alcohol, AR.

4. Sulfuric acid, $11.6M$. Add 640 ml of concentrated H_2SO_4 to 360 ml of distilled water.

5. Alcoholic sulfuric acid reagent. Mix 100 ml $11.6M$ of sulfuric acid with 50 ml of absolute ethanol.

6. Recrystallized phenylhydrazine hydrochloride. Phenylhydrazine hydrochloride may be twice recrystallized from absolute ethanol and dried in a desiccator over calcium chloride. Baker's phenylhydrazine hydrochloride can be used without recrystallization.

7. Phenylhydrazine-alcoholic sulfuric acid reagent. Dissolve 50 mg of recrystallized phenylhydrazine hydrochloride in 50 ml of alcoholic sulfuric acid reagent. This reagent must be freshly prepared before use.

8. β-Glucuronidase (bacterial), 1,000 units/ml in distilled water. Prepare fresh before use or keep frozen.

9. Phosphate buffer, $0.5M$, pH 6.8. Prepare by dissolving 34 g of KH_2PO_4 in 500 ml of water. Adjust pH to 6.8 by the addition of 50 to 75 ml of $1M$ NaOH and dilute to 1 liter in a volumetric flask.

10. Urinary 17-hydroxysteroid standards.

A. Stock standard, 1 ml = 100 μg. Transfer 25 mg of cortisol‖ (hydro-

‖ **Sigma** Chemical Co., St. Louis, Mo. 63118.

cortisone alcohol) to a 250 ml volumetric flask and dilute to volume with absolute ethanol.

B. Working standard, 1 ml = 5 μg. Dilute 5 ml of the stock standard solution to 100 ml with distilled water.

PROCEDURE

A 24-hr urine specimen is collected in a bottle containing 10 ml of toluol as a preservative. The total volume is measured and approximately 100 ml is retained for analysis. To a 100 ml glass-stoppered cylinder transfer 10 ml of urine, 1 ml of the β-glucuronidase enzyme, 2 ml of 0.5M phosphate buffer, and 1 ml of chloroform.

Prepare the reagent blank and standard samples in the same manner by using 10 ml of water and 10 ml of working standard instead of urine. Mix well and incubate at 37°C for 18 to 24 hr. Add to each cylinder 50 ml of chloroform and mix by repeated inversion for 30 sec. Allow to stand until the organic and aqueous phases have separated and remove the aqueous supernatant by aspiration. Add 10 ml of 0.1M NaOH to each cylinder and shake for 30 sec. Allow the two phases to separate and remove the alkali layer by aspiration. In the same manner wash the chloroform extracts twice with 10 ml of water. Transfer two 20-ml aliquots of the chloroform extracts from the blank and the urine sample to 50 ml glass-stoppered cylinders. Transfer one 20-ml aliquot of the standard to a 50 ml cylinder. One aliquot of the blank and the unknown serve as the control blank and the unknown control. The other aliquots serve as the reagent blank, standard, and unknown.

Add 5 ml of the alcoholic sulfuric acid reagent to the cylinders containing the control blank and the urine controls. Add 5 ml of phenylhydrazine-alcoholic sulfuric acid to the cylinders containing the reagent blank, standard, and unknowns. Stopper the cylinders tightly, shake vigorously for 30 sec, and allow to stand for 15 to 20 sec. Transfer the supernatants from each cylinder to a small cuvette and incubate in a 60°C water bath for 45 min.

The unknown controls are read against the control blank and the standard and the unknown are read against the reagent blank at a wavelength of 410 nm.

CALCULATION

$$\frac{U - C}{S} \times 0.005 \times V = \text{17-OH corticosteroids (mg/24 hr)}$$

where U = absorbance of unknown, C = absorbance of control, S = absorbance of standard, 0.005 = concentration of standard (mg/ml), and V = volume of urine in ml.

The normal values for 17-OH corticosteroids in urine are: adults, 10 to 15 mg/24 hr; children 0 to 1 year, 0.5 mg/24 hr; and 1 to 5 years, 1 to 2 mg/24 hr.

FLUOROMETRIC DETERMINATION OF CORTISOL

Cortisol is also called hydrocortisone or 17-hydroxycortisone. In this method [78–80] the cortisol is extracted with methylene dichloride and an aliquot of the extract treated with a mixture of concentrated sulfuric acid and ethanol, producing a fluorescence which is compared with that of standards similarly treated. Although the method is not absolutely specific for cortisol, it is sufficiently accurate for most clinical purposes.

REAGENTS

1. Methylene dichloride, AR.
2. Absolute ethyl alcohol, AR.
3. Sodium hydroxide, $0.1M$. Dissolve 4 g of sodium hydroxide in 1 liter of water, or dilute a more concentrated solution, if available.
4. Fluorescent reagent. To 150 ml of absolute ethyl alcohol which has been cooled in an ice bath, add slowly with constant mixing 350 ml of concentrated sulfuric acid with continued cooling. The solution should remain colorless. If it turns brown, either the sulfuric acid was added too rapidly or the ethyl alcohol needs to be redistilled before use. Usually a colored solution will give a very high reagent blank.
5. Cortisol standards.

   A. Stock solutions. Dissolve 50 mg of cortisol (alcohol) in ethyl alcohol to make 50 ml. This stock solution contains 1 mg/ml. Dilute 1 ml of the stock to 100 ml with water. This standard contains 10 μg/ml. These solutions are stable for several months in the refrigerator.

   B. Working standard. Just before use dilute 1 ml of the 10 μg/ml standard with 9 ml of water to give a working standard containing 1 μg/ml. This solution is not stable.

PROCEDURE

Blood. Serum or heparinized plasma is used for the determination. Serum is preferred since some lots of heparin can cause significant error due to the production of nonspecific fluorescence. It is important that the serum or plasma be separated from the cells as soon as possible after drawing the specimen and that they be kept cold until centrifuged. There is a pronounced diurnal variation in the plasma level of cortisol and the time of specimen collection is important. Usually two specimens are taken, one at 8 AM and the other at noon, since the difference between the two may be of significance in some cases. The separated plasma or serum is stable for several days in the refrigerator.

Urine. The urine is collected without preservative and kept under refrigeration. Because of the diurnal variation, a 24-hr specimen is necessary. If the analysis can not be made promptly after the collection period, an aliquot may be preserved for several days by freezing.

Extraction. Two milliliters of serum, plasma, urine, standard (1 μg/ml), or water (blank) are extracted with 15 ml of methylene dichloride in glass-stoppered test tubes, 16 × 125 mm. The extraction is done either by vigorous shaking by hand for a few minutes or by rotating the tubes on a rotator for 20 min. After extraction, the layers are allowed to separate completely (with centrifugation if necessary) and the upper aqueous layer is removed as completely as possible by suction. The urine extracts require further washing with two 5-ml portions of 0.1M sodium hydroxide and then with 5 ml of water. Ten-milliliter aliquots of the ethylene dichloride are then transferred to separate clean, dry, glass-stoppered tubes.

Development of fluorescence. Since the fluorescence developed increases with time, all the readings must be made at the same time interval after the addition of the reagent; 13 min is a convenient interval. With this interval, a blank, standard, and six sample extracts can be run in this order. At timed intervals of 1 min, 5 ml of the fluorescent reagent is added to the 10 ml of extract and the tube stoppered and shaken vigorously for 20 sec. The tube is allowed to stand until the reagent has been added to all the tubes of the series. The upper layer is removed as completely as possible by suction, and the lower layer transferred to a fluorometer cuvette. The samples are read against the blank exactly 13 min after the addition of the reagent, the standard being set at 100 scale divisions. The exact settings used will depend upon the particular fluorometer. The exciting light should have a maximum at about 450 nm, and the secondary filter should exclude light below about 520 nm. For the Turner fluorometer# suggested filters are: primary, 3 plus 48; secondary, 2A-15 plus 4-94.

CALCULATION

Plasma. Since 2 ml of plasma and 2 ml of standard containing 1 μg/ml are treated similarly.

$$\frac{\text{reading of sample}}{\text{reading of standard}} \times 100 = \text{cortisol (in } \mu\text{g/dl)}$$

G. K. Turner Assoc., Palo Alto, Calif. 94303.

The extraction of smaller quantities of serum does not always lead to consistent results, so that, for high samples, the fluorometer sensitivity should be reduced rather than using smaller samples.

Urine. Since 2 ml of urine and 2 ml of standard containing 1 µg/ml are treated similarly,

$$\frac{\text{reading of sample}}{\text{reading of standard}} \times \text{total urine volume (in ml)} = \text{cortisol (in µg/24 hr)}$$

NORMAL VALUES AND INTERPRETATION OF RESULTS

The normal range by this method may be taken as 7 to 30 µg/dl for morning samples. The levels are lower in adrenal insufficiency and higher in Cushing's syndrome and other instances of adrenal hyperplasia. Normally a sample taken at 9 PM will have a level 25 to 50 percent lower than one taken at 9 AM. In Cushing's syndrome not only is the 9 AM level higher than normal, but the evening specimen will show very little if any decrease.

The normal urinary excretion is in the range of 50 to 120 µg/24 hr.

Spironolactone treatment may result in an increase in nonspecific fluorescence in the extract, leading to erroneously high results. Urinary steroids determined by colorimetric methods are not affected by the drug. The administration of oral amphetamine or parenteral methamphetamine has been stated to cause increases in plasma cortisol levels, particularly in the morning specimen. This should be kept in mind in interpreting the results. It has also been claimed that large doses of alcohol or heavy smoking will cause increases in cortisol levels in the plasma. Spuriously high plasma levels are found in patients taking oral contraceptives.

Estrogens [81–84]

The main estrogens found in the urine are estradiol, estrone, and estriol. (See the introductory section on the steroid hormones.) These steroids have three double bonds in ring A (benzenoid structure) which gives them slightly acidic properties. Estrogens are excreted chiefly as glucuronides and must be hydrolyzed before extraction. This may be carried out either enzymatically or by heating with acid. The enzymatic hydrolysis is very lengthy. The acid treatment is usually used for the routine determination in pregnancy urine, although it may cause a loss of 10 to 20 percent of the estrogens. After hydrolysis, the estrogens are extracted with an immiscible solvent. The extract may then be purified somewhat by washing and then treated with a solution of hydroquinone in 12.6M sulfuric acid. On incubation a yellow color is developed which is a measure of the total estrogens present (Kober reaction). This is the only estrogen determination done routinely in the clinical laboratory.

Estrogens have been isolated from the ovaries, placenta, adrenal cortex, and testes. In males, postmenopausal women, and prepubertal females the total amount of estrogens excreted in the urine is very low. There is a cyclic variation of estrogens excretion during the menstrual cycle, with low levels during menstruation, a significant increase at midperiod (ovulatory phase), and a lesser increase in the latter part of the cycle (luteal phase).

In nonpregnant women the excretion of estrogens in the urine is relatively small. Their determination, particularly of the relative amounts of the three estrogens, is a rather complicated procedure. In pregnancy the amount of excreted estrogens increases markedly. A simple test for the total amount of excreted estrogens (most of the increase is due to estriol) is routinely used to monitor high-risk pregnancy patients. In these cases the excretion increases from about 5 mg/day in the 20th week of gestation to over 30 mg/day at term. The failure of the estrogen excretion to increase as expected or a fall in excretion in the later weeks of pregnancy is a grave prognostic sign in regard to the viability of the fetus.

Estrogens have been determined using bioassay procedures, fluorometric analyses, competitive protein binding, and radioimmunoassay (RIA). Except in cases of pregnancy, where there is a marked increase in the estrogen excretion, the routine determination of estrogens is not practical using common chemical methods. The introduction of radioimmunoassay techniques has brought these analyses within the realm of the specialized laboratory; details are given in Chapter 15.

Estrogen Determination in Pregnancy Urine

In this method the estrogens are extracted with ethyl acetate without hydrolysis. This gives slightly low recovery, but a correction can be made for this. An aliquot of the extract is evaporated and treated with concentrated sulfuric acid in the presence of hydroquinone. The yellow color produced is measured at 514 nm. Measurements are made at two other wavelengths to correct for nonspecific chromogens. Usually an aliquot of a 24-hr urine specimen is analyzed. However, in most instances good results can be obtained from a random specimen by also measuring the urinary creatinine and expressing the final results in terms of the estrogen-creatinine ratio.

REAGENTS

1. Hydrochloric acid, $1M$. Dilute 84 ml of concentrated acid to 1 liter with water.
2. Sodium chloride, AR crystals.

3. Ethyl acetate, reagent grade.

4. Hydroquinone solution.   Dissolve 2 g of purified hydroquinone in 100 ml of absolute ethanol.

5. Hydroquinone-sulfuric acid.   Add 68 ml of concentrated sulfuric acid to about 25 ml of water, cool to room temperature, add 2 g of purified hydroquinone, and dilute to 100 ml.

6. Estriol standards.

   A. Stock standard.   Dissolve 50 mg of estriol in absolute ethanol and dilute to 100 ml. This gives a stock solution containing 500 μg/ml.

   B. Working standard.   Dilute 10 ml of the stock to 100 ml with absolute ethanol to give a working standard containing 50 μg/ml.

PROCEDURE

If a 24-hr urine specimen is used, it should be collected without preservative and kept under refrigeration. The sample is well mixed and its volume measured. The specific gravity is determined, and if the urine is too concentrated, an aliquot is diluted with water to give a specific gravity of about 1.010. The dilution need not be exactly to this specific gravity, but if a 24-hr urine is used, the exact dilution made must be recorded for the calculations. If a random specimen is used with the results to be based on the estrogen-creatinine ratio, the dilution factor need not be recorded provided the estrogen and creatinine are determined on the same dilution. Add 2 ml of urine sample to an extraction tube (19 × 125 mm screw-capped tubes with Teflon-lined caps are preferable). Add 2 ml of 1M hydrochloric acid and about 2 g of sodium chloride crystals and mix well. Add 3 ml of ethyl acetate, cap tightly, and shake vigorously for about ½ min. Centrifuge to separate the layers completely and transfer 0.5 ml-aliquot of the ethyl acetate layer to a glass-stoppered tube (16 × 125 mm). Add 0.2 ml of the alcoholic hydroquinone solution and evaporate the solvent at about 50°C with the aid of a stream of nitrogen. To other properly labeled tubes add 0, 0.1, and 0.2 ml of the working estriol standard. To each tube add 0.2 ml of alcoholic hydroquinone and evaporate to dryness. It is absolutely necessary that all the solvents be completely evaporated from the tubes as they may also react with the color reagent.

Add 2.0 ml of the hydroquinone-sulfuric acid reagent to each tube, stopper tightly, and heat in a boiling water bath for 40 min. Swirl the tubes several times during the beginning of the heating period to aid in the solution of the residue. Then cool the tubes in running tap water and add 1.7 ml of water to each tube. Mix well and allow to stand in the bath until cooled. Read the absorbance of the standards and samples against the blank at 472, 514, and 556 nm.

CALCULATION

A corrected absorbance is calculated for each series of readings at the two wavelengths as follows:

$$A \text{ (corr.)} = 2 \times A_{514} - (A_{472} + A_{556})$$

where A_{514} refers to the measured absorbance at 514 nm and similarly for the other two wavelengths. Then,

$$\frac{A \text{ (corr.) of sample}}{A \text{ (corr.) of standard}} \times \mu g \text{ estrogen in standard} \times 3.6 = \text{estrogen in diluted urine (in } \mu g/ml)$$

The estrogen standards used, as given above, are equivalent to 5 and 10 μg of estrogen; the standard giving the absorbance closest to that of the sample is used. The factor 3.6 is derived from the fact that we must multiply by 6 (since only $\frac{1}{6}$ of the ethyl acetate extract is used, 0.5 ml out of 3 ml), divide by 2 (since 2 ml of urine was extracted), and multiply by 1.2 (to correct for incomplete extraction of the estrogens).

If the creatinine has been determined on the same urine dilution, the estrogen-creatinine ratio is found.

$$E/C = \frac{\mu g/ml \text{ of estrogen}}{mg/ml \text{ of creatinine}}$$

If a 24-hr sample was used,

$$\mu g/ml \text{ estrogen in diluted urine} \times F \times V = mg \text{ estrogen}/24 \text{ hr}$$

where F is the dilution factor to correct for any dilution of the urine and V is the total 24-hr urine volume in liters. If X ml of urine was diluted to a final volume of Y ml, then $F = Y/X$.

NORMAL VALUES AND INTERPRETATION OF RESULTS

Both the E/C ratio and the 24-hr estrogen excretion increase from very low values at the beginning of pregnancy to much higher values at term. The increase is only slight until about the 20th week of gestation. A sudden drop in the value is a grave prognostic sign and indicates fetal distress. A continued drop may indicate the need for induced delivery. A decline to very low values in the later weeks of gestation usually indicates fetal death.

Progestational Hormones [85–87]

These hormones are secreted by the corpus luteum of the ovary during the menstrual cycle. They are also produced by the adrenal cortex in both males and females. The principal steroid of these hormones is progesterone, a 21-carbon steroid partially responsible in pregnancy for cellular changes in the endometrium necessary for attachment and growth of the embryo.

Progesterone is found in extremely small amounts in the blood and its determination requires special techniques and equipment. The methods used include gas-liquid chromatography, double isotope derivatives, competitive protein binding, and radioimmunoassay; details for the radioimmunoassay technique are given in Chapter 15.

The chief metabolite of progesterone found in the urine is pregnanediol. Its concentration has been used as a measure of ovarian function. Pregnanediol is present in the urine in a concentration of about 2 to 7 mg/day during the latter half of the menstrual cycle. Smaller amounts are found in the earlier part of the cycle, in the urine of men, and in postmenopausal women. The methods for determination of pregnanediol utilize column chromatography and are tedious and complicated. In pregnancy, decreased levels of pregnanediol are often found in threatened abortion and intrauterine deaths involving the placenta. Increased levels are found in cases of corpus luteum cysts and in some cases of adrenal cortical tumors.

Aldosterone [88–90]

Aldosterone is an adrenal cortical hormone found in the urine in concentrations of 3 to 21 μg/day. This is the most active sodium-retaining hormone of the adrenal cortex. Its concentration varies with the state of the electrolyte balance. Sodium depletion raises the level, whereas administration of sodium lowers it. These changes are controlled by the extracellular fluid volume rather than by the serum or total body sodium. Administration of potassium also increases the aldosterone level, whereas potassium deficiency lowers it.

An increased production of aldosterone causes sodium retention, potassium elimination, alkalosis, and often hypertension. This increase is referred to as aldosteronism and occurs as primary and secondary forms.

In primary aldosteronism the increased aldosterone excretion is due to adrenal tumors (adenoma and carcinoma). Laboratory findings include normal 17-ketosteroids and 17-OH corticosteroid excretion, low serum potassium, alkalosis, low specific gravity of urine, increased 24-hr volume of urine, and elevation of serum sodium. The clinical findings include hypertension, weakness, polyuria, and tetany.

In secondary aldosteronism the increased aldosterone excretion may be due to salt depletion, hemorrhage, large doses of ACTH, cardiac failure, cirrhosis, pregnancy, lower nephron nephrosis, and postsurgical syndrome.

Many attempts have been made to find an adequate assay method for aldosterone. The compound is present in the urine in minute amounts representing less than 1 percent of the total corticosteroid metabolites in the urine. Extremely elaborate methods of separation and purification are required. Earlier methods used at least one separation by column chromatography, followed by two or more separations by paper or thin layer chromatography before the final determination. This procedure is very tedious and complicated, requires several working days for completion, and can be done only in a well-equipped laboratory.

Aldosterone may be determined more simply by competitive protein binding or radioimmunoassay (see Chapter 15).

Adrenal Cortex Function Tests [91–93]

The adrenal cortex secretes its hormones upon stimulation by the pituitary hormone corticotropin (ACTH, for adrenocorticotropic hormone). The amount of ACTH released by the pituitary is generally controlled by the concentration of cortisol in the circulation. Consequently the higher the concentration of cortisol, the lower the amount of ACTH released by the pituitary, and vice versa. Adrenal cortex hypofunction may be due to either failure of the pituitary to secrete the necessary ACTH or failure of the adrenal cortex to respond to the ACTH produced by the pituitary. The administration of exogenous ACTH will aid in differentiating between these two conditions. If the administered ACTH causes a marked rise in the excretion of ketosteroids and ketogenic steroids (in normal individuals the urinary output of ketosteroids may be increased two to three times and that of ketogenic steroids three to four times), the original low level is due to the failure of the pituitary to secrete ACTH. If the exogenous ACTH causes little rise in the steroid excretion, the adrenal gland has failed to respond to pituitary stimulation.

The first ACTH stimulation test that was used measured the decrease of eosinophils in the circulation following the administration of ACTH (Thorn test). If this test is properly controlled and interpreted, it can be useful as a screening test.

Stimulation Test

In this test, ACTH is administered to study the responsiveness of the adrenal cortex to its stimulation. A 24-hr urine specimen is collected prior to the test and assayed for 17-ketosteroids and 17-hydroxysteroids. On the day of the test, collection of the 24-hr urine specimen is started at 8 AM and an intravenous drip of saline solution containing 20 units of ACTH is begun. This should be allowed to run for 8 hr. Intramuscular injection of 80 units of ACTH on two successive days can be utilized as an alternative. The response to the stimulation test is related to the mass of the cortical tissue, the number of units of ACTH given, and the sensitivity of the adrenal cortex to stimulation. The test 24-hr urine specimen is assayed for 17-ketosteroids and 17-hydroxysteroids.

More recent procedures use the intravenous administration of 25 units of ACTH and the measurement of plasma cortisol before and 1 hr after the administration of ACTH.

Suppression Test

The theoretical basis for this test is the feedback mechanism. Cortisone derivatives decrease the output of adrenal cortical hormone (ACTH) by suppressing the anterior pituitary gland. Compounds such as α-fluorohydrocortisone, dexamethasone (Decadron), and prednisolone can be used for this purpose.

A 24-hr urine specimen is obtained prior to the test and is assayed for 17-ketosteroids and 17-hydroxysteroids. The test is performed by administering 1 mg of 1-α-fluorohydrocortisone, 0.5 mg of Decadron, or 2.5 mg of prednisolone daily for each 3 mg of 17-hydroxycorticoids in the urine. The dose is divided and given every 6 hr after meals for 3 days. A 24-hr urine specimen is collected again and assayed for 17-ketosteroids and 17-hydroxysteroids. The response of the adrenal is evaluated by the changes found in these urinary steroids.

The oral administration of dexamethasone has been used as a screening test for the diagnosis of Cushing's syndrome. One milligram of the compound is given at about noon and the plasma cortisol level is measured the next morning to evaluate suppressibility of the pituitary gland.

Metyrapone Inhibition Test

This test is used to assess pituitary function in regard to ACTH production [94]. Metyrapone inhibits the action of an enzyme that converts 11-deoxycortisol into cortisol. Thus the production of cortisol is inhibited and 11-deoxycortisol (compound S) is secreted by the adrenals in place of cortisol. Since the feedback mechanism controlling the production of ACTH is influenced chiefly by cortisol, the decrease in cortisol level will tend to stimulate the pituitary to secrete more ACTH with consequent production of more ketosteroids and ketogenic steroids if both the pituitary and adrenals are active.

Usually 750 mg of metyrapone is administered orally every 4 hr for 48 hr. If the pituitary and adrenals are both active, the result will be similar to the administration of ACTH, a marked increase over the basal level in urinary steroid excretion. In patients with pituitary insufficiency no increase is noted.

Catecholamines of the Adrenal Medulla [95–97]

The adrenal medulla is functionally distinct from the adrenal cortex and secretes an entirely different type of hormones. The main hormones secreted by the medulla are adrenaline (epinephrine) and noradrenaline (norepinephrine). These are collectively known as catecholamines (catechol is a dihydroxybenzene and the two compounds have this basic structure with an amine group attached at the end of a two-carbon chain that is also attached to the benzene ring).

The simplest determination is that of the total free (unconjugated) catecholamines. They are separated from the urine by adsorption on alumina at a pH of about 8.5. The alumina is then washed free of interfering substances and the amines eluted with acid. They are then oxidized to fluorescent compounds at a pH of about 6.5. One method uses ferricyanide as oxidizing agent (catalyzed by zinc ions), followed by treatment with strong sodium hydroxide to develop the fluorescence. Ascorbic acid is used to destroy the excess oxidizing agent and stabilize the fluorescent compound. A semiquantitative estimate of the amount of catecholamines present in urine may be made by visually comparing the fluorescence of the resulting solution under ultraviolet light with that seen in another urine sample that has been similarly treated after the addition of a known amount of standard. The fluorescence can also be quantitated in a fluorometer. Other methods of separation from the urine use a column of alumina or ion exchange resin instead of the batch method mentioned.

The adrenaline and noradrenaline can be determined separately either by making the fluorescent measurements with two different exciting wavelengths or by oxidizing the aliquots of the sample at two different pH values and making measurements on the two solutions at the same wavelength. Internal standards of the two compounds are also run. In either case, a pair of simultaneous equations have to be solved. In the catecholamine determinations an internal standard is often used to correct for quenching. This is said to be less necessary when an ion exchange resin is used for the preliminary separation. Some of the catecholamines also exist in the urine in the conjugated form. These may be hydrolyzed by heating with acid, but the usual determination is of the unconjugated amines. Some other amines, metadrenaline and normetadrenaline, are also present in the urine. If an ion exchange column is used to separate the adrenaline and noradrenaline after these have been eluted from the column with acid, the "met" compounds can be eluted with ammonia. These compounds may then be determined by oxidizing them to vanillin with metaperiodate. The resulting vanillin is determined by its ultraviolet absorption. Although the absorpton peak is at 350 nm, the measurements are usually made at 360 nm, as this reduces the interference from other compounds.

Tumors developing in the adrenal medulla and in the extraadrenal chromaffin tissue are called pheochromocytomas. These tumors produce hypertension, either paroxysmal or sustained, due to their excess production of the pressor amines.

Epinephrine is partially bound to albumin in the serum. In the urine, catecholamines are found either free or conjugated to glucuronides. Of the free catecholamines in the urine, 80 percent are norepinephrine and the remainder epinephrine.

Determination of Total Catecholamines

In this method catecholamines are adsorbed from alkaline urine into activated alumina, eluted with sulfuric acid, and reacted with reagents that produce a fluorescent derivative. The fluorescence is compared with that of standards treated in the same way, using a photofluorometer for quantitation, or compared visually with a long-wavelength ultraviolet lamp (360 nm) for semiquantitation.

A 24-hr urine specimen preserved with 1 ml of concentrated H_2SO_4 is required. The patient must abstain from using any drugs for 72 hr before the specimen is collected.

REAGENTS

1. Aluminum oxide, chromatography grade. Aluminum oxide is washed by shaking 200 g in a liter of hot $1M$ HCl, rinsing free from acid with several changes of distilled water, and drying.

2. Ethylene diamine tetracetic acid (EDTA), disodium salt.

3. Sodium hydroxide, $5M$ solution.

4. Acetic acid, $0.20M$ solution.

5. Phenolphthalein, 1 g/dl solution in ethanol.

6. Sodium bicarbonate, $0.24M$ solution.   Store in dark.

7. Potassium ferricyanide, $0.008M$ solution.   Store in dark.

8. Ascorbic acid, 2 g/dl solution.   Stable for 36 hr in refrigerator.

9. Alkaline ascorbate. Mix 9 parts $5M$ NaOH and 1 part 2% ascorbic acid solution. This mixture should be centrifuged briefly and prepared just before use.

10. Norepinephrine standards.

A. Stock standard. Levophed bitartrate* containing 100 mg/dl norepinephrine base.

* Winthrop Laboratories, Inc., New York, N.Y. 10016.

B. Working standard, 20 μg/ml. Dilute 0.1 ml of the stock standard to 5 ml in distilled water.

Separate 500 ml Erlenmeyer flasks are labeled unknown (U), standard 1 (S1), standard 2 (S2), and standard 3 (S3). To flask U add one-tenth of a 24-hr urine specimen (if the total volume is over 2,000 ml, use 200 ml of urine and make appropriate correction in the calculation). To flasks S1, S2, and S3 add 100 ml of a normal urine specimen, and dilute to 200 ml with water. To S1 add 0.3 ml of the working standard, to S2 add 0.7 ml of the working standard, and to S3 add 1.1 ml of the standard. The actual amounts of norepinephrine added to the three flasks are 6, 14, and 22 μg, respectively. If the normal urine is assumed to have a level of 4 μg/dl, the actual amounts of the amine in the three flasks are 10, 18, and 26 μg, respectively. This procedure is used to compensate for the quenching by impurities extracted from the urine.

To each flask add 2 g of aluminum oxide, 0.5 ml of the phenolphthalein solution, and 0.5 g of EDTA. While mixing, add 5M NaOH dropwise until a pink color persists. Continue mixing for about 2 min, then allow the aluminum oxide to settle. The granular aluminum oxide will settle very rapidly but the flocculent precipitate of phosphates will largely remain suspended. Pour off the supernatant fluid and phosphates as completely as possible. Wash the aluminum oxide by adding about 200 ml of distilled water, mix, allow the aluminum oxide to settle, and pour off the water.

Transfer the aluminum oxide to a 15 ml centrifuge tube. This can best be accomplished by adding 2 to 3 ml of water, making a slurry, and sucking this slurry up into a pipet. Centrifuge for 3 to 4 min at 1,500 to 2,000 rpm. Decant water and drain the centrifuge tube and wipe dry.

Add 5 ml of 0.2M acetic acid to each tube and suspend the aluminum oxide by thorough mixing with a glass rod. The catecholamines adsorbed on the aluminum oxide are eluted at this stage, so thorough mixing is required.

Centrifuge 2 to 3 min at 2,000 rpm. Pipet exactly 3 ml of supernatant to a Pyrex tube containing 3.0 ml of 0.24M NaHCO$_3$ and mix. Add 0.5 ml of 0.008M potassium ferricyanide and mix again. After 2 min add 1 ml of the alkaline ascorbate solution and mix well.

Centrifuge briefly to clarify, transfer supernatant to fluorometer cuvettes, and read against water blank, or transfer to flat-bottomed nonfluorescent tubes (bacteriology culture tubes are satisfactory) and examine tubes for green fluorescence in darkened room under ultraviolet light. This is best accomplished by holding the light under the tubes in a rack in such a manner that each tube is illuminated to approximately the same intensity. Compare intensity of fluorescence of the unknown to that of the standards. If one-tenth of the 24-hr volume has been used, the standards correspond to 100, 180, and 260 μg/day.

COMMENTS

Under these conditions, preparations from normal urine show faint green fluorescence, whereas those from urine having an abnormally high catecholamine content show an intense bright green fluorescence. Normal urine can be readily distinguished from urine containing 180 μg or more of catecholamines per 24 hr.

Quantitative Determination of Total Catecholamines

For a quantitative procedure the standards are added to aliquots of the urine sample (internal standard). Set up 3 flasks labeled A, B, and C. To each add the same aliquots of the urine sample (usually one-tenth of the total volume). To B add 1 ml of the working standard and to C add 2 ml of the working standard. Dilute each to 200 ml with water and carry through the procedure exactly as for the qualitative test. Also set up a blank by treating 3 ml of the 0.2M acetic acid similarly to the eluates from the aluminum oxide.

Read the samples in a fluorometer using a primary filter at 400 nm and a secondary filter at 530 nm. Set fluorometer to zero reading with the blank, then insert cuvette containing solution B and set to 100 or other convenient value. Insert cuvette with solution A and read (A_1). Check zero again, then insert cuvette with solution C and set to 100 or other convenient high value, again take reading on A (A_2).

CALCULATION

$$\frac{A_1}{B - A_1} \times 200 = \text{concentration of sample (in } \mu\text{g/day)}$$

or

$$\frac{A_2}{C - A_2} \times 400 = \text{concentration of sample (in } \mu\text{g/day)}$$

where A_1 is the reading of tube A when B is set to a definite value and A_2 is the reading of tube A when C is set to a definite value.

The standards are based on the use of a urine aliquot of one-tenth of the 24-hr output. If a different aliquot is used, an appropriate correction must be made in the calculations.

Determination of Epinephrine and Norepinephrine

The total catecholamines as determined by the above method include both epinephrine and norepinephrine. As an aid in diagnosis it may sometimes be helpful to determine the two substances separately. This may be accomplished by carrying out the oxidation to fluorescent compounds at two different pH values and making the fluorescent measurements on both tubes. Both epinephrine and norepinephrine are oxidized at pH 6.5. At pH 3.5 only a small proportion of the norepinephrine but almost all the epinephrine is oxidized. By treating similarly standards of the two compounds and using simultaneous equations, the amounts of the two compounds in the unknown can be calculated. The method is not done routinely and is not included here.

Normal Values and Interpretation of Results

As a screening test for pheochromocytoma, a 24-hr excretion of up to 100 μg of total free catecholamines is considered normal. From 100 to 200 μg is borderline and above 200 μg may be considered as suggestive of the disease.

A number of drugs may interfere with the determination of the catecholamines, usually giving erroneously high results. Quinine, quinidine, and the B complex vitamins may produce interfering fluorescence. Drugs such as methenamine compounds which liberate formaldehyde in the urine will also interfere. The excessive use of epinephrine inhalation or the use of isoproterenol or L-dopa can also produce high results. The foregoing drugs produce interfering chemical effects. Other drugs may produce elevated catecholamines by their physiological action; these include caffeine, aminophylline, and ethanol. It is advisable to limit the intake of these drugs for several days before the collection of the urine specimen.

Vanillylmandelic Acid (VMA)

Only a small fraction of the catecholamines are excreted as such in the urine; mainly they are excreted as metabolites, chiefly 4-hydroxy-3-methoxymandelic acid which is known as vanillylmandelic acid or VMA [92–102]. Since the amount of VMA excreted is much greater than that of the catecholamines (10 to 20 times), its determination is simpler, particularly since colorimetric methods are available. In the usual method the VMA is extracted from the urine with ethyl acetate or other solvent, and after some purification is reacted with diazotized *p*-nitroaniline to give a colored compound. The excess reagent and some interfering material is removed by extraction and the color read at two wavelengths—around 430 and 540 nm (different procedures may call for slightly different wavelengths). Comparison is made with a standard similarly treated. The difference in optical density at the two wavelengths is taken as proportional to the amount of VMA present. As usually performed this method is not very satisfactory, but it is sufficient in most cases to distinguish the normals from the abnormals. An alternative procedure involves the extraction of the VMA with ethyl acetate or other solvent, reextraction into potassium carbonate solution, and oxidation with metaperiodate. The resulting vanillin is read at 360 nm.

The amount of VMA excreted is a measure of endogenous secretion of catecholamines. Urinary VMA is highly elevated in the urine of patients with pheochromocytomas.

In the method described, acidified urine is treated with activated magnesium silicate (Florisil) to remove interfering chromogenic substances. VMA is extracted from the urine with ethyl acetate and subsequently reextracted into alkaline solution. An aliquot is treated with diazotized p-nitroaniline, and reextracted in butanol. The absorbance of the butanol layer is measured at 540 and 450 nm.

A 24-hr urine specimen preserved with 10 ml of concentrated HCl is collected. For 48 hr before the collection of the specimen, the patient must abstain from all medications, coffee, bananas, and substances containing vanilla.

REAGENTS

1. Hydrochloric acid, $2M$, saturated with sodium chloride.   Dilute 86 ml of concentrated acid to 500 ml with water and add 20 g of NaCl to saturate.

2. Hydrochloric acid, $0.10M$, saturated with sodium chloride.   Dilute 4.5 ml of concentrated acid to 500 ml with water and add 20 g of NaCl to saturate.

3. p-Nitroaniline.   Dissolve 0.5 g in 10 ml concentrated HCl and dilute to 500 ml with water.

4. Sodium nitrite, $0.29M$.   Dissolve 2 g in water to make 100 ml. Prepare fresh.

5. Diazotized p-nitroaniline.   Prepare fresh just before use by mixing 2.5 ml of nitroaniline solution, 7.5 ml of water, and 0.25 ml of sodium nitrite solution.

6. Potassium carbonate, $0.36M$.   Dissolve 5 g of anhydrous potassium carbonate (K_2CO_3) or 4.6 g of anhydrous sodium carbonate (Na_2CO_3) in water to make 100 ml.

7. Ethyl acetate, AR.

8. n-Butanol, AR, redistilled.

9. VMA standards.

   A. Stock standard.   Dissolve 10 mg of 4-hydroxy-3-methoxymandelic acid in 10 ml of water. This solution contains 1,000 μg/ml. It is stable when kept in refrigerator.

B. Working standard. Dilute stock standard 1:10 in water. This solution contains 100 μg/ml. Prepare weekly and keep refrigerated.

10. Florisil† (activated magnesium silicate), 60 to 100 mesh size.

PROCEDURE

To 10 ml of filtered urine adjusted to pH 2.5 to 5.5, add 0.5 g of Florisil. Shake vigorously for 30 sec and centrifuge. Using 40 ml round-bottom, narrow-neck centrifuge tubes, prepare solutions in duplicate as follows: unknown, 1 ml of the supernatant fluid, 3.0 ml of water, and 5 drops of 2M HCl; blank, 4.0 ml of water and 2 drops of 2M HCl; standard, 3.8 ml of water, 0.2 ml of working standard, and 2 drops of 2M HCl. Add 10 ml of ethyl acetate to each tube. Shake vigorously for 1 min. Add 2.5 ml of reagent 2 and again shake vigorously for 1 min. Centrifuge for 5 min at 1,500 rpm in order to separate the two layers. Pipet 8 ml of the upper layer (ethyl acetate) to another 40 ml centrifuge tube and add 5 ml of potassium carbonate. Shake vigorously for 1 min. Centrifuge for 5 min at 1,500 rpm, remove the ethyl acetate (upper layer) by suction, and discard. Transfer 4 ml of the K_2CO_3 extract to 12 ml glass centrifuge tubes and add 1 ml diazotized p-nitroaniline. Shake vigorously for 30 sec and let stand for 1 min. Add 5 ml of butanol, shake again for 30 sec, and centrifuge. Remove the upper layer (butanol extract) carefully.

Read absorbance of the extracted butanol layer in a colorimeter at 450 and 540 nm against the blank. Calculate the corrected absorbances (A_c) for the standard and unknowns as follows:

corr. absorbance (A_c) = absorbance at 540 nm − absorbance at 450 nm

CALCULATION

$$\frac{A_c \text{ of sample}}{A_c \text{ of standard}} \times 0.02 \times V = \text{mg VMA/24 hr}$$

where 0.02 is the concentration in mg of the standard (0.2 ml of 100 μg/ml) and V is the 24-hr urine volume in ml.

NORMAL VALUES AND INTERPRETATION OF RESULTS

The urinary excretion of VMA in the normal adult ranges from 2 to 14 mg/24 hr with a mean concentration of 5.0 mg. A marked increase is found in functioning pheochromocytoma, with concentrations reaching as high as 150 mg/day.

† Floridin Co., Pittsburgh, Pa. 15235.

Drugs such as epinephrine, norepinephrine, and L-dopa could be excreted as VMA or related compounds, but usually the dosage of these drugs is not high enough to cause a marked increase in the VMA excretion. Monoamine oxidase inhibitors may inhibit the conversion of catecholamines to VMA and thus cause a decrease in excretion. A similar effect is said to occur after the administration of imipramine. The presence of phenolsulfonphthalein or bromsulfophthalein in the urine may give an interfering color. Bananas, chocolate, coffee, tea, and foods containing vanilla extract also give increased color not due to VMA and thus a false positive reaction. On the other hand, meprobamate, dextroamphetamine, ephedrine, and isoproterenol apparently do not affect the VMA determination.

Serotonin and 5-Hydroxyindoleacetic Acid (5-HIAA)

Serotonin is derived from the amino acid tryptophan and excreted largely as the metabolite 5-hydroxyindoleacetic acid (5-HIAA) [103–105]. The tests for excessive serotonin production are usually based on determination of 5-HIAA in the urine.

In the body tryptophan is first oxidized and then decarboxylated to form serotonin (5-hydroxytryptamine). Serotonin is produced by the chromaffin cells of the intestines, and therefore under normal circumstances 90 to 95 percent is localized in the gastrointestinal mucosa. Small amounts are found in the spleen due to platelet disintegration. The small amount found in the blood is essentially due to the serotonin transported by the platelets. Normally only about 1 to 3 percent of dietary tryptophan is used for the production of serotonin. In patients with carcinoid tumors that have metastasized to the liver or lymph nodes, as much as 60 percent of the daily tryptophan is diverted by the tumor into the serotonin pathway, leaving less tryptophan available for the formation of niacin and protein. Under these conditions a clinical syndrome is produced called the *carcinoid syndrome*. This is characterized by skin changes such as flushing and cyanosis, cardiac signs and symptoms pointing toward involvement of the right side of the heart and lung, diarrhea, and occasionally asthma.

The methods described are based on the photometric measurement or visual comparison of the color complex formed by 5-HIAA with nitrous acid and 1-nitroso-2-naphthol. In a simple screening test the urine is treated with 1-nitroso-2-naphthol in the presence of nitrous acid. The excess reagent is removed by shaking with ethyl acetate and the pink to purple color evaluated visually. Although not very accurate, the test is relatively simple and the distinction between normal subjects and those with carcinoid tumor is definite, the former giving only a very faint pink or yellow color and the latter a very strong purple color. For a quantitative test the urine is first treated with dinitrophenylhydrazine to remove interfering keto acids. The excess reagent is removed by chloroform extraction, the pH adjusted, and the hydroxyindoles extracted with ether. These are then reextracted into a buffer and the reaction with nitrosonaphthol applied.

1. Chloroform, AR.
2. Ether, anhydrous, peroxide-free, AR.
3. Sodium chloride, AR.
4. Ethylene dichloride, AR.
5. Ethyl acetate, AR.
6. 2,4-Dinitrophenylhydrazine.   Dissolve 1 g in about 160 ml of $2M$ HCl by heating in boiling water bath. Allow to cool, dilute to 200 ml with $2M$ HCl, and filter.
7. Phosphate buffer, $0.5M$, pH 7.0.   Dissolve 27.6 g of $NaH_2PO_4 \cdot H_2O$ and 42.7 g of Na_2HPO_4 in about 800 ml of water. Check the pH and adjust if necessary with small amounts of $1M$ NaOH or HCl as required, then dilute to 1 liter.
8. 1-Nitroso-2-naphthol reagent, 100 mg/dl in 95% ethanol.   Dissolve 100 mg in 95% ethyl alcohol and dilute to 100 ml.
9. 5-HIAA standards.

   A. Stock standard, 1 mg/ml.   Dissolve 20 mg of 5-hydroxyindole-3-acetic acid monodicyclohexylammonium salt or 10 mg of 5-HIAA in 10 ml of $0.5M$ phosphate buffer.

   B. Working standard, 40 μg/ml.   Dilute 1 ml of stock standard to 25 ml with water.
10. Sulfuric acid, $1M$.   Carefully add 14 ml of concentrated H_2SO_4 to about 200 ml water, mix, cool, and dilute to 250 ml.
11. Hydrochloric acid, $2M$.   Carefully add 43 ml of concentrated HCl to about 200 ml of water, mix, cool, and dilute to 250 ml.
12. Sodium nitrite, $0.36M$.   Dissolve 2.5 g of sodium nitrite in water and dilute to 100 ml.
13. Nitrous acid reagent.   To 5 ml of $2M$ H_2SO_4 add 0.2 ml of $0.36M$ $NaNO_2$. This reagent should be prepared just before use.

QUALITATIVE TEST

Since 5-HIAA accounts for most of the 5-hydroxyindole material in urine, direct application of the nitrosonaphthol color test provides a simple diagnostic means for malignant carcinoid. This qualitative test is not specific for 5-HIAA and measures all 5-hydroxyindoles. Positive tests should be followed by the quantitative determination.

PROCEDURE

To a test tube containing 0.8 ml of water and 0.5 ml of 1-nitroso-2-naphthol reagent, add 0.2 ml of urine and mix. Add 0.5 ml of nitrous acid reagent and mix again. Let stand at room temperature for 10 min, add 5 ml of ethylene dichloride, and shake once more. Allow the phases to separate. If turbid, the tubes should be centrifuged. A positive test is indicated by a purple color in the top layer. Normal might show as a slight pink or yellow color. Assuming an average 24-hr urine volume of 1,000 ml, a purple color will be seen at levels of 5-HIAA excretion as low as 30 mg/24 hr. At high levels the color is more intense and is almost black at levels above 300 mg/24 hr.

QUANTITATIVE METHOD

Acidified urine is treated with 2,4-dinitrophenylhydrazine for removal of interfering keto acids and extracted with chloroform for removal of indoleacetic acid. The 5-HIAA is then extracted into ether, reextracted into a small volume of phosphate buffer, and reacted with nitrous acid and 1-nitroso-2-naphthol.

A 24-hr urine specimen preserved with 10 ml of concentrated HCl is required. It is important that patients refrain from taking nonessential medication for 72 hr prior to urine collection. The specificity of this method is very good.

PROCEDURE

To 8 ml of urine in a 50 ml glass-stoppered tube add 8 ml of 2,4-dinitrophenyl-hydrazine reagent. Mix and let stand 15 to 30 min to permit complete reaction and precipitation. Add 30 ml of chloroform, shake for a few minutes, and centrifuge.

Remove chloroform layer and replace with fresh 30 ml of chloroform and repeat extraction, centrifugation, and removal of chloroform layer. Remove a 12-ml aliquot of the aqueous layer and transfer to a 40 ml glass-stoppered centrifuge tube containing 5 g of NaCl and 30 ml of ether. Shake for 5 min and centrifuge. Transfer a 25-ml aliquot of the ether layer to another 40 ml glass-stoppered centrifuge tube containing 3.0 ml of $0.5M$ phosphate buffer (pH 7.0). Shake for 5 min, centrifuge, and remove ether by aspiration. Transfer 2 ml of aqueous phase into a 15 ml glass-stoppered centrifuge tube containing 1 ml of 1-nitroso-2-naphthol reagent and mix. Add 1 ml of nitrous acid reagent, mix, and incubate at 37°C for 5 min.

Add 8 ml of ethyl acetate and shake to remove yellow pigment in urine extract. After separation of phases, remove ethyl acetate by aspiration and repeat extraction with another 8-ml portion of ethyl acetate. Remove ethyl acetate, transfer aqueous layer to cuvette, and read at 540 nm against reagent blank.

Standards are prepared by treating 1, 2, and 4 ml of working standard, diluted to 8 ml with water, exactly as the urine samples. These standards contain 40, 80, and 160 μg of 5-HIAA, respectively, or, when diluted to 8 ml, 5, 10, and 20 μg/ml. The reagent blank is prepared by treating 8 ml of water in the same manner.

CALCULATION

$$\frac{\text{absorbance of unknown}}{\text{absorbance of standard}} \times \text{conc. of standard (mg/ml)} \times \text{urine volume (liters)} = \text{mg 5-HIAA/day}$$

NORMAL VALUES AND INTERPRETATION OF RESULTS

Normally very small amounts of the serotonin metabolite 5-hydroxyindoleacetic acid are found in the urine. The normal excretion is 2 to 9 mg/24 hr. In the presence of a carcinoid tumor the level may rise to 25 to 1,000 mg/24 hr. A number of drugs are said to cause lowered levels of 5-HIAA excretion; these include monoamine oxidase inhibitors, methyldopa, imipramine, and p-chlorophenylalanine. It is not clear whether the decrease is sufficient to interfere with the diagnosis of a carcinoid tumor. Methenamine and phenothiazines when excreted in the urine may result in some chemical interference with the test, leading to low values. The metabolites of some drugs such as acetanilide, phenacetin, mephenesin, and methocarbamol when excreted in the urine may give a chemical interference leading to erroneously high values. It is also stated that ingestion of certain foods high in serotonin may lead to elevated levels of HIAA in the urine. Foods said to contain large amounts of serotonin include eggplant, avocados, bananas, pineapples, and plums.

References

1. White, A., Handler, P., and Smith, E. L. *Principles of Biochemistry*, 4th ed. New York: McGraw-Hill, 1968.
2. Dorfman, R. I. [Ed.]. Methods in Hormone Research, 2nd ed. Vol. 1, *Chemical Determinations*. New York: Academic, 1968.
3. Loraine, J. A., and Bell, E. T. [Eds.]. *Hormone Assays and Their Clinical Application*, 3rd ed. Baltimore: Williams & Wilkins, 1971.
4. Porter, R. [Ed.]. *Gas Chromatography in Biology and Medicine*. London: Churchill, 1969.
5. Duczfalusy, E., and Duczfalusy, A. [Eds.]. *Steroid Assay by Protein Binding*. Stockholm: Karolinska Institute, 1970.
6. Peron, F. G., and Caldwell, B. V. [Eds.]. *Immunological Methods in Steroid Determinations*. New York: Appleton-Century-Crofts, 1971.
7. Inghar, S. H., and Woeber, K. A. In R. H. Williams [Ed.], *The Thyroid Gland. Textbook of Endocrinology*. Philadelphia: Saunders, 1968.

8. Langley, L. L. *Thyroid Gland. Review of Physiology.* New York: McGraw-Hill, 1971.
9. Recant, L., and Riggs, D. S. *J. Clin. Invest.* 31:789, 1952.
10. Sterling, K., and Brenner, M. A. *J. Clin. Invest.* 45:153, 1966.
11. Anderson, B. G. *J.A.M.A.* 203:135, 1968.
12. Silver, S. *Radioactive Isotopes in Medicine and Biology,* 2nd ed. Philadelphia: Lea & Febiger, 1963.
13. Sisson, J. C. *J. Nucl. Med.* 6:853, 1965.
14. Elzinga, K. E., Carr, E. A., and Beierwaltes, W. H. *Am. J. Clin. Pathol.* 36:125, 1961.
15. Scholer, J. F. *J. Nucl. Med.* 3:41, 1962.
16. Hamolski, M. W., Stein, M., and Freedberg, A. S. *J. Clin. Endocrinol. Metab.* 17:33, 1957.
17. Boerner, W., Noll, E., Ruppert, G., Rauh, E., and Nauman, P. *Nucl. Med.* 9(Suppl.): 799, 1971.
18. Braverman, V. A., and Williams, C. M. *J. Nucl. Med.* 12:55, 1971.
19. Bauer, J. D., Ackermann, P. G., and Toro, G. *Clinical Laboratory Methods,* 8th ed. St. Louis: Mosby, 1974.
20. Zak, B., Willard, H. H., Meyer, G. B., and Boyle, A. J. *Anal. Chem.* 24:1345, 1952.
21. Leffler, H. H., and McDougald, C. H. *Am. J. Clin. Pathol.* 45:344, 1966.
22. O'Neal, L. W., and Simms, E. S. *Am. J. Clin. Pathol.* 23:493, 1953.
23. Austin, E., and Koepke, J. A. *Am. J. Clin. Pathol.* 45:344, 1966.
24. Gambino, S. R., Schreiber, H., and Covolo, G. *First Technicon International Symposium.* New York: Mediad, 1966.
25. Strickler, H. S., Saier, E. L., and Grauer, R. *First Technicon International Symposium.* New York: Mediad, 1966.
26. Pileggi, V. J., Leen, N. D., Golub, O. J., and Henry, R. J. *J. Clin. Endocrinol. Metab.* 21:1272, 1961.
27. Pileggi, V. J., and Kessler, G. *Clin. Chem.* 14:339, 1968.
28. Kessler, G., and Pileggi, V. J. *Fifth Technicon International Symposium.* New York: Mediad, 1970. Vol. 1.
29. Yee, H. Y., Bowdell, J., and Jackson, D. *Clin. Chem.* 17:665, 1971.
30. Clark, F. *J. Clin. Pathol.* 20:344, 1966.
31. Leonards, J. R. *Clin. Chem.* 16:922, 1970.
32. Murphy, B. P., and Pattee, C. J. *J. Clin. Endocrinol. Metab.* 24:187, 1964.
33. Kessler, G., and Pileggi, V. J. *Clin. Chem.* 16:382, 1970.
34. Murphy, B. P., Pattee, C. J., and Gold, A. *J. Clin. Endocrinol. Metab.* 26:247, 1966.
35. Guaron, A. *J. Nucl. Med.* 10:532, 1969.
36. Murphy, B. P., and Jachan, C. *J. Lab. Clin. Med.* 66:161, 1965.
37. Murphy, B. P. *Semin. Nucl. Med.* 1:301, 1971.
38. Nusynowitz, M. L., and Waliszowski, J. *Am. J. Clin. Pathol.* 56:523, 1971.
39. Foster, L. B., and McFadden, D. L. *Clin. Chem.* 18:741, 1972.
40. Sterling, K., and Brenner, M.A. *J. Clin. Invest.* 45:153, 1966.
41. Malvaux, P. *J. Clin. Endocrinol. Metab.* 28:459, 1968.
42. Clark, I. F., and Horn, D. B. *J. Clin. Endocrinol. Metab.* 25:39, 1965.

43. Ingbar, S. H., Braverman, L. E., Dauber, N. A., and Lee, G. V. *J. Clin. Invest.* 44:1679, 1965.
44. Lee, N. D., Henry, R. J., and Golub, O. J. *J. Clin. Endocrinol. Metab.* 24:486, 1964.
45. Mincey, E. K., Thorson, S. C., and Brown, J. L. *Clin. Biochem.* 4:286, 1971.
46. Murray, I. P. C., and Parkin, J. *Med. J. Aust.* 1:1190, 1972.
47. Gladding, T. C. *J. Tenn. Med. Assoc.* 65:442, 1972.
48. Hoffman, K., and Yajima, H. *Recent Prog. Horm. Res.* 18:41, 1962.
49. Utiger, F. D. *J. Clin. Invest.* 44:1277, 1965.
50. Odell, W. D., Wilber, J. F., and Paul, W. E. *J. Clin. Endocrinol. Metab.* 25:1179, 1965.
51. Hershman, J. M., and Pittman, J. A., Jr. *Ann. Intern. Med.* 74:481, 1971.
52. Werner, S. C., Goodwin, L. D., and Quimby, E. H. *J. Clin. Endocrinol. Metab.* 9:342, 1947.
53. McKenzie, J. M. *Endocrinology* 63:372, 1958.
54. McKenzie, J. M. *Recent Prog. Horm. Res.* 23:1, 1967.
55. Purves, H. D., and Adams, D. P. *Br. Med. Bull.* 16:128, 1960.
56. Adams, D. D. *J. Clin. Endocrinol. Metab.* 18:699, 1958.
57. Boothby, W. M., and Sandiford, I. *J. Biol. Chem.* 54:767, 1922.
58. Hamolsky, M. W., and Freedberg, A. S. *N. Engl. J. Med.* 262:129, 1960.
59. Gavrod, L. P. *Br. Med. Bull.* 16:2, 1960.
60. Doniach, D., and Roitt, I. M. In P. G. H. Gell and R. R. A. Loombs [Eds.], *Clinical Aspects of Immunology,* 2nd ed. Philadelphia: Davis, 1968.
61. Rose, N. R., and Witebsky, E. *J. Immunol.* 76:417, 1956.
62. Squires, R. B., and Langhorne, W. H. *Am. J. Clin. Pathol.* 46:189, 1966.
63. Chaney, W. C. *J.A.M.A.* 82:2013, 1924.
64. Gilson, W. E. *N. Engl. J. Med.* 260:1027, 1959.
65. Heftman, E. *Steroid Biochemistry.* New York: Academic, 1965.
66. Dorfman, R. I., and Ungar, F. *Metabolism of Steroid Hormones.* New York: Academic, 1965.
67. Horton, R., Kato, T., and Sherins, R. *Steroids* 10:245, 1967.
68. Kato, T., and Horton, R. *Steroids* 12:631, 1968.
69. Mayes, D., and Nugent, C. A. *J. Clin. Endocrinol. Metab.* 28:1169, 1968.
70. Winter, J., and Grant, D. *Anal. Biochem.* 40:440, 1971.
71. Callow, N. H., Callow, R. K., and Emmons, C. W. *Biochem. J.* 32:1312, 1938.
72. Callow, N. H., and Callow, R. K. *Biochem. J.* 34:276, 1940.
73. Few, J. D. *J. Endocrinol.* 22:31, 1961.
74. Larsen, K. *Acta Endocrinol.* (Kbh) 57:228, 1968.
75. Silber, R. H., and Busch, R. D. *J. Clin. Endocrinol. Metab.* 16:1333, 1956.
76. Porter, C. C., and Silber, R. H. *J. Biol. Chem.* 185:1201, 1950.
77. Weichselbaum, T. E., and Margrat, H. W. *J. Clin. Endocrinol. Metab.* 15:970, 1955.
78. Mattingley, D. *J. Clin. Pathol.* 15:374, 1962.
79. Verjee, Z. H. M. *Clin. Chim. Acta* 33:268, 1971.
80. van de Vies, J. *Acta Endocrinol.* (Kbh) 38:399, 1961.
81. Dickey, R. P., Besch, P. K., Vorys, N., and Ullery, J. C. *Am. J. Obstet. Gynecol.* 94:591, 1966.

82. van Baelen, H., Heyns, W., and DeMoor, P.   *J. Clin. Endocrinol. Metab.* 27:1056, 1967.
83. Greene, J. W., Duhring, J. C., and Smith, K.   *Am. J. Obstet. Gynecol.* 92:1030, 1965.
84. Scommegna, A., and Chattoraj, S. C.   *Am. J. Obstet. Gynecol.* 99:1087, 1967.
85. van der Molen, H. J., and Groen, D. J.   *J. Clin. Endocrinol. Metab.* 25:625, 1965.
86. Johansson, E. D. B.   *Acta Endocrinol.* (Kbh) 61:592, 1969.
87. Klopper, A., Michie, E. A., and Brown, J. B.   *J. Endocrinol.* 12:209, 1955.
88. Grundy, H. M., Simpson, S. A., Tait, J. F., and Woodford, M.   *Acta Endocrinol.* (Kbh) 11:19, 1952.
89. Conn, J. W.   *J. Lab. and Clin. Med.* 45:6, 1955.
90. Conn, J. W.   *N. Engl. J. Med.* 273:1135, 1965.
91. Williams, R. H. [Ed.].   *Textbook of Endocrinology.* Philadelphia: Saunders, 1968.
92. Thorn, G. W., Forsham, P. H., and Emerson, K.   *The Diagnosis and Treatment of Adrenal Insufficiency,* 2nd ed. Springfield, Ill.: Thomas, 1951.
93. Lauler, D. P., and Thorn, G. W.   In T. R. Harrison [Ed.], *Principles of Internal Medicine,* 5th ed. New York: Blakiston, 1966.
94. Cope, C. L.   *Proc. R. Soc. Med.* 58:551, 1965.
95. Henry, R. J., and Sobel, C.   *Arch. Intern. Med.* 100:196, 1957.
96. Sobel, C., and Henry, R. J.   *Am. J. Clin. Pathol.* 27:240, 1957.
97. Jacobs, S. L., Sobel, C., and Henry, R. J.   *J. Clin. Endocrinol. Metab.* 21:305, 1961.
98. Mahler, D. J., and Humoller, F. L.   *Clin. Chem.* 8:47, 1962.
99. Georges, R. J.   *J. Med. Technol.* 21:126, 1964.
100. Sunderman, F. W., Jr.   *Am. J. Clin. Pathol.* 42:481, 1964.
101. Pisano, J. J., Crout, J. R., and Abraham, D.   *Clin. Chim. Acta* 7:285, 1962.
102. Hernandez, A.   *Lab. Dig.* 29:10, 1966.
103. Udenfriend, S., Titus, E., and Weissbach, H.   *J. Biol. Chem.* 216:499, 1955.
104. Udenfriend, S., Weissbach, H., and Brodie, B. B.   In D. Glick [Ed.], *Methods of Biochemical Analysis,* vol. 6. New York: Interscience, 1958.
105. Goldenberg, H.   *Clin. Chem.* 13:697, 1967.

14. Enzymes

Enzymes are complex, naturally occurring compounds that catalyze many biological reactions; i.e., they speed up reactions that might otherwise proceed very slowly [1–8a]. Some enzymes are simple proteins; others are conjugated proteins containing metal ions, coenzymes, or both. Coenzymes serve as intermediate carriers of electrons or specific functional groups (e.g., hydrogen atoms, amino groups), and aid in their transfer from one substrate to another. Most enzymes are very specific in their action in that they will catalyze only a definite type of chemical reaction or act on a particular compound (substrate). The enzymes have a complex molecular structure such that the active sites just match the corresponding configuration on the substrate whose change it will catalyze. The formation of the enzyme-substrate complex is necessary for the catalytic action of the enzyme.

Enzyme Nomenclature

Many enzymes are named by adding the suffix -*ase* to the name of the substrate on which they act: urease hydrolyzes urea, lipase hydrolyzes lipids, phosphatases act on organic phosphates, etc. Because of the increasing number of enzymes being isolated and studied, the Commission on Enzymes of the International Union of Pure and Applied Chemistry and the International Union of Biochemistry in 1964 adopted a number of rules for the naming of enzymes. These rules were revised in 1972 [1]. In this system the enzymes are divided into six main classes according to the type of reaction they catalyze. These are: (1) oxidoreductases, (2) transferases, (3) hydrolases, (4) lyases, (5) isomerases, and (6) ligases. This grouping provides the basis for assigning a distinctive number to each enzyme. Each enzyme's number consists of four numbers separated by periods. Thus the enzyme usually known in the clinical laboratory as glutamic oxalacetic transaminase was assigned the number E.C.2.6.1.1. The first digit (2) indicates that the enzyme belongs to the group of transferases (a chemical grouping is transferred from one molecule to another). The second digit (6) indicates that the group transferred contains a nitrogen atom (as distinguished from other functional groups such as sulfur, phosphate, aldehyde). The third digit (1) indicates that the nitrogen grouping is an amine group (as distinguished from oxime or other N-containing groups). The last digit (1) is merely the serial number of the enzyme in the particular class.

The commission also suggested a systematic and a trivial name for each enzyme. The former attempts to identify the substrate and describe the action of the enzyme as closely as possible. For glutamic oxalacetic transaminase the systematic name is L-aspartate:2-oxoglutarate aminotransferase, and the trivial name is aspartate aminotransferase. (Note that this is not the same as the

common name.) The reaction involved is: oxoglutarate + aspartate ⇌ glutamate + oxalacetate (oxoglutarate is also known as α-ketoglutarate). Actually,
in most determinations of this enzyme the substrate used is oxoglutarate +
aspartate, so that the E.C. name is more correct than the common name. It is not
likely, however, that the E.C. names will, in the near future, supersede the common names used in the past in the clinical laboratory, since the results are for the
use of physicians who are familiar with the common names rather than for
other biochemists. However, the commission names and numbers will be given
for most enzymes discussed here since there will be increasing reference to them in
the clinical chemistry literature. One suggestion should be adopted, that is, the
use of the salt form of the name rather than the acid form, e.g., glutamate
rather than glutamic, and lactate rather than lactic, since under the conditions of
the determinations it is usually the ionic form that reacts.

Enzyme Reactions and Substrate Concentration

The general reaction involving an enzyme may be written as follows:

$$E + S \rightleftharpoons ES \rightarrow E + S'$$

where E represents the enzyme, S the substrate on which the enzyme acts, ES
the postulated enzyme-substrate complex, and S' the reacted substrate that may
represent a changed molecular species for S or the splitting of S into two or more
different molecules. The enzyme thus represents a true catalyst in that it is not
changed in the reaction.

Enzymes are usually present in amounts too small to be measured directly in
the same sense that one might determine the concentration of other blood
constituents. They are determined by their effect on the rate of change of the
substrate. Like any chemical reaction, the rate at which the reaction $E + S \rightarrow ES$
proceeds depends upon the concentrations of E and S. Theoretically, the effective
concentration of E does not change during the reaction. If the concentration of
S is sufficiently high so that it does not change appreciably during the time in
which the reaction rate is measured, then this rate will depend only on the concentration of E. Enzymes are thus determined by measuring the rate at which
the enzyme-catalyzed reaction proceeds, usually by measuring either the amount
of S used up or the amount of S' produced during a given period of time. As the
reaction proceeds, the substrate will be decomposed or changed and the rate will
decrease (Fig. 14-1). Initially the total amount decomposed increases linearly
with the time (curve is a straight line), but as the substrate concentration
decreases the rate also decreases as indicated by the fact that the slope of the
curve begins to decrease (the dashed line in the figure illustrates a constant rate).

Figure 14-1. Change in amount of substrate decomposed with time for a constant enzyme concentration. See text.

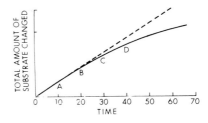

Figure 14-2. Change in initial velocity of enzyme reaction with substrate concentration. See text.

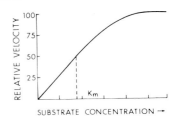

Thus the amount decomposed per unit time, which is the measure of enzyme activity, would not be the same if the measurements were taken at different times (10, 20, 30, and 40 units of time as indicated by the points A, B, C, and D). Therefore in the measurement of enzyme activity the substrate concentration should be great enough so that all measurements would fall on the linear portion of the curve.

This is illustrated in another way by the curve of Figure 14-2. Here the velocity is the initial velocity of the reaction. This is plotted against the substrate concentration. At low substrate concentration, the velocity is proportional to the substrate concentration, as might be expected. At high substrate concentration, the velocity becomes constant and independent of the substrate concentration. This is the concentration one would choose for making measurements, as then the velocity is dependent only on the enzyme concentration. We note from the original equation that before the substrate can be changed, the complex (ES) must be formed. This then decomposes into E and S'. At high concentrations of S practically all the enzyme will at all times be in the form of the ES complex, and the rate at which this decomposes will give the rate at which S' is formed.

The different enzymes are often characterized by the *Michaelis-Menten constant, K_m*. This is defined as the substrate concentration at which the reaction velocity is one-half the maximum. In the actual determination of enzyme activities the substrate concentration used should be of the order of 100 times K_m.

Enzyme determinations are made by measuring the reaction rate—the rate of decomposition of the substrate or the rate of formation of a product. This may be done by measuring the amount of material (substrate or product) present at zero time (on a separate aliquot if necessary) and after a fixed interval of time. The rate of change can then be calculated. This is known as the *fixed point method*. Usually the initial concentration of the substrate is assumed to have a fixed value initially or the initial concentration of the product is assumed to be zero, so that one reagent blank is sufficient for a set of determinations, unless a serum blank is required to compensate for any absorbance by the serum itself. In the other general method, the *kinetic method,* the absorbance is measured at a number of definite intervals (such as 1 min) and the average absorbance change per minute calculated. As illustrated in Figure 14-1, the reaction rate may fall off after an interval of time, due not only to a depletion of the substrate, but also to an inhibiting effect of excess products of the reaction. In the fixed point assay one cannot always be certain that measurements are being taken at points such as A or B of the figure, rather than at points such as C or D. In the kinetic method, measurements are made at a number of intervals and one can readily tell if the reaction rate is constant. Naturally if the measurements can be made only after the addition of a color reagent which usually stops the reaction, only one measurement can be made and the kinetic method cannot be used.

Other factors that may influence the initial reaction rate are illustrated in Figure 14-3. A lag period (curve B) often results when the final absorbance measurement is the end result of several coupled reactions as in the ultraviolet method for creatine phosphokinase (CPK) (see below). An initial greater reaction rate (curve C) is often due to the fact that there will be small amounts of other substrates in the serum added or in the solutions used. These may also react to give a change in absorbance. When these small amounts of impurities have been effectively used up, the reaction involving the desired substrate will proceed regularly. In the illustration the lines are exactly parallel after about 3 min. Thus in a kinetic method measuring the absorbance at, say, 3, 4, 5, and 6 min, one would find the same change in absorbance for each case. If, however, a fixed point method was used in which an additional reagent was added at 10 min to stop the reaction and develop the color, different results would be obtained in the three cases. If it were necessary to use such a colorimetric method, one could compensate for any initial changes in reaction rate by running the reaction under exactly the same conditions on two aliquots, except that one would be incubated for, say, 3 min and the other for 8 min. Then the difference in ab-

Figure 14-3. Absorbance (as a measure of the amount of product formed) is plotted against time. Curve A represents the ideal case in which the absorbance increases exactly linearly with time from the beginning. In curve B there is an initial lag phase and then a constant rate. Curve C represents the case where initially the reaction appears to proceed more rapidly, then levels off to a constant rate. See text.

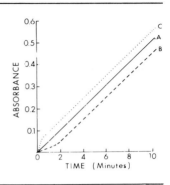

sorbance between the two samples would be a measure of the enzyme activity no matter whether there was an initial lag or accelerated phase or not. Manually this would mean additional pipettings, incubation timing, and photometer readings, which would increase the time required and more particularly the sources of errors. The method has been applied to some automated systems where these factors are more carefully controlled.

Thus the kinetic methods for enzyme determinations would be preferred when possible. Almost all the ultraviolet methods can be used in this way. For strictly manual methods, the kinetic method has the disadvantages that if the measurements are to be made over a short period of time, the cuvette compartment of the spectrophotometer must be kept at constant temperature, it may be tedious to make several readings for each sample, and usually only one sample can be run at a time. These disadvantages are not found with the use of one of the many automated or semiautomated instruments that are available. When manual kinetic methods are used with only short incubation times, the spectrophotometer must be capable of giving accurate absorbance readings to 0.001 unit. This was not possible with some of the older instruments, but digital readout accessories can now be obtained for many of these instruments to give the required accuracy.

Effect of Temperature and pH on Enzyme Reactions

All enzyme rates are temperature and pH sensitive. Figure 14-4 illustrates the change of enzyme activity with pH in a hypothetical example. Note that here the optimal pH is at about 6.0, and that at higher or lower pH values the activity is less. Most of the enzymes in the body are actually operative at a pH near 7.4 (except those involved in the digestive processes). However, the maximum activity in vitro may be at a different pH and this pH is usually used in the

Figure 14-4. Relation between enzyme activity and pH and temperature. See text.

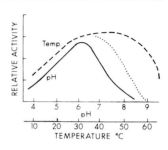

determination of enzymes. For this reason the substrate solutions are always buffered to maintain a constant pH.

The effect of temperature on the activity of enzymes is also illustrated in Figure 14-4. In general, a rise of 10°C will approximately double the rate of most chemical reactions. This also holds true for the enzyme-catalyzed reactions. The rate at 35°C will be approximately twice that at 25°C. The actual ratio of the rate at 35°C to that at 25°C, often designated by the symbol Q_{10}, may be used to characterize the change in rate with temperature. The value will vary with different enzymes. It may also be somewhat different for the same enzyme under different conditions (such as different substrate, substrate concentration, or pH).

At temperatures somewhat above 40°C, heat will begin to denature or deactivate most enzymes. Heating to 65°C for about 30 min will almost completely inactivate most enzymes. The dashed and dotted curves in the figure illustrate the changes in activity that might occur with temperature changes. The dashed curve might illustrate the relative activity when the enzyme is just added to the substrate at the indicated temperature. When, however, the enzyme is kept at the elevated temperature for some time before the reaction rate is measured, results like the dotted curve might be obtained. Even a temperature of 45°C for some time may cause some inactivation.

In the body most enzymes will presumably be functioning at a temperature near 37°C. This may not be the temperature of optimum activity, particularly in vitro where the conditions are different from in vivo. It should be obvious that in the determination of enzyme activity, the reaction temperature must be kept constant at a definite value. For purposes of comparison it would be preferable that all enzyme determinations be made at the same temperature. A number of different temperatures are still used. Many of the earlier methods used 37°C, possibly because this was the temperature at which the enzymes functioned in vivo, as well as for other reasons as given below. Originally the Enzyme Commission of the IUPAC and IUB first recommended a temperature of 25°C. Objections were raised to this in that the ambient temperature in many laboratories may at times rise above 25°C. This would require some sort of cooling bath, an added complication. The recommendation was then changed to 30°C, which is the temperature suggested for use in this volume. A temperature of 32°C has also been used by some.

The use of 37°C has the advantage that the reactions are more rapid at this temperature. Thus for a given amount of enzyme, a greater amount of product will be formed and there will be a greater absorbance difference. Theoretically this could give somewhat greater precision. This advantage was more relevant when the simpler colorimeters were used which could not be read to closer than 0.005 absorbance unit. With modern spectrophotometers and digital readouts the absorbance can be readily determined to 0.001 unit or less, and the greater absorbance difference is not so necessary. On the other hand, there are certain disadvantages to the use of 37°C. In many procedures, particularly those involving the single point method, the reaction is started by pipetting a fixed amount of one solution into another solution in the reaction tube or cuvette. Both solutions must be warmed to 37°C before mixing. It has been found that even pipetting a warm solution with a pipet that is at room temperature will cause some variation in the final temperature of the mixture. This objection probably applies more to manual methods than to automated ones. More important, some enzymes are slowly inactivated at 37°C under the conditions at which the assays are actually carried out. This will particularly be true when the diluted serum and buffer are incubated before the substrate is added to start the reaction. Thus a lower temperature is preferable.

Since the amount of substrate that a given amount of enzyme will decompose in a unit time will depend upon the temperature, the apparent amount of enzyme present at the two temperatures will be different if the temperature factor is not taken into account. This is illustrated in Figure 14-5 which shows the relative activity of two enzymes with temperature. The relative activity is the relative amounts of the substrate changed in a given time by the same absolute amounts of the enzyme at the different temperatures.

Since at present the same enzyme may be determined by different workers using different temperatures, it is convenient to have some rule for correlating the activities and units when measured at different temperatures. Suppose that one had determined the enzyme activities at 30°C and wished to compare the results with those obtained by others using the same method but making the determinations at 37°C. The reaction rate is higher at 37°C than at 30°C, so

Figure 14-5. Relative activity of two enzymes at different temperatures.

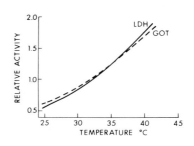

for the same actual amount of enzyme one would obtain a greater number of units of activity (higher absorbance, more micromoles of substrate decomposed) at the higher temperature.

Thus

units at 30°C = (units at 37°C) × F

where the factor F would be less than 1 since we are calculating the amount of absorbance change which would have occurred if the reaction had been run at the lower temperature rather than the higher one. Conversely,

units at 37°C = (units at 30°C) × F

where the factor F is now greater than unity. Thus for any two temperatures,

units at t_1 = (units at t_2) × F

where F is greater than 1 when $t_1 > t_2$ and F is less than 1 when $t_1 < t_2$.

Several points should be emphasized at this time. The same method must be compared at the two temperatures, since different methods might use different substrates or substrate concentrations or different pH and would not necessarily give the same results at the same temperature even though the results are both expressed in international units. This is discussed further elsewhere.

Second, the rate of change is different for different enzymes and may even be different for the same enzyme using different methods. One must choose the appropriate factor for the enzyme involved. Also the factors as given in the nomogram following are only approximate. They are useful in giving approximate comparisons for different temperatures but cannot be expected to be extremely accurate. When compiling material for the production of the nomogram, different workers were found to give different values for the rate change for the same enzyme; thus any value used can only be an average.

Figure 14-6. A nomogram for obtaining the proper conversion factor for comparing results at 30°C with those obtained at other temperatures. On the right are scales A and B for temperatures above and below 30°C. On the left are scales A′ and B′ for factors less than or greater than unity. On the diagonal scale are given values of Q_{10} used to characterize the enzyme rate change.

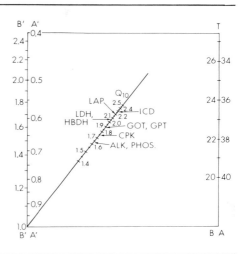

The nomogram (Fig. 14-6) is designed for obtaining the proper conversion factor for comparing results at 30°C with those obtained at other temperatures. To use the nomogram, align a ruler with the appropriate temperature on the right and the correct value of Q_{10} on the diagonal scale, and read the correct factor on the left. The choice of which of the F scales to use is made by the rules given below.

If the problem is stated as

units at 30°C = (units at $T°$) × F

then the two scales on the same sides of the vertical lines are used. If scale A is used (for temperatures above 30°C), scale A′ is used for the factor. If scale B is used (for temperatures below 30°C), scale B′ is used for the factor. Conversely, if the problem is stated as

units at $T°$ = (units at 30°C) × F

then the A and B′ scales are used together, or the A′ and B scales.

EXAMPLES

For LDH, $Q_{10} = 2.08$,

    units at 30°C = (units at 35°C) × F

    F = 0.71 (scales A and A′)

For an enzyme with $Q_{10} = 1.60$,

  units at 37°C = (units at 30°C) $\times$ F

  F = 1.39 (scales A and B')

The nomogram may also be used to convert the units between two different temperatures, neither one of which is 30°C.

For $Q_{10} = 1.60$,

  units at 25°C = (units at 30°C) $\times$ F

  F = 0.79 (scales B and A')

  units at 30°C = (units at 37°C) $\times$ F'

  F' = 0.72 (scales A and A')

  units at 25°C = (units at 37°C) $\times$ F $\times$ F' = (units at 37°C) $\times$ 0.57'

NOTE ON THE CONSTRUCTION OF THE NOMOGRAM

Suppose the ratio of the activity of a given enzyme at 35°C to that at 25°C were exactly 2.00; then the rate of increase per degree would be approximately 7.2%. That is, the ratio (activity at 26°C)/(activity at 25°C) = 1.072. Accordingly the ratio (activity at 35°C)/(activity at 25°C) = $(1.072)^{10}$ = 2.00. The nomogram is constructed on this basis using the equation:

$$\log \frac{\text{activity at } t_1}{\text{activity at } t_2} = \frac{t_1 - t_2}{10} \log Q_{10} \tag{1}$$

Theoretically one should use a somewhat more complicated form based on the Arrhenius equation

$$V = k\, e^{-\frac{E}{RT}} \tag{2}$$

where V is the velocity of a reaction, k is a constant that depends upon the units of measurement, E is a measure of the energy change in the reaction, R is a fundamental constant, and T is the absolute temperature (°C + 273). For a given reaction, one would have for the velocity at two different temperatures:

$$\frac{V_1}{V_2} = e^{-k\left(\frac{1}{T_1} - \frac{1}{T_2}\right)} \quad \text{or} \quad \log \frac{V_1}{V_2} = -k\left(\frac{T_2 - T_1}{T_1 T_2}\right) \tag{3}$$

If Q_{10} is taken at V_{308}/V_{298} (the two absolute temperatures are equal to 35 and 25°C), it can be shown that

$$\log \frac{V_1}{V_2} = \left[\frac{298 \times 308}{T_1 \times T_2} \right] \frac{T_1 - T_2}{10} \log Q_{10} \tag{4}$$

For the range of from 25 to 37°C, the expression in the brackets is always close to unity. If it is set equal to 1, equation 1 results.

Enzyme Units

The units of enzyme activity are all defined in terms of the amount of S decomposed or the amount of S′ formed in a given time under certain specified conditions of pH, substrate concentration, and temperature. If only the temperature were varied, we would find that at 37°C approximately 1.4 times as much S′ would be formed than at 32°C. In other words, if we measured the activity of the same amount of enzyme at these two temperatures, we would obtain 140 units at 37°C and 100 units at 32°C (measured in terms of the amount of S′ generated). Thus if our usual standard temperature were 37°C and we had inadvertently made a measurement at 32°C, we would have to multiply the result obtained at 32°C by 1.4 to compare the activity of this sample with our usual results.

As the enzyme determinations measure the rates of reaction, concentration of enzymes is usually expressed in units. The actual units used in reporting enzyme concentrations are still in a state of flux. In all of the older procedures arbitrary units were employed, and different units were used for the same enzyme in various procedures. For example, in the determination of the enzyme alkaline phosphatase, one method (Bodansky) uses a substrate of β-glycerophosphate and defines the unit as 1 mg of phosphate phosphorus liberated from the substrate at pH 8.6 by 100 ml of serum during 60 min incubation at 37°C. Another method (original King-Armstrong method) uses a substrate of phenylphosphate and defines the unit as 1 mg of phenol liberated from the substrate at pH 9.6 by 100 ml of serum in 30 min incubation at 37°C. A third method (Bessey-Lowry-Brock) uses p-nitrophenylphosphate as a substrate and defines the unit as 1 mmol of p-nitrophenol formed at pH 10.2 by 1,000 ml of serum in 60 min incubation at 37°C. One could not expect these units to be directly comparable. It is best to run the same serum by the different methods and compare the results. For a rough correlation one can look up the range of normal values given for two methods and compare them. One compilation gives the range of normal values for the King-Armstrong method as 3.7 to 13.1 units and for the Bessey-Lowry-Brock method as 0.8 to 2.9 units. The ratios for the lower and upper limits are $3.7/0.8 = 4.6$ and $13.1/2.9 = 4.5$, so that one can say that approximately 4.5 K-A units equals 1.0 B-L-B unit.

It has been proposed that all enzyme concentrations be expressed in international units (IU). One international unit is defined as 1 μmol of substrate used/min/liter of serum under specified conditions of pH and temperature. These units have not come into general use in clinical chemistry because the older units are firmly established. In some instances use of the international unit would facilitate better comparison between methods. For example, the normal ranges for three alkaline phosphatase methods (Shinowara-Jones-Reinhart, King-Armstrong, and Bessey-Lowry-Brock) are given as 2.2 to 8.6, 3.7 to 13.1, and 0.8 to 2.9 of their respective units. These all seem quite different, but when converted into international units, the ranges are essentially the same: 12.0 to 46.7, 13.0 to 46.0, and 13.3 to 48.3 IU. The Enzyme Commission [1] has recommended a new measure of enzyme activity, the katal. This is defined as the amount of activity that will convert one mole of substrate per second. This recommendation has not met with general approval and the katal is as yet rarely used.

The rate of reaction involving enzymes is markedly influenced by temperature, pH, concentration of substrate, and a number of other factors. Accordingly, all the details of a given procedure must be followed exactly in order to obtain accurate results. The time as well as the temperature of incubation must be closely controlled. With some enzymatic procedures the reaction rate is not constant with time, so one cannot assume that incubation for 60 min will utilize exactly twice the amount of substrate as incubation for 30 min. One cannot always compensate for increased activity by halving the incubation time and then multiplying the number of units found by 2. This may be true for some reactions but not all and should not be assumed to be true unless specifically stated in the procedure.

NAD-NADH Enzyme System

A number of enzyme systems involve the conversion of nicotinamide adenine dinucleotide (NAD) to its reduced form (NADH), or vice versa. The reduced form, NADH, has much greater absorption at 340 nm than does the oxidized form, and consequently the reactions may be followed by measuring the change in absorption at this wavelength. In addition, many other enzyme reactions that do not directly involve the NAD-NADH change can be coupled with another reaction that does, so that the change in NAD becomes a measure of the enzyme reaction. Thus a number of enzyme reactions can be determined by measurements at a wavelength of 340 nm. This wavelength is in the near ultraviolet range and most clinical photoelectric colorimeters do not operate at this wavelength. Good spectrophotometers that will operate in the ultraviolet range are rather expensive. However, there are instruments on the market that are relatively inexpensive and are specifically designed to read only at wavelength 340 nm.* This allows considerable simplification because these instruments can give accurate absorbance readings at this wavelength.

In the determination of enzymes by the ultraviolet methods using the change in absorbance that occurs with the intraconversion of NAD and NADH (or NADP and NADPH), the calculations are usually based on an extinction coefficient for NADH of 6.22×10^6 cm^2/mol at 340 nm. This means that a solution containing 1 μmol (10^{-6} mols)/ml of NADH would have a (theoretical) absorbance of 6.22 when read at 340 nm in a cuvette with exactly 1 cm light path. Thus if a solution having a volume of V ml when read in a cuvette with a light path of D cm has an absorbance of A units,

$$\frac{A \times V \text{ (in ml)}}{6.22 \times D \text{ (in cm)}} = \text{NADH (in } \mu \text{ mol)}$$

From the change in absorbance, the volume of sample used, and the time of incubation, one can then readily calculate the number of micromoles of NADH formed or oxidized per milliliter of sample per minute. This supposes that the reading is made at exactly 340 nm and that the light path D is accurately known. Since, as indicated in Figure 14-7, NADH has a fairly broad absorption peak at 340 nm, small errors in the wavelength will have only a slight effect on the result. Square cuvettes having an accurately known light path (usually 1.00 cm) are readily obtained. With round cuvettes, the actual light path distance may not be so accurately known. If a very narrow light beam strikes the cuvette exactly perpendicular to the surface and passes through the center of the cuvette, the light path may be taken as being equal to the internal diameter of the cuvette. With a wider light beam or if the cuvette is not exactly centered in the beam, the effective light path may be somewhat different from the internal diameter of the cuvette. Theoretically the way to check this as well as the absorbance scale of the photometer and the wavelength scale would be to measure the actual absorbance of solutions containing accurately known amounts of NADH. This is not practical because of the difficulty in obtaining solutions containing known amounts of NADH. Most commercial preparations of NADH are not pure and solutions of this material are not stable. With a spectrophotometer, the absorbance scale and cuvette path may be checked by using a stable solution having an absorbance curve similar to that of NADH in the region from 320 to 380 nm. The substance most often used is potassium dichromate whose absorption curve is also given

* Calbiometer, Calbiochem, San Diego, Calif. 92112; Coenzometer, Macalaster-Bicknell Co., New Haven, Conn. 06507.

Figure 14-7. A portion of the absorbance curves for NADH and $K_2Cr_2O_7$ in acid solution.

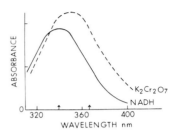

in Figure 14-7. The reagent grade of this salt is quite pure, or a very pure material may be obtained from the National Bureau of Standards (NBS, SRM #136c). The material is dried at 110°C, carefully weighed out, and dissolved in 0.005M sulfuric acid using distilled water free of all organic matter. According to Rand [8a], a solution of exactly 50.0 mg of potassium dichromate made up to exactly 1 liter with the sulfuric acid and read at 350 nm against a blank of 0.005M sulfuric acid should have an absorbance of 0.535 unit. In preparing such a standard it may be more convenient not to attempt to weight out exactly 50.0 mg but to weigh out accurately an amount close to 50 mg and use this in the calculations. Thus if exactly 52.7 mg of the dichromate were weighed out and made up to volume and this solution gave an average absorbance of 0.605, then the effective light path would be calculated as:

$$\frac{50.0 \times 0.605}{52.7 \times 0.535} = 1.07 \text{ cm}$$

This value would be used for D in the previous formula. The wavelength should be established as accurately as possible, but since the solution has a fairly broad absorption band at 350 nm, slight errors in this setting will have little effect. With the inexpensive instruments that read only at 340 nm, the problem of calibration is not so simple. Theoretically the potassium dichromate solution should have an absorbance of about 0.500 at 340 nm when measured with a narrow-band spectrophotometer, but this would not necessarily be true with an instrument having a wide-band filter for 340 nm. Presumably these filters are chosen to give the proper readings for NADH absorption but would not give the correct readings with other solutions having different absorption curves. Here one is forced to assume that the instrument is properly calibrated, though sometimes the cuvettes may be checked in a spectrophotometer.

Some kits for this type of enzymatic determinations, particularly kits of

European origin, suggest reading at 366 nm. At this wavelength a solution of NADH will have approximately 53 percent of the absorbance at 340 nm. As can be seen from Figure 14-7, one is reading on the slope of the absorption curve and the wavelength setting becomes more critical. Some of the instruments designed for reading at this wavelength have a mercury lamp as the source of light. Such a source has strong emission lines near 254, 313, 366, and 405 nm. By means of proper filters, the other lines can be filtered out and a nearly mono-chromatic light at 365 to 366 nm remains. In this way light of the same wave-length is always obtained without regard to any wavelength dial settings. Using a spectrophotometer at this wavelength, one must consider the band width and the accuracy of the wavelength setting since these may result in errors at this wavelength. With these single-wavelength instruments it is also difficult to check the cuvettes and absorbance scale. Since the sensitivity is also nearly twice as great at 340 nm as at 366 nm, it is preferable to use the former wavelength.

Stability of Enzymes

Most enzymes in biological fluids are stable at refrigerator temperature for 24 hr and at room temperature for a shorter time. For storage for longer periods, serum or plasma should usually be frozen. The stability of many common clinical en-zymes at various storage temperatures is shown in Table 14-1.

Standardization of Enzymes

The standardization of enzyme determinations may present some problems, par-ticularly in automated methods. In a strictly manual method one can theoretically add a definite amount of serum to a buffered substrate solution at the required temperature, incubate for exactly the specified period of time, then stop the reaction by the addition of another chemical, and measure the amount of S′ formed by a chemical reaction. This is the fixed time or single point method. The determination is made by measuring the amount of S′ formed at a given fixed time after the addition of the serum (point A, for example, in Fig. 14-1). In some instances, as in the determination of alkaline phosphatase by the Bodansky method or of amylase by a saccharogenic method, the added serum will initially contain a small amount of the S′ of the reaction (inorganic phosphate or glucose) and a correction must be made for this. If a chemical reaction must be used to determine S′, this is the only method that can be used since the addition of the reagents (e.g., to determine phosphate) stop the enzymic reaction so that it cannot proceed further.

If S′ is itself colored or absorbs in the ultraviolet, or if some other substance involved in the enzymic reaction has this property, the amount of S′ present can

Table 14-1. Stability of Some Common Clinical Enzymes at Various Storage Temperatures (no more than 10% change during specified time)

Enzyme	Room Temp., 25°C	Refrigerator, 0–4°C	Frozen, −25°C
Aldolase	2 days	2 days	Unstable[a]
HBDH	Unstable	3 days	Unstable[a]
Amylase	1 month	6 months	2 months
Arginase	4 hr	3 days	5 months
Ceruloplasmin	1 day[b]	2 weeks	2 weeks
Cholinesterase	1 week	1 week	1 week
CPK (nonactivated)	2 hr	6 hr	Unstable[a]
CPK (activated)	2 days	1 week	1 month
Glutamate dehydrogenase	1 day	2 days	1 day
GOT	3 days	1 week	1 month
GPT	3 days	1 week	Unstable
γ-GT	2 days	1 week	1 month
ICDH	5 hr	3 days	3 weeks
Lactate dehydrogenase	1 week	2 days[c]	2 days[c]
Leucine aminopeptidase	1 week	1 week	1 week
Lipase	1 week	3 weeks	3 weeks
Malate dehydrogenase	4 hr	8 hr	1 day
Phosphatase, acid	4 hr	3 days[d]	3 days[d]
Phosphatase, alkaline	2 days[e]	2 days[e]	1 month
Sorbitol dehydrogenase	4 hr	1 day	3 days

[a] Change caused by thawing.
[b] Variable.
[c] Isoenzyme pattern will change, particularly on freezing and thawing.
[d] With added acetic acid.
[e] Change may be due to change in pH on standing. HBDH, α-hydroxybutyrate dehydrogenase; CPK, creatine phosphokinase; GOT, glutamate-oxalacetate transaminase; GPT, glutamate-pyruvate transaminase; γ-GT, γ-glutamyl transpeptidase; ICDH, isocitrate dehydrogenase.

often be measured spectrometrically without interfering with the reaction. In this way a number of points on the curve of Figure 14-1 can be determined on one sample. One can measure the absorbance at 1-min intervals and thus determine that one is measuring the reaction rate on the linear portion of the curve. This is the kinetic or time rate method of measurement. It is used particularly in the enzyme-coupled ultraviolet methods to be discussed later. This method is particularly applicable to these ultraviolet reactions since there is often a time lag or delay before the reaction begins to proceed at the proper rate. If the time lag is a few minutes, then by measuring the absorbance at 1-min intervals between 5 and 10 min after the addition of the serum, one can obtain the true reaction rate, whereas measurements at zero time and 10 min would not give the proper result.

In the continuous-flow automated methods, the reaction time, although supposedly fixed, may vary somewhat from time to time and must be measured in some way. Thus by noting the time required for a colored solution pumped through the tubing to travel from the point where the serum enters the substrate to the point where a chemical is added to stop the enzymic reaction, one can obtain the reaction time. Usually, however, not all this flow path will be at a uniform temperature and this will introduce errors. Sometimes a dialysis step is used in which S′ dialyzes through the membrane to react with some chemical to give the color. This eliminates the interference the serum proteins may cause with the color reagent, but adds some further uncertainty as to the exact reaction time. In the discrete sample analyzers, some of these difficulties are eliminated in that the reaction time is more definite, but there may still be some temperature variation.

Often the automated methods are standardized against pooled or lyophilized serum which in turn has been carefully analyzed by a good manual method. Some difficulties arise here. The standardization should extend to fairly high levels in the abnormal range. It is difficult to obtain pooled serum in this range. One may either "spike" the serum with concentrated enzyme solution or powder obtained commercially or use one of the commercially lyophilized serums. The difficulty here is that these serums may be "spiked" with enzymes that are not of human origin and may not always react with the substrate in exactly the same manner as does the human enzyme. This will often introduce an error of unknown magnitude into the standardization. There is no simple solution to these difficulties, but the laboratory worker should be aware of them. The problem is particularly acute in connection with the automated continuous-flow instruments such as the Autoanalyzer SMA 12/60, as the units are not strictly comparable to the international units. Thus one cannot analyze the calibrated serum for, say, LDH and GOT by a good manual ultraviolet spectrophotometric method and use this value to standardize the instrument. One should, however, analyze the control serums at frequent intervals by a good manual method to note if the enzyme content has remained constant.

Throughout this section, reference to coenzymes nicotinamide adenine dinucleotide (NAD) and nicotinamide adenine dinucleotide phosphate (NADP) will be made. These compounds are also known as diphosphopyridine nucleotide (DPN) and triphosphopyridine nucleotide (TPN), respectively, and the two names for each compound may be used interchangeably.

The abbreviations given after the Enzyme Commission (E.C.) names are those suggested by Baron et al. [2] and are based on the E.C. names. Abbreviations given just after the common names are those in use in this country at present.

Isoenzymes

In recent years many investigators have demonstrated that enzymes with the same catalytic function but originating from different organs can be separated into several different fractions that show the same specificity. These multiple enzyme fractions have been called isoenzymes or isozymes.

Enzymes are normally found in the blood stream in small amounts as a result of the normal processes of cellular destruction. During pathological processes involving cell destruction the concentration of enzymes in blood can increase significantly. The origin of this increase may be one or more of several organs or tissues. Although several tissues may be rich in the same enzyme activity, the isoenzyme pattern of each tissue appears to be unique. By means of isoenzyme separation the identification of the specific tissue or organ injured and responsible for the enzyme increase becomes possible. This has become very useful in differential diagnosis.

Isoenzymes can exhibit differences in electrophoretic mobility, stability to heat denaturation, resistance to chemical inhibiting agents, and activity toward different substrates and coenzyme analogs. All these properties have been utilized for separation and identification of isoenzymes. Since the isoenzymes are protein in nature, they may be separated from each other by electrophoresis similar to that for serum proteins. This may be carried out on cellulose acetate or agar, agarose, starch, or polyacrylamide gel. After electrophoresis the supporting medium is incubated with the appropriate substrate and other chemicals such as are used in the colorimetric determination of the enzyme. At the positions where the isoenzyme is present, the color will develop and stain the medium. Thus the positions of the various isoenzymes will be visible as colored bands. These may be evaluated visually or scanned as in serum protein electrophoresis.

The isoenzymes may also differ in their susceptibility to heat inactivation. Heating to a high temperature for a long period of time will inactivate all enzymes. Under intermediate conditions one isoenzyme may be inactivated much more rapidly than another. Thus after carefully controlled heating, the amount of enzyme activity remaining will be a measure of the amount of heat-stable enzyme present. Similarly the isoenzymes may vary in their inactivation or inhibition by certain chemicals. Under standard conditions the amount of inactivation will be a measure of the amount of the more readily inactivated enzyme. The inactivation is never complete or absent. One isoenzyme may be

inactivated to about 10 or 15 percent and another by 85 to 90 percent; this difference is usually sufficient to distinguish between them.

It has been estimated that some 100 enzymes, including most of those discussed
in this chapter, exist as isoenzymes. These will be discussed further under the
different enzymes.

Oxidoreductases

As the name implies the oxidoreductases catalyze the complex mechanisms of
biological oxidations and reductions. The oxygen-activating systems acting with
hydrogen-activating systems must have hydrogen acceptors, or oxidation-reduc-
tion intermediaries, in order to bring about cellular oxidations.

Illustrated below are some of the fundamental problems of biological oxidation
and reduction. In the first example, the enzyme lactate dehydrogenase (LDH)
acts upon lactic acid to activate two H atoms which are transferred to the oxidized
form of an acceptor, NAD.

$$
\begin{array}{ccc}
\underset{\text{lactic acid}}{\overset{\displaystyle \text{COOH}}{\underset{\displaystyle \text{CH}_3}{\mid \atop \text{HCOH} \atop \mid}}} + \text{NAD}^+ \underset{\text{LDH}}{\rightleftharpoons} & \underset{\text{pyruvic acid}}{\overset{\displaystyle \text{COOH}}{\underset{\displaystyle \text{CH}_3}{\mid \atop \text{C}=\text{O} \atop \mid}}} + \text{NADH} + \text{H}^+
\end{array}
$$

The second example demonstrates the reduction of acetaldehyde to ethanol by
NADH, catalyzed by the enzyme alcohol dehydrogenase (ADH).

$$
\underset{\text{acetaldehyde}}{\overset{\displaystyle \text{CHO}}{\underset{\displaystyle \text{CH}_3}{\mid \atop \mid}}} + \text{NADH} + \text{H}^+ \underset{\text{ADH}}{\rightleftharpoons} \underset{\text{ethanol}}{\overset{\displaystyle \text{CH}_2\text{OH}}{\underset{\displaystyle \text{CH}_3}{\mid \atop \mid}}} + \text{NAD}^+
$$

The oxidoreductases, which constitute class 1 in the E.C. system, are usually
designated as dehydrogenases except when molecular oxygen acts as an acceptor,
when they are known as oxidases. The dehydrogenases include lactate (LDH),
hydroxybutyrate (HBDH), isocitrate (ICDH), glucose 6-phosphate (G-6-PDH),
malate (MDH), sorbitol (SDH), 6-phosphogluconate (6-PGDH), glutamate
(GDH), and alcohol (ADH) dehydrogenases, and diaphorase. The oxidases include ceruloplasmin (CLP), glutathione reductase (GR), xanthine oxidase
(XOD), soluble monoamine oxidase (MAO), and catalase.

Lactate Dehydrogenase (LDH)

E.C.1.1.1.27   L-lactate:NAD$^+$ oxidoreductase [9–16]

This enzyme catalyzes reversibly the conversion of lactate to pyruvate. The reaction may be written:

$$CH_3—HCOH—COO^- \rightleftharpoons CH_3—CO—COO- + 2(H)$$
$$\text{lactate} \qquad\qquad \text{pyruvate}$$

For this reaction to proceed in a biological system there must be a hydrogen acceptor to combine with the hydrogen atoms removed from the lactate. One such acceptor is the coenzyme nicotinamide adenine dinucleotide (NAD) (also known as diphosphopyridine nucleotide, DPN). The abbreviated formula for this compound is given below. In all such abbreviated structural formulas each angle in a ring represents a carbon atom unless another atom is shown. If there are not four bonds (lines) attached to each carbon, the other valences of the carbon are assumed to be satisfied by hydrogen atoms, unless some other atom is given in the formula.

nicotinamide adenine dinucleotide (NAD)
(oxidized form)

The reaction may be written:

$$\text{lactate} + \text{NAD} \xrightleftharpoons{\text{LDH}} \text{pyruvate} + \text{NADH}$$

The reduced form, NADH, is sometimes written as $NADH_2$ to indicate that there are two hydrogen atoms involved, although the actual reaction is

$$NAD^+ + 2(H) \rightleftharpoons NADH + H^+$$

as indicated below.

NAD+ (oxidized)   NADH (reduced)

NADH has a marked absorption at 340 nm whereas NAD has very little absorption at this wavelength. Thus the reaction may be followed by measuring the change in absorbance at 340 nm corresponding to an increase or decrease in the amount of NADH present.

In some reactions the acceptor nicotinamide adenosine dinucleotide phosphate (NADP) is used instead of NAD. As indicated in the asterisked note in the structural formula for NAD, NADP contains an additional phosphate group. This does not change the absorbance at 340 nm so NADP may be used the same way as NAD. (NADP is also known as triphosphopyridine nucleotide, TPN.)

The lactate-pyruvate reaction will proceed in either direction depending upon which substance is used as a substrate. LDH-P designates a system in which pyruvate is used as substrate with the reaction proceeding to the left and the absorbance at 340 nm decreasing as the reaction proceeds. LDH-L indicates a system in which lactate is used as substrate with the absorbance at 340 nm increasing as the reaction goes on. Using this reaction and ultraviolet measurement, the results may be calculated in terms of international units using the known absorbance of NADH. A solution containing 1 μmol of NADH/ml will have a calculated absorbance of 6.22 at 340 nm when cuvettes with a 1 cm light path are used. For 1 μmol of NADH in a total volume of 3 ml (the usual volume for these reactions), the calculated absorbance is 2.07. Thus the change in absorbance divided by 2.07 (or multiplied by 1/2.07 = 0.48) will give the actual number of micromoles of NADH transformed during the reaction time in a 3 ml volume. Knowing the actual reaction time and the amount of serum used, one can readily calculate the micromoles of NADH transformed per minute per liter of serum, that is, the activity in international units.

The results are sometimes reported in terms of spectrophotometric units. One such unit is defined as the amount of enzyme activity that will cause a change in absorbance of 0.001/min (due to change in NADH concentration) with a total reaction volume of 3 ml and measurements made at 340 nm with 1 cm cuvettes. Then, by calculation: spectrophotometric units × 0.48 = international units.

LDH may also be determined by colorimetric methods using the lactate-pyruvate reaction. In the Sigma† method the pyruvate is made to react with dinitrophenylhydrazine. The reaction product gives a reddish brown color when made alkaline. In this method the system used is usually LDH-P and the decrease in pyruvate color reaction is measured. In another method the NADH formed is caused to react with a tetrazolium compound to give a colored product. The reduction of the tetrazolium salt by NADH requires the presence of an intermediate electron carrier. In the LDH method in Technicon's SMA systems‡ the enzyme diaphorase is used for this purpose. In other methods phenazine methosulfate (PMS) may be used. These methods have the disadvantage that solutions of NAD or NADH are very unstable. For this reason the use of preweighed vials containing the NAD compound is preferred. These are kept frozen and made up as needed. Also available commercially are reagent tablets containing all the products necessary for the reaction; these are dissolved as needed.

SPECTROPHOTOMETRIC METHOD

This is a very simple and rapid method which requires a spectrophotometer capable of reading in the ultraviolet range (340 nm) [17]. It involves the forward

† Sigma Chemical Co., St. Louis, Mo. 63118.
‡ Technicon Instruments, Tarrytown, N.Y. 10591.

reaction, lactate to pyruvate. The reaction is followed by measuring the increase in absorbance due to conversion of NAD to NADH.

1. Buffered substrate. Dissolve 6.2 g of sodium pyrophosphate and 2 ml of lactic acid in 150 ml of hot distilled water. Cool, add 1.1 g of NAD, and dissolve. Adjust the pH to 8.8 with $1M$ NaOH or HCl as necessary and dilute to 280 ml. This solution contains 50 mmol of buffer, 77.5 mmol of lactic acid, and 5.25 mmol of NAD/liter. Pipet 2.8-ml aliquots of this substrate into 13×75 mm test tubes, stopper tightly, and store at $-20°C$. These are stable for 6 months.

PROCEDURE

Place tubes containing frozen substrate in a constant temperature bath and allow contents to come to bath temperature. Add 0.2 ml of serum, mix gently, and transfer to a cuvette with a 10 mm light path. Read the absorbance of the test (A_1) at 340 nm against a blank containing 2.8 ml of buffered substrate and 0.2 ml of water. Immediately start a timer set for 5 min. At the end of the 5 min incubation, read absorbance of the test against the blank (A_2).

CALCULATION

In this method 1 unit of LDH activity causes an increase in absorbance of 0.001/min/ml of serum under specified conditions.

$$\frac{A_2 - A_1}{5} \times \frac{1}{0.2} = \text{increase in absorbance/min/ml of serum}$$

This figure multiplied by 1,000 gives LDH units/min/ml.

The cuvette compartment of the photometer must be thermostatted and its temperature carefully controlled. If the measurements are not always made at exactly the same temperature, the results must be corrected to some standard temperature for comparison. For a rough comparison of values at different temperatures, refer to the nomogram in Figure 14-6.

COLORIMETRIC PROCEDURE

This method utilizes lactate with added NAD as substrate [18–20]. The amount of lactate changed is measured in terms of the amount of NADH produced, which in turn is estimated by its action in reducing a tetrazolium salt to an intensely colored form. The presence of an intermediate electron carrier is a requisite for the reduction of the tetrazolium salt by NADH. The compound phenazine methosulfate (PMS) is used for this purpose.

1. Tris buffer.   Dissolve 12.1 g of tris-(hydroxymethyl)aminomethane in about 60 ml of water. Adjust pH to 8.2 with $3M$ HCl (27 ml of concentrated HCl diluted to 100 ml). Dilute to 100 ml. Store in refrigerator.

2. Lactate substrate, $0.1M$, pH 5.5.   Add 5 ml of L(+)-lactic acid (20% solution) to about 60 ml of water. Adjust pH to 5.5 with $1M$ NaOH and dilute to 120 ml. Add a few drops of chloroform as preservative and store in refrigerator.

3. Control reagent; oxalate-EDTA.   Dissolve 0.2 g of potassium oxalate and 0.2 g of ethylenediaminetetraacetic acid (EDTA) disodium salt in 100 ml of water.

4. INT-PMS reagent.   Dissolve 200 mg of 2-p-iodophenyl-3-nitrophenyl-5-phenyltetrazolium chloride (INT) and 50 ml of phenazine methosulfate (PMS) in 100 ml of water.

5. NAD color reagent.   Dissolve 100 ml of NAD in 20 ml of the INT-PMS reagent. This mixture is stable several weeks under refrigeration.

6. HCl, $0.1M$.   Dilute 9.0 ml of concentrated HCl to 1 liter with water.

PROCEDURE

To each of two test tubes labeled sample and control add 0.2 ml of buffer and 0.1 ml of serum and mix. Add 0.5 ml of substrate to the sample tube and 0.5 ml of control reagent to the control tube. Mix thoroughly and warm to 37°C. With careful and exact timing add 0.2 ml of NAD color reagent, mix immediately, and return to 37°C bath. Incubate for exactly 5 min and add 5 ml of $0.1M$ HCl to stop the enzymatic reaction. Mix thoroughly and read absorbance of sample against control at 520 nm within 20 min. The intensity of the color is linear with concentration.

For standardization a carefully and accurately assayed serum with elevated LDH concentration can be diluted with a serum with low LDH concentration and a calibration curve made by treating serum mixtures exactly as described for unknown samples, or a mixture of normal and abnormal control serums commercially available may be used for this purpose.

Unknowns are read from the standardization curve or calculated using the standard whose absorbance is closest to the sample.

NORMAL VALUES AND INTERPRETATION OF RESULTS

The normal range for LDH as estimated by the colorimetric method is 90 to 200 mU/ml.

Marked increases in LDH activity have been found after myocardial infarction. This increase begins 48 to 72 hr after onset of pain and remains elevated for about 12 days. The increase of the enzyme level somewhat parallels the extent of cardiac damage. Increased LDH activity has also been reported in liver disease, especially toxic jaundice, and infectious mononucleosis. Significant increases have been found in leukemia, pulmonary infarction, sickle cell anemia, pernicious anemia, malignant lymphoma, and trauma of striated muscle.

Because LDH activity is increased in pulmonary infarction while the glutamate-oxalacetate transaminase (GOT) activity remains normal, the determination of these two enzymes may be used for the differential diagnosis of this condition.

LDH ISOENZYMES

With the use of electrophoretic techniques human LDH has been found to consist of five isoenzymes, all of identical molecular weight and all made up of four subunits (peptide chains) [8, 16, 21–27]. There are two types of subunits, designated H and M, and the five isoenzymes are the possible tetrametric combinations of these types. During electrophoresis the isoenzymes are numbered LDH-1 to LDH-5 in decreasing order of mobility toward the anode. The fastest-moving components, LDH-1 and LDH-2, are mostly found in heart tissue; the slowest-moving, LDH-4 and LDH-5, are predominantly found in liver tissue; and LDH-3 is found in other tissues. It has been found that tissue extracts from different organs or types of tissue have different isoenzyme patterns, different proportions of the five isoenzymes. Thus a disease causing tissue necrosis in a given organ should result in a large amount of the isoenzymes in the proportion existing in that organ being liberated into the blood. This would change the pattern normally found in serum and theoretically indicate the origin of the excess enzyme. In many cases the differences are not great enough for absolute diagnosis, but are usually suggestive. The instance in which the most clear-cut distinction is found is between heart and liver disease. Cardiac muscle contains an LDH isoenzyme with a large proportion of LDH-1, whereas the isoenzyme from liver contains a large proportion of LDH-5. Thus an increase in the amount of LDH-1 is indicative of cardiac disease and a large increase in LDH-5 would suggest liver disease. Both these conditions may cause increases in total LDH but they may be distinguished by the change in LDH isoenzyme pattern.

The electrophoretic separation of LDH isoenzymes can be carried out on agar, starch, or polyacrylamide gel, or cellulose acetate strips as supporting medium.

The sample is applied to the medium and electrophoresis allowed to proceed for the required time. After electrophoresis, the isoenzymes are located on the strip by a special staining process. The strip is overlayed with a solution containing lactate, DPN, phenazine methosulfate, and a tetrazolium compound such as nitroblue tetrazolium (NBT). The reaction is the same as described in the colorimetric procedure for LDH.

In the presence of phenazine methosulfate, NADH reduces the NBT to an intensely colored compound. The strip will be stained blue, the amount of stain depending upon the concentration of LDH isoenzyme at that point. Five bands corresponding to the five isoenzymes (LDH-1 to LDH-5) are usually obtained. The relative proportion of the different isoenzymes can be estimated either visually or quantitatively by using a densitometer. The method of choice will depend upon the equipment available. If the apparatus for cellulose acetate protein electrophoresis is available, it can readily be adapted to LDH isoenzyme electrophoresis.

LDH isoenzymes can be differentiated by means of other techniques such as heat inactivation, solvent precipitation, variations in substrate concentration, substrate affinity, coenzyme affinity, and selective chemical inhibition.

Using the different thermostability properties of LDH isoenzymes, it is possible to obtain a rough evaluation of LDH isoenzymes by incubating the serum at 65°C for 30 min. LDH-4 and LDH-5 are heat labile and are inactivated at this temperature whereas LDH-1 is relatively heat stable and most of its activity remains after incubation at this temperature. The total LDH concentration of the serum minus the heat-labile fraction will give the heat-stable isoenzyme. The normal values for the heat-labile fraction range from 10 to 25 percent of total. In patients with liver disease this value increases to 33 to 80 percent of total. For the heat-stable fraction the range of 20 to 40 percent of total is considered normal. Patients with myocardial infarction can have heat-stable values in the range of 45 to 80 percent total.

Babson [25] has recommended a simple and very useful procedure for the differentiation of LDH isoenzymes. In this method LDH assays are performed in $2M$ lactate substrate and in $0.02M$ lactate-$2M$ urea. The ratio of the two absorbances is used as an index to determine the predominant isoenzyme. This may be expressed as:

$$\frac{\text{absorbance of } 2M \text{ lactate}}{\text{absorbance of } 0.02M \text{ lactate-}2M \text{ urea}}$$

The normal index range suggested for clinical purposes is 0.8 to 1.3. Ratios of less than 0.8 indicate increases in LDH-1; ratios greater than 1.3, increases in LDH-5. Expressing LDH isoenzyme activity in this fashion completely eliminates the use of enzyme units and is independent of the instrument used.

Urea inhibits LDH-5 activity at lactate concentrations below $0.1M$; it has little effect on the heart LDH-1.

A sixth LDH isoenzyme designated as LDH-X has been reported. This is found in human postpubertal testes and seminal fluid and has an intermediate mobility between LDH-3 and LDH-4. It is supposedly also a tetramer but composed of four identical C subunits, which are different from the H and M subunits found in the other LDH isoenzymes.

Seminal plasma from healthy young male adults has been shown to contain a large concentration of LDH enzyme activity and a fairly constant isoenzyme percentage distribution, with some correlation between LDH_X activity and sperm concentration. Eliasson [27] found a concentration of LDH_X ranging between 20 and 45 percent of total LDH activity in human seminal plasma. LDH_X activity in seminal plasma has been suggested as a possible index for male fertility, but except for the correlation with sperm concentration, there doesn't seem to be a significant correlation with other tests for the evaluation of seminal fluid qualities. In order to preserve the LDH properties in seminal fluid, it should be stored at $-20°C$, as there is a marked decrease of activity at $4°C$.

LDH IN URINE [28, 29]

Urine contains substances that inhibit LDH activity and will interfere with its assay. In order to determine urinary LDH activity, dialysis of the urine must be performed. This also eliminates factors that might cause high values.

An accurately timed, clean voided 8-hr overnight urine specimen is collected without preservative. The analysis must be performed within 6 hr after collection of the specimen to prevent bacterial growth which can cause false elevations or false low values due to enzyme inactivation.

The LDH procedures described for serum may be used for urine simply by substituting clear dialyzed urine.

One unit of urinary LDH activity is defined as an increase in absorbance of 0.001/ml urine/min. Then:

total urinary LDH activity/8 hr = units/ml × volume of 8 hr urine (in ml)

NORMAL VALUES AND INTERPRETATION OF RESULTS

Normal urinary LDH activity by this method is 550 to 2,050 units/8-hr specimen. Wacker and Dorfman [28] have reported significant elevated values in cases involving carcinoma of the kidney or bladder. Considerably elevated values have been reported in several other diseases involving the urinary system including malignant hypertension, glomerulonephritis, lupus nephritis, acute tubular necrosis, and posssibly pyelonephritis. In cases of carcinoma of the kidneys or bladder, increases of as much as 5,000 percent have been reported.

Spurious elevations may result from diagnostic instrumentation of the urinary tract (cystoscopy with retrograde pyelography), hemolyzed blood in urine, menstrual contamination, and bacterial growth. Low results may be due to incomplete urine collection, the use of unclean glassware, or loss of activity due to the specimen's standing too long before assay.

LDH IN CEREBROSPINAL FLUID [30, 31]

Cerebrospinal fluid is collected in a sterile chemically clean test tube without anticoagulant. Enzyme activity is stable at room temperature for 4 hr or at refrigerator temperature (4°C) for 2 weeks.

LDH can be determined by either of the methods described, taking into consideration that the enzyme concentration is approximately one-fifth that of normal serum.

The normal values are 10 to 40 units. The concentration of LDH in spinal fluid is increased in head injury, degenerative disease of the central nervous system, and convulsive disorders. In patients with brain or meningeal tumors, the level is usually normal.

α-Hydroxybutyrate Dehydrogenase (HBDH)

This enzyme is considered to be the same as the LDH_1 isoenzyme of lactate dehydrogenase and has no separate E.C. number [32–36].

HBDH catalyzes the reversible reduction of α-ketobutyrate to α-hydroxybutyrate in the presence of reduced nicotinamide adenine dinucleotide (NADH). The reaction may be followed by the decrease in absorbance at 340 nm or by the addition of dinitrophenylhydrazine to give a brown color which is determined colorimetrically. Thus the decrease in the amount of color in the test sample in comparison with a blank will give the amount of substrate used, and hence the enzyme activity. HBDH is found in greatest concentration in the heart muscle, kidney, brain, and erythrocytes.

α-ketobutyrate + NADH ⇌ α-hydroxybutyrate + NAD

In the colorimetric procedure, serum is incubated with α-ketobutyrate substrate and reduced NADH. The reaction is stopped by the addition of 2,4-dinitrophenylhydrazine, which reacts with the unchanged α-ketobutyrate to form 2,4-dinitrophenylhydrazone. In the presence of excess sodium hydroxide this compound gives a brown color, which is determined colorimetrically, the intensity of the color being inversely proportional to the enzyme activity.

REAGENTS

1. Sörensen phosphate buffer, $0.067M$, pH 7.4. Dissolve 7.61 g of anhydrous disodium phosphate (Na_2HPO_4) and 1.78 g of dibasic potassium phosphate (KH_2PO_4) in water and dilute to 1 liter. Check pH and adjust to 7.4 if necessary.

2. α-Ketobutyrate stock solution, $0.1M$. Dissolve 1.02 g of α-ketobutyrate in phosphate buffer, adjust pH to 7.4 if necessary, and dilute to 100 ml. This is stable for 1 month at $-18°C$.

3. α-Ketobutyrate substrate. Dilute stock solution so that approximate absorbance differences shown in Table 14-2 are obtained when the calibration curve is prepared. Usually a dilution of 1:100 suffices. The substrate should be stored in 20 ml glass bottles at $-18°C$. This is stable for 1 month.

4. NADH solution. Dissolve 10 mg/ml in phosphate buffer. Prepare on the day the test is run.

5. 2,4-Dinitrophenylhydrazine, 40 mg/dl. Dissolve 400 mg of reagent grade 2,4-dinitrophenylhydrazine in 85 ml of concentrated hydrochloric acid and dilute to 1 liter with water. This reagent is stable indefinitely.

6. Sodium hydroxide, $0.4M$. Dissolve 16 g of reagent grade sodium hydroxide pellets in water and dilute to 1 liter.

Note: All glassware must be chemically clean. Washing with chromic acid, followed by thorough rinsing in distilled water, is recommended. Clear serum, free from hemolysis, should be used since red cells contain more than 100 times the serum enzyme activity.

CALIBRATION CURVE

Set up five test tubes and add reagents as shown in Table 14-2. To all tubes add 1 ml of dinitrophenylhydrazine reagent, mix thoroughly, and let stand for 20 min. Add 10 ml of $0.4M$ NaOH and mix well by inversion. Allow to stand at

Table 14-2. Calibration Curve for Colorimetric Determination of HBDH

Tube No.	Substrate (ml)	Phosphate Buffer (ml)	HBDH Activity (IU/liter of serum)	Net Decrease in Absorbance Compared with Tube No. 1
1	1.0	0.2	0	0
2	0.8	0.4	53	0.090
3	0.6	0.6	114	0.180
4	0.4	0.8	183	0.270
5	0.2	1.0	310	0.395

room temperature no less than 10 min and no more than 30 min. Read absorbance in a photometer at 490 nm against water. Plot absorbance against corresponding units of HBDH, starting with the highest absorbance reading in tube no. 1.

PROCEDURE

A test and a blank are required for each serum tested. For each series of tests a substrate blank and substrate NADH blank should be set up in duplicate. Set up a series of 10 × 175 mm test tubes, label, and add reagents as described in Table 14-3.

Place all tubes in a water bath at 37°C for about 5 min, add 0.1 ml of NADH to the serum test, and incubate all tubes for exactly 60 min at 37°C. Stop the reaction by adding 1 ml of dinitrophenylhydrazine reagent to each tube and mix thoroughly by gentle shaking. Remove tubes from the water bath and allow to stand at room temperature for 20 min. Add to all tubes 10 ml of 0.4M sodium hydroxide and mix by inversion. Allow to stand at room temperature no less than 10 min or more than 30 min and read absorbance at 490 nm against water.

CALCULATION

Subtract serum test absorbance from absorbance of serum blank. Subtract absorbance of substrate NADH blank and add this difference to the absorbance obtained for the serum. This value is converted into international units of HBDH per liter of serum by the calibration curve. If the activity of serum exceeds 310 units/liter, the test should be repeated using 1 volume of serum diluted with 4 volumes of phosphate buffer and multiplying the results from Table 14-1 by 5.

NORMAL VALUES AND INTERPRETATION OF RESULTS

The normal range of HBDH by this method is 56 to 125 IU/liter of serum. No significant changes occur with age after 10 years, but slightly higher values have

Table 14-3. Colorimetric Determination of HBDH

Reagent	Serum Test (ml)	Serum Blank (ml)	Substrate Blank (ml)	Substrate NADH Blank (ml)
α-Ketobutyrate substrate	1.0	1.0	1.0	1.0
Serum	0.1	0.1	—	—
Phosphate buffer	—	0.1	0.2	0.1
NADH	—	—	—	0.1

been obtained in younger children. Serum HBDH is increased in myocardial infarction, megaloblastic anemia, and Duchenne dystrophy. Following myocardial infarction, HBDH activity increases about 12 hr after infarction and reaches a peak 48 to 72 hr later. It remains elevated for 13 days.

Normal HBDH levels are found in coronary insufficiency, congestive heart failure, pulmonary infarction, pericarditis, chronic liver disease, obstructive jaundice, and rheumatic fever.

Isocitrate Dehydrogenase (ICDH)

E.C.1.1.1.42 $threo$-D_s-isocitrate:$NADP^+$ oxidoreductase (decarboxylating) [37–40]

This enzyme catalyzes the conversion of isocitrate to oxalosuccinate which is immediately decarboxylated to α-ketoglutarate and CO_2.

isocitrate $+$ $NADP^+$ $\rightleftharpoons$ $NADPH$ $+$ H^+ $+$ ketoglutarate $+$ CO_2

The enzyme is activated by the presence of manganous ions. The amount of α-ketoglutarate formed is proportional to the amount of enzyme activity present. The reaction may be followed by the increase in absorbance at 340 nm, or the amount of α-ketoglutarate formed may be reacted with 2,4-dinitrophenylhydrazine to form a highly colored hydrazone that can be measured colorimetrically. It is essential to include a serum blank with each determination as the amount of color contributed by the serum is significant and variable due partly to its bilirubin content. Since this enzyme is very sensitive to metals, all reagents should be made up with deionized water.

ICDH is widely distributed in the body and is mainly present in liver, heart, and skeletal muscle. Nonhemolyzed serum must be used since red cells contain high concentrations of the enzyme. It occurs in two major isoenzyme forms, a fast electrophoretic fraction found in the liver and a slow fraction found in heart muscle. The liver fraction is heat stable at 56°C whereas the heart fraction is heat labile and inactivated at this temperature.

REAGENTS

1. Sodium chloride, 0.15M. Dissolve 8.55 g of sodium chloride in distilled water and dilute to 1 liter.

2. Manganese chloride, 0.1M in 0.15M NaCl. Dissolve 200 mg of manganese chloride ($MnCl_2 \cdot 4H_2O$) in 100 ml of 0.15M NaCl.

3. Tris buffer, $0.1M$, pH 7.5.   Dissolve 12.1 g of tris-(hydroxymethyl)amino-methane in 850 ml of water in a 1 liter volumetric flask. Adjust pH to 7.5 with $1M$ HCl (about 80 ml will be required) and dilute to volume.

4. d-l-Isocitrate, $0.1M$ in $0.15M$ NaCl.   Dissolve 260 mg of trisodium iso-citrate§ in 10 ml of $0.15M$ NaCl.

5. NADP§, $0.004M$ in $0.15M$ NaCl.   Dissolve 30 mg of NADP in 10 ml of $0.15M$ NaCl (0.85% NaCl). Store frozen.

6. Ethylenediaminetetraacetic acid (EDTA), 5 g/dl.   Dissolve 5 g of EDTA (disodium) in 100 ml of water.

7. 2,4-Dinitrophenylhydrazine (DNPH), $0.001M$ in $1M$ HCl.   Dissolve 19.8 mg of DNPH in 100 ml of $1M$ HCl. Stable for 1 month at room temperature.

8. α-Ketoglutaric acid (KGA), $0.001M$.   Dissolve 150 mg of α-ketoglutaric acid (97.5% pure) in 1 liter of water.

9. Sodium hydroxide, $0.4M$.   Dissolve 16 g of sodium hydroxide pellets in 1 liter of water.

10. Hydrochloric acid, $1M$.   Add 83 ml of concentrated HCl (AR) to a 1 liter volumetric flask containing about 500 ml of water. Mix, allow to cool to room temperature, and dilute to volume.

PROCEDURE

Pipet 0.5 ml of serum into each of two test tubes labeled sample and blank. To both tubes add 0.4 ml of $MnCl_2$, 0.5 ml of tris buffer, and 0.2 ml of isocitrate. Place tubes in water bath at 37°C for a few minutes. Add 0.4 ml of NADP solution to the sample tube and 0.4 ml of NaCl solution to the blank tube. Incubate both tubes at 37°C for 1 hr. Then add 1 ml of DNPH solution to each tube, followed by 0.2 ml of EDTA solution, and let stand at room temperature for a minimum of 25 min (up to 1 hr). Add 10 ml of NaOH solution and mix well. Allow to stand 5 to 10 min. Read in a photometer at a wavelength of 410 nm, setting the instrument to 100% T (or zero absorbance) with the blank.

CALIBRATION CURVE

A standard curve for the hydrazone of KGA is prepared as described in the procedure, except that no isocitrate or NADP is required and incubation is omitted. Any serum may be used to set up the calibration curve.

Prepare a series of test tubes as described in Table 14-4. To each tube add 1 ml of DNPH solution and mix. Add immediately 0.2 ml of EDTA solution to all tubes, mix, and allow to stand at room temperature for a minimum of 25 min. Add 10 ml of NaOH to all tubes and mix well. Allow to stand 5 to 10 min. Read at

§ Sigma Chemical Co., St. Louis, Mo. 63118.

Table 14-4. Calibration Curve for Colorimetric Determination of ICDH

Tube No.	Serum (ml)	MnCl$_2$ (ml)	Tris Buffer (ml)	0.001M KGA (ml)	NaCl (ml)	KGA (mμmol)
Blank	0.5	0.4	0.5	0	0.6	0
1	0.5	0.4	0.5	0.1	0.5	100
2	0.5	0.4	0.5	0.2	0.4	200
3	0.5	0.4	0.5	0.3	0.3	300
4	0.5	0.4	0.5	0.4	0.2	400
5	0.5	0.4	0.5	0.5	0.1	500
6	0.5	0.4	0.5	0.6	0	600

410 nm, setting to 100% T (or zero absorbance) with the blank. Plot % T or absorbance against corresponding millimicromoles of KGA per milliliter of serum. When using a Beckman DU spectrophotometer at 410 nm, the curve is linear up to 400 mμmol KGA.

Units of activity are defined as the nanomols of KGA formed by the enzyme in 1 ml of serum at 37°C in 1 hr.

CALCULATION

ICDH activity in mμmol/ml serum = mμmol KGA read from curve

$$\times \frac{1}{\text{ml serum used}} \times \frac{60}{\text{incubation time (min)}}$$

For values exceeding 600 mμmol, either a smaller aliquot of serum may be used or samples can be incubated for a shorter time. This must be taken into consideration in the final calculation.

NORMAL VALUES AND INTERPRETATION OF RESULTS

The normal values by this method are 238 to 686 KGA units/ml, with an average of 420 ± 138. Significant elevated values have been reported in viral hepatitis and in about 50 percent of patients with metastatic carcinoma involving the liver. Dawkins et al. [41] suggested that in the absence of liver disease, an increase in serum ICDH may indicate active placental degeneration during pregnancy. The determination of ICDH has been suggested for the differential diagnosis of intrahepatic from extrahepatic obstructive jaundice. ICDH is also increased in patients with megaloblastic anemia. In myocardial infarction and muscular dystrophy, ICDH is usually normal.

Glucose 6-Phosphate Dehydrogenase (G-6-PDH)

E.C.1.1.1.49   D-glucose 6-phosphate:NADP$^+$ oxidoreductase [42–45]

This enzyme catalyzes the oxidation of glucose 6-phosphate to 6-phosphoglucono-lactone which immediately hyrolyzes to 6-phosphogluconate.

glucose 6-phosphate $+$ NADP$^+$ $\rightleftharpoons$ 6-phosphogluconate $+$ NADPH $+$ H$^+$

Another enzyme present in the red cells catalyzes the further oxidation of 6-phosphogluconate to ribulose-5-phosphate and CO_2:

6-phosphogluconate $+$ NADP$^+$ $\rightleftharpoons$ ribulose-5-phosphate $+$ CO_2 $+$ NADPH $+$ H$^+$

In the most commonly used assays the two reactions are not distinguished and the overall reaction is measured. The reaction may be followed as usual by the change in absorbance at 340 nm, as the NADP is reduced to NADPH. Since many of the reagents are unstable, it is recommended that if only an occasional assay is run, the reagents be purchased in lyophilized kit form such as obtainable from Calbiochem.‖

There are also some colorimetric methods which can be used as screening tests for this enzyme. The additional reactions are:

NADPH $+$ PMS $\rightleftharpoons$ NAD $+$ reduced PMS

reduced PMS $+$ dye (colored) $\rightleftharpoons$ PMS $+$ reduced dye (colorless)

Phenazine methosulfate (PMS) acts as an intermediate electron carrier between NADPH and the dye. A number of dyes can be used such as methylene blue or 2-6-dichlorophenolindophenol. Both are blue in color and are reduced to a colorless form by the reaction. The reaction must be carried out anaerobically to prevent the reoxidation of the dye by oxygen from the air. This is usually accomplished by layering mineral oil over the solution in the test tube. In the screening test, the time taken to cause complete disappearance of the blue color of the dye is measured. Blood from subjects with low levels of the enzyme requires a much longer time for decolorization than blood from normals. The reaction may also be used in a quantitative form by measuring the decrease in color in a photometer.

The procedure described in this section requires a photometer capable of

‖ Calbiochem, La Jolla, Calif. 92037.

reading at 340 nm with cuvettes with a 10 mm light path. It is preferable, though not absolutely essential, that a recording spectrophotometer with a thermostatted cuvette chamber be used. Martinek et al. [43] have suggested how round cuvettes can be adapted to this use, provided the band width of the spectrophotometer is no more than 20 nm.

REAGENTS

1. Tris buffer, $1.0M$, pH 8.0. Dissolve 12.1 g of tris-(hydroxymethyl)amino-methane in about 30 ml of water and 40 ml of $1M$ HCl. Check the pH and add additional $1M$ HCl to bring the pH to 8.0, then dilute to 100 ml.

2. Magnesium chloride, $0.1M$. Dissolve 2.06 g of $MgCl_2 \cdot 6H_2O$ in water to make 100 ml.

3. NADP, $2mM$. The sodium salt is preferred. As purchased, the material may contain varying amounts of water and other impurities. The label usually states the composition. Usually the sodium salt will contain 4 molecules of water of crystallization and will have a molecular weight of 837. If it is stated to be 96% pure, the equivalent molecular weight would be $837/0.96 = 872$. A $2mM$ solution would then contain 1.74 mg/ml. To make 10 ml of a $2mM$ solution, weigh out 17.4 mg and dissolve in 10 ml of water. The NADP can be purchased in pre-weighed vials containing 5, 10, or 25 mg.#

4. Glucose 6-phosphate, $6mM$. The disodium salt trihydrate has a molecular weight of 358. This would require 21.5 mg for 10 ml of the solution. If the salt is not pure, the amount required can be calculated as above.

PREPARATION OF HEMOLYSATE

The assay is carried out on a hemolysate of the cells. About 3 ml of heparinized blood is centrifuged strongly and the plasma and buffy coat removed as completely as possible. The cells are then washed several times with 5 volumes of ice-cold $0.15M$ saline. After the last wash, 0.2 ml of packed cells is added to 3.8 ml of ice-cold water. The solution is mixed well and allowed to stand in the refrigerator for 10 min and then centrifuged at 5,000 g for 10 min in a refrigerated centrifuge. The clear supernatant is poured off and stored in ice water until used. The hemoglobin content of the lysate is determined by adding 0.1 ml of the solution to 5 ml of the regular hemoglobin reagent and reading in the usual manner. Since this is a 1:51 dilution instead of the usual 1:251 dilution, the hemoglobin value as given by the calibration curve is multiplied by 0.203.

Sigma Chemical Co., St. Louis, Mo. 63118.

PROCEDURE

To each of two cuvettes add, with mixing after each addition, 0.25 ml of tris buffer, 0.25 ml of magnesium chloride solution, 0.25 ml of the NADP solution, and 0.05 ml of the hemolysate. To one cuvette, labeled blank, add 1.70 ml of water; to the other, labeled test, add 1.45 ml of water. Incubate both tubes at 37°C for 10 min. Add 0.25 ml of the glucose 6-phosphate solution to the test and read test against blank at 340 nm. If a recording spectrophotometer is used, record the absorbance change for about 5 min until a linear curve is obtained. With a regular photometer, take one reading immediately after the addition of the glucose 6-phosphate and one exactly 5 min later, incubating at 37°C between readings.

CALCULATION

From the readings, determine the absorbance change per minute. Then, if A is the absorbance change per minute,

$$\frac{A \times 2.5}{6.22 \times 0.05} = A \times 8.03 = \text{IU/ml}$$

where 2.5 is the total volume (in ml) in the cuvette and 0.05 is the amount (in ml) of hemolysate added. The results are usually expressed as units per gram of hemoglobin, so the above is further multipled by 100/Hb, where Hb is the hemoglobin content of the hemolysate (in g/dl).

NORMAL VALUES AND INTERPRETATION OF RESULTS

The normal values by this method are 8.6 to 17.4 IU/g of hemoglobin in hemolysate.

Decreased G-6-PDH is the most commonly found deficiency of the red cells. Individuals having low concentrations may experience little difficulty under favorable conditions, but will have sensitivity to drug-induced hemolytic anemia. After the ingestion of fava beans or a number of drugs such as primaquine or acetylphenylhydrazine, an excessive hemolysis of the erythrocytes occurs. This condition is known as favism if due to the fava bean or as primaquine defect if due to drugs. Viral hepatitis and diabetic acidosis have also been shown to induce hemolytic anemia in subjects deficient in G-6-PDH [46]. Severe deficiency of this enzyme is extremely rare.

Malate Dehydrogenase (MDH)

E.C.1.1.1.37   L-malate:NAD$^+$ oxidoreductase [47–50]

The enzyme catalyzes the reaction:

oxalacetate + NADH + H$^+$ $\rightleftharpoons$ malate + NAD$^+$

The reaction is followed by the decrease in absorbance at 340 nm that results when NADH is oxidized to NAD$^+$. Since the oxalacetate is not very stable, in some methods it is generated in situ by the addition of glutamate-oxalacetate transaminase (GOT), α-ketoglutarate, and aspartate (see reaction for determination of GOT later in this chapter).

The reaction may be run in the reverse direction and the oxalacetate formed determined with dinitrophenylhydrazine or a diazotized dye intermediate as in the methods for GOT.

The enzyme is present in the liver, heart, skeletal muscle, brain, and red blood cells.

In general the changes in malate dehydrogenase with disease are not specific. The level in serum may be normal or slightly elevated in cirrhosis and obstructive jaundice. It is elevated in acute hepatitis, renal injury, and Duchenne muscular dystrophy. After myocardial infarction MDH increases rapidly and elevated levels are found within 4 to 6 hr. The enzyme level returns to normal within 3 to 4 days following infarction.

Sorbitol Dehydrogenase (SDH)

E.C.1.1.1.14   L-iditol:NAD$^+$ 5-oxidoreductase [51–55]

This enzyme catalyzes the reaction:

fructose + NADH + H$^+$ $\rightleftharpoons$ sorbitol + NAD$^+$

In the reaction the keto group of fructose is reduced to form the polyhydric alcohol sorbitol. The enzyme will also oxidize some other isomeric alcohols such as iditol. Like all the dehydrogenases, SDH is determined at 340 nm by following the absorbance change in the conversion of NADH to NAD$^+$ or vice versa. This enzyme can also be determined by running the reaction in the reverse direction and measuring the fructose by the color formed with resorcinol.

SDH is found mainly in the liver, with smaller amounts also present in prostate gland, kidney, heart, and skeletal muscle. This enzyme is fairly specific for liver cell damage because normally it is not present in serum. Unlike other liver enzymes, SDH is not elevated in the serum in organ disease other than liver cell damage. SDH is markedly elevated in acute viral hepatitis or toxic necrosis of the liver. It is usually normal or very slightly elevated in nonhepatic diseases.

6-Phosphogluconate Dehydrogenase (6-PGDH)

E.C.1.1.1.44   6-phospho-D-gluconate:NADP$^+$ 2-oxidoreductase   (decarboxylating) [56–60]

This enzyme catalyzes the reaction:

6-phosphogluconate + NADP $\rightleftharpoons$ ribulose-5-phosphate + CO_2 + NADPH

This is one of the steps in the metabolism of glucose in the direct oxidative (HMP shunt) pathway. The reaction also requires the presence of Mg^{2+} or other divalent ion as activator. It is usually determined by measurement of the change in ultraviolet absorbance at 340 nm. It may be determined along with glucose 6-phosphate dehydrogenase in erythrocyte hemolysates in the study of red cell enzyme deficiencies.

The enzyme is normally present in serum only in very small amounts. The level is said to be elevated in liver disease but its determination has not been used to any extent for the diagnosis of hepatic disorders. The level of the enzyme in vaginal fluid has been studied as a possible indicator of uterine cancer. Some investigators have stated that a large proportion of women with cancer of the cervix will have increased enzyme levels in the vaginal fluid even in the early stages of the disease. This determination has been advocated as an aid in the diagnosis or early detection of such cancer. Other workers feel that the determination of the enzyme is of little if any diagnostic value and certainly should not replace the cytological methods as a screening test. Part of the difficulty may be in the proper collection of the specimens and obtaining a representative sample. Vaginal fluid will vary considerably in solid content and the amount of enzyme in a given volume may not be a good indication of the total amount present. The ratio of the enzyme activity to the potassium content of the fluid has been used to compensate for the varying dilution of the fluid. The use of cervical mucus has also been suggested as more suitable. The determination of the enzyme for this purpose has not gained wide acceptance as it was found that any condition causing vaginal or uterine irritation would also increase the amount of enzyme present.

Glutamate Dehydrogenase (GlDH)

E.C.1.4.1.3   L-glutamate:NAD(P)$^+$ oxidoreductase (deaminating) [61–65]

This enzyme catalyzes the reaction:

oxoglutarate + NADH + NH_3 + H^+ $\rightleftharpoons$ glutamate + NAD$^+$ + H_2O

The enzyme activity is measured by the decrease in absorbance at 340 nm due to the oxidation of NADH to NAD^+. A colorimetric method can also be used; this involves the addition of phenazine methosulfate and INT similar to that mentioned under isocitrate dehydrogenase.

The enzyme is found primarily in the liver, heart, and kidneys. Normally its activity in serum is very low. Moderate elevations are found in acute viral hepatitis and toxic necrosis of the liver. The serum level is usually normal in nonhepatic diseases. In liver disease the relative increase in serum levels is usually less than that of GOT. The ratio of glutamate dehydrogenase to GOT is usually higher in hepatitis and other diseases involving extensive diffuse parenchymal damage and lower in conditions involving localized but severe damage to individual cells, such as necrosis.

Alcohol Dehydrogenase (ADH)

E.C.1.1.1.1   alcohol:NAD^+ oxidoreductase [66–69]

This enzyme catalyzes the reaction:

$$\text{alcohol} + NAD^+ \rightleftharpoons \text{aldehyde} + NADH + H^+$$

The enzyme is not specific for any particular primary or secondary alcohol, but in most biological systems the alcohol of interest is ethyl alcohol. The enzyme has been used to determine alcohol in biological fluids. In this procedure, semicarbazide is usually added to combine with the acetaldehyde so that the reaction can go to completion. The same reaction may be used to determine the enzyme using ethanol as substrate. In carrying out the assay it has been found that more consistent results are obtained if the substrate, buffer, and material to be assayed are first mixed and then the reaction initiated by the addition of the NAD, although starting the reaction by the addition of the sample has proved fairly satisfactory. The results are usually calculated in terms of micromoles of NAD transformed (or alcohol oxidized) per minute per milliliter of sample.

The enzyme is present in the body in greatest amount in the liver parenchymal cells. Only small quantities are found in other organs such as gastric and intestinal mucosa and the kidneys. The determination of alcohol dehydrogenase in serum has thus been suggested as a test for the differential diagnosis of liver disease. The serum of normal subjects does not contain appreciable amounts of the enzyme. The serum level is increased in most patients with acute parenchymal liver disease. Similar to ornithine carbamyl transferase and sorbitol dehydrogenase, ADH is elevated almost exclusively in acute liver damage in comparison with damage to other organs. In patients with extrahepatic biliary obstruction, the enzyme level in the blood usually is not elevated. In some cases of liver disease the level of alcohol dehydrogenase in the serum has remained elevated for some time after the levels of aspartate amino transferase and alkaline phosphatase have returned to normal.

Diaphorase

E.C.1.6.4.3   NADH:lipoamide oxidoreductase, lipoamide dehydrogenase [70–72]

This enzyme catalyzes the reconversion of methemoglobin to hemoglobin in the presence of NADH. The reaction is:

$$MHb + H^+ + NADH \rightarrow Hb + NAD^+$$

The reaction can be followed using a methemoglobin ferrocyanide complex as a substrate, by measuring spectrophotometrically at 575 nm the reduction of the substrate complex.

In the congenital deficiency of this enzyme the reconversion rate is slower than in normal individuals and methemoglobin accumulates in the erythrocytes. In cases of methemoglobinemia the concentration of this enzyme may be measured as an aid in diagnosis. The parents of affected individuals have about half the concentration of the enzyme in the erythrocytes as normal individuals.

The enzyme may be determined by converting the hemoglobin in a hemolyzed sample with sodium nitrate to methemoglobin. The hemolysate, which contains both the enzyme and the methemoglobin, is incubated with NADH and the dye 2,6-dichlorobenzenoneindophenol (DCBIP). The oxidation of the dye is followed spectrophotometrically at 600 nm as a blue color is formed. The rate of formation of the color is a measure of the enzyme concentration.

Oxidases—General Comments

Oxidases are enzymes that catalyze the oxidation of a substrate by molecular oxygen (O_2). Oxidases are often metalloproteins containing iron, copper (e.g., ceruloplasmin), or molybdenum. Others, such as xanthine oxidase, are flavoproteins.

The reaction of the oxidases may be expressed as:

$$2S\text{-}H_2 + \tfrac{1}{2} O_2 \rightarrow 2S + H_2O$$

In the reaction only oxygen can serve as an acceptor for the hydrogen, and water is the product.

Ceruloplasmin (*CLP*)

E.C.1.16.3.1   Ferroxidase [73–78]

Ceruloplasmin is a blue copper protein found in serum associated with α_2-globulin. Although variations in its concentration are found in various pathological conditions, its exact metabolic role is unknown. In vitro it catalyzes the oxidation of aromatic diamines. The reaction is:

p-phenylenediamine $+ 2 O_2 \rightarrow$ Bandrowski base $+ 4 H_2O$

CLP is measured by its catalysis of the oxidation of p-phenylenediamine to form purple-blue compounds. The amount of color formed as measured in a spectrophotometer is taken as a measure of the activity of the enzyme. In an older method the activity was reported in arbitrary units by comparison with a standard solution of the dye Pontacyl Violet 6R. It was later found that the compound known as Bandrowski base is the blue substance formed by the oxidation of the p-phenylenediamine substrate. Since this substance has been prepared in pure form, the ceruloplasmin assay may be reported in international units, that is, in terms of micromoles of Bandrowski base formed per minute per liter of serum. A blank is usually run containing the substrate and serum together with sodium azide to inhibit the enzyme. This compensates for the nonenzymatic oxidation of the substrate during incubation. Bandrowski base is used to standardize the enzyme activity.

REAGENTS

1. Acetate buffer, $1.2M$, pH 5.2 ± 0.05.   Dissolve 163 g of crystalline sodium acetate trihydrate and 20 ml of glacial acetic acid in water and dilute to 1 liter. Check pH and adjust if necessary. Store in the refrigerator.

2. Sodium azide, 20 mg/dl.   Dissolve 10 mg of sodium azide in water and dilute to 50 ml. Prepare fresh daily as required.

3. Buffered p-phenylenediamine (PPD), 0.1 g/dl.   Dissolve 10 mg of recrystallized p-phenylenediamine dihydrochloride* in 10 ml of acetate buffer. Prepare fresh daily.

PPD is recrystallized by dissolving in a minimum amount of hot distilled water, decolorizing with charcoal, filtering while hot, and recrystallizing from the clear filtrate. The white crystals of PPD are dried and stored in a vacuum over calcium chloride.

* No. 207, Eastman Kodak Co., Rochester, N.Y. 14650.

PROCEDURE

Serum or heparinized plasma may be used. This should be separated from the cells as soon as possible and frozen if the test is not done the same day.

Place 1 ml of freshly prepared buffered PPD in each of three test tubes (15 × 125 mm) labeled blank, test 1, and test 2 (duplicate analysis), and incubate in water bath at 37°C for about 5 min. Add 0.1 ml of serum to each tube; then immediately add 5 ml of sodium azide solution to the blank tube. Mix by inversion and incubate in water bath at 37°C for 15 min. The level of the reaction mixture in the test tubes should be below the water level in the bath.

After incubation, add 5 ml of sodium azide solution to test 1 and 2 (not to the blank) to stop the reaction. Measure absorbance of test 1 and 2 against the blank at 540 nm. Dilutions should be made for specimens having very high activity.

CALIBRATION USING BANDROWSKI BASE

Bandrowski base† is a pure crystalline compound formed from the oxidation of p-phenylenediamine (PPD) in aqueous ammoniacal solution by hydrogen peroxide. This compound has been shown to have an absorption curve identical to the pigment formed by the oxidase activity of ceruloplasmin on p-phenylenediamine. Measurements of absorbance at 540 nm adhere strictly to Beer's law up to concentrations of 50 μg/ml standard solution.

Weigh accurately 5 mg of Bandrowski base, transfer to a 100 ml volumetric flask, add about 75 ml of acetate buffer, and agitate in a mechanical shaker for 2 hr. Dilute to 100 ml with acetate buffer. This solution contains 50 μg/ml base. It should be used for standardization within 1 hr after preparation. Prepare six test tubes (15 × 125 mm) and label as shown in Table 14-5.

Table 14-5. Calibration Curve for Colorimetric Determination of CLP

Tube No.	Sodium Azide (ml)	Serum (ml)	Bandrowski Base Solution (ml)	Acetate Buffer (ml)	Concentration (μg)	IU
Blank	5	0.1	0	1.0	0	0
1	5	0.1	0.1	0.9	5	10.5
2	5	0.1	0.2	0.8	10	21.0
3	5	0.1	0.4	0.6	20	42.0
4	5	0.1	0.6	0.4	30	63.0
5	5	0.1	1.0	0	50	105.0

† Nutritional Biochemicals Corp., Cleveland, Ohio. 44128.

Use any clear unhemolyzed serum. Mix all tubes and measure absorbance of standard tubes at 540 nm against blank.

One international unit is defined as the formation of 1 μmol of Bandrowski base/min/liter of serum. Since in our procedure 0.1 ml of serum is used and the incubation period is 15 min, the formation of 1 μg of base under these conditions would correspond to:

$$\frac{1}{318.4} \times \frac{1,000}{0.1} \times \frac{1}{15} = 2.1 \text{ IU}$$

where 318.4 is the molecular weight converting micrograms to micromoles, 1,000 converts to liters of serum, and $\frac{1}{15}$ converts to the amount formed per minute. Accordingly, tubes 1 to 5 correspond to 10.5, 21, 42, 63, and 105 IU. A calibration curve can be obtained by plotting the absorbance against the IU.

NORMAL VALUES AND INTERPRETATION OF RESULTS

The normal range of ceruloplasmin is 35 to 65 IU or 15 to 35 mg/dl. Concentrations below the lower limit have been reported only in patients with Wilson's disease, in the neonatal period, in the nephrotic syndrome, and occasionally in unaffected relatives of patients with Wilson's disease. There may be a deficiency of ceruloplasmin in patients with kwashiorkor or tropical sprue, and in certain infants with a syndrome of anemia and hypoproteinemia.

Increased concentrations of ceruloplasmin have been noted in pregnancy, subacute or chronic infection, myocardial infarction, hepatic cirrhosis, hyperthyroidism, aplastic or refractory anemia, Hodgkin's disease, acute leukemia, and patients receiving estrogen therapy, including oral contraceptives. The ceruloplasmin activity generally parallels the copper content of the serum.

Glutathione Reductase (GR)

E.C.1.6.4.2   NAD(P)H:oxidized glutathione oxidoreductase [79–83]

This enzyme catalyzes the reaction:

GSSG + NADPH + H$^+$ → 2 GSH + NADP$^+$

where GSSG represents the oxidized (disulfide) form of glutathione and GSH the reduced form. The reaction proceeds to completion and is irreversible.

The enzyme maintains glutathione in the reduced state in the presence of NADPH. It may be determined by incubating a dialyzed hemolysate under aerobic conditions with the oxidized form of glutathione and NADPH. The reduced glutathione formed may be determined colorimetrically by reaction with sodium nitroprusside or by the change from NADPH to NADP, which is followed by measuring the changes in absorption at 340 nm.

This enzyme is not directly concerned with the erythrocyte glucose metabolism but is essential for the functional integrity of the cell membrane and the conversion of methemoglobin to hemoglobin.

Xanthine Oxidase (XOD)

E.C.1.2.3.2   xanthine:oxygen oxidoreductase [84–88]

This enzyme catalyzes the oxidation of the purine xanthine to uric acid with the formation of hydrogen peroxide. The reaction is:

$$\text{xanthine} + H_2O + O_2 \rightarrow \text{uric acid} + H_2O_2$$

It has been determined fluorometrically by methods in which the hydrogen peroxide oxidizes 2-amino-4-hydroxypteridine or homovanillic acid to a fluorescent compound. Theoretically it could be determined by measuring the increase in uric acid absorption at 284 nm but the uric acid present in the serum would interfere. A method using this principle involves first separating the enzyme present in serum from the uric acid by Sephadex filtration. In experimental animals it has been found that xanthine oxidase is a much more sensitive indicator of liver damage due to exposure to carbon tetrachloride than is alanine aminotransferase. It is also said to be a good indicator of hepatic damage in the toxemia of pregnancy. However, it has not been routinely used as an indicator of hepatic cellular damage, possibly because no simple method of determination is available.

Monoamine Oxidase (MAO)

E.C.1.4.3.4   amine:oxygen oxidoreductase (deaminating) [89–93]

Monoamine oxidase catalyzes the oxidation of a number of amines by the following reaction:

$$\text{amine} + H_2O + O_2 \rightleftharpoons \text{aldehyde} + NH_3 + H_2O_2$$

The enzyme is present in a variety of body tissues. It is concerned with the oxidation of metepinephrine and normetepinephrine to vanillylmandelic acid (VMA) and with the oxidation of serotonin to 5-hydroxyindoleacetic acid (HIAA). MAO inhibitors have been used in the treatment of depression and hypertension as they apparently increase the amount of serotonin in the central nervous system. The concentration of the enzyme in serum has not been studied extensively.

A simple procedure for its estimation involves the use of p-dimethylaminobenzylamine as substrate. This is oxidized in the presence of the enzyme to form p-dimethylaminobenzaldehyde. The reaction is followed by measuring the absorption of the latter compound at 355 nm. A number of fluorometric methods have been devised in which either the oxidized amine is fluorescent or an added material is oxidized to a fluorescent compound by the H_2O_2 in the presence of peroxidase. A number of methods using ^{14}C have also been devised but these are used chiefly for the determination of the enzyme in tissue extracts.

The serum level of MAO is said to be increased in patients with chronic congestive heart failure, thyrotoxicosis, and diabetes, but the determination of the serum level has not been used to any extent in the diagnosis of these conditions. It was stated that the increase in the serum MAO level often occurred before other symptoms of diabetes. Monoamine oxidase in serum was also found to be elevated in chronic liver disease. The increase was not related to hepatocellular necrosis but was most closely associated with hepatic fibrosis.

Catalase

E.C.1.11.1.6   hydrogen-peroxide:hydrogen-peroxide oxidoreductase [94–97]

This enzyme catalyze the reactions:

$$2H_2O_2 \rightarrow 2H_2O + O_2 \tag{1}$$

$$ROOH + AH_2 \rightarrow H_2O + ROH + A \tag{2}$$

The role of this enzyme in vivo is not yet clear—whether it is to decompose hydrogen peroxide (equation 1) or to catalyze a peroxidation reaction (equation 2). The enzyme, with H_2O_2, is capable of oxidizing in vitro several substrates (AH_2), which include methanol, ethanol, formic acid, thiols, and phenols.

Catalase is usually determined by measuring the rate of decomposition of hydrogen peroxide after addition of material containing the enzyme. The excess of hydrogen peroxide after a definite period of incubation is measured either by titration with permanganate, by a colorimetric reaction with dichromate, or by determining volumetrically the amount of oxygen liberated.

The clinical interest in this enzyme is due to the rare genetically controlled condition of acatalasia. This condition was first discovered in Japanese individuals and most cases have been found among them. Some cases have been reported in the United States, Scandinavia, and Switzerland. Many of the Japanese subjects and most of the non-Japanese subjects are asymptomatic and the presence of acatalasia was discovered only by testing for the presence of the enzyme.

Transferases

This group of enzymes transfers a group of atoms or a radical from one molecule to another. They constitute class 2 in the E.C. system. They are classified on the basis of the type of grouping transferred. One of the most common ones, for example, is aspartate aminotransferase (GOT), the reaction being:

aspartate + oxoglutarate $\rightleftharpoons$ oxalacetate + glutamate

In the reaction, an amine group ($-NH_2$) is transferred from the aspartate to the oxoglutarate molecule forming oxalacetate and glutamate. Another type of transferase transfers a phosphate grouping from one molecule to another. These are generally known as kinases; an example is hexokinase, which catalyzes the reaction:

glucose + ATP $\rightleftharpoons$ glucose 6-phosphate + ADP

In this reaction a phosphate grouping is transferred from ATP to the glucose molecule. The kinases participate in the phosphorylation of a large number of compounds.

Glutamate-Oxalacetate Transaminase (GOT)

E.C.2.6.1.1   L-aspartate: 2-oxoglutarate aminotransferase: aspartate aminotransferase (AST) [98–101]

This enzyme catalyzes the transfer of the amino group from glutamate to the ketoacid, oxalacetate. It is present in greatest concentration in cardiac muscle, liver, skeletal muscle, and kidney. In blood it is normally present in relatively low concentrations. The reaction is:

glutamate + oxalacetate $\rightleftharpoons$ aspartate + α-ketoglutarate (oxoglutarate)

In some methods the reaction is run in the opposite direction and the amount of oxalacetate formed is determined. In the ultraviolet method, the reaction mixture contains, in addition to aspartate and ketoglutarate, an excess of the enzyme malate dehydrogenase and NADH. Any oxalacetate formed immediately reacts in accordance with the equation:

$$\text{oxalacetate} + \text{NADH} + \text{H}^+ \rightleftharpoons \text{malate} + \text{NAD}^+$$

This reaction is catalyzed by the enzyme malate dehydrogenase. The course of the reaction is followed by the decrease in absorbance at 340 nm produced by the oxidation of NADH. In other determinations the oxalacetate formed is determined colorimetrically. The Reitman-Frankel method uses the reagent dinitrophenylhydrazine which reacts with the oxalacetate to form a compound that gives a reddish brown color with the reagent; the initial concentration of the substrate cannot be very high or the initial color will be too strong. In other methods the oxalacetate reacts with a diazotate dye intermediate to give a red color which is measured photometrically.

Glutamate-Pyruvate Transaminase (GPT)

E.C.2.6.1.2   L-alanine:2-oxoglutarate aminotransferase:alanine aminotransferase (ALT)

The reaction with this enzyme is similar to that given for GOT except that the aspartate is replaced by alanine and the product is pyruvate instead of oxalacetate. The reaction is:

$$\text{pyruvate} + \text{glutamate} \rightleftharpoons \text{alanine} + \alpha\text{-ketoglutarate}$$

The reaction is usually run in the opposite direction and the amount of pyruvate formed is determined. In the ultraviolet method, LDH and NADH are present in the solution and the pyruvate reacts to form lactate as in the system LDH-P in accordance with the equation:

$$\text{pyruvate} + \text{NADH} + \text{H}^+ \xrightleftharpoons{\text{LDH}} \text{lactate} + \text{NAD}^+$$

The reaction is followed by measuring the decrease in absorbance at 340 nm due to the oxidation of NADH.

In the Reitman-Frankel method the reagent used is again dinitrophenylhydrazine with the formation of the reddish brown color with pyruvate.

Determination of GOT and GPT

GOT AND GPT BY COLORIMETRIC METHOD

The blood serum is added to a buffered solution of α-ketoglutarate and aspartate or alanine, and the resulting oxalacetate or pyruvate formed after incubation is measured colorimetrically by reaction with dinitrophenylhydrazine.

REAGENTS

1. Phosphate buffer.   Mix 420 ml of 0.1M disodium phosphate (26.81 g of $Na_2HPO_4 \cdot 7H_2O$/liter) and 80 ml of 0.1M potassium dihydrogen phosphate (13.61 g of KH_2PO_4/liter). The pH should be 7.4.

2. Pyruvate solution, 2mM.   Dissolve 22 mg of sodium pyruvate in 100 ml of phosphate buffer. This solution is used for the preparation of the standard curve.

3. α-Ketoglutarate-aspartate substrate for GOT.   Place 29.2 mg of α-ketoglutaric acid and 2.66 g of aspartic acid in a small beaker. Add 1M sodium hydroxide until solution is complete. Adjust to a pH of 7.4 with sodium hydroxide, transfer quantitatively to a 100 ml volumetric flask with phosphate buffer, and dilute to the mark with the buffer.

4. α-Ketoglutarate-alanine substrate for GPT.   Place 29.2 mg of α-ketoglutaric acid and 1.78 g of alanine in a small beaker. Add 1M sodium hydroxide until solution is complete. Adjust to a pH of 7.4 with sodium hydroxide, transfer quantitatively to a 100 ml volumetric flask with phosphate buffer, and dilute to the mark with the buffer.

5. 2,4-Dinitrophenylhydrazine solution.   Dissolve 19.8 mg of 2,4-dinitrophenylhydrazine in 100 ml of 1M hydrochloric acid (84 concentrated acid diluted to 1 liter with distilled water).

6. Sodium hydroxide solution, 0.4M.   Dissolve 16 g of sodium hydroxide in distilled water to make 1 liter.

The concentrations of the substrate solutions and the dinitrophenylhydrazine solution are critical. Under usual circumstances it is advisable not to attempt to make up the reagents (except the sodium hydroxide solution) but to obtain commercial preparations.‡

STANDARD CURVE

Since the reaction does not yield a linear result, it is necessary to set up an empirical standard curve to calculate the results. Set up a number of tubes as shown in Table 14-6 and add the amounts of pyruvate solution, substrate, and water as

‡ Sigma Chemical Co., St. Louis, Mo. 63118.

Table 14-6. Standard Curve for Colorimetric Determination of GOT and GPT

Tube No.	Pyruvate[a] (ml)	Substrate[b] (ml)	H_2O (ml)	GOT (units)	GPT (units)
1	0	1.0	0.2	0	0
2	0.1	0.9	0.2	24	28
3	0.2	0.8	0.2	61	57
4	0.3	0.7	0.2	114	97
5	0.4	0.6	0.2	190	—

[a] Pyruvate solution for standard curve.
[b] Ketoglutarate-aspartate substrate for GOT.

indicated. Add 1 ml of dinitrophenylhydrazine solution to each tube. Mix and allow to stand for 20 min. Then add 10 ml of $0.4M$ sodium hydroxide solution to each tube. Mix and after 10 min read in a photometer at 505 nm (500 to 520 nm). Set to zero with the blank tube and record absorbance reading for each tube. Plot the results against the corresponding units for GOT and GPT. Connect the points by a smooth curve.

PROCEDURE

Pipet 1 ml of the desired substrate into a test tube and place in 37°C water bath for 5 min. Prepare one extra tube for blank. Pipet 0.2 ml of serum into tube, mix by swirling, and begin timing. Incubate exactly 60 min for SGOT (serum GOT) and 30 min for SGPT (serum GPT). At the end of this time add 1 ml of dinitrophenylhydrazine solution, mix, and remove from bath. Allow to stand 20 min at room temperature (a longer time is not detrimental). Then add 10 ml of $0.4M$ sodium hydroxide, mix by inversion, and allow to stand for 10 min. Read as for standards, subtracting reading of blank from that of sample and reading units of enzyme from the prepared standard curve. If activities are too high to read on the curve, repeat with a dilution of serum (1:5 or 1:10).

NORMAL VALUES AND INTERPRETATION OF RESULTS

The normal values are 8 to 30 units for SGOT and 5 to 25 units for SGPT. Elevations in transaminase activity occur in myocardial infarction, infectious mononucleosis, and infectious hepatitis. Other conditions that often show some elevation are cirrhosis with active hepatic necrosis, biliary obstruction, acute interstitial pancreatitis, metastatic carcinoma of the liver, extensive traumatic injuries, and prolonged shock. Although the SGOT level is always increased in acute myocardial infarctions, the SGPT level does not always increase proportionately.

GOT USING DIAZONIUM SALT

In this method the oxalacetate formed is determined by reaction with a stabilized diazonium salt to form a red dye that is measured photometrically [102–105].

REAGENTS§

1. Substrate buffer.   Dissolve 0.731 g of α-ketoglutaric acid, 2.66 g of L-aspartic acid, 33.5 g of K_2HPO_4, 1 g of KH_2PO_4, 10 g of polyvinylpyrrolidone, || 1 g of sodium salt of ethylenediaminetetraacetic acid, and 5 ml of $1M$ NaOH in water, and dilute to 1 liter. Check the pH and adjust to 7.4 if necessary. Store in the refrigerator.

2. Control buffer.   Dissolve 28.6 g of K_2HPO_4, 4.9 g of KH_2PO_4, 10 g of polyvinylpyrrolidone, 1 g of disodium ethylenediaminetetraacetic acid, and 5 ml of $1M$ NaOH in water to make 1 liter. Check pH and adjust to 7.4 if necessary. Store in the refrigerator.

3. Color reagent.   Dissolve 0.2 g of 6-benzamido-4-methoxy-m-toluidine diazonium chloride# in 100 ml of $0.01M$ HCl (0.85 ml of concentrated HCl diluted to 1 liter). Store in the refrigerator. This is stable for only a few weeks. Discard when markedly colored or turbid.

4. Acid diluent.   Dilute 1.7 ml of concentrated HCl and 5 ml of nonionic detergent (such as Non-ionox*) in water to make 1 liter.

PROCEDURE

Pipet 1-ml aliquots of substrate buffer to test tubes, one tube for each sample plus one tube for the reagent blank. Place tubes in a 37°C water bath to come to temperature. At timed intervals add 0.2 ml of serum to the respective tubes, except the blank, and replace in the water bath. After exactly 20 min incubation and while the tubes are still in water bath, add 1 ml of color reagent to each tube, including the blank, and incubate for another 10 min. Remove from bath and add 10 ml of acid diluent. Mix tubes thoroughly and read samples against reagent blank at 540 nm. Also set up a similar series of tubes, using the control buffer

---

§ Reagents may be obtained as Transac from General Diagnostics Division, Warner-Lambert Pharmaceutical Co., Morris Plains, N.J. 07950.
|| Polyvinylpyrrolidone, K-30, Oxford Laboratory, Foster City, Calif. 94404, or through local laboratory supply houses.
Obtainable as Azoene Fast Violet B from Alliance Color and Chemical Co., Newark, N. J. 07105, and other suppliers.
* Curtis Scientific Corp., St. Louis, Mo. 63043.

instead of the substrate buffer. Carry through the same procedure and read tubes against control reagent blank. These are the serum blanks. For each serum sample subtract absorbance reading obtained with control buffer from that obtained with substrate buffer before converting to enzyme units.

CALCULATION

The method is best standardized by using dilutions of a control serum (such as Versatol E) with a high enzyme assay. For example, if the control serum contained 350 Karmen units, dilutions are made as follows:

Control serum (ml)	0.1	0.2	0.3	0.4	0.5	0.7	1
Saline solution (ml)	0.9	0.8	0.7	0.6	0.5	0.3	0
Units	35	70	105	140	175	245	350

These mixtures are run exactly as in the previous procedure to obtain a calibration curve. A graph is constructed with each new lot of diazonium salt. Serums with enzyme activity greater than 350 units should be accurately diluted with saline solution (1:5) and the analysis repeated. Final results from the calibration chart are multiplied by 5.

Normal values and interpretation of results are the same as for the previous method. In the GOT methods that use the coupling of a dye with the oxalacetate formed, erroneously high results may be obtained when ketosis is present since the ketone bodies present in the serum will also react with the dye reagent.

Creatine Phosphokinase (CPK)

E.C.2.7.3.2   ATP:creatine N-phosphotransferase; creatine kinase (CK) [106–110]

This enzyme catalyzes the reaction:

creatine + ATP $\rightleftharpoons$ phosphocreatine + ADP

ATP = adenosine triphosphate; ADP = adenosine diphosphate; these compounds aid in the transfer of phosphate groups in a large number of biological reactions. In the determination of CPK, the solution also contains an organic sulfur compound to activate the reaction; glutathione, cystine, or mercaptoethanol has been used in different procedures.

If the reaction is run in the forward direction indicated above, two methods have been used to determine the enzyme level. In the colorimetric method the reaction is stopped after incubation by the addition of an inhibitor such as an organic mercury compound. The phosphocreatine formed is allowed to hydrolyze spontaneously to creatine and inorganic phosphate. The amount of the latter is then determined by the usual method for serum inorganic phosphate (Chapter 8), with correction made for the amount of inorganic phosphate originally present in the serum sample. The units by this method are expressed in terms of milligrams of P transferred per hour per milliliter of serum. Some laboratories have not found this method very satisfactory.

In the ultraviolet method the solution also contains some of the compound phosphoenolpyruvate (PEP) which in the presence of added enzyme, pyruvic kinase, reacts as follows:

$$ADP + PEP \rightleftharpoons ATP + \text{pyruvate}$$

If LDH and NADH are also present, the pyruvate will react to form lactate with a corresponding decrease in the amount of NADH. The reactions can be followed by the change in absorbance at 340 nm. In this case the change will be proportional to the amount of pyruvate formed, which in turn will be proportional to the amount of ADP formed in the first reaction and thus to the CPK activity.

The reaction may also be run in the reverse direction and the amount of creatine or ATP formed determined. The creatine may be measured colorimetrically by reaction with diacetyl and α-naphthol or orcinol. A manual method has been developed using naphthol, and an automated method using orcinol.

The ATP formed may be measured by using a series of coupled reactions. If the appropriate enzymes and substrates are present in the solution, the following reactions will take place:

$$\text{glucose} + ATP \rightleftharpoons ADP + \text{glucose 6-phosphate (catalyzed by hexokinase)}$$

$$\text{glucose 6-phosphate} + NAD^+ \rightleftharpoons \text{6-phosphogluconic acid} + NADH + H^+ \text{ (catalyzed by}$$
$$\text{glucose 6-phosphate}$$
$$\text{dehydrogenase)}$$

Again one measures the change in absorbance at 340 nm. All the ultraviolet methods may be expressed in international units. The colorimetric method for creatine may be calibrated in terms of these units.

We will present here a colorimetric procedure determining the liberated creatine with diacetyl and α-naphthol.

REAGENTS

1. Tris buffer, $0.1M$.   Dissolve 1.21 g of tris-(hydroxymethyl)aminomethane and 0.54 g of magnesium acetate tetrahydrate in about 50 ml of water. Adjust to

a pH of 7.2 by the addition of $1M$ acetic acid (6 ml of glacial acetic acid diluted to 100 ml); about 9 ml should be required. Dilute to 100 ml with water.

2. Adenosine diphosphate (ADP), $0.012M$. Dissolve 60 mg of the trisodium salt of ADP in 10 ml of water. This solution is stable for about 1 week in the refrigerator. The solid material is said to decompose slowly even when stored below $0°C$, so it is advisable to purchase it in relatively small quantities.

3. Cysteine hydrochloride, $0.048M$. Dissolve 85 mg of L(+)-cysteine hydrochloride monohydrate in 10 ml of water. Store in the refrigerator.

4. Stock substrate. Dissolve 30 mg of creatine phosphate sodium salt in 9 ml of water and 1 ml of buffer. Store in the refrigerator.

5. Working substrate. Just before use, mix together 2 ml of buffer, 2 ml of stock substrate, 2.3 ml of water, 0.8 ml of cysteine solution, and 0.4 ml of $0.1M$ NaOH.

6. Alkaline solution. Dissolve 80 g of anhydrous sodium carbonate and 30 g of NaOH in water to make 500 ml. Discard if it becomes turbid.

7. Diacetyl stock solution. Dissolve 1 ml of diacetyl in 100 ml of water.

8. Diacetyl working solution. Just before use, dilute 1 ml of the stock to 25 ml with water.

9. α-Naphthol. Just before use, dissolve 200 mg of α-naphthol in 20 ml of the alkaline solution.

10. Silver nitrate. Stock solution, $0.5M$: Dissolve 8.49 g of silver nitrate in water to make 100 ml. This is stable if kept in a glass-stoppered bottle in the dark. Working solution, $0.005M$: As required, dilute 1 ml of the stock to 100 ml with water.

11. Alkaline EDTA. Dissolve 1 g of the disodium salt of EDTA and 10 ml of $1M$ NaOH in water to make 1 liter.

12. Standard solution. Dissolve 29.8 mg of creatine monohydrate in 100 ml of the alkaline EDTA. This solution contains 2 μg creatine/ml. This standard should be prepared fresh weekly and kept in the refrigerator.

PROCEDURE

Set up a series of tubes as shown in Table 14-7. Add the first four substances as indicated. Mix and incubate at $37°C$ for 30 min. Remove from bath and add, with mixing after each addition, the remaining reagents listed in the table. Allow to stand in the dark at room temperature for 30 min and read in photometer at 520 nm against water.

Serum is added to sample and sample blank, as the presence of serum affects the color development. Any clear, nonhemolyzed, nonicteric serum can be used for the standard and standard blank. For routine work on clear serum without

Table 14-7. Colorimetric Determination of CPK

Reagent	Sample (ml)	Sample Blank (ml)	Standard (ml)	Standard Blank (ml)
Working substrate	0.40	0.40	0.40	0.40
Water	—	0.10	0.10	0.10
Serum	0.05	0.05	0.05	0.05
ADP solution	0.10	—	—	—
Standard solution	—	—	0.05	—
Silver nitrate, 0.005M	0.5	0.5	0.5	0.5
Alkaline EDTA	3.75	3.75	3.70	3.75
α-Naphthol solution	1.0	1.0	1.0	1.0
Diacetyl working solution	0.5	0.5	0.5	0.5

visible hemolysis and without elevated bilirubin, the sample blank can be omitted and the standard blank used as a blank for all the serums. If the serum is icteric, lipemic, or contains visible hemolysis, a separate serum blank must be used for each such serum.

Exact timing is necessary only for the serum samples, although the timing for the other tubes should be close to 30 min.

CALCULATION

$$\frac{\text{absorbance of sample} - \text{absorbance of sample blank}}{\text{absorbance of standard} - \text{absorbance of standard blank}} \times 66.7 = \text{IU CPK}$$

The factor 66.7 is derived as follows: The standard contains 2 μmol of creatine/ ml and is treated exactly as an equal volume of serum. Thus it corresponds to 3 μmol of creatine liberated by 1 ml of serum in 30 min. Multiplying 2 by 1,000 to convert to μmol/liter and dividing by 30 to convert to mol/liter/min gives international units.

$(2 \times 1,000)/30 = 66.7$

NORMAL VALUES AND INTERPRETATION OF RESULTS

The normal range by this method is 20 to 50 IU for males and 10 to 37 IU for females. Values below the lower limit have no clinical significance. Elevated CPK values have been reported after onset of myocardial infarction; elevations occur about 6 hr after the onset of symptoms and reach peak levels in about 36 hr. There is usually a rapid return to normal levels by the fourth day. Extremely elevated values have been reported in Duchenne's muscular dystrophy; somewhat lower values have been reported in muscular dystrophy affecting the limbs and girdle. Following strenuous exercise, elevations of up to three times the normal CPK values have been observed. These elevations disappear within 24 to 48 hr. Significant elevations of CPK have been reported in cases of hypothyroidism.

γ-Glutamyl Transpeptidase (GGTP)

E.C.2.3.2.2   (γ-glutamyl)-peptide:amino-acid γ-glutamyl transferase [111–114]

This enzyme catalyzes the transfer of the glutamyl group from one peptidase to another. The substrate used in the procedure may be glutamylnitroanilide or glutamylnaphthylamide with the addition of glycylglycine, as the peptidase to which the glutamyl group is transferred. The reaction is as follows:

glutamylnitroanilide + glycylglycine ⇌ nitroaniline + glutamylglycineglycine

The liberated nitroaniline gives a yellow color in dilute alkaline solution which is measured photometrically. The amount of color is proportional to the amount of substrate decomposed and thus to the enzyme level. Usually a reagent blank is run to compensate for the small amount of spontaneous hydrolysis of the substrate.

REAGENTS

1. Buffer, $0.05M$, pH 8.6   Dissolve 5.26 g of 2-amino-2-methyl-1,3-propanediol in about 900 ml of water. Adjust to a pH of 8.6 with $1M$ HCl (about 30 ml will be required) and dilute to 1 liter.

2. Substrate.   Dissolve 250 mg of L-γ-glutamyl-*p*-nitroanilide, 872 mg of glycylglycine, and 672 mg of magnesium chloride hexahydrate in 300 ml of the buffer with constant stirring at 50 to 60°C. Store at room temperature. (Storing in the refrigerator may cause some material to crystallize out.) The solution is not stable for more than a few weeks. It should be made up fresh whenever it has turned appreciably yellower than when fresh.

3. Sodium hydroxide, $0.0075M$.   Dilute 7.5 ml of $1M$ NaOH to 1 liter.

PROCEDURE

For each serum sample set up two tubes. To one, labeled test, add 1 ml of substrate; to the other, labeled serum blank, add 1 ml of water. Place tubes in water bath at 37°C and allow to come to bath temperature. Add 0.1 ml of serum to each tube, incubate at 37°C for 45 min, then add 5 ml of $0.0075M$ sodium hydroxide, and mix well. For each series of samples set up a reagent blank by adding 0.1 ml of water to 1 ml of substrate. Incubate for 45 min, then add 5 ml of $0.0075M$ NaOH. Read each serum test against the corresponding serum blank and read the reagent blank against water at 405 nm. Subtract the absorbance of the reagent blank from each serum reading. Then, if the readings are made in cuvettes with an exactly 1 cm light path,

corrected absorbance of sample $\times$ 137 = IU/liter

The factor of 137 is calculated as follows: The international unit is defined as the amount of enzyme which will cause the decomposition of 1 μmol of substrate/min under the specified conditions. Accordingly, 0.1 ml of serum containing 1 IU/liter will in 45 min result in the formation of $(0.1 \times 45 \times 1)/1,000 = 0.0045$ μmol of p-nitroaniline. This amount is contained in a final volume of 6.1 ml. The concentration of nitroaniline is thus $0.0045/6.1 = 0.000738$ μmol/ml. A solution containing exactly 1 μmol/ml of nitroaniline has a calculated absorbance of 9.9 when the measurement is made with a 1 cm light path cuvette. The above solution will have an absorbance of $9.9 \times 0.000738 = 0.00730$. This would be the final corrected absorbance obtained with the procedure for a serum containing 1 IU/liter. Thus corrected absorbance/0.0073 = corrected absorbance $\times$ 137 = IU/liter.

If other cuvettes are used in which the light path cannot be accurately determined (round cuvettes), the standardization may be effected as follows. Dissolve 77.7 mg of p-nitroaniline† in water to make 100 ml (this is nearly a saturated solution and it may be necessary to warm to aid solution, or add one drop of concentrated hydrochloric acid). This solution will contain 5.63 μmol/ml. Dilution of 1 ml of this to 25 ml with the buffer, will give a solution containing 0.225 μmol/ml. To separate tubes add 0.2, 0.4, 0.6, 0.8, and 1.0 ml of the diluted nitroaniline and buffer to make 1.0 ml. To each tube add 0.1 ml of water and 5 ml of $0.0075M$ NaOH. Mix and read at 405 nm against a water blank. The solutions contain 0.045, 0.090, 0.135, 0.180, and 0.225 μmol/ml corresponding to 10, 20, 30, 40, and 50 IU/liter. A calibration curve may be constructed from these results, and if the curve is linear an appropriate factor may be calculated.

NORMAL VALUES AND INTERPRETATION OF RESULTS

The normal values by this mehod are 4 to 23 IU/liter for males and 3.5 to 13 IU/liter for females. The enzyme level is usually normal in infectious mononucleosis

† Eastman p-179, Eastman Kodak Co., Rochester, N.Y. 14650.

and only slightly elevated in hepatitis and cirrhosis of the liver. It is markedly elevated in obstructive jaundice, intrahepatic cholestasis, and metastatic carcinoma of the liver. Occasionally it may be slightly elevated in myocardial infarction, but it is normal in muscular dystrophy and bone diseases.

Ornithine Carbamoyl Transferase (OCT)

E.C.2.1.3.3.   carbamoyl phosphate:L-ornithine carbamoyl transferase [115–118]

This enzyme catalyzes the reaction:

ornithine + carbamoylphosphate $\rightleftharpoons$ citrulline + phosphate

This reaction is one of the steps in the formation of urea from ammonia in the body. The reaction is usually followed by adding urease to destroy any urea present (which would also react with the color reagents) and determining the citrulline by reaction with diacetyl monoxime, phenazone, and acid, similar to the colorimetric reaction used for urea.

OCT is found almost exclusively in the liver cells, and the serum levels in normal individuals are very low. Serum levels are markedly elevated in patients with acute viral hepatitis and other forms of hepatic necrosis. Only small elevations occur in obstructive jaundice, cirrhosis, or metastatic carcinoma. Slight elevations may sometimes occur after myocardial infarction, especially if there is liver involvement. Serum levels are usually normal in muscular dystrophy and bone disease.

Congenital deficiency of OCT may lead to a clinical condition in children, in which high protein diets cause a significant increase in blood ammonia due to the inability to convert ammonia to urea. This condition may subsequently lead to hepatic coma.

Galactose l-Phosphate Uridylyl Transferase (GPUT)

E.C.2.7.7.10   UTP-glucose:α-D-galactose-l-phosphate uridylyl transferase:hexose-l-phosphate uridylyl transferase [119–122]

This enzyme catalyzes the reaction:

galactose 1-phosphate + uridine diphosphate glucose $\rightleftharpoons$ glucose 1-phosphate
+uridine diphosphate galactose

The reaction is important in the metabolism of galactose through which it is converted into glucose. A deficiency of this enzyme is the basis for an inborn error of metabolism with the inability to utilize galactose. Testing for this enzyme has been used in the study of this genetic trait, especially in the detection of the heterozygous carriers of the disorder. The level of the enzyme in these individuals is about half that of normal subjects but higher than in the individuals homozygous for this trait.

GPUT governs the metabolism in all body cells; it is usually determined in a red cell hemolysate. In one method of determination the disappearance of the UDP-glucose is measured by the reaction:

$$\text{UDP-glucose} + 2\,\text{NAD}^+ \rightleftharpoons \text{UDP glucuronic acid} + 2\text{H}^+ + 2\,\text{NADPH}$$

The difference between the UDP-glucose in a blank without added galactose 1-phosphate and the system with the added substrate is measured, the hemolyzed red cells being added to both tubes. The procedure is not as simple as many other red cell enzyme tests.

There is also a fluorometric method in which the above reactions are carried out and after the reaction is stopped by the addition of phosphate buffer, the fluorescence of the NADPH is measured. The difficulty with this method is that the fluorescence of the product is influenced by a number of variables and no convenient primary standard is available.

Hexokinase (HK)

E.C.2.7.1.1   ATP:D-hexose 6-phosphotransferase [123–125]

This enzyme is involved in the metabolism of glucose; it catalyzes an important first step as follows:

$$\text{ATP} + \text{glucose} \rightleftharpoons \text{glucose-6-phosphate} + \text{ADP}$$

Although this reaction occurs in almost all cells that metabolize glucose, it is determined only in red cells or, more specificly, red cell hemolysates. In the determination the above reaction is coupled with the further reaction:

$$\text{glucose-6-phosphate} + \text{NADP}^+ \rightleftharpoons \text{6-phosphogluconate} + \text{NADPH} + \text{H}^+$$

This reaction is catalyzed by added enzyme glucose 6-phosphate dehydrogenase. In the determination in erythrocyte hemolysates there is always an excess of the enzyme 6-phosphogluconate dehydrogenase so that the following reaction also takes place:

6-phosphogluconate + NADP$^+$ $\rightleftharpoons$ ribulose-5-phosphate + NADPH + H$^+$ +CO$_2$

In this reaction two molecules of NADP are reduced for every molecule of glucose utilized. The reaction is usually followed by the change in absorbance at 340 nm. In vitro hexokinase is not very active and good instrumentation is necessary to obtain acceptable results. Hexokinase is one of the enzymes whose activity decreases most rapidly during aging of the red cells. A number of hexokinase isoenzymes have been discovered in red cells, but their significance has not been established.

Pyruvate Kinase (PK)

E.C.2.7.1.40   ATP:pyruvate 2-O-phosphotransferase [126–129]

This enzyme catalyzes the phosphorylation of ADP to ATP by phosphoenolpyruvate (PEP):

PEP + ADP $\rightarrow$ pyruvate + ATP

The rate of the reaction is usually determined by coupling with the following reaction:

pyruvate + NADH + H$^+$ $\rightleftharpoons$ lactate + NAD$^+$

This reaction is catalyzed by added LDH and is followed by measuring the change in absorbance at 340 nm from the conversion of NADH to NAD$^+$. The enzyme is important in erythrocyte metabolism and is usually determined in red cell hemolysates. Leukocytes also contain relatively large amounts of the enzyme but this is relatively constant and does not change in those conditions resulting in low levels in the red cells. Thus in preparation of a hemolysate, care must be taken to exclude all leukocytes. Pyruvate kinase deficiency is the most likely single cause of nonspherocytic hemolytic anemia. In this disease the enzyme level may be less than 25 percent of the normal value. The deficiency is apparently genetically controlled, with heterozygotes having an enzyme level of from 40 to 70 percent of normal without clinical symptoms. The determination of PK in red cells is thus used in the differential diagnosis of the various types of hemolytic anemia. Three isoenzymes of human pyruvate kinase have been found in different tissues—one chiefly in liver and erythrocytes, the second in kidney, and the third in liver, kidney, muscle, brain, and leukocytes. Pyruvate kinase is an important enzyme in glucose metabolism and thus must occur in most cells. The significance of the isoenzymes for diagnosis has not yet been established.

Transketolase (TK)

E.C.2.2.1.1   sedoheptulose-7-phosphate:D-glyceraldehyde-3-phosphate   glyceraldehyde transferase [130–133]

Transketolase (TK) is an enzyme involved in the hexose monophosphate shunt in the metabolism of glucose. It requires the presence of the coenzyme thiamine pyrophosphate (TPP) together with magnesium ions as a cofactor. The enzyme catalyzes the transfer of a ketol group from one phosphoketopentose to an aldehyde acceptor. One such reaction is

xylulose-5-phosphate + ribose-5-phosphate ⇌ sedoheptulose-7-phosphate
+ glyceraldehyde-3-phosphate

The enzyme is not found in any appreciable amount in serum and is usually determined in erythrocytes. Since the enzyme requires the presence of the coenzyme thiamine pyrophosphate, in instances of thiamine deficiency there will be a decrease in the apparent activity of the enzyme. This is one of the main indications for the determination of this enzyme. Usually the determinations are made with and without the addition of excess TPP. A marked increase in activity in the tube containing the added TPP indicates that the original sample was deficient in thiamine. The analysis is made on hemolyzed (by freezing) whole blood and the results calculated in units per 100 ml of packed cells by use of an initial hematocrit determination. Since the amount of enzyme in the serum is very small, it is not necessary to separate or wash the cells, and whole blood is hemolyzed directly by freezing.

The determinations are made using a reaction similar to that given above and measuring the amount of sedoheptulose formed. After incubation, the proteins are precipitated with trichloracetic acid and sedoheptulose determined in the filtrate by first heating with sulfuric acid and then adding cysteine to develop the color. The results are compared with standards containing known amounts of sedoheptulose and the results calculated in micromoles per milliliter per minute. Determinations are made on samples with and without the addition of excess TPP and the percentage of increase in activity over the sample without added TPP is calculated. Normal individuals were found to have activities of 42 to 86 μmol/min/ml of packed cells. After the addition of excess TPP the increase in activity was no more than 17 percent. In cases of thiamine deficiency the levels may be as low as 15 to 20 μmol/min/ml and the increase in activity after added TPP may be over 50 percent.

Hydrolases

These enzymes hydrolyze the substrate, splitting it into two or more molecules with the addition of water. They constitute class 3 in the E.C. system. The hydrolases are classified on the basis of the type of bond on which they act. One type hydrolyzes carboxylic ester bonds; an example is lipase which hydrolyzes glycerol esters of fatty acids. Another type hydrolyzes phosphate ester bonds; examples are the phosphatases and 5′-nucleotidase. Another type of hydrolase acts on glucoside bonds; an example is amylase which hydrolyzes the glucoside bonds of starch. A fourth type acts on single peptide bonds; an example is leucine aminopeptidase which hydrolyzes the leucine molecule from a peptide. The hydrolases also include the general peptide hydrolases which hydrolyze a large number of peptide bonds breaking down proteins into smaller molecules as does pepsin. A final type of hydrolase that may be mentioned is those that remove amino groups replacing them by hydroxyl groups; an example is urease.

Phosphatases—General Comments

These enzymes catalyze the hydrolysis of monoesters of orthophosphoric acid. Their action is of considerable importance in the transport of sugar and phosphate in the intestine, kidney, placenta, and bones. Acid and alkaline phosphatase enzymes are relatively nonspecific in their substrate requirements. As indicated by their names, they have maximum activity at acid and alkaline pH. Acid phosphatase is most active at a pH of 4 to 6 and alkaline phosphatase at a pH of 8 to 10. Another phosphatase, 5′-nucleotidase, hydrolyzes the specific substrate nucleotide pentose-5′-nucleotidase.

Alkaline Phosphatase (ALP)

E.C.3.1.3.3 orthophosphoric monoester phosphohydrolase (alkaline optimum) [134–137]

This enzyme is found in practically all human tissues, with the highest concentration found in bile and osteoblasts. Its use in clinical diagnosis is mainly in connection with bone and liver diseases.

Several different methods are used for the determination of this enzyme. The method by Bodansky uses the substrate β-glycerophosphate at a pH of 8.6. The reaction is followed by measuring the phosphate split off from the substrate. A correction is applied for the amount of inorganic phosphate originally present in the serum sample. The unit of activity is defined as that producing 1 mg of P/hr from 1 ml of serum. A modification of this method (Shinowara, Jones, and Rein-hard) uses a pH of 9.3 but the units are the same.

The method of King-Armstrong and its modifications use a substrate of phenyl phosphate at a pH of 9.0. The phenol liberated may be determined by the use of the Folin reagent. Later modifications use the reaction of phenol with ferricyanide and aminoantipyrine to give a red color. The unit is defined as the activity pro-ducing 1 mg of phenol in 30 min/dl of serum.

The Bessey-Lowry-Brock method uses a substrate of p-nitrophenyl phosphate at a pH of 10.1. The liberated nitrophenol gives an intense yellow color when the solution is made strongly alkaline after incubation; thus the determination is simple. This is probably the most commonly used manual method. The units are given as millimoles of nitrophenol formed per hour per liter of serum.

The phenolphthalein phosphate method uses phenolphthalein phosphate as substrate. When the solution is made alkaline after incubation, the phenolph-thalein formed imparts its well-known red color to the solution. The units used are given in terms of micromoles of phenolphthalein liberated per minute per liter of serum.

BESSEY-LOWRY-BROCK METHOD [136]

p-nitrophenylphosphate + H_2O
   (colorless in acid
      and alkali)

p-nitrophenol + H_3PO_4
   (yellow in alkali)

REAGENTS

1. Alkaline buffer solution.   Dissolve 7.5 g of glycine and 0.095 g of magnesium chloride in about 750 ml of distilled water. Add 85 ml of $0.1M$ sodium hydroxide solution and dilute to 1 liter with distilled water. The pH of the solution should be 10.5.

2. Stock substrate solution.   Dissolve 0.4 g of disodium p-nitrophenyl phos-phate in $0.001M$ hydrochloric acid (100 ml of distilled water plus 0.1 ml of $1M$ hydrochloric acid) to make 100 ml.

3. Alkaline buffered substrate.   Mix equal volumes of alkaline buffer solution and stock substrate solution. This should be kept frozen.

4. Sodium hydroxide, $0.02M$.   Mix 200 ml of $0.1M$ sodium hydroxide with 800 ml of distilled water.

5. Stock p-nitrophenol, 10 mM.   Dissolve 139.1 mg of p-nitrophenol in distilled water and dilute to 100 ml in volumetric flask.

Place two tubes containing 1 ml of alkaline buffered substrate in a 37°C water bath and warm to temperature. Pipet 0.1 ml of serum to one tube, mix, and start timing. Add 0.1 ml of water to the other tube as a blank. Exactly 30 min after the addition of serum, add 10 ml of 0.02M sodium hydroxide to each tube. Mix by inversion and read absorbance of the sample tube at 410 nm, setting to zero with the blank tube. To each tube add 2 drops of concentrated hydrochloric acid and mix. Read the sample tube again, using the blank tube as a reference. Subtract the absorbance obtained from the earlier reading on the sample. This is the corrected absorbance. Obtain units from the calibration curve.

CALIBRATION CURVE

Dilute 0.5, 1, 2, 3, and 5 ml of stock p-nitrophenol standard to 100 ml with distilled water in separate volumetric flasks. To 1 ml of each dilution add 0.1 ml of water and 10 ml of 0.02M sodium hydroxide solution. Read in a photometer with water as a reference. The dilutions correspond to 1, 2, 4, 6, 8, and 10 units (Bessey-Lowry). Plot the absorbance against units for the calibration curve.

NORMAL VALUES AND INTERPRETATION OF RESULTS

The normal range for adults is 0.8 to 2.3 units (Bessey-Lowry) and for children, 2.8 to 6.7 units. Alkaline phosphatase may be elevated in liver disease. In bone disease it is increased in those conditions in which bone regeneration is taking place. It is not increased when there is bone destruction unless there is simultaneous formation of new bone or osteoid tissue. In rickets the increase is very marked, roughly paralleling the severity of the disease. Values from 20 to 60 units may be found. In osteomalacia (adult rickets) there is some increase, but not as marked as in children. In Paget's disease, values over 30 units are not unusual. In hyperparathyroidism the increase may be present, but it is less marked (15 units). In bone tumors the findings are variable. Serum alkaline phosphatase is usually normal in multiple myeloma, with never more than a very slight increase.

ALP USING PHENOLPHTHALEIN MONOPHOSPHATE [138]

phenolphthalein monophosphate + $H_2O \rightarrow$ phenolphthalein + H_3PO_4
(red in alkali)

This method has an advantage over p-nitrophenyl phosphate because the absorption peak of phenolphthalein is much farther from those of bilirubin and hemoglobin. Hence the interference from these compounds is significantly reduced. Also, for a given enzyme concentration, much more color (greater absorbance) is produced by phenolphthalein. This increases the sensitivity of the method.

REAGENTS

1. Stock substrate solution. Dissolve 0.39 g of monohydrated dicyclohexylamine salt of phenolphthalein monophosphate in a mixture of 73.2 g of 2-amino-2-methyl-1-propanol and 21.9 ml of concentrated hydrochloric acid. This solution is $0.065M$ in phenolphthalein monophosphate and $7.8M$ in the buffer. The pH is 10.15. It is stable indefinitely when stored in the refrigerator and should be warmed to room temperature before use. This stock solution is diluted 1:26 before use.

2. Color stabilizer (phosphate buffer, $0.1M$, pH 11.2). Dissolve 9.3 g of $Na_3PO_4 \cdot 12H_2O$ and 20.3 g of $Na_2HPO_4 \cdot 17H_2O$ (or 10.8 g of anhydrous salt) in water and dilute to 1 liter with water. The solution is stable at room temperature.

PROCEDURE

Dilute 1 drop (0.04 ml) of stock substrate solution with 1 ml of water and warm to 37°C. Add 0.1 ml of serum and mix. After exactly 20 min, add 5 ml of color stabilizer and read absorbance at 550 nm against a reagent blank without serum. The absorbance due to the serum has been found to be negligible.

CALIBRATION CURVE

Dissolve 79.6 mg of phenolphthalein in 50 ml of alcohol in a 100 ml volumetric flask and dilute to 100 ml with water. This stock solution is stable. Dilute 1 ml of stock solution to 50 ml with water in a volumetric flask. This working standard contains 15.9 μg or 0.05 μmol of phenolphthalein/ml. To 1, 2, 3, and 4 ml of working standard add, respectively, 5.14, 4.14, 3.14, and 2.14 ml of color stabilizer. This brings the volume of each standard up to that of the test samples. Read against a blank at 550 nm. If 1 IU is defined as the amount of enzyme that will liberate 1 μmol of product in 1 min for each liter of serum, then the 0.05 μmol standard corresponds to:

$$\frac{0.05 \times 1,000}{0.1 \times 20} = 25 \text{ IU}$$

where the factor 1,000 converts to liters, 0.1 ml serum is used, and the incubation period is 20 min.

Thus the standards correspond to 25, 50, 75, and 100 IU. The absorbance may be plotted against the units to obtain a calibration curve. There may be some deviation from linearity with a filter photometer because of the narrow absorption peak of phenolphthalein.

NORMAL VALUES AND INTERPRETATION OF RESULTS

The normal range given for this method is 9 to 35 IU. It may be noted that the method of Bessey, Lowry, and Brock, in which p-nitrophenyl phosphate is used, defines the unit as millimoles per hour. By multiplying their values by 1,000 (to change to micromoles) and dividing by 60 (to change from per hour to per minute), one obtains the range for this method of approximately 13 to 48 IU. This does not differ greatly from the phenolphthalein monophosphate method.

ALKALINE PHOSPHATASE ISOENZYMES

Alkaline phosphatase is composed of several isoenzymes [139–141]. These are probably derived from the liver, bone, and intestinal mucosa. Electrophoresis on starch, agar gel, or acrylamide gel has been used to fractionate ALP isoenzymes. Depending on the technique used, a number of bands can be demonstrated by staining the gel. The main bands found in serums from patients with bone and liver disease can be differentiated by their electrophoretic mobility. ALP originating in the bones migrates more slowly than that originating in the liver, and this can be used to distinguish between elevations in total alkaline phosphatase due to liver disease and to bone disorders. As yet this is not as commonly used as LDH electrophoresis. Another source of an alkaline phosphatase isoenzyme is the placenta. The placental alkaline phosphatase is usually responsible for the increase in serum alkaline phosphatase found during pregnancy. The placental isoenzyme may be distinguished from the others by its greater resistance to heat inactivation. The determination of heat-stable alkaline phosphatase is used in the study of some pregnancy disorders. The failure of the isoenzyme to increase at the normal rate during pregnancy may indicate some difficulty in the development of the placenta and fetus.

HEAT INACTIVATION OF ALP ISOENZYMES

One method for distinguishing between the alkaline phosphatase isoenzymes from bone and liver is by heat inactivation. With tissue extracts, the bone isoenzyme is about 90 percent inhibited by heating at 55°C for exactly 16 min, whereas the isoenzyme from liver is only about 50 percent inhibited. Thus if a serum with an elevated alkaline phosphatase has more than 70 percent of the activity lost on heating as described, the rise is probably due to bone disease. If the percentage inhibition is less than 65 percent, the increased alkaline phosphatase is probably due to liver. Unfortunately, although the bone enzyme seems to be fairly constant in degree of inhibition, the liver and the intestinal isoenzymes are more variable, thus detracting from the value of the method. Winkelman et al. [144] reported an extensive study of the comparison of the alkaline phosphatase isoenzymes as determined by heat inactivation, phenylalanine and urea inhibition, and agarose film electrophoresis. They could not find any consistent correlation between the clinical diagnosis and the known organ involvement with the results of the various isoenzyme studies. The heat inactivation method gave a considerable overlap between the bone and liver groups. They concluded that at best the heat inactivation or the other methods could serve only as a guide and not as a positive diagnostic test.

The placenta is the source of an isoenzyme that is very stable toward heat. Even heating at 55°C for 30 min will cause no appreciable inhibition. The placental alkaline phosphatase has been suggested as a measure of placental function. Some of the albumin solutions used for intravenous administration are prepared from cord blood which contains a high level of placental alkaline phosphatase. After the administration of such a solution, the patient may have an elevated alkaline phosphatase level. If this is suspected as the origin of an unusually elevated alkaline phosphatase level, it can be checked by heating a portion of the serum to 55°C for 30 min. If an appreciable amount of enzyme activity remains, the serum contains some placental alkaline phosphatase.

Acid Phosphatase (ACP)

E.C.3.1.3.2   orthophosphoric monoester phosphohydrolase (acid optimum)

This enzyme is found in greatest concentration in the prostate, erythrocytes, and seminal fluid. The acid phosphatase in erythrocytes is selectively inhibited by formaldehyde; that in serum, by fluoride ions, and prostatic phosphatase, by $0.02M$ tartrate solution.

The enzyme can use three of the substrates mentioned earlier for alkaline phosphatase—β-glycerophosphate, phenyl phosphate, and p-nitrophenyl phosphate—with a buffer at a pH of about 5.0. The most common method uses p-nitrophenyl phosphate.

Bessey-Lowry-Brock method [136]

REAGENTS

1. Acid buffer solution.   In a 1 liter volumetric flask dissolve 8.91 g of citric acid in 180 ml of $1M$ sodium hydroxide. Add 100 ml of $0.1M$ hydrochloric acid,

dilute to volume with distilled water, and mix. The pH of this solution should be 4.8.

2. Stock substrate solution.   This is the same as used for alkaline phosphatase.

3. Acid buffered substrate.   Mix equal parts of acid buffer solution and stock substrate solution. This solution is not stable and should be kept frozen if possible. A convenient method is to pipet 1-ml portions to test tubes that are then tightly stoppered (culture tubes with screw caps are excellent) and kept frozen. The tubes required are thawed just before use.

4. Sodium hydroxide, $0.1M$.   Dissolve 4 g of sodium hydroxide in distilled water to make 1 liter.

PROCEDURE

Place two tubes containing 1 ml of buffered substrate in a 37°C water bath for 5 min to warm to this temperature (a slightly longer time may be required if the tubes are initially frozen).

Pipet exactly 0.2 ml of serum to one tube, mix, and start timing. Add 0.2 ml of water to the other tube as a reagent blank. Exactly 30 min after the addition of serum, add 4ml of $0.1M$ sodium hydroxide to each tube. Read sample tube at 410 nm, setting to zero with the reagent blank. To correct for any color contributed by serum, add 0.2 ml of serum to 5 ml of $0.1M$ sodium hydroxide and read in a photometer, setting to zero absorbance with $0.1M$ sodium hydroxide. Subtract the absorbance obtained from that for the sample determination. This is the corrected absorbance. Read units of activity from the calibration curve using the corrected absorbance.

CALIBRATION CURVE

The same stock p-nitrophenol solution used in the alkaline phosphatase method is used. Dilute 1, 2, 4, 6, 8, and 10 ml of stock nitrophenol solution, respectively, to 100 ml in separate volumetric flasks with distilled water. To 0.2 ml of each of the dilutions add 5 ml of $0.1M$ sodium hydroxide and read absorbance at 410 nm against a water blank. The dilutions correspond to 0.2, 0.4, 0.8, 1.2, 1.6, and 2.0 Bessey-Lowry units. Plot the absorbance against the units to obtain the calibration curve.

NORMAL VALUES AND INTERPRETATION OF RESULTS

The normal values range from 0.13 to 0.63 unit (Bessey-Lowry) for men and from 0.01 to 0.56 unit for women. The determination of acid phosphatase is usually carried out in connection with malignant disease of the prostate gland. Small increases are found in conditions such as Paget's disease in which there are very high levels of alkaline phosphatase. Very high values are often found in metastasizing carcinoma of the prostate.

PROSTATIC ACID PHOSPHATASE

Serum prostatic acid phosphatase activity is inhibited by the presence of tartrate, whereas acid phosphatase from other sources is not. By assaying for total acid phosphatase activity in the presence and absence of tartrate, one can determine by difference the activity due to prostatic secretion [145, 146].

REAGENTS

1. Acid buffered substrate.   Same as for previous method.
2. Sodium hydroxide, $0.1M$.   Same as for previous method.
3. Tartrate, $0.2M$.   Dissolve 3.002 g of tartaric acid in about 50 ml of water in a 100 ml volumetric flask. Add 35 ml of $0.1M$ sodium hydroxide. Adjust pH to 4.9 and dilute to the mark with water.

PROCEDURE

Prepare four test tubes as shown in Table 14-8. Pipet substrate, water, and tartrate in tubes 1, 2, and 3, mix, and place in water bath at 37°C for 5 min. Add 0.2 ml of serum to tubes 2 and 3, mix, and start timer. Incubate for exactly 30 min. While tubes 2 and 3 are incubating, prepare tube 4. At the end of the incubation period, add sodium hydroxide to tubes 1, 2, and 3. Read absorbance of tubes 2 and 3 against tube 1 and read tube 4 against water at a wavelength of 410 nm. Subtract absorbance of tube 4 from that of tubes 2 and 3 and convert to units by using the calibration curve from acid phosphatase. Subtract units in tube 3 from tube 2 to obtain units of tartrate-inhibited acid phosphatase.

Table 14-8. Determination of Prostatic Acid Phosphatase

Reagent	Tube No. 1	Tube No. 2	Tube No. 3	Tube No. 4
Step 1				
Substrate (ml)	1.0	1.0	1.0	0
Water (ml)	0.4	0.2	0	0.2
Tartrate (ml)	0	0	0.2	0
Step 2				
Serum (ml)	0	0.2	0.2	0.2
Step 3				
NaOH (ml)	3.8	3.8	3.8	4.8

COMMENTS

Prostatic acid phosphatase is very unstable. Blood should be refrigerated immediately after drawing. Centrifuge after standing 1 hr and separate serum. Do not use if hemolyzed. Keep serum at 0° to 5°C at all times or add 0.02 ml of 3.5M acetic acid to each 2 ml of serum to stabilize the enzyme.

NORMAL VALUES AND INTERPRETATION OF RESULTS

The normal value may range up to 0.15 unit. Prostatic acid phosphatase assay is most valuable in cases of prostatic carcinoma, where it is found to be elevated even when the total acid phosphatase may be normal.

ACID PHOSPHATASE USING α-NAPHTHYL PHOSPHATE [147–150]

The α-naphthol liberated by enzymatic hydrolysis is measured by coupling with diazotized 5-nitro-o-anisidine.

α-Naphthyl phosphate has been shown to be the substrate of choice for prostatic acid phosphatase; therefore this method is highly specific for its determination.

REAGENTS‡

1. α-Naphthyl phosphate substrate, 2.7mM in 0.07M citrate buffer at pH 5.2.

A. Dissolve 5.30 g of citric acid monohydrate and 13.18 g of sodium citrate dihydrate in water and dilute to 1 liter. Check and adjust pH to 5.2 if necessary.

B. Dissolve 30 mg of α-naphthyl phosphate in 50 ml of citrate buffer. This solution is stable for 1 week under refrigeration.

2. Diazonium salt. Dissolve 20 mg of diazotized 5-nitro-o-anisidine (Fast Red Salt B) in 100 ml of 0.1M HCl. This solution is stable for 1 week under refrigeration.

3. Standards. Dissolve 86.5 mg of α-naphthol in 10 ml of ethanol and dilute to 100 ml with water. Dilute 1 ml of this solution to 10 ml with pooled serum. Pipet 0, 1, 2, 3, and 4 ml of this solution into correspondingly labeled test tubes and add 4, 3, 2, 1, and 0 ml of pooled serum. These mixtures are equivalent to 0, 5, 10, 15, and 20 IU of acid phosphatase, respectively, and are stable for several days under refrigeration. Alternatively, mixtures of quality control serums of known activities may be used as standards; these have the further advantage of standardizing the enzymatic reaction in addition to the color reaction.

4. Sodium hydroxide, 0.1M. Dilute 1M NaOH 1:10 in a volumetric flask.

PROCEDURE

To 1 ml of substrate warmed to 37°C add 0.2 ml of serum, mix, and return immediately to water bath. Exactly 30 min later, remove from water bath and add

‡ Reagents are available from Warner-Chilcott Laboratories Division, Warner-Lambert Pharmaceutical Co., Morris Plains, N.J. 07950.

1ml of diazonium salt, mix, and immediately add 5 ml of 0.1M NaOH. Mixing is accomplished by blowing the NaOH forcibly through a 5 ml serological pipet.

Read absorbance at 590 nm against a water blank. The control is run the same way, except the substrate is not warmed to 37°C and the diazonium salt is added immediately after the serum. Enzyme standards are treated exactly the same as unknowns, and α-naphthol standards are treated the same as the controls.

CALCULATION

The absorbances of the α-naphthol standards minus the zero unit standard or of the enzyme standards minus controls are plotted against the known units. The activity of the unknowns can be read directly from the curve after subtracting the control reading.

The international unit is defined as the amount of acid phosphatase in 1 liter of serum that will liberate 1 μmol of α-naphthol in 1 min under the conditions of assay.

NORMAL VALUES

The normal serum level for men and women by this method is 1 to 1.9 IU/liter. The only interfering substance encountered was bilirubin, which forms a chromogen with the diazonium salt. The resulting increase in control color is equivalent to about 0.3/mg/dl bilirubin.

5′-Nucleotidase (5′NT)

E.C.3.1.3.5   5′-ribonucleotide phosphohydrolase [151–154]

This enzyme is a phosphatase that specifically catalyzes the hydrolysis of nucleotides with a phosphate radical attached to the 5′-position of the pentose. The reaction is:

adenosine 5′-phosphate + H_2O ⇌ adenosine + H_3PO_4

The enzyme is strongly inhibited by nickel ions at a final concentration of about 0.01M. Thus if the determination is run in the presence of nickel ions and in their absence, the difference in activity will be that caused by the nucleotidase. The reactions are usually followed by measuring the amount of inorganic phosphate produced in the two reactions. In another method the substrate and serum are incubated in the presence of a large excess of glycerophosphate which inhibits the action of alkaline phosphatase on the adenosine phosphate. The adenosine formed is determined by the addition of the enzyme adenosine deaminase and the ammonia formed determined by the phenol-hypochlorite reaction or by measuring the decrease in absorbance of adenosine at 265 nm.

5'-Nucleotidase determination is very helpful in cases where there is questionable hepatobiliary disease and there is an elevation in alkaline phosphatase. If 5'NT is also elevated, it will confirm the diagnosis of liver involvement. However, if its level is normal, it may be an indication of osteoblastic rather than liver disease.

Increased levels of nucleotidase are found in most patients with obstructive jaundice and hepatic metastases. Increases may be only slight to moderate in hepatitis and cirrhosis. Usually the level of nucleotidase is not elevated in bone diseases.

The determination of this enzyme may be useful in confirming liver disease in children, where borderline alkaline phosphatase elevations may overlap the higher range of normal found in childhood.

Amylase (Diastase) (AMY)

E.C.3.1.1.1.   1, 4-α-D-glucan-glucanohydrolase; α-amylase

Amylase is a digestive enzyme which hydrolyzes starch into smaller molecular units. Starch is a long-chained polymer composed of glucose molecules. α-Amylase attacks the chain at random positions, breaking it down into smaller units and eventually into maltose and some glucose. These will react with the copper reagents such as used in the Folin-Wu or Somogyi methods for blood glucose.

The reaction may be followed by measuring the amount of reducing sugars formed or by measuring the decrease in the amount of starch present. The former methods are called saccharogenic, the latter amyloclastic.

In the saccharogenic method (Somogyi) the buffered starch substrate is allowed to react with the enzyme-containing material. After incubation, the excess starch and serum proteins are precipitated with barium hydroxide and zinc sulfate or other precipitating reagents and the amount of reducing sugars in the filtrate determined by one of the copper reduction methods. A correction is applied for the amount of reducing sugars originally present in the sample. The units are given in terms of milligrams of reducing sugars (calculated as glucose) formed in 30 min by 100 ml of serum. The other methods for amylase are usually calibrated in terms of the saccharogenic method and units.

The decrease in the amount of starch present (amyloclastic methods) may be determined in a number of ways. Starch gives an intense blue color with a dilute solution of iodine. As the starch molecule is broken down, the color formed changes from deep blue to purple-red to reddish brown, and finally with almost complete hydrolysis of the starch very little color is formed. In the original Somogyi method, the time required for the disappearance of the blue color formed with iodine when the sample is added to a definite amount of starch is measured. Aliquots of the incubation mixture are withdrawn at definite times and added to a dilute solution of iodine. The endpoint is taken as the time when the blue color is no longer formed. This method has the disadvantage that the disappearance of the blue color is not sharp and there is some subjective error in judging the endpoint. The activity is inversely proportional to the time required. The proportionality constant is chosen so that the units are the same as in the saccharogenic method.

In a modification of this method, the reaction is allowed to proceed until only a fraction of the starch is decomposed. Iodine is then added and the amount of blue color present is compared in a photometer with the deeper color formed in an aliquot of the original mixture without incubation. The decrease in absorbance divided by the original absorbance would be a measure of the percentage of starch decomposed.

In another method special starch preparations are used which have considerable turbidity. As the starch is broken down into smaller molecules, the turbidity decreases. Measurement of the decrease in turbidity in a photometer would give a measure of the amount of starch decomposed. This method is not widely used.

More recently another method has been introduced. Starch is treated with a dye substance with which it forms a very stable complex. As the starch is hydrolyzed by the added enzyme, some of the dye is liberated. After incubation, the excess starch and combined dye are precipitated by alcohol or other means. The amount of free dye found in the supernatant or filtrate is then a measure of the amount of starch that has been hydrolyzed by the enzyme and thus of the enzyme activity. This method requires standardization against control serums whose amylase activity has been determined by another method, usually the saccharogenic method.

SACCHAROGENIC METHOD OF SOMOGYI [155, 156]

REAGENTS

1. Buffer solution.   Dissolve 2.25 g of anhydrous monobasic potassium phosphate (KH_2PO_4), 2.4 g of anhydrous disodium phosphate (Na_2HPO_4), 2.5 g of sodium chloride, 2 g of sodium fluoride, and 0.3 g of propyl p-hydroxybenzoate§ in distilled water and dilute to 1 liter.

2. Starch solution.   Best results are obtained with a starch purified as follows:

§ No. 2992, Eastman Kodak Co., Rochester, N.Y. 14650.

Dissolve 4.5 g of sodium hydroxide in about 1,000 ml of water in a 2,000 ml beaker. Heat to 50° to 55°C. Discontinue heating and introduce about 200 g of cornstarch while agitating with a mechanical stirrer. Continue agitation for 1 to 2 hr; allow starch to settle overnight. Decant the yellow liquid, suspend the starch in about 1,800 ml of distilled water, and stir the mixture well before allowing it to settle again overnight. After the water is decanted, wash the starch twice more with water in a similar manner. After final removal of the water, allow the starch to dry in the air. For the preparation of the starch solution, suspend 15 g of purified starch in 100 ml of buffer solution; heat 900 ml of buffer solution to boiling and add to the starch suspension with vigorous agitation. Further heating is unnecessary.

3. Barium hydroxide, 0.15M, and zinc sulfate, 0.175M. These are the same as used for the preparation of the Somogyi blood filtrate (Chapter 5).

4. Somogyi high-alkalinity copper reagent. Dissolve 28 g of disodium phosphate and 40 g of potassium sodium tartrate in 700 ml of water. Add 100 ml of 1M NaOH and 80 ml of 0.4M copper sulfate and dilute to 1 liter.

PROCEDURE

Warm 5 ml of starch solution to 37°C in a water bath, add 1 ml of serum, and incubate the mixture for 30 min. Add 2 ml of 0.15M barium hydroxide and, after mixing, 2 ml of 0.175M zinc sulfate. Mix well and allow to stand for 30 min. Filter and determine the sugar in the filtrate using the Somogyi method (Chapter 5) except that the high-alkalinity copper solution is used and the samples are boiled for 20 min rather than 15 min. Since the dilution of the serum is 1:10, the calculations are exactly the same as for the blood glucose method. Also determine the blood glucose in the serum by the regular Somogyi method. Subtract the blood glucose value (mg/dl) found for the original serum from that found for the serum after incubation with starch (also expressed as mg/dl). The result is the diastase activity of the serum in Somogyi units. If the value found is over 500 units, the incubation should be repeated using a 1-ml aliquot of serum diluted with 0.15M sodium chloride (dilute 1:2 or 1:4 as required), and the results are multiplied by the appropriate dilution factor.

AMYLOCLASTIC METHOD OF CARAWAY [157, 158]

In this method the amount of starch and the incubation time are adjusted so that only a portion of the starch is hydrolyzed when the reaction is stopped by the addition of iodine. The difference between the amount of blue color formed in the incubated sample and a blank prepared by the addition of the serum after incubation will be a measure of the amount of starch hydrolyzed and hence of the amylase activity of the sample.

REAGENTS

1. Buffered starch substrate, pH 7.0.   Dissolve 4.30 g of benzoic acid and 13.30 g of disodium phosphate (Na_2HPO_4) in about 250 ml of water and heat to boiling. Mix 0.200 g of soluble starch with 5 ml of cold water in a small beaker. Add the starch suspension to the boiling solution with stirring. Rinse out the beaker with additional water to transfer all the starch to the boiling solution. Boil for an additional minute. Allow the solution to cool and dilute to 500 ml. Store in the refrigerator. This solution is fairly stable but there will be a tendency for molds to form. It may be preferable to use sterile water and sterile containers in the preparation and the solution should be transferred aseptically to small sterile bottles. Usually containers that have been well washed and dried by heating in an oven will be sufficiently sterile. Not all brands of soluble starch will be equally satisfactory. Harleco's Starch Powder—Smith and Roe||and Harleco's starch reagent labeled "prepared according to Caraway" have been found satisfactory, as is Merck's Soluble Starch—Lintner.# Starch labeled "prepared according to Somogyi" and some other starches may not be satisfactory for this method as they may give a turbidity in the final solution, which should be perfectly clear.

2. Iodine solutions.

A. Stock solution.   Dissolve exactly 3.567 g of reagent potassium iodate and 45 g of potassium iodide in about 800 ml of water. Slowly add with stirring 9 ml of concentrated hydrochloric acid. Cool and dilute to 1 liter. Store in a brown bottle in the refrigerator. This solution is stable.

B. Working solution.   Dilute the stock 1:10 with water. Store in brown bottle in the refrigerator. This solution should be prepared fresh each month.

PROCEDURE

Add 5 ml of buffered starch substrate to each of two 50 ml volumetric flasks, labeling one test and the other blank. If they are available, it may be convenient to use large test tubes graduated at 50 ml, such as NPN digestion tubes. Place the two flasks in a water bath at 37°C for 5 min. Add 0.10 ml of serum to test flask and mix well. Replace in bath and incubate for exactly 7.5 min. Remove from bath; add about 35 ml of water and 5 ml of working iodine solution. Remove blank from bath, add 0.100 ml of serum, then at once 35 ml of water and 5 ml of

|| Harleco, Div. American Hospital Supply, Philadelphia, Pa. 19143.
Merck Chemical, Div. Merck & Co., Rahway, N.J. 07065.

iodine solution. Dilute contents of flasks to 50 ml and mix well. Read solutions in colorimeter at 660 nm against water.

CALCULATION

Caraway defined the amylase unit as the amount of enzyme that will hydrolyze 10 mg of starch in 30 min to a colorless stage. In the procedure the substrate theoretically should be completely hydrolyzed by 800 units of amylase activity in 100 ml of serum. Then:

$$\frac{\text{absorbance of blank} - \text{absorbance of test}}{\text{absorbance of blank}} \times 800 = \text{amylase units/dl of serum}$$

If the reading is equivalent to over 400 units, an aliquot of the serum should be diluted with 0.16M sodium chloride and the test repeated, multiplying the result obtained by the appropriate dilution factor. Several samples can be run in one series by making the additions at timed intervals. A serum blank is required for each sample since the serum proteins decrease the amount of color produced by the starch-iodine complex and this will be different for each serum. Urine or other body fluids can be run in a similar manner. At least once each day it is advisable to run a reagent blank, carrying out the above procedure on one flask without the addition of any serum. The resulting absorbance should be constant from day to day. A decrease in absorbance would indicate deterioration of the starch solution.

The upper limit of normal by this method is 160 to 180 units/dl; thus the results are comparable with those by the Somogyi method. For 2-hr urine specimens the upper limit of normal is about 300 units/hr.

URINARY AMYLASE

The same methods just described for serum may be used, merely by substituting urine. A clean-voided, accurately timed specimen ranging from 1 to 24 hr is required.

NORMAL VALUES AND INTERPRETATION OF RESULTS

The normal range of amylase by methods calibrated in Somogyi units is 60 to 180 units. Serum amylase is greatly increased in acute nonhemorrhagic pancreatitis early in the course of the disease. Levels can rise as high as 2,000 to 3,000 units. Increased levels may also be found in patients with perforated gastric or duodenal ulcers. In chronic pancreatitis the levels may range from 250 to 400 units. The injection of morphine causes a temporary rise in serum amylase levels.

Other narcotic analgesics may have a similar effect. Some increase in serum amylase activity has been reported following the ingestion of relatively large amounts of alcohol. Thiazide diuretics have also been reported to cause increases in serum amylase of up to 200 percent.

Lipase (*LIP*)

E.C.3.1.1.3   triacylglycerol acyl-hydrolase [159–161]

This enzyme hydrolyzes fats into fatty acid and glycerol. The fatty acids liberated are determined by titration with standard sodium hydroxide or extracted in petroleum ether and determined colorimetrically by the change in the color of a buffered indicator solution. Fat emulsions such as olive oil, coconut oil, and tributyrin have been used as substrates, olive oil being the preferred one.

Lipase is found in greatest concentration in the pancreas, gastrointestinal mucosa, and white blood cells.

TITRIMETRIC METHOD

In this method the fatty acids liberated in the reaction are titrated with standard $0.05M$ sodium hydroxide using thymolphthalein as indicator. The amount of sodium hydroxide used to neutralize the fatty acids liberated is equivalent to the units of lipase in the sample.

REAGENTS

1. Olive oil emulsion.   To 100 ml of distilled water add 200 mg of sodium benzoate and 7 g of gum arabic (acacia). Mix in a blender at low speed until dissolved. With the blender at low speed, slowly add 100 ml of pure olive oil. Mix for an additional 10 min at high speed. This reagent should be kept at refrigerator temperature. Freezing or exposure to excessive heat will destroy the emulsion. A creamy layer on top of the emulsion may form during storage; shake the reagent thoroughly before using. Discard if excessive separation occurs after mixing. Olive oil should be purified as follows. To 300 ml of pure olive oil add 60 g of aluminum oxide while stirring. Stir at 10-min intervals for 1 hr. Let the aluminum oxide settle and filter through Whatman No. 1 filter paper. The olive oil may be checked by mixing 5 ml of purified oil with 5 ml of ether and 5 ml of 95% ethanol and titrating with thymolphthalein as indicator. If the titration requires more than 0.5 ml of $0.05M$ sodium hydroxide, repeat the purification.

2. Buffer.

A. Stock solution.   In a 500 ml volumetric flask dissolve 48.55 g of tris-(hydroxymethyl)aminomethane and dilute to volume.

B. Working solution.   Dilute 50 ml of stock buffer and 21.5 ml of 1M HCl with distilled water to 200 ml in a volumetric flask. Adjust the pH to 8.0 using 1M HCl and refrigerate.

3. Sodium hydroxide, 0.05M.   In a 100 ml volumetric flask dilute 5 ml of standardized 1M sodium hydroxide to volume with water.

4. Thymolphthalein indicator.   Dissolve 1 g of thymolphthalein in 95% ethanol and dilute to 100 ml.

5. Ethanol, 95%.

PROCEDURE

Into each of two tubes labeled blank and test, pipet 2.5 ml of water, 3 ml of olive oil emulsion, and 1 ml of working buffer solution. Add 1 ml of serum to the tube labeled test and stopper both tubes. Shake vigorously for a few seconds. Place both tubes in a water bath at 37°C and incubate for 6 hr. Immediately after starting the incubation, pipet 1 ml test serum into a 50 ml Erlenmeyer flask, label it blank, and store it in the refrigerator. At the end of the incubation period pour the contents of the blank tube into the cold blank flask containing test serum and pour the contents of the test tube into a clean Erlenmeyer flask labeled test. Rinse both tubes with 3 ml of ethanol. Add washings to the respective flasks. Mix the contents of the flasks by rotation and add 4 drops thymolphthalein. With the use of an accurate buret, titrate both flasks with 0.05M sodium hydroxide to a light but distinct blue color (the test and blank must be titrated to the same color intensity).

CALCULATION

Subtract the milliliters of 0.05M sodium hydroxide taken by the blank from that taken by the unknown. The difference will be the units of lipase per milliliter of serum.

NORMAL VALUES AND INTERPRETATION OF RESULTS

Normal values found by this method range from zero to 1 unit. Elevated serum lipase values are found in acute pancreatic disease. In this condition the elevation usually occurs 24 hr after onset of disease. The increase in lipase may not parallel that of amylase, which may show an elevation earlier in the disease, but the elevation of the lipase may persist for a longer period. In chronic pancreatitis the lipase level may be normal. Moderate increases are found in some cases of pancreatic carcinoma. Occasionally elevated values are found in kidney diseases, intestinal obstruction, and duodenal ulcers.

Colorimetric method [162, 163]

This method for lipase requires only 30 min incubation time and eliminates the subjective errors encountered in determining the titration endpoint. It requires careful technique to obtain good results. The fatty acids liberated by the hydrolysis of the substrate are extracted with petroleum ether and an aliquot of the extract evaporated to dryness. A buffered indicator solution is added to the residue and the change in color of the indicator solution due to the fatty acids measured photometrically.

REAGENTS

1. Olive oil. A good quality olive oil (reagent grade) purified as described in the titrimetric procedure. The blank value is checked by adding 0.05 ml of the oil to 3 ml of the methyl red reagent, shaking well, and centrifuging to remove the excess oil (similar to the corresponding steps in the procedure). The absorbance of the solution in 10 mm light path cuvettes should not be more than 0.08 greater than the reagent alone. If it is, the oil should be further purified.

2. Buffer. To about 800 ml of water in a 1 liter volumetric flask, add 2.42 g of tris-(hydoxymethyl)aminomethane, 3.5 g of deoxycholic acid, and 0.2 g of sorbic acid; dissolve and dilute to 1 liter.

3. Substrate. Homogenize 5.0 ml of the purified olive oil with 100 ml of buffer in a high-speed blender. The blending should be done in short intervals of from 30 to 60 sec with periods of cooling between to avoid overheating of the solution and consequent hydrolysis of some of the oil. The blending is continued until no film of oil is seen after the foam settles. The pH is adjusted to 8.5 by the addition of $1M$ NaOH or HCl. This solution is stable for at least 2 months in the refrigerator.

4. Absolute ethanol, reagent grade.

5. Ethanol, 95%, reagent grade.

6. Sulfuric acid, $0.14M$. Dilute 4 ml of concentrated sulfuric acid to 500 ml with water.

7. Petroleum ether, reagent grade.

8. Acid-alcohol mixture. Mix 3 volumes of absolute ethanol and 2 volumes of $0.14M$ sulfuric acid. This solution should be prepared fresh as needed.

9. Methyl red, 0.2%. Dissolve 200 mg of methyl red (reagent grade, free acid) in 100 ml of 95% alcohol. The indicator dissolves rather slowly, requiring a day or more. This can be speeded up by using a magnetic stirrer.

10. Methyl red reagent (buffered indicator solution). Add 10 ml of $1M$ NaOH to 1 liter of 95% alcohol. After mixing, add sufficient 0.2% methyl red solution (usually 10 to 13 ml is required) to bring the absorbance to 0.095 to 0.100 in 10

mm cuvettes when read against alcohol at 500 nm. Then add 1 ml of 1M sodium acetate (13.6 g of the trihydrate diluted to 100 ml with water) while mixing with a magnetic stirrer. With continued mixing add 1M HCl dropwise until a faint orange-red tinge remains. Measure the absorbance of the solution and add further acid carefully until the absorbance is 0.200 ± 0.005 in 10 mm cuvettes. If too much acid has been added, add 1M NaOH carefully to bring the solution to the proper absorbance. The color of the solution may fade slightly at first but it should then remain stable for at least 1 month when kept in a brown bottle at room temperature. The method as originally described used cuvettes having exactly 10 mm light path for standardization of the reagent. Other cuvettes having an approximately equivalent light path such as Coleman 12 mm cuvettes could be used, provided all measurements are made in the same cuvettes.

11. Standard solution. Dissolve 14.3 mg of reagent grade stearic acid in 100 ml of petroleum ether or hexane. The latter is preferred because of its higher boiling point. This solution contains 0.5 μmol/ml.

PROCEDURE

Screw-capped culture tubes (16 × 100 mm) with Teflon-lined caps are used throughout. Aliquots of the serum and substrate are warmed to 37°C in a water bath. To one tube labeled test, add 1.0 ml of substrate and 0.050 ml of serum, mix, cap, and incubate at 37°C for exactly 30 min. To a second tube labeled blank, add 1.0 ml of substrate and incubate for 30 min. This will serve as a serum blank and a separate tube must be set up for each sample. A portion of the serum is also incubated separately at 37°C for 30 min. After incubation, add 3.3 ml of the acid-alcohol mixture to all the tubes, then 0.050 ml of the incubated serum to the blank tube. Add 4 ml of petroleum ether to all the tubes. Cap the tubes, shake vigorously for 2 min, and centrifuge at 2,000 rpm for 5 min. Carefully pipet and transfer to another tube a 2.0-ml aliquot of the upper petroleum ether layer. Care must be taken not to pick up the slightest trace of the aqueous layer. Evaporate the petroleum ether at 50°C with the aid of a gentle stream of air. Set up three standards by carefully pipetting 0.25, 0.50, and 0.75 ml of the standard solution to corresponding tubes and evaporating the solvent. Several sets of standards can be prepared at one time, since the evaporated residue is stable for several weeks in the refrigerator when kept in tightly stoppered tubes.

Add 3 ml of methyl red reagent to each tube of standards, samples, and blanks. Cap the tubes tightly and shake vigorously for 1 min. Centrifuge samples and blanks for a short time at 2,000 rpm to remove suspended oil. If the room temperature is too high, difficulty may be experienced in completely centrifuging down the oil. This can be remedied by cooling the tubes in the refrigerator for several minutes before centrifuging. The standards are not centrifuged, but care must be taken that all the material is dissolved. Read standards, samples, and blanks at 502 nm against alcohol.

CALCULATION

The absorbances of the three standards are plotted against the concentrations on linear graph paper. Since an aliquot of half the original petroleum ether extract was used, the standards are equivalent to 0.25, 0.50, and 0.75 μmol in the sample. The concentration of each sample and its respective blank is read from the calibration curve. Then:

(μmol in the sample $-$ μmol in blank) $\times$ 20 $\times$ 2 = μmol/ml/hr = units of lipase

If the sample reading is higher than that of the highest standard, the sample tube and its corresponding blank are each diluted with an equal volume of the reagent and read against alcohol; the result of the calculation is multiplied by 2 to take the dilution into account. If necessary, a second dilution can be made with the appropriate change in the calculation.

The normal range by this method is 2.0 to 7.5 units.

Leucine Aminopeptidase (LAP)

E.C.3.4.11.1 α-aminoacyl-peptide hydrolase (cytosol) [164–169]

This enzyme catalyzes the hydrolysis of amino acids containing alpha amino groups. In vitro it catalyzes the following reaction:

L-leucyl-β-naphthylamide + H_2O → leucine + β-naphthylamine

It is found in highest concentration in the pancreas and liver.

In the determination of this enzyme the substrate used is the synthetic compound leucyl-β-naphthylamide. The naphthylamine formed by hydrolysis is treated with nitrous acid to form the diazo compound; the excess nitrous acid is removed by ammonium sulfamate or urea. The diazotized compound is reacted with N-(1-naphthyl)ethylenediamine to form a blue azo dye which is proportional to the amount of β-naphthylamine liberated. This is measured spectrophotometrically at 560 nm.

β-naphthylamine + $NaNO_2$ + dye base → blue azo dye

The usual units are given as micrograms of naphthylamine liberated by 0.02 ml of serum in 2 hr incubation.

LAP may also be determined using leucyl-*p*-nitroanilide as substrate. This has the advantage that an immediate yellow color is produced on the addition of alkali, but the disadvantage that high bilirubin levels in the serum will result in high blank values.

REAGENTS*

1. Phosphate buffer, 0.2M, pH 7.0.

    A. Dissolve 28.4 g of anhydrous disodium phosphate (Na_2HPO_4) in water and dilute to 1 liter.

    B. Dissolve 27.2 g of anhydrous potassium dihydrogen phosphate (KH_2PO_4) in water and dilute to 1 liter.

    C. Mix 7 parts of solution A and 3 parts of solution B. Check and adjust pH if necessary.

2. L-Leucyl-β-naphthylamine hydrochloride, 0.0012M, pH 7.1. Dissolve 400 mg in water with gentle warming and dilute to 1 liter.

3. Buffered substrate solution, 0.1M, pH 7.0. Mix equal volumes of reagents 1C and 2. Make fresh every month.

4. Trichloracetic acid, 2.5M. Dissolve 40 g of trichloracetic acid in water and dilute to 100 ml.

5. Ammonium sulfamate, 0.04M. Dissolve 0.5 g of ammonium sulfamate in water and dilute to 100 ml.

6. Sodium nitrite, 0.15M. Dissolve 100 mg in water and dilute to 100 ml in a volumetric flask. Make fresh daily.

7. N-(1-Naphthyl)ethylenediamine dihydrochloride, 50 mg/dl. Dissolve 50 mg in 95% ethyl alcohol and dilute to 100 ml with the alcohol. This solution is stable for 30 days.

8. β-Naphthylamine standard solution, 0.036 mg/ml. Dissolve exactly 22.6 mg of β-naphthylamine hydrochloride in water and dilute to 500 ml in a volumetric flask. Free base = $(22.6 \times 0.797)/500$. The factor 0.797 converts β-naphthylamine hydrochloride to the free base.

PROCEDURE FOR SERUM

Dilute serum 1:50 with distilled water in a 50 ml volumetric flask. Label three tubes 1, 2, and 3 for reagent blank, sample, and sample blank, respectively. Add 1 ml of water to tube 1 and 1 ml of diluted serum (0.02 ml of serum) to tubes 2 and 3. Then add 1 ml of buffered substrate to tubes 1 and 2 and 1 ml of phosphate buffer to tube 3. Mix well and incubate for 2 hr at 37°C. Terminate hydrolysis

* All reagents and complete procedures may be obtained from Sigma Chemical Co., St. Louis, Mo. 63118. Reagents 2 and 7 are available from Dajac Laboratories, Philadelphia, Pa. 19124.

by adding 1 ml of trichloracetic acid to all tubes. Mix well and allow to stand about 5 min for complete precipitation of the protein. Centrifuge and transfer 1 ml of supernatant from each tube to correspondingly marked tubes. Add 1 ml of sodium nitrite to all tubes, mix well, and let stand 3 min. Add 1 ml of ammonium sulfamate, mix well, and let stand 2 min. Add 2 ml of N-(1-naphthyl)ethylenediamine dihydrochloride to all tubes, mix well, and let stand 10 min. Read absorbance of all tubes against water at a wavelength of 560 nm. Subtract absorbance of tubes 1 and 3 from tube 2. Convert corrected absorbance to micrograms of β-naphthylamine from calibration curve. Calculate units as follows:

units = μg β-naphthylamine $\times$ 3

where the factor 3 corrects for the 1-ml aliquots diluted to 3 ml. To convert these units to those described by the original authors (Klett units), multiply by the factor 12. If the activity is too high to read, repeat the test with a more dilute sample. Units of serum LAP are defined as the micrograms of β-naphthylamine liberated by 0.02 ml of serum after 2 hr of incubation.

CALIBRATION CURVE

Using the β-naphthylamine standard solution, prepare a series of test tubes as shown in Table 14-9.

Mix thoroughly all tubes containing working standard and pipet 1 ml from each into correspondingly labeled test tubes. Add 1 ml of phosphate buffer and 1 ml of trichloracetic acid to all tubes. Mix thoroughly, withdraw 1-ml aliquot from each

Table 14-9. Calibration Curve for Determination of LAP

Tube No.	Stock Standard Solution (ml)	Water (ml)	Working Standard (μg/ml)	β-Naphthylamine (μg) in Aliquot Used
1	6.0	0	36	12
2	5.0	1.0	30	10
3	4.0	2.0	24	8
4	3.0	3.0	18	6
5	2.0	4.0	12	4
6	1.0	5.0	6	2
7	0.5	5.5	3	1
8	0	6.0	0	0

tube, and put into correspondingly labeled tubes or cuvettes. Proceed with color development as for the serum assay. Read absorbance of all tubes against reagent blank (tube 8) in a photometer at a wavelength of 560 nm. Plot absorbance against micrograms of β-naphthylamine as given in Table 14-9.

PROCEDURE FOR URINE

Collect 24-hr urine specimen without preservative and refrigerate (4°C). Measure and record total volume in milliliters. Because of the occasional presence of chromogens in the urine, it is necessary to dialyze a portion of the specimen before proceeding with the assay.

Dialyze a 50-ml aliquot of urine overnight against running tap water (same procedure as for LDH). After urine is dialyzed, it may be stored, if necessary, at 4°C for 7 days without significant loss of LAP activity.

Dilute 1 ml of dialyzed urine with 9 ml of water (1:10). Use 1 ml of this dilution and follow procedure for serum LAP.

CALCULATION

$$\text{units} = \mu\text{g } \beta\text{-naphthylamine (from curve)} \times 3 \times 10 \times \frac{1}{1,000} \times \text{vol. (ml)}$$

$$= \mu\text{g } \beta\text{-naphthylamine} \times 30 \times \text{vol. (liter)}$$

The factor 3 corrects for the 1-ml aliquots diluted to 3 ml, the factor 10 corrects for the 1:10 dilution of urine, and the factor 1,000 converts micrograms to milligrams.

Units of urinary LAP are defined as the milligrams of β-naphthylamine liberated by a 24-hr urine specimen under specified conditions of assay after 2 hr of incubation.

NORMAL VALUES AND INTERPRETATION OF RESULTS

The normal levels of serum LAP by this method are 5 to 20 units in adults and 5 to 24 units in infants. The values for urine are 2 to 18 units/24 hr. Serum LAP activity is moderately increased in diseases of the liver (viral hepatitis, infectious mononucleosis, neoplasm, obstructive jaundice) and in carcinoma of the pancreas, where the increase is greater. There is a temporary elevation of LAP in acute pancreatitis. Patients with elevated serum LAP always show elevated urine levels; however, elevated urine LAP levels may be found with normal serum levels in diseases such as carcinoma of the colon and rectum, lymphomas, and leukemias.

Cystine Aminopeptidase (CAP; Oxytocinase)

E.C. 3.4.11.2   α-aminoacyl-peptide hydrolase (microsomal) [170–176]

This enzyme acts similarly to a number of other aminopeptidases; it hydrolyzes a cystine group from one end of a peptide.

cystine-peptide + H_2O ⇌ cystine + peptide

It has been determined in a number of different ways. The simplest method uses the synthetic substrates cystine-*bis*-*p*-nitroanilide or *S*-benzyl-cysteine-4′-nitro-anilide. In either case, the nitroanilide is split off and produces a yellow color in alkaline solution which is measured colorimetrically. This enzyme is produced by the placenta.

Two forms of this enzyme have been identified by starch gel electrophoresis. They have been designated cystine aminopeptidase 1 (CAP_1), and cystine amino-peptidase 2 (CAP_2). CAP_1 activity increases significantly in early pregnancy and declines as the pregnancy progresses whereas CAP_2 shows its highest activity in the third trimester.

The determination of CAP has been used as a pregnancy test. It has been claimed that with this technique pregnancy can be diagnosed as early as 4 weeks after conception.

The serial determination of the enzyme may be of value in detecting deviations from the normal pattern in pregnancy and in evaluating the treatment of abnormal pregnancies. Technically its determination should be simpler than that of the total estrogens in a 24-hr urine specimen and would serve a similar diagnostic purpose. Sufficient work has not yet been done to evaluate the relative values of estrogen and the enzyme determinations.

Cholinesterase: "True" Acetylcholinesterase (ACE)

E.C.3.1.1.7   acetylcholine: hydrolase [177–180]

Pseudocholinesterase (PCE)

E.C.3.1.1.8   acylcholine:acyl-hydrolase

There are two different enzymes known as cholinesterase—the true cholinesterase, found in red cells and nerve tissue, and the pseudocholinesterase, found in serum, the liver, and other organs. The true cholinesterase is more specific and is active only against acetylcholine and very closely related compounds. Pseudocholinesterase is active against a larger range of substrates and a number of these have been used for its determination. With acetylcholine the reaction is:

acetylcholine + H_2O → choline + acetic acid

A simple way of determining the amount of reaction is to measure the change in pH produced by the liberated acetic acid. In one method (Michel) this is done by a direct measurement of the pH with a glass electrode before and after incubation. The units are given in terms of actual change in pH under the specified conditions (change in pH units produced by incubating 0.1 ml of serum or packed cells for 1 hr). Another method measures the pH change by the change in color of an added indicator; the color change is measured spectrophotometrically. A number of different indicators have been used, including phenol red, bromthymol blue, and m-nitrophenol. These methods may be calibrated in terms of international units by measuring the change in color when known amounts of acetic acid (in micromoles) are added to the buffered indicator. The results may also be given in other units such as micromoles of acetic acid formed per milliliter of serum in 30 min incubation.

One colorimetric method uses acetylthiocholine as substrate. With this substrate the reaction is:

acetylthiocholine + H_2O → thiocholine + acetic acid

thiocholine (SH group) + dithiobis(nitrobenzoic acid) → color

The liberated thiocholine is measured by reaction with the reagent dithiobis-(nitrobenzoic acid). This gives a yellow color with the liberated thio group with maximum absorbance at 405 nm which is proportional to the concentration of the enzyme. This may be calibrated in terms of international units by using a stable thiol compound such as glutathione as standard. For pseudocholinesterase a number of other substrates have also been used. With benzoyl choline as substrate, the liberated benzoic acid is measured by its ultraviolet absorption at 240 nm. With naphthyl acetate as substrate, the liberated naphthol is measured photometrically after reaction with a diazotized dye intermediate as a color reagent.

METHOD USING ACETYLCHOLINE AS SUBSTRATE

In this method the change in pH due to the liberation of acetic acid is measured using m-nitrophenol as indicator. The indicator is yellow in alkaline solution and colorless in acid. The amount of yellow color decreases as acetic acid is liberated.

This decrease in color is a measure of the acetic acid formed and hence of the enzyme activity.

1. Buffered nitrophenol.   Dissolve 6.65 g of anhydrous disodium phosphate (Na_2HPO_4) and 0.43 g of potassium dihydrogen phosphate (KH_2PO_4) in about 200 ml of distilled water. Dissolve 0.30 g of m-nitrophenol† in about 200 ml of distilled water (with the aid of slight heating if necessary). Mix the two solutions and adjust to pH 7.8 with 1.0M sodium hydroxide solution; then dilute to 1 liter.

2. Acetylcholine chloride solution, 0.83M.   Dissolve 15 g of acetylcholine chloride in water to make 100 ml. Store in the refrigerator.

3. Sodium chloride solution, 0.15M.   Dissolve 9 g of sodium chloride in water and make up to 1 liter.

To each of two tubes add 0.1 ml of sodium chloride solution and 0.1 ml of serum. Heat one tube to 60°C in a water bath for 3 min. This is the blank tube; the heating inactivates the enzyme. The other tube is the sample tube. To each tube add 2.5 ml of buffered nitrophenol and 0.1 ml of acetylcholine solution. Mix and incubate at 25°C for 30 min. Read absorbance of both tubes at 420 nm, setting to zero with water. Read sample exactly 30 min after addition of acetylcholine. Subtract the absorbance of the sample tube from the absorbance for the blank.

Dilute 58 ml of glacial acetic acid to 1 liter with distilled water. Titrate against standard alkali solution with phenolphthalein indicator and adjust to exactly 1M. Dilute 1, 2, 3, 4, and 5 ml of the 1M acid to 50 ml in volumetric flasks. These solutions will correspond to 20, 40, 60, 80, and 100 units, respectively.

Pipet 2.5 ml of buffered nitrophenol and 0.1 ml of pooled inactivated serum (heated to 60°C for 3 min) to each of six tubes. Do not use hemolyzed, icteric, or turbid serum. To one tube add 0.1 ml of distilled water; this is the blank. To the other tubes add 0.1 ml of diluted standard acetic acid solutions. Mix and read absorbance at 420 nm, setting to zero absorbance with water. Subtract the absorbance of each standard tube from that for the blank. Plot the values obtained against the units of standard. The unknown samples are read from this curve. A new standard curve must be made up for every new batch of reagents. For values higher than 120 units, dilute the serum with an equal volume of sodium

† No. 1340, Eastman Kodak Co., Rochester, N.Y. 14650.

chloride solution and repeat the test, taking 0.1 ml of diluted serum and multiplying the result obtained by **2**.

NORMAL VALUES AND INTERPRETATION OF RESULTS

The normal values by this method are 40 to 80 units. Low levels have been found in anemia, tuberculosis, hypoproteinemia, uremia, and shock; increased values in hyperthyroidism and diabetes. Low serum pseudocholinesterase levels are also found in most liver diseases, although the decrease may only be slight in infectious mononucleosis, cirrhosis, or metastatic carcinoma.

The organic phosphorus insecticides are potent cholinesterase inhibitors and exposure to these may result in low levels of the enzymes in both the serum and the red cells. Another important use of the determination of pseudocholinesterase is in measuring the susceptibility to succinyl choline. In major surgery succinyl choline is often given as a muscular relaxant. This compound is slowly hydrolyzed by the pseudocholinesterase. If the level of the enzyme is low, the inactivation of the succinyl choline is slow and the patient may experience a period of dyspnea in the recovery room. The enzyme level may also be low in a small percentage of individuals who have a genetically determined low pseudocholinesterase activity.

METHOD USING ACETYLTHIOCHOLINE AS SUBSTRATE

In this method the enzyme releases thiocholine from the substrate, which reacts with 5, 5-dithiobis(2-nitrobenzoic acid) (DTNB) to produce a yellow color [181, 182]. Samples are incubated for 3 min at 37°C and the reaction is stopped by adding quinidine sulfate. The difference in absorbance between the sample and the blank is measured at 412 nm.

REAGENTS

1. DTNB-buffer. To a 250 ml volumetric flask add: 100 ml of 0.1M HCl, 25 mg of DTNB,‡ 1.66 g of sodium chloride, and 62.5 ml of 0.2M tris-(hydroxymethyl)aminomethane. Dissolve and dilute to volume with distilled water. The pH of this solution should be 7.40 at 37°C. It is stable for 2 weeks at refrigerator temperature.

2. Acetylthiocholine iodide substrate, 0.18M. Add 5.20 mg of acetylthiocholine iodide to 100 ml volumetric flask and dilute to volume with distilled water. This solution is stable for 1 week in the refrigerator.

3. Quinidine sulfate, 500 mg/dl. Dissolve 500 mg of quinidine sulfate in 100 ml of distilled water.

‡ Aldrich Chemical Co., Milwaukee, Wis. 53233.

PROCEDURE

To each of two tubes, labeled test and blank, add 4.0 ml of DTNB-buffer and allow to equilibrate in a 37°C water bath. Add 0.02 ml (20 μl) of serum to both tubes and immediately add 1.0 ml of quinidine sulfate to the blank tube. Add 0.5 ml of the substrate to both tubes and mix. Exactly 3 min after addition of substrate, stop enzyme reaction in test sample by adding 1.0 ml of quinidine sulfate. Remove tubes from 37°C bath and measure absorbance of both tubes at 412 nm, setting to zero with water. The readings of each pair of tubes (sample and blank) should be made within 1 min of each other. Subtract the absorbance of the blank tube from that of the sample to obtain corrected absorbance. Determine cholinesterase units from calibration curve.

CALIBRATION CURVE

The cholinesterase unit of activity is expressed as micromoles of sulfhydryl (SH) groups liberated in 3 min from 1 ml of serum.

Prepare a stock solution of glutathione containing 24.6 mg in 100 ml of distilled water. This solution contains 0.80 μmol/ml buffer.

Prepare six test tubes and label as shown in Table 14-10. Mix the contents of each tube and measure the absorbance at 412 nm, setting to zero with water. Subtract the absorbance of the blank from that of the standards to obtain corrected absorbances. Plot corrected absorbance for each standard against corresponding cholinesterase units.

NORMAL VALUES

The normal range obtained by this method is 7.8 to 16.6 units for males and 5.8 to 13.8 units for females.

Table 14-10. Calibration Curve for Determination of Cholinesterase

Tube No.	Stock Glutathione (ml)	DTNB Substrate (ml)	Water (ml)	Concentration (μmol)	Cholinesterase Units (μmol SH/3 min/ml serum)
Blank	0	4.0	1.5	0	0
1	0.10	4.0	1.4	0.08	4
2	0.20	4.0	1.3	0.16	8
3	0.30	4.0	1.2	0.24	12
4	0.40	4.0	1.1	0.32	16
5	0.50	4.0	1.0	0.40	20

Pepsin (PS)

E.C.3.4.4.1 [183, 184]

Pepsin is an enzyme, found in gastric juice, that hydrolyzes the peptide linkages of proteins, breaking them down into smaller units and eventually into amino acids. It is secreted as the precursor pepsinogen which is changed by acid into the active form. Any pepsin in the plasma is probably present as pepsinogen. Pepsin is excreted in the urine as uropepsin which may also be activated by acid. Pepsin has been determined by using a protein such as hemoglobin or dried serum as a substrate. After activation by incubating with acid, the sample is incubated with the protein substrate. The proteins are then precipitated with trichloracetic acid (TCA) and the liberated phenolic amino acids determined with the Folin-Ciocalteu reagent. A blank is run by mixing the sample and the substrate after incubation and immediately precipitating with TCA. The difference in the amount of color with the phenol reagent between the test and blank is taken as a measure of the enzyme activity. The results are usually reported in arbitrary units by comparison with the color produced by a standard solution of phenol.

Uropepsin has been determined in urine using casein—as a diluted solution of homogenized milk—as substrate. The enzyme converts the milk casein into the insoluble paracasein. The endpoint is taken as the first appearance of small particles of the insoluble paracasein in the solution. The time required to reach the endpoint is measured. The results are reported in terms of arbitrary units; a unit is defined as ten times the number of milliliters of urine which would be required to reach the endpoint in exactly 100 sec under the conditions of the test.

Trypsin (TR)

E.C.3.4.21.4 [185–188]

Chymotrypsin A and B

E.C.3.4.21.1

These proteolytic enzymes have no systematic names. Trypsin is secreted as trypsinogen and is activated to trypsin by the enzyme enterokinase of the intestinal mucosa. The chymotrypsins are secreted as chymotrypsinogens which are activated by the presence of trypsin. Since the enzymes will hydrolyze a number of peptides, simple methods will not distinguish between them, although for many clinical purposes this is not necessary. However, by using a proper substrate they can be distinguished, since trypsin preferentially attacks peptides at points involving the carboxyl groups of arginine or lysine, and the chymotrypsins preferentially attack bonds involving aromatic amino acids. In one method the substrate N-tosylarginine methyl ester was used as the substrate for trypsin assay and N-acetyltyrosine methyl ester as the substrate for chymotrypsin. For both enzymes the hydrolysis of the substrate produced free acid. The reaction is carried out in a buffered solution of phenol red, and the production of the acid changes the intensity of the phenol red color as a consequence of the pH change. Similar assays have been carried out using benzoyl derivatives and measuring the liberated benzoic acid by its absorbance at 253 nm in the ultraviolet. This is less convenient, particularly with the analysis of duodenal samples, which have variable absorption in the ultraviolet. The analysis of the trypsins in duodenal contents is used to measure the function of the pancreas in secreting these enzymes. Usually the specimen is collected by duodenal intubation and samples taken before and after stimulation by pancreozymin or secretin.

Guanase

E.C. 3.5.4.3   guanine aminohydrolase; guanine deaminase [189–193]

This enzyme catalyzes the reaction:

guanine $+$ H_2O $\rightarrow$ xanthine $+$ NH_3

Like several other of the enzymes dealing with ammonia metabolism, this enzyme is found chiefly in the liver. The measurement of the serum activity of this enzyme has been used as an index of hepatic cell necrosis.

The reaction has been followed by measuring the change in guanine or xanthine concentration by ultraviolet spectrophotometry. The absorption maximum of guanine is at 245 nm and of xanthine at 268 nm. However, guanine has some absorption at 268 nm and xanthine some at 245 nm so that to use one wavelength for measurement requires the preparation of standards containing both compounds. Xanthine oxidase may be added to the reaction mixture and the xanthine formed oxidized to uric acid, which is then measured at 290 nm. These methods all require an ultraviolet spectrophotometer.

Simpler methods measure the amount of ammonia formed. Any free ammonia present can be corrected for by the use of a serum blank. In one method guanine is used as a substrate; after incubation the reaction is stopped by the addition of $0.33M$ sulfuric acid and the proteins are precipitated by the addition of sodium tungstate. The ammonia is determined in the filtrate by the use of the Berthelot reaction with phenol and hypochlorite. In another method azaguanine is used as a substrate. This compound is also deaminated by the enzyme. The reaction apparently proceeds somewhat more smoothly than with guanine. This method uses smaller amounts of serum, but the sensitivity is increased by adding the Berthelot reagents directly to the incubated mixture. The reaction is stopped by the addition of the phenol reagent. A small correction is made for the slight inhibiting effect of the protein present on the Berthelot reaction. In both methods the colorimetric readings are compared with those made on ammonia standards. The results are reported in international units as micromoles of ammonia formed per minute per liter of serum. With azaguanine as substrate the normal range was found to be 0.6 to 2.0 IU.

Increased serum guanase activity is generally found in association with acute hepatocellular disease. This increase generally parallels the rise in aspartate aminotransferase (GOT). Normal values are usually found in patients with primarily extrahepatic obstructive processes without secondary hepatocellular involvement. Since guanase is generally absent from cardiac, muscular, or pancreatic tissue, it should be helpful in the differential diagnosis of jaundice.

Adenosine Deaminase

E.C.3.5.4.4   adenosine aminohydrolase [194–197]

This enzyme catalyzes the reaction:

adenosine + H_2O → inosine + NH_3

In one method of assay the change from adenosine to inosine is followed by measuring the decrease in absorbance at 265 nm. This method has the disadvantages that the optimum amount of substrate cannot be used and changes in turbidity or other substances absorbing in the ultraviolet may interfere. The simpler method uses the determination of the amount of ammonia formed. This has been done by the use of Nessler's reagent, but the methods utilizing the Berthelot reaction are preferred. For a kinetic ultraviolet method the ammonia is determined by means of the reaction:

2-oxoglutarate + NH_3 + NADH → glutamate + H_2O + NAD

The results are usually expressed as micromoles of ammonia formed per minute per liter of serum.

There have been some reports that the determination of serum adenosine deaminase was of considerable value in the diagnosis of various types of cancer;

other investigators have not been able to confirm this but have felt that the determination may be of some value in the diagnosis of liver disease. The adenosine deaminase serum level was found to be increased in most cases of infectious hepatitis, most cases of hepatic cirrhosis, and many cases of biliary cirrhosis and drug jaundice. For liver disease it was stated that the determination of the adenosine deaminase level is as valuable as the determination of the ratio GOT:GPT, and is thus a useful test for liver disorders.

The use of adenosine deaminase determinations in some types of cancer has also been studied using agar gel electrophoresis to separate the isoenzymes, but the method has not yet proved of much value.

Arginase

E.C.3.5.3.1   L-arginine amidinohydrolase [198–200]

This enzyme catalyzes the reaction:

L-arginine + H_2O → L-ornithine + urea

This reaction is one of the steps in the formation of urea in the body. Arginase is found almost exclusively in the liver; thus its determination has been suggested for the diagnosis of liver disease. The methods for its determination have not proved very satisfactory. Those based on the determination of the urea formed suffer from the disadvantage that the urea already present in the serum may be an appreciable fraction of the total. In one procedure the original urea present is separated by means of Sephadex filtration. The urea formed during the incubation is then determined with diacetyl monoxime and thiosemicarbazide. In another method the urea is first removed from the serum by the action of urease. The urease is then inactivated by phenylmercuric nitrate (which does not affect the arginase) and the serum incubated with the arginine substrate. The urea formed is then determined by the same reaction as for the other procedure given above. In another method the endogenous urea is not separated but the amount of urea is determined in aliquots before and after incubation. The proteins are precipitated with trichloracetic acid and the urea determined using the reaction between urea and p-dimethylaminobenzaldehyde.

In another approach the decrease in the amount of arginine is measured. The arginine is determined by the Sakaguchi reaction with hydroxyquinoline and sodium hypobromite. However, this reaction has been shown to be rather unspecific, and a later method in which the ornithine formed is determined by a reaction with ninhydrin is preferable. Arginine reacts only slightly and this can be compensated for by means of a serum blank. In all the methods the presence of manganous ions is necessary for activation of the enzyme.

The normal levels for serum arginase have been given as 0.4 to 8.0 units (micromoles of urea formed per minute per liter of serum). Elevated levels have been found in patients with some inflammatory liver conditions. In liver disease the arginase activity has been found to parallel roughly the serum activity of aspartate aminotransferase, but does not appear to be a very satisfactory indicator of the extent of hepatic damage.

β-Glucuronidase

E.C.3.2.1.31   β-D-glucuronide glucuronohydrolase [201–205]

This enzyme catalyzes the hydrolysis of a number of β-D-glucuronides to an alcohol and a D-glucuronate. The enzyme is found in several different tissues, such as liver, spleen, lung, and endocrine tissues. Many substances are conjugated to glucuronides in the liver or elsewhere, but the exact role of glucuronidase in their hydrolysis is not clear.

In the determination of this enzyme several substrates have been used which are hydrolyzed to fluorescent substances (such as α-naphthyl and methylumbelliferyl glucuronides). These fluorescent methods are very sensitive, but the usual method is colorimetric, using phenolphthalein glucuronide which is hydrolyzed to phenolphthalein giving a red color in alkaline solution.

The serum levels of glucuronidase have been found to be elevated in a number of conditions including hyperthyroidism (decreased in hypothyroidism), diabetes, and atherosclerosis. The degree of increase in serum does not appear to be clearly related to the duration or severity of any of these conditions and the measurement of the enzyme has not been used for diagnosis or prognosis. Some studies have been made on the relation between the level of the enzyme in vaginal fluid and the presence of carcinoma of the uterus and, more especially, of the cervix. The correlation has not been very satisfactory. An elevation of serum glucuronidase levels has also been found in women with impending eclampsia, even before the appearance of the other symptoms, but elevations have also been found in a significant proportion of normal pregnancies.

A marked decrease in the levels of β-glucuronidase has been found in the serum and urine of patients suffering from a genetically controlled deficiency of the enzyme. The disease is characterized by short stature, progressive skeletal deformities, and mental retardation.

Lysozyme

E.C.3.2.1.17   mucopeptide *N*-acetylmuramoylhydrolase;
(muramidase) [206–210]

This enzyme probably hydrolyzes the 1, 4 links between *N*-acetylmuramic acid and 2-acetamido-2-deoxyglucose residues in some mucopolysaccharides and mucopeptides. The usual method for the determination of this enzyme is based on the fact that the enzyme will attack the cell walls of the microorganism *Micrococcus lysodeikticus*, lysing the cells and thus causing a decrease in the turbidity of a cell suspension. The material to be assayed is added to a buffered suspension of the cells and the rate of decrease in the turbidity when measured at 450 nm is noted. The method has also been automated. The results may be expressed in arbitrary units. One unit is equal to a decrease of 0.001 in absorbance/min, when the reaction is carried out at 30°C at a pH of 7.0, and read in cuvettes with a 1 cm light path and containing a total volume of 3 ml. A standard can also be used that consists of crystallized lysozyme obtained from chicken egg white. Using such standards the enzyme concentration may be expressed as micrograms per milliliter. This is one of the few instances in which the enzyme concentration may be expressed in mass units. Other methods have also been used, particularly for measuring smaller amounts; they include immunochemical methods and a simple diffusion method in which the clearing of an agar suspension of the microorganism is noted around wells containing the sample material.

One of the chief clinical uses of lysozyme assays is in the classification of the leukemias. The serum lysozyme level is greatly increased in patients with the monocytic forms of acute leukemia and moderately elevated in the acute and chronic forms of granulocytic leukemia. It is usually not elevated in acute or chronic lymphocytic leukemia. It has also been suggested that the determination of the serum lysozyme level be used to follow the course of the disease under antileukemic therapy. Some instances of elevated lysozyme levels have been noted in other diseases such as tuberculosis, sarcoidosis, and azotemia.

The determination of the urinary lysozyme concentration has been used for the diagnosis of kidney disease. Normally only very small amounts of lysozyme are found in the urine of normal individuals (less than 2 μg/ml). In patients with acute or chronic glomerulonephritis with severe lesions as judged by renal biopsy, the urinary excretion of lysozyme is usually markedly elevated (up to 80 μg/ml). In those patients in whom the lesions were less severe, the urinary excretion of the enzyme was less. The serum level of the enzyme is also sometimes increased in renal disease. If the level rises above a threshold value of about 50 μg/ml, some lysozyme may be excreted in the urine regardless of the state of the kidneys.

Thus an excretion of lysozyme in the urine with a serum level below the threshold value is a good indication of renal damage.

L-*Asparaginase*

E.C.3.5.1.1   L-asparagine amidohydrolase [211, 212]

This enzyme catalyzes the hydrolysis of the amine group from asparagine to form aspartate and ammonia. The reaction is not reversible.

asparagine $+ H_2O \rightarrow$ aspartate $+ NH_3$

The enzyme may be determined by measuring the ammonia produced in the reaction by the phenol-hypochlorite method (Berthelot reaction) or by nesslerization.

Asparaginase is not normally found in human blood. The administration of this enzyme has been found useful in the treatment of lymphoblastic leukemia. During this form of therapy it is necessary to determine the enzyme level in serum to arrive at an effective dose. The enzyme is obtained from *Escherichia coli* and guinea pig serum.

The clinical use of this enzyme determination is not for diagnostic purposes, but rather to determine the level when it is used as a drug.

Trehalase

E.C.3.2.1.28 α, α-trehalose glucohydrolase [213–216]

This enzyme catalyzes the reaction:

trehalose $+ H_2O \rightarrow 2$ glucose

It is the only enzyme found in human serum that hydrolyzes a disaccharide. The disaccharide trehalose is found in some yeasts and fungi and in the hemolymph of many insects. The sugar is not found in appreciable quantities in human serum or tissue, but the enzyme that hydrolyzes it is found in small amounts in serum and the liver and in larger amounts in the cortex of the kidneys. It has been postulated that the enzyme has some function in the transport of glucose in the kidneys, but this has not been demonstrated. Since the enzyme is found chiefly in the kidneys, its determination in serum or urine has been suggested for the diagnosis of some types of kidney disease.

The determination of the enzyme is relatively simple. The sample of serum or urine is incubated with the substrate trehalose in phosphate buffer at a pH of between 5 and 6, for from 6 to 18 hr (depending upon the exact method used). The glucose formed is then measured by any of the standard methods for this sugar. With serum a blank is run to compensate for any glucose originally present. The results are calculated as micrograms of glucose produced per hour per milliliter of serum or urine, or as micromoles of trehalose decomposed per minute per milliliter.

It is claimed that the serum trehalase is derived chiefly from the liver and that it is decreased in cirrhosis and increased in jaundice, but in view of the wide range of values found in normal subjects (by one method, 25 to 900 units), its diagnostic value in serum would appear to be slight. Urinary trehalase appears to vary independently of the serum level and is very probably derived from the kidney cortex. The level in urine is normally low (in terms of units per milliliter, less than one-tenth of that in serum) and the value of its determination in kidney disease has not been established.

Lyases

The lyases, which constitute class 4 in the E.C. system, remove groups from their substrate, without hydrolysis, leaving double bonds, or conversely add groups to double bonds. Thus the enzyme aldolase catalyzes the reaction:

fructose-1,6-diphosphate $\rightleftharpoons$ dihydroxyacetone phosphate + glyceraldehyde-3-phosphate

It is somewhat difficult to see the relation between this reaction and the definition of lyases, but it may be noted that fructose has one double bond C=O group, and the two products each have one C=O group, so the net result is the formation of an additional double bond.

Aldolase (ALD)

E.C.4.1.2.13   fructose-1,6-diphosphate:D-glyceraldehyde-3-phosphate lyase [224–228]

This enzyme takes part in the intermediary breakdown of glucose at the level of fructose-1,6-diphosphate and converts it into dihydroxyacetone phosphate and glyceraldehyde-3-phosphate. It catalyzes the reaction shown above. Under the influence of the enzyme triosephosphate isomerase, the dihydroxyacetone phosphate is changed into glyceraldehyde-3-phosphate so that the final reaction product is two molecules of this compound.

For determination of ALD, the additional enzyme, glycerol phosphate dehydrogenase, and NADH may be added so that the following reaction occurs:

glyceraldehyde-3-phosphate $+$ NADH $+$ H$^+$ $\rightleftharpoons$ NAD $+$ glycerol phosphate

The reaction is followed by measuring the decrease in absorbance at 340 nm as the concentration of NADH decreases.

In the colorimetric method, given below, hydrazine is usually added to combine with the glyceraldehyde phosphate and dihydroxyacetone. After incubation, the reaction is stopped by precipitating the proteins with tungstic acid or trichloracetic acid. The filtrate is then hydrolyzed with alkali and treated with dinitrophenylhydrazine to form osazones, which give a characteristic reddish brown color. The intensity of the color produced is proportional to the amount of trioses. This is related to the aldolase activity.

The results have been reported in dihydroxyacetone units; 1 unit corresponds to the formation of 10 μg of dihydroxyacetone/ml of serum in 1 hr. In the Sibley-Lehninger procedure the unit corresponds to 1 mm^3 of fructose 1,6-diphosphate/hr at 37°C.

REAGENTS

1. Fructose 1,6-diphosphate, 0.05M.   Dissolve 250 mg of fructose 1,6-diphosphate sodium salt§ in 10 ml of water. This is stable for approximately 2 weeks at 0° to 5°C.

2. Hydrazine sulfate solution, 0.56M, pH 7.4.   Dissolve 7.28 g of hydrazine sulfate in 50 ml of water. Adjust pH to 7.4 with 1M NaOH and dilute to 100 ml with water.

3. Tris buffer, 0.05M, pH 7.4.   Dissolve 6.05 g of tris-(hydroxymethyl)aminomethane in about 800 ml of water in a 1 liter volumetric flask. Add 1M HCl (84 ml of concentrated acid diluted to 1 liter) until a pH of 7.4 is reached (about 40 ml required) and dilute to 1 liter.

4. 2,4-Dinitrophenylhydrazine, 100 mg/dl in 2M HCl.   Dissolve 100 mg of 2,4-dinitrophenylhydrazine in 100 ml of 2M HCl. Filter and store in the dark.

5. Sodium hydroxide, 0.75M.   Dissolve 30 g of reagent grade sodium hydroxide in water and dilute to 1 liter.

6. Trichloracetic acid (TCA), 0.62M.   Dissolve 20 g of TCA in water and dilute to 200 ml.

7. Aldolase calibration solution (1.5 mM dihydroxyacetone).   Dissolve 27 mg of dihydroxyacetone|| in 200 ml of tris buffer. This is used for preparing the calibration curve.

§ Sigma Chemical Co., St. Louis, Mo. 63118.
|| Schwarz/Mann Div., Becton, Dickinson & Co., Orangeburg, N.Y. 10962.

PROCEDURE

To test tubes labeled test and blank add 1.4 ml of tris buffer, 0.2 ml of hydrazine sulfate solution, and 0.2 ml of serum. Warm in water bath at 37°C for several minutes and then add 0.2 ml of the substrate (fructose 1,6-diphosphate solution) to test only. Incubate both tubes at 37°C for 30 min; then add 2.0 ml of 0.62M TCA to each tube. Add 0.2 ml of substrate to blank tube. Mix tubes well and centrifuge. Add 1 ml of supernatant from each tube to separate tubes, again labeled test and blank. To each tube add 1.0 ml of 0.75M sodium hydroxide and mix. Allow to stand for 10 min. Add 1.0 ml of the dinitrophenylhydrazine solution and mix. Place in a 37°C bath for exactly 10 min. Remove from bath, add 7.0 ml of 0.75M NaOH, and allow to stand at room temperature for 5 min. Read absorbance of test against blank at 540 nm. Obtain enzyme units from calibration curve. If value exceeds 125 IU, repeat the test using a fivefold dilution of the serum and multiply value obtained from the curve by 5.

CALIBRATION CURVE

To a series of six test tubes, add the amounts of dihydroxyacetone solution and tris buffer as given in Table 14-11. Then add to each tube, with mixing after each addition, 0.1 ml of hydrazine sulfate solution, 1.0 ml of TCA solution, 2.0 ml of 0.75M NaOH, and 2.0 ml of dinitrophenylhydrazine solution. Incubate at 37°C for 10 min. Add 14 ml of 0.75M NaOH to each tube and read absorbance at 540 nm against tube 1 (blank). The readings should be made within 15 min after the addition of the alkali. Plot absorbances of the five standards against the corresponding units of aldolase for a calibration curve.

Table 14-11. Calibration Curve for Determination of ALD

Tube No.	Dihydroxyacetone Solution (ml)	Tris Buffer (ml)	Units[a]	
			IU	S-L
1	0	0.9	0	0
2	0.1	0.8	25	33
3	0.2	0.7	50	67
4	0.3	0.6	75	100
5	0.4	0.5	100	134
6	0.5	0.4	125	167

[a]IU = international units = μmol/min/liter of serum. S-L = Sibley-Lehninger units.

CALCULATION

When cuvettes having exactly 1 cm light path are used, the units may be cal-culated as follows:

absorbance $\times$ 225 = S-L units

or

absorbance $\times$ 167 = IU/liter

NORMAL VALUES AND INTERPRETATION OF RESULTS

The normal levels by this method are 11 to 40 IU or 15 to 53 S-L units. Significant elevations are found in acute infectious hepatitis, usually up to 20 times the aver-age normal. Normal levels are usually found in portal cirrhosis and obstructive jaundice.

A significant increase is also found in progressive muscular dystrophy (primary myopathy). No elevation is found in muscular dystrophy secondary to alterations of the nerves or nerve centers.

Increases in aldolase activity have been reported after myocardial infarction, advanced prostatic carcinoma, hepatoma, large pulmonary infarcts, hemorrhagic pericarditis, erythroblastosis fetalis, acute pancreatitis, and hemolytic anemia.

ALDOLASE ISOENZYMES

Two aldolase isoenzymes have been identified in human blood, one found pri-marily in skeletal muscle and the other in the liver. They have been differentiated by their action on the two substrates fructose 1,6-diphosphate and fructose-1-phosphate.

The aldolase of muscle origin shows more than 30 times the activity with fructose 1,6-diphosphate than it does with fructose-1-phosphate. The liver aldo-lase and that found in normal serum are equally active against both substrates.

Isomerases

The isomerases act to change a molecule into an isomer, another molecule with the same empirical formula but different arrangement of the atoms. Thus the enzyme glucose phosphate isomerase changes glucose 1-phosphate into fructose 1-phosphate; the two molecules have the same number of carbon, hydrogen, and oxygen atoms, but in glucose the carbonyl group is on the end atom (aldehyde) and in fructose, the carbonyl group is on the second atom in the chain (ketone). The isomerases constitute class 5 in the E.C. system.

Phosphohexose Isomerase (PHI; Glucose Phosphate Isomerase)

E.C.5.3.1.9.   D-glucose-6-phosphate ketol-isomerase [217–222]

This enzyme catalyzes the interconversion of glucose 6-phosphate and fructose 6-phosphate. It has been determined by using fructose 6-phosphate as substrate and measuring the glucose 6-phosphate formed by a series of coupled reactions involving glucose 6-phosphate dehydrogenase (G-6-PDH) and sometimes additionally with 6-phosphogluconate dehydrogenase (6-PGDH). The reactions are:

$$\text{fructose-6-phosphate} \xrightleftharpoons{\text{PHI}} \text{glucose-6-phosphate}$$

$$\text{glucose-6-phospsate} + \text{NAD}^+ \xrightleftharpoons{\text{G-6-PDH}} \text{6-phosphogluconate} + \text{NADH} + \text{H}^+$$

For increased sensitivity the following may be added:

$$\text{6-phosphogluconate} + \text{NADP}^+ \xrightleftharpoons{\text{6-PGDH}} \text{ribose-5-phosphate} + \text{NADPH} + \text{H}^+ + \text{CO}_2$$

The NADH produced in the second reaction is proportional to the glucose 6-phosphate produced in the first. The increased absorbance at 340 nm is used as the measure of PHI activity. The results are reported in international units; 1 IU is defined as the reduction of 1 μmol of NAD/min at 30°C. The normal levels by this method range from 19 to 87 IU/liter.

The method of Bodansky uses glucose-6-phosphate as substrate and the fructose-6-phosphate formed is determined by the Seliwanoff reaction, as a measurement of the enzyme activity. In this reaction fructose 6-phosphate is reacted with resorcinol in hydrochloric acid to produce a red chromogen which is measured spectrophotometrically. The enzyme unit is defined as the reciprocal of that concentration of serum, expressed as milliliters of serum per milliliter of reaction mixture, that would cause the formation of 25 μg of fructose, as fructose 6-phosphate, per milliliter of reaction mixture under the specified conditions of the test.

METHOD OF BODANSKY

REAGENTS

1. Buffered substrate, 0.0024M G-6-P, pH 7.4.   Dissolve 425 mg of glucose 6-phosphate, disodium salt, hydrate# and 6.1 g of tris-(hydroxymethyl)aminomethane in about 300 ml of water. Add 125 ml of 0.1M HCl, check pH and adjust to 7.4, and dilute to 500 ml. This solution is stable for 6 months at 5°C.

Sigma Chemical Co., St. Louis, Mo. 63118.

2. Sodium chloride, $0.15M$. Dissolve 9.0 g of sodium chloride in 1 liter of distilled water.

3. Resorcinol, 100 mg/dl ethanol. Dissolve 100 mg of resorcinol in 100 ml of 95% ethyl alcohol. Store in amber bottle at 5°C.

4. Hydrochloric acid, $10M$. Carefully add 250 ml of concentrated HCl to 50 ml of water and mix well.

5. Trichloracetic acid (TCA), $0.31M$. Dissolve 10 g of TCA in about 150 ml of water and dilute to 200 ml.

6. Ethyl alcohol, 95%, undenatured.

7. Fructose-6-phosphate calibration solutions.

A. Stock solution, 1 mg/ml. Dissolve 138 mg of the disodium salt of fructose 6-phosphate* (equivalent to 100 mg of fructose 6-phosphate) in water in a 100 ml volumetric flask and dilute to volume.

B. Working solution, 100 μg/ml. Dilute 10 ml of the stock solution to 100 ml with water, in a 100 ml volumetric flask.

PROCEDURE

Add 2.0 ml of buffered substrate and 0.4 ml of sodium chloride solution to two test tubes labeled test and blank. Place the test in a 37°C water bath to warm to bath temperature; leave the blank at room temperature. Add 0.10 ml of serum to test and incubate for exactly 30 min. After incubation, stop reaction by adding 2.5 ml of TCA, mixing rapidly and thoroughly. Add 2.5 ml of TCA and 0.1 ml of serum to blank tube and mix thoroughly. Allow to stand at room temperature for about 5 min and centrifuge both tubes.

Into correspondingly labeled tubes, pipet 2.0 ml of clear supernatant. Pipet a series of standards to be run with each set of unknowns, as shown in Table 14.12.

Table 14-12. Determination of PHI by the Bodansky Method

Tube No.	Working Standard Solution (ml)	Water (ml)	Fructose (μg)
1	0	2.0	0
2	0.25	1.75	25
3	0.50	1.50	50
4	1.00	1.00	100
5	1.50	0.50	150
6	2.00	0	200

* Sigma Chemical Co., St. Louis, Mo. 63118.

To all tubes—tests, blanks, and standards—add 2.0 ml of resorcinol solution and 6.0 ml of 10M HCl. Mix by inversion and place in water bath at 80°±1°C for 15 min. Cool and read absorbance at 490 nm within 30 min after the heating period.

Read the standard tubes, tests, and corresponding blanks against tube no. 1 from the standard curve.

Prepare a calibration curve by plotting absorbances of the various standards against the corresponding micrograms of fructose. Determine amount of fructose in the tests and blanks from calibration curve and subtract the blanks from the corresponding tests, to obtain the micrograms of fructose formed. Convert to Bodansky units by referring to Table 14.13. The table applies only when 0.04 ml (i.e., 0.1/2.5) of serum being tested is used per milliliter of reaction mixture. If a value exceeds 300 units, it should be repeated using a dilution of the serum and multiplying the value obtained from the table by the dilution factor.

NORMAL VALUES AND INTERPRETATION OF RESULTS

The normal PHI activity by this method ranges from 8 to 40 units. Elevations have been reported in all types of carcinoma and in viral and acute hepatitis. This enzyme has been reported elevated in the serum of the greater percentage of patients with cancer of the stomach, colon, pancreas, head, neck, esophagus, lung, breast, and prostate. Elevated values have also been reported in myelocytic leukemia. Determination of this enzyme has been proposed to follow the course of neoplasia, the growth or regression of tumor, and for following the effectiveness of therapy.

Triosephosphate Isomerase (TPI)

E.C.5.3.1.1.   D-glyceraldehyde-3-phosphate ketol-isomerase [223]

This enzyme catalyzes the interconversion of glyceraldehyde-3-phosphate and dihydroxyacetone phosphate.

glyceraldehyde-3-phosphate ⇌ dihydroxyacetone phosphate

The rate of formation of dihydroxyacetone phosphate is measured by linking it to the oxidation of NADH to NAD in the presence of the enzyme α-glycerophosphate dehydrogenase.

$$\text{dihydroxyacetone phosphate} + \text{NADH} + \text{H}^+ \xrightarrow{\ \alpha\text{-GPDH}\ } \alpha\text{-glycerophosphate} + \text{NAD}^+$$

The decrease in absorbance caused by the oxidation of NADH to NAD at 340 nm is used as a measure of enzyme activity. A screening test has also been devised in which the loss of fluorescence of NADH under ultraviolet light serves as an indicator of enzyme activity.

The enzyme is determined in red blood cells hemolysate and the results are reported in terms of international units per gram of hemoglobin. The normal adult level is 1440 ± 160 IU/gHb.

TPI is markedly inhibited by ammonium sulfate.

Several cases of TPI deficiency resulting in nonspherocytic congenital hemolytic anemia have been reported.

In Tay-Sachs disease (inborn error of lipid metabolism), aldolase isoenzymes of muscle origin are occasionally found to be reduced and those of liver origin may be absent or markedly decreased. Low levels are also found in heterozygous carriers of the disease.

Aminolevulinate Dehydratase

E.C.4.2.1.24   5-aminolevulinate hydro-lyase [229–234]

This enzyme catalyzes the condensation of two molecules of 5-aminolevulinic acid to form one molecule of porphobilinogen with the elimination of two molecules of water. The activity of this enzyme, which is found chiefly in the erythrocytes, is inhibited by traces of heavy metals such as mercury or lead. A decrease in the enzyme activity has been suggested as an indicator of poisoning by such metals, particularly lead. This determination could serve to complement other tests for lead poisoning such as the determination of aminolevulinic acid and porphyrin excretion in the urine and the determination of the lead content of blood or urine.

The determination of the enzyme is relatively simple. The washed cells are lysed, or, since the enzyme is almost all in the cells, whole blood can be lysed directly, and added to 5-aminolevulinic acid in a buffered solution. The porphobilinogen formed is determined with a modified Ehrlich's reagent (*p*-dimethylaminobenzaldehyde in perchloric and acetic acids). Since pure porphobilinogen is not available as a standard, one either uses the literature value for the molar extinction coefficient for porphobilinogen or expresses the activity in arbitrary units in terms of the change in absorbance per minute under specified conditions. After incubation, the proteins are precipitated with trichloracetic acid before determining the porphobilinogen. A blank is generally run but this is usually very low. The results are usually calculated in terms of units per milliliter or deciliter of packed red cells.

Table 14-13. Conversion of Micrograms of Fructose Formed into Bodansky Units of Phosphohexose Isomerase Activity

Micro-grams	Bodansky Units	Micro-grams	Bodansky Units	Micro-grams	Bodansky Units	Micro-grams	Bodansky Units
1	1	36	36	71	84	106	161
2	2	37	37	72	86	107	163
3	3	38	38	73	88	108	166
4	4	39	39	74	90	109	170
5	5	40	40	75	92	110	172
6	6	41	41	76	93	111	176
7	7	42	42	77	94	112	179
8	8	43	43	78	97	113	182
9	9	44	44	79	99	114	186
10	10	45	45	80	100	115	190
11	11	46	46	81	101	116	194
12	12	47	47	82	103	117	198
13	13	48	48	83	106	118	202
14	14	49	50	84	107	119	206
15	15	50	51	85	109	120	208
16	16	51	52	86	111	121	212
17	17	52	54	87	113	122	216
18	18	53	55	88	115	123	220
19	19	54	56	89	117	124	224
20	20	55	58	90	120	125	229
21	21	56	60	91	122	126	234
22	22	57	61	92	123	127	240
23	23	58	62	93	126	128	245
24	24	59	64	94	128	129	252
25	25	60	66	95	131	130	257
26	26	61	67	96	133	131	264
27	27	62	69	97	136	132	272
28	28	63	70	98	139	133	278
29	29	64	72	99	141	134	286
30	30	65	74	100	144	135	293
31	31	66	76	101	147	136	300
32	32	67	78	102	149	137	307
33	33	68	80	103	153	138	314
34	34	69	81	104	155	139	324
35	35	70	84	105	158	140	331

The inverse correlation between enzyme activity and blood lead level (i.e., the higher the blood lead level, the lower the enzyme activity) is fairly good between blood lead levels of approximately 10 and 50 μg/dl. At higher blood levels of lead, the further decrease in enzyme activity is much slower so that exact decrease in blood enzyme level is not as satisfactory for distinguishing between levels of, say, 50 and 100 μg/dl as it is between 25 and 50 μg/dl, but this latter range is often the one of most interest. An increased level of blood alcohol will cause a temporary decrease in the enzyme activity, but this should offer no difficulties in most situations. Because of the different units used by the various methods, no normal values are given here. One must use those given for the particular method, or preferably determine one's own normal values. The method of Burch and Siegel appears to be a reliable one.

References

1. *Enzyme Nomenclature.* Recommendations (1972) of the International Union of Pure and Applied Chemistry and the International Union of Biochemistry. Amsterdam: Elsevier, 1973.
2. Baron, D. N., Moss, D. W., Walker, R. G., and Wilkinson, J. H. *J. Clin. Pathol.* 24:656, 1971.
3. Guilbaut, G. G. *Enzymic Methods of Analysis.* London: Pergamon, 1970.
4. Boyer P. D. [Ed.]. The Enzymes, 3rd ed. Vol. 2, *Kinetics and Mechanism.* New York: Academic, 1970.
5. Laidler, K. J. *The Chemical Kinetics of Enzyme Action.* Oxford: Oxford University Press, 1958.
6. Edsall, J. T. *Enzymes and Enzyme Systems.* Cambridge, Mass.: Harvard University, Press, 1951.
7. Harper, H. A. *Review of Physiological Chemistry,* 10th ed. Los Altos, Calif.: Lange, 1965.
8. Wilkinson, J. H. *Isoenzymes,* 2nd ed. Philadelphia: Lippincott, 1970.
8a. Rand, R. N. *Clin. Chem.* 15:839, 1969.
9. Bergmeyer, H. U., Bernt, E., and Hess, B. *Methods of Enzymatic Analysis.* New York: Academic, 1965. P. 736.
10. Batsaki, J. G., Briere, R. O., and Markel, S. F. *Diagnostic Enzymology.* Chicago: ASCP Manual Commission on Continuing Education, 1970.
11. Henry, R. J. *Clinical Chemistry, Principles and Technics.* New York: Hoeber Med. Div., Harper & Row, 1965.
12. Amador, E., and Wacker, W. E. C. In D. Glick [Ed.], *Methods of Biochemical Analysis,* vol. 13. New York: Interscience, 1965.
13. Babson, A. L. *Clin. Chim. Acta* 1:121, 1967.
14. Caband, P. G., and Wroblewski, F. *Am. J. Clin. Pathol.* 30:234, 1958.
15. Snodgras, P. J., Wacker, W. E. C., Eppinger, E. C., and Vallee, B. L. *N. Engl. J. Med.* 261:1259, 1959.
16. Berger, L., and Broida, D. *Sigma Technical Bulletin No. 500.* St. Louis: Sigma Chemical Co., 1964.

17. Wacker, W. E. C., Ulmer, D. D., and Vallee, B. L.   *N. Engl. J. Med.* 255:449, 1956.
18. Amador, E., Dorfman, L. E., and Wacker, W. E. C.   *Clin. Chem.* 9:391, 1963.
19. Babson, A. L., and Phillips, G. E.   *Clin. Chim. Acta* 12:210, 1965.
20. Briere, R. O., Preston, J. A., and Batsakis, J. G.   *Am. J. Clin. Pathol.* 45:544, 1966.
21. Wieland, T., and Pfleiderer, G.   *Biochem. Z.* 329:112, 1957.
22. Vessell, E. S., and Bearn, A. G.   *J. Clin. Invest.* 37:672, 1958.
23. Wroblewski, F., and Gregory, K.   *Ann. N.Y. Acad. Sci.* 94:912, 1961.
24. Strandjord, P. E., Clayson, K. J., and Freier, E. F.   *J.A.M.A.* 182:1099, 1962.
25. Babson, A. L.   *Clin. Chim. Acta* 16:121, 1967.
26. Blanco, A., and Zinkham, W. H.   *Science* 139:601, 1963.
27. Eliasson, R., Haggman, K., and Wiklund, B.   *Scand. J. Lab. Invest.* 20:353, 1967.
28. Wacker, W. E. C., and Dorfman, L. E.   *J.A.M.A.* 181:972, 1962.
29. Dorfman, L. E., Amador, E., and Wacker, W. E. C.   *J.A.M.A.* 184:123, 1963.
30. Fleisher, G. A., Wakim, K. G., and Goldstein, N. P.   *Proc. Staff Meet. Mayo Clin.* 32:188, 1957.
31. Wroblewski, F., Decker, B., and Wroblewski, R.   *Am. J. Clin. Pathol.* 28:269, 1957.
32. Rosalki, S. B.   *J. Clin. Pathol.* 15:566, 1962.
33. Preston, J. A., Batsakis, J. G., and Briere, R. O.   *Am. J. Clin. Pathol.* 41:237, 1964.
34. Elliot, B. A., and Wilkinson, J. H.   *Lancet* 1:698, 1961.
35. Rosalki, S. B., and Wilkinson, J. H.   *J.A.M.A.* 189:61, 1964.
36. Knottinen, A., and Haolmen, P. I.   *Am. J. Cardiol.* 101:525, 1962.
37. Taylor, T. H., and Friedman, M. E.   *Clin. Chem.* 6:209, 1960.
38. Bowers, G. N., Jr., and MacDuffee, R. C.   *Clin. Chem.* 5:369, 1959,
39. Wolfson, S. K., and Williams-Ashman, H. G.   *Proc. Soc. Exp. Biol. Med.* 96:231, 1957.
40. Okumura, M., and Spellberg, M. A.   *Gastroenterology* 39:305, 1960.
41. Dawkins, M. R., MacGregor, W. G., and McLean, A. E.   *Lancet* 2:827, 1959.
42. Beutler, E.   *Red Cell Metabolism,* New York: Grune & Stratton, 1971.
43. Martinek, R. G., Jacobs, S. L., and Hammer, F. E.   *Clin. Chim. Acta* 36:75, 1972.
44. Brewer, G. J., and Dern, R. J.   *Clin. Res.* 12:215, 1964.
45. Brewer, G. J., Tarlov, A. R., and Alving, A. S.   *J.A.M.A.* 180:386, 1962.
46. Burka, E. R., Weaver, Z., and Marks, P. A.   *Ann. Intern. Med.* 64:817, 1966.
47. Wacker, W. E. C., Ulmer, D. D., and Vallee, B. L.   *N. Engl. J. Med.* 255:449, 1956.
48. Bergmeyer, H. U., and Bernt, E.   In H. Bergemeyer [Ed.], *Methods of Enzymatic Analysis.* New York: Academic, 1965. P. 757.
49. Schwartz, P. L., and Burns, H. J., Jr.   *Clin. Chem.* 16:579, 1970.
50. Bing, R. J., Castellanos, A., and Siege, A.   *J.A.M.A.* 164:647, 1957.
51. Blakely, R. L.   *Biochem. J.* 49:259, 1951.
52. Holzer, H., Haan, J., and Schneider, S.   *Biochem. Z.* 326:451, 1955.
53. Asada, M., and Galambos, J. T.   *Gastroenterology* 44:578, 1963.
54. DeRitis, F., Giusti, G., Piccinino, F., and Cacciatore, L.   *Bull. W.H.O.* 32:59, 1965.
55. Wiesner, I. S., Rawnsley, H. M., Brooks, F. P., and Senior, J. R.   *Am. J. Digest. Dis.* 10:147, 1965.
56. Hohorst, H. J.   In H. U. Bergmeyer [Ed.], *Methods of Enzymatic Analysis.* New York: Academic, 1965. P. 141.
57. Wooten, I. D. P., and Sheppard, Y.   *J. Obstet. Gynaecol. Br. Commonw.* 74:270, 1967.

543

58. Cameron, C. B., and Husain, O. A. N.   *Br. Med. J.* 1:1529, 1965.
59. Hoffman, R. L., and Merritt, J. W.   *Am. J. Obstet. Gynecol.* 92:650, 1965.
60. Sanner, T., Bentzen, H., Kolstad, P., Norbye, K., and Pihl, A.   *Acta Obstet. Gynecol. Scand.* 49:371, 1970.
61. Frieden, C.   *J. Biol. Chem.* 234:809, 1959.
62. Carlson, A. S., Siegelman, A. M., and Robertson, T.   *Am. J. Clin. Pathol.* 38:260, 1962.
63. Schmidt, E., and Schmidt, F. W.   *Klin. Wochenschr.* 40:962, 1962.
64. Filippa, G.   *Enzymol. Biol. Clin.* 3:97, 1963.
65. Jung, K.   *Clin. Chim. Acta* 36:231, 1972.
66. Schmidt, E., Schmidt, F. W., Horn, H. D., and Gerlach, U.   In H. U. Bergmeyer [Ed.], *Methods of Enzymatic Analysis.* New York: Academic Press, 1965. P. 658.
67. Mezey, E., Cherrick, G. R., and Holt, P. R.   *N. Engl. J. Med.* 279:241, 1968.
68. Mezey, E., Cherrick, G. R., and Holt, P. R.   *J. Lab. Clin. Med.* 71:798, 1968.
69. Wolfson, S. K., Spencer, J. A., Starkel, R. L., and William-Ashman, H. G.   *Ann. N.Y. Acad. Sci.* 75:260, 1968.
70. Huennekens, F. M., Caffery, R. W., Basford, R. D., and Gabrio, B. W.   *J. Biol. Chem.* 227:261, 1957.
71. Hegesh, E., Calmanovici, N., and Avrom, M.   *J. Lab. Clin. Med.* 72:339, 1968.
72. Scott, E. M.   *J. Clin. Invest.* 39:1176, 1960.
73. Houchin, O. D.   *Clin. Chem.* 4:519, 1958.
74. Ravin, H. A.   *Lancet* 2:726, 1956.
75. Ravin, H. A.   *J. Lab. Clin. Med.* 58:161, 1961.
76. Scheinberg, I. H., and Morrell, A. G.   *J. Clin. Invest.* 36:1193, 1957.
77. Rice, E. W.   *Anal. Biochem.* 3:452, 1962.
78. Scheinberg, I. H., Morrell, A. G., Harris, R. S., and Berger, A.   *Science* 126:925, 1957.
79. Asnis, R. E.   *J. Biol. Chem.* 213:77, 1955.
80. Manso, C., and Wroblewski, F.   *J. Clin. Invest.* 37:214, 1958.
81. Manso, C., Sugiura, K., and Wroblewski, F.   *Cancer Res.* 18:682, 1958.
82. Aldridge, W. N., and Johnson, M. K.   *Biochem. J.* 73:270, 1959.
83. Beutler, E.   *J. Clin. Invest.* 48:1957, 1969.
84. Haining, J. L., and Legan, J. S.   *Anal. Biochem.* 21:337, 1967.
85. Guilbault, G. G., Brignac, P., Jr., and Zimmer, M.   *Anal. Chem.* 40:190, 1968.
86. Rambber, C. R. H.   *J. Lab. Clin. Med.* 74:828, 1969.
87. Al-Khalidi, U. A. S., and Geha, R. S.   *Clin. Chim. Acta* 14:833, 1966.
88. Hajj, S., Nasrallah, S. M., Shaamma, M. H., and Al-Khalidi, R.   *Enzymol. Biol. Clin.* 8:1, 1967.
89. Dietrich, R. D., and Erwin, V. G.   *Anal. Biochem.* 30:395, 1969.
90. Tabakoff, B., and Avoivisauos, S. G. A.   *Anal. Chem.* 44:427, 1972.
91. Drujan, B. S., and Diaz Borges, J. M.   *Z. Anal. Chem.* 243:662, 1968.
92. Guilbault, G. G., Kuan, S. S., and Brignac, P. J., Jr.   *Anal. Chim. Acta* 47:503, 1969.
93. Tipton, K. F.   *Anal. Biochem.* 28:318, 1969.
94. Wyngaarden, J. B., and Howell, R. R.   In J. B. Stanbury, J. B. Wyngaarden, and D. S. Fredrickson [Eds.], *The Metabolic Basis of Inherited Disease,* 2nd ed. New York: Blakiston Div., McGraw-Hill, 1966. P. 1343.
95. Werner, E., and Reider, R.   *Z. Klin. Chem.* 1:115, 1963.

96. Sinha, A. K.   *Anal. Biochem.* 47:389, 1972.
97. Kirk, J. E.   *Clin. Chem.* 9:763, 1963.
98. Cabaud, P., Leeper, R., and Wroblewski, F.   *Am. J. Clin. Pathol.* 26:1101, 1956.
99. Karmen, A.   *J. Clin. Invest.* 34:131, 1955.
100. Reitman, S., and Frankel, S.   *Am. J. Clin. Pathol.* 28:56, 1957.
101. Annino, J. S.   *Tech. Bull. Regist. Med. Technol.* 36:203, 1966.
102. Babson, A. L., Shapiro, P. O., Williams, P. A. R., and Phillips, G. E.   *Clin. Chim. Acta* 7:199, 1962.
103. Babson, A. L.   *Clin. Chem.* 6:394, 1960.
104. Amador, E., and Wacker, W. E. C.   *Clin. Chem.* 8:343, 1962.
105. Wolff, P. L., Langton, C., Potolsky, A. I., and Williams, D. J.   *Clin. Chem.* 17:341, 1971.
106. Hughes, B. P.   *Clin. Chim. Acta* 7:597, 1962.
107. Nielsen, L., and Ludvigsen, B.   *J. Lab. Clin. Med.* 62:159, 1963.
108. Menache, R., and Gaist, L.   *Clin. Chim, Acta* 26:591, 1969.
109. Nuttal, F. Q., and Wedlin, D. S.   *J. Lab. Clin. Med.* 68:324, 1966.
110. Rosalki, S. B.   *J. Lab. Clin. Med.* 69:696, 1967.
111. Szasz, F.   *Clin. Chem.* 15:124, 1969.
112. Jacobs, W. L. W.   *Clin. Chim. Acta* 31:175, 1971.
113. Zein, M., and Discombe, G.   *Lancet* 2:748, 1970.
114. Lum, G., and Gambino, R.   *Clin. Chem.* 18:110, 1972.
115. Ceriotti, G., and Gazaniga, A.   *Clin. Chim. Acta* 16:436, 1967.
116. Reichard, H., and Reichard, P.   *J. Lab. Clin. Med.* 52:709, 1958.
117. Knottinen, A.   *Clin. Chim. Acta* 21:29, 1968.
118. Lorentz, K., and Wrabetz, W.   *Z. Klin. Chem. Klin. Biochem.* 9:220, 1971.
119. Beutler, E., and Baluda, M. C.   *Clin. Chim. Acta* 13:369, 1966.
120. Beutler, E., and Mitchell, M.   *J. Lab. Clin. Med.* 72:527, 1968.
121. Copenhaver, J. H., Bausch, L. C., and Fitzgibbons, J. F.   *Anal. Biochem.* 30:327, 1969.
122. Beutler, E.   *Red Cell Metabolism.* New York: Grune & Stratton, 1971. P. 79.
123. Beutler, E.   *Red Cell Metabolism.* New York: Grune & Stratton, 1971. P. 38.
124. Altay, C., Alper, C. A., and Nathan, D. G.   *Blood* 36:219, 1960.
125. Kaplan, J. C., and Beutler, E.   *Science* 159:215, 1968.
126. Brunetti, P., and Nenci, G.   *Enzymol. Biol. Clin.* 4:51, 1964.
127. Tanphaichitr, V. S., and Van Eys, J.   *Clin. Chim. Acta* 41:41, 1972.
128. Tanaka, K. R., Valentine, W. N., and Miwa, S.   *Blood* 19:267, 1962.
129. Bigley, R. H., Stenzel, P., Jones, R. T., Campos, J. O., and Koler, R. D.   *Folia Haematol.* (Leipz.) 91:61, 1969.
130. Dreyfus, P. M.   *N. Engl. J. Med.* 267:596, 1962.
131. Abkarian, M., and Dreyfus, P. M.   *J.A.M.A.* 203:77, 1968.
132. Massod, M. F., McGuire, S. L., and Werner, K. R.   *Am. J. Clin. Pathol.* 55:165, 1971.
133. Hoffman, I., Knapp, A., Reitz, K., and Milner, C. L.   *Clin. Chim Acta* 33:415, 1971.
134. Bodansky, M.   *J. Biol. Chem.* 101:93, 1933.
135. King, E. J., and Armstrong, M. D.   *Can. Med. Assoc. J.* 31:376, 1934.
136. Bessey, O. A., Lowry, O. H., and Brock, M. J.   *J. Biol. Chem.* 164:321, 1946.

137. Huggins, C., and Talalay, P. *J. Biol. Chem.* 159:399, 1945.
138. Babson, A. L., Greeley, S. J., Coleman, C. M., and Phillips, G. E. *Clin. Chem.* 12:482, 1966.
139. Kerkhoff, J. F. *Clin. Chim. Acta* 22:231, 1968.
140. Kreisher, J. H., Close, V. A., and Fishman, W. H. *Clin. Chim. Acta* 11:122, 1965.
141. Posen, S., Neale, F. C., and Clubb, J. S. *Ann. Intern. Med.* 62:1234, 1965.
142. Stolbach, L. L., Krant, M. J., and Fishman, W. H. *N. Engl. J. Med.* 281:757, 1969.
143. Fishman, W. H., and Ghosh, N. K. *Adv. Clin. Chem.* 10:255, 1967.
144. Winkelman, J., Nadler, S., Demetriou, J., and Pileggi, V. J. *Am. J. Clin. Pathol.* 57:625, 1972.
145. Fishman, W. H., and Lerner, F. *J. Biol. Chem.* 200:89, 1953.
146. Jacobsson, K. *Scand. J. Clin. Lab. Invest.* 11:358, 1960.
147. Babson, A. L., and Phillips, G. E. *Clin. Chim. Acta* 13:264, 1966.
148. Babson, A. L., and Read, P. A. *Am. J. Clin. Pathol.* 32:88, 1959.
149. Sudhof, H., Meumann, G., and Oloffs, J. *Dtsch. Med. Wochenschr.* 89:217, 1964.
150. Van Gorkon, W. J. *Ned. Tijdschr. Geneeskd.* 106:297, 1962.
151. Eschar, J., Rudzk, C., and Zimmerman, H. J. *Am. J. Clin. Pathol.* 47:589, 1967.
152. Belfield, A., Ellis, G., and Goldberg, D. M. *Clin. Chem.* 16:396, 1970.
153. Rieder, S. V., and Otero, M. *Clin. Chem.* 15:727, 1969.
154. Bodansky, O., and Schwartz, M. K. *Adv. Clin. Chem.* 11:277, 1968.
155. Somogyi, M. *J. Biol. Chem.* 134:315, 1940.
156. Somogyi, M. *J. Biol. Chem.* 142:579, 1942.
157. Caraway, W. T. *Am. J. Clin. Pathol.* 32:97, 1959.
158. McNair, R. D. In R. P. MacDonald [Ed.], *Standard Methods of Clinical Chemistry*, vol. 6. New York: Academic, 1970.
159. Tietz, N. W., Borden, T., and Stapleton, J. D. *Am. J. Clin. Pathol.* 31:148, 1959.
160. Cherry, I. S., and Crandall, L. A. *Am. J. Physiol.* 100:266, 1932.
161. Saifer, A., and Perle, F. *Clin. Chem.* 7:178, 1960.
162. Massion, C. G., and Seligson, D. *Am. J. Clin. Pathol.* 48:307, 1967.
163. Massion, C. G. *Lab. Med.* 2:26, 1961.
164. Rutenburg, A. M., Goldbarg, J. A., and Pineda, E. P. *N. Engl. J. Med.* 259:469, 1966.
165. Goldbarg, J. A., and Rutenburg, A. M. *Cancer* 11:283, 1958.
166. Shay, H., Sun, D. C., and Siplet, H. *Am. J. Dig. Dis.* 5:217, 1960.
167. Green, M. N., Tsou, K. O., Bressler, R., and Seligman, A. M. *Arch. Biochem.* 57:458, 1955.
168. Pineda, E. P., Goldbarg, J. A., Banks, B. M., and Rutenburg, A. M. *Gastroenterology* 38:698, 1960.
169. Arst, H. E., Manning, R. T., and Depl, M. H. *Am. J. Med. Sci.* 238:598, 1959.
170. Peeters, J. A. B. M. *Clin. Chem.* 18:563, 1972.
171. Tovey, J. E., Dawson, P. J. G., and Fellowes, K. P. *Clin. Chem.* 19:756, 1973.
172. Tovey, J. E., Sykes, J. R., Robinson, D. A., and Hurry, D. J. *Clin. Biochem.* 5:71, 1972.
173. Chapman, L., Silk, E., Skupny, A., Tooth, E. A., and Barnes, A. *J. Obstet. Gynaecol. Br. Commonw.* 78:435, 1971.
174. Van Oudheusden, A. P. M. *Clin. Chim. Acta* 32:140, 1970.

175. Page, E. W., Titus, M. A., Mohun, G., and Glending, M. B. *Am. J. Obstet. Gynecol.* 82:1090, 1961.
176. Melander, S. *Acta Endocrinol.* [Suppl.] (Kbh) 96:94, 1965.
177. Rappaport, F., Fischl, J., and Pinto, N. *Clin. Chim. Acta* 4:227, 1959.
178. de la Huerga, J., Yesinick, C., and Popper, H. *Am. J. Clin. Pathol.* 22:1126, 1956.
179. Seligson, D. [Ed.]. *Standard Methods of Clinical Chemistry*, vol. 3. New York: Academic, 1961. P. 93.
180. Wetstone, H. J., Tennant, R., and White, B. V. *Gastroenterology* 33:41, 1957.
181. Garry, P. J., and Routh, J. I. *Clin. Chem.* 11:91, 1965.
182. Ellman, G. L., Courtney, D. K., Andres, V., and Featherstone, R. M. *Biochem. Pharmacol.* 7:88, 1961.
183. King, J. *Practical Clinical Enzymology.* Princeton: Van Nostrand, 1965.
184. Mirsky, I. A., Block, S., Osher, J., and Broh-Kahn, R. H. *J. Clin. Invest.* 27:818, 1948.
185. Vandermeers, M. C., Vandermeers-Piret, J., and Christophe, J. *Clin. Chem.* 18:1514, 1972.
186. Vandermeers, A., Lelotte, H., and Christophe, J. *Anal. Biochem.* 42:437, 1971.
187. Ozawa, Y., Suzuki, K., and Mogi, K. *Anal. Biochem.* 43:15, 1971.
188. Schwert, G. W., and Tanenaka, Y. *Biochim. Biophys. Acta* 26:570, 1955.
189. Caraway, W. T. *Clin. Chem.* 12:187, 1966.
190. Ellis, G., and Goldberg, D. M. *Clin. Chim. Acta* 37:47, 1972.
191. Knights, E. M., Jr., Whitehouse, J. L., Hue, A. C., and Santos, C. S. *J. Lab. Clin. Med.* 65:355, 1965.
192. Coodley, E. L. *Am. J. Gastroenterol.* 50:55, 1968.
193. Mandel, E. E., and Macalincag, L. R. *Am. J. Gastroenterol.* 54:253, 1970.
194. Martinek, R. G. *Clin. Chem.* 9:620, 1963.
195. Ellis, G., and Goldberg, D. M. *J. Lab. Clin. Med.* 76:507, 1970.
196. Goldberg, D. V., Fletcher, M. J., and Watts, C. *Clin. Chim. Acta* 14:720, 1966.
197. Nishihara, H., Akedo, H., Hiroko, O., and Hattori, S. *Clin. Chim. Acta* 30:251, 1970.
198. Jergovic, I., Zuzic, I., Fiser-Herman, M., and Straus, B. *Clin. Chim. Acta* 30:765, 1970.
199. Loeb, W. F., and Stuhlman, R. A. *Clin. Chem.* 15:162, 1969.
200. Roman, W., and Ruys, J. *Proceedings of the Seventh International Congress of Clinical Chemistry*, vol. 2. Basel: Karger, 1970. P. 121.
201. Greenberg, L. J. *Anal. Biochem.* 14:265, 1966.
202. Fishman, W. H. In H. U. Bergmeyer [Ed.], *Methods of Enzymic Analysis.* New York: Academic, 1965. P. 869.
203. Weissman, G., and Segal, R. L. *Proc. Soc. Exp. Biol. Med.* 134:812, 1970.
204. Miller, B. F., Keyes, F. P., and Curreri, P. W. *Science* 152:775, 1966.
205. Glaser, J. H., and Sly, W. S. *J. Lab. Clin. Med.* 82:969, 1973.
206. Harrison, F. J., and Barnes, A. D. *Clin. Sci.* 38:533, 1970.
207. Terry, J. M., Baliney, J. D., and Swingler, M. C. *Clin. Chim. Acta* 35:317, 1971.
208. Wieczorek, A., Czajka, M., and Kowalezyk, H. *Arch. Immunol. Ther. Exp.* (Warsz.) 15:829, 1967.
209. Perillie, P. E., and Finch, S. C. *N. Engl. J. Med.* 283:456, 1970.

210. Prockop, D. J., and Davidson, W. D. *N. Engl. J. Med.* 270:269, 1964.
211. Schwartz, M. K. In B. P. Colowick [Ed.], *Methods in Enzymology*, vol. 17-B. New York: Academic, 1971.
212. Boiron, M., and Levy, D. *Pathol. Biol.* (Paris) 17:1129, 1969.
213. Van Handel, E. *Clin. Chim. Acta* 29:349, 1970.
214. Grossman, I. W., and Sacktor, B. *Science* 161:57, 1968.
215. Van Handel, E. *Science* 163:1075, 1969.
216. Demelier, J. F., Bark, C., Labat, J., and Courtois, J. E. *Proceedings of the Seventy-fifth International Congress of Clinical Chemistry*, vol. 2. Basel: Karger, 1970. P. 188.
217. Ratliff, C. R., Culp, T. W., and Hall, F. F. *Am. J. Gastroenterol.* 56:199, 1971.
218. German, A., Galli, A., Maison, C., and Galli, J. *Clin. Chim. Acta* 22:551, 1968.
219. Bodansky, O. *J. Biol. Chem.* 202:829, 1953.
220. Bodansky, O. *Cancer* 7:1191, 1954.
221. Berger, L., and Broida, D. *Technical Bulletin No. 650*. St. Louis: Sigma Chemical Co., 1973.
222. Schwartz, M. K., and Bodansky, O. *Am. J. Med.* 40:231, 1966.
223. Beutler, E. *Red Cell Metabolism*. New York: Grune & Stratton, 1971. Pp. 47–49, 115.
224. Baron, D. N., Moss, D. W., Walker, R. G., and Wilkinson, J. H. *J. Clin. Pathol.* 24:656, 1971.
225. Friedman, M. M., and Lapan, B. *J. Lab. Clin. Med.* 51:745, 1958.
226. Sibley, J. A., and Lehninger, A. L. *J. Natl. Cancer Inst.* 9:303, 1948.
227. Sibley, J. A. *Ann. N.Y. Acad. Sci.* 75:399, 1958.
228. Pinto, P. V. C., Van Dreal, P. A., and Kaplan, A. *Clin. Chem.* 15:339, 1969.
229. Bonsignore, D., Calissano, P., and Carstasegna, C. *Med. Lav.* 56:199, 1965.
230. Nakao, K., Wada, O., and Yano, Y. *Clin. Chim. Acta* 19:319, 1968.
231. Hernberg, S., Nikkanen, J., Melin, G., and Lilius, H. *Arch. Environ. Health* 21:140, 1970.
232. Weissberg, J. B., Lipshuta, F., and Oski, F. A. *N. Engl. J. Med.* 284:565, 1971.
233. Moore, M. R., Beattie, A. D., Thompson, G. G., and Goldberg, A. *Clin. Sci.* 40:81, 1971.
234. Burch, H. G., and Siegel, A. L. *Clin. Chem.* 17:1038, 1971.

15. Competitive Protein Binding and Radioimmunoassay

The basic principle of both competitive protein binding assay (CPB) and radio-immunoassay (RIA) is the quantitative combination of the substance to be measured with a protein that binds the substance. The three major requirements for all such assays are: availability of a radioactive labeled form of the substance to be measured, a protein that will specifically bind the substance, and a method for separating the protein-bound material from the material that is not bound [1].

The difference between CPB and RIA lies in the source and nature of the specific binding protein involved. In CPB the binding agents are naturally occurring proteins (mainly glycoproteins of molecular weight less than 100,000 [2]) that are found in the blood. RIA utilizes antibodies—immunoglobulins with an average molecular weight of 150,000 [3]. Macromolecules foreign to an organism become antigens when present in that organism's blood stream, triggering the immune response to produce antibodies that are specific in their structure to bind with that antigen. In contrast to the binding proteins used in CPB, it is not necessary to find a naturally occurring antibody for the substance one wishes to assay by RIA. Antibodies for use in RIA may be induced in various suitable animals by injection of the substance at repeated intervals. Dilutions of the animals' serum may then, after suitable testing and characterization, be used to measure the antigen in an in vitro assay.

The basis of both CPB and RIA is competition of radioactive labeled molecules with a known standard or an unknown sample of unlabeled molecules for binding sites on the binding protein or antibody. The radioactivity bound is inversely proportional to the amount of nonradioactive substance present. These methods involve the use of radioactive atoms as tracers. Some of the same compound as the one to be assayed is made radioactive. For example, the iodine-containing thyroid hormone, thyroxine, is made radioactive by replacing some of the normal iodine atoms with ^{125}I. Radioactive iodine atoms may also be incorporated in compounds not normally containing this element, in sufficient amounts for use as a tracer without affecting their physical or chemical properties significantly. In other organic compounds a small proportion of the carbon or hydrogen atoms are replaced by ^{14}C or ^{3}H. The addition of the radioactive atoms does not change the chemical properties of the molecules so that they react the same as the nonradioactive molecules.

The theoretical basis for the methods derives from the law of mass action, and is best explained by the numerical example given below and illustrated in Table 15-1 and Figure 15-1. X will designate ordinary molecules of the substance being determined (in the sample or standards) and X* will designate similar molecules containing radioactive atoms. In the example we will use small numbers of molecules for simplicity, although actually very large numbers are involved (1 nano-

Figure 15-1. Theoretical basis for RIA procedures.

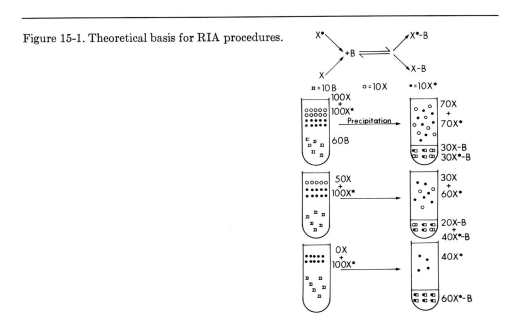

Table 15-1. Illustration of Radioimmunoassay

| | Amounts After Binding | |
Original Amounts	Bound	Free
100 X + 100 X* + 60 B	30 XB + 30 X*B	70 X + 70 X*
50 X + 100 X* + 60 B	20 XB + 40 X*B	30 X + 60 X*
0 X + 100 X* + 60 B	60 X*B	40 X*

gram of a compound will contain 10^{10} to 10^{12} molecules). We will also have a sub-stance B which will combine with or bind the X molecules to form a stable X-B complex. A standard curve is first set up by adding 100, 50, and 0 molecules of X to three tubes as indicated in the table. (Actually more than three points are usually used for the preparation of the standard curve but three will serve for illustrative purposes.) Then to each tube 100 molecules of X* are added and the solutions mixed. Sixty molecules of the binding substance B are then added to each tube. Since the binding substance does not distinguish between X and X*, in the first tube, where there are equal numbers of X and X* molecules, an equal number of each will be bound and 30 molecules of XB and 30 of X*B are formed. The rest of the X and X* molecules (70 of each) will remain unbound. Similarly in the second tube, where one-third of the molecules are X and two-thirds X*, one-third of the bound complexes will be XB and two-thirds X*B, with the rest of the X and X* molecules remaining unbound. In the third tube, where only X* is present, there will be after the addition of B, 60 molecules of X*B and 40 of X*. The free (F) and bound (B) fractions can now be separated by one of several methods and the radioactivity in the separate fractions determined. Since the amount of radioactivity will be proportional to the amount of X* present, we find that by plotting the radioactivity in either fraction (%B or %F) or their ratio (B/F) against the amount of X originally added, a smooth curve can be drawn which is then used to calculate the amount of X in samples similarly treated.

There are many ways of plotting radioimmunoassays. The logit of B/B_0 vs the log of the concentration has the particular advantage of converting most curves to straight line graphs. For this reason the logit plot is used in computer and desktop programmable calculator programs for RIA. Figure 15-2 shows actual data from the same HFSH assay plotted in four different ways: B/B_0 vs the $\log_{10}$ of the concentration; % of initial B/F vs the concentration; logit $B/B_0 = (\log_e \left[\dfrac{B/B_0}{1 - B/B_0} \right])$ vs $\log_{10}$ of the concentration; and % of initial B/F vs $\log_{10}$ of the concentration. If the B/F ratio of radioactivity were 40:60 in a sample determination, this would mean that the sample has the same concentration of X as the second standard in the above series. A discussion of the relative advantages of the different methods of plotting is given by Robard. [3a]

In our example we have assumed that each of the 60 molecules of B binds 1 molecule of X. We can see that if we had used 30 molecules of B, each of which bound 2 molecules of X, the result would be exactly the same as given. Thus the actual number of X molecules bound by each B molecule is not important as long as it remains constant. We have also assumed that each of the X* molecules added was radioactive. If to each tube in the initial step we had added not 100 X* but 90 X + 10 X*, it is easy to calculate that the amount of X*B in the three bound fractions would be, respectively, 3, 4, and 6. This would have no effect on the plotting of the standard curve since the percentage of the total radioactivity in each reaction would be the same. Thus it does not matter what proportion of

Figure 15-2. Comparison of different methods
of plotting RIA standard curves.

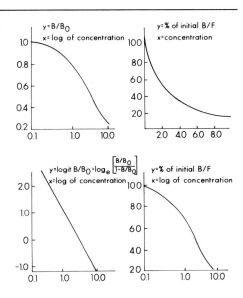

the molecules designated by X* in the illustration is actually radioactive, as long as this proportion is constant for all the determinations of a given set of standards and samples.

It is important that no more B molecules be added that can combine with all the X* molecules present. As an extreme example we can see from the table that if we had added 200 B molecules to each tube, then all of the X* molecules in each tube would be bound and we could not distinguish among the tubes. It is also important that the B molecules be specific in their binding capacity. If, for example, the B molecules bound not only the X molecules present in the sample but also some other molecules C that were also present, our method would determine not X but X + C. If the binding of the C molecules were weaker than for X molecules, the error would be less, but still might be significant. In such cases, if a suitable specific binding agent cannot be found, it is necessary to make a preliminary separation of X and C before applying the method.

RIA offers greater sensitivity and specificity than CPB, but in some cases this may be balanced by the greater availability of plasma binding proteins. Antibody production is not always a simple matter, and since the immune response is highly individual, serums from different animals and even serums taken from the same animal at different times must be separately evaluated. Binding proteins tend to have greater uniformity in their properties within a species. CPB always requires a deproteinization step; this is not usually the case with RIA. The greater specificity of antibodies often allows RIA performance without a prior separation of components in the sample. Binding proteins are more likely to react with a variety of ligands. Murphy [2] has assayed as many as eight different substances with corticosteroid binding globulin (CBG). This, of course, necessitates complicated chromatographic procedures before the desired substance can be measured.

Both CPB and RIA procedures offer an advantage over previous methods, particularly in the field of endocrinology, in that they are more sensitive, allowing the hormone to be measured in a much smaller sample. The volume of serum or plasma that would have been needed to assay most steroid hormones, for example, would have been prohibitive with previously available techniques. Therefore such tests were usually done on 24-hr collections of urine. These samples were bulky and inconvenient, and in some cases necessitated overnight hospitalization of the patient in order to ensure complete and accurate collection. RIA, on the other hand, may be run on small volumes of serum (sometimes as little as 0.01 ml), greatly cutting expense and inconvenience both for the patient and the laboratory. Blood specimens also allow more accurate assessment of the results of stimulation tests, the effects of which may be blunted by collection of the sample over a prolonged period.

Radioimmunoassays for enzymes have a distinct advantage over catalytic methods, in that they measure the actual protein concentration of the enzyme rather than its activity. Enzyme RIAs are also more sensitive and specific than the catalytic methods [4]. Care must be taken in interpretation, however, since RIA measures immunochemical determinants, such as molecular fragments of ACTH or parathormone (PTH), which are not biologically active.

Yalow and Berson et al. [5], during the course of an investigation of the metabolism of ^{131}I-labeled insulin in diabetics, discovered that virtually all patients who had been treated with insulin had circulating insulin-binding antibodies. This discovery took place in the early 1950s, and in 1960 the same investigators published a procedure for the radioimmunoassay of insulin [6]. The numerous RIA procedures now in use are all based on that discovery.

The Antibody

The immune response has been described as consisting of four phases [7]: (1) an afferent phase in which macrophages concentrate and process the antigen into an immunogenic complex, (2) a recognition phase in which appropriate antigen-sensitive cells interact with the processed antigen, (3) a stimulatory phase in which the antigen-sensitive cells are stimulated to differentiate and replicate to produce a large number of immunologically committed plasma cells and lymphocytes, and (4) an efferent phase in which antibodies and sensitized cells are produced and interact with antigen to produce an inflammatory response.

The antibody-producing cells and the concentration of antibody in the serum increase logarithmically for a time. As the antibody increases, however, it binds and neutralizes the antigen present, thus reducing the stimulus for further antibody production. Proliferation of the antibody-producing cells ceases, and the concentration or titer of antibody in the blood gradually falls. Reintroduction of a small amount of the antigen at this point usually causes a rapid and dramatic increase in antibody titer. This "booster" effect may be repeated at regular intervals.

Not all molecules are antigenic or immunogenic—that is, capable of stimulating antibody production. The substance must be a macromolecule or part of a macromolecule. Almost all proteins are immunogenic. Polypeptides may or may not be effective antigens, depending in large part on their molecular weight. Molecules with molecular weights greater than 10,000 tend to have excellent antigenic properties, whereas those with molecular weights of 1,000 to 5,000 are often only weakly antigenic [8, 9]. Under certain conditions, however, peptides with molecular weights of 1,000 or less have been antigenic [10].

In many cases, haptens, molecules that are not antigenic, may be made so by conjugating them covalently to protein carriers [11]. In producing a hapten with good antigenic properties, it is important that the site of linkage be as far distant from the site of antigenicity of the proposed antigen as possible. Usually an antibody is directed against, and combines specifically with, not an entire molecule but a certain portion of that molecule. If this site of antibody recognition is bound to or covered by the carrier protein, antibodies will be formed not to that molecule but to the carrier protein molecule. In the case of closely related compounds such as the steroid hormones, much greater cross reactivity occurs between compounds whose structural differences exist near the site of linkage to the carrier than in those with such differences more distant from the binding site. [12].

In many cases where the structural differences are minimal, or necessarily near the site of binding to the carrier protein, the same antiserums may be used to assay more than one hormone after appropriate chromatographic separation of the hormones involved. This is the case for testosterone and dihydrotestosterone [13], and for the estrogens: estrone, estradiol, and estriol [14]. Such separation procedures are tedious and time consuming, however, and separate, specific antiserums for each hormone will be a distinct advantage when they can be obtained.

The carrier proteins linked to the desired antigen will produce populations of antibodies directed against themselves in the antiserums obtained. In many cases this is no problem, if the carrier protein is a substance unlikely to be encountered in the sample being assayed or to cross react with the labeled antigen. However, when necessary or desirable, the antiserums can be pretreated with the carrier protein prior to use in the assay, to remove antibodies to that protein. In the case of an antibody produced with estradiol covalently linked to bovine serum albumin (BSA), for example, the antiserum need only be mixed with an excess of BSA to precipitate the anti-BSA globulins, leaving the antiestrogen antibodies in solution [15].

Although methods of antibody induction have been described [16] using injections of antigen (angiotensin II) absorbed onto fine microparticles of charcoal (diameter <30 nm), covalent bonding to protein carriers is by far the most common method of hapten conjugation. Of the methods for covalent bonding, peptide bond formation is often the most desirable due to the strength of the bond, and the fact that it is biologically stable and thus unlikely to undergo lysis subsequent to injection [17]. Among the more common coupling reactions are the carbodiimide condensation reaction [17, 18], the glutaraldehyde (Schiff's base) reaction [17, 19, 20], esterification of hydroxyl groups with succinic anhydride [17, 21, 22], and formation of oxime derivatives of ketone groups by using (o-carboxymethyl) hydroxylamine [17, 23, 24].

Immunization

Whether an unconjugated antigen or a hapten is used, the actual process of immunization is similar. The first step is to choose an animal species for immunization. Commonly used for this purpose are guinea pigs, rabbits, chickens, monkeys, goats, sheep, cattle, and horses. The smaller species obviously are the only choice in most laboratories, where animal quarters are limited. On the other hand, where space is available, as with the large commercial producers of antiserums, enough antiserum may be obtained from one large animal to supply many laboratories for years. This is important due to the individuality of the immune response; of ten animals immunized it is possible that only one or two will produce a suitable antiserum.

Another primary consideration in the choice of an animal is its degree of relation to the species from which the antigen is obtained. Since a hormone from a sheep is more likely to share structural similarity with the same hormone produced by a goat than that produced by a guinea pig, the immune response to that hormone is likely to be much stronger and more specific if the sheep hormone is injected into guinea pigs than if it is injected into goats. The evolutionary cul-de-sac occupied solely by rabbits makes them a good choice for general-purpose antibody production.

The antigen used for immunization must be as pure as possible to avoid producing antibodies to contaminating substances if there is a possibility that such antibodies will be cross reactive with the labeled antigen. In cases where the highly purified antigen is very rare or expensive, it is possible to use the highly purified substance for the first or primary injection, and a less pure preparation for subsequent injections, since the specificity of the antiserum is in large part determined by the primary injection [12]. The dose of antigen most frequently used for injection is between 0.2 and 2.0 mg [12], although good results have been obtained with as little as 20 to 100 μg [25] or as much as 5 to 15 mg [26]. Ordinarily antibody production will increase progressively with increased doses of antigen over a certain range, but a plateau is then reached wherein larger doses fail to produce significantly higher antibody concentrations [27]. The choice of optimum dosage of a particular antigen for a particular animal is very often a matter of trial and error. Another variable is the frequency of injection. Often injections are given at 7- or 10-day intervals until the antibody titer reaches a plateau. The titer is then maintained by booster injections given monthly or every 6 weeks.

Antibodies have been produced to some polypeptides such as human growth hormone (HGH) by simple injection of the pure substance [28]. More commonly, the antigen, in aqueous solution, is emulsified with Freund's Complete Adjuvant [29, 30], a preparation (available commercially) with a mineral oil base containing waxes, killed mycobacteria, and an emulsifying agent. The effect of injecting the antigen in the form of a thick emulsion is to delay absorption and destruction of the substance in vivo, thus making it available for antibody production over a longer period of time. The killed mycobacteria act as an irritant to enhance the immune response of the animal. Many investigators prefer to use complete adjuvant for the primary injection, and incomplete adjuvant (without mycobacteria) subsequently. This reduces antibody production directed against the bacterial proteins and also minimizes severe allergic reactions and lesions at the injection site which may threaten the well-being of the animal.

Injections can be given intraperitoneally, subcutaneously, intradermally, intravenously (only when adjuvant is not used), or directly into the lymph nodes or spleen. Footpads on the smaller animals are a popular site. A single immunization may be divided into several aliquots for injection into multiple sites, a technique that can often greatly increase the effectiveness of the dose.

Blood is taken from the animals 7 to 10 days following an immunization and the serum is evaluated. Small animals from which relatively large volumes of blood are obtained must often be given iron supplements to avoid anemia.

Evaluation and Characterization of Antibody

After the antiserum is obtained, it must be carefully tested to determine its suitability for use in an immunoassay.

The first and simplest characteristic to test is the titer or concentration of antibody in the serum. This determines the dilution at which it will be used in the assay. Much confusion has resulted from the fact that some authors have used the term "titer" to express the concentration of the working dilution of the antibody that is added to the other ingredients in the assay tube, whereas others have used titer to refer to the final concentration of antibody in the assay tube during incubation. Obviously it is the latter that is the important concentration. The working dilution is chosen so that a convenient amount may be added to the assay tube to result in the optimal final concentration. The word *titer* will hereafter be used to denote final concentration, but care should be exercised in studying other sources to be sure exactly how the term is used.

The titer of an antiserum is dependent on many factors: the size and molecular configuration of the antigen itself; the frequency, duration, dose, and site of immunization [27]; the "foreignness" of the antigen to the antibody-producing animal; and the genetic and immunological characteristics of the individual animal. It is this last factor that makes it necessary to test each serum separately even though ten rabbits of equal size may have been injected with the same compound on identical schedules.

The titer may be determined by incubating different dilutions of the antiserum with labeled antigen (X^*). The percentage of X^* bound by each dilution is called the zero point (B_0), or initial binding, of the antibody at that dilution. The optimal dilution or titer chosen is often that at which B_0 is approximately equal to 50%. This is extremely variable. If one wishes to increase sensitivity, a higher dilution of antibody is usually chosen, whereas lesser dilution often increases the range of values covered by the assay curve. If the initial binding is too high, however, there is likely to be an excess of antibody present in the system. In this case unlabeled antigen introduced into the system would be bound by unoccupied binding sites rather than acting to displace labeled antigen. If, on the other hand, the initial binding is too low, the resulting assay curve would be too shallow and the results less accurate.

Theoretically, titer is not of extreme importance in determining the value of an antibody. An antiserum used at a final concentration of 1:500 can give results as valid as those given by an antiserum used at 1:500,000. The obvious advantage in the latter is that it will assay 1,000 times as many tubes for the same initial volume of serum. This is an advantage in all situations, but in a case where the antiserum will be used commercially or by a large number of investigators it may be critical.

Specificity may be defined as the degree of exclusiveness with which the antibody binds the substance being measured. If the antibody binds not only the desired antigen, but also another substance present in the assay tube, this substance also will displace labeled antigen and give a falsely high value for the substance being measured. This is true even if the interfering substance is bound less strongly.

Lack of specificity can arise from immunization with impure antigen, cross reaction with molecularly similar compounds that share antigenic sites with the antigen in question, or the presence of naturally occurring antibodies in the antiserum that react with some substance in the assay mixture. Unlike the situation in bioassay, cross reaction in RIA is more likely to occur between compounds of similar molecular structure, regardless of the similarity of their biological action, than between compounds with greater molecular differences but with similar biological properties. Thus an antibody to human growth hormone is more likely to cross react with human placental lactogen than with rat growth hormone.

Specificity is tested by incubating the antibody with different amounts of substances that might, by their structure, be expected to interfere, or that are expected to be present in significant amounts in the samples to be tested. If, for example, dihydrotestosterone (DHT) displaces as much label as testosterone (T) from an antitestosterone antibody, or even exhibits a 50% or 25% cross reaction (in a 50% cross reaction two molecules of DHT must be present to cause the same amount of displacement as one molecule of T; in a 25% cross reaction four molecules of DHT are required), the antibody cannot be used to measure testosterone without previous removal of DHT from the sample. If the degree of cross reaction is only 0.1%, 1,000 times as many molecules of DHT would be required to show the same displacement as T. In this case the cross reaction may be safely ignored, as concentrations of DHT this high are not known to occur in human serum. Likewise even a 50% cross reaction of anti-HGH with human chorionic somatomammotropin (HCS) can be disregarded if the antibody is going to be used to assay HGH only in nonpregnant patients, since no HCS will be present in the serum of such patients. In some cases an antibody may be made more specific by pretreatment with the interfering compound. To assay for follicle stimulating hormone (FSH), it is necessary first to expose the antibody to suitable amounts of HCS, which combines with any binding sites cross reactive with this compound [31].

If a suitable specific antibody cannot be produced, separation of interfering compounds before assay is necessary. The method of separation is dependent on the chemical composition of the compounds involved. Sometimes a simple ethanolic deproteinization will suffice; more frequently complex thin layer or column chromatography is required. In any case such separations are time consuming and introduce added sources of error. A specific antibody is always preferable. Occasionally the use of a heterologous system may confer advantages in special situations as regards specificity.

Not all lack of specificity arises from cross reaction. In some systems, such as the glucagon RIA, circulating enzymes destroy tracer and render it unbindable, unless Trasylol or some other effective protease inhibitor is added to the assay mixture.

The *sensitivity* of an assay may be defined as the smallest amount of antigen that can be distinguished from no antigen [32]. It is in a way a property of the assay system rather than of the antibody alone, since it can be increased by lowering antibody concentration, by using minimal doses of labeled antigen, by using low-ionic-strength buffer systems, and by preincubating the unlabeled antigen with antibody before addition of labeled antigen. This last technique is called *disequilibration*. However, changing assay conditions can increase sensitivity only to a certain degree; ultimately the sensitivity of a given assay depends on the affinity of the antibody for the antigen. This is variable not only from antiserum to antiserum, but for different populations of antibody molecules within an antiserum.

Sensitivity is potentially greater for RIA than for CPB or other clinical procedures. It is most essential when measuring compounds that are present only in minute quantities in the blood. The steroid hormones, for example, are present in the range of nanograms (1 ng = 0.000,000,001 g) per deciliter. With substances present in larger concentration, it is more desirable to develop an assay able to detect a larger range of values than one capable of distinguishing extremely small quantities.

Storage and Handling of Antibody

If properly stored, antibodies are stable over long periods. After the initial testing of a new antiserum, it may be pooled with other antiserums exhibiting similar properties. The entire batch is then diluted with the buffer (often phosphosaline or borate) that will be used in the assay. To this buffer extra protein—e.g., bovine serum albumin (BSA), bovine gamma globulin (BγG), or gelatin—is usually added to coat the glass and prevent adsorption of the antibody. Some prefer to store the antibody diluted in Rivanol which precipitates all serum proteins except the γ-globulins [33]. Sodium azide is often added to the diluent to prevent bacterial growth. The degree of dilution is dictated by the titer of the antibody, so that a convenient amount may be used for further dilution to the working concentration. Once diluted, the antibody is divided into small aliquots, to avoid repeated freezing and thawing. These aliquots are usually stored frozen until needed for assay, then thawed at room temperature or 37°C. Antiserum that is shipped is usually lyophilized (freeze-dried). Once the working dilution is made, it is often better stored at 4°C than refrozen. Most antiserums are stable for several days at this temperature; others must be prepared at the time of use.

Antibody molecules are proteins that can be denatured by violent shaking, vortexing, or other agitation. It is necessary to mix solutions thoroughly, but gently enough to avoid foaming. High temperatures should also be avoided.

The Radioactive Tracer

The radioactive tracer used in RIA is a highly purified form of the antigen in which one or more of the atoms have been replaced by the radioactive isotope of that atom. It is essential that the molecule is not damaged in the course of this substitution, or it may lose its immunoreactivity and fail to be bound by the antibody.

Most polypeptide hormones have iodine-containing tyrosine residues and can be conveniently labeled with an isotope of iodine. ^{125}I has a longer half-life, is less expensive, and has a lower gamma energy than ^{131}I; thus it is the iodine isotope preferable for use in a clinical laboratory.

Both ^{125}I and ^{131}I are gamma emitters. They have short half-lives and often become too "cold" for use in a matter of weeks. The high-energy radiation and the oxidation reaction used to iodinate may cause damage at the time of iodination, requiring purification before use. Continued damage may also occur upon storage, requiring repurification, although this is not usually necessary with kits bearing definite expiration dates [34]. The high-energy radiation of ^{131}I easily penetrates most materials except lead, and special precautions must be taken in its handling. These isotopes can be counted in a crystal scintillation counter (gamma counter), equipment commonly found in most clinical laboratories. ^{3}H and ^{14}C are beta emitters and are most often used in labeling steroid hormones. They have much longer half-lives and are stable over much longer periods of time than the gamma emitters. The weaker radiation cannot penetrate glass or plastic, making its handling much safer, as long as direct contact is avoided; if it gets through the skin, however, it causes far greater tissue damage than the gamma emitters. This same characteristic makes liquid scintillation counting more expensive, more time consuming, and subject to greater error.

Sophisticated equipment for either gamma or beta counting that includes two or more channels for simultaneous counting of different isotopes makes possible the double-isotope variation of RIA. With these systems more than one substance can be measured at the same time. For example, insulin and human growth hormone may be assayed in the same tube by labeling one with ^{125}I and the other with ^{131}I, incubating with both antibodies, and counting each isotope separately [36].

The iodination procedure should be carried out in an area designated for radioactive work only, preferably behind a screen of lead bricks and under a fume hood. Disposable gloves should be worn. The procedure itself is not difficult, but great speed is necessary to prevent excessive radiation damage to the molecules.

Radioactive sodium iodide (Na^{125}I) is oxidized by chloramine-T* in the presence of the substance to be labeled, in phosphate buffer. After a brief reaction time (10 sec to 2 min) [12], the reaction is stopped by the addition of sodium metabisulfite. The reaction time will determine the specific activity. It is desirable to have as high a specific activity as possible so that only a small amount of tracer need be added to the system; too high a specific activity, however, is likely to cause immunoreactivity damage [36].

The reaction mixture must then be immediately purified to remove free ^{125}I and unreacted chloramine-T to minimize further molecular damage. This may be done by Sephadex gel filtration, microgranules of silica (QUSO), cellulose absorption, dialysis, or ion exchange resin [12, 37].

After purification, the labeled peptide is pooled and diluted and then frozen in small aliquots. Only enough material for a single assay should be thawed at a time. In using commercially prepared tracer or RIA kits, careful attention should be paid to expiration dates and outdated material should be properly disposed of.

The radioactivity present in working dilutions of tracer and in RIA kits is much less than that present during the iodination procedures. However, special care should always be taken in handling these materials.

All radioactive materials should be stored in one designated area and should be clearly marked with a radiation symbol with the words CAUTION: RADIOACTIVE MATERIAL. The laboratory bench should be covered with disposable material, preferably absorbent on the upper side and waterproof underneath, to protect the area from contamination in case of spillage. If a spill does occur, it should be cleaned with large quantities of soap and water. Disposable plastic

---

* Eastman #1022, Eastman Kodak Co., Rochester, N.Y. 14650.

gloves should be worn during the clean-up process. The paper towels used in cleaning up spills should be flushed down the toilet.

Radioactive solutions should never be pipetted by mouth or allowed to come in contact with the skin. Eating, drinking, and smoking should be prohibited in the area designated for radioactive work.

Excess or outdated solutions containing radioactivity may be disposed of by flushing down the laboratory drain with large quantities of water. The handling of larger amounts of radioactivity requires special licensing by the Atomic Energy Commission (AEC), and that organization will supply further instructions for use and disposal of this material.

Standards and Quality Control

The material used as a standard in RIA must share identical immunological properties with the substance being measured. Standards for most substances being assayed by clinical laboratories are available commercially. In establishing a new assay, it is necessary to compare curves derived from the dilution of human serum containing a high concentration of the substance to be assayed with dilutions of the purified material. If the curves are not parallel, the immunoreactive characteristics may have been damaged during purification, and the material is not suitable for use in the assay.

In a research situation, one is sometimes faced with the lack of purified material for use as standards. This was the case when the RIA for human prolactin (HPR) was developed [38]. HPR had not yet been isolated and purified. Ovine prolactin, which had been used as a standard in the bioassay, failed to exhibit parallelism with human prolactin. A human serum extremely high in HPR was therefore diluted and used as a standard. As soon as a purified form of HPR was available, this was compared to the human standard and the actual weight value of the arbitrary units determined.

Once a standard is shown to exhibit a curve parallel to that of the samples, the values obtained from the system should be compared to those obtained by bioassay and other methods. However, sometimes the biological potencies of a substance do not correlate well with their value as measured by RIA. The demonstration of parallelism of standards and samples is enough to establish the *accuracy* of the assay (defined as "the extent to which a given measurement or measurements of a substance agrees with the standard measured value of that substance") [12].

The *precision* of an assay has been defined as "the extent to which a given set of measurements of the same sample agrees with the mean of that set" [32]. Precision is greatest when the slope of an assay curve is steep and nearly linear.

This means that precision is greatest in the central range of values covered by the curve; therefore it is best to choose sample sizes and dilutions so that the unknowns will fall in this range, rather than at the extremes of the curve. To some extent precision is dependent on the concentration of antibody, being greater when concentration is relatively high. To measure extremely small quantities, therefore, it is sometimes necessary to sacrifice a certain degree of precision for sensitivity, since sensitivity is greatest at low antibody concentrations. However, when the substance being measured is present in sufficient quantities, it is far more important to obtain maximum precision than maximum sensitivity.

Reproducibility, or *replicability,* is the extent to which assay values for the same sample may be duplicated, both within the same assay and between assays. This should be extensively tested when a new assay is first set up. Poor replicability may result from incomplete separation or the presence of nonspecific interference, but the most common cause is variation in individual techniques and this may be greatly reduced by careful training and practice. Once an assay is established, it is good practice to run high and low quality control pools with each assay to monitor long-term reproducibility of the assay. Freeze-dried quality control material specifically designed for use in RIA is available† for angiotensin I, cortisol, digoxin, digitoxin, gastrin, insulin, testosterone, T_3, and vitamin B_{12}.

Nonspecific binding is a problem in many RIA systems. It can easily be tested for by running duplicate blank tubes with each assay. To these tubes are added buffer and label but no antibody. The percentage of free label in these tubes should be within 5% of the total counts added.

Binding of significant quantities of label in the absence of antibody can be a serious source of error in the assay system. In some cases this is due to binding of label to the test tube walls. This may be reduced by increasing the protein in the system (i.e., adding BSA to the buffer) or by using plastic or siliconized tubes. Because nonspecific binding to various serum proteins sometimes occurs, a system assaying serum or plasma directly must have the same amount of serum or plasma in each tube. If dilutions of the sample are made, they must be made with a serum or plasma known to contain no measurable amounts of the substance being measured. This nonreactive serum or plasma must also be added to the standards. Nonreactive plasma is sometimes obtained from patients deficient in the substance being tested. Plasma from a noncross-reacting species may also be used, or the plasma can be stripped of the substance by various extraction methods prior to use. To be sure that it is completely nonreactive, the plasma or serum is first assayed in comparison to a water blank.

† A. R. Smith, Los Angeles, Calif. 90051.

Nonspecific binding may also result from substances present on the tubes or pipets used. In these cases careful washing of all glassware with acid or organic solvents or both prior to use may be required.

In addition to nonspecific binding, binding interference is sometimes a problem in RIA. Temperature, age of sample, hemolysis, heparin, and ethylenediamine-tetraacetic acid (EDTA) may all cause interference in some assays but not in others. EDTA causes interference in some angiotensin II assays, but in growth hormone and other assays it is added to the buffer to prevent interference. Each possible factor must then be tested individually for each antiserum. Commercially prepared antibodies and those in kits have usually been tested for such interfering substances and should carry warnings in the instructions.

Serum complement may sometimes cause error in radioimmunoassay systems by interfering with the antigen-antibody reaction. Complement has been reported both to slow down and to accelerate precipitating antibody reactions. Interference is thus primarily seen in double-antibody separation systems. Addition of EDTA to bind calcium, which is necessary for complement activity, usually is a sufficient remedy. Alternatively the complement can be destroyed by heating the antibody, carrier protein, and second antibody to 56°C for 20 min prior to use in the assay. Great care must be taken not to exceed either the temperature or the time limit to avoid denaturing the antibody.

Incubation of the Assay

The factor of prime importance in the actual incubation of the assay is that all tubes, standards, and samples contain the same components in the same concentrations and that all are treated alike at every stage of the assay.

All assays are incubated with a buffer solution, usually borate or phosphosaline. Sodium azide is often added to prevent bacterial contamination. The pH of the system is usually in the range of 7.0 to 8.6. Within this range most assays do not seem to be adversely affected by slight changes in pH. Incubation volume is determined by sample size, antibody titer, and convenience. Concentrations of antibody and tracer have already been discussed.

Antibody and tracer are either added separately or mixed together prior to being added to the assay tubes. The latter method saves time and reduces significantly the chances of pipetting error, but is not feasible if the label is bound so strongly by the antibody that subsequent presence of unlabeled antigen will not displace it. This is not often the case; most antibody-antigen reactions are reversible. The theory is the same that is used in the delayed addition of tracer. In this variation, unlabeled antigen is first given time to bind to the antibody, then tracer is added to react with the unoccupied binding sites. This procedure sometimes confers increased sensitivity.

The time needed for the assay to come to equilibrium is variable and is dependent on the temperature of incubation and the characteristics of the antibody. Some assays have been developed requiring as much as 4-day incubation in the cold. These are obviously impractical for clinical purposes; moreover, some hormones such as ACTH and glucagon are subject to incubation damage over long periods of time. Most antibodies work well if incubated at 37°C for 1 or 2 hr, at room temperature for 4 hr, or overnight in the refrigerator (4°C). In most cases these incubation conditions will all give comparable results. This is a distinct advantage in that the assays may be set up at any time of the day and incubated overnight if there is not enough time to complete the separation the same day. However, it must be stressed again that standards and samples for a given assay must be treated identically.

RIA Variations

So far we have discussed radioimmunoassays in which antibody, cold antigen, and labeled antigen are incubated together, allowed to bind, and then separated by one of the methods yet to be discussed. It is appropriate at this time that mention be made of two types of immunoassay that share the basic components of antibody and antigen and yet are different in concept.

Noncompetitive Binding Assays Using Labeled Antibodies

Sometimes called *immunoradiometric assay*, this variaton on RIA involves radioactive labeling of the antibody molecules, rather than the antigen [39]. The labeled antibody reacts with the antigen present in the tube, and the antibody-antigen complex is then separated from the free antibody. The amount of radioactivity present in the bound complex is directly proportional to the amount of antigen added to the tube. An excess of antibody is used, so that virtually all the antigen present is bound. This gives a greater potential sensitivity to the assay. Use of labeled antibodies also reduces background radioactivity and nonspecific binding, allowing greater precision. The major disadvantage of the method is that the specific population of binding antibodies must first be laboriously isolated from all other IγG molecules in the antiserum prior to labeling.

Enzyme Immunoassay

In this technique [40, 41] an enzyme is labeled with a drug molecule, such as morphine. The labeled enzyme is then bound by an antimorphine antibody. In this complex, the enzyme is rendered inactive. When free morphine is present, it competes with the enzyme-morphine for antibody binding sites and displaces the enzyme from the antibody. The enzyme becomes active again and reacts with the bacterial substrate present in the tube. The enzyme activity is directly related to the concentration of free morphine in the sample.

This technique requires no handling of radioactive materials, no separation procedure, and no complicated expensive instrumentation. The enzyme-substrate reaction involves the measurement of changes in absorbance on a spectrophotometer. It is simple, accurate, and very sensitive, with the advantage of requiring little time for performance of assay. Results of an individual test may be completed a few minutes after the sample is obtained. So far the major application of the method has been in the drug abuse area (morphine, amphetamine, barbiturate, cocaine metabolite, methadone, and opiates assays are available in kit form under the trade mark EMIT‡). The advantage of speed in emergency and legal situations in this field is obvious. The method also holds great promise for many areas of clinical chemistry.

Heterologous Assay

Another variation on the basic RIA is the heterologous assay. In the usual homologous assays, samples, standards, labeled antigen, and antigen used for immunization are all from the same species. This is not the case in the heterologous assay, where the radioactive tracer and the antigen used to produce the antibody are from different species; neither need be identical to the substance being measured. We have seen that the identity of standard and unknown is essential, but identity of unknown and labeled antigen is not. If a material is bound strongly enough by the antibody, but not too strongly to be displaced by the sample, it can be used as a radioactive tracer. It is not necessary that the material behave in an identical manner to the substance being measured as long as its behavior is consistent from tube to tube. The antibody also may be raised against some other substance as long as it cross reacts strongly enough with the desired substance to give a good curve and none of the original antigen is present to interfere. Thus in the RIA for human prolactin already mentioned [38], a human serum was diluted to provide standards, but radioactive porcine prolactin was used as a tracer and the antibody was antiovine (sheep) prolactin. However, when the purified material is available, a homologous system is usually preferable.

‡ Syva, Palo Alto, Calif. 94304.

Separation of Bound from Unbound Material

After equilibrium has been established between free and bound antigen, it is necessary to separate them without disruption of the antigen-antibody bond. An ideal method must be simple, quick, and capable of producing complete separation. Most methods are based on the difference in size of the free antigen molecule and the antigen-antibody complex.

Separation may be accomplished by gel filtration on Sephadex. This gives a clean separation, but is somewhat impractical for use in a clinical laboratory because of the time involved in preparation and use of the individual columns [42].

Electrophoresis and chromatoelectrophoresis separate the fractions by means of differential migration in an electrical field, enhanced in the latter by use of a support medium with high affinity for the free antigen. Although these methods give a highly effective separation, they are difficult, time consuming, and require expensive electrophoretic equipment and coldroom facilities.

The double-antibody technique, or immunoprecipitation, is much more suitable for use in a clinical situation. In this method a second antibody is produced against the γ-globulin of the animal species that produced the first antibody. The second antibody binds the antibody-antigen complex, forming a larger complex. In many cases this larger complex can be centrifuged out of solution directly. Serum from the animal against which the second antibody is directed is often added to provide a bulkier precipitate. One disadvantage is that relatively long incubations are sometimes necessary for the second antibody reaction to reach equilibrium. This is sometimes circumvented by preprecipitating the primary antibody with the secondary antibody, and then allowing the insoluble complex to react with the antigen [12].

The second antibody is used in much higher concentrations than is the primary antibody; therefore it is most frequently produced in the larger animal species.

Once the precipitate is formed, it is centrifuged and the supernatant aspirated or decanted. This step requires great care to avoid disturbing the precipitate. Some laboratories count the radioactivity in the bound precipitate, whereas others count the free portion in the entire supernatant or in an aliquot of the supernatant.

Nonspecific precipitation of the antigen-antibody complex may be accomplished by the addition of various salts and solvents. Because of the nonspecific nature of the precipitating agent, the danger of also precipitating the free antigen is much greater than in the double-antibody method. The conditions and concentration of precipitant must be carefully regulated to avoid this. The advantage over double antibody is that precipitation occurs immediately, and the time-consuming second incubation is eliminated. Precipitation of steroids and other small antigens with equal volumes of saturated ammonium sulfate is a widely used method [43]. For some systems a mixture of salt and solvent is used to achieve precipitation [43]. As in double-antibody precipitation, γ-globulin or normal animal serum is often added as a carrier (or coprecipitant) to increase the size of the precipitate.

Solid phase adsorption of the free antigen is an extremely popular method. It has the advantages of speed and ease of performance. Small particles of dextran-coated charcoal, talc, QUSO, Florisil, Fuller's earth, or other substances are added to the assay tube. The small, free antigen molecules are adsorbed onto the surface of the particles, leaving the bound complex in solution. The particles may then be centrifuged and either the free or bound fraction counted. If the adsorbing particles are added in solution, constant stirring is required to maintain a uniform suspension. Careful adjustment of plasma protein is necessary for the success of this method of separation. If too much protein is present, adsorption of the free antigen may be inhibited, whereas too low a protein concentration allows adsorption of the antigen-antibody complex [12].

Solid phase antibody is coming into increasing use, especially in the clinical laboratory. The antibody is polymerized, coupled to discs or beads, bound to the test tube wall, or otherwise rendered insoluble. Separation is rapid and almost foolproof, involving simple washing techniques. The method requires more antibody than the other techniques but for most commercial suppliers this is no problem.

In all the separation techniques it is important to keep the assay at the temperature at which equilibrium was reached. Changes of temperature can change the equilibrium constant and reverse the binding reaction. Therefore, if an assay has been incubated in the refrigerator, it should be placed in an ice bath upon removal; if a precipitating solution is used, this solution should also be cold. In most cases a refrigerated centrifuge is required and the tubes should be returned to the ice bath following centrifugation.

Table 15-2 lists commercial suppliers of antiserums and radioimmunoassay kits, as well as the isotope and method of separation for each kit.

RIA Procedures

It is not possible to give detailed procedures for each substance that can be measured by RIA, as the details of the procedure will vary according to the antibody used. When antibodies are obtained from commercial sources, detailed procedures

Table 15-2. Sources of Radioimmunoassay Kits

Antigen	Supplier	Isotope	Method of Separation[a]
Adrenocorticotropic hormone (ACTH)	Amersham/Searle Arlington Heights, Ill. 60005	^{125}I	D.C.C.
	Wellcome Reagents Beckenham, Kent, Eng.	^{125}I	D.C.C.
Aldosterone	Antibodies, Inc.[b] Davis, Calif. 95616	—	—
	Calbiochem[b] San Diego, Calif. 92112	—	—
	Diagnostic Products Los Angeles, Calif. 90029	^3H	D.C.C.
	Endocrine Sciences[b] Encino, Calif. 91316	—	—
	ICN Medical Diagnostic Products Portland, Ore. 97208	^3H	$(NH_4)_2SO_4$
	New England Nuclear Worcester, Mass. 01608	^3H	D.C.C.
	Pharmacia[b] Piscataway, N.J. 08854	—	—
	Research Plus Steroid Labs[b] Denville, N.J. 07834	—	—
	Sorin Saluggia, Vercelli, Italy	^3H	D.C.C.
	Wien Laboratories Succasunna, N.J. 07876	^3H	D.C.C.
Amphetamines	ICN Medical Diagnostic Products[b]	—	—
Angiotensin I	Clinical Assays Cambridge, Mass. 02142	^{125}I	2nd Ab
	ICN Medical Diagnostic Products	^{125}I	D.C.C.
	Isolab, Inc. Akron, Ohio 44321	^{125}I	2nd Ab
	Mallinckrodt Chemicals St. Louis, Mo. 63160	^{125}I	Anion resin strip
	New England Nuclear	^{125}I	D.C.C.
	Schwarz-Mann Orangeburg, N.Y. 10962	^{125}I	D.C.C.
	Sorin	^{125}I	D.C.C.
	Squibb New Brunswick, N.J. 08903	^{125}I	D.C.C.

[a] D.C.C., dextran-coated charcoal; 2nd Ab, second antibody; S.P. Ab, solid phase antibody; EtOH, ethanol precipitation.
[b] Company supplies antiserum rather than complete assay kit.

Table 15-2 (*continued*)

Antigen	Supplier	Isotope	Method of Separation[a]
Angiotensin II	Sorin	^{125}I	D.C.C.
Barbiturates	ICN Medical Diagnostic Products[b]	—	—
Bradykinin	Calbiochem[b]	—	—
Carcinoembryonic antigen (CEA)	Cordis Laboratories[b] Miami, Fla. 33137	—	—
	Roche Diagnostics Nutley, N.J. 07110	^{125}I	Zirconyl phosphate gel
	Spectrum[b] Los Angeles, Calif. 90054	—	—
Ceruloplasmin	Spectrum[b]	—	—
Cortisol	Bioware, Inc.[b] Wichita, Kan. 67208	—	—
	Cordis Laboratories[b]	—	—
	Endocrine Sciences[b]	—	—
	ICN Medical Diagnostic Products	^{3}H	Florisil
	Micromedic Systems Philadelphia, Pa. 19105	^{125}I	2nd Ab
	Schwarz-Mann	^{3}H	D.C.C.
	Serono Laboratories Boston, Mass. 02116	^{3}H	D.C.C.
	Wien Laboratories	^{3}H	D.C.C.
Cyclic adenosine monophosphate (c-AMP)	Amersham/Searle	^{3}H	D.C.C.
	Collaborative Research Waltham, Mass. 02154	^{125}I	2nd Ab
	Diagnostic Products	^{3}H	D.C.C.
	ICN Medical Diagnostic Products	^{125}I	$(NH_4)_2SO_4$
	Schwarz-Mann	^{125}I	D.C.C.
Cyclic guanosine monophosphate (c-GMP)	Collaborative Research	^{125}I	2nd Ab
	Diagnostic Products	^{3}H	D.C.C.
Digitoxin	Beckman Instruments Fullerton, Calif. 92634	^{125}I	2nd Ab
	Bioware[b]	—	—
	Clinical Assays	^{3}H	D.C.C.
	Diagnostic Products	^{3}H	D.C.C.
	ICN Medical Diagnostic Products	^{3}H	D.C.C.
	Isolab	^{3}H	D.C.C.
	Kallestad Laboratories Minneapolis, Minn. 55416	^{3}H	D.C.C.
	Mallinckrodt	^{125}I	Anion resin strip

Table 15-2 (*continued*)

Antigen	Supplier	Isotope	Method of Separation[a]
	Meloy Laboratories Springfield, Va. 22157	^3H	D.C.C.
	New England Nuclear	^3H	D.C.C.
	Schwarz-Mann	^3H	D.C.C.
	Schwarz-Mann	^{125}I	D.C.C.
	Sorin	^3H	D.C.C.
	Sorin	^{125}I	D.C.C.
	Wien Laboratories	^3H	D.C.C.
Digoxin	Beckman Instruments	^{125}I	2nd Ab
	Bioware[b]	—	—
	Clinical Assays	^3H	D.C.C.
	Curtis Nuclear Los Angeles, Calif. 90058	^{125}I	S.P. Ab
	Diagnostic Products	^3H	D.C.C.
	ICN Medical Diagnostic Products	^3H	D.C.C.
	Isolab	^3H	D.C.C.
	Kallestad Laboratories	^3H	D.C.C.
	Mallinckrodt	^{125}I	Anion resin strip
	Meloy Laboratories	^3H	D.C.C.
	New England Nuclear	^3H	D.C.C.
	Schwarz-Mann	^3H	D.C.C.
	Schwarz-Mann	^{125}I	D.C.C.
	Serono Laboratories	^3H	D.C.C.
	Wellcome Reagents	^3H	D.C.C.
	Wien Laboratories	^3H	D.C.C.
Estrogens Estrone (E1)	Amersham/Searle[b]	—	—
	Endocrine Sciences[b] Encino, Calif. 91316	—	—
	Endocrine Services[b] Tarzana, Calif. 91356	—	—
	ImChem[b] Fort Collins, Colo. 80521	—	—
	Micromedic Systems	^{125}I	2nd Ab
	New England Nuclear	^3H	D.C.C.
	Wien Laboratories	^3H	D.C.C.
Estradiol (E2)	Amersham/Searle[b]	—	—
	Calbiochem[b]	—	—
	Endocrine Sciences[b] Encino, Calif. 91316	—	—

Table 15-2 (*continued*)

Antigen	Supplier	Isotope	Method of Separation[a]
	Endocrine Services[b] Tarzana, Calif. 91356	—	—
	ImChem[b]	—	—
	Meloy Laboratories	^{3}H	D.C.C.
	New England Nuclear	^{3}H	D.C.C.
	Research Plus Steroid Labs[b]	—	—
	Wien Laboratories	^{3}H	D.C.C.
Estriol (E3)	Endocrine Services[b] Tarzana, Calif. 91356	—	—
	ImChem[b]	—	—
	Micromedic Systems	^{125}I	2nd Ab
	Research Plus Steroid Labs[b]	—	—
Follicle stimulating hormone (FSH)	Calbiochem	^{125}I	2nd Ab
	ICN Medical Diagnostic Products	^{125}I	2nd Ab
	Serono Laboratories	^{125}I	2nd Ab
Gastrin	Amersham/Searle[b]	—	—
	Sorin	^{125}I	D.C.C.
Hepatitis-associated antigen (HAA)	Abbott North Chicago, Ill. 60064	^{125}I	S.P. Ab
	Virgo Reagents[b] Bethesda, Md. 20014	—	—
Human chorionic gonadotropin (HCG)	Calbiochem	^{125}I	2nd Ab
	Serono	^{125}I	2nd Ab
	Wellcome	^{125}I	2nd Ab
Human chorionic somatomammotropin (HCS)	*See* Human placental lactogen (HPL)		
Human growth hormone (HGH)	Abbott	^{125}I	2nd Ab
	Beckman Instruments	^{125}I	2nd AB
	Collaborative Research	^{125}I	2nd Ab
	Curtis Nuclear	^{125}I	S.P. Ab
	ICN Medical Diagnostic Products	^{125}I	2nd Ab
	Kallestad	^{125}I	2nd Ab
	Schwarz-Mann	^{3}H	2nd Ab
	Serono Laboratories[b]	—	—
	Sorin	^{125}I	2nd Ab
	Wellcome	^{125}I	2nd Ab
Human luteinizing hormone (HLH)	Calbiochem	^{125}I	2nd Ab
	Serono Laboratories	^{125}I	2nd Ab
	Wellcome	^{125}I	2nd Ab
Human placental lactogen (HPL)	Amersham/Searle	^{125}I	EtOH
	Antibodies, Inc.[b]	—	—

Table 15-2 (*continued*)

Antigen	Supplier	Isotope	Method of Separation[a]
(Human chorionic somatomammotropin, HCS)	Collaborative Research	^{125}I	2nd Ab
	Micromedic Systems	^{125}I	2nd Ab
	New England Nuclear	^{125}I	D.C.C.
	Pharmacia	^{125}I	EtOH
	Schwarz-Mann	^{125}I	2nd Ab
	Sorin	^{125}I	2nd Ab
Immunoglobin E (IgE)	Bioware[b]	—	—
	Pharmacia	^{125}I	S.P. Ab
Insulin	Amersham/Searle	^{125}I	2nd Ab
	Arnel Products[b] Brooklyn, N.Y. 11235	—	—
	Collaborative Research	^{125}I	2nd Ab
	ICN Medical Diagnostic Products	^{125}I	2nd Ab
	New England Nuclear	^{125}I	2nd Ab
	Pharmacia	^{125}I	S.P. Ab
	Schwarz-Mann	^{3}H	2nd Ab
	Schwarz-Mann	^{125}I	2nd Ab
	Serono Laboratories	^{125}I	2nd Ab
	Sorin	^{125}I	2nd Ab
	Sorin	^{125}I	D.C.C.
	Wellcome	^{125}I	2nd Ab
Lupus erythematosus	Virgo Reagents	^{3}H	$(NH_4)_2SO_4$
	Virgo Reagents	^{14}C	$(NH_4)_2SO_4$
Lysergic acid diethylamide (LSD)	Collaborative Research	^{125}I	2nd Ab
Morphine	ICN Medical Diagnostic Products[b]	—	—
	Roche Diagnostics	^{3}H	$(NH_4)_2SO_4$
	Roche Diagnostics	^{125}I	$(NH_4)_2SO_4$
Ouabain	Clinical Assays	^{3}H	D.C.C.
Parathormone (PTH)	Amersham/Searle[b]	—	—
	Calbiochem[b]	—	—
Phenobarbital	ICN Medical Diagnostic Products	^{3}H	$(NH_4)_2SO_4$
Progesterone	Amersham/Searle[b]	—	—
	Calbiochem[b]	—	—
	Endocrine Sciences[b] Encino, Calif. 91316	—	—
	ICN Medical Diagnostic Products	^{3}H	$(NH_4)_2SO_4$
	ImChem	^{125}I	2nd Ab
	Micromedic Systems	^{125}I	2nd Ab
	New England Nuclear	^{3}H	D.C.C.
	Pharmacia[b]	—	—

Table 15-2 (*continued*)

Antigen	Supplier	Isotope	Method of Separation[a]
	Research Plus Steroid[b]	—	—
	Sorin	^3H	D.C.C.
Prolactin	Amersham/Searle[b]	—	—
Prostaglandin	Clinical Assays, Inc.	^3H	D.C.C.
Renin activity	Schwarz-Mann	^3H	D.C.C.
	Schwarz-Mann	^{125}I	D.C.C.
Testosterone	Amersham/Searle[b]	—	—
	Bioware[b]	—	—
	Calbiochem[b]	—	—
	Endocrine Sciences[b] Encino, Calif. 91316	—	—
	ICN Medical Diagnostic Products	^{125}I	$(NH_4)_2SO_4$
	ImChem	^{125}I	2nd Ab
	Micromedic Systems	^{125}I	2nd Ab
	New England Nuclear	^3H	D.C.C.
	Pharmacia[b]	—	—
	Research Plus Steroid[b]	—	—
	Serono Laboratories	^3H	2nd Ab
	Serono Laboratories	^{125}I	2nd Ab
	Sorin	^3H	D.C.C.
	Sorin	^{125}I	D.C.C.
	Wien Laboratories	^3H	D.C.C.
Thyrocalcitonin	Amersham/Searle[b]	—	—
Thyroid stimulating hormone (TSH)	Amersham/Searle[b]	—	—
	Beckman Instruments	^{125}I	2nd Ab
	Calbiochem	^{125}I	2nd Ab
	ICN Medical Diagnostic Products	^{125}I	2nd Ab
Thyroxine (T$_4$)	Pantex Inglewood, Calif. 90307	^{125}I	2nd Ab
	Wien Laboratories[b]	—	—
Triiodothyronine (T$_3$)	Mallinckrodt	^{125}I	Anion resin strip
	Pantex	^{125}I	2nd Ab
	Wien Laboratories	^{125}I	D.C.C.
Vitamin B$_{12}$	Pharmacia	^{57}Co	S.P. Ab
	Schwarz-Mann	^{57}Co	D.C.C.

will be supplied by the manufacturer and should be followed as closely as possible. The procedures given here are intended as samples and are, in fact, applicable to a wide range of substances with minor variations. In spite of the variations encountered in RIA, such as the buffer used, the incubation volume, amount of sample, concentration of antibody and of radioactive tracer, and length and temperature of incubation, all radioimmunoassays share a basic similarity in that they consist of the following procedural steps:

1. Incubation of the samples and standards with antibody and radioactive tracer
2. Separation of free from antibody-bound tracer
3. Counting of radioactivity in either the free or bound fraction
4. Calculation of the unknowns from the standard curve

Human Growth Hormone by RIA

The basic procedure for radioimmunoassay of human growth hormone (HGH) [32, 44–46], has been used successfully for several polypeptide hormones, including human prolactin (HPR) [38], luteinizing hormone (HLH), follicle stimulating hormone (FSH), thyroid stimulating hormone (TSH), insulin [6], and glucagon. A similar procedure for lysergic acid diethylamide (LSD) has also been described [47].

EQUIPMENT

Gamma counter, refrigerated centrifuge, vortex mixer, automatic syringes or pipets (1 and 3 ml), automatic pipet or repeating dispensor (100 μl), micropipets, glass ampules (1 ml), 10 × 75 mm disposable glass tubes, 15 × 150 mm plastic gama counter tubes, pH meter.

REAGENTS

1. Bovine serum albumin (BSA) phosphosaline buffer, 0.01M.   To 900 ml of distilled water, add 1.43 g of $NaH_2PO_4 \cdot H_2O$, 8.18 g of NaCl, 9.62 g of EDTA, and 1.0 g of NaN_3. Stir until dissolved, then adjust pH to 7.6 by adding small quantities of 10M NaOH until the desired pH is reached. Bring the volume to 966.7 ml, and add 33.3 ml of 30% BSA. Store at 4°C.

2. Guinea pig anti-HGH (GPA-HGH).

    A. Stock solution.   Dilute pure antiserum 1:1,000 with phosphosaline buffer (PSB). Store 1-ml aliquots frozen in sealed ampules.

    B. Working solution.   Dilute stock solution to desired working dilution with PSB. Add enough normal guinea pig serum to give a concentration of 1:300. Titer will vary between lots of antiserum; information on the correct working dilution will be supplied with the antiserum.

3. HGH-^{125}I.   Instructions on both stock and working dilutions vary according to amount of radioactivity in the shipment and will be supplied by the manufacturer. Working solutions should contain between 12,000 and 20,000 CPM/100

μl in PSB. Careful attention should be paid to the expiration date, and outdated HGH-^{125}I promptly discarded.

4. Goat anti-guinea pig γ-globulin (GAGP).

   A. Stock solution.   Store frozen in sealed glass ampules.

   B. Working solution.   Dilute stock with PSB to desired working concentration, usually 1:10 to 1:50.

5. Serum from hypophysectomized patients, or other serum free from HGH.

6. HGH standards.

   A. Stock standard, 10 μg/ml.   Weigh out exactly 1.000 mg of pure HGH,§ dissolve in PSB solution, and dilute to 100 ml.

   B. Working standard.   Dilute stock standard with PSB to give solutions containing 0.05, 0.1, 0.25, 0.5, 1.0, 2.5, and 5 ng/100 μl. Aliquot approximately 300 μl into glass ampules, seal, and store frozen. Thaw one ampule of each concentration for each assay.

7. Normal guinea pig serum (NGPS). Store in freezer.

PROCEDURE

Write assay protocol and number tubes. Tubes 1 and 2 are blanks, and tubes 3 to 18 are standards beginning with zero HGH in increasing order of concentration of HGH. Each concentration is pipetted into duplicate tubes. The rest of the tubes contain samples, including quality control pools, set up in duplicate.

Add PSB to the tubes. The blanks (tubes 1 and 2), zeroes (tubes 3 and 4), and all sample tubes receive 200 μl of PSB, and the standards (tubes 5 to 18) receive 100 μl of PSB. Using a micropipet, add 100 μl of the appropriate standard to tubes 5 to 18. All tubes now contain a volume of 200 μl. Add 100 μl of HGH-free serum to tubes 1 to 18, and pipet 100 μl of unknown serum in duplicate into the remaining tubes and gently mix all tubes. Add 100 μl of first antibody (GPA-HGH) working solution to all tubes except blanks. To blanks add 100 μl of 1:300 NGPS in PSB. Add 100 μl of HGH-^{125}I working solution to all tubes and to six additional tubes labeled TC (total counts). Stopper the TC tubes. Mix all tubes by vortexing *gently*. Cover the tubes and incubate at room temperature for 24 hr. If more convenient, the incubation can be carried out for 72 hr in the refrigerator at 4°C (e.g., over the weekend).

Add 100 μl of working solution of second antibody (GAGP) to all tubes except TC. Cover and refrigerate. Incubate overnight in refrigerator at 4°C. Add 3 ml of cold distilled water to all tubes except TC. Centrifuge in the cold at 3,000 rpm for 30 min. Remove supernatant by aspiration or decantation very carefully so as to avoid dislodging the almost invisible precipitate.

Place all tubes (including TC) in plastic gamma counter tubes and count for 1

§ Obtained from the National Pituitary Agency, Bethesda, Md. 20014.

min. The mean of the blank tubes should be dialed into the background subtraction of the gamma counter before counting the remaining tubes.

CALCULATION

Compute blank:

$$\frac{\text{mean counts of tubes 1 and 2}}{\text{mean TC}} \times 100 = \% \text{ Binding (B) of blank}$$

The blank should be less than 5%. Next calculate the % binding of each standard, including zero:

$$\frac{\text{mean counts of standard}}{\text{mean TC}} \times 100 = \% \text{ B}$$

Calculate the % binding (B) for the unknowns:

$$\frac{\text{mean counts of samples}}{\text{mean TC}} = \% \text{ B}$$

Plot the % B for each standard against the concentration of that standard as in Figure 15-3. Using the average % B of each sample, read the concentration in nanograms per 100 microliters from the standard curve. Multiply by 10 to convert to nanograms per milliliter. Samples having % B higher than that of the lowest concentration of standard (0.05 ng/100 μl) should be reported as < 0.5 ng/ml. Samples having % B lower than that of the highest standard should be diluted 1:10 with HGH-free serum and reassayed.

NORMAL VALUES AND INTERPRETATION OF RESULTS

Normal basal HGH levels in adults and children are less than 3 ng/ml. There can be great fluctuation in HGH levels due to activity, stress, hypoglycemia, and other factors; consequently a single determination is seldom of diagnostic significance. Stimulation or suppression tests should be used to evaluate growth hormone secretion [34].

Two commonly used stimulation tests are arginine infusion and insulin-induced hypoglycemia. A normal response to either test involves the attainment of a level of 6 ng/ml or greater in any sample obtained during the test procedure. If the HGH response to insulin is deficient, this should not be considered abnormal unless the blood glucose has fallen at least 50% from the baseline value. About 10 to 15 percent of the subjects are nonresponders to each stimulus without having HGH deficiency. As a result, diagnosis of HGH deficiency requires demonstration

Figure 15-3. Human growth hormone RIA standard curve.

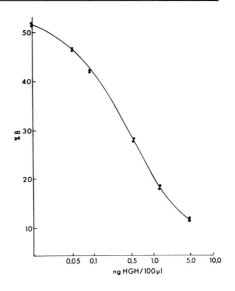

of failure to respond to at least two HGH stimuli. (L-Dopa is a third HGH stimulus used clinically.)

Hypopituitary patients will show little or no response to arginine and insulin HGH stimulation tests. In children, dwarfism may result if HGH replacement therapy is not given.

Glucose suppression tests are used for the diagnosis of acromegaly. In normal individuals 100 g of glucose given orally suppresses HGH to less than 3.0 ng/ml within 60 min. Acromegalics and hyperpituitary children (in whom the condition leads to gigantism) do not exhibit such suppression.

Hepatitis-Associated Antigen by RIA

The procedure given here is the one used with Abbott Radio-Pharmaceuticals Ausria™-I Kit [48]. It is the most sensitive test available for hepatitis-associated antigen (also called Australia antigen), which is of great clinical significance in the diagnosis of hepatitis and the elimination of hepatitis carriers from blood bank donors.

The procedure utilizes a solid phase separation technique [49–51]. Antibody-coated tubes are used to incubate samples [52]. After the antigen in the serum sample is bound to the antibody coating the tube, additional radioactively labeled antibody [39] is added to form an antibody-antigen-antibody "sandwich."

Solid phase separation kits are also available for the radioimmunoassay of insulin, immunoglobin E, digoxin, and human growth hormone (HGH).△

EQUIPMENT

Gamma counter, automatic pipets (100 µl), automatic syringe or repeating dispensor (2 ml), aspiration device such as a cannula or aspiration tip with a vacuum source and a trap for retaining the aspirate.

REAGENTS

1. Rinse solution. Dilute an appropriate aliquot of the Tromethamine Buffer concentrate supplied with the kit 1:400 with distilled water. Prepare only as much as will be used within 48 hr. Store at 2° to 4°C after dilution.

2. Hepatitis-associated antibody-^{125}I solution. Use as supplied.

PROCEDURE

Add 100 µl of serum or plasma to antibody-coated tubes. One tube should be used for each unknown serum, and seven negative and three positive controls should be run with each group. Tap tubes gently so that the sample evenly coats the bottom of the tube. Cover the tubes to prevent evaporation and place on a level surface. Incubate overnight at room temperature.

Aspirate the contents of the tube. Wash with 2 ml of the rinse solution and aspirate. Repeat the rinse procedure four more times. Remove all remaining rinse solution with the last aspiration, to avoid dilution of the labeled antibody.

Carefully add 100 µl of hepatitis-associated antibody-^{125}I to the bottom of each tube, making sure the solution evenly covers the bottom of the tube. Cover and incubate the tubes on a level surface for 90 min at room temperature.

Aspirate the contents of each tube, then rinse and aspirate five times as previously described. Count the tubes in a well-type gamma scintillation counter for 1 min.

CALCULATION

The presence or absence of hepatitis-associated antigen is determined by relating CPM of the unknown sample to CPM of the negative control mean times the factor 2.1.

△ A second kit, Ausria-II-125 (TM, Abbott Radio-Pharmaceuticals) is currently available. This kit utilizes antibody coated beads instead of coated tubes, and has the advantage of a three hour incubation period at 45°C in place of the overnight incubation necessary with Ausria-I. Slightly greater sensitivity can also be achieved in the Ausria-II-125 procedure.

Samples whose count rate is higher than the mean cutoff value established with respect to the negative control are considered positive; those whose count rate is lower than the cutoff value are considered negative.

Calculation of the negative control mean: Add the CPM of each negative control and divide by the number of negative controls. Discard those values that fall outside the range 0.5 to 1.5 times the mean and recalculate the mean without using these aberrant values. If more than one aberrant value is consistently obtained, problems with the technique should be investigated.

Calculation of the cutoff value: Multiply the negative control mean by the factor 2.1. All samples having CPM higher than this number should be considered positive.

CAUTION

All samples and controls in this procedure should be considered potentially infectious. Great care should be used in handling and disposal of all materials.

Digoxin by RIA

This procedure, developed by Horgan and Riley [53], is typical of many commercially available RIA kits in that it is very rapid. The separation is by dextran-coated charcoal [54, 55], one of the most popular methods used for this purpose. This technique can also be used in procedures for aldosterone, angiotensin I, angiotensin II, cortisol, digitoxin, estrogens, gastrin, human growth hormone [56], human placental lactogen, insulin [57], progesterone, renin, triiodothyronine (T_3), testosterone, and vitamin B_{12}, among others [58].

EQUIPMENT

Polystyrene assay tubes, 10 × 75 mm, gamma counter, refrigerated centrifuge, micropipets (100 μl, 50 μl, and 10 μl), adjustable automatic pipet (1 ml), vortex mixer.

REAGENTS

1. Digoxin, pure crystalline (Sandor Pharmaceutical). Dilute in buffer to concentrations of 50, 100, 150, 200, and 250 pg/50 μl. Aliquot small quantities and store at −20°C.

2. Digoxin-^{125}I (Schwarz-Mann). Store at −20°C. Dilute with buffer to about 5,000 CPM/100 μl prior to use.

3. Diluent buffer. To 1 liter of 0.01M sodium phosphate (pH 7.4) add 9 g of sodium chloride, 5 g of bovine serum albumin, and 1 g of sodium azide. Store at 4°C.

4. Antidigoxin serum. Store lyophilized in 0.1-ml aliquots. Prepare 1:10,000 stock dilution and store 100-μl aliquots at −20°C. Dilute each aliquot with 3 ml of buffer as required for working solution.

5. Normal human (digoxin-free) serum. Store in small aliquots at −20°C.

6. Dextran-coated charcoal. For stock suspension add 3 g of Norit A Charcoal to 30 ml of buffer containing 75 mg of dextran T70.‖ Store at 4°C; dilute 1:10 with diluent buffer as needed.

PROCEDURE

Add buffer to the assay tubes as follows: blanks, 1,000 μl; zero standard, 900 μl; other standards, 850 μl; unknowns, 900 μl. Pipet 50 μl of standards and samples into duplicate tubes. Add 50 μl of normal serum to blanks and all standards. Add 10 μl of digoxin-^{125}I to all tubes and 100 μl of 1:30,000 antidigoxin to all tubes except the blanks.

Incubate 30 min at room temperature. Quickly squirt 500 μl of cold dextran-coated charcoal suspension into each tube, agitating the charcoal suspension constantly to avoid settling. Let tubes stand 5 min after last addition. Centrifuge at 1,500 × g for 10 min at 4°C. Decant supernatant into another tube and count either precipitate (free counts) or supernatant (bound counts) for 1 min. Digoxin values (in micrograms per liter or nanograms per milliliter of plasma) are calculated from a standard curve of % bound vs μg/liter (or ng/ml) digoxin on semilogarithmic graph paper, as illustrated in Figure 15-4.

NORMAL VALUES AND INTERPRETATION OF RESULTS

Serum or plasma digoxin levels are usually measured in patients receiving oral digoxin compounds.

Digoxin levels in nontoxic patients taking 0.25 mg/day of digoxin range from 0.8 to 1.6 ng/ml; those taking 0.50 mg/day range from 0.9 to 2.4 ng/ml. Serum digoxin levels lower than these may indicate inadequate absorption of the orally administered drug.

Toxic patients show serum digoxin levels ranging from 2.1 to 8.7 ng/ml, depending on the dose [59].

Testosterone by RIA [60]

Testosterone is a steroid hormone for which a completely specific antibody has not yet been produced [61–63]. In common with many other steroids (dihydro-testosterone [64], aldosterone, progesterone, estrone, estradiol, and estriol [21], for

‖ Pharmacia Fine Chemicals Inc., Piscataway, N.J. 08854.

Figure 15-4. Digoxin RIA standard curve.

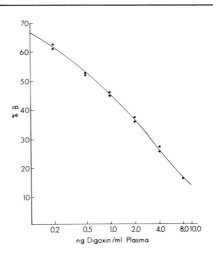

ng Digoxin /ml Plasma

example), its radioimmunoassay depends on prior extraction and separation to avoid cross reaction.

EQUIPMENT

Liquid scintillation counter (beta counter), centrifuge, automatic pipets, 10×75 mm glass test tubes, extraction tubes with Teflon-lined caps, Pasteur pipets, vortex mixer, and rotating or horizontal extraction device.

REAGENTS

1. Borate buffer, $0.065M$, pH 8.0.   Dissolve 2 g of reagent grade boric acid in 500 ml of distilled water. Add 400 μl of $10M$ NaOH. Store at 4°C.

2. Bovine γ-globulin (BγG).   Dissolve 250 mg in 10 ml of $0.08M$ saline containing 0.1% sodium azide. Store at 4°C.

3. Bovine serum albumin (BSA).   Dissolve 1 g in 10 ml of distilled water containing 0.1% sodium azide. Store at 4°C.

4. Stock 1,2-^{3}H-testosterone.   Dilute to approximately 280,000 DPM/100 μl with methanol (spectro grade). Store at 4°C.

5. Recovery tracer.   Dilute stock 1,2-^{3}H-testosterone to 6,000 DPM/100 μl by adding 200 μl of stock to 10 ml of borate buffer. Store at -10°C and mix well before use.

6. Testosterone standards.

A. Stock standard. Prepare stock standards of 1 $\mu g/ml$ and 1 ng/ml in redistilled ethanol.

B. Working standards. Prepare working dilutions of 10 $pg/100$ μl and 1 $pg/100$ μl just before use by properly diluting stock standards.

7. Stock antiserum. Reconstitute and store as instructed by antiserum supplier.

8. Dilute antiserum. Prepare working solution fresh for each assay by adding to 8 ml of borate buffer: 50 μl of stock solution 1,2-^{3}H-testosterone, 100 μl of BSA solution, and 200 μl of BγG solution. Mix gently but well. Pipet 250 μl of this mixture to duplicate blank tubes; then add to the remaining solution sufficient stock antibody to achieve the desired final dilution. This will vary between lots of antiserum.

9. Ammonium sulfate. A saturated solution of reagent grade $(NH_4)_2SO_4$ is prepared in distilled water. Crystals should be present in the bottom of the container to assure complete saturation of the solution. Store at 4°C.

10. Scintillation fluid. Any commercial fluid that will allow incorporation of 250 μl of aqueous solution is satisfactory. The authors have used routinely Packard's Insta Gel# with excellent results.

11. Alumina (Al_2O_3). Wash 200 g of alumina with hot 6M HCl in a 2 liter Erlenmeyer flask. Mix well and decant the fines. Repeat until acid remains colorless, then rinse at least ten times with distilled water, decanting the fines after each rinse. Rinse three times with methanol, transfer to a Büchner funnel, and rinse twice with dichloromethane. Dry in a 100°C oven and store in a tightly capped jar.

12. Testosterone-free plasma. Obtain a pool of plasma from normal nonpregnant females not taking oral contraceptives. Place 500 ml of the pool in a 3 liter Florence flask and add 1.5 liters of spectro grade ether. Agitate on a mechanical shaker for 30 min, decant ether, and repeat extraction twice. Leave plasma overnight in a glass pan or other vessel with a large surface area to allow the last traces of ether to evaporate. Measure plasma volume and replace water lost from evaporation. Mix thoroughly and aliquot in quantities sufficient for a single assay.

13. Hexane, spectro grade.

14. Methanol, spectro grade.

15. Ethyl acetate, spectro grade.

16. Methylene chloride, spectro grade.

17. Absolute ethanol, spectro grade.

Packard Instruments, Downers Grove, Ill. 60515.

PROCEDURE

Prepare microcolumns by packing the tip of disposable Pasteur pipets with glass wool until the flow rate with hexane is 1.5 ml/5 min. Slurry the washed Al_2O_3 with dichloromethane and transfer to the column to a height of 4 cm. Allow columns to run dry. Immediately before applying sample, wash 2 to 3 times with hexane: ethyl acetate 9:1, delivering the solvent with a long-tipped Pasteur pipet to disperse the packing completely and remove air bubbles.

EXTRACTION PROCEDURE

Pipet 50, 100, 250, 500, 750, and 1,000 pg of testosterone standard to methanol-washed extraction tubes, and dry under a stream of air at 4°C. Pipet 100 μl of serum or plasma from each male unknown and 500 μl from each female unknown into washed extraction tubes. Add 400 μl of distilled water to each male sample to equalize the volumes. Add 500 μl of testosterone-free plasma to each standard tube and to a zero standard tube. Pipet 100 μl of recovery tracer to each tube and to duplicate scintillation vials labeled TRC (total recovery counts). Mix all tubes on a vortex. Add 4 ml of hexane:ethyl acetate 9:1 and extract 10 min on a horizontal mechanical shaker, or extract by mixing each tube for at least 60 sec on a vortex mixer. Centrifuge at about 2,500 rpm for 5 min. Transfer the extraction solvent from each tube to a microcolumn and allow to drain. Add 1.6 ml of hexane:ethyl acetate 9:1 and allow to drain. Add, consecutively, 5 aliquots of 1.6 ml of 0.5% ethanol in hexane. Discard. (If desired, the first five rinses may be saved and assayed for dihydrotestosterone. If this is to be done, dihydrotestosterone standards are also extracted, and dihydrotestosterone recovery tracer is added to all standards and samples prior to extraction.)

Elute the testosterone with two 1.6-ml aliquots of 1.1% ethanol in hexane. Collect in a clean tube and evaporate to dryness under air at 40°C.

ASSAY PROCEDURE

Reconstitute each eluate with 3 ml of hexane:ethyl acetate 9:1. Pipet 500 μl from each aliquot tube into duplicate assay tubes and evaporate to dryness. Pipet a 1-ml aliquot of each eluate into a scintillation vial and dry. Add 10 ml of scintillation fluid to these vials and to the two vials marked TRC, mix well, and count for 5 min to determine recovery.

To each assay tube add 250 μl of dilute antibody reagent. Mix well on a vortex mixer to ensure solution of the dried sample but at a low enough speed to avoid foaming. Incubate overnight at 4°C.

Add 250 μl of cold, saturated $(NH_4)_2SO_4$ to each tube and mix well. Allow to stand 10 min in an ice bath, then centrifuge at 4°C for 10 min at 4,000 rpm. Pipet a 250-μl aliquot from each tube into a scintillation vial, add 10 ml of scintillation fluid, mix well, and count for 5 min.

CALCULATION

Plot % F (percent free testosterone) for each standard on the arithmetic scale of semilog graph paper against the amount of testosterone present in each standard after extraction and chromatography on the logarithmic scale, as in Figure 15-5. Read the amount of testosterone in picograms per tube for each unknown. Divide by the % recovery. The number obtained is reported in nanograms per deciliter for males; female values must be divided by 5 and then reported as nanograms per deciliter.

NORMAL VALUES AND INTERPRETATION OF RESULTS

Normal testosterone levels for adult males range from 350 to 1,000 ng/dl; for adult females, from 15 to 70 ng/dl. Normal male prepubertal children have testosterone levels of less than 100 ng/dl; normal female children, less than 20 ng/dl.

Testosterone levels greater than 60 ng/dl in prepubertal females or greater than 120 ng/dl in adult females are usually reflected by hirsutism or virilization or both. Adrenogenital syndrome, adrenal virilizing tumors, ovarian tumors, and severely polycystic ovaries can all cause elevated testosterone secretion. The administration of dexamethasone suppresses plasma testosterone levels, indicating the presence of adrenogenital syndrome. Adrenal disease raises the levels of urinary and serum 17-ketosteroids in addition to testosterone.

Testosterone values less than 350 ng/dl in the adult male are an indication of hypogonadism. In primary hypogonadism, luteinizing hormone (LH) levels are elevated; in secondary hypogonadism, LH values are low.

Carcinoembryonic Antigen (CEA) RIA

Carcinoembryonic antigen is a glycoprotein normally occurring in fetal but not in adult gastrointestinal tissues [65]. CEA is also produced by certain types of carcinoma cells, particularly those of entodermal origin. The presence of CEA in the serum of patients with such tumors is an aid in the diagnosis and management of the carcinoma.

The procedure is that of Roche Diagnostics' CEA-Roche Test Kit [66], currently the only RIA kit for CEA available. The method requires extraction and dialysis, since proteins and electrolytes present in plasma interfere with the assay. Separation of bound from free antigen is accomplished by precipitation of the bound portion with zirconyl phosphate gel (Z-gel) [67].

Figure 15-5. Testosterone RIA standard curve.

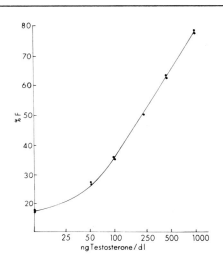

ng Testosterone / dl

EQUIPMENT

Gamma counter, dialysis bath, dialysis bags, 45°C water bath, horizontal head centrifuge, 10, 25, 50, 100, and 500 μl dispensing pipets, dispensing bottles, pH meter, vortex mixer, test tubes 1.5 × 150 mm or 1.2 × 120 mm, either glass or plastic.

REAGENTS (* indicates those supplied with CEA-Roche kit)

1. Deionized water.

2. $1.2M$ cold perchloric acid. Store at 4°C. Make fresh from 70% stock solution at least once a month.

3. Normal saline solution, $0.15M$.

4. Ammonium acetate buffer, $0.01M$, pH 6.8. Dissolve 12.8 ml of glacial acetic acid in 20 liters of deionized water. Adjust pH to 6.8 with concentrated ammonium hydroxide and dilute to 22.5 liters with deionized water. Store at room temperature.

5. Ammonium acetate buffer, $0.1M$, pH 6.25. Dissolve 2.85 ml of glacial acetic acid in 400 ml of deionized water. Add concentrated ammonium hydroxide to bring pH to 6.25 and dilute to 500 ml with deionized water. Store at room temperature.

6. CEA antiserum* (goat anti-CEA). Use as supplied. Store at 4°C.

7. CEA standard.* Use as supplied. Store at 4°C.

8. EDTA buffer stock solution* with bovine serum albumin and 0.17% sodium azide as preservative. Store at room temperature. Dilute 1:10 immediately prior to use with deionized water. The final pH should be 6.5 ± 0.1.

9. ^{125}I-CEA.* Use as supplied. Store at 4°C.

10. Zirconyl phosphate gel (Z-gel)* in ammonium acetate. Use as supplied. Store at room temperature. Do *not* freeze.

SAMPLE REQUIREMENT

One 10 ml lavender stoppered vacutainer tube containing EDTA and potassium sorbate. Do not use clotted blood, plasma containing heparin, or obviously hemolysed plasma. Separate plasma from cells within 6 hr and store frozen or at 4°C for up to 2 weeks. Blood collection tubes containing EDTA and sorbate are not routinely available commercially. The sorbate acts chiefly as a fungistatic agent. If the plasma is separated as soon as possible after collection and kept frozen until assayed, the use of the sorbate is not essential.

PROCEDURE

Pipet 500 μl of plasma into duplicate 15 × 150 mm or 12 × 120 mm test tubes. Add 2 ml of normal saline to each tube and vortex gently. Add 2.5 ml of $1.2M$ cold perchloric acid to each tube with a dispensing bottle. Do not pipet by mouth. Immediately mix each tube for 30 sec on a vortex mixer. Centrifuge tubes at 1,000 rpm for 20 min. Decant supernatant into prewetted dialysis bags, tie, and place in a tank containing 50 times the total volume of the extracts of deionized water. Dialyse at least 3 hr. Change the water three more times, allowing at least 3 hr between changes.

Place bags in 50 volumes of $0.01M$ ammonium acetate for 3 to 24 hr. Transfer contents of bags to clean 15 × 150 mm or 12 × 120 mm test tubes and check pH. If pH of extracts is not 6.5 ± 0.2 discard and repeat extraction. The extracts should be assayed the same day they are removed from dialysis bags.

Prepare a standard inhibition curve as follows: Add 5 ml of a 1:10 dilution of the stock EDTA buffer to 5 pairs of 15 × 150 mm or 12 × 120 mm test tubes labeled with the appropriate concentration of standard. Add the proper amount of standard to duplicate tubes. Use 0, 10, 25, 50, and 100 μl of the standard to obtain concentrations of 0, 1.25, 3.125, and 12.5 ng of CEA activity. Vortex the standard tubes and treat the same as tubes containing plasma extracts.

Add 25 μl of CEA antiserum to each tube (standards and samples). Vortex and place tubes in a 45°C (± 1°C) water bath for 30 min. Remove tubes and add 25 μl of ^{125}I-CEA to each tube. Vortex and return to bath for 30 min. Remove from bath and immediately add 2.5 ml of Z-gel. Shake Z-gel vigorously before beginning additions and agitate from time to time to assure even distribution to all tubes.

Mix and immediately centrifuge tubes at 1,000 rpm for 5 min. Gently decant supernatant into an isotope waste container, add 5 ml of $0.1M$ ammonium acetate to each tube, vortex and centrifuge at 1,000 rpm for 5 min. Again decant supernatant and blot tube tops with a paper towel.

Count tubes for 1 min in a gamma counter to determine the amount of bound ^{125}I-CEA in each tube. Draw a standard curve by plotting the % ^{125}I-CEA bound in each standard tube.

Read sample values from curve and multiply by 2 to convert to nanograms of CEA per mililiter of plasma. Values above 20 ng/ml obtained by this method are not quantitatively accurate and these samples must be reassayed by a direct (nonextracted) method for which directions are supplied with the kit.

NORMAL VALUES AND INTERPRETATION OF RESULTS

Virtually all healthy adults have circulating CEA levels of less than 2.5 ng/ml [68]. Highly elevated levels (greater than 20 ng/ml) are found in many, but by no means all, patients with gastrointestinal tract cancers [69]. Elevated levels of CEA have also been reported in patients with nonmalignant diseases such as ulcerative colitis [68], benign tumors, emphysema, cirrhosis, and diverticulitis [70], and in healthy volunteers who smoke [70]. Because of this, and because a significant percentage of patients with known gastrointestinal malignancies fail to have elevated (greater than 2.5 ng/ml) CEA levels, the test is not recommended as a general screening test for cancer [66].

When used in conjunction with other tests and procedures, CEA titers can be of value in confirming a diagnosis of cancer [71]. Where elevated CEA titers do exist in cancer patients, this test can be useful in monitoring the progress of treatment by comparison of the patients' initial or baseline CEA levels with subsequent values [70]. A dramatic decrease in CEA is usually seen following successful surgery or x-ray therapy for cancer [72]. Failure to show a decrease of CEA to normal levels following surgery might indicate the presence of unsuspected metastases [71], while a sudden increase in CEA levels of presumably cured cancer patients can give early indications of a recurrence of the disease.

Normal and Abnormal Values Found in Common RIA Procedures

Normal and abnormal values obtained by some of the more common radioimmunoassay procedures are listed in Table 15-3, along with brief comments on interpretation and a reference for each procedure.

Table 15-3. Normal Values of Hormones and Other Substances Assayed by RIA

Substance	Normal Values	Comments	References
Adrenocorticotropic hormone (ACTH)	10–80 pg/ml (AM)	Depressed—adrenal tumor Elevated—Cushing's syndrome, ectopic ACTH syndrome	Berson, S. A., and Yalow, R. S.: *J. Clin. Invest.* 47:2725, 1968.
Aldosterone	1–8 ng/dl (supine) 5–25 ng/dl (upright) 2–5 × baseline (low Na diet) 2–5 × baseline (ACTH stimulation)	Failure to respond to ACTH—adrenal insufficiency Depressed—Addison's disease	Ito, T., Woo, J., Janing, R., and Horton, R.: *J. Clin. Endocrinol. Metab.* 34:106, 1972.
Angiotensin I	100–300 ng/dl (supine) 200–600 ng/dl (upright)	Depressed—primary aldosteronism Elevated—renal hypertension	Haber, E, Koerner, T, Page, L. B., Kliman, B., and Purnode, A.: *J. Clin. Endocrinol. Metab.* 29:1349, 1969.
Cortisol	7–18 µg/dl (AM) 2–9 µg/dl (PM)	Depressed—Addison's disease Depressed—hypopituitarism No diurnal variation—Cushing's syndrome	Van de Vies, J.: *Acta Endocrinol.* 38:399, 1961.
Digitoxin	Not present unless on medication	Therapeutic range: 9–25 ng/ml Toxic range: 26–39 ng/ml	Smith, T. W.: *J. Pharmacol. Exp. Ther.* 175:352, 1970.
Digoxin	Not present unless on medication	Therapeutic range: 0.9–2.4 ng/ml Toxic range: 2.1–8.7 ng/ml	Smith, T. W., Butler, V. P., and Haber, E.: *N. Engl. J. Med.* 28:1212, 1969.
Estrone (E1)	< 20 ng/dl, female luteal 20–50 ng/dl, female ovulatory peak < 10 ng/dl, male	Elevated—pregnancy Depressed or failure to peak—ovarian failure	Wu, C. H., and Lundy, L. E.: *Steroids* 18:91, 1971.
Estradiol (E2)	< 20 ng/dl, female luteal 20–60 ng/dl, female ovulatory peak < 10 ng/dl, male	Elevated—pregnancy Depressed or failure to peak—ovarian failure	Wu, C. H., and Lundy, L. E.: *Steroids* 18:91, 1971.

Table 15-3 (*continued*)

Substance	Normal Values	Comments	References
Estriol (E3)	< 10 ng/dl, female luteal 10–30 ng/dl, female ovulatory peak	Elevated—pregnancy Depressed or failure to peak—ovarian failure	Wu, C. H., and Lundy, L. E.: *Steroids* 18:91, 1971.
Follicle stimulating hormone (FSH)	5–10 mIU/ml, male 5–15 mIU/ml, female luteal 50–70 mIU/ml, female ovulatory peak 40–70 mIU/ml, female postmenopausal	Elevated—pituitary tumor, ovarian failure Depressed or failure to peak—pituitary failure	Midgley, A. R.: *J. Clin. Endocrinol. Metab.* 27:1711, 1967.
Gastrin	< 300 pg/ml	Elevated—stomach or duodenal ulcer, pernicious anemia	Yalow, R. S., and Berson, S. A.: *Gastroenterology* 58:1, 1970.
Human chorionic gonadotropin (HCG)	2,000–40,000 IU/liter	Detectable level—confirms pregnancy Low level—threatened pregnancy	Wide, L.: *Acta Endocrinol.* (Supp 70), 1962.
Human growth hormone (HGH)	Baseline < 3 ng/ml	Elevated—acromegaly Failure to stimulate with arginine or insulin—hypopituitarism	Utiger, R. D., Parker, M. L., and Daughaday, W. H.: *J. Clin. Invest.* 41:254, 1962.
Human luteinizing hormone (HLH)	< 11 mIU/ml, male < 25 mIU/ml, female > 3 × baseline, female ovulatory peak > 25 mIU/ml, female postmenopausal	Elevated—pituitary tumor, ovarian failure Depressed or failure to peak—pituitary failure	Schalch, D. S., Parlow, A. F., and Boon, R. C.: *J. Clin. Invest.* 47:665, 1968.
Human placental lactogen (HPL)	4.8–8.8 µg/ml at 32 wk of pregnancy	Elevated—multiple pregnancy Depressed—retarded fetal growth, placental insufficiency	Saxena, B. N., Refetoff, S., Emerson, K., and Selenkow, H. A.: *Am. J. Obstet. Gynecol.* 101:874, 1968.
Insulin	5–20 µU/ml (adult fasting)	Elevated—hypoglycemia, insulinoma	Yalow, R. S., and Berson, S. A.: *J. Clin. Invest.* 39:1157, 1960.

Hormone	Normal range	Clinical interpretation	Reference
Parathormone (PTH)	0–200 mIU/ml	Highly elevated—hyperparathyroidism	Lequin, R. M., Hackeng, W. H. L., and Schapman, W.: *Acta Endocrinol.* 63:655, 1970.
Prostaglandin F	200–300 pg/ml (luteal)	Highly elevated—pregnancy	Caldwell, B. V., Burstein, S., Brock, W. A., and Speroff, L.: *J. Clin. Endocrinol. Metab.* 33:171, 1971.
Prolactin	< 20 ng/ml; 20 × baseline, pregnancy and lactation	Elevated—pituitary tumor	Jacobs, L. S., Mariz, I. K., and Daughaday, W. H.: *J. Clin. Endocrinol. Metab.* 34:484, 1972.
Testosterone	350–1,000 ng/dl, male; 20–70 ng/dl, female	Elevated (female)—adrenal or ovarian tumor; Depressed (male)—hypogonadism	Furuyama, S., Mayes, D. M., and Nugent, C. A.: *Steroids* 16:415, 1970.
Thyrocalcitonin	20–400 pg/ml	Highly elevated—medullary carcinoma of the thyroid	Tashijan, A. H., Howland, B. A., Melvin, K. E., and Hill, C. W.: *N. Engl. J. Med.* 283:890, 1970.
Thyroid stimulating hormone (TSH)	< 10 μU/ml	Elevated—hypothyroidism; < 10 μU/ml in hypothyroid patients—hypopituitarism	Odell, W. D., Wilber, J. F., and Paul, W. E.: *J. Clin. Endocrinol. Metab.* 25:1179, 1965.
Thyroxine (T_4)	1.5–4.0 ng/dl	Elevated—hyperthyroidism; Depressed—hypothyroidism	Chopra, I. J.: *J. Clin. Endocrinol. Metab.* 34:938, 1972.
Thyroid binding globulin (TBG)	2.0–4.8 ng/dl	Elevated—pregnancy, acute liver disease; Depressed—major illness, nephrosis, thyrotoxicosis	Levy, R. P., Marshall, J. S., and Velayo, N. L.: *J. Clin. Endocrinol. Metab.* 32:372, 1971.
Triiodothyronine (T_3)	60–190 ng/dl	Elevated—pregnancy, hyperthyroidism; Depressed—hypothyroidism	Chopra, I. J., Ho, R. S., and Lam, R.: *J. Lab. Clin. Med.* 80:729, 1972.

References

1. Murphy, B. E. P.   *Nature* (Lond.) 201:679, 1964.
2. Murphy, B. E. P.   In W. D. Odell and W. H. Daughaday [Eds.], *Principles of Competitive Protein Binding*. Philadelphia: Lippincott, 1971.
3. Parker, C. W.   In W. D. Odell and W. H. Daughaday [Eds.], *Principles of Competitive Protein Binding*. Philadelphia: Lippincott, 1971.
3a. Robard, D.   *Clin. Chem.* 20:1255, 1974.
4. Felber, J. P.   *Metabolism* 22:1089, 1973.
5. Berson, S. A., Yalow, R. S., Bauman, A., Rothschild, M. A., and Newerly, K.   *J. Clin. Invest.* 35:170, 1956.
6. Yalow, R. S., and Berson, S. A.   *J. Clin. Invest.* 39:1157, 1960.
7. Parker, C. W., and Vavra, J. D.   *Prog. Hematol.* 6:1, 1969.
8. Sela, M.   *Adv. Immunol.* 5:29, 1966.
9. Kabat, E. A., and Bexen, A. E.   *Arch. Biochem. Biophys.* 78:306, 1958.
10. Boyd, G. W., Adamson, A. R., Fritz, A. E., and Peart, W. S.   *Lancet* 1:213, 1969.
11. Landsteiner, K., and Van der Scheer, J.   *J. Exp. Med.* 63:325, 1936.
12. Skelley, D. S., Brown, L. P., and Besch, P. K.   *Clin. Chem.* 19:146, 1973.
13. Caldwell, B. V.   Unpublished data.
14. Wu, C. H., and Lundy, L. E.   *Steroids* 18:91, 1971.
15. Hollander, F. C., and Schuurs, A. H. W. M.   *Scand. J. Clin. Lab. Invest.* [Suppl.] 29:126. Abstr. 14.7, 1972.
16. Boyd, G. W., and Peart, W. S.   *Lancet* 2:129, 1968.
17. Abraham, G. E., and Grover, P. K.   In W. D. Odell and W. H. Daughaday [Eds.], *Principles of Competitive Protein Binding*. Philadelphia: Lippincott, 1971.
18. Khorana, H. G.   *Chem. Rev.* 53:145, 1959.
19. Habeeb, A. F. S. A., and Hiramoto, R.   *Arch. Biochem. Biophys.* 126:16, 1968.
20. Avrameas, S., and Ternynck, T.   *Immunochemistry* 6:53, 1969.
21. Mikhail, G., Wu, C. H., Ferin, M., and Vande Wiele, R. L.   *Acta Endocrinol.* [Suppl.] (Kbh) 64:147, 341, 1970.
22. Abraham, G. E., Swerdloff, R., Tulchinsky, D., and Odell, W. D.   *J. Clin. Endocrinol. Metab.* 32:619, 1971.
23. Feiser, L. F.   *Experientia* 6:312, 1950.
24. Erlanger, B. F., Borek, R., Beiser, S. M., and Lieberman, S.   *J. Biol. Chem.* 228:713, 1957.
25. Vaitukaitis, J., Robbins, J. B., Vieschlag, E., and Ross, G. T.   *J. Clin. Endocrinol. Metab.* 33:988, 1971.
26. Lindner, H. R., Perel, E., Friedlander, A., and Zeitlin, A.   *Steroids* 19:357, 1972.
27. Odell, W. D., Abraham, G. A., Skowsky, W. R., Hescox, M. A., and Fisher, D. A.   In W. D. Odell and W. H. Daughaday [Eds.], *Principles of Competitive Protein Binding*. Philadelphia: Lippincott, 1971.
28. Moudgal, N. R., and Li, C. H.   *Endocrinology* 68:704, 1961.
29. Freund, J.   *Annu. Rev. Microbiol.* 1:291, 1947.
30. Freund, J.   *Am. J. Clin. Pathol.* 21:645, 1951.
31. Saxena, B. B., Demura, H., Gandy, H. M., and Peterson, R. E.   *J. Clin. Endocrinol. Metab.* 28:519, 1968.

32. Midgley, A. R., Jr., Niswender, G. D., and Regar, R. W. *Acta Endocrinol.* [Suppl.] (Kbh) 63:142, 163, 1969.
33. Horejsi, J., and Smetana, R. *Acta Med. Scand.* 155:509, 1971.
34. Nichols Institute for Endocrinology. *Radioimmunoassay Manual.* Wilmington, Calif., 1973.
35. Glick, S. M., and Kagan, A. *J. Clin. Endocrinol. Metab.* 27:133, 1967.
36. Berson, S. A., and Yalow, R. S. *Science* 152:205, 1966.
37. Greenwood, F. C. In W. D. Odell and W. H. Daughaday [Eds.], *Principles of Competitive Protein Binding.* Philadelphia: Lippincott, 1971.
38. Jacobs, L. S., Mariz, I. K., and Daughaday, W. D. *J. Clin. Endocrinol. Metab.* 34:484, 1972.
39. Miles, L. E. M. In W. D. Odell and W. H. Daughaday [Eds.], *Principles of Competitive Protein Binding.* Philadelphia: Lippincott, 1971.
40. Rubenstein, K. E., Schneider, R. S., and Ullman, E. F. *Biochem. Biophys. Res. Commun.* 47:864, 1972.
41. Schneider, R. S., Lindquist, P., Tong-in-Wong, E., Rubenstein, K. E., and Ullman, E. F. *Clin. Chem.* 19:821, 1973.
42. Daughaday, W. H., and Jacobs, L. S. In W. D. Odell and W. H. Daughaday [Eds.], *Principles of Competitive Protein Binding.* Philadelphia: Lippincott, 1971.
43. Odell, W. D., Wilber, J. F., and Paul, W. E. *J. Clin. Endocrinol. Metab.* 25:1179, 1965.
44. Berson, S. A., Yalow, R. S., Glick, S. M., and Roth, J. *Metabolism* 13:1135, 1964.
45. Daughaday, W. H. *Postgrad. Med.* 46:84, 1969.
46. Takahashi, Y., Kipnis, D. M., and Daughaday, W. H. *J. Clin. Invest.* 47:2079, 1968.
47. Taunton-Rigby, A., Sher, S. E., and Kelley, P. R. *Science* 181:165, 1973.
48. Abbott Radio–Pharmaceuticals Technical Bulletin. *Hepatitis Associated Antibody ^{125}I-Ausria®.* Chicago: Abbott, 1972.
49. Abraham, G. E. *J. Clin. Endocrinol. Metab.* 29:866, 1969.
50. Goodfriend, T. L., Ball, D. L., and Updike, S. *Immunochemistry* 6:487, 1969.
51. Moore, P. H., Jr., and Axelrod, L. R. *Steroids* 20:199, 1972.
52. Catt, K., and Tregear, G. W. *Science* 158:1570, 1967.
53. Horgan, E. D., and Riley, W. J. *Clin. Chem.* 19:187, 1973.
54. Donald, R. A. *J. Endocrinol.* 41:499, 1968.
55. Herbert, V. In M. Margoulies [Ed.], *Proceedings, International Symposium on Proteins and Polypeptide Hormones.* Amsterdam: Excerpta Medica, 1969.
56. Jacobs, H. S. *J. Clin. Pathol.* 22:710, 1969.
57. Keane, P. M., Pearson, J., and Walker, W. H. C. *Diabetologia* 4:339, 1969.
58. Butler, V. P. *Lancet* 1:186, 1971.
59. Smith, R. W., Butler, V. P., and Haber, E. *N. Engl. J. Med.* 281:1212, 1969.
60. *Endocrine Sciences Technical Bulletin: Antiserum No. T3-125.* Encino, Calif.: 1972.
61. Furuyama, S., Mayes, D. M., and Nugent, C. A. *Steroids* 16:415, 1970.
62. Horton, R., Kato, T., and Sherins, R. *Steroids* 10:245, 1967.
63. Kinouchi, T., Pages, L., and Horton, R. *J. Lab. Clin. Med.* 82:309, 1973.
64. Ito, T., and Horton, R. *J. Clin. Endocrinol. Metab.* 31:326, 1970.
65. Gold, P. *Hosp. Prac.* 7:79, 1972.
66. Roche Diagnostics Technical Bulletin. *CEA-Roche Test Kit.* Nutley, N.J., 1973.

67. Chu, T. M., and Reynoso, G.   *Clin. Chem.* 18:918, 1972.
68. Rule, A. H., Straus, E., Vandevoorde, J., and Janowitz, H. D.   *N. Engl. Med.* 287:24, 1972.
69. Reynoso, G., Chu, T. M., Holyoke, D., Cohen, E., Nemoto, T., Wang, J. J., Chuang, J., Guinan, P., and Murphy, G. P.   *J.A.M.A.* 220:361, 1973.
70. *The Bulletin of Laboratory Medicine.* Biochemical Procedures, March 1974.
71. LoGerfo, P., LoGerfo, F., Herter, F., Barker, H. G., and Hansen, H. J.   *Am. J. Surg.* 123:127, 1972.
72. Holyoke, D., Reynoso, G., and Chu, T. M.   *Ann. Surg.* 176:559, 1972.

16. Hemoglobin and Hemoglobin Derivatives

Hemoglobin

The various methods for the determination of blood hemoglobin have been reviewed in the literature [1] and only a few of them will be mentioned here. The hemoglobin methods are frequently standardized against an iron determination, and some methods have been introduced in which the iron determination is sufficiently rapid to serve as a routine method for hemoglobin. In several of these the hemoglobin iron is split off by treatment with sodium hypochlorite [2, 3] and the iron determined with thiocyanate or one of the newer color reagents [4]. In another method the blood may be diluted with water and the total iron determined by atomic absorption spectrophotometry [5]. Most methods are based on the absorption of hemoglobin derivatives in solution, usually around 540 nm. Pyridine forms a stable chromogen with reduced hemoglobin [6, 7] but the odor of this reagent is objectionable and it is rarely used routinely. Oxyhemoglobin is formed when the diluted blood is shaken with a solution of dilute alkali (ammonium hydroxide, $0.06M$, or sodium carbonate, $0.01M$) [8–11]; this technique has been used extensively. The most commonly used method is some variation of that introduced by Drabkin [12], in which the hemoglobin is changed to cyanmethemoglobin by the action of cyanide and ferricyanide. This compound has a relatively broad absorption peak near 540 nm and is sufficiently stable so that standard reference solutions can be prepared for standardization.

Hemoglobin by Conversion to Cyanmethemoglobin

The procedure is relatively simple using the modified reagent introduced by Van Kampen and Zijlstra [13].

REAGENT

Dissolve 200 mg of potassium ferricyanide [$K_3Fe(CN)_6$], 140 mg of dipotassium phosphate (K_2HPO_4), and 50 mg of potassium cyanide (KCN) in water to make 1 liter. Add 0.5 ml of Sterox SE* and mix thoroughly. If Sterox SE is not available, other surfactants suggested by Van Assendelft are Quolac Nic-218† and Nonic 218.‡ The reagent is relatively stable at room temperature. It has been shown that freezing will cause serious deterioration of the reagent [14, 15].

PROCEDURE

To 5 ml of the reagent add 20 μl of well-mixed whole blood. Mix the solution and allow to stand for about 5 min. Read in a photometer at 540 nm. Since the absor-

---

* Harleco, Philadelphia, Pa. 19143.
† Unibasic, Inc., Arnold, Md. 21012.
‡ Pennsalt Chemical Corp., Philadelphia, Pa. 19102.

bance of the reagent is negligible at 540 nm, a blank of water is generally used, particularly if the standardization is carried out in the same way. The concentration of hemoglobin is usually obtained from a calibration chart. Occasionally in bloods containing very high levels of γ-globulins, a turbidity may result in the solution. This can be eliminated by adding a few drops of ammonia ($1M$). The conversion of reduced and oxidized hemoglobin to cyanmethemoglobin takes only a few minutes. When appreciable amounts of carboxyhemoglobin are present, the conversion of the carbon monoxide derivative takes longer and the solution should be allowed to stand for 30 min before reading. It has been suggested [16] that the conversion can be speeded up by heating to 56°C for 2 to 3 min.

CALIBRATION

Standard ampules containing definite amounts of cyanmethemoglobin are commercially available. These usually contain about 60 mg of cyanmethemoglobin (the exact amount being stated on the label). Since the usual dilution is 1:251, if the solution contained exactly 60 mg of cyanmethemoglobin, this would correspond to a blood sample containing $251 \times 60 = 15.06$ g/dl hemoglobin. Although there has been some controversy, it seems fairly well established that the quarter millimolar extinction coefficient for cyanmethemoglobin is 11.0 [17] based on a molecular weight of 16,114. Determination by various methods agree very well [18–21]. On this basis it can be calculated that if the cyanmethemoglobin is measured in a 1:251 dilution of blood at 540 nm in cuvettes having an exact light path of 1 cm, absorbance $\times 36.8 = $ g/dl of hemoglobin in the original sample.

Hemoglobin by Iron Determination

The above procedure may also be standardized by analyzing a number of samples for hemoglobin by the above method, and for iron, with the percentage of iron in hemoglobin being taken as 0.347%. We present the relatively simple method of Connerty and Briggs [2]. If the reagent is available, the elegant method of Klein et al. [4] might be simpler. In the method presented, sodium hypochlorite is used to split off the iron from hemoglobin. The iron is then determined with thiocyanate [22–24].

REAGENTS

1. Sodium hypochlorite, 5%. The commercial chlorine bleaches such as Clorox are used.

2. Ammonium thiocyanate solution, 6.5M. Dissolve 50 g of ammonium thiocyanate in water to make 100 ml. The solution may be freed of traces of iron by

extracting it three times with 25-ml portions of a mixture of 1 part ether and 3 parts isoamyl alcohol. The organic layer is discarded. Store the solution in the refrigerator.

3. Potassium nitrate, $1M$. Dissolve 10 g of potassium nitrate in water to make 100 ml.

4. Thiocyanate reagent. Mix 20 ml of the ammonium thiocyanate solution and 1 ml of the potassium nitrate solution and dilute to 100 ml with water.

5. Sulfuric acid, $9M$. Cautiously add 100 ml of concentrated sulfuric acid to 100 ml of water and cool.

6. Iron standards.

A. Stock standard, 100 μg/ml. Dissolve 100 mg of reagent grade iron wire in a mixture of 5 ml of 70% perchloric acid, 2 ml of concentrated hydrochloric acid, and 5 ml of water in a small beaker covered with a watch glass. When dissolved, transfer quantitatively to a 1 liter volumetric flask and dilute to the mark with water.

B. Working standard, 8 μg/ml. Dilute 8 ml of the stock standard and 40 ml of the $9M$ sulfuric acid to 100 ml with water.

7. Modified Drabkin's solution for determination of hemoglobin. See previous section.

PROCEDURE

To avoid the difficulty inherent in pipetting exact duplicate samples of whole blood, a portion of blood is hemolyzed and the solution used for analysis of hemoglobin and iron since only the relation between the two determinations is desired. About 5 ml of whole blood (oxalate or EDTA) is mixed well with 20 ml of water and filtered by suction.

Pipet 2 ml of the hemolysate into a 50 ml volumetric flask, add 5 ml of 5% sodium hypochlorite, and mix. Allow to stand for a few minutes and add about 25 ml of distilled water and 10 ml of the $9M$ sulfuric acid. A small amount of flocculant precipitate is formed. Mix and dilute almost to 50 ml with water. Add a small drop of isoamyl or caprylic alcohol to dispel foam and dilute to 50 ml. Mix well and filter. To 10 ml of the filtrate add 2.5 ml of the thiocyanate reagent and mix. Read absorbance at 480 nm against a blank made up of 2 ml of sulfuric acid, 8 ml of water, and 2.5 ml of the thiocyanate reagent. Prepare a standard as follows: Dilute 0.25 ml of the hypochlorite solution with 10 ml of water. Mix 5 ml of this solution with 5 ml of the working standard and add 2.5 ml of the thiocyanate reagent. Mix and read against the blank.

CALCULATION

The standard used contains 40 μg of iron (5 ml of 8 μg/ml). This is compared with 10 ml of a 2:50 dilution of the hemolysate, or the equivalent of 0.4 ml of hemolysate. Thus:

$$\frac{\text{absorbance of sample}}{\text{absorbance of standard}} \times \frac{40}{0.4} = \mu\text{g Fe/ml hemolysate}$$

$$\mu\text{g/ml} \times \frac{100}{1,000} = \text{mg Fe/dl hemolysate}$$

Hemoglobin contains 0.347% Fe.

$$\frac{\text{Fe}}{0.00347} = \text{Fe} \times 288 = \text{hemoglobin}$$

$$\frac{\text{Fe (mg/dl)}}{0.00347 \times 100} = \text{Hb (in g/dl)}$$

The hemoglobin in the hemolysate is also determined using the Drabkin's reagent. For blood the usual dilution is 0.02 ml to 5.0 ml, or 1:251. For greater accuracy larger aliquots are used. If, for example, exactly 1 ml of the hemolysate is diluted to exactly 50 ml with Drabkin's solution and then 0.2 ml more of the diluting solution is added making a total volume of 50.2 ml, the absorbance of the solution at 540 nm will correspond to that of a blood diluted 1:251 and having a hemoglobin content exactly five times that of the hemolysate used. (The hemolysate will generally contain between 2.5 and 3 g/dl of hemoglobin.) Similarly 0.5 ml of the hemolysate diluted in the same way will correspond to a blood sample having a hemoglobin content two and one-half times that of the hemolysate.

NORMAL VALUES AND INTERPRETATION OF RESULTS

The normal level of blood hemoglobin may be taken as 14 to 17 g/dl for adult males and about 1 g/dl less for females. It varies with age and may be lower in children and the aged. Decreased blood hemoglobin levels may be the result of excessive blood loss or decreased erythrocyte production. For the ramifications of the diagnosis of hematological disorders, a hematology text should be consulted.

Carboxyhemoglobin by Carbon Monoxide Determination

Blood gases including carbon monoxide can be determined with a gas chromatograph [28] (see Chapter 9). The gas can also be extracted and determined by infrared spectrophotometry [29], but this requires special apparatus. It has also been determined by diffusion methods using the Conway dish [30] or other apparatus. The CO [31, 32] diffuses into a solution of a palladium salt which reacts with the CO and is finally determined colorimetrically. The simplest and most rapid methods are the spectrophotometric. The diluted sample is treated with dithionite which reduces the oxyhemoglobin but does not immediately affect the carboxyhemoglobin. The two components, reduced hemoglobin and carboxyhemoglobin, are then determined by reading at two wavelengths [21, 33]. Alternatively a portion of the diluted sample may be oxygenated and the spectra of the oxygenated and the original samples compared for a calculation of carboxyhemoglobin [34–36]. An instrument is available to measure simultaneously reduced hemoglobin, oxyhemoglobin, and carboxyhemoglobin. We present a simple two-wavelength method for the measurement of carbon monoxide.

REAGENTS

1. Ammonium hydroxide, $0.1M$. Dilute 7 ml of concentrated ammonium hydroxide to 1 liter with water.

2. Sodium dithionite (sodium hydrosulfite, $Na_2S_2O_4$). Fresh crystals should be used as the material tends to decompose on standing, especially when exposed to moisture.

3. Carbon monoxide cylinder for preparation of standards. Can be purchased as small lecture cylinders from most laboratory supply houses.

4. Oxygen cylinder for preparation of standards. Any cylinder containing 90% or more oxygen is satisfactory.

PROCEDURE

Heparinized blood is preferable and the sample should be obtained as soon as possible after exposure since the carbon monoxide level falls fairly rapidly when the patient is removed from the source. The tightly stoppered tube of blood may be kept in the refrigerator for several hours if necessary, but preferably the analysis should be made as soon as possible after collection.

To 10 ml of the $0.1M$ ammonia add 0.05 ml of blood and mix well. Allow to stand for several minutes for complete hemolysis. If the sample is not completely clear, it should be centrifuged and the supernatant used. Transfer an aliquot of the diluted blood to a cuvette and add about 5 mg of fresh sodium dithionite. Dissolve by mixing by inversion. Then make readings as rapidly as possible at 480, 538, 555, and 578 nm against a blank of the ammonia solution with as narrow a slit width as possible. A large excess of dithionite should be avoided as this may produce some turbidity causing an error in the readings. The ratios A_{555}/A_{480}

and A_{538}/A_{578} are calculated and the percentage of carbon monoxide calculated from the calibration curves. As with the oxygen saturation method, only two points are needed for the calibration curve, 0 and 100% carboxyhemoglobin. The latter is obtained by equilibrating a portion of the gas with carbon monoxide. This may be done similarly to the procedure for oxygen saturation (Chapter 9) except that the saturation should be carried out in a good fume hood. The 0% CO sample is prepared by equilibrating the blood with pure oxygen (to remove traces of CO). This blood should be obtained from a nonsmoker with minimal exposure to CO. The two standards are carried through the procedure exactly the same as the samples. Values obtained will depend somewhat on the spectrophotometer, but they should be close to the values given below.

	0% CO	100% CO
A_{555}/A_{480}	3.1	1.9
A_{538}/A_{578}	1.1	1.5

The two points for each ratio are then plotted for 0 and 100%, and a straight line drawn between them. This will give two different calibration curves and the average value obtained can be used. For levels of CO under about 25%, the 0% point is the most important and should be checked often, as these lower values will be the most frequently found.

NORMAL VALUES AND INTERPRETATION OF RESULTS

The amount of carbon monoxide in the blood of an individual depends upon previous exposure. Under the most favorable conditions, the carboxyhemoglobin level will be of the order of 1% of the total hemoglobin (some endogenous CO is formed by the decomposition of hemoglobin). In smokers the level may be from 4 to 6% and may be as high as 8% in very heavy smokers. Taxi drivers, traffic policemen, and others having considerable exposure to automobile exhaust fumes may have levels up to 10%. Rather mild symptoms sometimes occur with levels of 10 to 15% saturation. At 20 to 25% there will be severe headaches, ready fatigue, and impairment of judgment. Higher levels cause increasingly severe symptoms and death may result at levels of 60 to 70%

Methemoglobin

Although two-wavelength methods have been tried for the determination of methemoglobin, these have not proved very successful. The usual method is that of Evelyn and Malloy [37]. This is based on the fact that the addition of cyanide to a diluted blood sample will abolish the absorption peak of methemoglobin at 632 nm. The decrease in absorption after the addition of cyanide is thus a measure of the amount of methemoglobin present. To determine the percentage of methemoglobin, ferricyanide is added to another aliquot of the dilution to change all the hemoglobin present to methemoglobin. The difference in absorption after the addition of cyanide will be a measure of the total hemoglobin pigment. The method for sulfhemoglobin given later also determines methemoglobin, but the Evelyn and Malloy method is much simpler for routine use. (In the European literature methemoglobin is often referred to as hemiglobin.) [21]

REAGENTS

1. Stock phosphate buffer, $2M$. Dissolve 36.1 g of anhydrous disodium phosphate (Na_2HPO_4) and 34.6 g of monobasic potassium phosphate (KH_2PO_4) in water to make 250 ml. Adjust the pH so that a 1:20 dilution will have a pH of 6.8.

2. Sterox SE, 5%. Dilute 5 ml of Sterox SE to 100 ml with water.

3. Working buffer. Dilute 1 ml of the stock phosphate buffer and 6 ml of the Sterox solution to 50 ml with water.

4. Potassium ferricyanide, AR fine crystals.

5. Potassium cyanide, AR fine crystals.

PROCEDURE

To 10 ml of the working buffer add 0.2 ml of well-mixed blood. Mix and allow to stand for complete hemolysis. If the solution is not perfectly clear after hemolysis, centrifuge and use the supernatant. If only the relative amount of methemoglobin is desired, the blood need not be measured very accurately. If the diluent and blood are accurately measured, the total hemoglobin can be measured to give the absolute amount of methemoglobin present. Add a portion of the hemolysate to a cuvette and read at 632 nm against a blank of the working buffer. Record the reading as A_1. Add a few milligrams of potassium cyanide, mix by inversion to dissolve, and read again at the same wavelength (A_2). If A_1 and A_2 are the same, there is no methemoglobin present and the following portion of the procedure need not be performed. To another aliquot of the original hemolysate add a few milligrams of potassium ferricyanide and mix to dissolve. Transfer to a cuvette and read at 632 nm against the working buffer (A_3). Add a few crystals of potassium cyanide and mix. Read again at the same wavelength (A_4). If the same cuvettes are used for the different determinations, they must be thoroughly rinsed between determinations.

CALCULATION

The percentage of methemoglobin is then calculated as follows:

$$\% \text{ methemoglobin} = \frac{A_1 - A_2}{A_3 - A_4} \times 100$$

If the hemoglobin content of the original blood sample is known, the absolute amount of hemoglobin may be calculated. If the original dilution was made exactly, the solution remaining after reading A_4 may be accurately diluted 1:5 with the usual Drabkin's solution and the cyanmethemoglobin is read at 540 nm. The result as obtained from the usual hemoglobin calibration curve is multiplied by 1.02 to take into account the fact that the dilution used here is 1:255 instead of the usual 1:251.

RANGE OF VALUES AND INTERPRETATION OF RESULTS

Traces of methemoglobin may be found in normal individuals. Levels of up to 0.5 g/dl have been found in hospital patients who have not been administered drugs which might cause methemoglobinemia. Increased levels of methemoglobin may be found after the administration of a number of drugs such as nitrites, antipyrine, phenacetin, acetanilide, and sulfa drugs. Levels up to 30% may not produce any noticeable symptoms. At somewhat higher levels dyspnea may be noted. The lethal level is generally above 70% saturation.

Sulfhemoglobin

The determination of sulfhemoglobin is less precise than the methods for the other hemoglobin derivatives since absorption of pure sulfhemoglobin is not known exactly. The method of Nicol and Morrell [38] is accurate but requires a good spectrophotometer with a narrow band width, accurate wavelength calibration, and the use of cuvettes with exactly 1 cm light path. The hemoglobin is first converted into the carbon monoxide derivative and the absorption measured at several wavelengths before and after the addition of dithionite. The percentages of methemoglobin and sulfhemoglobin are then calculated by the use of empirically determined factors.

Method of Nicol and Morrell

REAGENTS

1. Phosphate buffer, 0.1M, pH 6.0.   Dissolve 11.8 g of monopotassium phosphate (KH_2PO_4) and 1.9 g of disodium phosphate (Na_2HPO_4) in water and dilute to 1 liter. Check the pH and adjust to 6.0 if necessary.

2. Sodium dithionite (sodium hydrosulfite, $Na_2S_2O_4$), fresh crystals.

3. Carbon monoxide gas. Obtained from a small lecture cylinder of the gas.

PROCEDURE

Dilute 1 ml of oxalated or citrated blood with water to make 10 ml. Mix well and allow to stand until hemolysis is complete. Centrifuge strongly to remove stroma. Dilute 1 ml of the clear supernatant to 10 ml with the phosphate buffer. A moderate stream of carbon monoxide is bubbled through the solution for 3 min. The gas should pass through the solution in the form of rather large bubbles. This must be done in a good fume hood. Measure the absorption, as soon as possible after gassing, in 1 cm cuvettes at 569 and 638 nm against a blank of the phosphate buffer. Add a few crystals of fresh dithionite, mix by inversion, and read at 569, 614, and 638 nm.

CALCULATION

The percentage of methemoglobin is calculated from the formula:

$$\% \text{ methemoglobin} = (\Delta A_{569} + \Delta A_{638}) \times 10.4$$

where ΔA_{569} is the difference between the two readings at 569 nm and ΔA_{638} the difference at 638 nm, both differences being considered as positive in the formula.

The percentage of sulfhemoglobin is calculated from the formula:

$$\% \text{ sulfhemoglobin} = (A_{614} - 0.036 \times A_{569}) \times 6.4$$

where A_{614} is the absorbance at 614 nm and A_{569} is the absorbance at 569 nm, both after the addition of dithionite. Note that in the second formula no differences are used.

To determine the absolute amounts of methemoglobin or sulfhemoglobin, one must know the total amount of hemoglobin present. If this is determined by the usual cyanmethemoglobin method, the sulfhemoglobin is converted only very slowly into the cyanmethemoglobin derivative and the solution must be allowed to stand for at least 30 min for complete conversion.

Screening Test

The following approximate method, based on the work of Van Kampen and Zijlstra [21, 36], is satisfactory as a screening test. It is based on the fact that the addition of cyanide will abolish the absorption of methemoglobin at 620 nm but will not affect the absorption of sulfhemoglobin which has an absorption peak near this wavelength. The measurement of the absorption at 620 nm is corrected for the absorption of other hemoglobin compounds at this wavelength by an empirical factor. The derivation of the formula is not given here. The constants used are somewhat different from those in the original reference since we have used the absorption coefficient of sulfhemoglobin as given by Nicol and Morrell [38].

REAGENTS

1. Sterox SE, 2%.§   Dilute 2 ml of Sterox SE to 100 ml with water.
2. Potassium cyanide, fine crystals.

PROCEDURE

Add 0.1 ml of blood to 10 ml of the Sterox solution. Allow to stand for several minutes until hemolysis is complete. Centrifuge the solution if it is not sparkling clear. To about 3 ml of the clear solution in a cuvette add a few crystals of potassium cyanide and mix. Read at 577 and 620 nm in 1 cm light path cuvettes against a blank of the Sterox solution.

CALCULATION

The percentage of sulfhemoglobin is given by the formula:

$$\% \text{ sulfhemoglobin} = 1.3 \left(F \frac{A_{620}}{A_{577}} - 1.0 \right)$$

where A_{620} and A_{577} are the measured absorbances at the two wavelengths and F is the ratio of the absorbance of pure oxyhemoglobin at 577 nm to that at 620 nm. This is approximately 35 but it should be determined experimentally for the spectrophotometer used by diluting a sample of normal oxygenated blood with the Sterox solution and measuring the absorbance at the two wavelengths.

RANGE OF VALUES AND INTERPRETATION OF RESULTS

Normally only very small amounts of sulfhemoglobin are found in blood. It may be increased after the ingestion of drugs such as those that cause methemoglobinemia, but it rarely exceeds 10%.

Plasma Hemoglobin

The determination of plasma hemoglobin is important in the diagnosis of transfusion reactions, paroxysmal nocturnal hemoglobinuria, and other conditions in

§ Harleco, Philadelphia, Pa. 19143.

which there is intravascular hemolysis. A number of methods have been tried for the determination. In spite of the high absorbance of oxyhemoglobin [39, 40], cyanmethemoglobin [41], and other hemoglobin derivatives [42, 43] at the Soret band near 415 nm, the direct spectrometric measurements have not proved sufficiently sensitive. Most methods have depended upon the catalytic action of hemoglobin derivatives on hydrogen peroxide to produce a color with a chromogen such as benzidine, o-tolidine, guaiac, pyrogallol, or other substances. One difficulty encountered with these methods is that the plasma itself interferes with the color reaction [44].

Two methods have been used to overcome this. In one, the hemoglobin is converted to hematin and extracted from the plasma with diethyl ether, then re-extracted into an aqueous solution for determination [45, 46]. In the other method, the plasma is first treated with hydrogen peroxide before the addition of the chromogen which apparently inactivates the inhibiting factors [47, 48]. The first method is more sensitive and yields lower normal values. The second method, although probably not as accurate, is much more rapid and thus more suitable when the determination of relatively high levels of hemoglobin are present. We present the second method, since we feel that the first method will usually yield slightly high values in normal individuals unless extraordinary precautions are taken to prevent any hemolysis during the collection of the sample. The second method is, as explained, more suitable for higher levels of hemoglobin.

COLLECTION OF SAMPLE

This is the most critical part of the whole determination since any hemolysis during collection of the sample will invalidate the results. About 0.05 ml of heparin solution (any commercial preparation containing about 10 mg/ml heparin) is added to a dry 5 ml glass syringe. The plunger is moved back and forth to distribute the heparin solution on the walls of the syringe, and the plunger moved up to the tip to expel excess heparin. A sterile 20 gauge needle is attached and the cubital vein is punctured after a short application of the tourniquet. The blood is allowed to fill the syringe by venous pressure only. If there is difficulty in entering the vein, if aspiration is required, or if more than a few air bubbles appear in the syringe, the puncture is not satisfactory and a fresh puncture should be made in another vein with a different syringe. As soon as the syringe is filled, the needle is removed and about 1 ml of air drawn into the syringe. The tip is closed with the fingertip and the syringe gently inverted a few times to mix the blood with the anticoagulant. The blood is then slowly and gently expelled along the walls into a dry centrifuge tube (preferably of plastic).

The blood is centrifuged for 10 min at about 1,000 g. The plasma is then carefully removed with a Pasteur pipet to a level about 5 mm above the buffy coat, with care taken not to remove any cells. The separated plasma is recentrifuged. There should be only a very slight trace, if any, of cells at the bottom of the tube. The plasma from the second centrifugation is again pipetted off and stored in a plastic tube. The frozen plasma may be kept for several months.

REAGENTS

1. Benzidine, 10 g/dl.   Five grams of benzidine base (not the hydrochloride which is less sensitive) is dissolved in 50 ml of glacial acetic acid. This solution may be dark on preparation and darken further on storage in the refrigerator, but this does not seem to affect the determination. Benzidine base is said to be carcinogenic and care should be taken in handling. It may be more difficult to secure than the hydrochloride since not all suppliers handle it, but it is available.||

2. Hemoglobin standards.

   A. Stock standard, 10 g/dl.   About 10 ml of heparinized blood is obtained from a normal individual. The blood is centrifuged strongly and the plasma removed and discarded. The packed cells are washed three times with 3 volumes of isotonic saline (0.85 g/dl), mixing each time by inversion and then centrifuging. After the final washing add about 1.1 volume of water for each volume of cells, then add 0.5 volume of carbon tetrachloride and mix well by inversion. Allow to stand in the refrigerator for about 10 min and centrifuge. Remove the upper clear hemolysate and determine the hemoglobin content by the usual method for blood. Dilute to exactly 10.0 g/dl. This solution is stable for months when frozen.

   B. Working standard, 100 mg/dl.   Dilute 1 ml of the stock standard to 100 ml with water and saturate with carbon monoxide from a small lecture cylinder. This solution is stable in the refrigerator for at least 2 months.

3. Standard-benzidine mixture.   One milliliter of the working standard solution of hemoglobin is diluted to 100 ml with water to make a solution containing 1 mg/dl. Mix equal volumes of this and the benzidine solution (reagent 1). This should be prepared on the day used.

4. Blank-benzidine mixture.   Mix 1 volume of water with 1 volume of the benzidine solution. This solution is stable for 1 week in the refrigerator.

5. Hydrogen peroxide, 3%.   The commercial 3% solution is satisfactory, or the 30% reagent may be diluted 1:10.

6. Acetic acid solution, 3.4M.   Dilute 20 ml of glacial acetic acid, reagent grade, to 100 ml with water.

|| Sigma Chemical Co., St. Louis, Mo. 63118; K & K Laboratories, Inc., Plainview, N.Y. 11803.

PROCEDURE

Disposable plastic tubes are preferred. Set up tubes labeled standard, sample, and blank. Each separate determination requires its own standard and blank tubes. Add the amount of sample and reagents as given in Table 16-1.

Read sample and standard against blank at 500 nm. If the color of the sample is too intense, it may be diluted 1:2 or 1:3 with the acetic acid solution before reading; then the appropriate correction must be made in the calculation.

CALCULATION

Since the 1 mg/dl standard is diluted with an equal volume of benzidine and 0.5 ml of this is compared with 0.25 ml of plasma:

$$\frac{\text{absorbance of sample}}{\text{absorbance of standard}} \times 1 = \text{hemoglobin (in mg/dl)}$$

If a high level is suspected, such as found after transfusion reaction or hemoglobinuria, a 1:5 or 1:10 dilution of plasma may be used initially. For very high levels a greater dilution may be required.

NORMAL VALUES AND INTERPRETATION OF RESULTS

This method gives normal levels of up to 2.5 mg/dl. (The other method mentioned earlier gives normal levels of under 1 mg/dl.) Increased levels are noted after heavy exercise. Surgical trauma may raise the level to as high as 35 mg/dl. A transfusion reaction after surgery can be suspected only if the level is above 50 mg/dl.

Table 16-1. Determination of Plasma Hemoglobin

Sample or Reagent	Sample (ml)	Standard (ml)	Blank (ml)
Plasma or diluted plasma	0.25	0.25	0.25
Hydrogen peroxide	—	0.25	0.25
Mix and allow to stand for 10 min			
Standard-benzidine mixture	—	0.5	—
Blank-benzidine mixture	0.5	—	0.5
Hydrogen peroxide	0.25	—	—
Mix and allow to stand for 15 min			
Acetic acid solution	10.0	10.0	10.0

Plasma Methemalbumin

The plasma methemalbumin may also be elevated in conditions that cause intravascular hemolysis. It is determined by measuring the change in absorbance of the diluted plasma upon the addition of dithionite. In the method to be presented, hematin combined with serum albumin is used as a standard [49, 50].

REAGENTS

1. Phosphate buffer, $1M$, pH 7.4.   Dissolve 14.2 g of disodium phosphate (Na_2HPO_4) in water to make 100 ml. Dissolve 13.6 g of monopotassium phosphate (KH_2PO_4) in water to make 100 ml. Mix 3 volumes of the sodium phosphate solution with 1 volume of the potassium phosphate solution. Check the pH and adjust to 7.4 by the addition of more potassium phosphate solution if the mixture is too alkaline or more sodium phosphate if the solution is too acid.

2. Sodium hydroxide, $1M$.   Dissolve 4 g of sodium hydroxide in water to make 100 ml.

3. Human serum albumin, 4 g/dl.   Dissolve 4 g of human serum albumin in water to make 100 ml.

4. Hematin standard.   Weigh out accurately about 5 mg of hematin,# dissolve in the minimum amount of sodium hydroxide solution, and dilute to 100 ml with the human serum albumin solution. (Hematin, a derivative of hemin, must not be confused with hematein or hematine, a stain obtained from logwood.)

5. Sodium dithionite (sodium hydrosulfite, $Na_2S_2O_4$), fresh crystals.

PROCEDURE

Heparinized blood is collected with precautions to avoid hemolysis and the plasma is separated. (See previous method for plasma hemoglobin.) The plasma is stable for several days if frozen. Mix 2 ml of plasma with 1 ml of the phosphate buffer. Centrifuge strongly and read cleared plasma at 469 nm against a water blank. Add 5 mg of sodium dithionite, mix, and allow to stand for 5 min; read again at the same wavelength. Calculate the difference in absorbance. Treat a standard similarly. The absorbance change of the standard is fairly constant and the value obtained may be used for other tests without running a standard each time, but it should be checked occasionally.

NORMAL VALUES AND INTERPRETATION OF RESULTS

In intravascular hemolysis if the amount of hemoglobin liberated exceeds the capacity of the haptoglobulin to bind hemoglobin, the free hemoglobin may be converted to hematin producing methemalbumin.

Calbiochem, San Diego, Calif. 92037.

The normal levels by this method are 0 to 0.6 mg/dl calculated as hematin. High levels may be found in conditions resulting in intravascular hemolysis and in hemorrhagic pancreatitis and the determination has been used as a test for this disease [51, 52]. The results by this method may be compared with those obtained by that of Shinowara [53] by multiplying the latter units by 0.04.

Variant Hemogloblins [54, 55]

The normal adult hemoglobin (Hb-A) molecule consists of four globulin peptide chains. Two of these, called alpha chains, each contain 141 amino acids in the peptide sequence. The other two chains, called beta chains, each contain 146 amino acids in the peptide sequence. The normal hemoglobin A is thus designated $\alpha_2\beta_2$. Each peptide chain has attached to it one heme grouping containing one atom of iron. The entire molecule thus contains four iron atoms and has a calculated molecular weight of 64,458. In the earlier discussion of the determination of total hemoglobin, mention was made of the quarter millimolar extinction of cyanmethemoglobin, that corresponding to one atom of iron and having an equivalent weight of 16,114.

A large number of different hemoglobins have been identified and characterized. The more commonly occurring ones are usually designated by letters, such as hemoglobin-A (Hb-A), the usual normal type (which has been divided into three subtypes, A_1, A_2, and A_3), hemoglobin-S (Hb-S), that found in individuals with sickle-cell disease, and fetal hemoglobin (Hb-F). Since the formation of the various hemoglobins is genetically controlled, an individual may be either homozygous or heterozygous for a given hemoglobin trait. When it is desired to indicate the genotype, the hemoglobin may be referred to as, for example, SS in an individual homozygous for the sickle-cell trait and SC if the individual is heterozygous for the abnormal hemoglobins Hb-S and Hb-C. In the latter instance, both types of hemoglobins will be found in the red cells.

These differ in the exact amino acid sequence in one type of chain. Fetal hemoglobin (Hb-F) contains two alpha chains and two slightly different chains known as gamma chains. The gamma chains also contain 146 amino acids per chain but the sequence is slightly different from that in the beta chains. Fetal hemoglobin is then characterized as $\alpha_2\gamma_2$. In the early development of the embryonic hemoglobin two other types of chains are found. These have delta and epsilon chains and are known as $\alpha_2\delta_2$ and $\alpha_2\varepsilon_2$. These latter chains are slightly different in the amino acid sequence from the alpha, beta, and gamma chains. The embryonic hemoglobins usually disappear before birth, but the fetal hemoglobin may persist for several months after birth and, in exceptional cases, into adult life.

The nomenclature of the large number of different hemoglobins that have been found cannot be discussed here. The various hemoglobins have been differentiated by chemical properties, electrophoresis, and Sephadex chromatography. None of these gives a complete separation. Often the chemical properties are too similar. Electrophoretic separation depends only on the size and charge of the molecule. All the hemoglobin molecules are the same size, and the charge depends upon the presence of certain amino acids in the peptide chain. Two hemoglobins could have slightly different arrangements of the amino acids in a given chain, which might give them slightly different biological properties, but have the same charge and hence the same electrophoretic mobility. We present only a few tests for distinguishing certain important hemoglobins—an alkali denaturation test for fetal hemoglobin, a solubility test for sickle cell hemoglobin, a simple microchromatographic method for separating some types—and a few general remarks about hemoglobin electrophoresis.

Fetal Hemoglobin by Alkali Denaturation

Fetal hemoglobin is more resistant to denaturation by alkali than other hemoglobins and this is the basis for most methods of determination. We present two methods; one, that of Singer, is more suitable for relatively large proportions of fetal hemoglobin (over 5%), and the other is more suitable for the determination of the small amounts of fetal hemoglobin found in normal adults. In both methods the hemoglobin is subjected to the action of alkali for a timed period, and then ammonium sulfate is added which precipitates the denatured hemoglobins but not the fetal hemoglobin. The amount of hemoglobin in the filtrate as compared with the amount in the original sample represents the proportion of the fetal hemoglobin.

METHOD OF SINGER [56]

REAGENTS

1. Potassium hydroxide, 0.0833M. This may be made by careful dilution of a stronger solution (e.g., 8.33 ml of 1M diluted to 100 ml), or by dissolving 5.5 g of potassium hydroxide in water to make 1,000 ml. This is then standardized against 0.1M acid. The 0.0833M hydroxide solution should be kept in a tightly stoppered plastic bottle in the refrigerator.

2. Saturated ammonium sulfate. Mix 390 g of ammonium sulfate with 500 ml of water in a flask. Shake vigorously for about 15 min or allow to stand for several hours with frequent shaking. Allow excess salt to settle out and use the clear supernatant.

3. Half-saturated ammonium sulfate reagent. Mix together 400 ml of saturated ammonium sulfate, 400 ml of water, and 1.7 ml of concentrated hydrochloric acid.

4. Sodium chloride, 0.15M. Dissolve 8.7 g of sodium chloride in water to make 1 liter.

5. Carbon tetrachloride, AR.

PROCEDURE

Prepare a hemolysate with 2 or 3 ml of blood as described in the earlier section in this chapter, Plasma Hemoglobin, reagent 2. Into properly labeled test tubes, pipet 3.2 ml of the KOH solution and place in a water bath at 20°C for 5 to 10 min. Add 0.2 ml of hemolysate to the tubes and immediately start a stopwatch. Mix by rinsing the solution up and down in the pipet five or six times, then shake gently for 10 sec. Exactly 1 min after the hemolysate has been added, quickly add 6.8 ml of the half-saturated ammonium sulfate and mix at once by inversion. Filter the solution through a 9 cm Whatman No. 44 filter paper. The filtrate should be completely clear. While the solution is filtering, add 0.02 ml of the hemolysate to 5 ml of water and mix well. This tube will be a measure of the total hemoglobin, whereas the filtrate will be a measure of the fetal hemoglobin. Read the filtrate and the total hemoglobin tube against water at 540 nm.

CALCULATION

$$\frac{\text{absorbance of filtrate}}{\text{absorbance of total Hb}} \times 20.3 = \% \text{ fetal hemoglobin}$$

The factor arises from the fact that the filtrate is equivalent to a dilution of 1:51 (0.2 ml to 10.2) whereas the total hemoglobin tube is a dilution of 1:251 (0.02 ml to 5.02). Thus $(51/251) \times 100 = 20.3$. Preferably the determination should be done in duplicate if sufficient hemolysate is available.

METHOD OF PEMBREY ET AL. [57, 58]

REAGENTS

1. Sodium hydroxide, 1.2M. Dissolve 26 g of sodium hydroxide in water to make 500 ml. Standardize against 1M standard acid. This solution should be checked frequently for exact molarity.

2. Drabkin's solution (for this test). Dissolve 50 mg of potassium cyanide and 200 mg of potassium ferricyanide in water to make 1 liter.

3. Saturated ammonium sulfate. Same as in the previous method.

PROCEDURE

The hemolysate is prepared and diluted to 10 g/dl exactly as described for Plasma Hemoglobin, reagent 2.

Add 0.6 ml of the hemolysate to 10 ml of the Drabkin's solution and mix. Pipet 5.6 ml of this solution to a test tube. Add 0.4 ml of the sodium hydroxide solution, starting a stopwatch and mixing at once by inversion. Exactly 2 min after the addition of the hydroxide, add 4 ml of saturated ammonium sulfate with vigorous mixing. Allow the mixture to stand for about 5 min, then filter through a double thickness of Whatman No. 6 filter paper. A tube equivalent to total hemoglobin is prepared by adding 2 ml of the saturated ammonium sulfate and 1.4 ml of the cyanmethemoglobin solution to a 50 ml volumetric flask and diluting to the mark with water. The filtrate and the total hemoglobin solution are read in a photometer at 415 nm against a water blank.

CALCULATION

The filtrate represents a dilution of 5.6 ml of the cyanmethemoglobin solution to 10 ml, and the total hemoglobin represents a dilution of 1.4 ml of the cyanmethemoglobin solution to 50 ml (or 5.6 to 200); thus:

$$\frac{\text{absorbance of filtrate}}{\text{absorbance of total Hb}} \times \frac{10}{200} \times 100 = \frac{\text{absorbance of filtrate}}{\text{absorbance of total Hb}} \times 5 = \% \text{ fetal hemoglobin}$$

This procedure, like the previous one, should be performed at near 20°C.

NORMAL VALUES

At birth the blood contains 55 to 85% fetal hemoglobin. This falls to about 45% within a few days and then to 10 to 12% at 2 weeks. It falls further to the adult level of 1 to 2% by the end of the second year. In some abnormal conditions the fetal hemoglobin will persist at higher concentrations even in adult life.

Sickle Cell Hemoglobin by Dithionite Tube Test

This test [59] is based on the fact that at low oxygen tension (presence of dithionite) certain hemoglobins that cause sickling within the cell will form large molecular aggregates in a buffered solution. This renders the solution cloudy or opaque whereas with ordinary hemoglobins the solution remains transparent. With Hb-S the aggregation is prevented by urea and the solution remains clear, but with other sickling hemoglobins the solution will be cloudy. A number of variations based on this principle have been introduced [60–62]; the one described in this section is widely used as a screening test. It has been reported that this method seems to give better results (fewer false positives) than a number of commercial tests based on the same principle [63] A complete discussion of the properties of Hb-S may be found in the book by Marayama and Nalbandian [64].

1. Stock phosphate buffer, 2.8M. Dissolve 160.5 g of monopotassium phosphate (KH_2PO_4) and 281 g of dipotassium phosphate (K_2HPO_4) in water to make 1 liter. The large amount of salts may dissolve slowly. The simplest way to make the solution is to add about 500 ml of water to a 1 liter volumetric flask along with a magnetic stirring bar. With stirring, the salts are added gradually to the flask. More water is added to rinse down the neck of the flask and stirring continued until all the salts have dissolved. The stirring bar is then removed and the solution diluted to 1 liter. This solution is stable at room temperature.

2. Saponin solution, 1 g/dl. Dissolve 5 g of saponin in water to make 500 ml. A good grade of saponin which is very light tan in color and not dark brown should be used. Store in the refrigerator. Stable for several months.

3. Working solution. Transfer 800 ml of the stock buffer to a 1 liter volumetric flask; dissolve 20 g of sodium dithionite ($Na_2S_2O_4$) in this solution. Add 60 ml of the saponin solution and dilute to 1 liter with water. Store in the refrigerator.

4. Urea-working solution. Just before use dissolve 12 g of urea in 100 ml of the working solution.

Pipet 2 ml of the working solution (reagent 3) to a 12 × 75 mm test tube. Test tubes of this particular size must be used; otherwise one will not obtain consistent results, particularly with samples from anemic patients. Add 0.02 ml of blood (the volume is not critical), mix, and allow to stand for 5 min at room temperature. View the tube against a background of black diagonal lines ruled on white cardboard. If the lines are not visible through the solution, the test is considered positive. If the solution is clear so that the lines are easily visible, the test is negative. If the test is positive with this reagent, repeat the test using the urea-working solution. This procedure has also been automated [65].

INTERPRETATION OF RESULTS

If the first test with the regular working solution is positive, the blood may contain sickle cell hemoglobin or nonsickling hemoglobin such as C (Harlem), C (Georgetown), Bart's, or possibly Alexandra. If the second test with the urea solution is negative, this indicates that the hemoglobin present is either S or G

(Harlem). The diagnosis of sickle cell hemoglobin should always be checked by hemoglobin electrophoresis. This will be discussed in more detail later in this chapter.

Hemoglobins S and C by Microchromatographic Screening Test

This method [66], used to distinguish between some types of hemoglobin, is particularly adapted to screening cord blood at birth. The other methods are not always satisfactory for this blood because of the large amount of Hb-F present.

REAGENTS

1. Tris solution, 0.05M.   Dissolve 6.055 g of reagent grade tris [tris (hydroxymethyl) aminomethane] in water to make 1 liter.

2. Tris maleate solution, 0.05M.   Dissolve 11.86 g of tris maleate* in water to make 1 liter.

3. Tris maleate buffer, 0.05M, pH 6.8.   Mix equal volumes of the tris and tris maleate solutions. Check the pH and adjust to 6.8 by the addition of more of the proper solution as necessary (the tris maleate is the more acid). Note the exact volumes used to give the proper pH for use in subsequent preparation of additional buffer.

4. Tris maleate buffer, 0.05M, pH 7.4.   Mix 4 volumes of tris solution and 3 volumes of the tris maleate. Check pH and adjust to 7.4. Again note the total volumes for use in subsequent preparation.

5. Preparation of column.   Add 1 g of CM-Sephadex (C-50; capacity 4.5 mEq/g)† to 100 ml of the pH 6.8 buffer and swirl gently. Allow to stand overnight in a stoppered flask. The Sephadex should give a swollen volume of about 50 ml. Allow the resin to settle completely and remove the supernatant buffer by suction. Add 100 ml more of buffer, mix gently by inversion, and allow to stand for several hours. When resin has settled, remove supernatant by suction and add 50 ml of fresh buffer. The resin may be kept as a 50% suspension in the buffer. Place a small pledget of cotton or glass wool in the constriction of a Pasteur pipet. Add a slurry of the resin until a column 6 to 8 cm long is formed. The columns may be stored for some time by closing the outlets with lengths of plugged tubing.

PROCEDURE

Mix a 0.1-ml sample of heparinized whole blood with 0.15 ml of water and allow to hemolyze for 1 or 2 min at room temperature. Clamp the column in a

---

* Sigma Chemical Co., St. Louis, Mo. 63118.
† Pharmacia Fine Chemicals, Piscataway, N.J. 08854.

vertical position and carefully remove the excess buffer over the resin with a pipet. Add the hemolyzed blood sample to the top of the column and carefully layer some of the buffer over it to fill the tube. Connect the tube with tubing to reservoir of the pH 6.8 buffer at a sufficient height so that there is a column of liquid of 40 to 50 cm above the top of the column. A glass siphon tube with a stopcock or pinch clamp on tubing just above the attachment to the column may be used. Develop the column with 40 to 50 ml of the pH 6.8 buffer. The flow rate should be about 30 to 40 ml/hr. Most of the fetal hemoglobin washes off the column during the first 10 ml or so of the buffer. At the end of the 50 ml of buffer flow, the Hb-F will have been completely washed off the column and the Hb-A will occupy a zone in the lower part of the column. (With short columns some of the Hb-A may wash through but it should be still visible at the bottom of the column.) If there is no strongly fixed band of the hemoglobin near the top of the column, this indicates the absence of hemoglobins other than A and F. A red band, in the upper 1 cm or so of the column, indicates the presence of Hb-S or Hb-C. Now develop the column with 5 to 6 ml of the pH 7.4 buffer. If Hb-A is present as indicated by a band in the lower part of the column after the first development, this will be washed out by the pH 7.4 buffer. If the upper zone is also washed off by the buffer, this indicates the presence of AS hemoglobin. If the A band is washed off but the upper zone remains, this indicates the presence of AC. If the first column development fails to show the Hb-A band in the lower column but there is a band at the top of the column, the development with the pH 7.4 buffer will indicate three possibilities. If the upper zone is not washed off by the buffer, the hemoglobin present is CC. If the upper zone is washed off, the hemoglobin present is SS. If it is only partially washed off, the hemoglobin is SC. This last possibility may be difficult to see, but if the effluent is collected from the pH 7.4 buffer, a red color would indicate that the zone is being partially washed off.

Hemoglobin Electrophoresis

Cellulose Acetate Electrophoresis

Hemoglobin electrophoresis is most often carried out on cellulose acetate [67–73]. As with protein electrophoresis, the strips made by different manufacturers may use slightly different conditions, buffers, and methods of staining and clearing. The exact technique may vary with the type of apparatus used. The manufacturer's directions should be followed. One commonly used buffer is a tris borate buffer with a pH of 8.4. This may be prepared by dissolving 10.2 g of tris(hydroxymethyl)aminomethane, 0.6 g of ethylenediaminetetraacetic acid, and 6.4 g of boric acid in water to make 1 liter. The pH should be checked and adjusted if necessary. The strips are usually stained with Ponceau S. A staining solution may be prepared by dissolving 0.5 g of the dye and 5 g of trichloroacetic acid in water to make 100 ml.

The hemolysate for application to the strip may be prepared by the method given for the determination of fetal hemoglobin. For capillary samples, the hemolyzing solution furnished by the manufacturer may be used or the following solution can be made up. Dissolve 0.1 g of a purified saponin in 100 ml of water and add 0.6 g of potassium cyanide dissolved in 20 ml of water. Store at room temperature. For blood samples obtained in capillary tubes, add 1 drop of blood to 3 drops of the hemolyzing solution and mix, or collect the blood in a microhematocrit tube and centrifuge. Cut off a section of the tube containing the packed cells 1.6 cm in length and add to 6 drops of the hemolyzing solution. Mix vigorously for 30 sec to remove cells from tube and hemolyze. The capillary methods do not yield as consistent results as the macrohemolysate but may often be the only type of samples available.

The directions of the manufacturer are followed for the application of the sample and electrophoresis. After electrophoresis, the strips are stained and cleared. They may be scanned in a densitometer for quantitation or the presence of the different hemoglobins noted by visual inspection. Since the migration of the different hemoglobins may vary slightly with the experimental conditions or the particular type of cellulose acetate strip used, hemoglobin controls should be run with each set of determinations. Hemoglobin controls may be obtained from several suppliers.‡ After a sample has been obtained that gives a definite distinct pattern, the blood hemolysate may be frozen and also used as a control. On cellulose acetate the more common hemoglobins will appear on the strip in the following order of increasing migration distance from the point of application: A_2+C, S, F, A_1, A_3 (Fig. 16-1).

Agar Gel Electrophoresis

Further characterization of abnormal or unusual hemoglobins may be made by electrophoresis on agar gel [67, 74, 75], starch gel [76–78], or polyacrylamide gel [79–81]. The method using agar gel [67] is relatively simple. The electrophoresis is carried out on slides coated with a gel of agar in pH 6.0 buffer.

‡ Gelman Instruments, Ann Arbor, Mich. 48106; Helena Laboratories, Beaumont, Tex. 77704; Hyland Laboratories, Diagnostic Division, Costa Mesa, Calif. 92626; Schering Diagnostics, Port Reading, N.J. 07064.

Figure 16-1. Hemoglobin electrophoresis pattern using cellulose acetate. Upper scan shows pattern from a hemoglobin control; lower scan, that of a normal individual.

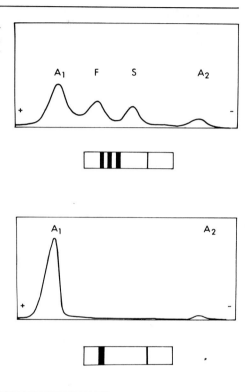

1. Stock citrate buffer, $0.5M$, pH 6.0. Dissolve 147 g of sodium citrate ($Na_3C_6H_5O_7 \cdot 2H_2O$) in about 700 ml of water in a 1 liter volumetric flask. Adjust the pH to 6.0 by the addition of a citric acid solution (30 g/dl) (slightly over 80 ml will be required), and dilute to 1 liter. Store in the refrigerator.

2. Working buffer solution. Dilute the stock 1:10 with distilled water, check the pH, and adjust if necessary to between 6.0 and 6.1.

3. Benzidine staining solution. Prepare just before use by mixing together the following: 5 ml of benzidine solution (0.2 g of benzidine base in 100 ml 95% ethanol), 10 ml of acetic acid solution (3 ml of glacial acetic acid diluted to 100 with water), 1 ml of sodium nitroferricyanide solution [1 g of sodium nitro-

ferricyanide (sodium nitroprusside) in water to make 100 ml], and 1 ml of hydrogen peroxide (commercial 3% hydrogen peroxide or the 30% solution diluted 1:10).

PREPARATION OF SLIDES

A preliminary coating is made on the slides (2.5 × 7.5 cm) by first dissolving about 0.1 g of a good grade of agar in 100 ml of hot water. Stand the cleaned slides on one end and from a pipet pour some of the agar solution over one side. Allow to drain, then wipe off any excess agar solution accumulated at the lower end. Allow the slides to air dry completely. Make an identifying mark on the uncoated side with a wax pencil. These slides may be stored indefinitely. Just before use, coat the required number of slides with buffered agar. Dissolve 1.0 g of agar in 100 ml of the working buffer by warming in hot water just enough to dissolve the agar. Cool to about 50°C and from a pipet carefully add 2 ml of the agar to the previously thinly coated side of the slide with the slide on a perfectly horizontal flat surface. Allow to cool without disturbing.

APPLICATION OF SAMPLE AND
ELECTROPHORESIS

Dip the end of a rounded toothpick into a hemolyzed blood sample (which should contain between 2 and 3 g/dl of hemoglobin). Insert into agar gel on a slide at a distance of 2 to 3 cm from the anodal end. Several samples or controls can be added across one slide. Place slides with agar side down with ends resting on small sponges or wicks to connect with the buffer solutions and electrodes. The electrophoresis is carried out in a small closed container with the buffer solutions surrounded by crushed ice, or the slides may be cooled by placing above them a small plastic bag filled with finely crushed ice. Alternatively the electrophoresis may be carried out in a coldroom without additional cooling. Apply a potential of 50 to 90 v to yield 15 to 20 ma for 30 to 60 min. After electrophoresis, remove the slides and stain with freshly prepared benzidine stain for 3 to 5 min. The different hemoglobins may be identified by visual inspection in comparison with controls which must be run at the same time. On this gel the order of migration of the common hemoglobins is (in order of increasing migration): F, A, S, C. Hematology texts may be consulted for details of interpretation [25–27].

Porphyrins

The determination of porphyrins is included in this chapter since these substances are either precursors in the synthesis of hemoglobin or may be produced by side reactions in the metabolic steps leading to hemoglobin synthesis. A number of different porphyrins have been isolated, differing slightly in chemical structure. They may be conveniently classified as *coproporphyrins* (so called because one of the members of this group was first isolated from feces) and *uroporphyrins* (first isolated from urine). Both types occur in urine and feces. The two classes may be separated by differential extraction. At a pH of 4.8 the coproporphyrins are almost completed extracted by ethyl acetate but only a small fraction of the uroporphyrins are extracted. If the pH is lowered to 3.0, the uroporphyrins may be extracted with *n*-butanol. The separated porphyrins are reextracted from the organic solvents with dilute hydrochloric acid and estimated either by fluorometry or spectrophotometry. The two types of porphyrins may also be separated by chromatography on alumina [83].

We present an extraction method with the final determination done spectrophotometrically [83–85], together with the modifications necessary for fluorometry [86]. The fluorometric method is much more sensitive but requires the use of accurate standards. The spectrophotometric method requires the use of a good spectrophotometer with a band width of not more than 2 nm.

Spectrophotometric Determination

COLLECTION OF SPECIMENS

Since porphyrins are unstable in solution at a pH below 6 and are sensitive to light, a 24-hr urine specimen should be collected in a brown bottle containing 3 to 4 g of anhydrous sodium carbonate [84]. During the collection, contact with metallic caps or other metal should be avoided. Random urine specimens can be collected without preservative if they are analyzed without delay, but a 24-hr specimen is preferable to give an accurate measurement of the amount excreted.

REAGENTS

1. Ethyl acetate, reagent grade.
2. *n*-Butanol, reagent grade.
3. Petroleum ether (30°C–60°C), reagent grade.
4. Hydrochloric acid, concentrated.
5. Hydrochloric acid, $1.5M$. Dilute 125 ml of the concentrated acid to 1 liter with water.
6. Acetate buffer, pH 4.8.

    A. Saturated sodium acetate. Mix together 250 g of sodium acetate ($NaC_2H_3O_2 \cdot 3H_2O$) and 200 ml of water. Warm to dissolve most of the salt, then allow to cool. Some excess salt should crystallize out.

    B. Glacial acetic acid, reagent grade.

To prepare the buffer, mix together as required 4 volumes of solution A with 1 volume of solution B.

7. Acetic acid, $0.08M$. Dilute 5 ml of glacial acetic acid to 1 liter with water.

8. Sodium acetate solution, $0.12M$. Dissolve 1.6 g of sodium acetate trihydrate in water to make 100 ml.

9. Iodine solution, $0.2mM$. Prepare a stock solution by dissolving 1 g of iodine crystals in 100 ml of 95% ethanol. Store this in a glass-stoppered brown bottle in the refrigerator. Just before use, dilute 1 ml of the stock solution with 200 ml of water.

10. Coproporphyrin standard (if required). Dissolve 5.0 mg of pure coproporphyrin in 10 ml of $1.5M$ hydrochloric acid. Gentle warming may aid in solution. The coproporphyrin may be obtained§ as preweighed vials each containing 5 mg of coproporphyrin. Ten milliliters of the $1.5M$ hydrochloric acid is added to the vial and the coproporphyrin dissolved with the aid of heat. This standard is stable when kept in the refrigerator. With a good spectrophotometer and 1 cm cells, the concentrations can be calculated from the known absorptivities of the compounds without the use of a standard [87].

SPECIAL APPARATUS

Long-wavelength ultraviolet light (365 nm) for checking fluorescence of samples.

PROCEDURE FOR COPROPORPHYRIN

Mix the total urine collected, measure, and record the volume. Since some porphyrins may be absorbed on the precipitate formed in alkaline urine, the urine should not be filtered before use. A well-mixed aliquot, including any suspended material, should be used. Examine a portion of the urine under ultraviolet light in the dark. If a noticeable red fluorescence is visible, the urine should be diluted 1:10 before analysis.

Transfer a 100-ml aliquot of the urine (or diluted urine) to a separatory funnel. Add 50 ml of acetate buffer and 150 ml of ethyl acetate. Shake for 1 min and allow to stand until the layers separate completely. Withdraw the lower aqueous layer and save for the analysis of uroporphyrin. Wash the ethyl acetate layer with several 5-ml portions of $0.12M$ sodium acetate. Shake gently for 1 min, allow the layers to separate, and withdraw the lower aqueous wash layer and add to the solution previously reserved for the uroporphyrin analysis. Make several washings until no red fluorescence shows under ultraviolet. Add all the washings to the uroporphyrin sample.

Gently shake the ethyl acetate with 10 ml of the $0.2M$ iodine solution for 30

§ Sigma Chemical Co., St. Louis, Mo. 63118.

sec. Allow the layers to separate and quickly draw off and discard the iodine solution. The iodine solution should not be in contact with the ethyl acetate for more than 5 min. This treatment is necessary to oxidize some coproporphyrinogen to coproporphyrin in fresh urine samples. In urine that has been standing for 24 hr or more, exposure to air has already caused the oxidation and the iodine treatment is not necessary.

Extract the ethyl acetate with several 2.5-ml portions of 1.5M hydrochloric acid. Shake gently for 1 min and allow the layers to separate; then withdraw the lower acid extract into a 25 ml graduated cylinder. Repeat extraction until the last extract shows no red fluorescence under ultraviolet. Combine all the acid extracts in the cylinder and record the volume or dilute with the hydrochloric acid to a convenient volume such as 15 ml. Centrifuge if not perfectly clear and read the solution in a spectrophotometer having a half-band width of not more than 2 nm, with cuvettes having a 1 cm light path against a blank of the hydrochloric acid solution. Read at 380 nm (A_1) and 430 nm (A_3) and at 1-nm intervals between 399 and 405 nm. For A_2 take the maximum reading obtained (peak of absorption curve).

CALCULATION

First find A_{corr}:

$$A_{corr} = 2\,A_2 - (A_1 + A_3)$$

Then:

μg of coproporphyrin in extract $= A_{corr} \times 0.81 \times V_2$

where V_2 is the total volume of the acid extract. Since this is the extract from 100 ml of urine, the above gives the micrograms per deciliter of urine.

$$\mu\text{g of coproporphyrin in total urine sample} = \frac{\mu\text{g/dl} \times V_1}{100}$$

where V_1 is the total urine volume. If the original urine sample was diluted 1:10 for analysis, the above figures must be multiplied by the dilution factor of 10.

PROCEDURE FOR UROPORPHYRIN

To the combined extracts saved for uroporphyrin analysis, add concentrated hydrochloric acid dropwise until the pH is reduced to 3.0 $\pm$ 0.2 (use narrow-range indicator paper or preferably a pH meter). Transfer to a separatory funnel and extract with 20 ml of n-butanol, shaking well for 1 min. Allow the phases to separate completely, then withdraw the lower aqueous layer into a flask and the upper butanol layer into another flask. Add the aqueous solution back to the separatory funnel and extract twice more with 10-ml portions of butanol. Then transfer the combined butanol extracts to a separatory funnel and wash twice with 10-ml portions of 0.08M acetic acid. Discard the washing. To the washed butanol add 30 ml of petroleum ether and 1 ml of concentrated hydrochloric acid. Shake well and allow phases to separate. Draw off lower layer into a 25 ml graduated cylinder. Extract the organic layer with an additional 2 ml of the hydrochloric acid and add this to the first acid extract. Finally extract the organic layer with 5 ml of 1.5M hydrochloric acid. Do not add this to the acid extract but save for the spectrophotometer blank. Note the volume of the combined first two acid extracts, or it may be diluted to a convenient volume with 1.5M hydrochloric acid and mixed well. Read in the spectrophotometer against the blank. Read at 380 nm (A_1) and 430 nm (A_3) and at 1-nm intervals between 404 and 410 nm, taking the maximum absorbance obtained for A_2.

CALCULATION

$A_{corr} = 2 \times A_2 - (A_1 + A_3)$

μg uroporphyrin in total extract $= A_{corr} \times 0.90 \times V_3$

where V_3 is the volume of the total acid extract.

$$\frac{\mu g \text{ uroporphyrin in extract} \times V_1}{100} = \text{total } \mu g \text{ of uroporphyrin excreted}$$

If the urine was diluted 1:10 before analysis, the above figures are multiplied by 10.

Modifications for Fluorometric Determination [86]

A sample of only 5 ml of urine is used, or a 5-ml aliquot of a diluted urine if the original urine showed red fluorescence under ultraviolet light. For the coproporphyrin the 5 ml of urine is mixed with 5 ml of acetate buffer and 10 ml of water and extracted with 75 ml of ethyl acetate. Save the aqueous layer for uroporphyrin determination. Wash the ethyl acetate (saving the washings for uroporphyrin) and treat with iodine (if necessary) as given above. Extract the coproporphyrin from the ethyl acetate with 4-ml portions of 1.5M hydrochloric acid; four extracts are sufficient. These are combined in a 25 ml stoppered cylinder and diluted to 25 ml with 1.5M hydrochloric acid. The uroporphyrin extraction proceeds just as given in the previous section, the combined extracts being diluted up to a convenient volume such as 15 ml.

For fluorometric determination a primary filter having a maximum transmittance at 405 nm (such as Corning #5113) and a secondary filter with transmittance in the red (such as Corning #2412 or Wratten #45) are used. The fluorescence of the samples is compared with that of the standard. The concentration of the standard used will depend upon the sensitivity of the fluorometer. With Turner Model 111 with the high-intensity attachment, a standard of 0.05 μg/ml of coproporphyrin is satisfactory. This is made by diluting the standard given earlier 1:10 with the 1.5M hydrochloric acid. The extracts may also have to be diluted since the fluorescence may not be linear much above 0.05 μg/ml.

If 5 ml of urine was extracted and the final acid extract made up to 25 ml and compared with a 0.05 μg/ml standard:

$$\mu g/\text{ml in urine} = \frac{\text{reading of sample}}{\text{reading of standard}} \times 0.925 \times 0.05 \times \frac{25}{5}$$

Since the fluorescence of the coproporphyrin is about 8% higher in the acid saturated with ethyl acetate than in the standard, a factor of 0.925 is included to correct for this.

The procedure for uroporphyrin is similar to that given for the spectrophotometric method. Since standards of uroporphyrin are not readily available, the fluorescence is read against the coproporphyrin standard and the result multiplied by a factor of 0.75 to correct for the difference in fluorescence of the two compounds.

Normal Values and Interpretation of Results [83]

The normal range for 24-hr excretion is up to 160 μg for coproporphyrin and up to 26 μg for uroporphyrin. The excretion of the porphyrins is greatly increased in certain metabolic diseases (porphyrinurias). Coproporphyrin excretion may also be increased whenever there is active blood regeneration, in infectious hepatitis, and in obstructive jaundice.

Urobilinogen

Urobilinogen is a breakdown metabolite of bilirubin and its determination could well be included in Chapter 10 on liver function tests. However, it is also related to the decomposition products of hemoglobin and so is included in this chapter. Urobilinogen produces a red color with a modified Ehrlich's reagent (*p*-dimethylaminobenzaldehyde). The sample is first reacted with ferrous sulfate in alkaline solution to reduce any urobilin present to urobilinogen. In the method presented, an extraction with petroleum ether is used to reduce the amount of interfering substances in the final reacting solution [87–89].

REAGENTS

1. Ferrous sulfate, 0.7M.   Dissolve 20 g of ferrous sulfate ($FeSO_4 \cdot 7H_2O$) in water to make 100 ml. This solution should be made up fresh as required.

2. Sodium hydroxide, 2.5M.   Dissolve 10 g of sodium hydroxide in water, cool, and dilute to 100 ml.

3. Sodium acetate, saturated.   Same as reagent 6A used in the porphyrin determination.

4. Glacial acetic acid, reagent grade.

5. Acetic acid, 0.08M.   Dilute 5 ml of glacial acetic acid to 1 liter with water.

6. Petroleum ether (30°C–60°C), reagent grade.

7. Modified Ehrlich's reagent.   Dissolve 0.7 g of p-dimethylaminobenzaldehyde in 150 ml of concentrated hydrochloric acid and add 100 ml of water.

8. Standards.   Since pure urobilinogen standards are not available, a dye solution is used instead [90].

A. Stock standard.   Dissolve 100 mg of Pontacyl Carmine 2B|| in 0.08M acetic acid and dilute to 500 ml with this acid. Pipet 25 ml of this solution to a 1 liter volumetric flask. Add 95 mg of Pontacyl Violet 6R|| and dilute to the mark with the dilute acetic acid.

B. Working standard.   Dilute exactly 20.4 ml of the stock to 100 ml with the 0.08M acetic acid.

Procedure for Urine

Collect a 24-hr specimen in a brown bottle containing 5 g of anhydrous sodium carbonate and 50 ml of petroleum ether. Refrigerate during collection. Measure to total volume of urine collected and mix well.

Pipet 50 ml of urine to an Erlenmeyer flask, add 25 ml of the ferrous sulfate solution, and mix. Add 25 ml of the sodium hydroxide solution and mix. Stopper the flask and allow to stand for 1 hr in the dark. Transfer mixture to two 50 ml centrifuge tubes and centrifuge strongly.

Add 10 ml of the supernatant to a 125 ml separatory funnel. Add 1 ml of glacial acetic acid and 40 ml of water. Add 35 ml of petroleum ether and shake vigorously for at least 5 min. Allow the layers to separate completely and re-

|| Harleco, Div. American Hospital Supply Corp., Philadelphia, Pa. 19193.

move and discard lower aqueous layer. To the petroleum ether layer add 3 ml of modified Ehrlich's reagent and shake well. Then add 9 ml of saturated sodium acetate and shake vigorously. After the layers have separated, draw off the lower layer into a graduated cylinder. Repeat the extraction with Ehrlich's reagent and sodium acetate solution until no more color is extracted from the ether layer. Add all of the extracts to the same cylinder. Note total volume of extract and mix well. Prepare a blank by adding 3 ml of modified Ehrlich's reagent and 9 ml of saturated sodium acetate to 35 ml of petroleum ether in a separatory funnel. Shake, allow layers to separate, and withdraw lower layer for use as blank in reading sample. Read sample against blank at 565 nm.

For standards add 5, 10, and 20 ml of the working standard solution to tubes and add 0.08M acetic acid to make 20 ml. Read these standards against a blank of the dilute acetic acid at 565 nm. These standards may be taken as equivalent to concentrations of 0.15, 0.30, and 0.60 mg/dl of urobilinogen. Compare the absorbance of the sample with that of the standard giving an absorbance reading closest to that of the sample.

CALCULATION

$$\text{mg urobilinogen/dl urine} = \frac{\text{absorbance of sample}}{\text{absorbance of standard}}$$

$$\times \text{conc. of standard} \times \frac{\text{vol. of colored extract (ml)}}{10} \times \frac{100}{50}$$

where the concentration of standard is taken as the equivalent concentration given by the dye standard whose absorbance is used in the calculation. Then:

$$\text{mg/dl} \times \frac{\text{total urine vol. (ml)}}{100} = \text{mg urobilinogen excreted in urine/day}$$

If the absorbance of the sample is higher than that of the highest standard, the extraction and colorimetric determination are repeated using 5 ml of the supernatant instead of 10 and multiplying the result obtained by 2.

NORMAL VALUES AND INTERPRETATION OF RESULTS

The normal excretion by this method is 0.05 to 2.5 mg/day in adults. The amount excreted is low in obstructive jaundice since the urobilinogen is formed from bile in the intestines. Antibiotics resulting in changes in the intestinal flora may also result in low levels. The excretion will be increased in infectious hepatitis and in diseases causing intravascular hemolysis.

Procedure for Feces

It is recommended that a 72-hr stool collection be used. A random specimen gives little information since the portion of the daily output represented by the specimen is unknown. The reagents and standards are the same as for the urine procedure.

Homogenize the entire stool sample with the minimum amount of water in a blender and transfer to a tared flask. It may be necessary to homogenize the sample in several portions. Mix the total homogenized specimen and washings from the blender and determine the total weight.

Transfer a sample of the homogenate weighing 5.0 g to a mortar. Measure 145 ml of water into a flask. Use portions of this water in the various steps so that in the whole procedure 145 ml of water is used. Add 5 to 10 ml of water to the sample in the mortar and grind to a paste. Add about 40 ml of additional water, grind further, then allow to stand for a few minutes. Add 50 ml of ferrous sulfate solution to a flask and add the supernatant fine material from the mortar to the flask. Add additional water to the mortar, grind further, and again decant the supernatant into the flask. Repeat again and use the last of the 145 ml of water to rinse out the mortar and add to the flask.

Add slowly with shaking 50 ml of 2.5M sodium hydroxide solution and mix well. Allow the flask to stand in the dark for 2 to 3 hr or until the supernatant is relatively colorless. Filter through Whatman No. 2 paper or equivalent. If the filtrate is clear and not highly colored, proceed with the analysis using an aliquot of the filtrate. If the filtrate is highly colored, a further purification is necessary. To 50 ml of the filtrate add 25 ml of the ferrous sulfate solution, mix, and add 25 ml of the sodium hydroxide solution. Allow to stand in the dark for 2 hr and then filter. Occasionally the second filtrate will also be dark, and a third treatment of 50 ml of the second filtrate with 25 ml of ferrous sulfate and 25 ml of sodium hydroxide solution may be necessary. If a second treatment was needed, the calculated value is multiplied by 2 since an additional 1:2 dilution was made; if a third treatment was necessary, the calculated value is multiplied by 4.

Transfer 10 ml of the filtrate to a separatory funnel and add 40 ml of water and 0.5 ml of glacial acetic acid. Mix and add 35 ml of petroleum ether. Shake vigorously for at least 5 min. Allow the layers to separate and draw off and discard the lower aqueous layer. Wash the petroleum ether layer by shaking with 10 ml of water and discard the aqueous washings. The color is developed exactly as in the procedure for urine. Add 3 ml of modified Ehrlich's reagent to the petroleum ether, shake well, add 9 ml of saturated sodium acetate, and again shake well. Draw off the lower aqueous layer into a graduated cylinder. Repeat the extraction with 3 ml of Ehrlich's reagent and 9 ml of saturated sodium acetate until no more color is extracted. Combine the extracts, note the volume, and mix. Read at 565 nm against a blank prepared as in the urine procedure. The same standards are used as in the urine procedure. As with the urine, if the sample gives a reading higher than the highest standard, repeat the color development using 5 ml of filtrate instead of 10 and multiply the calculated result by 2.

CALCULATION

$$\text{mg urobilinogen}/100\text{ g stool} = \frac{\text{absorbance of sample}}{\text{absorbance of standard}}$$

$$\times \text{ conc. of standard} \times \frac{\text{vol. of colored extract (ml)}}{10} \times \frac{250}{5}$$

where conc. of standard represents the equivalent concentration of urobilinogen represented by the standard used. Then:

$$\frac{\text{total wt stool homogenate (g)}}{100} \times \text{mg}/100\text{ g}$$

$$\times \frac{24}{\text{total hr collection period}} = \text{mg urobilinogen excreted in stool}/\text{day}$$

The calculated amount must be multiplied by a factor of 2 if only 5 ml of filtrate instead of 10 was used, and by a factor of 2 or 4 if one or two additional treatments with ferrous sulfate and sodium hydroxide were necessary.

NORMAL VALUES AND INTERPRETATION OF RESULTS

The normal range of 24-hr excretion of urobilinogen in feces is 60 to 200 mg in males and 10 to 150 mg in females. The amount of urobilinogen in the stool is governed by the amount of bile reaching the gut. In liver damage such as cirrhosis and hepatitis, the fecal urobilinogen is decreased. Increases are found in diseases resulting in intravascular hemolysis such as hemolytic anemia and paroxysmal hemoglobinuria.

References

1. Martinek, R. G. *Am. J. Med. Technol.* 32:37, 1970.
2. Connerty, H. V., and Briggs, A. R. *Clin. Chem.* 8:151, 1962.

3. Fine, J.   *J. Clin. Pathol.* 14:561, 1961.
4. Klein, B., Weber, B. K., Foreman, J. A., and Searcy, R. L.   *Clin. Chim. Acta* 26:77, 1969.
5. Zenttner, A., and Mensch, A. H.   *Am. J. Clin. Pathol.* 48:225, 1967.
6. Kumlien, A., Paul, K. G., and Ljungberg, S.   *Scand. J. Clin. Lab. Invest.* 12:381, 1960.
7. Rimington, C.   *Br. Med. J.* 1:177, 1960.
8. Sheard, C., and Sanford, A. H.   *J. Lab. Clin. Med.* 14:558, 1929.
9. Hill, D. K., and Pincock, A. C.   *Lancet* 2:754, 1941.
10. King, E. J., et al.   *Lancet* 2:563, 1948.
11. Sunderman, F. W., Copeland, B. E., McFate, R. P., Martins, W. E., Naumann, H. N., and Stevenson, G. F.   *Am. J. Clin. Pathol.* 25:489, 1955.
12. Drabkin, D. L., and Austin, J. H.   *J. Biol. Chem.* 112:51, 1935.
13. Van Kampen, E. J., and Zijlstra, W. G.   *Clin. Chim. Acta* 6:538, 1961.
14. Walters, M. I.   *Clin. Chem.* 14:682, 1968.
15. Rice, E. W., and Lapara, C. Z.   *Clin. Chem.* 11:531, 1965.
16. Rice, E. W.   *Clin. Chim. Acta* 18:89, 1967.
17. Van Assendelft, O. W., Zijlstra, W. G., Buursma, A., and Van Kampen, E. J.   *Clin. Chim. Acta* 22:281, 1968.
18. Itano, H. A., Fogarty, W. M., Jr., and Alford, W. C.   *Am. J. Clin. Pathol.* 55:135, 1971.
19. Tentori, L., Vivaldi, G., and Salvati, A. M.   *Clin. Chim. Acta* 14:276, 1966.
20. Morningstar, D. A., Williams, G. Z., and Suurarinen, P.   *Am. J. Clin. Pathol.* 46:603, 1967.
21. Van Kampen, E. J., and Zijlstra, W. G.   *Adv. Clin. Chem.* 8:141, 1965.
22. Baginski, E. S., Foa, P. P., Suchocka, S. M., and Zak, B.   *Microchem. J.* 14:293, 1969.
23. Martinek, R. G.   *Am. J. Med. Technol.* 30:376, 1968.
24. Rice, E. W.   *J. Lab. Clin. Med.* 71:318, 1968.
25. Williams, W. J., Beutler, E., Erslev, A. J., and Rundles, R. W.   *Hematology.* New York: Blakiston-McGraw Hill, 1972.
26. Miale, J. B.   *Laboratory Medicine: Hematology,* 4th ed. St. Louis: Mosby, 1972.
27. Wintrobe, M. M.   *Clinical Hematology,* 6th ed. Philadelphia: Lea and Febiger, 1967.
28. Blackmore, D. J.   *Analyst* (Lond.) 95:439, 1970.
29. Boiteau, H. K., and Gelot, S.   *Eur. J. Toxicol.* 5:217, 1972.
30. Trinder, P., and Harper, F. E.   *J. Clin. Pathol.* 15:82, 1962.
31. Ciuhandu, R., and Rusu, V.   *Z. Klin. Chem. Klin. Biochem.* 6:204, 1968.
32. Allen, T. H., and Root, W. S.   *J. Biol. Chem.* 216:319, 1955.
33. Klenshoj, N. C., Feldstein, M., and Sprague, A. L.   *J. Biol. Chem.* 183:297, 1955.
34. Cummins, B. T., and Lawther, P. J.   *Br. J. Ind. Med.* 22:139, 1965.
35. Beeckmans, J. M.   *Br. J. Ind. Med.* 24:71, 1967.
36. Van Assendelft, O. W.   *Spectrophotometry of Hemoglobin Derivatives.* Springfield, Ill.: Thomas, 1970.
37. Evelyn, K. A., and Malloy, A. T.   *J. Biol. Chem.* 126:655, 1938.
38. Nicol, A. W., and Morrell, D. B.   *Clin. Chim. Acta* 22:157, 1968.
39. Hunter, F. T.   *The Quantitation of Mixtures of Hemoglobin Derivatives by Photoelectric Spectrophotometry.* Springfield, Ill.: Thomas, 1951.

40. Shinowara, G. Y. *Am. J. Clin. Pathol.* 24:696, 1954.
41. McCall, K. B. *Anal. Chem.* 28:189, 1956.
42. Hunter, F. T., Grove-Rassmussen, M., and Sputter. I. *Am. J. Clin. Pathol.* 20:429, 1950.
43. Fielding, H. E. *Am. J. Clin. Pathol.* 35:578, 1961.
44. Creditor, M. C. *J. Lab. Clin. Med.* 41:307, 1953.
45. Vanzetti, G., and Valente, D. *Clin. Chim. Acta* 11:442, 1965.
46. Jacobs, S. L., and Fernandez, A. A. *Stand. Methods Clin. Chem.* 6:107, 1970.
47. Cockrell, J. M., and Naumann, H. N. *Surg. Forum* 7:105, 1957.
48. Naumann, H. M. In F. W. Sunderman and F. W. Sunderman, Jr. [Eds.], *Hemoglobin: Its Precursors and Metabolites*. Philadelphia: Lippincott, 1964. P. 40.
49. Chong, G. C., and Owen, J. A. *J. Clin. Pathol.* 20:211, 1967.
50. Murray, K., Knight, M., and Owen, J. A. *J. Clin. Pathol.* 26:446, 1973.
51. Joseph, W. L., Stevens, G. H., and Longmire, W. P., Jr. *J. Surg. Res.* 8:206, 1968.
52. Kelly, T. R., Klein, R. L., Porquez, J. M., and Homer, G. M. *Ann. Surg.* 175:15, 1972.
53. Shinowara, G. Y., and Walters, M. I. *Am. J. Clin. Pathol.* 40:113, 1963.
54. Murayama, M., and Nalbandian, R. M. *Sickle Cell Hemoglobin: Molecule to Man*. Boston: Little, Brown, 1973. Chap. 1.
55. Murayama, M. *Ann. Clin. Lab. Sci.* 1:1, 1971.
56. Singer, K., Chernoff, A., and Singer, L. *Blood* 6:413, 1951.
57. Betke, K., Marti, H. R., and Schlicht, L. *Nature* (Lond.) 184:1877, 1959.
58. Pembrey, M. E., McWade, P., and Weatherall, D. J. *J. Clin. Pathol.* 25:738, 1972.
59. Nalbandian, R. M., Nichols, B. M., Camp, F. R., Jr., Lusher, J. M., Conte, N. F., Henry, R. L., and Wolf, P. L. *Clin. Chem.* 17:1028, 1971.
60. Clark, K. G. A. *J. Clin. Pathol.* 25:700, 1972.
61. Matusik, J. E., Powell, J. B., and Gregory, D. M. *Clin. Chem.* 17:1081, 1971.
62. Kelly, S., and Desjardins, L. *Clin. Chem.* 18:934, 1972.
63. Schmidt, R. M., and Wilson, S. M. *J.A.M.A.* 225:1225, 1973.
64. Murayama, M., and Nalbandian, R. M. *Sickle Cell Hemoglobin: Molecule to Man*. Boston: Little, Brown, 1973. Chaps. 2 and 3.
65. Nalbandian, R. M., Nichols, B. M., Camp, F. R., Jr., Lusher, J. M., Conte, N. F., Henry, R. L., and Wolf, P. L. *Clin. Chem.* 17:1033, 1971.
66. Schroeder, W. A., Jakway, J., and Powars, D. *J. Lab. Clin. Med.* 82:303, 1973.
67. Schmidt, R. M., and Brosious, E. M. *Laboratory Methods of Hemoglobin Detection*. Atlanta, Ga.: U.S. Dept. Health, Education and Welfare, Public Health Service, Center for Disease Control, 1973.
68. Evans, D. I. K. *J. Clin. Pathol.* 24:877, 1971.
69. Hicks, E. J., Griep, J. A., and Nordschow, C. D. *Lab. Med.* 4:27, 1973.
70. Kind, E. J. G. *Can. J. Med. Technol.* 34:83, 1972.
71. Penalver, J. A., and Abren de Miani, M. S. *Clin. Chim. Acta* 30:657, 1970.
72. Sheena, A. H., Fox, F. A., Hayha, M., and Stevens, K. M. *Am. J. Clin. Pathol.* 50:142, 1968.
73. Klepach, G. L., and Climie, A. R. *Harper Hosp. Bull.* 25:130, 1967.
74. Thompson, R. B., Odom, J., Bell, W. N., and Grower, M. K. *Am. J. Med. Technol.* 32:269, 1966.

75. McDonald, C. D., Jr., and Huisman, T. H. J.   *Clin. Chim. Acta* 8:639, 1963.
76. Ricco, G., Gallo, E., Fiorina, L., and Prato, V.   *Acta Haematol.* 38:306, 1967.
77. Aksov, M., and Erdem, S.   *Clin. Chim. Acta* 12:696, 1965.
78. Chernoff, A. I., and Pettit, N. M.   *J. Lab. Clin. Med.* 63:290, 1964.
79. Smith, E. W., and Evatt, B. L.   *J. Lab. Clin. Med.* 69:1018, 1967.
80. Neerhout, R. C., Kimmel, J. R., Wilson, J. F., and Lahey, M. E.   *J. Lab. Clin. Med.* 67:314, 1966.
81. Ferris, T. G., Easterling, R. E., and Budd, R. E.   *Nature* 208:1103, 1965.
82. Harper, H. A.   *Review of Physiological Chemistry*, 10th ed. Los Altos, Calif.: Lange, 1965. Pp. 50 ff.
83. Fernandez, A. A., Henry, R. J., and Goldenberg, H.   *Clin. Chem.* 12:463, 1966.
84. Schwartz, S., Zieve, L., and Watson, C. J.   *J. Lab. Clin. Med.* 37:843, 1951.
85. Schwartz, S., Berg, M. H., Bossenmaier, I., and Dinsmore, H.   *Methods Biochem. Anal.* 8:221, 1960.
86. Talman, E. L.   *Stand. Methods Clin. Chem.* 2:137, 1958.
87. Balikov, B.   *Stand. Methods Clin. Chem.* 2:192, 1958.
88. Schwartz, S., Sborov, V., and Watson, C. J.   *Am. J. Clin. Pathol.* 14:598, 1944.
89. Watson, C. J., Schwartz, S., Sborov, V., and Bertie, E.   *Am. J. Clin. Pathol.* 14:605, 1944.
90. Watson, C. J., and Hawkinson, V.   *Am. J. Clin. Pathol.* 17:108, 1947.

17. Miscellaneous Methods

Vitamin A and Carotene

The carotenes (alpha and beta forms) are precursors of vitamin A and are usually determined together in serum. The proteins are precipitated with ethanol and the carotenes and vitamin A extracted with hexane, cyclohexane, or similar solvent. The carotenes are determined directly by their absorbance at 450 nm. The yellow color extracted by the solvent is not entirely due to the two carotenes—other pigments such as xanthophylls may also be present. The determination should more properly be labeled as that of carotenoids. However, β-carotene is usually used as standard and the results expressed in terms of this standard. Vitamin A in the extract has been determined by several methods. It may be determined fluorometrically as a true micromethod [1–4]. Most commonly it is determined by a colorimetric method. Ethylene dichlorhydrin [5] and antimony trichloride [6–8] have been used as the color reagent, but more recent methods use trifluoroacetic acid [9–12] which has several advantages over antimony trichloride. The former reagent is much less sensitive to traces of moisture and is much easier to remove from glassware.

In theory the vitamin esters should be hydrolyzed before analysis, but many methods omit the hydrolysis step as the difference in the results is not great, relative to the wide range of values found in normal individuals. A method using trifluoroacetic acid without hydrolysis is presented here.

REAGENTS

1. Trifluoroacetic acid (TFA)-chloroform. Mix 1 volume of trifluoroacetic acid and 2 volumes of chloroform. Prepare fresh as required and protect from moisture.

2. Absolute ethanol, reagent grade.

3. Hexane. The spectro grade should be used. This is usually sold as hexanes since it contains small amounts of isomers other than n-hexane.

4. Chloroform, spectro grade.

5. Carotene standards.

A. Stock standard, 500 μg/ml. Dissolve 50 mg of β-carotene in 10 ml of chloroform and dilute to 100 ml with hexane. The stock solution is fairly stable if kept in a tightly stoppered brown bottle in the refrigerator.

B. Working standard, 10 μg/ml. Dilute 1 ml of the stock solution to 50 ml with hexane. Prepare this fresh as needed. If an accurate spectrophotometer with 1 cm precision cuvettes and a narrow band width is available, the carotene concentration of the working standard can be checked making use of the fact that its absorptivity in hexane at 451 nm is $E^{1\%}_{1\ cm} = 2{,}518$. Thus a solution containing 1 μg/ml should have an absorbance of 0.252 under these conditions. If all the measurements are made on the spectrophotometer, the carotene standard is not needed except for the determination of the correction factor for vitamin A as mentioned later.

6. Vitamin A (retinol) standards.

A. Stock standard, 40 mg/dl.   Dissolve 45.9 mg of retinyl acetate in chloroform to make 100 ml. The vitamin A acetate (retinyl acetate) is preferred as a standard since it is more stable in solution and gives the same amount of color on a molar basis as the free alcohol. Since vitamin A is sensitive to light, the standard solutions should preferably be diluted in low-actinic (red) volumetric flasks and stored in the refrigerator.

B. Working standard, 4 μg/ml.   Just before use, dilute 1 ml of the working standard to 100 ml with chloroform.

EXTRACTION OF CAROTENE AND VITAMIN A

Pipet 3 ml of serum to a screw-capped or glass-stoppered test tube and add with constant agitation 3 ml of absolute ethanol. Add 6 ml of hexane, stopper, and shake vigorously for 10 min. Centrifuge (with stopper in place) for a few minutes at slow speed. Pipet off as much as possible of the upper hexane layer to a second tube and reserve for analysis.

CAROTENE DETERMINATION

Tranfer a portion of the hexane extract to a cuvette and read at 451 nm against a blank of hexane. If standards are used, add to a series of cuvettes 0.5, 1.0, 2.0, and 3.0 ml of the working standard and dilute each to 10 ml with hexane. Read the standards similarly to samples. Since each 6 ml of hexane contains the carotene from 3 ml of serum (a 1:2 dilution), the standards that contain 5, 10, 20, and 30 μg of carotene in 10 ml are equivalent to 100, 200, 400, and 600 μg/dl in the serum.

CALCULATION

$$\frac{\text{absorbance of sample}}{\text{absorbance of standard}} \times \text{conc. of standard} = \mu\text{g/dl carotene}$$

Use in the calculations the standard having the absorbance nearest that of the sample, or plot a calibration curve and read results from this curve.

If it is desired to use the known absorptivity for calculations, since:

absorbance/2518 = g/dl carotene (by definition)

or

absorbance $\times$ 3.971 $\times$ 10^{-4} = g/dl carotene then:

$A \times 397.1 = \mu g/dl$ carotene

Taking into account the 1:2 dilution of the serum:

absorbance of sample (451 nm, 1 cm light path) $\times$ 794 $= \mu g/dl$ carotene

VITAMIN A DETERMINATION

Pipet 4 ml of the hexane extract to a cuvette and evaporate the solvent on a water bath at about 60°C with the aid of a stream of dry nitrogen or carbon dioxide. Dissolve the residue in 0.5 ml of chloroform. Set the instrument wavelength to 620 nm and adjust the zero absorbance with a chloroform blank. Quickly add 3 ml of the TFA-chloroform reagent to the sample, mix at once, and read in the photometer. The color develops rapidly, reaching a maximum in about 5 sec, and then fades. Record the maximum absorbance. For a calibration curve add to cuvettes 0.1, 0.2, 0.3, and 0.4 ml of the working standard (4 $\mu g/ml$) and dilute to 0.5 ml with chloroform. Treat these similarly to the sample. These standard tubes contain 0.4, 0.8, 1.2, and 1.6 μg. Taking into account the 1:2 dilution of serum in preparation of the extract, the standards are equivalent to 20, 40, 60, and 80 $\mu g/dl$ of vitamin A in the serum.

CALCULATION

$$\frac{\text{absorbance of sample}}{\text{absorbance of standard}} \times \text{conc. of standard} = \mu g/dl \text{ vitamin A}$$

Use the standard giving an absorbance closest to that of the sample for the calculations or plot a calibration curve and read the values from the curve.

CORRECTION FOR CAROTENE

Carotene will give some color with the TFA-chloroform reagent, so a correction must be made. This correction is adequate for normal levels of carotene, but may be somewhat in error in extreme hypercarotenemia (such as might be caused by ingestion of large quantities of leafy vegetables and carrots).

Pipet 1 ml of the diluted carotene standard (10 $\mu g/ml$) to a cuvette and evaporate the solvent. Dissolve the residue in 0.5 ml of chloroform and treat exactly as in the vitamin A determination. The absorbance obtained is compared with that from the 80 $\mu g/dl$ vitamin A standard. The carotene standard used contains 10 μg of carotene in the cuvette, and the vitamin A standard used (see above) contains 1.6 μg of vitamin A. Thus the correction factor is:

$$\frac{\text{absorbance of carotene standard}}{\text{absorbance of vitamin A standard}} \times \frac{1.6}{10} = F$$

The amount of carotene actually present multiplied by this factor gives the correction to be applied to the vitamin A determination. As an example, if $F = 0.05$ and the carotene level $= 160$ μg/dl, a correction of $0.05 \times 160 = 8.0$ μg/dl would be subtracted from the measured vitamin A level.

NORMAL VALUES AND INTERPRETATION OF RESULTS

The normal range of serum carotene is from 40 to 250 μg/dl, although levels as low as 30 μg/dl have been found in normal individuals on a normal diet. The level may be higher if a diet rich in leafy vegetables and carrots is consumed. Low values are found in steatorrhea or malabsorption syndrome and in dietary deficiencies.

The normal range of vitamin A is 25 to 70 μg/dl in adults, with somewhat lower values in infants. Low levels are generally associated with malnutrition and inadequate diet. High levels may be found in individuals with an excessive intake of vitamin A preparations.

Ascorbic Acid (Vitamin C)

Ascorbic acid may be determined by oxidation to dehydroascorbic acid and reaction with dinitrophenylhydrazine, which in the presence of strong acid gives a red color for photometric estimation. Though this is an accurate method, it is lengthy and simpler methods are generally satisfactory for most clinical purposes [13–15]. Ascorbic acid has been determined by reaction with certain diazotized amines [16, 17], but the usual simple methods involve the reduction of the colored dye 2,6-dichlorophenolindophenol from a blue to a colorless form. The standard solution of the dye may be titrated with the solution containing the ascorbic acid to the disappearance of the dye color [18, 19] or the ascorbic acid may be added to an excess of the dye solution and the amount of decolorization determined photometrically [20, 21]. This method is fairly satisfactory using a metaphosphoric acid filtrate of serum, but in urine, sulfhydryl groups present will interfere. This is usually overcome by the addition of a mercury compound to combine with the sulfur groups.

REAGENTS

1. Metaphosphoric acid, 3 g/dl. Dissolve 3 g of reagent grade metaphosphoric acid in water to make 100 ml. (This compound as purchased is a mixture of metaphosphoric acid and its sodium salt.) This solution is stable for only about 1 week in the refrigerator.

2. Sodium citrate solution, $0.15M$. Dissolve 4.37 g of sodium citrate dihydrate in water to make 100 ml.

3. Mercuribenzoate solution, 0.2 g/dl. Dissolve 200 mg of the sodium salt of mercuribenzoic acid in water to make 100 ml.

4. Stock indophenol solution, $0.025M$. Dissolve 100 mg of sodium dichlorophenolindophenol in sufficient water to make 100 ml. This solution is stable for several months in the refrigerator.

5. Working indophenol solution, $0.0025M$. Just before use as required, dilute 5 ml of the stock solution to 50 ml with water. If necessary, the concentration of the dye can be adjusted so that the reagent blank as given below will have an absorbance between 0.65 and 0.70.

6. Ascorbic acid standards.

A. Stock standard, 40 mg/dl. Dissolve 40.0 mg of ascorbic acid in 40 ml of water and dilute to 100 ml with the metaphosphoric acid solution. This is stable for about 1 month in the refrigerator.

B. Working standard, 0.80 mg/dl. As required, just before use, add 1 ml of the stock standard to 10 ml of water and dilute to 50 ml with the metaphosphoric acid solution.

PROCEDURE FOR PLASMA

Ascorbic acid is stable in plasma for only about 30 min, so the analysis must be started immediately after collection of the specimen. The protein-free filtrate (see below) is stable for several hours in the refrigerator.

Add 2 ml of plasma (heparin or oxalate) to 3 ml of the metaphosphoric acid solution and mix well. Centrifuge strongly to obtain a clear supernatant. To one tube labeled standard add 2.0 ml of the working standard and 0.5 ml of the sodium citrate solution. To a second tube labeled sample add 2 ml of the protein-free supernatant, 0.2 ml of the sodium citrate solution, and 0.3 ml of water. To a third tube labeled blank add 1.2 ml of the metaphosphoric acid solution, 0.5 ml of water, and 0.8 ml of the sodium citrate solution. To each tube add 1 ml of the working dye solution and read after 30 sec at 520 nm against a water blank. Then to each tube add a few crystals of ascorbic acid to decolorize completely and read again. For each tube subtract the reading obtained after the addition of the excess ascorbic acid from that obtained initially to give the corrected absorbance. (This serves to correct for any turbidity that might develop in the solutions.)

CALCULATION

The decrease in color from that of the blank produced by 2 ml of standard is compared with that produced by 2 ml of supernatant equivalent to 0.4 ml of plasma. The standard contains 0.80 mg/dl and in the calculation is equivalent to 2 mg/dl for the sample (0.80/0.40). Then:

$$\frac{\text{corr. absorbance of blank} - \text{corr. absorbance of sample}}{\text{corr. absorbance of blank} - \text{corr. absorbance of standard}} \times 2 = \text{mg/dl ascorbic acid}$$

PROCEDURE FOR URINE

Add 2 ml of urine to 3 ml of the metaphosphoric acid solution, mix well, and centrifuge. To 3 ml of the supernatant add 1 ml of the mercuribenzoate solution and mix well. Allow to stand for about 10 min, then centrifuge, if necessary, to give a clear solution. For sample tube use 2 ml of this supernatant plus 0.5 ml of the sodium citrate solution. Use the same blank and standard as given for plasma, add the indophenol dye, and read in the same way. The calculations are the same except that a factor of 2.67 is used instead of 2.00 to compensate for the extra dilution of the urine.

Occasionally one may find very high plasma or urine levels in which the sample tube is almost completely decolorized initially. In this case use a smaller aliquot of the supernatant, making up the required volume with the proper amount of the metaphosphoric acid solution diluted 1:2 with water.

NORMAL VALUES AND INTERPRETATION OF RESULTS

The normal plasma level of ascorbic acid is 0.5 to 2.0 mg/dl. The plasma level is not a good indicator of scurvy since it is subject to considerable variation with food ingestion. A fasting level below 0.4 mg/dl would suggest the possibility of ascorbic acid deficiency. Due to the popularity of large doses of ascorbic acid as a prophylactic against the common cold, one may find very high levels in the blood and urine.

The daily output of ascorbic acid in the urine varies with the intake. Due to the instability of the vitamin in the urine, the determination of a 24-hr output is not practical. In ascorbic acid deficiency the level may be well below 1 mg/dl in a random specimen.

Lactic Acid

Lactic acid has been determined by a number of colorimetric reactions, the most common one being that devised by Barker and Sommerson in which the isolated lactic acid is dehydrated by strong sulfuric acid to acetaldehyde which then reacts with p-hydroxybiphenyl to produce a purple color [22, 23]. Although this method has been commonly used, it is long and tedious, particularly for only an occasional determination. The enzymatic method [24–28] is much more rapid and simple. It is based on the reaction:

$$\text{lactate} + \text{NAD}^+ \xrightleftharpoons{\text{LDH}} \text{pyruvate} + \text{NADH} + \text{H}^+$$

The extent of the reaction and hence the amount of lactic acid present are determined by the increase in absorbance at 340 nm as the NAD is converted to NADH. At a pH near neutrality the equilibrium in the reaction lies strongly to the left. In order to produce the complete conversion of lactate to pyruvate, the reaction is run at an alkaline pH and hydrazine is added to combine with the pyruvate formed, to prevent the reverse reaction.

The procedure has the disadvantage that some of the reagents are unstable. Satisfactory results may be obtained by using one of the kits made for the enzymatic determination of lactic acid.* If a large number of determinations are to be made, there will be a significant saving if the reagents are made up in the laboratory as outlined here.

REAGENTS

Unless otherwise specified, store all reagents in the refrigerator at 4°C.

1. Perchloric acid, 0.6M. Dilute 5 ml of 70% reagent grade perchloric acid with water to make 100 ml.

2. Sodium hydroxide, 2M. Dissolve 80 g of NaOH in water and dilute to 1 liter. Store at room temperature.

3. Potassium carbonate, 5M. Dissolve 69 g of anhydrous potassium carbonate in water to make 100 ml. This solution need not be stored in the refrigerator.

4. Indicator solution, methyl orange, 50 mg/dl. Dissolve 50 mg of methyl orange in water to make 100 ml.

5. Glycine-hydrazine buffer (glycine, 1M; hydrazine, 0.4M; pH 9.5). To a 100 ml volumetric flask add 7.5 g of glycine, 5.2 g of hydrazine sulfate, and 0.2

---

* Sigma Chemical Co., St. Louis, Mo. 63118; Boehringer-Mannheim Corp., New York, N.Y. 10017; Calbiochem, San Diego, Calif. 92112.

g of disodium ethylenediaminetetraacetic acid. Suspend in about 30 ml of water and dissolve by the addition of 51 ml of the $2M$ sodium hydroxide solution; then dilute to 100 ml with water. This buffer is stable for about 1 week in the refrigerator.

6. Nicotinamide adenine dinucleotide (NAD), $0.066M$. The dry crystals of NAD are stable for nearly a year if kept tightly stoppered at 4°C. A solution of 40 mg/ml in water is prepared as required. This is stable for only a day or two in the refrigerator.

7. Ammonium sulfate, $2.2M$. Dissolve 29 g of ammonium sulfate in water to make 100 ml.

8. Lactate dehydrogenase (LDH). The LDH from rabbit muscle is used. (Sigma type II, PK-free is satisfactory.) This is obtained as a suspension in ammonium sulfate solution. Obtain a solution containing about 5 mg protein/ml, or dilute a more concentrated solution with $2.2M$ ammonium sulfate to give this concentration. This suspension is stable for 1 year in the refrigerator.

9. Lactic acid standards.

A. Stock standard, $0.1M$. Dissolve 0.9 g of anhydrous L(+)-lactic acid or 0.96 g of lithium lactate (the salt is more stable and less hygroscopic than the free acid) in water to make 100 ml. Note that the enzyme acts only on L(+)-lactate and this must be used and not the synthetic DL form.

B. Working standard, $0.2mM$. Dilute stock 1:500 to give a solution containing 0.20 mmol/liter (0.20 μmol/ml).

COLLECTION OF SAMPLE

The blood should be obtained without stasis and without exposure to air. The skin must be thoroughly cleansed before the sample is taken. If a tourniquet is required for the venipuncture, it should be released before the sample is drawn. There is a considerable difference between the lactic acid content of arterial and venous blood, the ratio being about 3:5. Venous blood is usually used as it is much more convenient to obtain. The sample should be placed in ice at the bedside and taken at once to the laboratory for analysis.

PRECIPITATION OF PROTEINS

Pipet 1 ml of whole blood into 7.2 ml of chilled perchloric acid solution and mix at once. Centrifuge for 5 min and pipet 4.0 ml of supernatant to a chilled flask for neutralization. Add 0.01 ml of the methyl orange indicator solution, and with vigorous agitation in an ice bath add 0.05 ml of the potassium carbonate solution. After the evolution of CO_2 has slowed down, add additional potassium carbonate solution in small increments until the indicator color changes to a salmon pink. Usually about 0.08 ml of the potassium carbonate solution is required. (The potassium carbonate neutralizes the acidity and removes the perchlorate ion as insoluble potassium perchlorate.) Allow the solution to stand for about 5 min in the ice bath, then pipet or decant off the supernatant for analysis. This solution is stable for at least 2 days in the refrigerator.

PROCEDURE

Set up tubes for blank, standard, and sample. One blank and one standard will be suitable for a series of samples run during the same period.

To each tube add 1.3 ml of the glycine-hydrazine buffer and 0.2 ml of the NAD solution. Add 0.3 ml of working standard to one tube labeled standard, 0.3 ml of the neutralized filtrate to the tube labeled sample, and 0.3 ml of water to tube labeled blank. Then add 1.2 ml of water to each tube to make a total volume of 3.0 ml. Transfer to cuvettes and read samples and standards against blank at 340 nm. Add 0.05 ml of the LDH suspension to each tube and mix with a small plastic paddle. Allow to stand at room temperature for about 10 min. Then read samples and standards against the blank at frequent intervals and record the maximum absorbance obtained. The maximum should be reached in less than 20 min. If a longer time is required, this indicates that the activity of the LDH preparation used is too low.

CALCULATION

Subtract the initial absorbance for each sample or standard from the corresponding final maximum absorbance to obtain the absorbance change. Since 0.3 ml of the working standard containing 0.2 μmol/ml is compared with an equal volume of the filtrate,

$$\frac{\text{absorbance change of sample}}{\text{absorbance change of standard}} \times 0.2 \times 8.2 = \mu\text{mol/ml lactic acid} = \text{mmol/liter}$$

where 8.2 is the dilution factor for the preparation of the filtrate.

If cuvettes with exactly 1 cm light path and a spectrophotometer with a narrow band width is used, the amount of lactic acid can be calculated from the known absorptivity of NADH without the use of a standard. Since 1 μmol/ml of NADH has an absorbance of 6.22,

$$\text{absorbance change of sample} \times F = \frac{8.2 \times 3.05}{6.22 \times 0.30} = 13.40$$

where 8.2 is again the dilution factor used in the preparation of the filtrate which is further diluted in the cuvette from 0.3 ml to 3.05. Thus:

absorbance change $\times$ 13.40 = μmol/ml lactic acid = mmol/liter

If it is desired to check this method using the standard, since the samples are diluted an additional 1:8.2 and the standards are not, when using the above factor of 13.4 for the calculation, the working standard used should be equivalent to 0.2 $\times$ 8.2 = 1.64 mmol/liter of lactic acid.

The lactic acid concentration is often expressed in mg/dl which may be obtained as follows:

$$mmol/liter \times \frac{90.1 \times 100}{1,000} = mg/dl$$

where 90.1 is the molecular weight of lactic acid.

ALTERNATIVE PREPARATION PROCEDURE

An alternative procedure for the collection of the sample and preparation of the filtrates is as follows. Place 7.5 of the perchloric acid solution in a 25 ml Erlenmeyer flask, stopper tightly, and weigh to the nearest 10 mg. Cool the flask and contents in ice and immediately after withdrawing the blood sample into a syringe, expel any air bubbles from the syringe, remove the needle, and eject about 1.0 ml of blood directly into the perchloric acid solution in the tared flask, restopper, and mix well. Then wipe off all moisture from the outside of the flask and reweigh. The increase in weight in grams divided by 1.06 will represent the volume of blood added. The remainder of the procedure is the same except that the dilution factor for the blood will be different from the 8.2 used above in the calculations. The factor will be instead:

$$\frac{7.5 + ml\ of\ blood\ added}{ml\ of\ blood\ added}$$

If it is desired to determine pyruvate as well, blood can be added to a second flask for this determination, using a different ratio of blood to perchloric acid (see directions for pyruvate determination in the next section).

NORMAL VALUES AND INTERPRETATION OF RESULTS

The normal levels for lactic acid in adults are 0.55 to 0.95 mmol/liter (5.0 to 8.6 mg/dl) for arterial blood and 1.00 to 1.50 mmol/liter (9.0 to 13.5 mg/dl) for venous blood. Some of the earlier colorimetric methods yielded higher values.

The level is increased by strenuous exercise but is not greatly increased by moderate walking. It may be lower in bed patients than in ambulatory ones. Lactic acid acidosis may occur during anesthesia, in hyperthermia, and in shock.

Pyruvic Acid

Pyruvic acid may be determined by the same reaction as used for lactic acid when it is run in the opposite direction. The amount of pyruvic acid present is determined by the decrease in absorbance at 340 nm caused by the conversion of NADH to NAD. The reaction is run at a lower pH [25–29]. For an occasional sample one of the commercial kits may be used (see lactic acid).

REAGENTS

1–4. These are the same as the first four reagents used in the determination of lactic acid.

5. Triethanolamine buffer, $0.4M$, pH 7.6.   Dissolve 18.6 g of triethanolamine hydrochloride in about 200 ml of water, add 18 ml of the $2M$ sodium hydroxide and 3.7 g of the disodium salt of ethylenediaminetetraacetic acid, and dilute with water to 250 ml.

6. Reduced nicotinamide adenine dinucleotide (NADH), $0.06M$.   Dissolve 14 mg of the disodium salt of NADH in 3.0 ml of water. This solution is stable for 48 hr in the refrigerator.

7. Lactate dehydrogenase (LDH).   See method for lactic acid.

8. Pyruvate standards.

A. Stock solution, $0.1M$.   Dissolve 1.12 g of lithium pyruvate in water and dilute to 100 ml. The lithium salt is used as it is more stable than the free acid.

B. Working solution, $0.1mM$.   Dilute stock solution 1:1,000 to obtain a solution containing 0.1 mmol/liter (0.1 μmol/ml).

As with the lactic acid method, the reagents should be stored in the refrigerator unless otherwise specified.

PROCEDURE

The preparation of the protein-free filtrate and neutralization is carried out similarly to that for lactic acid except that a different dilution is used. Add 2 ml of whole blood to 6.3 ml of the perchloric acid solution; this gives a dilution of 1:4.15. When the alternative method mentioned for lactic acid is used, add 6.5 ml of perchloric acid to the flask, then add about 2 ml of blood from the syringe, and mix. Again calculate the exact dilution factor from the weight of blood used. After centrifugation, 4 ml of the supernatant is neutralized with the potassium carbonate solution just as for the lactic acid procedure. The neutralization must be complete to the pink color but an excess of carbonate should be avoided.

Set up cuvettes for samples and standard by adding 2 ml of the sample extract or working standard to the cuvette, followed by 1 ml of buffer and 0.04 ml of NADH solution. For a blank add 3 ml of buffer and 0.03 ml of the indicator solution to a cuvette. Read standard and samples against the blank, setting the blank at an absorbance of 0.200 rather than zero. Since the absorbance is decreasing with increasing pyruvate, the initial absorbance of the solutions of standard and samples will be high and the spectrophotometer should be readable at least to 1.5 absorbance units. If the absorbance is too high, the amount of NADH added is decreased slightly. Then to each tube add 0.01 ml of the LDH suspension and mix with a plastic paddle. Proceed with the readings as outlined for lactic acid except that here the readings are continued until the minimum reading is obtained. In each reading set the blank to the same point, 0.200 absorbance, before reading. Calculate the maximum change in absorbance between the initial and final readings.

CALCULATION

If the standard is used for the calculations, since 2 ml of the working standard (0.1 mmol/liter) and 2 ml of the blood filtrate are treated similarly,

$$\frac{\text{change in absorbance of sample}}{\text{change in absorbance of standard}} \times 0.1 \times 4.15 = \text{mmol/liter pyruvic acid}$$

where 4.15 is the dilution factor for the protein-free filtrate. If the concentration is to be calculated from the use of a 1 cm light path cuvette and the known molar absorptivity of NADH, the calculations are similar for lactic acid allowing for the difference in dilutions. Thus the factor F for pyruvate would be:

$$F = \frac{4.15 \times 3.05}{6.22 \times 2.00} = 1.02$$

since 2 ml of the filtrate is diluted to 3.05 ml in the cuvette. Thus with sufficient accuracy:

change in absorbance at 340 nm (1 cm) $\times$ 1.02 = μmol/ml = mmol/liter pyruvic acid

The pyruvic acid concentration may also be expressed in milligrams per deciliter by multiplying by the factor 88.1/10 where 88.1 is the molecular weight of pyruvic acid.

The fasting levels of blood pyruvate in normal adults are 0.095 to 0.190 mmol/liter (0.84 to 1.67 mg/dl) for arterial blood and 0.080 to 0.160 mmol/liter (0.71 to 1.41 mg/dl) for venous blood. It may also be increased in those conditions causing increases in lactic acid, but usually not to the same extent as lactic acid.

Gastric Acidity [30–32]

The determination of the amount of acid (almost entirely hydrochloric acid) secreted by the parietal cells of the stomach is an important diagnostic tool in the study of disorders of this organ. For many years the main method used has been the titration with standard alkali (0.1M sodium hydroxide) of the stomach contents removed via a stomach tube. Since the gastric contents, even in the absence of food, contain other substances besides the hydrochloric acid, such as pepsin and other enzymes, inorganic ions, and mucus, the titration curve is not as simple as it is with pure hydrochloric acid. Conventionally the titration was made to a pH of about 3.5 (endpoint with Toepfer's reagent), then the titration continued to a pH of about 8.4 (endpoint with phenolphthalein). The alkali required for the first titration was considered a measure of free acidity and the titration to the second endpoint a measure of total acidity. The results were conventionally reported in degrees of acidity, calculated as the milliliters of 0.1M sodium hydroxide required to neutralize 100 ml of gastric juice. This is numerically equal to milliequivalents (or millimoles) per liter and this latter expression is to be preferred. Some have preferred to make the titration using a pH meter to detect the endpoint. Theoretically this should be more satisfactory, but practically gives little additional information.

The preferred method would appear to be the measurement of the actual pH of the stomach contents as obtained. This gives a more accurate measurement of the effective gastric acidity but requires an accurate pH meter and may be somewhat more troublesome than the simple titration. Since the titration method is still commonly used, we present this method briefly as well as the use of the glass electrode for gastric acidity.

Gastric Acidity by Titration

REAGENTS

1. Sodium hydroxide, 0.1M. See Chapter 20 for the preparation of standard alkali.

2. Toepfer's reagent. Dissolve 0.5 g of p-dimethylaminobenzaldehyde in 100 ml of 95% ethanol.

3. Phenolphthalein indicator.   Dissolve 1 g of phenolphthalein in 100 ml of 95% ethanol.

COLLECTION OF SPECIMEN

Since it is generally agreed that for best results the stomach tube must be positioned fluoroscopically, the procedure will not be given here as it is beyond the scope of this book. Similarly the gastric secretions may be stimulated by the administration of histamine or other drug; this also must be done by a physician. Usually four 15-min basal specimens are obtained followed by six (if possible) 15-min specimens after stimulation. The samples are best collected by continuous suction using an apparatus that limits the suction to about 35 cm of water.

PROCEDURE

The volume of each specimen as received is measured and if it contains a large amount of mucus, it may be filtered through gauze. If available, a 10-ml aliquot is used for the titration; otherwise 5 ml or smaller measured sample is used. Add 2 drops of Toepfer's reagent. If a red color is present, titrate with the sodium hydroxide solution to the endpoint (salmon-pink color). Record the titration, add 1 drop of phenolphthalein indicator, and continue the titration to the endpoint for phenolphthalein (pink color). The first titration is considered the free acidity, and the total amount of alkali used to reach the phenolphthalein endpoint is considered the total acidity.

CALCULATION

$$\text{free or total acidity in mmol/liter} = \frac{\text{ml titration} \times 1{,}000}{\text{ml gastric contents used} \times 10}$$

If the sodium hydroxide solution used is not exactly $0.100M$, a correction must be applied (multiply by actual molarity/0.1). For the free acid one can also calculate the actual millimoles secreted:

$$\text{mmol/liter} \times \frac{\text{vol. of sample (ml)}}{1{,}000} = \text{mmol secreted}$$

NORMAL VALUES AND INTERPRETATION OF RESULTS

The actual amount of acid secreted may be more important than the concentration in the stomach contents. This is discussed further in the section on pH measurement by the glass electrode. For the simple titration the basal free acid is generally taken as 0 to 40 mmol/liter and the free acid after histamine stimulation as 10 to 130 mmol/liter. Before stimulation, the total acidity is 10 to 20 mmol/liter greater than the free acidity.

Gastric Acidity Using the Glass Electrode

This requires a good glass electrode and a pH meter capable of reading to 0.001 pH unit. These are generally available from a number of manufacturers. Those with digital readouts are more convenient.

REAGENTS

1. pH standard, 7.0. This standard may be purchased from a number of manufacturers of buffers or can be prepared as a phosphate buffer (see Chapter 20).

2. Potassium tetraoxalate, $0.05M$. Dissolve 3.151 g of pure potassium tetraoxalate in water to make 250 ml. Preferably the grade of this salt obtained from the National Bureau of Standards should be used. Since this is relatively expensive for routine use, the pH of a solution of an ordinary reagent grade may be compared with the NBS grade for use as a secondary standard. If the material is not available commercially, it may be readily prepared as follows. Dissolve 50 g of potassium oxalate in 150 ml of warm distilled water. Dissolve 100 g of oxalic acid in 220 ml of hot distilled water. Mix the two solutions and allow to stand in the refrigerator for several hours. Filter off the crystals on a Büchner funnel and wash with several small portions of ice cold water and then several times with methanol. Allow to dry in the air and store in a closed container. The NBS salt when made up as directed will have a pH of 1.675 at 20°C, 1.679 at 25°C, and 1.683 at 30°C.

OTHER APPARATUS

Although not absolutely necessary, a small magnetic stirrer is convenient. This is used with a small beaker and a small stirring bar (flea).

CALIBRATION

Rinse the electrodes well with distilled water, wipe off gently, and place in the pH 7.00 buffer. Allow to stand for a few minutes with occasional swirling. Adjust meter reading to 7.00 with the calibrating dial. Remove electrodes, rinse well with distilled water, wipe off gently, and place in the tetraoxalate buffer. Adjust the pH reading to the pH of the buffer as given above by use of the temperature-compensating knob or scale-expansion resistance if the instrument has one.

PROCEDURE

Measure the volume of the specimen, mix well, and place an aliquot in a small beaker just large enough to admit the two electrodes. Mix well with a magnetic

Table 17-1. Conversion of Measured pH to True Hydrogen Ion Concentration

pH	Factor	pH	Factor
3.50	1209	1.40	1259
3.00	1211	1.35	1263
2.80	1212	1.30	1269
2.60	1214	1.25	1274
2.40	1216	1.20	1278
2.20	1218	1.15	1284
2.00	1224	1.10	1289
1.90	1227	1.05	1294
1.80	1231	1.00	1297
1.70	1237	0.95	1303
1.60	1244	0.90	1309
1.50	1252		

For pH values greater than 3.5, use the value for 3.5. Interpolate in the table as necessary. The factor is actually $1,000\gamma/$ where γ is the activity coefficient obtained from Moore's tables [31].

stirrer, taking care that the stirring bar does not hit the glass electrode. Turn off stirrer and make a measurement of pH. Stir again and make several measurements. Find the average of the measurements and obtain the actual concentration of hydrogen ions from Table 17-1.

CALCULATION

The pH measurement is related to the thermodynamic activity of the hydrogen ions rather than their actual concentration. At pH values over 4 the differences are small, particularly in solutions containing only hydrogen ions and negative ions. At the pH values encountered here the differences are larger, particularly in the presence of other positive ions such as sodium and potassium. A complete table for calculation with correction for different amounts of sodium and potassium ions is given by Moore [31]. Since these ions are usually not measured routinely in the gastric aspirate, we present only an abbreviated method that assumes an average sodium plus potassium concentration of 50 mmol/liter.

Convert the measured pH value into concentration units in accordance with the definition: $pH = -\log C$ or $C = $ antilog $(-pH)$. The concentration will then be expressed in mols/liter. Multiply this by the factor given in the table (the factor depending upon the measured pH). The resulting figure will then be the actual concentration of hydrogen ions in millimoles per liter in the gastric fluid.

EXAMPLE

Measured pH $= 1.75$; factor from table $= 1234$.

$C =$ antilog $(-\text{pH}) =$ antilog $(-1.75) =$ antilog $(-2 + 0.25) = 10^{-2} \times$ antilog $(0.25) = 1.778 \times 10^{-2}$. Then the actual hydrogen ion concentration $= C \times$ factor $= 1.778 \times 10^{-2} \times 1234 = 21.9$ mmol/liter.

Usually four 15-min fractions are collected for the basal (fasting) specimens and at least four (preferably six) 15-min specimens after stimulation by histamine. The volume of each specimen is measured along with the pH. Then the amount of hydrogen ions in each specimen is calculated as follows:

$$\text{mmol/liter} \times \frac{\text{vol. of specimen (ml)}}{1,000} = \text{mmol in specimen}$$

The millimoles in all four basal specimens are added to obtain the total output in millimoles per hour. If only three specimens are collected, add the results for these three and multiply by 4/3. If more than four specimens are obtained after stimulation, add the millimoles for the four specimens having the highest values to obtain the maximum output after stimulation. The acid output in the basal state is referred to as BAO and the maximum output after stimulation as MAO.

NORMAL VALUES AND INTERPRETATION OF RESULTS

The BAO in normal individuals ranges from 0 to 2.8 mmol/hr with an average of 1.4. The results will vary somewhat in the same individual at different times and any values in the above range are considered normal. The mean value for the MAO in normal individuals is about 12 mmol/hr with an upper limit of about 22. There is considerable overlap between normal individuals and those with gastric or duodenal ulcers. In the absence of complicating factors, usually no peptic ulcer is present if the MAO is below 15 mmol/hr.

Gastric Acidity by Tubeless Analysis

This method estimates the amount of gastric acidity without the use of a stomach tube. A relatively inert, easily detectable substance combined with an ion exchange resin is ingested by the patient. In the presence of gastric acidity some of the substance is released from the resin by the hydrogen ions, absorbed into the blood stream, and excreted by the kidneys. The amount of the substance excreted in the urine is then a rough measure of the amount of acidity in the stomach. The greater the acidity, the more of the substance is released from the resin and the more excreted in the urine. Quinine was first used as the absorbed substance and detected in the urine by its fluorescence [33]. Later a blue dye Azure A (3-amino-7-dimethylaminophenazathionium chloride) was used [34]. This may be readily detected in the urine.

During the process of absorption and excretion some of the dye may be reduced to the colorless form. This is reoxidized by acidifying and boiling with the addition of a small amount of copper ions as catalyst. The comparison is usually made with the color standards furnished by the suppliers of the resin-dye combination.† Even with no acidity present some of the dye may be released from the resin by the presence of other ions such as sodium, potassium, magnesium, or calcium.

The test is usually performed by collecting the urine for a definite period of time after ingestion of the dye-resin. The amount excreted in the urine will depend upon the emptying time of the stomach, the rate of absorption by the intestines, and the ability of the kidneys to excrete the dye readily. Thus in cases of malabsorption or reduced kidney function, erroneous results may be obtained. Caffeine sodium benzoate is usually given along with the resin to stimulate the flow of gastric juice, but this may not prove a sufficient stimulus in many instances. It is nonetheless a relatively simple screening test that can be done on an ambulatory basis, if the possible causes of errors are kept in mind in the interpretation of the results. Complete directions are furnished by the manufacturers of the kit.

It has been estimated that even with good technique the test will yield about 3 to 5% false positives (dye excretion in the absence of gastric acidity) and an approximately equal number of false negatives (no dye excretion although some gastric acidity is present) [35–38]. It has been suggested that, when kidney function is grossly impaired or complete collection of urine specimens is difficult, the dye be determined in the serum [39], but this adds to the complexity of the test.

Xylose Excretion Test

Xylose is a pentose sugar that is readily absorbed in the upper intestines, is not metabolized, and is rapidly excreted by the kidneys chiefly by glomerular filtration. The amount excreted in 5 hr after a standard dose has been used as a measure of intestinal absorption [40–42]. If the excretory power of the kidneys is grossly impaired, the amount excreted will be low irrespective of the amount absorbed. To distinguish between decreased excretion due to decreased absorption

† Diagnex Blue, E. R. Squibb & Sons, Division Olin Mathieson Chemical Corp., New York, N.Y. 10022.

and that due to decreased kidney function, the blood level of xylose may also be determined 1 or 2 hr after the ingestion of the sugar. Although the blood level by itself is not very satisfactory as a diagnostic test, it does serve to distinguish between the two types of decreased excretion.

A dose of 25 g of xylose has been used but this sometimes causes nausea and vomiting or diarrhea which interfere with the test. A lower dose of 5 g has proved satisfactory [43]. Some investigators [42, 44, 45] have adjusted the dose on the basis of body weight, particularly for children, but this is ordinarily not done for adults. The determination of xylose by the o-toluidine reaction has been described in Chapter 5. The usual method for the determination of xylose for the present test is based on the reaction with p-bromoaniline to give a red color [46].

Xylose by p-*Bromoaniline Method* [47]

The xylose is given dissolved in about 250 ml of water and an additional 250 ml of water is given to insure adequate urine flow. The bladder is emptied and the urine excreted during the next 5 hr is collected and the total volume noted. A blood sample is taken 2 hr after the administration of the sugar and analyzed for xylose by the method given. Only the D(+)-xylose should be used, the L(−)-isomer being less readily absorbed. The former is usually sold as D-xylose. The material, as sold, usually is labeled for laboratory and experimental use only, not for drug use, and it is stated that xylose has not yet been released by the FDA for oral administration and thus may be used only for investigative purposes [48]. Although the test has been used for many years with few reports of deleterious effects, the physician should be aware of the above proscription when ordering the test.

REAGENTS

1. Somogyi reagents for protein-free filtrate: zinc sulfate and barium hydroxide solutions (see Chapter 5 on the preparation of protein-free filtrates). If these reagents are not used for other procedures, it may be more convenient to use 0.3M sodium hydroxide instead of the barium hydroxide solution given. Dissolve 12 g of sodium hydroxide in water and dilute to 1 liter. This solution is then titrated against the zinc sulfate solution and the two solutions adjusted to neutralize each other exactly (see Chapter 5 under the preparation of the barium hydroxide solution).

2. Acetic acid saturated with thiourea. Add 10 g of thiourea to about 250 ml of glacial acetic acid, shake well for several minutes and then allow to stand for several hours with frequent shaking, or stir with a magnetic stirrer for 30 min. The stability of the reagent is better if some excess thiourea remains in contact with the solution which is then decanted as needed.

3. *p*-Bromoaniline reagent.   Add 2 g of *p*-bromoaniline to 100 ml of the acetic acid saturated with thiourea. Since this reagent is not very stable, it should be prepared as needed.

4. Xylose standards.

A. Stock standard, 10 mg/ml.   Dissolve exactly 1.0 g of pure D(+)-xylose in benzoic acid solution (1 g/liter) and dilute to 100 ml with the benzoic acid solution.

B. Working standards.   Dilute 1, 2, and 4 ml of the stock solution to 200 ml with benzoic acid solution to give standards containing 5, 10, and 20 mg/dl of xylose (0.05, 0.10, and 0.20 mg/ml).

PROCEDURE

Blood or serum.   Prepare a 1:10 protein-free Somogyi filtrate as given in Chapter 5, replacing the barium hydroxide by sodium hydroxide if desired. The specimens should be refrigerated if the filtrate is not made within 1 hr after collection. The filtrate is stable for a longer time.

Urine.   For clear, relatively dilute urines, 1:50 and 1:100 dilutions are made with water. If the urine is dark, contains blood, or is turbid after centrifugation, a 1:10 filtrate may be made as for blood and the filtrate further diluted 1:5 and 1:10. The urine specimens are stable for several days if kept refrigerated.

Prepare duplicate tubes for each sample, standard, and reagent blank. To pairs of tubes add 1 ml of sample (blood filtrate or diluted urine), standard (run all three standards), or water (reagent blank). Add 5 ml of the bromoaniline reagent to each tube. (An immediate pink color when the reagent is added to a urine dilution usually indicates that the concentration of xylose in the aliquot is too high and a further dilution of the urine should be made.) Heat one set of the duplicate tubes at 70°C in a water bath or heating block for 10 min. (The water level or top of block should be above the level of solution in the tubes.) Keep the other set of tubes in ice water during this time. Then allow all the tubes to stand at room temperature in the dark for 70 min. Read in a photometer at 520 nm within 30 min. For each pair of tubes, read the heated tube against the unheated tube. Subtract from these absorbance values the absorbance difference found for the two reagent blank tubes. These will give the corrected absorbances for samples and standards.

CALCULATION

For blood or serum:

$$\frac{\text{corr. absorbance of sample}}{\text{corr. absorbance of standard}} \times \text{conc. of standard} = \text{mg/dl xylose}$$

where the three standards made up to contain 5, 10, and 20 mg/dl are equivalent to 50, 100, and 200 mg/dl in blood, taking into account the 1:10 dilution in the filtrate. Usually for blood only the 50 mg/dl standard need be run and used for the calculations.

For urine:

$$\frac{\text{corr. absorbance of sample}}{\text{corr. absorbance of standard}} \times \text{conc. of standard} \times D = \text{mg/dl xylose}$$

where the standards are now taken as 5, 10, and 20 mg/dl and the results multiplied by the total dilution factor D of the urine (usually 50 or 100). The standard having the absorbance nearest that of a given sample is used for the calculation with that sample. Then:

$$\text{mg/dl in urine} \times \frac{V}{1,000} = \text{g xylose excreted}$$

$$\% \text{ excreted} = \frac{\text{g xylose excreted} \times 100}{\text{g xylose ingested}}$$

where V is the total volume of the 5-hr urine specimen in milliliters and 1,000 converts milligrams to grams. Preferably the total volume of urine should be over 150 ml for good results. If a 25-g dose is used, it may be necessary to make further dilutions of the urine and one should probably begin with 1:100 and 1:200 dilutions.

The color developed by the standards will vary enough from day to day to make it necessary to run standards with each group of samples and not rely on a calibration curve.

Xylose by o-Toluidine Method

As mentioned in Chapter 5 under the o-toluidine method for glucose, xylose can also be determined by this reagent even in the presence of glucose. If the o-toluidine reagent is routinely used for glucose determinations, this might also be the simplest reagent for xylose. This method should be satisfactory since the actual blood levels are not considered to be of great diagnostic value. Since the determinations are made directly on serum, the xylose standards are prepared by diluting the stock 1:20 and 1:10 to give standards of 50 and 100 mg/dl. Usually the urine need be diluted only 1:5 and 1:10. The results can be obtained in a much shorter time than with the previous method and if an additional dilution of urine must be run, this can be done within 15 min. The calculations from the readings at two wavelengths may seem a bit more complicated but this is not necessarily true.

As mentioned in the discussion on this section in Chapter 5, if only the urine is analyzed and it is essentially free of glucose (tested with glucose oxidase test strip), one can determine the xylose directly by reading sample and standard at 430 nm.

RESULTS AND INTERPRETATION

On a 5-g dose the normal amount excreted is 20 to 40% of the ingested dose in 5 hr; with a 25-g dose, 16 to 33%. With the lower dose, the 1-hr blood level is 10 to 25 mg/dl; on the higher dose, 30 to 70 mg/dl. The 2-hr sample is generally about 20% lower. Decreased excretion is found in various types of malabsorption or steatorrhea.

Ketone Bodies

The substances usually determined as ketone bodies are acetone, acetoacetic acid, and β-hydroxybutyric acid [49–55]. These may be determined separately by chemical separation but are usually determined together by conversion to acetone. The acetone is determined by a colorimetric method with dinitrophenylhydrazine. (Hydroxybutyric acid is not strictly a ketone but is readily converted to one and is usually determined along with the other two compounds.) The conversion to acetone may be accomplished by refluxing and distillation or by autoclaving. The procedure presented uses heating in an oil bath followed by treatment with dichromate and further heating for the conversion to acetone. This is somewhat simpler than the other methods.

The results have usually been reported as milligrams per deciliter of acetone equivalents. It would seem more logical to express the results as micromoles per liter since the other two compounds are converted into equimolar amounts of acetone, whereas on a weight basis 1 mg of acetone is equivalent to about 1.8 mg of acetoacetic acid and 2.2 mg of hydroxybutyric acid.

REAGENTS

1. Barium hydroxide and zinc sulfate for preparation of Somogyi protein-free filtrate. See Chapter 5.

2. Sulfuric acid, 6M. Cautiously add 335 ml of concentrated sulfuric acid to about 500 ml of water. Cool and dilute to 1 liter with water.

3. Potassium dichromate, 0.012M. Dissolve 3.5 g of potassium dichromate in water and dilute to 1 liter.

4. Sodium sulfite, 0.4M. Dissolve 5.0 g of sodium sulfite in water to make 100 ml.

5. Sodium hydroxide, 0.25M. Dissolve 10 g of sodium hydroxide in water and dilute to 1 liter.

6. Dinitrophenylhydrazine, 0.1M in 5M sulfuric acid. Add 28 ml of concentrated sulfuric acid to about 60 ml of water. Cool and dissolve in it 200 mg of dinitrophenylhydrazine. Dilute to 100 ml with water. Transfer to a separatory funnel and extract three times with 10-ml portions of carbon tetrachloride; discard the carbon tetrachloride extract. This removes interfering colored material. This reagent is stable for at least 1 month in a tightly stoppered bottle.

7. Carbon tetrachloride, reagent grade.

8. Acetone standards.

A. Dilute exactly 1.0 ml of reagent grade acetone to 1 liter with water. This standard contains 785 mg or 13.5 mmol/liter. Prepare fresh monthly.

B. Working standard, 270 μmol/liter. As required, dilute 2 ml of the stock solution to 100 ml with water.

SPECIAL APPARATUS

1. Screw-capped vials, 20 × 150 mm, of borosilicate glass, with Teflon-lined caps.

2. Heating bath at 110°C and 140°C. A large beaker filled with mineral oil heated on a hot plate is satisfactory. The temperature of the bath should be kept within ±1° at 110°C and within ±2° for the higher temperature. With an adjustable hot plate the temperature can be regulated with sufficient accuracy after a little experimentation to obtain the proper control settings. If desired, one may use polyethylene glycol (PEG-20,00) instead of mineral oil. This material is a solid at room temperature and melts below 100°C. It may be more convenient to use because it is water soluble and thus readily rinses off the outside of the tubes. If this material is used, the heating should be done in a fume hood as the vapors are not entirely innocuous.

PROCEDURE FOR BLOOD

The blood specimen (heparinized) is collected in the morning in the fasting state. Prepare a 1:10 Somogyi filtrate of the blood as given in Chapter 5, using 2 ml of blood, 10 ml of water, and 4 ml each of the barium hydroxide and zinc sulfate solutions. Filter or centrifuge strongly to obtain a clear supernatant. Also dilute 1 ml of the working standard with 9 ml of water to obtain a dilution of the standard equivalent to that for the blood.

To separate screw-capped tubes add 5 ml of sample supernatant, 5 ml of diluted working standard, and 5 ml of water (blank). If high values are expected, it may be advisable to set up another sample tube using 1 ml of filtrate or supernatant and 4 ml of water. The reading for this tube is multiplied by an extra factor of 5 in the calculations. To each tube add 1 ml of the 6M sulfuric acid and mix. Cap tubes tightly and heat in bath at 110°C for 10 min. Cool tubes in an ice bath and mix by inversion. Add 1 ml of the potassium dichromate to each tube, cap tightly, and mix. Heat at 140°C for 45 min. Cool, add 1 ml of the sodium sulfite solution to each tube, and mix. To each tube add 1 ml of the dinitrophenylhydrazine solution and 4 ml of carbon tetrachloride. Cap and shake vigorously for 5 min. Allow layers to separate completely (centrifuge briefly if necessary) and remove upper aqueous layer and discard. To each tube add 6 ml of the 0.25M sodium hydroxide solution and shake vigorously for 3 min. Allow the layers to separate and remove upper aqueous layer as completely as possible. Transfer carbon tetrachloride layer to cuvettes and read samples and standard against blank at 420 nm. For low levels of ketones, some increases in sensitivity may be obtained by reading at 375 nm instead of 420 nm.

CALCULATION

Since the working standard and the blood sample are diluted similarly and carried through the same procedure:

$$\frac{\text{absorbance of sample}}{\text{absorbance of standard}} \times 270 = \mu\text{mol/liter of total ketone bodies}$$

μmol/liter $\times$ 0.0058 = mg/dl (as acetone)

If the filtrate has been diluted an additional amount for high values, the appropriate factor must be entered into the calculation.

PROCEDURE FOR URINE

The determination is preferably made on a 24-hr urine specimen collected under refrigeration without preservative. Mix the entire sample and measure the volume. Prepare a 1:20 filtrate of the urine by using 1 ml of urine, 15 ml of water, and 2 ml each of the barium hydroxide and zinc sulfate solutions. Treat a 5-ml aliquot of filtrate exactly the same as for blood filtrate, using the same standard. Here also, if high values are expected, it may be advisable to run further dilutions of the filtrate at the same time.

CALCULATION

Since the urine is diluted 1:20 in comparison with a 1:10 dilution for the standard, the calculations are similar to those for blood, except that a factor of 270 is used instead of 135. The result is multiplied by the appropriate dilution factor, if a further dilution of the urine filtrate has been made. Then:

μmol/liter $\times$ urine volume (in liters) $= \mu$mol excreted in 24 hr

NORMAL VALUES AND INTERPRETATION OF RESULTS

The normal levels in blood are 80 to 240 μmol/liter (0.46 to 1.40 mg/dl) and for 24-hr urine excretion 300 to 1,100 μmol (17.5 to 64.0 mg).

Increased levels in blood and urine are found in all types of metabolic acidosis, particularly in diabetes and after several days fasting. Levels of up to ten times the upper limit of normal have been found in these conditions, depending upon the degree of acidosis.

References

1. Kahan, J. *Int. J. Vitam. Nutr. Res.* 43:127, 1973.
2. Garry, P. J., Pollack, J. D., and Owen, G. M. *Clin. Chem.* 16:766, 1970.
3. Hansen, L. G., and Warwick, W. J. *Am. J. Clin. Pathol.* 51:538, 1969.
4. Kahan, J. *Scand. J. Clin. Lab. Invest.* 18:679, 1966.
5. Fegeler, F., and Quinkert, E. *Aerztl. Lab.* 6:14, 1960.
6. Roels, O. A., and Trout, M. *Am. J. Clin. Nutr.* 7:197, 1959.
7. Ditlefsen, E. M. L., and Stoa, K. F. *Scand. J. Clin. Lab. Invest.* 6:210, 1954.
8. Carr, F. H., and Price, E. A. *Biochem. J.* 20:497, 1926.
9. Bradley, D. W., and Hornbeck, C. L. *Biochem. Med.* 7:78, 1973.
10. Roels, O. A., and Trout, M. *Stand. Methods Clin. Chem.* 7:215, 1972.
11. Neeld, J. B., Jr., and Pearson, W. N. *J. Nutr.* 79:454, 1963.
12. Dugan, R. E., Frigerio, N. A., and Siebert, J. M. *Anal. Chem.* 36:114, 1964.
13. Roe, J. H. *Stand. Methods Clin. Chem.* 3:35, 1961.
14. Roe, J. H., and Keuther, C. A. *J. Biol. Chem.* 147:399, 1943.
15. Roe, J. H. *J. Biol. Chem.* 236:1611, 1961.
16. Michaelsson, G., and Michaelsson, M. *Scand. J. Clin. Lab. Invest.* 20:97, 1967.
17. Wilson, S. S., and Guillan, R. A. *Clin. Chem.* 15:282, 1969.
18. Harris, L. J., and Ray, S. N. *Lancet* 1:71, 1935.
19. Farmer, C. J., and Abt, A. F. *Proc. Soc. Exp. Biol. Med.* 34:146, 1936.
20. Roe, J. H. *Ann. N.Y. Acad. Sci.* 92:277, 1965.
21. Owen, J. A., and Iggo, R. J. *Biochem. J.* 62:675, 1956.
22. Barker, S. B., and Summerson, W. H. *J. Biol. Chem.* 138:535, 1941.
23. Barker, S. B. *Stand. Methods Clin. Chem.* 3:167, 1961.
24. Hohorst, H. J. In H. U. Bergmeyer [Ed.], *Methods of Enzymatic Analysis.* New York: Academic, 1963, P. 266.
25. Rosenberg, J. C., and Rush, S. F. *Clin. Chem.* 12:299, 1966.
26. Hadjivassiliou, A. G., and Rieder, S. V. *Clin. Chim. Acta* 19:357, 1968.
27. Fleischer, W. R. *Stand. Methods Clin. Chem.* 6:245, 1970.
28. Neville, J. E., Jr., and Gelder, R. L. *Am. J. Clin. Pathol.* 55:125, 1971.

29. Segal, S., Blair, A. E., and Wyngaarden, J. B.   *J. Lab. Clin. Med.* 48:177, 1956.
30. Moore, E. W., and Scarlatta, R. W.   *Gastroenterology* 49:178, 1965.
31. Moore, E. W.   *Gastroenterology* 54:501, 1968.
32. Burke, E. L.   *U.S. Navy Newsletter* 55:32, 1970.
33. Segal, H. L., Miller, L. L., Morton, J. J., and Young, H. Y.   *Gastroenterology* 16:380, 1950.
34. Segal, H. L.   *Ann. N.Y. Acad. Sci.* 57:308, 1953.
35. Segal, H. L., Miller, L. L., and Plume, E. J.   *Gastroenterology* 28:402, 1955.
36. Fentress, V., and Sandweiss, D. J.   *J.A.M.A.* 165:21, 1955.
37. Gilbert, S. S., Matzner, M. J., and Schwartz, I. R.   *Am. J. Gastroenterol.* 27:277, 1957.
38. Mortimer, D. C.   *Can. Med. Assoc. J.* 80:607, 1959.
39. Segal, H. L., and Plosscowe, R. P.   *Am. J. Digest. Dis.* 6:485, 1960.
40. Christiansen, P. A., Kirsner, J. B., and Ablaza, J.   *Am. J. Med.* 27:443, 1959.
41. Brien, F. S., Turner, D. A., Watson, E. M., and Geddes, J. H.   *Gastroenterology* 20:287, 1952.
42. Benson, J. A., Culver, P. J., Ragland, S., Jones, C. M., Drummey, G. D., and Bongas, E.   *N. Engl. J. Med.* 256:335, 1957.
43. Santini, R., Sheehy, T. W., and Martinez de Jesus, J.   *Gastroenterology* 40:772, 1961.
44. Clark, P. A.   *Gut* 3:333, 1962.
45. Jones, W. O., and de Sant' Agnese, P. A.   *J. Pediatr.* 62:50, 1963.
46. Roe, J. H., and Rice, B. W.   *J. Biol. Chem.* 173:507, 1948.
47. Reiner, M., and Cheung, H. L.   *Stand. Methods Clin. Chem.* 5:257, 1965.
48. Tietz, N. W.   In N. W. Tietz [Ed.], *Fundamentals of Clinical Chemistry.* Philadelphia: Saunders, 1970. P. 727.
49. Peden, V. H.   *J. Lab. Clin. Med.* 63:332, 1964.
50. Kaesar, H.   *Clin. Chim. Acta* 6:337, 1961.
51. Ahole, T., and Somersalo, O.   *Ann. Med. Exp. Biol. Fenn.* 41:237, 1963.
52. Britton, H. G.   *Anal. Biochem.* 15:261, 1966.
53. Procos, J.   *Clin. Chem.* 7:97, 1961.
54. Levine, V. E., and Taterka, M.   *Clin. Chem.* 3:646, 1957.
55. Goeschke, H.   *Clin. Chim. Acta* 28:359, 1970.

18. Toxicology

Drugs and Poisons

In recent years the multiplicity of potent drugs, newer insecticides and pesticides, and industrial compounds of a toxic nature has considerably increased the number of potential poisons. The acute or chronic poisoning with many of these agents may mimic a number of disease states or alter the symptoms of preexisting disease. In order to arrive at a correct diagnosis, the determination of the presence or absence of toxic compounds in body fluids may be necessary.

In general, the substances whose determinations are discussed in this chapter may be found in the body as the result of accidental or other exposure to toxic agents not normally found in the body (at least not in high concentrations), or they may be due to overdosage of drugs. In addition, the therapeutic levels of some drugs in the blood or other body fluids may be helpful in maintaining proper therapy.

A complete listing of methods for all the drugs and poisons that might be encountered would be much too lengthy for this book. We present only simple methods for the most common drugs and poisons. For some, we have given only qualitative or semiquantitative tests, since the more accurate methods are generally not practical for the clinical laboratory when only an occasional sample is encountered. For accurate analysis, the sample is best sent to a specialized toxicological laboratory.

A good reference to quantitative methods is the manual edited by Sunshine [1]; many of the methods require the use of a gas chromatograph, ultraviolet spectrophotometer, or atomic absorption spectrophotometer. The book by Kaye [2] gives more data on the qualitative identification of many poisons and useful information on the symptoms of poisoning.

Ethyl Alcohol

Ethyl alcohol in blood, urine, or other body fluid may be determined in a number of ways. In one of the older methods, the alcohol is separated from the non-volatile constituents by distillation. The distillate is then treated with potassium dichromate in a strong acid solution. The amount of dichromate reduction is measured by colorimetric means [3, 4]. Another method uses a Conway micro-diffusion apparatus [5, 6]. Alcohol may also be determined enzymatically using alcohol dehydrogenase, the primary reaction being:

$$C_2H_5OH + NAD^+ \rightleftharpoons CH_3CHO + NADH + H^+$$

Usually a substance such as semicarbazide is added to react with the acetaldehyde formed and enable the reaction to go to completion. The measurement may be made at 340 nm by following the change in absorbance caused by the reduction of NAD to NADH [7–9]. A colorimetric procedure may also be used with the NADH coupled to reduce a formazan salt [10, 11].

Probably the best method of determination if the apparatus is available is gas-liquid chromatography [12–14]. This allows the determination not only of ethyl alcohol but also of any other alcohols, aldehydes, or ketones that might be present.

Two methods will be discussed—one using the Conway microdiffusion dish and the other using the enzyme, alcohol dehydrogenase.

BLOOD ALCOHOL BY MICRODIFFUSION [5, 6]

REAGENTS

1. Acid dichromate.   Dissolve 3.70 g of reagent grade potassium dichromate in 15 ml of water. Add carefully with swirling, 280 ml of concentrated sulfuric acid. Cool to room temperature and dilute to 650 ml. This solution is stable in a glass-stoppered bottle if protected from contact with organic matter.

2. Saturated potassium carbonate.   Mix 50 g of anhydrous potassium carbonate and 40 ml of water. When cooled, there should be some excess salt undissolved.

3. Alcohol standards.

A. Stock standard.   Dilute exactly 2.0 ml of absolute ethyl alcohol measured at 20°C to 100 ml with water. This solution contains 1.580 g/dl of alcohol. If only the 95% (by volume) ethyl alcohol is available, 2.0 ml will contain 1.500 g of pure alcohol.

B. Working standard.   Dilute the stock standard 1:10 to give a solution obtaining 0.158 g/dl (or 0.150 g/dl if the 95% alcohol was used). Although it is customary in clinical chemistry to express such concentrations as 158 mg/dl, all the legal definitions of intoxication express the concentration in grams per deciliter, so this should be used in reporting.

SPECIAL APPARATUS

Conway microdiffusion dishes are available from most laboratory supply houses. Two different types are available. One type is made of glass with a cover of a ground glass plate; a tight fit is made by the use of a thin coating of high-vacuum stopcock grease on the plate. The other type of dish is made of plastic, and the seal is made by a turned-down edge of the cover dipping into an outer well containing a sealing liquid. In the method presented here, this liquid is the saturated potassium carbonate solution.

PROCEDURE

Accurately pipet 1 ml of the dichromate solution to the center well of a number of dishes. In the outer compartment of each dish, place 1 ml of the saturated potassium carbonate solution. Mix 1 ml of the blood sample with 1 ml of water and add 1 ml of this mixture to the outer compartment of one dish and cover immediately. Rotate gently to mix the blood sample and potassium carbonate solution. For standard and blank obtain some alcohol-free blood (outdated blood bank blood may be used). For the blank, mix 1 ml of this blood with 1 ml of water and add 1 ml of this mixture to the outer compartment of a second dish, seal at once, and mix. For the standard, mix 1 ml of the alcohol-free blood with 1 ml of the working standard solution and add 1 ml of the mixture to the outer compartment of a third dish. Seal and mix. Allow the dishes to stand undisturbed in a warm place for at least 2 hr. At the end of this time remove the cover and transfer the dichromate from the center wells to separate 10 ml volumetric flasks or glass-stoppered 10 ml graduated cylinders with the aid of a Pasteur pipet, rinsing the well several times with small portions of water and adding the rinsings to the flask. Dilute the flasks to 10 ml and mix well. Read the tubes at 600 nm against a reagent blank prepared by diluting 1 ml of the acid dichromate to 10 ml with water.

CALCULATION

$$\frac{\text{absorbance of sample}}{\text{absorbance of standard} - \text{absorbance of blood blank}} \times \text{conc. of standard} = \text{conc. of sample}$$

If the standard used is prepared from absolute alcohol, the diluted standard will be equivalent to 0.158 g/dl of alcohol. If the concentration of alcohol is higher than about 0.30 g/dl, all the dichromate will be reduced, and the test should be repeated with a greater dilution of the sample.

The dichromate will be reduced by other substances besides ethyl alcohol, such as methyl alcohol, formaldehyde, or paraldehyde. As a check, these should be tested for in aliquots of a trichloroacetic acid filtrate.

METHYL ALCOHOL, FORMALDEHYDE, AND PARALDEHYDE

REAGENTS

1. Trichloroacetic acid (TCA), $1.2M$. Dissolve 20 g of TCA in water and dilute to 100 ml.

2. Sulfuric acid, concentrated, reagent grade.

3. Sulfuric acid, $1.8M$. Dilute 10 ml of concentrated acid with water to make 100 ml.

4. Chromotropic acid sodium salt, 0.5 g/dl.   Dissolve 0.5 g of chromotropic acid sodium salt (4,5-dihydroxy-2,7-naphthalenedisulfonic acid disodium salt) in water to make 100 ml.

5. Potassium permanganate, $0.32M$.   Dissolve 5 g of potassium permanganate in water to make 100 ml.

6. Sodium bisulfite, $0.15M$.   Dissolve 2.8 g of sodium metabisulfite ($Na_2S_2O_5$) in water to make 100 ml.

7. Copper sulfate, $0.16M$.   Dissolve 4 g of cupric sulfate ($CuSO_4 \cdot 5H_2O$) in water to make 100 ml.

8. p-Phenylphenol, $0.06M$.   Dissolve 2 g of sodium hydroxide in 100 ml of water, then add 0.5 g of p-phenylphenol and mix until dissolved.

PROCEDURE

Prepare a protein-free filtrate by adding to 2 ml of blood, 4 ml of the trichloroacetic acid. Shake well and filter or centrifuge strongly. Pipet a 1-ml aliquot of the TCA filtrate to a test tube. Add 1 ml of the $1.8M$ sulfuric acid and mix. Add 0.5 ml of the potassium permanganate solution and allow to stand for 15 min. Add sodium bisulfite solution dropwise until the permanganate color has just disappeared, avoiding any excess of sulfite. Place tube in an ice bath and add 0.2 ml of the chromotropic acid solution followed by the cautious addition of 4 ml of concentrated sulfuric acid. Mix and place in a boiling water bath for 15 min. A purple color indicates the presence of methyl alcohol or formaldehyde. Treat a second 1-ml aliquot of the filtrate with the chromotropic acid and sulfuric acid without prior oxidation with permanganate. A purple color indicates the presence of formaldehyde in the distillate. To a third 1-ml aliquot of the filtrate add 0.05 ml of copper sulfate solution and 6 ml of concentrated sulfuric acid. Mix, then add 0.2 ml of the phenylphenol reagent with constant shaking. Place in a boiling water bath for 3 min and cool. If formaldehyde is absent, a pink color indicates the presence of paraldehyde in the filtrate.

BLOOD ALCOHOL BY ENZYMATIC DETERMINATION

The reactions involved have already been mentioned. The enzyme preparations used are unstable and it is preferable to purchase the lyophilized material in a kit form.* The procedure will differ slightly depending upon the kit used, but the general directions are given below. A protein-free filtrate is first made with the blood sample. Either perchloric acid (4.8 ml of the 70% acid diluted to 100 ml)

---

* Sigma Chemical Co., St. Louis, Mo. 63118; Calbiochem, San Diego, Calif. 92112; Boehringer-Mannheim Corp., New York, N.Y. 10017; Worthington Biochemicals, Freehold, N.J. 07728

or trichloroacetic acid (6.25 g of the acid dissolved in water to make 100 ml) may be used. A dilution of the blood is made by adding 1 ml of blood or serum to 9 ml of the precipitating reagent to give a 1:10 dilution. Some kits may suggest a 1:20 dilution, which would be prepared by using 0.5 ml of blood and 9.5 ml of reagent. After precipitation, the mixture is centrifuged in a stoppered tube to obtain a clear supernatant. This has proved to be a critical step in the procedure. A very clear supernatant is required which necessitates prolonged centrifuging at high speed. Since only about 0.2 ml or less of supernatant is needed, a quantity of the cloudy supernatant after one centrifugation may be transferred to microcentrifuge cups and centrifuged at very high speed in a microcentrifuge.

The substrate is prepared as directed and aliquots of the appropriate size (usually to make a total volume of 3 ml) are added to a number of cuvettes. A reading is taken for each cuvette at 340 nm against a water blank. Then 0.1 ml of sample supernatant, diluted standard (the same diluted standard as used in the previous methods [0.158 g/dl] further diluted 1:10 or 1:20 with the precipitating reagent) used for the blood sample, or precipitating solution (as blank) is added to the separate cuvettes containing the substrate. The mixtures are allowed to stand at 37°C for 20 min and a second reading taken at 340 nm. The initial reading is subtracted from the final reading for each tube. Subtract any increase in absorbance obtained for the blank from the absorbances obtained for the sample and standard. Since the sample and standard are treated similarly,

$$\frac{\text{corr. absorbance change for sample}}{\text{corr. absorbance change for standard}} \times \text{conc. of standard} = \text{conc. of sample}$$

If the diluted standard given in the first method is used, the concentration of alcohol will be 0.158 g/dl (or 0.150 g/dl if the 95% alcohol was used to prepare the standard).

If a good spectrophotometer with a narrow band width and cuvettes of exactly 10 mm light path are available, the theoretical factor may be calculated as:

$$F = \frac{V_1 \times D \times 100 \times 46.07}{6.22 \times V_2 \times 1,000,000}$$

Then:

$F \times$ corr. change in absorbance of sample $= $ g/dl alcohol in the sample

In the above equation, 6.22 is the absorbance of a solution containing 1 μmol of NADH/ml, V_1 the total volume of solution in the cuvette, D the dilution factor for the blood filtrate, V_2 the volume of filtrate used, and 46.07 the molecular weight of ethyl alcohol. The other factors convert to grams per deciliter from micrograms per milliliter. Although the alcohol level has been conventionally reported in grams per deciliter of whole blood, as with many other substances, it is probably the amount in plasma that is physiologically important. The ratio of the concentration of alcohol in plasma to that in packed cells has been found to be approximately 1.54 (varying from 1.26 to 1.98), and the ratio between the concentration in plasma and that in whole blood may vary between 1.10 and 1.35 [15].

ALCOHOL DETERMINATION IN URINE

The alcohol in urine may be determined by any of the methods given above for blood. The determination in a random specimen is of little value. Preferably the subject should void, and the urine obtained during the next 30 min should be used for the determination. This will give a value related to the average level in the blood during this time. The level in the urine will actually be more closely related to the plasma level than to the level in whole blood.

Since a specimen of urine is often easier to obtain than a blood sample, the suggestion is often made to determine the alcohol content of a urine sample as a measure of intoxication. This assumes a fairly constant ratio between the alcohol in blood and that in urine. In a number of patients in which the blood and urine alcohol were measured at the same time, the ratio of alcohol in urine to that in blood varied from 0.21 to 2.66 with a mean of 1.28. When this factor was used to calculate the blood level from the urine level, the results were high in 22 percent of the cases and low in 35 percent [16]. In another study [17] the calculated values agreed with the measured values only 44 percent of the time. Thus, although the urine concentration may be sufficient to determine whether the patient has ingested alcohol recently, it is not very accurate as a measurement of the state of intoxication.

INTERPRETATION OF RESULTS

Alcohol is rapidly absorbed. When taken without food, absorption will be complete in 60 to 90 min. After absorption, about 90 percent of the alcohol is metabolized and the rest excreted in the urine and breath. The rate at which alcohol is metabolized is relatively constant and independent of the amount of alcohol in the blood. In regard to driving after alcohol ingestion, the Committee on Alcohol and Drugs of the National Safety Council has recommended the following: Below .050 g/dl of blood, ethanol exerts no influence on driving; between 0.050 and 0.100 g/dl, alcohol may have some influence on driving ability, and above 0.100

g/dl the person is considered to be "under the influence" in respect to driving ability. The response to alcohol varies greatly among individuals and depends also on the form in which the alcohol was taken and whether it was accompanied by the ingestion of food. Usually some loss of skill and slurred speech are noted in most persons at levels between 0.100 and 0.200 g/dl. Disturbances of equilibrium may occur at levels between 0.150 and 0.300 g/dl, and death may result at levels over 0.500 g/dl.

A simple test that may be helpful in cases of suspected alcohol intoxication, particularly at higher levels of alcohol, is the determination of the serum osmolality [18, 19]. This is a simple and rapid test and is nondestructive so that the serum can afterward be used for other tests. Since alcohol is a substance of low molecular weight (46), a level of 184 mg/dl would correspond to a concentration of 40 mmol/liter or roughly 40 mosmol/kg. Thus a marked increase in serum osmolality would be presumptive evidence of intoxication by ethyl alcohol (or other low-molecular-weight compound such as methyl alcohol, acetone, isopropanol, or ether). The normal osmolality is calculated from the formula

$$\text{osm (calc.)} = 1.86\,[\text{Na}] + \frac{[\text{G}]}{18} + \frac{[\text{BUN}]}{2.8}$$

where Na is the concentration of sodium in the serum in mmol/liter, G is the concentration of glucose in mg/dl, and BUN is the concentration of urea nitrogen in mg/dl. Then the osmolality difference may be determined as:

$$\Delta\,\text{osm} = \text{osm (meas.)} - \frac{\text{osm (calc.)}}{0.93}$$

If the figures to calculate the normal osmolality are not available, a normal value of 286 mosmol/kg can be assumed and substituted for the last term in the equation [osm (calc.)/0.93]. Using the calculated difference, it was found that when ethyl alcohol was the only low-molecular-weight volatile substance in the blood, the following equation was roughly true:

$$\Delta\,\text{osm} \times 4.6 = \text{mg/dl alcohol in serum}$$

Most other drugs such as barbiturates and tranquilizers, because of their higher molecular weight and lower concentration in the serum, will have only a slight effect on the osmolality. Thus in a comatose patient with a serum osmolality of less than 310 mosmol/kg, it would be very unlikely that the condition is due to alcohol intoxication. On the other hand, if the osmolality were over 320 mosmol/kg, alcohol intoxication could be definitely suspected.

Arsenic, Mercury, and Bismuth

COMBINED SEMIQUANTITATIVE TEST

We present a qualitative and semiquantitative test for these metallic elements [20, 21]. It will detect relatively small quantities in urine, vomitus, or stomach contents. It is not applicable to blood which requires special procedures. The sample is boiled gently with dilute hydrochloric acid in the presence of a small strip of copper. Under the conditions of the test, these elements will plate out on the strip. They are detected by a change in color of the strip and further identified by confirmatory tests. An estimate of the quantity present can be made by comparing the copper strip from the sample with those from several different standards. If the presence of one of the elements is definitely suspected, two samples, a blank, and several standards are run. If it is not known which of the elements is present, one strip is run to determine the element present; then standards of the appropriate element can be run.

GENERAL REAGENTS

1. Copper strip or spiral. Strip 5 × 10 mm cut from a sheet of copper 0.02 in. in thickness, or spiral formed by winding 20 gauge copper wire tightly around a glass rod (about 5 mm in diameter) for ten turns, then removing from the rod.

2. Nitric acid, 2.5M. Dilute 15 ml of concentrated nitric acid to 100 ml with water.

3. Ethanol, 95%.

4. Hydrochloric acid, concentrated, reagent grade.

5. Standards.

A. Arsenic standards. A stock solution containing 1 mg/ml is prepared. Dissolve 1.32 g of arsenic trioxide in 10 ml of sodium hydroxide solution (40 g of NaOH to 100 ml with water) in a 1 liter volumetric flask. Add about 500 ml of water, neutralize with concentrated hydrochloric acid, and dilute to 1 liter. The working standard is prepared by diluting the stock 1:100 with water, giving a solution containing 10 μg/ml.

B. Mercury standards. A stock solution containing 1 mg/ml is prepared by dissolving 1.61 g of mercuric nitrate [$Hg(NO_3)_2$] and 2 ml of concentrated nitric acid in water to make 1 liter. A working standard containing 20 μg/ml is made up as required by diluting the stock 1:50 with water.

C. Bismuth standards. A stock solution is prepared by dissolving 100 mg of pure metallic bismuth in 5 ml of concentrated nitric acid and 5 ml of water, then diluting to 100 ml with water. This contains 1 mg/ml. A working standard is prepared as required by diluting the stock 1:20 with water, to give a solution containing 50 μg/ml.

CONFIRMATORY REAGENTS

6. Potassium cyanide, 10 g/dl. Dissolve 10 g of potassium cyanide in water to make 100 ml. (*Caution:* Poison!)

7. Cuprous iodide suspension. Dissolve 3 g of ferrous sulfate and 5 g of copper sulfate pentahydrate in 15 ml of water. Dissolve 7 g of potassium iodide in 50 ml of water. Add the iodide solution slowly to the copper solution with constant stirring. Allow to stand for 20 min, then filter off the cuprous iodide. Wash the precipitate well with water. Transfer the moist precipitate to a small brown bottle with the aid of a minimum amount of water. The suspension is stable for several months.

8. Nitric acid, 3.3M. Dilute concentrated nitric acid 1:5 with water.

9. Quinine-iodide reagent. Dissolve 1 g of quinine sulfate and 0.5 ml of nitric acid in water to make 100 ml. Then add 2 g of potassium iodide and dissolve.

10. Sodium sulfite solution, 0.4M. Dissolve 5 g of anhydrous sodium sulfite in water and dilute to 100 ml.

PROCEDURE

Place 20 ml of sample in a 150 ml Erlenmeyer flask. If sufficient material is available, set up several flasks to be used for confirmatory tests. Also set up a blank with 20 ml of water. If standards are run, add 1, 3, and 5 ml of the appropriate standards to 20 ml of water in separate flasks. To each flask add one copper strip or spiral. Add 4 ml of concentrated hydrochloric acid to each flask. Loosely stopper the flasks with inverted beakers or glass bulbs, and heat to gentle boiling for 20 min. Cool and decant the solution and discard. Wash the copper strip remaining in the flask several times with water, remove from the flask, and dry on filter paper. Compare the appearance of the strips from the samples to those from the standards and blank. A silvery deposit on the strip indicates mercury; a dull black deposit, arsenic; and a shiny black deposit, bismuth. Antimony, which is less likely to be present, will give a dark purple color. If the sample strip does not appear different from the blank, less than 5 μg of arsenic or 30 μg of mercury or bismuth was present in the sample aliquot. The confirmatory tests may be carried out on the strips or spirals.

CONFIRMATORY TESTS

Arsenic.   Place the strip in 2 ml of potassium cyanide solution. The black deposits of arsenic will dissolve, those of bismuth or antimony will not.

Arsenic or mercury.   Place a processed strip or spiral in a small borosilicate test tube. Heat the bottom of the tube very gently, inclining the tube at an angle so that the top remains cool. Cool the tube and examine under a microscope at low power. If arsenic is present, it will sublime to the upper part of the tube as small white octahedral crystals of arsenic trioxide. Mercury will sublime as minute shiny globules. Antimony and bismuth will not sublime.

Mercury [22].   Place a small square of filter paper in a watch glass. Add several drops of the cuprous iodide suspension to the filter paper. Place the copper strip on the cuprous iodide and cover with a second watch glass. Allow to stand for at least 30 min. Mercuric iodide is formed with the development of a rose to salmon pink color if mercury is present (compare color with some untreated cuprous iodide on another piece of filter paper).

Bismuth [23].   Place a strip or spiral in a small test tube. Add 1 ml of the sodium sulfite solution and 1 ml of the nitric acid solution and agitate gently for about 10 min. Add 1 ml of water and 1 ml of the quinine-iodide solution and mix. Remove the strip or spiral with the aid of a wooden applicator stick. A definite orange turbidity will develop in the presence of bismuth. Compare with a bismuth standard.

INTERPRETATION OF RESULTS

These tests will not detect the minute amount of these metals normally present in body fluids. Any positive test thus indicates possible poisoning.

ARSENIC BY THE GUTZEIT METHOD

This method [24–26] will detect smaller quantities of arsenic and give a better quantitative result than the previous method, but requires a preliminary acid digestion. Thus it is not as convenient when only an occasional test is run.

The organic matter is first destroyed by digestion with sulfuric, nitric, and perchloric acids. An aliquot of the digest is treated in a special apparatus with metallic zinc. The hydrogen formed by the action of the acid on the zinc carries along the arsenic as arsine (AsH_3) past a filter paper strip impregnated with mercuric bromide. The arsenic reacts with the mercuric salt to form a brown coloration. The length of the colored portion of the strip is proportional to the amount of arsenic present. By comparison with the color produced by standards, the amount of arsenic in the sample may be estimated.

1. Sulfuric acid, reagent grade.

2. Perchloric acid, 70%, reagent grade.

3. Nitric acid, concentrated, reagent grade.

4. Sodium bisulfite, granular, reagent grade.

5. Stannous chloride solution, 1.8M. Dissolve 40 g of stannous chloride ($SnCl_2 \cdot 2H_2O$) in water to make 100 ml. A slight turbidity is of no consequence.

6. Cupric sulfate solution, 0.4M. Dissolve 10 g of cupric sulfate ($CuSO_4 \cdot 5H_2O$) in water to make 100 ml.

7. Zinc metal, granular, special, arsenic-free.

8. Lead acetate solution, 1.8M. Dissolve 50 g of lead acetate [$Pb(C_2H_3O_2)_2 \cdot 3H_2O$] in water to make 100 ml. The turbidity present does not interfere with the test.

9. Glass wool with lead acetate. Immerse pledgets of glass wool, sufficient to fill loosely about two-thirds of the lower tube of the Gutzeit apparatus, in the lead acetate solution for a few minutes and then dry in air or in an oven at low heat.

10. Methanolic mercuric bromide. Dissolve 10 g of mercuric bromide in absolute methanol to make 100 ml. Store in a glass-stoppered brown bottle out of contact with light.

11. Filter paper strips with mercuric bromide. Dip 15 cm circles of Whatman No. 1 filter paper (or its equivalent) in the mercuric bromide solution, remove, and dry in air. Avoid contact of the fingers with the strips after impregnation; handle with plastic or Teflon-coated tweezers. After drying, store in an airtight container out of contact with light. Cut the strips to fit in the upper small tube of the apparatus with a short tail protruding for ease in removal. Cut strips can be purchased for treatment with mercuric bromide.† These are convenient if a large number of tests are run.

12. Arsenic standards. Use the same working standard (10 μg/ml) as given in the earlier method for arsenic.

Gutzeit apparatus. This consists of a lower reaction bottle, an intermediate constriction tube for the lead acetate, and an upper small tube for the impregnated strip. The apparatus may be purchased from a number of suppliers.‡

Kjeldahl digestion flasks, 300 or 500 ml.

† No. 1180-H, A. H. Thomas Co., Philadelphia, Pa. 19105; No. S-1095, Sargent-Welch Scientific Co., Skokie, Ill. 60076.

‡ No. 1180-A, A. H. Thomas Co.; No. S-1065, Sargent-Welch Scientific Co.; No. 022-285, Curtin-Matheson, Houston, Tex. 77001.

PROCEDURE

Place an aliquot of sample (100 ml of urine, 20 ml of blood, or 50 ml of vomitus or stomach contents) in a Kjeldahl flask. Add nitric acid (10 ml for urine, 20 ml for blood, 50 ml for vomitus or stomach contents), then add 5 ml of sulfuric acid, and mix. Heat over a low flame in a fume hood until the solution begins to darken. During the heating, swirl the flask occasionally to prevent charring. When the solution darkens, remove from the flame and add dropwise with swirling 1 ml of a mixture of 2 parts perchloric acid and 1 part nitric acid. Heat again until the mixture becomes dark, then add another 1 ml of the perchloric-nitric mixture as before. Heat further and repeat the addition, if necessary, until the solution does not darken but remains a clear yellow on further heating. Now heat more strongly until fumes of sulfur trioxide fill the flask. Remove from the flame and allow to cool. Cautiously add 5 ml of water and about 100 mg of sodium bisulfite, mix, and heat again to white fumes. Cool, add cautiously about 5 ml of water, then cool again. Transfer the mixture quantitatively to a 50 ml glass-stoppered cylinder with the aid of several small portions of water. Add 3 drops of the stannous chloride solution and dilute to 50 ml.

Prepare a blank solution by adding to a Kjeldahl flask twice the amount of all acids used in the digestion of the sample. Digest over a flame similarly to the sample. Since the solution contains no organic matter, the digestion can be carried out more rapidly by stronger heating from the beginning. Digest as for the sample, heating to fumes, adding bisulfite, and heating again to fumes. Cool, add 10 ml of water, cool further, and transfer quantitatively to a 100 ml cylinder. Add 6 drops of stannous chloride and dilute to 100 ml.

Prepare a number of the Gutzeit setups by adding a pledget of glass wool with lead acetate to the lower tube (do not pack the glass wool too tightly in the tube; it must permit the free flow of gas), and adding a mercuric bromide-treated strip to each upper small tube. To one of the lower reaction bottles add a 20-ml aliquot of the sample digest. To three other reaction bottles add 20-ml aliquots of the blank digest. Reserve one of these as a blank, to the other two add 1 and 2 ml of the working standard (10 μg/ml). To each reaction bottle add 1 ml of the copper sulfate solution and about 5 g of arsenic-free granulated zinc, stopper immediately with the stopper containing the two tubes, and allow to stand for 1 hr in a pan of water at about 25°C. At the end of this period examine the mercuric bromide strips. Arsenic, if present, will change the color of the strip to an orange-brown. The length of the color on the strip and its intensity will be a measure of the amount of arsenic present. By comparing the strips from the blank, the sample, and the two standards, the approximate amount of arsenic present can be estimated. The two standards, as given, represent 10 and 20 μg arsenic in two-fifths of the sample taken. If the amounts of samples specified are used, the standards correspond to 12.5 and 25 μg/dl for urine, 62.5 and 125 μg/dl for blood, and 25 and 50 μg/dl for stomach contents. If larger amounts are found, the test can be repeated by using a smaller aliquot of the sample digest and making the volume up to 20 ml with some of the blank digest.

NORMAL VALUES AND INTERPRETATION OF RESULTS

The normal level of arsenic excretion in urine is considered to be up to 100 μg/liter. The level in blood is normally in the range of 6 to 20 μg/dl. Higher levels indicate accidental or intentional poisoning or excessive occupational exposure.

Mercury

Mercury may be determined colorimetrically with dithizon after first digesting the sample with nitric acid and permanganate to destroy organic matter. The digestion must be done carefully to avoid loss of mercury by volatilization. The procedure is lengthy and will not be given here [27, 28]. A somewhat simpler method is to use a commercially available instrument§ which measures the ultraviolet absorption of mercury vapor in what is essentially a flameless atomic absorption method [29, 30]. After digestion, the mercury is reduced to the metallic state and a stream of air carries the mercury vapor through a plastic tube in the path of the light from a mercury lamp. The instrument is calibrated to read directly in micrograms of mercury in the sample.

Barbiturates

There are probably more than 50 different barbiturates on the market distributed under at least 150 brand names. Suspected barbiturate poisoning is one of the most common types that might be encountered in the laboratory. With unconscious patients, knowledge of the presence or absence of barbiturates in the body fluids is helpful in the treatment. Probably the best method for the detection and quantitation of the barbiturates is by means of gas chromatography [31–34]. The results can be obtained in a relatively short time and will give not only the amount of barbiturate present but also the type and perhaps even the particular compound. For this to be done on a routine basis, however, requires a simple gas chromatograph devoted solely to the analysis of these compounds. Barbiturates may also be quantitated by use of their ultraviolet spectra [35–38]. This does

§ Model MAS-50 Mercury Analyzer System, Coleman Instruments, Division of Perkin Elmer Corp., Maywood, Ill. 10153.

not always distinguish between the different barbiturates and requires an ultra-violet spectrophotometer capable of giving readings in the range of 230 to 290 nm.

We present a simple semiquantitative method for the estimation of barbiturates [39]. It is very rapid and gives a good estimate of amount but does not distinguish between the types of barbiturate. It does give sufficient information for emergency treatment. The barbiturate is extracted with chloroform and the extract shaken with a buffered mercuric nitrate solution. Any barbiturates present dissolve some of the mercury as barbiturate salts in the chloroform layer. The excess reagent is removed and the mercury present in the chloroform layer determined with diphenylcarbazone. The different barbiturates do not give the same amount of color on a molar basis; thus the amount determined cannot be an absolute amount. As with many methods for barbiturates, dilantin will interfere.

REAGENTS

1. Chloroform, reagent grade.
2. Buffered mercuric nitrate.

A. Mercuric nitrate, 0.06M. Dissolve 1 g of mercuric nitrate in a few milliliters of water containing 0.2 ml of concentrated nitric acid and dilute to 50 ml.

B. Phosphate buffer, 0.15M, pH 8.0. Dissolve 4.48 g of anhydrous disodium phosphate (Na_2HPO_4) and 0.25 g of monopotassium phosphate (KH_2PO_4) in water and dilute to 500 ml. Check the pH and adjust to 8.0 if necessary.

For the reagent, mix 2 ml of A and 50 ml of B. Filter if turbid. This solution is stable at room temperature.

3. Diphenylcarbazone reagent. Dissolve 100 mg of diphenylcarbazone in 100 ml of chloroform. Allow to age for several days in a stoppered flask at room temperature, then transfer to a brown bottle and store in the refrigerator.

4. Sulfuric acid, 1M. Dilute 5.6 ml of concentrated sulfuric acid to 100 ml with water.

5. Barbital standards.

A. Stock standard. Dissolve 55 mg of secobarbital sodium in about 100 ml of water, add 1 ml of approx. 0.1M hydrochloric acid (8.5 ml to 100 ml with water), and dilute to 200 ml with water. Alkaline solutions of the barbiturates tend to decompose on standing. This standard contains 25 mg/dl of the free barbiturate.

B. Working standard. Dilute the stock standard 1:5 and 1:10 as required, to obtain solutions containing 5 and 2.5 mg/dl. Alternatively, in the procedure add 1 ml of water to the tube, followed by 0.1 or 0.05 ml of the stock standard.

This gives the same equivalent working standards but obviates the frequent preparation of the dilute standards, which are not very stable.

PROCEDURE

Place 1 ml of sample (serum, urine, or gastric contents) in a glass-stoppered tube of 15 to 20 ml capacity. Add 1 ml of water to a second tube as blank, and add 1 ml of the two diluted standards to other tubes. To each tube add 2 drops of the sulfuric acid solution and 10 ml of chloroform. Stopper tubes tightly and shake vigorously for 20 sec. Allow to stand for several minutes for separation of the layers. Remove as much as possible of the upper layer by suction; if the upper layer forms a thick clot, it is not necessary to try to remove it. Filter the chloroform layer through 9 cm Whatman No. 31 paper or equivalent in another glass-stoppered tube. To each of the filtrates add 1 ml of the buffered mercuric nitrate reagent, stopper the tubes, and again shake vigorously for 20 sec. Allow the layers to separate, remove most of the upper layer by suction, and filter the chloroform layer through Whatman No. 31 paper. Pipet 5 ml of the second filtrate from the samples, standards, and blank into dry cuvettes. Add 1 ml of the diphenylcarbazone solution to each tube and mix. Allow to stand for 5 min; then read standards and samples against the blank at 550 nm.

CALCULATION

$$\frac{\text{absorbance of sample}}{\text{absorbance of standard}} \times \text{conc. of standard} = \text{mg / dl barbiturate as secobarbital}$$

The different barbiturates give different amounts of color on a weight basis. Thus serum containing 1 mg/dl of secobarbital will give about as much color as 1.2 mg/dl of amobarbital, 1.4 mg/dl of pentobarbital, and 2 mg/dl of phenobarbital. In general the long-acting barbiturates give less color on a weight basis than do the short-acting ones. The approximate serum levels that will be found in patients who are comatose but with active reflexes are: 1 mg/dl for secobarbital, 1.5 mg/dl for pentobarbital, 3.0 mg/dl for amobarbital, and 5 mg/dl for phenobarbital.

Bromide

In this simple method [40, 41] a trichloroacetic acid filtrate is treated with gold chloride which forms a yellow color with bromide ions. The method will not detect very low levels of bromide in serum but will detect levels somewhat below those that may cause symptoms of bromide intoxication. Iodide will also give the color but it is rarely found in blood in sufficiently high concentrations to interfere.

1. Trichloroacetic acid, $0.6M$.   Dissolve 10 g of trichloroacetic acid and 120 mg of sodium chloride in water and dilute to 100 ml.

2. Gold chloride, $0.0125M$.   Dissolve 0.5 g of gold chloride (the material sold as gold chloride is usually $HAuCl_4 \cdot 3H_2O$) in water and dilute to 100 ml. Store in a small, brown, glass-stoppered bottle.

3. Bromide standards.

   A. Stock standard, 50 mmol/liter (400 mg/dl).   Dissolve 1.286 g of sodium bromide or 1.488 g of potassium bromide, previously dried at 110°C for several hours, in water to make 250 ml.

   B. Working standard.   Dilute 5, 10, and 15 ml of the stock with water to make 50 ml. These standards will contain 5, 10, and 15 mmol/liter (40, 80, and 120 mg/dl).

To 2 ml of serum in a test tube add slowly with constant shaking 8 ml of the trichloroacetic acid solution. Also set up a blank with 2 ml of water and 8 ml of trichloroacetic acid, and standards with 2 ml of each standard plus 8 ml of trichloroacetic acid. Mix well. Allow the serum tube to stand for about 15 min, then centrifuge (the standards and blank need not be centrifuged). If the supernatant is not perfectly clear, filter through a small Whatman No. 1 filter paper. Pipet 4 ml of each sample and 4 ml of the standards and blank mixtures to separate tubes. To each tube add 1 ml of the gold chloride solution. Mix and allow to stand for 10 min. Read standards and samples against the blank at 440 nm.

$$\frac{\text{absorbance of sample}}{\text{absorbance of standard}} \times \text{conc. of standard} = \text{conc. of sample (mmol/liter or mg/dl)}$$

Although the bromide concentration has been conventionally expressed in milligrams per deciliter, the use of millimoles per liter allows a simpler comparison with the chloride concentration (see below).

This method does not detect very low levels of bromide but will detect levels somewhat below the point at which symptoms of toxicity may appear. Symptoms of toxicity such as mental confusion may occur at levels of 6 to 12 mmol/liter (50 to 100 mg/dl). At concentrations between 20 and 30 mmol/liter (160 to 240 mg/dl), more striking symptoms may occur such as emotional outbursts and incoordination. Concentrations over 40 mmol/liter (320 mg/dl) may be fatal. If the normal chloride level is taken as 100 mmol/liter, then symptoms first occur when about 10% of the chloride has been replaced by bromide. The symptoms become more severe when 25% of the chloride has been replaced and may be fatal if more than 40% of the chloride is replaced by bromide. The bromide is excreted at about the same rate as chloride. After any bromide intake is stopped, the level falls rather slowly, reaching one-half the initial value in about 12 days.

Carbon Monoxide—Carboxyhemoglobin

A spectrophotometric method for carbon monoxide has been described in Chapter 16 in the section on carboxyhemoglobin. We present here a chemical method using the Conway microdiffusion apparatus [42–44]. The carbon monoxide released from the blood diffuses into a solution of palladium chloride; this reduces the palladium chloride to metallic palladium and the change in absorbance of the solution is a measure of the amount of carbon monoxide present.

REAGENTS

1. Sulfuric acid, $1M$.   Dilute 5.6 ml of concentrated sulfuric acid to 100 ml with water.

2. Hydrochloric acid, $0.1M$.   Dilute 8.5 ml of concentrated acid to 1 liter with water.

3. Potassium iodide, $0.13M$.   Dissolve 1 g of potassium iodide in 50 ml of water. Prepare just before use as it is not stable.

4. Palladium chloride solution, $0.005M$.   Dissolve 88 mg of palladium chloride $(PdCl_2)$ in $0.1M$ hydrochloric acid and dilute to 100 ml with the acid.

PROCEDURE

Pipet 2 ml of the palladium chloride solution to the inner chamber of a Conway dish. Pipet 2 ml of the $1M$ sulfuric acid to the outer chamber. If the dishes are of the type that use a liquid seal, add the sulfuric acid solution to the sealing chamber as well. Add 1 ml of blood to the outer chamber, cover at once, and mix blood with the acid by gentle rotation of the dish. Also set up a blank using the same amounts of reagents but use 1 ml of water instead of a blood sample. Allow to stand at room temperature for 2 hr. If carbon monoxide is present, some small particles of reduced palladium metal will be seen on the surface of the solution in the center well after the diffusion. An experienced worker can roughly estimate the amount of carbon monoxide present from the appearance of the solution.

Taking care not to include any of the particles of metallic palladium floating on the surface, pipet 0.2 ml of the solution from the inner well of the sample dish to a cuvette and add 10 ml of the potassium iodide solution. Pipet 0.2 ml of the solution from the inner well of the blank dish and treat similarly. Read sample and blank cuvettes against a blank of potassium iodide solution at 490 nm.

CALCULATION

The 2 ml of palladium solution will react with 0.01 mmol of carbon monoxide, which in turn will react with 0.0025 mmol of hemoglobin (161 mg of hemoglobin in the 1-ml blood sample, or 16.1 g/dl).

$$\frac{\text{absorbance of blank} - \text{absorbance of sample}}{\text{absorbance of blank}} \times \frac{16.1 \times 100}{\text{Hb}}$$

$$= \% \text{ saturation with carbon monoxide}$$

where Hb represents the hemoglobin content of the blood sample in grams per deciliter. If the hemoglobin has not been measured, a value of 15 g/dl may be assumed.

As an alternative method of measurement, pipet 0.2 ml from the inner wells to tubes containing 6 ml of the hydrochloric acid solution. Mix; read samples and blank against the hydrochloric acid in the ultraviolet at 278 nm [45]. The calculations are the same as given above.

The precautions for the collection of the sample and the interpretation of the results have been given in Chapter 16.

Lead

Lead determination is used for detecting industrial exposure and for the detection of lead poisoning in young children, whose exposure to the metal is due to the ingestion of flakes of old lead-based paint. The older methods [46–48] of determination required wet digestion of the sample followed by colorimetric determination. This method used relatively large quantities of reagents to determine microgram amounts of lead, and the problem of lead contamination in the reagents was always present. In addition, the method was lengthy and tedious. It would be rarely worthwhile for a laboratory to use this method for only an occasional blood analysis for lead. The preferred method for the determination of lead in biological fluids is by atomic absorption [49–51]. For blood and urine an extraction method can be used in which the lead is extracted into methyl isobutyl ketone by the addition of pyrrolidine dithiocarbamate which forms a lead salt soluble in the organic solvent. The extract is then separated and aspirated directly into the flame. For urine there are other methods in which the lead is coprecipitated with lead-free bismuth [52] or thorium [53] and the precipitate dissolved in a small amount of acid for aspiration into the flame. Separation of lead by means of an ion exchange resin has also been used [54]. Newer methods use the sample cup technique [55–58] or flameless [59–62] carbon rod furnace for atomic absorption. These require that a potentiometric recorder be attached to the atomic absorption instrument as the peak absorption occurs over a very short time interval. They may also require some type of background corrector. The principles of the sample cup and carbon rod analyzer have been explained in Chapter 1. The sample cup method may not be quite as precise as the carbon rod method, but the former is less expensive and may be more suitable for only an occasional analysis.

Lead in Blood and Urine by Atomic Absorption [51, 63]

COLLECTION OF SPECIMENS AND SPECIAL PRECAUTIONS

The blood specimens should be collected in special lead-free heparinized Vacutainers (Becton-Dickinson No. L3200XF313). The blood may also be collected in a disposable plastic syringe and transferred at once to the tube. Since the lead in blood is chiefly in the erythrocytes, difficulties with coagulation may be eliminated by immediately mixing blood in the tube with the heparin to insure prompt anticoagulation. Urine should be collected in a plastic container that has been previously rinsed several times with dilute (1:5) nitric acid. To avoid contamination, disposable plastic pipets should be used whenever possible (do not use plastic for pipetting the methyl isobutyl ketone). All glassware should be soaked overnight in dilute nitric acid (1:3), then rinsed well with tap water and distilled water. Glass-stoppered 15 ml centrifuge tubes are suitable for blood extraction, and larger screw-capped culture tubes with Teflon-lined caps may be used for urine extraction.

REAGENTS

1. Ammonium pyrrolidine dithiocarbamate|| (APDC) Triton reagent.

A. Dissolve 2 g of APDC in water to make 100 ml. This solution is stable for only about 1 week in the refrigerator.

B. Triton X-100# solution. Dissolve 25 ml of Triton X-100 in water to make 500 ml. This solution is stable and larger quantities may be prepared and stored.

For use, mix equal volumes of solutions A and B.

|| K and K Laboratories, Plainview, N.Y. 11803.
Manufactured by Rohm and Haas, Philadelphia, Pa. 19105; may be obtained from most laboratory suppliers.

2. Methyl isobutyl ketone (MIBK), reagent grade, water saturated. To 1 liter of methyl isobutyl ketone add about 10 ml of water and shake. Allow the water to settle out completely and pipet quantities of the ketone as required, taking care not to include any water droplets. The water-saturated solvent must be used for all steps in the procedure whenever methyl isobutyl ketone is required.

3. Lead standards.

A. Stock solution. Dissolve 1.399 g of lead nitrate, which has been previously dried at 110°C for 1 hr, in water containing a few milliliters of concentrated nitric acid, and dilute to 1 liter. This stock contains 1 mg/ml of lead. Atomic absorption standards may be purchased from a number of suppliers. These may be labeled 1,000 ppm which is the same as 1 mg/ml (1 mg/ml = 1 g/1,000 ml = 1,000 g/1,000,000 ml = 1,000 ppm). This solution is stable for several months if kept tightly closed in a polyethylene bottle.

B. Working solution. Dilute 1 ml of the stock solution to 100 ml with water. This solution contains 1 mg/dl (10 μg/ml). It should be made up fresh as required.

PROCEDURE FOR WHOLE BLOOD

Transfer 5 ml of whole blood to an acid-washed glass-stoppered tube, add 2 ml of the APDC-Triton solution, and mix well. Add exactly 5 ml of MIBK, stopper tube, and shake well for 1 min. Centrifuge strongly to obtain a clean separation with an upper layer of MIBK (this layer will be colored yellow to brown by substances extracted from the blood). With care the upper layer can be aspirated directly into the flame of the atomic absorption apparatus, but it is preferable to remove as much of the upper layer as possible to a small disposable glass tube with the aid of a Pasteur pipet. If some of the lower layer is inadvertently added, centrifuge the small tube and decant the upper clear solution to another tube.

If one attempts to prepare a series of standards by adding aliquots of the working standard to 5-ml portions of water and treating similarly to the samples, a troublesome emulsion may form on shaking which may not separate on ordinary centrifugation. One method of avoiding this problem is to make up the standards in aliquots of pooled normal blood (outdated blood bank blood is satisfactory). To four extraction tubes add 5 ml of pooled normal blood. To the separate tubes add, respectively, 0, 0.15, 0.30, and 0.60 ml of the working standard. These will correspond to 0, 30, 60, and 120 μg/dl of added lead. Treat these exactly like the blood samples. An alternative procedure is to add to larger extraction tubes (capacity 35 to 40 ml such as would be used for urine) 0, 0.6, 1.2, and 2.4 ml of the working standard. Add water to make 20 ml, then add 4 ml of the APDC-Triton reagent and 20 ml of MIBK. Stopper, shake well, and centrifuge. With this larger volume one can usually obtain sufficient MIBK extract for use. Note that in this alternative, four times as much standard is added to each tube and these are then extracted with four times as much MIBK so that the concentration of lead in the extract is the same as in the other procedure for standards.

The measurement of the standards and samples in the atomic absorption instrument may differ slightly with the model used. Such parameters as air and acetylene flow rates, aspiration rate, and sensitivity setting will vary with the instrument but the following general principles should be applicable to most instruments. The instrument is turned on, the current to the cathode tube adjusted to the proper value, and the apparatus allowed to warm up for the time suggested in the instruction manual. The slit width and sensitivity are adjusted to the recommended values. The air-acetylene flame is lit and allowed to burn for 15 min or more. The wavelength is set to 283.3 nm. At the time the samples are to be read, water-saturated MIBK is aspirated into the flame. This will cause a very yellow flame. The acetylene flow is reduced until the desired clear blue flame is obtained. After this adjustment, MIBK must be aspirated at all times except when running samples or standards. The instrument is now set to zero while aspirating MIBK. The standards and sample extracts are then aspirated and readings taken. Usually there is sufficient sample to obtain several readings with each tube. If the instrument does not give readings directly in absorbance units, the units are obtained from the percent absorption readings.

CALCULATION

The average readings obtained from the zero standard are subtracted from the average readings for the other standards (this will be noticeable when the standards are made up in blood). The corrected values are then plotted against the concentrations (30, 60, and 120 μg/dl) and the values for the samples obtained from this curve.

An alternative procedure for the extraction step, which eliminates some pipetting, particularly of the blood samples, is as follows [64]: Four to five milliliters of blood is collected in one of the Vacutainers. If more than 5 ml is collected, some must be removed to give a volume of well-mixed blood of about 4.5 ml. The volume of blood in the tube is then carefully marked on the outside of the tube at the top of the blood level. The blood is then poured into a 30 ml polyethylene bottle with a good screw cap. The blood remaining in the tube is rinsed out with 2 ml of the APDC-Triton reagent and this is added to the bottle, followed by a rinse of 1 ml of water. A volume of MIBK is then measured in the original blood tube to equal the volume of blood used. This is poured into the bottle, which is then capped and shaken vigorously for 1 min. The mixture is poured back into the original tube and centrifuged. The resultant upper MIBK layer is removed and run as above. This procedure eliminates the pipetting of the blood and the use of special extraction tubes (the plastic is much less liable to contamination than glass). Since blood is extracted by an equal volume of MIBK, the results are exactly the same as when 5 ml of blood is extracted by 5 ml of MIBK in the first method.

PROCEDURE FOR URINE

Pipet 25 ml of urine into an acid-washed extraction tube of about 40 ml capacity. Adjust the urine to a pH of between 6 and 7 (indicator paper) and add 3 ml of the APDC-Triton solution and mix. Add exactly 5 ml of MIBK, stopper, and shake well. For standards and blank use the aqueous standards given under the alternative standards for blood. Separate the MIBK layers and read in atomic absorption apparatus similar to that for the blood samples and standards. Sometimes a troublesome emulsion will form in the urine samples which will not be broken by strong centrifugation. If this occurs, the following manipulations will usually break the emulsion to allow sufficient extract to be pipetted from the top for reading. After centrifugation, carefully insert a disposable pipet through the upper layer containing the emulsion and remove as much as possible of the lower aqueous layer by suction. Transfer the remaining solution including the emulsion to a clean smaller tube. Immerse the tube containing the emulsion in hot water (about 90°C) for several minutes and cool to room temperature in tap water. Centrifuge strongly.

The calculations are similar to those for blood except that here the standards correspond to 0, 0.06, 0.12, and 0.24 mg/liter of urine. (The highest standard, for erample, will contain $0.6 \times 10 = 6$ μg in 5 ml of MIBK extract equivalent to that in 25 ml of urine, or $6 \times 1,000/25 = 240$ μg/liter or 0.24 mg/liter.)

NORMAL VALUES AND INTERPRETATION OF RESULTS

The normal levels for lead are 0.01 to 0.05 mg/dl in blood and 0.01 to 0.08 mg/dl in urine. Levels of over 0.08 mg/dl in blood and over 0.15 mg/dl in urine are evidence of lead intoxication. Intermediate values are equivocal and the determination should be repeated. In individuals with occupational exposure to lead, levels up to 0.07 mg/dl in blood and up to 0.15 mg/dl in urine may be accepted as normal; levels of over 0.10 mg/dl and 0.20 mg/dl, respectively, are considered evidence of lead intoxication.

Other tests that have been suggested for the diagnosis of lead poisoning are: decreased levels of activity of the enzyme δ-aminolevulinate dehydratase, and increased excretion of δ-aminolevulinic acid and porphobilinogen [65, 66]. The determination of the latter substances is readily carried out using a column chromatographic separation [67], but the test is still somewhat lengthy for screening purposes.

Lead by δ-Aminolevulinic Acid (ALA) Determination

A simple test has been proposed for the determination of urinary δ-aminolevulinic acid (ALA) without any prior separation [68]. This test might prove helpful in screening for lead poisoning since only a small amount of urine is needed. The ALA reacts with ethyl acetoacetate to give a colored compound which is extracted with ethyl acetate and measured in a spectrophotometer.

REAGENTS

1. Acetate buffer, 2M, pH 4.6.   Dissolve 57 ml of glacial acetic acid and 136 g of sodium acetate trihydrate in about 700 ml of water and dilute to 1 liter.

2. Ethyl acetate, reagent grade.

3. Ethyl acetoacetate, reagent grade.

4. Ehrlich's reagent, modified.   To about 30 ml of glacial acetic acid add 1.0 g of p-dimethylaminobenzaldehyde, 5 ml of 60% perchloric acid, and 5 ml of water (or 4 ml of 70% acid and 6 ml of water). Mix and dilute to 50 ml with glacial acetic acid.

5. δ-Aminolevulinic acid (ALA) standards.

    A. Stock standard, 50 mg/dl.   Dissolve 64 mg of δ-aminolevulinic acid hydrochloride* in water and dilute to 100 ml. This solution is stable for 2 months in the refrigerator.

    B. Working standards, 0.5 and 1.0 mg/dl.   Dilute 1 and 2 ml of the stock standard to 100 ml with water. Prepare fresh as needed.

PROCEDURE

For screening purposes random urine specimens are satisfactory, but preferably the first-voided morning specimen should be used. Urine specimens obtained after excessive fluid intake or voided after 8:00 PM should not be used. If the analysis cannot be performed immediately after collection, add about 0.1 ml of glacial acetic acid for every 20 ml of urine and store in the refrigerator.

For each sample or standard, set up two glass-stoppered tubes labeled test or standard and blank. To each pair add 1 ml of urine or 1 ml of the standards; preferably both standards should be set up. To all tubes add 1 ml of acetate buffer and mix. To test and standards tubes but not to blanks, add 0.2 ml of

* Sigma Chemical Co., St. Louis, Mo. 63118.

ethyl acetoacetate. Mix and heat all tubes in boiling water bath for 10 min. Cool to room temperature, add 3 ml of ethyl acetate, stopper, and shake each tube 50 times by hand. Centrifuge; then pipet 2 ml of the upper ethyl acetate layer from each tube to a second set of tubes correspondingly labeled. To each tube add 2 ml of modified Ehrlich's reagent and mix. Allow to stand for 10 min and read each test and standard against the corresponding blank at 530 nm.

CALCULATION

Since standard and samples are treated similarly,

$$\frac{\text{absorbance of sample}}{\text{absorbance of standard}} \times \text{conc. of standard} = \text{mg/dl ALA in urine}$$

The concentration of the standard used is that of the standard (0.5 or 1.0 mg/dl) having an absorbance nearest that of the sample.

NORMAL VALUES AND INTERPRETATION OF RESULTS

The normal range found by this method is up to 0.5 mg/dl. Values over 1.0 mg/dl are strongly indicative of lead poisoning, although the correlation is not completely satisfactory.

Salicylate

Salicylates are administered in relatively high doses in the treatment of rheumatic fever, and the monitoring of the blood level is often helpful in assessing the results of therapy. In addition, salicylate (aspirin) poisoning is a relatively common cause of accidental poisoning in small children.

Salicylates have been determined in serum plasma and urine by a number of different methods. Gas chromatography has been used [69], as well as several fluorometric methods [70–72] based on the fact that salicylates exhibit a strong fluorescence in highly alkaline solutions. The salicylate may be extracted from the acidified solution by an organic solvent such as ethylene dichloride, then reextracted into an aqueous solution and identified and quantitated by means of its ultraviolet spectrum [73, 74]. The extract may also be shaken with a solution of ferric nitrate to give the characteristic purple color formed by the reaction between iron and salicylate [75, 76]. In the simpler method of Trinder [77] presented in this section, the proteins are precipitated with a mercuric chloride reagent also containing ferric ions. After centrifugation, the supernatant contains the ferric salicylate complex which is then determined colorimetrically. This method may not be as precise as others but it is satisfactory for clinical purposes.

REAGENTS

1. Mercuric chloride reagent. Dissolve 40 g of mercuric chloride in about 900 ml of water. Add 10 ml of concentrated hydrochloric acid and 40 g of ferric nitrate $[Fe(NO_3)_3 \cdot 9H_2O]$, dissolve, and dilute to 1 liter with water.

2. Salicylate standards.

A. Stock standard, 200 mg/dl. Dissolve 0.580 g of sodium salicylate in water to make 250 ml. Store in the refrigerator.

B. Working standards. Dilute 5 and 10 ml of the stock to 50 ml with water. These standards contain 20 and 40 mg/dl, respectively.

PROCEDURE FOR SERUM OR PLASMA

To 1.8 ml of water add 0.2 ml serum or plasma. Also prepare a blank with 2 ml of water and standards with 0.2 ml of each standard and 1.8 ml water. To each tube add 2 ml of the mercuric chloride reagent. Mix well and allow to stand for a few minutes, then centrifuge strongly. Decant the clear supernatant into cuvettes and read samples and standards against the blank at 540 nm.

CALCULATION

Since the sample and standard are treated similarly,

$$\frac{\text{absorbance of sample}}{\text{absorbance of standard}} \times \text{conc. of standard} = \text{mg/dl salicylate in sample}$$

where the concentration of the standard will be 20 or 40 mg/dl, depending upon which is used. Preferably use the standard giving an absorbance nearest that of the sample.

PROCEDURE FOR URINE

Since the concentration is generally higher in urine than in serum, make a preliminary 1:5 dilution of the urine. Treat the diluted urine sample the same as serum as given above. To correct for any color due to the urine, set up a urine blank by adding 0.2 ml of the diluted urine to 2 ml of the reagent, then add 1.6 ml of water and 0.2 ml of 85% phosphoric acid. Read against a blank prepared from 0.2 ml of water instead of the diluted urine. Subtract any absorbance obtained from that found for the urine sample. Using the corrected absorbance for the urine, the calculations are as given above multiplied by the dilution of the urine.

INTERPRETATION OF RESULTS

In rheumatic fever, salicylate levels of 30 to 40 mg/dl may be desirable. Toxic symptoms may be evident at levels over 50 mg/dl.

Sulfonamides (Sulfa Drugs) [78]

Although these drugs are not used as often as they were before the introduction of the antibiotics, the determination of the level in the blood may occasionally be requested. Many of the drugs are fairly well absorbed from the intestines and therapeutic concentrations in the blood may easily be attained. Such drugs include sulfanilamide, sulfapyridine, sulfathiazole, sulfadiazine, and sulfamerazine. A determination of the level in blood or urine will indicate whether an effective amount has been absorbed. Some other sulfonamides such as sulfaguanidine, succinylsulfathiazole, and phthalylsulfathiazole are less readily absorbed. These are used chiefly for their bacteriostatic action in the intestinal tract and the level in the blood is of less significance.

The drugs are all determined by the same reaction: The free amino group or the molecule is diazotized and coupled with a naphthyl amine to produce a characteristic red-purple color. On a molecular basis all the sulfa drugs produce about the same amount of color; thus one of the group can be used as a standard for all, provided a suitable correction is made for the difference in molecular weight. For example, 172 ng of sulfanilamide (1 nanomole) will produce the same color as 255 ng of sulfathiazole (1 nanomole). Thus if sulfanilamide is used as a standard, to convert to sulfathiazole multiply the result obtained in terms of sulfanilamide by the factor $255/172 = 1.48$.

REAGENTS

1. Trichloroacetic acid, $0.9M$.   Dissolve 15 g of trichloroacetic acid in water to make 100 ml, or dilute the $1.8M$ solution prepared as given in Chapter 5 under protein-free filtrates with an equal volume of water. Store in the refrigerator.

2. Sodium nitrite, $0.014M$.   Dissolve 0.1 g of sodium nitrite in water to make 100 ml. This should preferably be made up fresh as needed.

3. Hydrochloric acid, $4M$.   Mix 50 ml of concentrated hydrochloric acid and 100 ml of water.

4. Ammonium sulfamate, $0.044M$.   Dissolve 0.5 g of ammonium sulfamate in water to make 100 ml.

5. Color reagent.   Dissolve 100 mg of N-(1-naphthyl)ethylenediamine dihydrochloride in 100 ml of water. Store in a brown bottle in the refrigerator.

6. Sulfanilamide standards.

A. Stock standard, 10 mg/dl.   Dissolve 100 mg of sulfanilamide in about 800 ml of water with the aid of gentle heating. Cool and dilute to 1 liter.

B. Working standard, 0.5 mg/dl.   Dilute 5 ml of the stock standard and 20 ml of $0.9M$ trichloroacetic acid to 100 ml with water. This standard is suitable for all sulfonamide drugs; the results have to be multiplied by the factors given for the other drugs besides sulfanilamide. The factors are: sulfapyridine, **1.45**; sulfathiazole, **1.48**; sulfadiazine, **1.45**; sulfamerazine, **1.54**; Gantrisin, **1.55**; sulfacetamide, **1.24**; succinylsulfathiazole, **2.06**.

PROCEDURE FOR BLOOD

In this procedure only the free (unconjugated) sulfonamide in the blood or urine is determined. The unconjugated form is the therapeutically active form of these drugs. Pipet 15 ml of water into a small flask, add 1 ml of oxalated blood, and mix. Allow to stand for a few minutes until hemolyzed. Add 4 ml of the trichloroacetic acid slowly with shaking. Shake well, allow to stand for about 15 min, then filter. Transfer to separate tubes 10 ml of the filtrate (sample), **10** ml of the working standard (standard), and 10 ml of water (blank). To each tube add 1 ml of sodium nitrite solution, mix, and allow to stand for 3 min. Add 1 ml of ammonium sulfamate solution to each tube, mix, and allow to stand for 2 min. Add 1 ml of the color reagent to each tube, mix, and allow to stand for 10 min. Read standard and samples against blank at **540 nm**.

CALCULATION

The working standard contains 0.5 mg/dl sulfanilamide. Equal volumes of the standard and a 1:20 dilution of blood are compared.

$$\frac{\text{absorbance of sample}}{\text{absorbance of standard}} \times 0.5 \times 20 = \text{mg/dl sulfanilamide}$$

When other sulfonamide drugs are determined, this result is multiplied by the appropriate factor as mentioned under standards.

PROCEDURE FOR URINE

The same procedure can be used for urine after a preliminary dilution. For readily absorbed sulfonamides, dilute the urine 1:100 with water, then make a further 1:20 dilution with trichloroacetic acid just as in preparing the blood filtrate. The procedure and calculations are the same with the extra dilution factor included. For the slightly absorbed sulfonamide drugs such as sulfaguanidine, make an original dilution of only 1:5, then proceed as above.

This procedure will determine only the free (unconjugated) drug. To obtain the total, proceed as follows: To 10 ml of the final dilution in a graduated tube add 0.5 ml of $4M$ hydrochloric acid and heat in a boiling water bath for 1 hr. Cool and dilute back to the original volume of 10 ml. Then proceed with the analysis. Since the amount in the urine will vary much more than in blood, it may be necessary to make different dilutions from those given above.

The reaction used is not specific for sulfonamides; any free aromatic amine may react. Substances that may be found in blood and may interfere include procaine, barbiturates, and p-aminobenzoic acid.

The optimum concentration of the sulfonamide drugs in blood is about 5 to 10 mg/dl and in urine 50 to 100 mg/dl.

p-Aminobenzoic Acid (PABA) [79]

This drug has been used in the treatment of rickettsial diseases. It may be determined by the same reactions as used for the sulfonamides. For convenience a sulfanilamide standard may be used with a conversion factor of 0.80 (see discussion under sulfonamides). A PABA standard may be made by dissolving 100 mg of the drug in about 90 ml of water, adding a few drops of $1M$ hydrochloric acid to aid in solution, and diluting to 100 ml. A working standard is prepared just as for the sulfonamides.

Since the blood levels are generally higher for PABA than for the sulfonamides, an aliquot of 3 ml of filtrate plus 7 ml of water is used instead of the 10 ml of filtrate as outlined under sulfonamides. The other directions are the same.

CALCULATION

$$\frac{\text{absorbance of sample}}{\text{absorbance of standard}} \times 0.5 \times 20 \times \frac{10}{3} = \frac{\text{absorbance of sample}}{\text{absorbance of standard}} \times 33.3 = \text{mg/dl PABA}$$

If a sulfanilamide standard is used, the results must be multiplied by a factor of 0.80. If a PABA standard is used, this factor is omitted.

Usually a PABA blood level in the range of 30 to 50 mg/dl is desired. The determination could also be made in serum, but the older criteria for desirable therapeutic levels are given in terms of whole blood, and the levels in serum would be somewhat higher (20 to 60% depending upon whether any of the drug enters the erythrocytes). If it were desirable to run a number of determinations on serum, a comparison could be made of the values in serum and whole blood obtained from a patient at the same time.

p-Aminosalicylic Acid (PAS) [80, 81]

This drug has been used in the treatment of tuberculosis. Since the drug is usually given over a rather long period of time, it may be helpful to monitor the level in the blood occasionally. In the determination a trichloroacetic acid filtrate is treated with Ehrlich's reagent to give a color that is fairly specific for PAS.

1. Trichloroacetic acid, $0.45M$. Dissolve 7.5 g of trichloroacetic acid in water to make 100 ml or dilute the $1.8M$ solution as prepared for trichloroacetic acid filtrate (Chapter 5) 1:4. Store the solution in the refrigerator.

2. Sodium hydroxide, $1M$. Dissolve 40 g of sodium hydroxide in water to make 1 liter.

3. Hydrochloric acid, $1M$. Dilute 84 ml of concentrated hydrochloric acid to 1 liter with water.

4. Ehrlich's reagent. Dissolve 3 g of p-dimethylaminobenzaldehyde in about 100 ml of water containing 7 ml of concentrated sulfuric acid and dilute to 1 liter.

5. p-Aminosalicylic acid standards.

A. Stock standard, 100 mg/dl. Dissolve 100 mg of p-aminosalicylic acid in the $0.45M$ trichloroacetic acid solution and dilute to 100 ml with this acid.

B. Working standard, 0.5 mg/dl. Dilute 0.5 ml of the stock to 100 ml with the trichloroacetic acid solution.

Add 0.5 ml of blood to 9.5 ml of the trichloroacetic acid solution, mix well, allow to stand for about 5 min, and centrifuge. Pipet 2 ml of the clear supernatant to a test tube. In another tube add 2 ml of trichloroacetic acid solution as a blank and to a third tube add 2 ml of the working standard. To each tube add 0.5 ml of sodium hydroxide solution, 1 ml of water, and 1 ml of Ehrlich's reagent, mixing after each addition. Allow to stand for about 5 min, then read samples and standard against blank at 470 nm.

The 0.5 mg/dl standard is compared with an equal volume of a 1:20 dilution of the blood; thus it is equivalent to $0.5 \times 20 = 10$ mg/dl in the blood.

$$\frac{\text{absorbance of sample}}{\text{absorbance of standard}} \times 10 = \text{mg/dl PAS in blood}$$

A desirable therapeutic blood level is in the range of 4 to 8 mg/dl. The determination could be made in serum (see remarks under PABA levels).

Isoniazid [82, 83]

This drug (isonicotinic acid hydrazide) is used in the treatment of tuberculosis. In long-term therapy the occasional monitoring of the blood levels may be desirable. The method uses a protein-free filtrate prepared with metaphosphoric acid. An aliquot of the filtrate is reacted with sodium pentacyanoamineferroate to give a yellow color which is fairly specific for isoniazid.

REAGENTS

1. Metaphosphoric acid, 20 g/dl.   Dissolve 20 g of metaphosphoric acid in water to make 100 ml. The material labeled metaphosphoric acid is actually a mixture of sodium metaphosphate and metaphosphoric acid containing about 35% free acid.

2. Diammonium phosphate [$(NH_4)_2HPO_4$]. 1.25M.   Dissolve 16.5 g of diammonium phosphate in water to make 100 ml.

3. Sodium pentacyanoamineferroate,† 0.2 g/dl.   Dissolve 200 mg of the salt in water containing 1.5 ml of concentrated ammonium hydroxide and dilute to 100 ml.

4. Isoniazid standards.

   A. Stock standard. 1 mg/ml.   Dissolve 100 mg of isoniazid in water and dilute to 100 ml. This solution is stable for about a week when stored in the refrigerator.

   B. Working standard. 0.01 mg/ml.   Dilute 1 ml of the stock to 100 with water. Prepare fresh as needed.

PROCEDURE FOR BLOOD

Pipet into a centrifuge tube 4 ml of water and 2 ml of serum. Mix and add 2 ml of the metaphosphoric acid solution. Mix and allow to stand for about 10 min, then centrifuge. To one tube add 4 ml of clear supernatant, to a second tube 4 ml of the working standard, and to a third tube 4 ml of water as blank. To each tube add 1 ml of the diammonium phosphate solution and 1 ml of the pentacyanoamineferroate solution and mix. Allow to stand for 10 min and then read standard and samples against blank at 420 nm.

CALCULATION

The working standard contains 0.01 mg/ml or 1 mg/dl; equal volumes of this and a 1:4 serum dilution are compared.

† K and K Laboratories, Plainview, N.Y. 11803.

$$\frac{\text{absorbance of sample}}{\text{absorbance of standard}} \times 4 = \text{mg/dl isoniazid in serum}$$

PROCEDURE FOR CSF

A determination can also be made in cerebrospinal fluid by using 4 ml of fluid, 2 ml of water, and 2 ml of metaphosphoric acid solution. The procedure is carried out just as for serum filtrate. Since the CSF is diluted 1:2, the multiplier in the equation is 2 rather than 4.

INTERPRETATION OF RESULTS

The therapeutic range of serum concentration for isoniazid is 0.5 to 1.2 mg/dl.

Phenothiazines [84, 85]

A number of drugs used as tranquilizers are derivatives of phenothiazine (promazine, chlorpromazine, prochlorperazine), as are many of the antihistamine drugs. The following test may be used to detect the presence of the drug in the urine. This detection may be helpful to the physician in determining whether the patient is taking the drug.

The test is based on the color produced with ferric chloride in the presence of an oxidizing agent. The intensity of the color gives some indication of the amount of drug in the urine.

REAGENT

Dissolve 1 g of ferric chloride ($FeCl_3 \cdot 6H_2O$) in 250 ml of water; add 100 ml of concentrated nitric acid and 50 ml of 70% perchloric acid. This solution is stable for a long time, if kept out of contact with organic matter.

PROCEDURE

Add 1 ml of the reagent to 1 ml of urine and note the color that develops within 20 sec.

INTERPRETATION OF RESULTS

A positive reaction is given by the development of a pink to violet color. A slight yellowish pink color will indicate an intake of about 5 to 20 mg/day of one of the drugs, a darker salmon pink an intake of 25 to 70 mg/day, a light lavender color an intake of about 100 mg/day, and a dark purple color over 125 mg/day. These colors are only approximate and will vary slightly with the particular drug and with the dilution of the urine.

Lithium

Lithium is an alkali metal similar to sodium and potassium. Normally it is not found in body fluids (for this reason it was chosen for use as an internal standard). Recently it has been introduced for the treatment of some types of mental disorders. The blood or serum level may be determined as an aid to therapy, particularly as the therapeutic index is not high. Lithium may be determined by either flame or atomic absorption photometry [86–90]. Flame photometry is accurate and convenient. However, many of the modern flame photometers use an internal standard of lithium, and these cannot be used for the determination of lithium without some changes. Some photometers have adjustments making it possible to determine sodium and potassium using a lithium internal standard and to determine lithium with a potassium internal standard. The older types of flame photometers which did not use an internal standard can be used for lithium determinations with the proper filter. The use of atomic absorption is also a convenient method for lithium provided an instrument is available and the frequency of use would justify the purchase of the hollow cathode tube.

For those flame photometers that use lithium as an internal standard for sodium and potassium but can be converted to lithium determinations, the procedure manual accompanying these instruments must be consulted. For flame photometers ordinarily not using an internal standard and for atomic absorption, some general details may be given.

To prepare a lithium standard solution, dry some lithium carbonate (reagent grade) for several hours at 200°C, then cool in a desiccator. Accurately weigh out 369.5 mg of the lithium carbonate and transfer it quantitatively to a 1 liter volumetric flask with the aid of about 200 ml of water. Add 10 ml of hydrochloric acid ($1M$) and swirl until all the carbonate is dissolved. A few additional drops of HCl may be added if needed. When all the material has dissolved, dilute to 1 liter and mix well. This solution contains 10 mmol/liter of lithium. A working standard of 1 mmol/liter is made by diluting exactly 10 ml of the stock to 100 ml.

The serum samples and standards are diluted 1:10 or 1:50 depending upon the particular instrument used. For the serum sample, 1 ml of serum is diluted to volume with water. For the standard, 1 ml of the working standard and 1 ml of a sodium and potassium standard containing 5 mmol/liter of potassium and 140 (or 150) mmol/liter of sodium are diluted to volume. For the blank 1 ml of the sodium-potassium standard is diluted to volume. The samples and standards are run on the flame photometer or atomic absorption apparatus in the usual way, comparing the reading of the standard with that of the unknown. With some instruments using only a 1:10 dilution the presence of the serum may cause slightly different aspiration rates for the sample and standard. This can be checked by comparing a dilution of a serum known to be lithium-free to which 1 ml of standard has been added with the same dilution of standard without added serum. If there appears to be a significant difference, then the sodium-potassium solution added to the standard and blank dilutions may be replaced by lithium-free serum. This latter is not difficult to obtain since lithium is present in appreciable amounts only in patients on lithium therapy. It has been stated that some instruments do not show linearity of response above 1.5 mmol/liter of lithium. This can be checked by running 1 and 2 mmol/liter standards and comparing. Levels above 1.5 mmol/liter are generally considered in the toxic range so that ordinarily there will not be many samples in this range. If the instrument does show some nonlinearity, the occasional high sample can be repeated using a greater dilution of the serum.

Drug Screening

In recent years significant interest has arisen in the rapid identification and determination of drugs of abuse and their metabolites in urine. A number of different types of drugs have been included: barbiturates, amphetamines, tranquilizers such as meprobamate, chlorpromazine, and prochlorperazine, antihistamines such as methapyrilene, chlorpheniramine, and diphenhydramine, opium alkaloids and their substitutes, along with many others. The screening methods are generally based on thin layer chromatography after a preliminary separation from the urine.

Using different extraction conditions, solvent systems for development, and spray solutions for the visualization of the spots, a large number of drugs can be detected. Even when the exact drug cannot be positively identified, the general class to which it belongs usually can be, e.g., long- or short-acting barbiturates and amphetamine derivatives. For many purposes the information gained by the screening procedure may be sufficient. At other times it may be desirable to make confirmatory tests. These can be additional colorimetric tests or thin layer chromatography techniques, but the best confirmatory tests are made with the gas chromatograph [91–94] or in more sophisticated systems with the gas chromatograph–mass spectrograph combination [95, 96]. In some instances the spots suspected of being due to a given drug may be scraped from the plate, the absorbent material eluted with a small amount of solvent [97], and the drug confirmed by microchemical tests or gas chromatography [98].

The drugs may be separated from the urine by extraction with an organic solvent, either at a pH near neutrality or at high and low pH values to separate the acidic neutral and the basic drugs [99–101]. A nonionic resin, either in a small column [102, 103] or on resin-impregnated paper [104, 105], has also been used. The drugs are then eluted or extracted from the resin with an organic solvent. The solvent is then evaporated to a small volume and applied to a chromatographic sheet or plate. Several different types of sheets have been used together with different developing solvents [106, 107]. After development, the plates may be viewed in ultraviolet light to locate certain spots and then sequentially sprayed with a number of different reagents to develop the characteristic colored spots. The different drugs are identified by means of the color produced with the reagents and the Rf values of the spots (see Chapter 3, section on thin layer chromatography). Usually standards containing several of the most commonly found drugs are run simultaneously to aid in identifying the unknowns since the Rf values and even the colors produced may vary with the exact conditions. A number of drug screening kits‡ are available containing all the reagents and chromatographic sheets for the procedure with separation either by solvent extraction or resin absorption.

The solvent extraction methods have the advantage that by extraction at two different pH values, the possible drugs present are separated into two classes— acid-neutral and basic—for separate chromatography. This decreases the number of possible drugs that may be found on one plate. The resin absorption methods may be somewhat shorter but may be more expensive for occasional use, unless the columns are prepared in the laboratory [103].

We present an extraction procedure based in general on the work of Broich and associates [99], with some modifications suggested by Baden et al [98].

Drug Screening by Chromatography

SPECIAL APPARATUS REQUIRED

1. Drying oven. This will be found in most laboratories, but it should have a temperature regulator so that the oven temperature does not exceed about 110°C (lower temperatures may be required for some applications). If glass plates are used, they can be laid, glass side down, on the oven shelves. For other types of sheets, the oven should contain some clean glass plates on which the sheets can be placed or provision should be made for suspending the sheets in the oven from small stainless steel clips.

2. Source of warm air for drying spots. A small hand hair dryer can be used, but the "heat guns" available from most laboratory supply houses are more

‡ Chromat-O-Screen, Eastman Kodak Co., Rochester, N.Y. 14650; Drug-Skreen, Brinkman Instruments, Inc., Westbury, N.Y. 11590; Seprachrom Drug System, Gelman Instruments, Inc., Ann Arbor, Mich. 48806; Toxi-X Drug Test System, ICL Scientific Co., Fountain Valley, Calif. 92708; QSD-3, Quantum Industries, Fairfield, N.J. 07006.

substantial. Those rated up to 300°F are satisfactory as high temperatures are not needed.

3. Ultraviolet lamp. A dual-purpose lamp giving both the long-wavelength (366 nm) and the short-wavelength (254 nm) ultraviolet light is preferable. Such a lamp can be used for other tests as well. A high intensity is not required, but one should be able to see the fluorescent spots without having to darken the room completely. These lamps are available from most laboratory supply houses.

4. Chromatographic developing chamber for plates or sheets. For glass plates a number of glass jars are available, but these are less satisfactory for plastic or other sheets. The Gelman Chromatographic chamber§ has been found satisfactory for almost all types of sheets and can be used with glass plates as well. It will not support sheets less than 10 cm long.

5. Reagent sprayer. Some means of spraying the developed chromatograms with the reagents is needed. All-glass sprayers operated by compressed air available from most supply houses are satisfactory, if a supply of compressed air is available. The use of an atomizer bulb as air source is less convenient. The sprayers that operate with compressed dichlorodifluoromethane propellant are convenient. It is difficult to regulate the rate of spraying and the tendency is to overspray. If considerable use is made of the sprayers, those operated by compressed air are less expensive.

6. Micropipets or capillary tubes for application of samples. Regular micropipets can be used, but these must be rinsed out well with solvent after each sample. It is more convenient to use disposable capillary tubes. The Drummond Micro-caps‖ (5 μl size) can be used. The Strumia microhematocrit tubes# (plain) are also satisfactory; they have a capacity of about 5 μl. The small capillaries have the advantage that a smaller spot can be more easily made even though the capillary may have to be filled several times to transfer all the liquid to the sheet.

EXTRACTION REAGENTS

1. Sulfuric acid, concentrated, reagent grade.
2. Ammonium hydroxide, concentrated, reagent grade.
3. Sodium bicarbonate, saturated solution. Add about 12 g of solid sodium bicarbonate to 100 ml of distilled water and shake.
4. Sodium bicarbonate, solid, reagent grade.
5. Chloroform, reagent grade.

§ Gelman Instrument Co., Ann Arbor, Mich. 48106.
‖ Drummond Scientific Co., Broomall, Pa., 19008.
Cat. No. 21805, Sherwood Medical Industries, St. Louis, Mo. 63103.

6. Chloroform-isopropanol, 4:1.   Mix 100 ml of isopropanol with 400 ml of chloroform.

7. Methanol, reagent grade.

8. Isopropanol, reagent grade.

SPRAY REAGENTS

1. Ninhydrin, 0.1 g/dl.   Dissolve 0.1 g of ninhydrin in 100 ml of methanol.

2. Mercurous nitrate, 1 g/dl.   Dissolve 1 g of mercurous nitrate in 1 ml of nitric acid and 5 ml of water; then dilute to 100 ml with water

3. Diphenylcarbazone, 10 mg/dl.   Dissolve 10 mg of diphenylcarbazone in 100 ml of chloroform. Store in a brown bottle in the refrigerator and prepare fresh every 2 weeks.

4. Vanillin, 1 g/dl.   Dissolve 1 g of vanillin in 100 ml of methanol and add 2 ml of concentrated sulfuric acid. Store in the refrigerator and make up fresh if the solution becomes darker in color.

5. Iodoplatinate solution.   Dissolve 1 g of potassium iodide in about 10 ml of water and add 1 ml of 5% platinum chloride solution; then dilute to 50 ml with water. The platinum chloride is purchased as a 5% solution.

6. Dragendorf reagent.   Add 1.3 g of bismuth subnitrate to 60 ml of water. Add 15 ml of glacial acetic acid and mix until the bismuth salt has dissolved. Dissolve 12 g of potassium iodide in 30 ml of water and add to the bismuth solution. Add 100 ml of water and 25 ml of additional glacial acetic acid and mix thoroughly.

7. Standards.   Solutions that contain definite amounts of known drugs should be chromatographed along with the samples. For the acid neutral fraction a satisfactory mixture is made by dissolving 20 mg each of phenobarbital, secobarbital, glutethimide, and meprobamate or diphenylhydantoin in 10 ml of methanol. For the basic group dissolve 20 mg each of amphetamine, morphine, propoxyphene, and quinine in 10 ml of methanol. The standards are fairly stable if kept in tightly stoppered bottles in the refrigerator. Since the sale of most of these drugs is governed by federal regulations, it may be difficult to secure them in pure form. Three standard mixtures may be purchased directly,* one containing 3 mg/dl each of phenobarbital, secobarbital, diphenylhydantoin, and glutethimide, a second containing 3 mg/ml each of propoxyphene, methadone, quinine, and metamphetamine; the third containing 3 mg/ml each of cocaine, meperidine, amphetamine, codeine, and morphine. Another source† supplies a large number

* Cat. No. 51910, Gelman Instrument Co., Ann Arbor, Mich. 48106.
† Theta Corporation, Chemical Division, Media, Pa. 19063.

of separate standards each containing 5 mg/ml of the drug. These are packaged in small vials with a rubber septum. If a 5 µl Hamilton syringe‡ is used to withdraw the required amount, the solution will remain stable for several months. Most kits provide their own standards.

EXTRACTION OF DRUGS

For each urine, two tubes are set up. These may be marked A (acid-neutral) and B (basic). Convenient extraction tubes are screw-capped culture tubes (20 × 150 mm) with Teflon-lined caps. Add 5 ml of urine to each tube. To tube A add 2 drops of concentrated sulfuric acid and 12 ml of chloroform. To tube B add 4 drops of concentrated ammonium hydroxide and sufficient solid sodium bicarbonate just to saturate the solution (about 60 mg) and 12 ml of the chloroform-isopropanol mixture. Cap the tubes and shake well for several minutes. Allow to stand until the layers separate (centrifuge the tubes for a short time if necessary). Remove the upper layer in tube A by disposable pipet and discard. Shake the chloroform layer briefly with 2 ml of saturated sodium bicarbonate solution. Then remove the upper bicarbonate layer and save for possible testing for salicylate (see below under special tests). Remove the upper layer in tube B and save for possible testing for morphine and codeine (fraction C). Filter the two solvent layers through small filter papers (Whatman no. 31) into appropriately marked beakers.

EVAPORATION OF THE SOLVENTS

Evaporate the solvent filtrates in the two beakers just to dryness on a steam bath in a fume hood. When about half the solvent in beaker B has been evaporated, add 1 drop of concentrated hydrochloric acid and continue the evaporation. The beakers should not be heated for more than a very short time after all the solvent has evaporated. Dissolve the residue in about 0.2 ml of chloroform (beaker A) or methanol (beaker B). Incline the beakers at an angle of 45° and further evaporate the solvent to a final volume of about 10 µl. The beakers can be held in the proper position by placing in a large pan containing a layer of sand and heating on a hot plate at low heat. Alternatively one can use a micro-preparation dish.§ This consists of a Teflon block with a number of wells with conical bottoms. Transfer the final solution to the wells with the aid of a Pasteur pipet and small portions of the solvent. Evaporate the solvent to the required volume with the aid of a stream of warm air from the heat gun.

‡ Hamilton Co., Reno, Nev. 89502.
§ Furnished with the Helena Fetal Maturity Kit or obtained separately as item No. 8003, Helena Laboratories, Beaumont, Tex. 77703.

Plastic sheets (20 × 20 cm)|| with a coating of silica gel and fluorescent indicator are used. These may be cut to a smaller size, if needed, with scissors or a paper cutter. For a few samples a 5 × 20 cm sheet may be used; one standard and three unknowns may be placed on this size sheet. With a sharp pencil place small dots about 1 cm apart in a line long one edge (short side in narrower strips), about 1.5 cm from the edge. Dip the microcapillary in the solution to be applied and allow to fill by capillary action. Then touch it briefly to the sheet at the marked position. Some of the liquid will flow into the strip as noted by the change to a whiter color. Allow the spot to dry somewhat and apply additional liquid. Too long contact with the sheet at one time will result in too large a spot. Since the spots tend to increase in size on migration, they should initially be kept as small as possible. For the samples several capillary fillings may be required. About half the total extract is added to the spot. For the standards, 5 µl is sufficient. For the larger strips the standard is applied at both end positions. Note that one sheet is prepared from the A extracts with the appropriate standard mixture and one sheet from the B extracts with the standard for this extract. After all the spots have been applied, remove the last traces of solvent by heating for a short time with a stream of warm air.

CHROMATOGRAPHIC DEVELOPMENT

The developing solvent for the samples from the A fraction is hexane-ethanol. The original directions called for a ratio of 97:3. Particularly with older sheets we found that a higher percentage of ethanol was needed to give a good separation. A ratio of 94:6 appears to give better results. It would be advisable to run the mixed standard by itself on a narrow strip with several different hexane: ethanol ratios to see which gives a wider separation of the spots. It is not necessary to spray these sheets; the ultraviolet viewing will be sufficient. For the B extract the developing solvent is a modified Davidow's reagent, ethyl acetate:methanol: ammonium hydroxide (85:10:1). This should be prepared just before use.

Pour the developing solution into the chromatographic chamber and wet the saturating pad with the solution. After allowing 10 or 15 min for saturation, carefully set the sheet in place. The sheet should dip in the solution so that the line of spots is about 0.5 cm above the top of the liquid. Close the chamber and allow the development to proceed until the solvent front has moved 8 to 10 cm. Remove the sheets from the chamber and immediately mark the position of the

|| Cat. No. 6060, Eastman Kodak Co., Rochester, N.Y. 14650.

solvent front with a pencil. Then evaporate the solvent in a stream of warm air or in an oven at about 90°C.

VISUALIZATION OF SPOTS

Fraction A—acid-neutral drugs. Expose the sheet to ammonia vapors for a few minutes by moving it back and forth over a beaker containing a small amount of concentrated ammonium hydroxide. Then air dry for a few minutes to remove excess ammonia vapors and view under short-wave ultraviolet light. Barbiturates and related compounds will be seen as dark spots against the orange background. A blue fluorescence will be given by salicylic acid and salicylamide. If only small amounts of salicylic acid are present, it may not appear here but can be detected by the extra test given below. Carefully outline the spots observed with light pencil marks and make a record of any special features.

First spray the sheet with the mercurous nitrate reagent and then lightly dry in a current of warm air. Barbiturates, dilantin, and other compounds noted under ultraviolet light will give clearly visible white spots with the reagent. Glutethimide may give only a faint spot, and meprobamate none. Now spray the sheet with the diphenylcarbazone reagent. Most of the compounds already mentioned will give blue or violet spots. There will be some blue background but this will fade gradually leaving the spots clearly visible, especially when viewed by transmitted light. For the detection of meprobamate, spray the lower part of the sheet with the vanillin reagent. (Only the lower part of the sheet need be sprayed since meprobamate will be found in this part.)

Note the colors of any spots formed and outline the position of the spots in pencil as some of the spots may fade on standing. The spraying should be done rather lightly, avoiding any excess. It is easy to add additional spray if deemed necessary but impossible to correct overspraying which may cause the spots to run.

Heat the sheet in an oven at about 110°C, observing it frequently. Too much heating may char the sheet, but before this occurs, a distinct yellow spot will be seen if meprobamate is present.

Ordinarily in drug screening there will be not more than one or two drugs found in any given urine, but the position and color of the spot are necessary to identify the drug. Calculate the Rf values for the various spots. Measure the distance from the application line to the solvent front. Then measure the distance from the center of each spot to the application line. The Rf value is defined as the ratio of the distance traveled by the particular drug to the distance traveled by the solvent front. The actual Rf values may vary somewhat from time to time

Table 18-1. Color Reactions and Rf Values of Acid–Neutral Drugs [9, 99]

Compound	Rf Value	Ultraviolet[a] (254 nm)	Mercurous Nitrate Reagent[b]	Diphenyl-carbazone Reagent[b]
Meprobamate	0.13	—	n	[c]
Phenobarbital	0.32	+	W	B
Diphenylhydantoin	0.34	+	W	BV
Barbital	0.39	+	W	B
Bromural	0.41	+	ft W	BV
Butabarbital	0.41	+	W	BV
Cyclobarbital	0.41	+	W	BV
Amobarbital	0.47	+	W	BV
Secobarbital	0.48	+	W	B
Glutethimide	0.53	+	ft W	BV
Hexabarbital	0.55	+	W	BV

[a] +, dark spot (UV absorption); —, no absorption.
[b] W, white; n, no change; B, blue; V, violet; ft, faint.
[c] Meprobamate may appear as a colorless spot against the blue background. A more extensive table of drugs reacting with the above reagents is given by Frings [118].

and will vary much more if a different solvent mixture is used. By comparing the Rf values of the known standards with that of the unknown, the drug can usually be identified (Table 18-1).

Fraction B—basic drugs.   Examine the dried sheet under ultraviolet light. If quinine or quinidine is present, these will be seen as fluorescent blue spots. Mark their position with a pencil. Quinine has been used to adulterate heroin and its presence may indicate the possible use of heroin. Then spray the sheet with the ninhydrin reagent and heat gently in a stream of warm air. Amphetamine, if present, may appear as a faint yellow spot. Then expose the sheet to long-wavelength ultraviolet light for 15 to 20 min. The amphetamine spot should now appear violet-red and metamphetamine will show up as a light violet spot. With a strong source of ultraviolet, the spots should appear fairly rapidly if the drug is present; with a weaker source, a longer time may be required. Spray the sheet with the iodoplatinate reagent. Blue or violet spots will be given by amphetamine, alkaloids including codeine and morphine, and many phenothiazine derivatives. Nicotine, usually found in the urine of smokers, will give a blue-green spot. Outline the position of the spots and note their color. Now spray the sheet with Dragendorf

Table 18-2. Color Reactions and Rf Values of Basic Drugs

Compound	Rf Value	Ninhydrin Reagent	Ultraviolet[a] (366 nm)	Iodo-platinate Reagent[a]	Dragendorf Reagent[a]
Hydromorphone	0.24	—	—	V	RO
Morphine	0.32	—	—	B	RO
Codeine	0.51	—	—	V	RO
Quinine	0.52	—	—	V	RO
Metamphetamine	0.58	—	wk V	RV	O
Quinidine	0.58	—	—	V	RO
Nalorphine	0.60	—	—	V	O
Amphetamine	0.64	wk Y	RV	RV	O
Prochlorperazine	0.64	—	—	V	BrO
Chlordiazepoxide	0.72	—	—	VO	O
Nicotine	0.78	—	—	BGr	RO
Chlorpromazine	0.83	—	—	V	BrO
Procaine	0.84	—	—	V	O
Demerol	0.84	—	—	V	RO
Diazepam	0.92	—	—	V	BrO
Methadone	0.94	—	—	V	O
Cocaine	0.96	—	—	V	O
Propoxyphene	0.99	—	—	V	O

Adapted from Broich et al. [99].
[a] Y, yellow; V, violet; R, red; Br, brown; O, orange; Gr, green; B, blue; wk, weak; —, no color. A more extensive table of drugs reacting with these reagents is given by Frings [118].

reagent. The spots will change to various shades of red, orange, or brown as indicated in Table 18-2. The Rf values for the various spots are computed similarly to the spots on the A sheet. The various possible drugs are identified by the use of the Rf values and colors as given in Table 18-2. The Rf values may vary somewhat from run to run but by comparison with the known standards the drugs can usually be identified. In the screening procedure usually only one or two drugs (other than nicotine) will be found in any one urine. For forensic purposes the presence of suspected drugs must be confirmed by other means such as gas chromatography, infrared or ultraviolet spectroscopy, or in some cases special chemical tests.

Tables 18-1 and 18-2 list the color reactions for many of the most commonly encountered drugs of abuse. The Rf values given are used only as a guide; the values obtained for the samples must be compared with those of the standards run under the same conditions. In these tables and elsewhere in this section, the drugs are referred to only by their generic names. The different brand names for a given generic drug can usually be obtained from the *Physicians' Desk Reference* [108] or the *Merck Index* [109].

Special Tests

SALICYLATES

The presence of salicylates in the bicarbonate wash from the A extract may be detected as follows: Acidify the bicarbonate wash to a pH of below 2 by the cautious addition of a few drops of concentrated hydrochloric acid. Then shake this with about 10 ml of ether. Separate the ether layer and evaporate just to dryness in a small beaker on a water bath. Dissolve the residue in about 0.2 ml of methanol. Spot a portion of the methanol solution on filter paper, dry, and examine under ultraviolet light. Salicylates should appear as a blue fluorescent spot. Add another portion of the methanol solution to a depression in a white spot plate. Add 1 or 2 drops of Trinder's reagent (see quantitative determination of salicylate earlier in this chapter). A blue to purple color indicates the presence of salicylate.

BARBITURATES

The quantitative test given earlier in this chapter may be used as a confirmatory test for barbiturates. As mentioned, this does not distinguish between the different barbiturates as does the thin layer chromatography procedure. In the latter, a spot corresponding to the position of the phenobarbital standard will indicate a long-acting barbiturate, a spot corresponding to the secobarbital standard may indicate a short-acting barbiturate, and an intermediate spot, a barbiturate intermediate in length of action.

OPIUM ALKALOIDS

Heroin is metabolized in the body to morphine and excreted as such. Both morphine and codeine are excreted chiefly as glucuronide conjugates which are not extracted by the organic solvents. If only a small amount of these alkaloids is present in the urine, there may not be a sufficient amount in the free form extracted to give a spot on the chromatogram. If the presence of these alkaloids is suspected, the fraction C reserved earlier may be tested. To this fraction add dropwise sulfuric acid (1.8M, prepared by diluting concentrated H_2SO_4 1:10) to a pH of about 1. Add an additional amount of the diluted sulfuric acid equal to half the volume of the solution and autoclave for 40 min at 18 to 20 psi. Cool, and neutralize with concentrated ammonium hydroxide to a pH of 8.5. Extract with chloroform-isopropanol as in the regular procedure for basic drugs. Evaporate the solvent, spot on a sheet, and develop the chromatogram as given previously. Dry and spray with iodoplatinate reagent. If morphine or codeine is present, the spots should be visible. This procedure cannot be used as a general screening test since the autoclaving would decompose a number of other drugs that might be present.

Phenothiazines can be detected directly in urine by a color reaction (see earlier in this chapter). The hallucinogenic drugs (LSD, mescaline, marihuana) are much more difficult to detect in urine or blood. Furthermore, one is handicapped by the fact that in most jurisdictions the possession of these drugs even for scientific purposes is illegal. The following tests taken from the literature may be helpful if they can be checked by the use of authentic specimens of the drugs.

MARIHUANA

The following method has been suggested for the detection of cannabis derivatives on the fingers of marihuana smokers. Wash the fingers with chloroform, evaporate the chloroform washing to a few microliters, and apply to a strip of Eastman 6060 chromatographic sheet that has been previously dipped in a saturated solution of silver nitrate in 95% ethanol for a few seconds and then air dried. Develop the sheet with toluene for about 5 cm, then remove and air dry. Then spray it with a freshly prepared solution of 0.1 g/dl of Fast Blue B salt in 70% ethanol. A positive test is indicated by a purple-red spot a few centimeters above the point of application. The application point may also be stained but this is not considered a positive test. The determination of cannabis derivatives in urine is still too difficult for routine use [110, 111].

LYSERGIC ACID DIETHYLAMIDE (LSD)

Although LSD is usually taken in very small doses, a fluorometric method has been suggested for its determination in plasma. In the procedure 5 ml of plasma is saturated with solid sodium chloride (about 1.5 g). The plasma is then made slightly alkaline with a few drops of 2M sodium hydroxide solution (8 g/dl) and extracted by shaking vigorously with 20 ml of hexane containing a few drops of isoamyl alcohol. Then the layers are separated and the LSD reextracted from the hexane layer by shaking with 2 ml of $0.004M$ hydrochloric acid (1 ml of $1M$ HCl diluted to 250 ml). The LSD is then determined fluorometrically. Two methods have been suggested. In one the excitation wavelength used is 335 nm and fluorescence is measured at 435 nm [112]. In the other the exciting wavelength may be 318 nm with emission at 413 nm if a good spectrofluorometer is available [113]. After measuring the fluorescence against a blank of the hydrochloric acid solution, the sample tube is irradiated at 254 nm for 3 hr. This destroys the fluorescent compounds and the residual fluorescence is measured again. The difference in fluorescence is proportional to the amount of LSD present. A sample of pure LSD must be available for comparison.

A procedure for the determination of LSD by radioimmunoassay (see Chapter 15) has been described by Taunton-Rigby et al. [114]. This method is faster, more specific, and more sensitive than previously used techniques. Picogram quantities of LSD can be measured in human urine, serum, bile, gastric fluid, and other biological fluids without prior extraction.

Identification of Drugs in Dosage Forms and Other Material

Spot Tests

Occasionally the laboratory may be asked to identify a capsule, tablet, powder, or other material that may possibly contain a drug. If the material is a capsule or tablet, the colored plates found in the *Physicians Desk Reference* [108] may be consulted. These illustrate the shape and color of most drugs commercially available in these dosage forms. Also given is a Drug Code Number which is now placed on almost all such products. If the capsule or tablet cannot be positively identified from the charts, or if the material is in a form other than a commercial preparation, some of the following spot tests may be helpful. Most of these tests are for a given drug or a closely related group of drugs. A negative test for one type of drug does not give any information as to what type should be tested for next. Usually, however, one has some idea of what type of drug is most likely present and this is investigated first. The tests given below are derived from a number of sources, principally Carlton and Quittner [115], Kay [2], and Fiorese et al. [116]. The last two references also give tests for a number of other alkaloids not mentioned here. There is also available commercially a kit# containing reagents for the identification of some of the most common drugs of abuse.

DIK-12 Drug Identification Kit, G. Frederick Smith Chemical Co., Columbus, Ohio 43223.

In theory the spot tests could be used to detect the drugs in an extract from urine or gastric contents, but they are not as sensitive as the thin layer chromatography methods. In addition to the spot tests there are in use some microcrystal tests [117] that are very sensitive but require a polarizing microscope for best results. These tests identify the drugs by the color, size, and shape of small crystals formed on a microscope slide by the reaction between the drug and various reagents.

PREPARATION OF SAMPLES

The tests can sometimes be run on the powder from a capsule or the crushed tablet. Often, however, the diluent or excipient in the commercial preparation may interfere with the test. A number of the spot tests use concentrated acids and these could react with the common diluents such as starch or lactose to produce interfering colors. In addition, if the drug is present in only a small amount, a much more sensitive test would result by separating the drug from the diluent. With tablets, the outer coloring may also interfere. Attempts should be made to remove this before crushing the tablet. One can agitate the tablets briefly with water or ethanol to remove the color. A quantity of the powdered material is first shaken with a small amount of methanol and the extract decanted to a small beaker. As some of the drugs may be present as sodium salts which are relatively insoluble in methanol, the residue is then extracted with methanol containing about 1 ml of concentrated hydrochloric acid/dl. The separate extracts are evaporated on a water bath to a small volume. Aliquots of the concentrated extract are then transferred to the depressions in a spot plate and evaporated to dryness. If the sample is vegetable matter suspected of containing cannabis, the dried material may be extracted by shaking with petroleum ether. The extract is then evaporated on a water bath and the test applied on the residue. Gastric washing or large amounts of urine may be extracted with chloroform or chloroform-isopropanol after adjustment of the pH as given in the earlier thin layer chromatography procedure. These extracts may also be evaporated and tested.

REAGENTS

A number of reagents are given below. In some cases the directions call for two different solutions (A and B) for a given test; these are added sequentially to the spot. In other instances the two solutions are mixed just before use, since the mixed reagent is not stable.

1. Concentrated sulfuric acid, reagent grade.

2. Marquis reagent. Prepare fresh as needed by adding 5 drops of 40% formaldehyde to 5 ml of concentrated sulfuric acid.

3. Mecke reagent. Dissolve 0.25 g of selenous acid in 25 ml of concentrated sulfuric acid.

4. Mandelin reagent. Dissolve 5 mg of ammonium vanadate in 5 ml of concentrated sulfuric acid.

5. Froehde reagent. Prepare fresh as needed by dissolving 5 mg of molybdic acid in 5 ml of concentrated sulfuric acid.

6. Nitric-sulfuric acid. Just before use, cautiously mix 2 ml of concentrated nitric acid and 2 ml of concentrated sulfuric acid.

7. Ninhydrin-sulfuric acid.

A. Prepare a saturated solution of ninhydrin in chloroform by shaking about 0.1 g of ninhydrin with 10 ml of chloroform.

B. Concentrated sulfuric acid.

8. Vanillin reagent. Dissolve 0.25 g of vanillin in 15 ml of concentrated sulfuric acid.

9. Cobalt-thiocyanate reagent.

A. Dissolve 4 g of cobaltous chloride in water to make 100 ml.

B. Dissolve 4 g of potassium thiocyanate in water to make 100 ml.

Just before use mix equal volumes of the two solutions.

10. Dille-Koppanyi reagent.

A. Mix 5 ml of isopropylamine in methanol to make 100 ml.

B. Dissolve 0.1 g of cobaltous acetate in 100 ml of methanol containing 0.2 ml of glacial acetic acid.

11. Duquenois reagent. Dissolve 0.5 g of vanillin in 20 ml of 95% ethanol and then add 0.25 ml of acetaldehyde. The reagent is relatively stable but should be discarded if it turns yellow.

12. Dimethylaminobenzaldehyde reagent.

A. Dissolve 12.5 mg of p-dimethylaminobenzaldehyde in 6.5 ml of concentrated sulfuric acid and 3.5 ml of water. The reagent is relatively stable but is best discarded when it turns yellow.

B. Dissolve 1 g of tartaric acid in water to make 100 ml.

13. Mercuric chloride reagent. Dissolve 2 g of mercuric chloride in 100 ml of 50% ethanol.

14. Iodoplatinate reagent. Dissolve 0.5 g of potassium iodide in 15 ml of water and add 0.5 ml of 5% platinum chloride solution.

PROCEDURE

The spot tests can in some instances be applied directly to the powdered material, particularly for a quick test, but an extraction of the sample will eliminate many interferences. To the powdered material or an aliquot of the extract evaporated to dryness in a spot plate depression add 1 or 2 drops of the reagent or two reagents sequentially as indicated. Use a small glass rod to mix the reagent with the residue. Note the immediate colors produced and the changes in color with time. The colors may vary somewhat from those given here and, if possible, known specimens should be treated similarly. In those reagents containing a high concentration of sulfuric acid, a yellow-brown color may be produced by interfering substances and should not ordinarily be called a positive test. A few of the tests given are done in small test tubes rather than on the spot plate. In these instances a portion of the extract is evaporated in the test tube with the aid of a stream of air and the test applied to the residue.

Table 18-3 gives the results of the spot tests applied to a number of different drugs.

Additional Spot Plate Tests

To a small sample of powdered material or extract residue add a few drops of each of the two Dille-Koppanyi reagents (reagent 10) and mix. A red, blue, or violet color indicates the presence of barbiturates, glutethimide, or diphenylhydantoin.

To a small amount of material on a spot plate add a few drops of the cobalt thiocyanate reagent (reagent 9) and mix. A blue color indicates the presence of cocaine, meperidine, methadone, or an opium alkaloid.

Additional Test Tube Tests

To a small portion of a sample in a small test tube add 2 ml of Duquenois reagent (reagent 11). Shake for about 1 min; then add 2 ml of concentrated hydrochloric acid and mix. Note the color changes during the next few minutes. Add 2 ml of chloroform and shake. Allow the layers to separate. The presence of cannabis is indicated by a green to blue to blue-violet color change.

To a small portion of the sample in a test tube add 1 ml of water, 1 ml of reagent 12A, and 1 ml of reagent 12B. Mix and allow to stand for 10 min. A lavender or blue-red color indicates the presence of LSD or related amines.

Place a small amount of sample in a test tube; add dropwise 1 ml of the mercuric chloride reagent (reagent 13). If a red color is produced almost immediately, atropine or homatropine is present.

Table 18-3. Color Reactions of Drug Spot Tests with Various Reagents[a]

Drug	1 Sulfuric Acid	2 Marquis Reagent	3 Mecke Reagent	4 Mandelin Reagent	5 Froehde Reagent	6 Nitric-Sulfuric Acid	7 Ninhydrin-Sulfuric Acid	8 Vanillin Reagent	14 Iodoplatinate Reagent
Amitriptyline	—	—	—	—	—	—	R-G-B	—	B
Amphetamines	VBr-R(H)	O-OBr	IR	O-R-Br	R-Y	—	—	—	B
Codeine	—	BV	G-B	G-B	G-R-Br	—	—	—	B
Diazepam	—	RV	—	—	—	—	—	—	—
Ethinamate	—	R	—	—	—	—	—	—	—
Heroin	—	IV-V	B-G	IG	IV-GrG	—	—	—	B
Imipramine	IY	—	—	—	—	G	V	—	—
Meperidine	—	—	O-Br	IBr	—	—	—	—	—
Meprobamate	—	—	—	—	—	—	—	R	—
Mescaline	Y-Br-O	—	Y-G-B-Br	GBr	Y-dBr	—	—	—	—
Methadone	Y-R	—	IR	dG-Br	O-GrG-G	—	—	—	B
Morphine	—	RV	BG	B-V-BBk	RV-B-f	R	—	—	—
Phenothiazines	—	V	—	R	—	—	—	—	B
Propoxyphene	—	V	—	—	—	—	—	—	—
Reserpine	—	—	B-G-Y	—	—	—	B	—	—
Salicylate	—	—	—	G	—	—	—	—	—
4-Methyl-2,5-dimethoxy-α-methylamphetamines (STP)	—	—	YG-O	—	—	—	—	—	—

[a] R, red; O, orange; Y, yellow; G, green; B, blue; V, violet; Br, brown; Gr, gray; Bk, black; l, light or faint color; d, dark or intense color. A combination of two symbols indicates a mixed color, e.g., BG, blue-green. When several colors are given in order, as R-B-V, this indicates that the initial color is followed by the others in turn, usually within a minute or so. f, color fades fairly rapidly. (H), color is produced only when the spot plate is warmed.

References

1. Sunshine, I. [Ed.]. *Manual of Analytical Toxicology.* Cleveland: Chemical Rubber Co., 1971.
2. Kaye, S. *Handbook of Emergency Toxicology.* 3rd ed. Springfield, Ill.: Thomas, 1970.
3. Sunshine, I. *Stand. Methods Clin. Chem.* 3:1, 1960.
4. Sunshine, I., and Nenad, R. *Anal. Chem.* 25:653, 1963.
5. Powell, F. J. N., and Ellam, E. *Clin. Chim. Acta* 10:472, 1964.
6. Feldstein, M., and Klendshoj, N. C. *Forensic Sci.* 2:39, 1957.
7. Jones, D., Gerber, L. P., and Drell, W. *Clin. Chem.* 16:402, 1970.
8. Brink, N. C., Bonnischsen, R. K., and Thoerell, H. *Acta Pharmacol. Toxicol.* 10:233, 1954.
9. Stiles, D., Batsakis, J. G., Kremers, B., and Briere, R. O. *Am. J. Clin. Pathol.* 46:608, 1966.
10. Leric, H., Kaplan, J. C., and Brown, C. *Clin. Chim. Acta* 29:523, 1970.
11. Roos, K. J. *Clin. Chim. Acta* 31:285, 1971.
12. Karnitis, L., and Porter, L. J. *J. Forensic Sci.* 17:318, 1972.
13. Gupta, R. N., Galdenzi, S., and Keane, P. M. *J. Forensic Sci.* 17:453, 1972.
14. Siek, T. J. *J. Forensic Sci.* 17:334, 1972.
15. Savory, J., Sunderman, F. W., Jr., and Roszel, N. O. *Clin. Chem.* 14:132, 1968.
16. Kaye, S., and Cardona, E. *Am. J. Clin. Pathol.* 52:577, 1969.
17. Morgan, W. H. D. *J. Forensic Sci.* 5:15, 1964.
18. Redetzki, H. M., Koerner, T. A., and Hughes, J. R. *Clin. Toxicol.* 5:343, 1972.
19. Glasser, L., Sternglanz, P. D., Combie, J., and Robinson, A. *Am. J. Clin. Pathol.* 60:695, 1973.
20. Kaye, S. *Am. J. Clin. Pathol.* 14:83, 1944.
21. Kaye, S. In I. Sunshine [Ed.], *Manual of Analytical Toxicology.* Cleveland: Chemical Rubber Co., 1971. Pp. 30, 39, 212.
22. Gettler, A. O., and Kaye, S. *J. Lab. Clin. Med.* 35:146, 1950.
23. Mehr, C. *Z. Anal. Chem.* 94:161, 1933.
24. Gettler, A. O. *Am. J. Clin. Pathol.* 2:76, 1939.
25. Leifheit, H. C. *Stand. Methods Clin. Chem.* 3:23, 1960.
26. Helwig, C. A. *Am. J. Clin. Pathol.* 13:96, 1943.
27. Kaye, S. *Bol. Asoc. Med. PR* 60:602, 1968.
28. Kaye, S. In I. Sunshine [Ed.], *Manual of Analytical Toxicology.* Cleveland: Chemical Rubber Co., 1971.
29. Hatch, N. R., and Ott, W. L. *Anal. Chem.* 40:2085, 1968.
30. Eathje, A. O. *Am. Ind. Hyg. Assoc. J.* 30:126, 1969.
31. Flanagan, R. J., and Withers, G. *J. Clin. Pathol.* 25:899, 1972.
32. Fiereck, E. A., and Tietz, N. W. *Clin. Chem.* 17:1024, 1971.
33. MacGee, J. *Clin. Chem.* 17:587, 1971.
34. Street, H. V. *Clin. Chim. Acta* 34:357, 1971.
35. Jatlow, P. *Am. J. Clin. Pathol.* 59:167, 1973.
36. Zak, B., and Williams, L. A. In F. W. Sunderman [Ed.], *Laboratory Diagnosis of Diseases Caused by Toxic Agents.* St. Louis: Green, 1970.

37. Bjerre, S., and Porter, C. J.   *Clin. Chem.* 11:137, 1965.
38. Richterich, R.   *Clin. Chim. Acta* 3:183, 1958.
39. Baer, D. M.   *Am. J. Clin. Pathol.* 44:114, 1965.
40. Wuth, O. J.   *J.A.M.A.* 88:2013, 1927.
41. Sunshine, I.   In I. Sunshine [Ed.], *Manual of Analytical Toxicology.* Cleveland: Chemical Rubber Co., 1971.
42. Gettler, A. O., and Friemuth, H. C.   *Am. J. Clin. Pathol.* 7:79, 1942.
43. Harper, P. V.   *J. Lab. Clin. Med.* 40:634, 1952.
44. Beringer, H., and Smith, R.   *Clin. Chem.* 5:127, 1959.
45. Williams, L. A., Linn, R. A., and Zak, B.   *Am. J. Clin. Pathol.* 34:324, 1960.
46. Bessman, S. P., and Layne, E. C., Jr.   *J. Lab Clin. Med.* 45:159, 1955.
47. Mecherly, P. A., Lilly, A., and Whitman, N. E.   *Am. Ind. Hyg. Assoc. Q.* 18:161, 1957.
48. Wintner, P.   *J. Med. Lab. Technol.* (Lond.) 21:281, 1964.
49. Mitchell, D. G., Ryan, F. J., and Aldous, K. M.   *At. Absorption Newslett.* 11:120, 1972.
50. Zintnerhofer, L. J. M., Jatlow, P. I., and Papiano, A.   *J. Lab. Clin. Med.* 78:644, 1971.
51. Hessel, D. W.   *At. Absorption Newslett.* 7:55, 1968.
52. Kopito, L., and Schwachman, H.   *J. Lab. Clin. Med.* 70:326, 1967.
53. Zurlo, N., Griffine, A. M., and Colombo, G.   *Anal. Chim. Acta* 47:203, 1969.
54. Lyons, M., and Quinn, F. E.   *Clin. Chem.* 17:152, 1971.
55. Ediger, R. D., and Goleman, R. L.   *At. Absorption Newslett.* 11:33, 1972.
56. Olsen, E. D., and Jatlow, P. I.   *Clin. Chem.* 18:1312, 1972.
57. Hildebrand, D. C., Koityrohann, S. R., and Pickett, E. E.   *Biochem. Med.* 3:437, 1970.
58. Delves, H. T.   *Analyst* (Lond.) 95:431, 1970.
59. Hwang, J. Y., Ulluci, P. A., and Mokeler, C. J.   *Anal. Chem.* 45:795, 1973.
60. Kubasik, N. P., Volosin, M. T., and Murray, M. H.   *Clin. Biochem.* 5:266, 1972.
61. Norval, E., and Butler, L. R. P.   *Anal. Chim. Acta* 58:47, 1972.
62. Rosen, J. F., and Trinidad, E. E.   *J. Lab. Clin. Med.* 80:567, 1972.
63. Lubran, M.   In F. W. Sunderman [Ed.], *Laboratory Diagnosis of Diseases Caused by Toxic Agents.* St. Louis: Green, 1970.
64. Davidow, B., Searle, B., and Chan, W.   In I. Sunshine [Ed.], *Manual of Analytical Toxicology.* Cleveland: Chemical Rubber Co., 1971.
65. Davis, J. R., and Andelman, S. L.   *Arch. Environ. Health* 15:53, 1967.
66. Haeger-Aronsen, G.   *Scand. J. Clin. Lab. Invest.* [Suppl.] 12:47, 1960.
67. Mauserall, D., and Granick, S.   *J. Biol. Chem.* 219:435, 1956.
68. Tomokuni, K., and Ogata, M.   *Clin. Chem.* 18:1534, 1972.
69. Thomas, B. H., Solomonraj, G., and Coldwell, B. B.   *J. Pharm. Pharmacol.* 25:201, 1973.
70. Lever, M., and Powell, J. C.   *Biochem. Med.* 7:203, 1973.
71. Chirigos, M. A., and Udenfriend, S.   *J. Lab. Clin. Med.* 54:769, 1959.
72. Saltzman, A.   *J. Biol. Chem.* 174:399, 1948.
73. Williams, L. A., Linn, R. A., and Zak, B.   *J. Lab. Clin. Med.* 53:156, 1959.

74. Stevenson, G. W. *Anal. Chem.* 32:1522, 1960.
75. Routh, J. I., Paul, W. D., Arrendondo, E., and Dryer, R. L. *Clin. Chem.* 2:432, 1956.
76. Brodie, B. B., Udenfriend, S., and Coburn, A. F. *J. Pharmacol.* 80:114, 1944.
77. Trinder, P. *Biochem. J.* 57:301, 1954.
78. Bratton, A. C., and Marshall, E. K. *J. Biol. Chem.* 128:537, 1939.
79. Smalley, A. E. *Bull. Inst. Med. Lab. Technol.* 14:109, 1949.
80. Reider, H. P. *Klin. Wochenschr.* 39:813, 1961.
81. Sardi, V. *Clin. Chem. Acta* 2:134, 1957.
82. Sardi, V., and Bonavita, V. *Clin. Chem.* 3:728, 1957.
83. Forrest, I. S., and Forrest, F. M. *Clin. Chem.* 6:11, 1960.
84. Forrest, F. M., Forrest, I. S., and Mason, A. S. *Am. J. Psychiatry* 116:549, 1959.
85. Brownstein, H., and Roberge, A. R. *Clin. Chem.* 12:844, 1966.
86. Levy, A. L., and Katz, E. M. *Clin. Chem.* 16:840, 1970.
87. Pybus, J. N., and Bowers, G. N., Jr. *Clin. Chem.* 16:139, 1970.
88. Little, B. R., Platman, S. R., and Fieve, R. R. *Clin. Chem.* 14:1211, 1968.
89. Amidisen, A. *Scand. J. Clin. Lab. Invest.* 20:104, 1967.
90. Pybus, J. N., and Bowers, G. N., Jr. *Stand. Methods Clin. Chem.* 6:189, 1970.
91. Cooper, R. G., Greaves, M. S., and Osen, G. *Clin. Chem.* 18:1343, 1972.
92. Dolon, S. J. *Anal. Instrum.* 10:185, 1972.
93. Mule, S. J. *J. Chromatogr.* 55:255, 1971.
94. Reid, R. W., Katzen, R., and Clinger, J. M. *Am. J. Clin. Pathol.* 53:462, 1970.
95. Holmstedt, G., and Lindgren, J. E. *Fresenius' Z. Anal. Chem.* 261:291, 1972.
96. Milne, G. W. A., Fales, H. M., and Axenrod, T. *Anal. Chem.* 43:1815, 1971.
97. Decker, W. J. *Clin. Toxicol.* 5:365, 1972.
98. Baden, M. M., Valanju, N. N., Verma, S. K., and Valanju, S. N. *Am. J. Clin. Pathol.* 57:43, 1972.
99. Broich, J. R. M., Hoffman, D. B., Andryauskas, S., Galante, L., and Umberger, C. J. *J. Chromatogr.* 60:95, 1971.
100. Kaistha, K. K., and Jaffe, J. H. *J. Chromatogr.* 60:83, 1971.
101. Sunshine, I. *Am. J. Clin. Pathol.* 40:576, 1963.
102. Bastos, M. L., Jukofsky, D., Saffer, E., Chedekel, M., and Mule, S. J. *J. Chromatogr.* 71:549, 1972.
103. Weissman, N., Lowe, M. L., Beattie, J. M., and Demetrious, J. A. *Clin. Chem.* 17:875, 1971.
104. Juselius, R. E., and Barnhart, F. *Clin. Toxicol.* 6:53, 1973.
105. Dole, V. P., Kim, W. K., and Eglitis, I. *J.A.M.A.* 198:349, 1966.
106. Ho, I. K., Loh, H. H., and Way, E. L. *J. Chromatogr.* 65:577, 1972.
107. Kaistha, K. K., and Jaffe, J. H. *J. Pharm. Sci.* 61:679, 1972.
108. *Physicians' Desk Reference.* Oradell, N. J.: Medical Economics. Published yearly.
109. *Merck Index*, 8th ed. Rahway, N. J.: Merck, 1967.
110. Salascheck, M., Matte, A., and Seifert, K. *J. Chromatogr.* 78:393, 1973.
111. Kisser, N. *Arch. Toxikol.* 29:331, 1972.
112. Noirfalise, N. *J. Pharm. Belg.* 23:387, 1963.
113. Upshall, D. G., and Wailling, D. G. *Clin. Chim. Acta* 36:67, 1973.
114. Taunton-Rigby, A., Sher, S. E., and Kelley, P. R. *Science* 181:165, 1973.

115. Carlton, R. F., and Quidtner, H.   In F. W. Sunderman, Jr. [Ed.], *Manual of Procedures for Laboratory Diagnosis of Diseases Caused by Toxic Agents*. Philadelphia: Association of Clinical Scientists, 1968.
116. Fiorese, F. F., Forster, G. F., and Brown, M. L.   In M. Stefanini [Ed.], *Progress in Clinical Pathology*, vol. 2. New York: Grune & Stratton, 1969.
117. Fulton, C. G.   *Modern Microcrystal Tests for Drugs*. New York: Interscience, 1969.
118. Frings, C. S.   *Crit. Rev. Clin. Lab. Sci.* 4:353, 1973.

19. Automation

In the past decade the number and types of apparatus for automated and semi-automated analysis in the clinical laboratory have markedly increased. It is difficult for any individual to assess the relative merits of the different instruments and decide which might be the most satisfactory for a given task. In this chapter we will attempt to give some information concerning the major instruments available at the time of writing. Obviously no individual has seen all the instruments in actual laboratory operation or has had practical experience with more than a few of them. The material presented is based on information furnished by the various manufacturers, a series of survey articles presented in *Lab World* by Alpert [1], and other information as referred to in the text.

In any chemical determination the following steps are generally required:

1. Sample pick-up and identification
2. Dilution and addition of reagents
3. Heating or incubation
4. Transfer to photometer cuvette
5. Choice of wavelength and photometric reading
6. Calculation of results
7. Printout or other permanent record of results

The above list does not include the calibration procedure, which in a fully automated system would require only the entry of the calibration values into the system with no further adjustments by the operator. This is possible with many systems if they are connected to a computer but ordinarily in most automated systems some manual adjustments are necessary.

The simplest semiautomated system includes one of the automatic pipettors and diluters and a spectrophotometer reading in concentration units with a printout. This is a relatively simple system that could be assembled from a number of separate components on the market. Such a system would involve a minimum of manual manipulations: pipetting and dilution of the sample, transfer to incubation bath, and transfer to cuvette for reading. Such a system is not included in the following discussion which deals in general only with complete systems, each by a single manufacturer.

The automated systems may be divided into several classes on the basis of their method of operation. One that has been in use for many years is the continuous-flow analyzers as exemplified by the various modules from Technicon,*

* Technicon Instruments Co., Tarrytown, N.Y. 10591.

the only manufacturer of this type of system. The different models of this manufacturer will be discussed later. A second type, more recently introduced, involves
the use of centrifugal force in the mixing of the reagent and sample and measurement of the color in the rotor. These centrifugal systems are produced by two
manufacturers and will be discussed in detail later. The most common type of
automated analyzer is the discrete sampler system in which the individual samples
are handled in a manner similar to manual operation; the samples are pipetted
into tubes or other containers, reagents added, the mixture incubated or heated as
required, further reagent added if necessary, and the measurement made at the
proper wavelength, the results then being calculated from the proper calibration
curve. All this is done automatically but in theory differs little from the manual
method.

A considerable number of different discrete analyzer systems are on the market,
differing in size (which to some extent determines the total number of samples
that can be processed in a given time), versatility (making different types of
determinations), and ease of operation. Many of these will be discussed in detail
later.

Discrete analyzer systems that follow the manual procedures most closely have
the advantage that new or improved methods can be more readily adapted to the
automated operation with these systems, since most new methods are tried out
initially manually. The available discrete systems vary somewhat in their degree
of automation. The simpler ones require that the sample be added with a
manually operated pipettor and sometimes that the reagent also be added
manually. Some of the automated and semiautomated systems are programmed
for kinetic determinations and are used almost exclusively for enzyme work, although some colorimetric reactions for other substances can be adapted to a
kinetic mode. The smaller discrete analyzers determine only one substance at a
time on a series of samples at rates as high as 120/hr and vary in the time required for a change-over to a different determination. The larger discrete analyzers simultaneously or sequentially determine a number of different substances
in one sample. These are more suitable for survey and screening tests.

In the discussion of the individual instruments we will depart somewhat from
the order in which they have been first mentioned and begin with those made
particularly for enzyme analyses. These generally use a single wavelength of 340
(or 366) nm which results in some simplifications in the instrumentation. Most
instruments also use kinetic rate measurements over short lengths of time and
thus have photodetector systems that read to 0.0001 absorbance unit or less.

Guilford 300 Series Automatic Enzyme Analyzer†

This system comprises a number of modules including a spectrophotometer and printout which can be used separately for other analyses. The complete system includes also a reagent dispenser, automated transfer to photometer cuvette, and printout of absorbance change. The temperature of the cuvette can be regulated to 30°, 32°, or 37°C. The serum or other samples must be added to the tubes with a pipet, usually of the Eppendorf type. Included also is a calculator for obtaining the printout in any desired units. The various modules can be used separately for nonautomated analyses. Since the system is not designed specially for automated work, it is not as convenient for large-scale automated procedures. It may be satisfactory in a smaller laboratory where the spectrophotometer can be used for other nonautomated tests.

Beckman Enzyme Activity Analyzer, System TR‡

This is a small, moderately priced system for the automated determination of enzymes by kinetic methods. A turntable with a capacity of 20 samples is supplied. With this 20 to 45 samples/hr can be processed depending upon the time required for the rate reaction to stabilize for a given enzyme. The reactions can be run at 25°, 30°, or 37°C, though for 25°C an external source of cooling water may be necessary if the ambient temperature is higher than this. The system has a digital readout and direct printout with identification number. Provision is also made for the attachment of a recorder if desired. A quartz iodine tungsten lamp is used with a narrow-beam grating spectrophotometer operating in the double-beam mode. The cuvette has an accurate light path of 1.00 cm so that the readout can be calculated directly in international units from known absorptivities. The operation with stat samples from the standby position is very simple and the results are usually obtained within 5 min. Although the manufacturer supplies reagents for only the most commonly determined enzymes (AST, SLT, LDH, CPK, HBDH, and alkaline phosphatase), adaptation can easily be made for any other enzyme reaction read at 340 or 405 nm—this would include a great majority of the enzymes.

Photovolt ERA-1 Enzyme Rate Analyzer§

This instrument is specifically designed for kinetic enzyme assays. It has a fixed mercury light source and thus takes measurements only at 340 nm. Only filters

† Guilford Laboratories, Inc., Oberlin, Ohio 44074.
‡ Beckman Instruments, Inc., Clinical Instruments Div., Fullerton, Calif. 92634.
§ Photovolt Corp., New York, N.Y. 10010.

are used with a very simple optical system. The instrument is usually operated in the double-beam mode. The electronic measuring system is designed to make measurements of small differences in absorbances over short time intervals. An absorbance difference of 0.02 unit causes a full scale deflection in the normal mode, but the initial absorbance can be set to any position in the range from zero to 2.5 absorbance units. An accessory is provided for maintaining the cuvette compartment at a constant temperature, although some users apparently prefer to make readings at the ambient temperature and then use a correction factor to the standard temperature. The instrument is designed to take readings at 15-sec intervals and then calculate the average absorbance change, which makes the operation quite rapid. The instrument does not have a precise cuvette and wavelength specification so that the calibration point as established by the measurement of an NADH solution at the factory must be used for the calculations unless the absorbance of an NADH solution in the instrument can be compared with that measured on a precise spectrophotometer.

Sherwood Digecon System||

This is a semiautomated instrument that can be used for measurement of absorbances directly as well as for kinetic rate measurements. The system is semiautomatic in that the serum (or other sample) and reagent must be introduced manually into the reaction chamber, using syringe-type pipets for the reagent and plunger-type automatic pipets for the sample. The bath temperature, wavelength, and reaction time or rate are previously set on the appropriate dials. Thus by pressing the start button the reaction is carried out, the spectrophotometer readings made, and the results calculated and printed out in the appropriate concentration or enzyme units. The cuvette and reaction chamber are automatically washed and dried and are then ready for the next sample. The determination may take only a few minutes, depending upon the particular reaction involved. To change to another determination it is only necessary to make changes in the appropriate settings for wavelength, bath temperature, reaction timer sequence, and factor for conversion from absorbance to concentration units. The printout gives the sequence number and the concentration in the appropriate units. Digital readouts show not only the printout figures, but also the actual absorbance change. The instrument has a very stable light source and a spectrophotometer having a range of 270 to 700 nm, a band pass of 5 nm, and an absorbance range of zero to 2.5 absorbance units. Although the instrument is

---

|| Sherwood Medical Industries, St. Louis, Mo. 63103.

only semiautomated, it can be adapted to almost any manual chemical test. It has a good spectrophotometer and stable light system for accurate measurements. The apparatus has two reaction chambers so that two samples can be set up sequentially and both read within a period of 3 min. A good technician can run from 30 to 40 samples/hr.

Beckman Discrete Sample Analyzer—DSA564#

This system is a completely automated discrete sample analyzer capable of running almost any test that has been performed manually. The photometer uses a number of interference filters covering the range of 340 to 700 nm, with a nominal band pass of 15 nm in the visible range and 20 nm at 340 nm. The filter carrier has provision for only four different filter pairs which can be changed automatically; when other tests requiring different filters are run, the filters must be changed manually. The sample may be automatically pipetted and diluted into one of four reaction cups. Thus either four different tests may be run simultaneously, or two determinations run each with a serum blank. The instrument is unique in that it has a provision for automatically preparing a protein-free filtrate from a sample and carrying the filtrate through the analytical procedure. This feature is not often used since almost all discrete automated procedures are designed for use with serum or plasma directly. This instrument has a rather sophisticated computer incorporated with it (PDP8E). The computer monitors all operations of the system, calculates the results, and generates the printout. It also performs the calibrations automatically and corrects for the baseline drift. Many different tests, including the most common ones, have been adapted to the instrument. The enzyme analyses are made by a two-point method as the instrument is not adapted for short-time kinetic measurements. Four different tests at a rate of 60/hr is equivalent to 240 tests/hr, a good output. The changeover from one set of tests to another is rapid, usually taking only about 10 min or less. The reactions are carried out in small disposable plastic containers which are discarded at the end of the run after an aliquot of the final solution has been pipetted into the colorimeter cuvette.

Abbott ABA-100*

This is a compact automated system for the determination of enzymes by kinetic methods and other chemical determinations by the usual endpoint method. The instrument uses single-beam optics but compensation for turbidity and other

Beckman Instruments, Inc., Clinical Instruments Div., Fullerton, Calif. 92634.
* Abbott Scientific Products Div., Abbott Laboratories, South Pasadena, Calif. 91030.

interference is made by reading at two different wavelengths. One wavelength is that of an absorbance peak and the other is a suitable wavelength where the absorbance by the substance being determined is small. The disposable 32-cell multiple cuvette is used for the reaction chamber and as cuvette. The use of the two wavelengths is said to compensate for the slight differences in the light path in the cuvettes. The spectrophotometer output can be read to 0.0001 absorbance unit. The absorbance may be displayed on a digital readout and a small internal computer calculates the result in terms of the desired units after calibration. The result is then printed out on a tape together with a sequential identification number. The calibration factor must be set manually for each run but this is simple. The samples (generally 5 or 10 μl) are automatically pipetted from the sample tube into the cuvette, followed by the reagent (generally 250 to 500 μl). The cuvettes pass through a water heating bath which can be set at either 30° or 37°C, and then to the spectrophotometer. Generally the speed of rotation of the cuvette carrier is such that readings are taken 5 min after the addition of the reagent. For a chemical reaction that is complete within this time the output is thus 32 samples in 10 min. During this time a second carrier with sample tubes can be prepared. This can be substituted for the other carrier within a few seconds. For reactions that do not go to completion in 5 min or for enzyme reactions in which several readings are made, a longer time is required, but even so an output of 120 samples/hr (including standards and controls) could be readily attained. The change to a different procedure with different reagents can be made within a very few minutes. The instrument uses true micro methods and the small amounts of reagents used make it economical for enzyme determination. A considerable number of tests have been developed for the instrument, and others are being added from time to time.

Instrumentation Laboratories Clinicard Analyzer 368†

This is not a completely automated instrument but has some features that make it more rapid and foolproof than a simple reagent kit system. The reagents are supplied in prepackaged small plastic blocks which also serve as cuvettes. For some enzyme reactions dry tablets are used which are added to the cuvette block together with the required amount of water. The sample is added manually with an aspirator dispensor. Each block contains three cuvette wells so that sample, blank, and standard can be run at the same time. After addition of reagent and sample, the cuvettes are placed in the appropriate heating blocks (two are provided at 37° and 90°C), then removed, and the absorbance read. All the steps

† Instrumentation Laboratories, Lexington, Mass. 02173.

are controlled by a mini-computer, which is programmed by an inserted punched card. A separate card is used for each determination. The card gives the proper computer instructions to regulate the amount of sample picked up by the aspirator, the time of incubation before an audible signal is sounded, the wavelength to be used for the spectrophotometer reading, and the calibration factors necessary for calculation of the results. The results are then displayed as a digital readout. If the result is off scale, an error signal will flash. A number of other punched cards are also available for monitoring the spectrophotometer and some of the electronic circuits. When these various cards are inserted in the instrument, certain specified figures should appear on the readout. If incorrect figures appear, this indicates a fault in that portion of the instrument as indicated by the card. The instrument is relatively easy to operate and, although not completely automated, may be satisfactory for stat or back-up work. For most tests the reagents are supplied in the cuvettes and only these can be used. Thus one cannot prepare his own reagents for even the simpler tests. The average cost is around $0.75 for a nonenzyme test, with higher cost for the enzyme tests. For an occasional stat work this cost is not so important, but for large-scale routine work it should be considered.

duPont ACA Automatic Clinical Analyzer‡

This is a fully automated instrument that has a number of distinctive features not found in any of the instruments mentioned in this chapter. The system is built around clear flexible plastic packs that are expendable; they serve as reagent containers, reaction vessel, and photometer cuvette. The sample is injected into the pack either directly or after passing through a small column to remove interfering substances. The column may be ion exchange resin, dextran polymer gel, or small porous glass beads coated with protein precipitating reagents. The pack also contains the reagents in separate compartments. Under processing, the compartments are broken down so that the reagents mix with the sample. All the reagents may be mixed at the same time or in sequence at different times. When the reaction is complete, the pack passes to the photometer module where the portion of the pack containing the liquid is compressed to form an optical cell of exact light path. After photometer reading, the pack is discarded. The pack for each test is coded so that the integral computer can regulate the treatment the pack receives, the wavelength for reading, and the calibration factor for calculation. An identification card is filled out and placed in the apparatus on the sample cup. This is followed by separate packs for all the tests desired on the

‡ Instrument Products Div., E. I. duPont de Nemours and Co., Wilmington, Del. 19898.

particular sample. At the end of the run for this sample a copy of the identification card together with the results of all the tests ordered is printed out. Since the reagent packs are all self contained, any test (for which reagents are available) can be run on any sample. The samples need not be placed in the instrument in any particular order, the only requirement being that reagent packs for all the tests desired on a given sample immediately follow this sample. Stat tests can be readily introduced at any time. The time required for the result to be printed out after the sample enters the instrument is about 7 min. The instrument is generally left in a standby state at all times and requires only the placing of the sample container and the reagent pack in position for an immediate stat run. Since the reagents are contained in the special packs, one cannot use other reagents and can use only the method specifically developed for the instrument. The cost per test is then relatively high compared with the continuous-flow methods, for example. The instrument is not well adapted for micro work, generally requiring 100 to 250 μl of serum per test.

Vickers Multi-Channel 300 System§

This is a complete automated discrete sampling system which in many instances can be adapted to micro methods using around 25 μl of serum per test. The basic unit is a two-channel automated analyzer, in which the sample and reagents are added to the reaction tubes at the appropriate times, mixed, incubated, and the color formed read after the required time. The output rate will be 150 of each of two different tests/hr or 150 tests/hr if a blank is needed for each test. A number of these basic units can be combined with a single sample container with a positive identification system and a single computer to process a much larger number of different tests. As many as ten basic units can be combined, giving a maximum of 10×300 tests/hr. Within certain limits not all these basic units need be installed at one time; a few can be used at first and more added as required. The complete system uses a PDP8/L computer and teletype printout of the results. As with other computer-controlled systems, the calibration is done automatically. The basic units can be operated separately without the computer interface but with a plain tape printout. In this case the calibrations must be done manually. This instrument is manfactured in England by Vickers, Ltd. and has not as yet been widely used in this country. Since it is a discrete sampling system closely simulating manual methods, reagents similar to those in manual methods are used and in most cases can be made up by the laboratory

§ Medi-Computer Corp., Englewood Cliffs, N.J. 07632.

if desired. Aqueous standards are usually used, except for enzymes. As with most discrete analyzers, serum is used directly for all tests.

AcuChem Microanalyzer||

This is a complete discrete sample automated analyzer which is a true micro system, performing up to 16 different analyses on 200 μl of serum. The sample aspiration and reagent dispensing are not done with syringe pumps as in many other systems but with a system of pneumatic valves and a sensor device that allows a definite volume of liquid to be held in the measuring tube. After aspiration, the sample is diluted with water and aliquots delivered into separate tubes for the different analyses. The tubes are on a rotating turntable. Some pass through a bath at 25°C, some through a bath at 37°C, and others through a bath at 90°C, depending upon the reaction involved. The reagents are added automatically as required during the cycle. The absorbances are measured in quartz flow-through cells of small volume using fiber optics. After the optical measurements, the tubes are washed and dried automatically and progress around in the turntable to the starting point for the next set of determinations. Each analysis is done at the rate of 60 samples/hr; with 15 channels this is equivalent to 900 tests/hr. The instrument uses a computer in a separate console for monitoring the operation, calculation of results, and forming a printout. The present modules in operation have capabilities for 15 channels but later production models will be expanded to at least 20 channels. The special racks carrying the sample tubes (in line and not on a turntable) have identification tables giving the rack number and the sample number in the rack. The tubes containing control or calibration samples are placed in special racks to identify them to the computer. The calibration is generally done on a high and a low standard and periodically a mid-range standard is added which the computer uses to make adjustments in the calibration curve as required. Presumably most of the reagents would be similar to those used in manual methods and could be made in the laboratory if desired. Because of the microsystem, only relatively small amounts of reagents are used.

Programachem 1040#

This instrument is a very rapid automated discrete analyzer system. It will analyze for only one constituent at a time but at an average rate of 350/hr. The sample and reaction tubes are on separate turntables rotating in opposite direc-

|| Ortho Diagnostic Instruments, Raritan, N.J. 08869.
Fisher Scientific Co., Pittsburgh, Pa. 15219.

tions to bring the separate tubes in close proximity for sampling. The reaction tubes are not reused but are automatically discarded and new ones must be added to the turntable as required. The changeover to a new test is very simple. All the reagents for the different tests are stored in the cabinet of the instrument. To make the changeover, the operator merely sets the wavelength and calibration dials to the correct values, then inserts a punched card into the instrument. This automatically selects the correct reagents and sample and reagent volumes. The instrument is now ready for the new analysis; one needs only to check the calibration factor with standards. For each test the internal computer automatically calculates the results and gives a printout recording the test procedure number (each different analytical procedure being assigned an identification number), the sample sequence number, and the analytical result in the desired units. The sample tubes may be kept in the carrier for several different analyses on the same samples. Provision is made so that if in a given run it is desired not to make the analysis on some particular samples, these need not be removed from the carrier but by setting switches the desired samples will not be picked up and processed for the run. With this instrument the usual reagents, similar to those for manual methods, are used and these could in most cases be prepared in the laboratory. Even with the purchased reagents, the cost per test (other than enzyme tests) is usually less than 10 cents. The instrument appears to have good capabilities for large-volume runs on a single analysis.

Hycel Mark XVII and Mark X*

These are multichannel discrete analyzer systems. As indicated by the designations, the former has 17 channels and the latter 10. The output capacity is 40 samples/hr or 680 and 400 tests/hr, respectively. The serum samples are automatically pipetted into a series of tubes which are carried along on a linear track. The reagents are added at the proper time and the tubes then pass through the desired heating or incubation cycle. As in all the parallel analysis systems, although the tubes for different tests on the same serum may be subject to different incubation or heating temperatures, the total elapsed time must be the same for all. The reaction mixture is then pipetted into the colorimeter cuvette and the measurements of absorbance made. The tubes are then washed and dried automatically to be ready for the next cycle. The results are calculated automatically from the photometer readings and entered through a recorder on a chart paper. This latter contains scales of values for each determination, with the normal ranges designated. A line drawn by the recorder pen at the proper position on the

* Hycel, Inc., Houston, Tex. 77036.

scale indicates the result of the analysis. Thus all the results obtained on one sample are shown on one chart together with the normal values. By proper manipulation of the control buttons, any desired determinations can be omitted from the series for a given serum without affecting the other samples. Calibration is done with serum containing known values of all the constituents being determined and adjusting the recorder input to indicate the proper values on the scale. Once the instrument is set up to do a given battery of tests, it is not simple to replace one test by another since this would require, among other things, a different chart paper. This instrument is the only one of the multichannel units that offers a wet ash determination of PBI as one of the channels.

Centrifugal Analyzers

In contrast to the systems already described, three discrete sampler instruments operate on an entirely different principle—the centrifugal principle. The sample and reagents are placed in wells in the inner part of the rotor and the centrifugal force mixes the sample and reagents and finally forces the mixture into an optical cell at the periphery of the rotor where the absorbance measurements are made. The three instruments that operate on this principle are the CentrifiChem,[†] GEMSAEC,[‡] and Rotochem.[§] There are certain differences between them but they all operate on similar principles. In the simplest operating mode the samples and reagents are pipetted into separate wells in the inner rotor (only this part is removable for loading; the outer part of the rotor which contains the optical cells remains fixed). The loaded rotor is locked in place and, assuming that all the parameters, such as optical wavelength to be used for measurement, calibration factors for calculation, and time of spinning, have already been set, one merely pushes the start button. The sample and reagents are mixed and forced into the outer portion of the rotor containing the optical cell. Here a light beam passing vertically through the cell measures the absorbance as the rotor revolves. The resulting light pulses are displayed on an oscilloscope for monitoring and fed to a computer that separates the pulses from the individual cells and calculates the results. Since the measurements on each cell are made several hundred times per minute, kinetic methods can be readily used.

The GEMSAEC and the Rotochem have automatic washing cycles at the end of the run for cleaning the rotors; in the CentrifiChem the wash cycle must be initiated manually. In all models the rotor is contained in a temperature-controlled chamber. Each instrument has a loader which automatically loads the

† Union Carbide Co., Biomedical Products Div., Tarrytown, N.Y. 10591.
‡ Electro-Nucleonics, Fairfield, N.J. 07006.
§ American Instrument Co., Div. of Travenol Laboratories, Silver Springs, Md. 20910.

samples and reagents into the appropriate rotor wells, usually within less than 4 min. The time of the rotation cycle depends upon the particular test involved, but is usually less than 5 min so that a complete set of 15 samples can be completed in less than 10 min.

In the CentrifiChem the operating program (wavelength, calibration factors, etc.) must be entered manually by punching the appropriate buttons. In the other two instruments more of the programming is stored in the computer or introduced by means of punched tapes as required. The sample sizes required for most determinations are generally in the range of 10 to 50 μl per test.

Continuous-Flow Analyzers||

An entirely different type of instrument is based on the continuous-flow principle as developed by Technicon Instruments. In all these instruments the sample and reagents are introduced into a continuous-flow system by means of peristaltic pumps. The rate of flow for the sample and different reagents is regulated by using different sizes of pump tubings; the pumps themselves operate at a constant speed. The sample and reagents are combined in the flow system by the use of T and Y connectors. The liquid flow is segmented by air bubbles to aid in mixing by passage through mixing coils. The reagents flow through the system continuously and the samples are automatically and consecutively picked up for definite periods of time, usually with a shorter water wash between samples. If it is desired to remove proteins or other interfering material, continuous dialysis is used. The sample and recipient stream flow in parallel on opposite sides of a cellophane membrane. The small crystalloids will dialyze through the membrane but the larger protein molecules will not. Incubation or heating is accomplished by passage of the solution through glass coils immersed in the heating bath at the proper temperature. The mixture then flows to a debubbler where the bubbles are removed and a portion of the liquid pumped through a flow-through cuvette for absorbance measurements. Interference filters are usually used for wavelength selection.

In the earlier single- and double-channel modules, the output was a series of peaks on a strip chart and the results had to be calculated manually by comparing the peak heights of a series of standards with those of the samples. In later modules this is done automatically and the results printed out on tape.

For multiple analysis the SMA 12/60 and SMA 6/60 models are used. (SMA stands for sequential multiple analyzer.) The principles are similar to those of the simpler analyzers. The sample is picked up in a continuous-flow system and

|| Technicon Instruments Corp., Tarrytown, N.Y. 10591.

split as required to send a portion of the sample or diluted sample to each of the 12 channels where the sample reacts with the reagent (after dialysis if required), passes through a delay coil for incubation or heating if required, then to a de-bubbler and a flow-through cuvette for absorbance measurement. As in the simpler models, the color developed for a given determination on a sample, if measured continuously on a recorder, would show a rise to a broad peak and then a fall. The flow through the separate colorimeters for the different tests is so regulated that the peak absorbance for the different channels occurs at successive intervals of time and the colorimeter is switched to the recording system during this time. Thus the recorder chart would show a succession of nearly straight lines corresponding to the peak absorbances for the different tests. On the chart paper each channel has a corresponding concentration scale printed on it; thus the successive lines indicate the concentrations in the different tests. Glass coils (phasing coils) may be inserted in any given channel to insure that, when a given colorimeter is connected to the recording system, the peak concentration is actually in the cuvette. This adjustment is facilitated (in the 12/60 model) by the continuous display of the colorimeter readings for each channel on an oscilloscope. The oscilloscope trace also indicates when the particular colorimeter is connected to the recorder. Calibration is made by first manually adjusting the baseline potentiometers with only reagents in the system, then running through the system a standard serum containing known values of all the constituents being determined and adjusting the recorder pen to indicate the standard values. The 12/60 will analyze serum for 12 different constituents at the rate of 60 samples/hr. The total serum required for the analysis is about 2.6 ml. There is some variation in the tests that can be run, but the usual configuration for the 12/60 is the determination of calcium, inorganic phosphate, glucose, BUN, uric acid, cholesterol, total protein, albumin, total bilirubin, alkaline phosphatase, lactate dehydrogenase, and GOT (not necessarily in that order). As the instrument is usually set up, there is no way in which a few tests can be omitted from a given sample analysis if desired. All will be recorded. If it is desired to run only a few tests on a number of samples, it is usually possible to set the reagent valves so that water instead of reagents is pumped through the channels not desired, but this does not decrease the amount of sample required. The 6/60 runs six different tests at a rate of 60 samples/hr. The usual tests programmed for this are glucose, BUN, CO_2, chloride, sodium, and potassium, the last two by flame photometer. Although there are some relatively completely automated flame photometers for the determination of sodium and potassium, none of the other automated multichannel instruments offers sodium and potassium as part of a group of tests.

The newer SMAC (sequential multiple analyzer computerized) system operates on the same principles as the older SMA 12/60 and 6/60 but with several innovations. The initial configuration is for 20 channels. Additional channels can be added later, not separately but as a group of ten. By using smaller volumes of reagents together with smaller flow-through photometer cells with fiber optics, the consumption of reagents and sample has been reduced considerably over that of the older models so that the 20 tests can be performed on less than 1 ml of serum.

Sodium and potassium, are determined by flow-through specific ion electrodes rather than by the flame photometer. The elimination of the flame photometer results in some simplification in the manifolds. The other determinations are made by modifications of the methods used in the earlier models. The results of the determinations are given as a numerical printout together with a small scale chart which indicates the relative position of each result with respect to a scale of the normal range. This enables the physician to see at a glance the approximate position of each result in relation to the normal values. If desired, adjustments can be made so that only certain specified tests will be printed out from the total tests available. The arrangement is such that even if the result of a given test were not requested, it would still be printed if the result were higher or lower than a designated normal range. Thus the instrument might be programmed to print out only a certain battery of tests that might be requested, for example, a liver profile. But if the result of any of the other tests, not in this group, were definitely abnormal, it would also be printed out, warning the physician of this abnormality. The instrument operates at the rate of 150 samples/hr and has built-in sample identification system, automatic startup, and calibration and shutdown procedures. Operation and troubleshooting are monitored by the computer. This instrument may prove to be one of the most versatile and satisfactory arrangements for large-scale screening.

Many of the larger instruments mentioned in the previous discussion have some type of computer included for calculation of results, and in some cases automatic calibration and drift correction. Most of the instruments are readily compatible with larger laboratory computers. For some of the smaller instruments, minicomputers may be available for automatic calibration procedures and give a numerical printout of the results, with identification numbers which might not otherwise be available. These computers are usually not made by the manufacturers of the automated instruments but must be obtained from other sources. Although the addition of computer facilities will increase the versatility and capacity of almost any automated system, a discussion of complete computerized systems is beyond the scope of this book.

Ames Clinilab#

Although the simple routine urinalysis tests for glucose, protein, pH, occult blood, ketones, bilirubin, and specific gravity have not been specifically treated in this volume, mention might be made of an automated instrument for these determinations. The Ames Clinilab automatically performs all these tests in urine samples, at the rate of 120 samples/hr. The tests are for the most part performed similarly to the manual strip methods with an optical scanning of the different strips. The specific gravity is measured by a falling drop method. The automated instrument is a great timesaver and is probably as accurate as manual methods, considering the inherent limitations of the strip tests.

Reference

1. Alpert, N. L. *Lab World*. 1. November 1972, p. 1272; 2. December 1972, p. 1408; 3. January 1973, p. 35; 4. February 1973, p. 36; 5. March 1973, p. 30; 6. April 1973, p. 32; 7. May 1973, p. 34; 8. June 1973, p. 27; 9. July 1973, p. 33; 10. August 1973, p. 32; 11. September 1973, p. 32; 12. October 1973, p. 46; 13. November 1973, p. 32; 14. December 1973, p. 40.

Ames Co., Div. Miles Laboratories, Elkhart, Ind. 46514.

20. Hydrogen Ion Concentration; Preparation of Buffers and Standard Acids and Bases

Hydrogen Ion Concentration and Buffers

It is customary to express the hydrogen ion concentration in biological fluids in terms of the pH. This is defined as the logarithm of the reciprocal of the hydrogen ion concentration (in moles per liter). It is a significant factor in biological systems and reactions. For biological fluids the major exception to expressing hydrogen ion concentration in terms of pH is gastric acidity, which is expressed in moles of hydrogen ion per liter. The concentration of standard acids and bases is also usually expressed in moles per liter (or equivalents per liter).

In many biological analyses it is necessary to maintain the pH (hydrogen ion concentration) relatively constant. The solutions used for this purpose are called buffers; they maintain the pH relatively constant in spite of the addition or formation of small amounts of acid or base. Accurately made buffer solutions of known pH are also used for the calibration of glass electrode pH meters. Buffers usually consist of a mixture of either a weak (very slightly ionized) acid and its salt or a weak base and its salt. For a weak acid the fundamental relation, the Henderson-Hasselbalch equation, may be written:

$$pH = pK'_a + \log \frac{[\text{salt}]}{[\text{acid}]} \tag{1}$$

where pK'_a is a measure of the effective ionization constant of the acid ($pK = -\log K$, K being the ionization constant as given below), and the brackets, as usual, denote concentrations (in moles per liter). If the ionization of the acid is written as

$$HAc \rightleftharpoons H^+ + Ac^-$$

the ionization equilibrium constant is

$$K'_a = \frac{[H^+] \cdot [Ac^-]}{[HAc]} \tag{2}$$

The Henderson-Hasselbalch equation may be readily derived from this with $pK'_a = -\log K'_a$. The concentration of the undissociated acid, [HAc], is usually taken as the total concentration of acid in the solution since the acid is only very slightly ionized and thus contributes very few Ac^- ions to the total, and the concentration of these ions, $[Ac^-]$, is taken to be that of the total added salt which is thus assumed to be completely ionized. The K'_a (or pK'_a) is written with a prime

725

to distinguish it from the "true" ionization constant which is derived from an equation similar to equation 2 except that the concentrations are replaced by the thermodynamic activities of the molecular species. Thus the pK'_a is theoretically equal to the pH of a solution containing equimolar concentrations of the acid and salt.

A similar equation holds for a mixture of a weak base and its salt, with K'_a replaced by K'_b, the ionization constant of the base, and pH replaced by pOH (defined similarly to pH). Since at 25°C pH + pOH is taken to be equal to 14,

$$pH = (14 - K'_b) - \log \frac{[salt]}{[base]} \tag{3}$$

The concentration of the base is taken as the total amount of free base and that of the salt taken as the total concentration of added salt of the base. The bases used for buffers are usually tertiary ammonium derivatives; using the system ammonium chloride-ammonium hydroxide as an example (although this is actually little used as a buffer), the salt concentration would be that of the added ammonium chloride, and the base concentration that of the ammonium hydroxide added. Theoretically, then, for equimolar concentrations of the base and salt the pH should be $14 - pK'_b$.

Thus, knowing pK'_a or pK'_b one could calculate the pH of any solution having known quantities of the acid (or base) and its salt. The calculations are only approximate since pK'_a and pK'_b as well as the ionization of the salt will be influenced by the total amount of acid and salt present. In the tables given later in this chapter for the preparation of buffers, it will be noted that the pH obtained from the tables for an equimolar mixture of acid and salt is not always exactly the same as the pK'_a given for that acid. Similarly in tables such as Tables 20-5 and 20-9 it will be seen that for a given pH the ratio of salt to acid will vary with the total amount of salt and acid present. In Table 20-5, for example, the ratio of salt to acid for a pH of 4.6 is only slightly different for the ionic strength = 0.2 buffer from what it is for the ionic strength = 0.05 buffer, indicating that the dilution of an acetate buffer will not result in any great change in pH. On the other hand, for the phosphate buffer (Table 20-9) at the pH of 6.8 the ratio of salt to acid for the ionic strength = 0.2 buffer is considerably different from that for the ionic strength = 0.05 buffer at the same pH. Calculations will show that the phosphate buffer in dilution from an ionic strength of 0.2 to one of 0.05 can change in pH by as much as 0.1 unit. Since this dilution effect varies from buffer to buffer, if a particular buffer must be diluted, the pH after dilution should always be checked. Sometimes it is convenient to prepare a concentrated stock buffer and dilute aliquots to make a working buffer as required. If this is done, the pH of the diluted aliquot should always be checked rather than that of the stock.

The pH of a buffer is also affected by the temperature. In general the pH will decrease slightly with increase in temperature, but this will vary with the particular buffer. In the following tables the approximate change in pH between 25° and 37°C is given in many cases. If the buffer is calibrated at the temperature of use, no problem will arise. Often in enzyme determinations the buffer is prepared and calibrated at 25°C and then used at 37°C. Since some enzyme reactions may be quite sensitive to changes in pH, this change in temperature might cause some variation in results. In many instances, however, in determining the optimal conditions for an enzyme reaction, the pH of the trial buffers used is measured at 25°C no matter what the reaction temperature. If one uses such a procedure in the laboratory, one can measure the pH of the buffers at 25°C without having to worry about any temperature effect.

For most purposes the buffers are prepared to a constant molarity of acid plus salt. Thus a $0.2M$ (0.2 mol/liter) acetate buffer will contain a total of 0.2 mol/liter of sodium acetate plus acetic acid, the varying pH being obtained by varying the proportion of the two constituents. In some instances, such as in electrophoresis, it is desirable to make up the buffers to a constant ionic strength rather than to constant molarity. The ionic strength is calculated by multiplying the concentration of each ion (in moles per liter) by the square of the charge, summing over all ions, and dividing the result by 2. Mathematically:

$$\text{ionic strength} = \Sigma \; \frac{c_i \, (x_i)^2}{2}$$

where c_i is the concentration of the i ion and x_i its charge, the summation being taken for all ions. For a uni-univalent molecule such as sodium chloride, the ionic strength is equal to the molarity, since for a solution of $0.1M$,

$$\text{ionic strength} = \frac{0.1 \times 1^2 + 0.1 \times 1^2}{2} = 0.1$$

This is not true for molecules containing polyvalent ions. For a $0.1M$ solution of Na_2SO_4,

$$\text{ionic strength} = \frac{2 \times 0.1 \times 1^2 + 0.1 \times 2^2}{2} = 0.3$$

since there are 2×0.1 mol of Na ions of charge 1, and 0.1 mol of SO_4 ions with a charge of 2 (note that since the charge is squared, its sign is irrelevant). Since the ionic strength is concerned only with ions, the unionized acid (or base) in a buffer solution does not contribute to the ionic strength; only the salt form does. Thus (Table 20-5) to make an acetate buffer having an ionic strength of 0.1, 0.1 mol of sodium acetate (100 ml of $1M$) is diluted to 1 liter with water and acetic acid to bring the pH to the desired value.

Preparation of Buffers

In the following tables and the discussion of their use, all references to water mean a good grade of distilled or deionized water. The directions for the preparation of the buffers may call for the use of acids (hydrochloric or acetic) or base (sodium hydroxide) in solutions of definite molarities (usually 1.0, 0.5, 0.2, or $0.1M$). The procedure for the preparation of standardized acids and bases is given in a section following and is not included separately for each buffer table. When making up the acid, it is best to prepare a quantity of $1.0M$ acid and make careful dilutions when lower strengths are required.

In the directions for the preparation of a number of the buffers, the added acid (or base), together with the added salt, is further diluted with water in the preparation of the final solution. For this the added acid need not be exactly, say, $0.1M$, as long as its exact molarity is known. If the directions call for the addition of A ml of a solution of B mol/liter (M) of acid and the concentration of the acid solution available is C mol/liter (M), the amount of this acid to be used is $(A \times B)/C$. For example, if the directions call for 25.00 ml of $0.200M$ acid and a $0.215M$ acid is available, $(25 \times 0.20)/0.215 = 22.3$ ml would be used. Of course, the stronger acid could be diluted to exactly $0.200M$ and 25 ml of it used, but the other procedure is often easier and more convenient.

Conventionally the procedures for the preparation of buffers call for the mixing of definite solutions of the acid and salt. This is very convenient if one wishes to make several solutions of different pH values in the same buffer system. Suppose one wishes to make up only one buffer—for example, 1 liter of a citrate buffer of pH 4.0. The directions (Table 20-3) call for the preparation of $0.1M$ solutions of citric acid and sodium citrate and then mixing 330 ml of the former with 170 ml of the latter and diluting to 1 liter (ten times the quantities shown in the table for 100 ml of buffer). It might be simpler to weigh out the amounts of citric acid and sodium citrate contained in the volumes of the solution given, dissolve these in water, and dilute to 1 liter. Thus in most of the tables these weights have been included. In order to distinguish clearly between the volumes of

Figure 20-1. pH ranges for various buffers. **The** numbers at the left refer to the buffers listed in the tables.

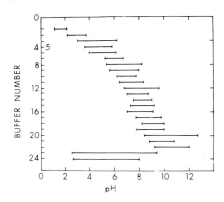

solutions and alternatively the weights of acids and salts to be used, the latter are given in parentheses in the body of the tables. For example, referring again to the pH 4.0 citrate buffer of Table 20-3, the figures given under citric acid for this pH are 33.0 (0.694). This means that 33.0 ml of the specified citric acid solution (or 0.694 g of citric acid) plus the proper amount of sodium citrate are measured out and the solution made up to 100 ml. Note also that in some tables the volume is made up to 100 ml and in others to 1 liter. In Tables 20-10, 20-11, 20-15, 20-16, 20-17, and 20-18, which use solutions of HCl and a base, the base solution can be titrated against the HCl solution using methyl orange as indicator to check if the two solutions are in the proper ratio. If, for example, the two solutions are both supposed to be $0.2M$, the resulting pH would be very nearly the same if the two solutions were both $0.19M$ or both were $0.21M$.

Preferably after being made up all the buffers should be checked against a standard buffer with a glass electrode pH meter. Most commercially available buffers are accurate. Usually only three or four buffers in the range of 3 to 10 pH units are required. Some special buffers that can also be used as standards are given later.

Figure 20-1 shows the approximate pH range for the various buffers described in the tables. It may be used to select the desired buffer for any particular need.

Table 20-1

HCl-KCl Buffer of Clark and Lubs [1] ionic strength $= 0.10$

Potassium chloride, 0.2M. Dissolve 14.91 g of reagent grade KCl in water to make 1 liter.
Hydrochloric acid, 0.2M.

Add the designated amounts of the KCl and HCl solutions to a 100 ml volumetric flask
and dilute to the mark with water.

pH	HCl ml	KCl ml	pH	HCl ml	KCl ml
1.1	47.3	2.7	1.7	11.9	38.1
1.2	37.6	12.5	1.8	9.4	40.6
1.3	29.8	20.2	1.9	7.5	42.5
1.4	23.7	26.3	2.0	6.0	44.0
1.5	18.8	31.2	2.1	4.7	45.3
1.6	15.0	35.0	2.2	3.8	46.2

This buffer has approximately the same pH between 25° and 37°C.

Table 20-2

Potassium Biphthalate = HCl Buffer [1, 2] 0.05M
(pK'_a = 2.8)

Potassium biphthalate, 0.2M. Dissolve 40.83 g of reagent grade potassium biphthalate,
which has been dried at 110°C for 2 hr, in water to make 1 liter.
Hydrochloric acid, 0.2M.

Add 25.0 ml of the potassium biphthalate solution or 1.021 g of the salt and the designated amount of the HCl solution to a 100 ml volumetric flask, add water to dissolve the
salt if added as such, and dilute to the mark with water.

pH	HCl ml	pH	HCl ml
2.2	46.6	3.2	14.8
2.4	39.6	3.4	9.95
2.6	33.0	3.6	6.0
2.8	26.5	3.8	2.65
3.0	20.4		

The pH at 37°C will be not more than 0.02 unit lower than at 25°C.
Note: For preparing 1 liter of solution 10.21 g of the salt and *twice* the quantities
given above of *1.0M* HCl can be used.

730

Table 20-3

Citrate Buffer [3, 4]   0.05M
(pK$'_a$ = 4.7)

Citric acid, 0.1M. Dissolve 21.02 g of citric acid monohydrate in water to make 1 liter. (*Note:* The monohydrate tends to lose water on standing and should be kept in a tightly closed container and a fresh bottle used when possible.)
Sodium citrate, 0.1M. Dissolve 29.41 g of trisodium citrate dihydrate in CO_2-free water and dilute to 1 liter.

Add the designated volumes of the two solutions to a 100 ml volumetric flask and dilute to the mark with water. Alternatively, add the designated weights of the acid and salt to a 100 ml volumetric flask, add water to dissolve, and dilute to the mark.

pH	Citric Acid ml	(g)	Sodium Citrate ml	(g)	pH	Citric Acid ml	(g)	Sodium Citrate ml	(g)
3.0	46.5	(0.977)	3.5	(0.103)	4.8	23.0	(0.483)	27.0	(0.794)
3.2	43.7	(0.918)	6.3	(0.183)	5.0	20.5	(0.431)	29.5	(0.868)
3.4	40.0	(0.841)	10.0	(0.294)	5.2	18.0	(0.378)	32.0	(0.941)
3.6	37.0	(0.778)	13.0	(0.382)	5.4	16.0	(0.336)	34.0	(1.000)
3.8	35.0	(0.736)	15.0	(0.441)	5.6	13.7	(0.288)	36.3	(1.124)
4.0	33.0	(0.694)	17.0	(0.500)	5.8	11.8	(0.248)	38.2	(1.231)
4.2	31.5	(0.662)	18.5	(0.544)	6.0	9.5	(0.198)	40.5	(1.271)
4.4	28.0	(0.588)	22.0	(0.647)	6.2	7.2	(0.151)	42.8	(1.259)
4.6	25.5	(0.536)	24.5	(0.721)					

Table 20-4

Acetate Buffer [5, 6]   0.2M

Sodium acetate, 0.2M. Dissolve 27.22 g of reagent grade sodium acetate trihydrate in water to make 1 liter.
Acetic acid, 0.2M.

Mix together the designated volumes of the two solutions. Alternatively, add the designated volume of acetic acid to a 100 ml volumetric flask, add the designated weight of sodium acetate and water if necessary to dissolve the salt, and dilute to the mark with water.

pH	Acetic Acid ml	Sodium Acetate ml	(g)	pH	Acetic Acid ml	Sodium Acetate ml	(g)
3.6	92.5	7.5	(0.204)	4.8	41.0	59.0	(1.606)
3.8	88.0	12.0	(0.327)	5.0	30.0	70.0	(1.905)
4.0	82.0	18.0	(0.490)	5.2	21.0	79.0	(2.150)
4.2	73.5	26.5	(0.721)	5.4	14.0	86.0	(2.341)
4.4	63.0	37.0	(1.007)	5.6	9.0	91.0	(2.477)
4.6	52.0	48.0	(1.306)	5.8	6.0	94.0	(2.559)

At 37°C these buffers are about 0.05 pH unit lower than at 25°C.

731

Table 20-5

Sodium acetate, 1.0M. Dissolve 126.1 g of reagent grade sodium acetate trihydrate in water and dilute to 1 liter.
Acetic acid, 1.0M.

A. For ionic strength = 0.05, add 50 ml of the sodium acetate solution or 6.804 g of the salt to a 1 liter flask, add the amount of acid given in column A, add water if necessary to dissolve the salt if added, and dilute to the mark with water.

B. For ionic strength = 0.10, add 100 ml of the sodium acetate solution or 13.61 g of the salt to a 1 liter volumetric flask. Then add the amount of acetic acid given in column B, add water if necessary to dissolve the salt, and dilute to 1 liter.

C. For ionic strength = 0.20, add 200 ml of the sodium acetic solution or 27.22 g of the salt to a 1 liter volumetric flask, add the amount of acetic acid given in column C, add water to dissolve the salt if necessary, and dilute to 1 liter.

	ml Acetic Acid				ml Acetic Acid		
pH	A	B	C	pH	A	B	C
3.6	600	—	—	4.8	37.7	73	141
3.8	378	728	—	5.0	23.8	46	89
4.0	238	459	—	5.2	15.0	29	56
4.2	150	289	560	5.4	9.5	18	35
4.4	95	183	353	5.6	6.0	12	22
4.6	60	115	223	5.8	3.8	7	12.5

Table 20-6

Potassium Biphthalate-Sodium Hydroxide Buffer [1, 2]   0.05M
(pK$_a'$ = 5.0)

Potassium biphthalate, 0.2M. Dissolve 40.83 g of reagent grade potassium biphthalate (previously dried for 2 hr at 120°C) in water and dilute to 1 liter.
Sodium hydroxide, 0.2M.

To a 100 ml volumetric flask add 25 ml of the biphthalate solution or 1.021 g of the salt. Add the designated amount of sodium hydroxide, plus water to dissolve the salt if added, and dilute to 100 ml.

	NaOH		NaOH
pH	ml	pH	ml
4.0	0.40	5.2	29.75
4.2	3.65	5.4	35.25
4.4	7.35	5.6	39.70
4.6	12.0	5.8	43.1
4.8	17.5	6.0	45.4
5.0	23.65	6.2	47.00

These buffers will be only about 0.02 pH unit lower at 37°C than at 25°C.
Note: For making up 1 liter of buffer 10.21 g of biphthalate and *1.0M* NaOH in *twice* the amounts given in the table above can be used.

Table 20-7

Maleic Acid–Sodium Hydroxide Buffer [4, 8]   0.05M
(pK'_a = 6.1)

Sodium hydroxide, 0.2M.
Monosodium maleate, 0.2M. Dissolve 23.22 g of maleic acid or 19.61 g of maleic anhydride (the latter can often be obtained in a purer form [reagent grade] than the acid) in 0.2M sodium hydroxide solution and dilute to 1 liter with the hydroxide solution.

Add to a 1 liter volumetric flask 50 ml of sodium maleate solution and the designated volume of NaOH solution. Dilute to the mark with water. Alternatively, add to a 1 liter volumetric flask 1.161 g of maleic acid or 0.981 g of maleic anhydride, 50 ml of the sodium hydroxide solution, plus the additional sodium hydroxide solution given in the table. Dissolve the salt and dilute to 1 liter with water.

pH	NaOH ml	pH	NaOH ml
5.2	7.2	6.2	33.0
5.4	10.5	6.4	38.0
5.6	15.3	6.6	41.6
5.8	20.8	6.8	44.4
6.0	26.9		

Table 20-8

Sorensen's Phosphate Buffer [9–12]   0.0667M
(pK'_a = 6.7)

Disodium phosphate, 0.0667M. Dissolve 9.47 g of reagent grade Na_2HPO_4 in water to make 1 liter. This salt will take up some moisture on exposure to air and should preferably be dried at 110°C for a few hours before use.
Monopotassium phosphate, 0.0667M. Dissolve 9.08 g of reagent grade KH_2PO_4 in water to make I liter.

Mix together the designated volumes of the two solutions. Alternatively, add the designated weights of the two salts to a 100 ml volumetric flask, dissolve in water, and dilute to the mark.

pH	Na_2HPO_4 ml	(g)	KH_2PO_4 ml	(g)	pH	Na_2HPO_4 ml	(g)	KH_2PO_4 ml	(g)
5.4	3.0	(0.0284)	97.0	(0.880)	7.0	61.1	(0.578)	38.9	(0.353)
5.6	5.0	(0.0473)	95.0	(0.862)	7.2	71.5	(0.677)	28.5	(0.259)
5.8	7.8	(0.0739)	92.2	(0.837)	7.4	80.4	(0.761)	19.6	(0.1778)
6.0	12.0	(0.1136)	88.0	(0.798)	7.6	86.8	(0.822)	13.2	(0.1198)
6.2	18.5	(0.1751)	81.5	(0.740)	7.8	91.4	(0.865)	8.6	(0.0780)
6.4	26.5	(0.251)	73.5	(0.667)	8.0	94.5	(0.894)	5.5	(0.0499)
6.6	37.5	(0.355)	62.5	(0.568)	8.2	97.0	(0.918)	3.0	(0.0272)
6.8	50.0	(0.473)	50.0	(0.454)					

At 37°C the values are only about 0.02 pH units less than at 25°C.
This is the original Sorensen's phosphate buffer. For most purposes the equivalent amount of K_2HPO_4 (11.61 g) could be used instead of Na_2HPO_4. Note that for the phosphate buffers the acid form is $K^+ + H_2PO_4^-$ and the salt form is $Na^+ + Na^+ + HPO_4^{2-}$.

Table 20-9
Phosphate Buffer [7]
($pK'_a = 6.7$) constant ionic strength

Disodium phosphate, 0.5M. Dissolve 70.99 g of reagent grade Na_2HPO_4 (previously dried at 110°C for 2 hr) in water and dilute to 1 liter.
Monopotassium phosphate, 0.5M. Dissolve 68.05 g of reagent grade KH_2PO_4 in water and dilute to 1 liter.

Add the designated volumes of the two solutions to a 1 liter volumetric flask and dilute to 1 liter with water. Alternatively, add the designated weights of the two salts to a 1 liter volumetric flask, dissolve, and dilute to 1 liter.

| | Ionic Strength = 0.05 | | | | Ionic Strength = 0.10 | | | | Ionic Strength = 0.20 | | | |
| | KH_2PO_4 | | Na_2HPO_4 | | KH_2PO_4 | | Na_2HPO_4 | | KH_2PO_4 | | Na_2HPO_4 | |
pH	ml	(g)	ml	(g)	ml	(g)	ml	(g)	ml	(g)	ml	(g)
5.6	—	—	—	—	—	—	—	—	333	(22.66)	22.4	(1.590)
5.8	—	—	—	—	159	(10.82)	13.8	(0.980)	303	(20.62)	32.4	(2.300)
6.0	74.2	(5.05)	8.6	(0.610)	142	(9.66)	19.5	(1.384)	265	(18.03)	44.8	(3.180)
6.2	64.6	(4.40)	11.8	(0.838)	121	(8.23)	26.4	(1.874)	222	(15.11)	59.4	(4.22)
6.4	53.4	(3.634)	15.5	(1.100)	98.2	(6.68)	34.0	(2.414)	176	(11.98)	74.6	(5.30)
6.6	42.0	(2.858)	19.3	(1.370)	75.6	(5.15)	41.4	(2.939)	133	(9.05)	89.2	(6.33)
6.8	31.4	(2.137)	22.8	(1.619)	55.4	(3.770)	48.2	(3.422)	95.2	(6.48)	102	(7.24)
7.0	22.4	(1.525)	25.8	(1.832)	39.0	(2.654)	53.6	(3.805)	65.8	(4.48)	111	(7.88)
7.2	15.4	(1.048)	28.2	(2.002)	26.4	(1.797)	57.8	(4.103)	44.2	(3.008)	119	(8.45)
7.4	10.3	(0.701)	30.0	(2.130)	17.6	(1.198)	60.8	(4.316)	29.2	(1.983)	124	(8.80)
7.6	6.7	(0.456)	31.0	(2.201)	11.5	(0.783)	62.8	(4.46)	18.9	(1.286)	127	(9.02)
7.8	4.4	(0.299)	31.8	(2.258)	7.4	(0.504)	64.2	(4.56)	12.1	(0.823)	129	(9.16)
8.0	2.8	(0.191)	32.4	(2.300)	—	—	—	—	—	—	—	—

Table 20-10

Imidazole Buffer [14] $0.05M$
(pK'_a = approx. 7)

Imidazole, 0.2M. Dissolve 13.62 g of imidazole[a] in water to make 1 liter. This material may absorb water from the air and should be dried in a desiccator before use.
Hydrochloric acid, 0.1M.

Add the designated amount of the HCl solution to 25.00 ml of the imidazole solution (or use 0.4305 g of the base) and dilute to 100 ml with water.

pH	HCl ml	pH	HCl ml
6.2	42.9	7.2	18.6
6.4	39.8	7.4	13.6
6.6	35.5	7.6	9.3
6.8	30.4	7.8	6.0
7.0	24.3		

[a] Obtainable from Sigma Chemical Co., St. Louis, Mo. 63118.

Table 20-11

Collidine Buffer [15] $0.05M$
(pK'_b = 6.6)

Collidine, 0.2M. Dissolve 26.4 g of s-collidine[a] (2,4,6-trimethylpyridine) in distilled water and dilute to 1 liter.
Hydrochloric acid, 0.1M.

To 25 ml of the collidine solution (or 0.660 g of the free base) add the designated volume of HCl solution and dilute to 100 ml with water.

pH	HCl ml	pH	HCl ml
6.4	45.7	7.6	19.0
6.6	42.8	7.8	14.0
6.8	40.0	8.0	10.0
7.0	35.7	8.2	7.0
7.2	30.5	8.4	4.1
7.4	25.0		

At 37°C the pH values are about 0.1 pH unit lower than at 25°C.
The above figures are calculated by graphical interpolation from the data given by Gomori [15]. Because collidine is a liquid, it may be difficult to weigh out exactly 26.4 g. Weigh out an amount close to this; then if A g was weighed out, use instead of 25 ml of the collidine solution, the amount: $(26.4/A) \times 25.0$ ml.

[a] Obtainable from Sigma Chemical Co., St. Louis, Mo. 63118.

Table 20-12

Sodium Barbital-Hydrochloric Acid Buffer [16, 17]   (See footnote[a])
(pK$'_a$ = 8.0)

Sodium barbital, 0.1M. Dissolve 20.60 g of sodium diethylbarbiturate in water to make 1 liter. The U.S.P. grade is satisfactory without further treatment.
Hydrochloric acid, 0.1M.

Mix together the designated volumes of the two solutions. Alternatively, add the designated weight of sodium barbital to a 100 ml volumetric flask, add about 50 ml of water to dissolve, add the designated amount of HCl, and dilute to 100 ml.

pH	Sodium Barbital ml	Sodium Barbital (g)	HCl ml	pH	Sodium Barbital ml	Sodium Barbital (g)	HCl ml
6.8	52.2	(1.076)	47.8	8.4	82.4	(1.697)	17.7
7.0	53.6	(1.105)	46.4	8.6	87.1	(1.795)	12.9
7.2	55.4	(1.142)	44.6	8.8	90.8	(1.870)	9.2
7.4	58.1	(1.198)	41.9	9.0	93.6	(1.930)	6.4
7.6	61.5	(1.268)	38.5	9.2	95.2	(1.963)	4.8
7.8	66.2	(1.365)	33.8	9.4	97.4	(2.007)	2.6
8.0	71.6	(1.476)	28.4	9.6	98.5	(2.030)	1.5
8.2	76.9	(1.584)	23.1				

At 37°C this buffer will be about 0.1 pH unit lower than at 25°C.

[a] This series of buffers is neither of constant molarity nor of constant ionic strength. The molarity and ionic strength vary from approximately 0.05 at pH 6.8 to approximately 0.10 at pH 9.6. For some purposes this is not a disadvantage and these buffers are easier to prepare than the subsequent ones made up with barbituric acid, which is difficult to dissolve.

Table 20-13

Barbital-Sodium Barbital Buffer [16, 18]   0.04*M*
(pK$'_a$ = 8.0)

Sodium barbital, 0.04M. Dissolve 8.24 g of sodium diethylbarbituric acid in water and dilute to 1 liter.
Barbital, 0.04M. Dissolve 7.36 g of diethylbarbituric acid in water and dilute to 1 liter. Heating is usually required to dissolve the acid in a reasonable length of time.

Mix together the designated volumes of the two solutions. Alternatively, add the designated weights to a 100 ml volumetric flask, dissolve in water (heating may be needed), and dilute to 100 ml.

pH	Sodium Barbital ml	Sodium Barbital (g)	Barbital ml	Barbital (g)	pH	Sodium Barbital ml	Sodium Barbital (g)	Barbital ml	Barbital (g)
7.0	10.0	(0.0825)	90.0	(0.663)	8.0	50.0	(0.412)	50.0	(0.368)
7.2	15.5	(0.1278)	84.5	(0.623)	8.2	65.0	(0.536)	35.0	(0.258)
7.4	22.5	(0.1854)	77.5	(0.571)	8.4	76.5	(0.630)	23.5	(0.1732)
7.6	30.0	(0.247)	70.0	(0.516)	8.6	83.0	(0.685)	17.0	(0.1251)
7.8	39.5	(0.326)	60.5	(0.446)	8.8	88.0	(0.726)	12.0	(0.0884)

At 37°C these buffers will be about 0.1 pH unit lower than at 25°C.

Table 20-14

Barbital-Sodium Barbital Buffer [5]   constant ionic strength
(pK$_a'$ = 8.0)

Barbital, 0.025M. Dissolve 4.60 g of diethylbarbituric acid in water to make 1 liter. Heating may aid in solution.
Sodium barbital, 0.5M. Dissolve 103.1 g of sodium diethylbarbiturate in CO_2-free water and dilute to 1 liter. This solution is not stable and tends to decompose on standing.
Sodium chloride, 0.5M. Dissolve 29.22 g of reagent grade NaCl in water and dilute to 1 liter.

Mix together the designated volumes of the three solutions and dilute to 1 liter. The formulation on a weight basis is given in a separate table.

| | Ionic Strength = 0.05 | | | Ionic Strength = 0.10 | | | Ionic Strength = 0.20 | | |
pH	Barb. Acid ml	Na Barb ml	NaCl ml	Barb. Acid ml	Na Barb. ml	NaCl ml	Barb. Acid ml	Na Barb. ml	NaCl ml
7.4	648	10	90	639	10	100	275	5	395
7.6	409	10	90	403	10	190	348	10	390
7.8	645	25	75	636	25	175	219	10	390
8.0	814	50	50	401	25	175	346	25	375
8.2	514	50	50	506	50	150	218	25	275
8.4	648	100	—	639	100	100	275	50	350
8.6	409	100	—	403	100	100	348	100	300
8.8	258	100	—	509	200	—	438	200	200
9.0	163	100	—	321	200	—	277	200	200

To make up this buffer on a weight basis, add the designated amounts of the three compounds to a 1 liter volumetric flask, dissolve in water, and dilute to volume. The use of heat may be necessary to dissolve the barbituric acid.

| | Ionic Strength = 0.05 | | | Ionic Strength = 0.10 | | | Ionic Strength = 0.20 | | |
pH	Barb. Acid (g)	Na Barb. (g)	NaCl (g)	Barb. Acid (g)	Na Barb. (g)	NaCl (g)	Barb. Acid (g)	Na Barb. (g)	NaCl (g)
7.4	(2.981)	(1.031)	(2.63)	(2.940)	(1.031)	(2.922)	(1.266)	(0.516)	(11.54)
7.6	(1.883)	(1.031)	(2.63)	(1.856)	(1.031)	(5.553)	(1.603)	(1.031)	(11.40)
7.8	(2.970)	(2.577)	(2.192)	(2.929)	(2.577)	(5.114)	(1.009)	(1.031)	(11.40)
8.0	(3.744)	(5.155)	(1.461)	(1.847)	(2.577)	(5.114)	(1.593)	(2.577)	(10.96)
8.2	(2.367)	(5.155)	(1.461)	(2.330)	(5.155)	(4.384)	(1.004)	(2.577)	(8.04)
8.4	(2.981)	(10.31)	—	(2.940)	(10.31)	(2.922)	(1.266)	(5.155)	(10.23)
8.6	(1.883)	(10.31)	—	(1.856)	(10.31)	(2.922)	(1.603)	(10.31)	(8.77)
8.8	(1.188)	(10.31)	—	(2.344)	(20.62)	—	(2.017)	(20.62)	(5.85)
9.0	(0.751)	(10.31)	—	(1.478)	(20.62)	—	(1.274)	(20.62)	(5.85)

Table 20-15

Tris Buffer [15, 18]   0.05M
(pK'_b = 5.76)

Tris, 0.2M. Dissolve 24.23 g of tris(hydroxymethyl)aminomethane in water and dilute to 1 liter.
Hydrochloric acid, 0.1M.

Mix 25.00 ml of tris solution with the designated volume of the HCl solution and dilute to 11 ml. Alternatively, dissolve 0.6057 g of tris in water in a 100 ml volumetric flask, add the designated amount of the HCl solution, and dilute to 100 ml.

pH	HCl ml	pH	HCl ml
7.2	45.0	8.4	17.5
7.4	42.0	8.6	12.8
7.6	38.9	8.8	9.0
7.8	34.0	9.0	6.3
8.0	29.0	9.2	3.7
8.2	23.3		

At 37°C these buffers are about 0.1 pH unit lower than at 25°C.
A very pure grade of tris is commercially available,[a] as is also a series of mixtures of tris and tris hydrochloride for the simple preparation of buffers of different pH values.

[a] Sigma Chemical Co., St. Louis, Mo. 63118.

Table 20-16

Tris Buffer [7, 19]   constant ionic strength
(K'_b = 5.76)

Tris, 1.0M. Dissolve 121.14 g of tris(hydroxymethyl)aminomethane in CO_2-free water and dilute to 1 liter.
Hydrochloric acid, 1.0M.

To a 1 liter volumetric flask add either the designated volume of tris solution or the designated weight of tris plus water to dissolve. Then add 50 ml of HCl for all buffers with ionic strength = 0.05, 100 ml of HCl for all buffers with ionic strength = 0.10, and 200 ml of HCl for all buffers with ionic strength = 0.20, and dilute to the mark.

pH	Ionic Strength = 0.05 Tris ml	(g)	Ionic Strength = 0.10 Tris ml	(g)	Ionic Strength = 0.20 Tris ml	(g)
7.0	53.6	(6.49)	107	(12.96)	214	(25.92)
7.2	55.7	(6.75)	111	(13.45)	222	(26.89)
7.4	59.1	(7.16)	118	(14.30)	235	(28.47)
7.6	64.4	(7.80)	128	(15.51)	257	(31.13)
7.8	72.9	(8.83)	144	(17.44)	290	(35.13)
8.0	86.1	(10.43)	169	(20.47)	342	(41.43)
8.2	107	(12.96)	208	(25.20)	421	(51.00)
8.4	141	(17.08)	270	(32.71)	550	(66.63)
8.6	194	(23.50)	367	(44.46)	738	(89.40)
8.8	279	(33.80)	524	(63.48)	—	—
9.0	414	(50.15)	761	(92.19)	—	—
9.2	627	(75.96)	—	—	—	—

Table 20-17

2-Amino-2-methyl-1,3-propanediol Buffer [15] $0.05M$
($pK'_b = 5.22$)

2-Amino-2-methyl-1,3-propanediol,[a] *0.2M.* Dissolve 21.0 g of the crystalline base in water to make 1 liter. This material is hydroscopic and should be dried in a desiccator before use.
Hydrochloric acid, 0.1M.

To a 100 ml volumetric flask add the designated quantity of the hydrochloric acid solution and either 25 ml of the 0.2M solution of the base or exactly 0.526 g. Add water if necessary to dissolve and dilute to 100 ml.

pH	HCl ml	pH	HCl ml
7.8	45.6	9.0	18.9
8.0	42.5	9.2	13.7
8.2	39.6	9.4	9.7
8.4	35.0	9.6	6.8
8.6	30.0	9.8	4.0
8.8	24.5		

At 37°C these buffers are approximately 0.1 pH unit lower than at 25°C.
The figures presented are calculated by graphical interpolation from the data of Gomori [15].

[a] Obtainable from Sigma Chemical Co., St. Louis, Mo. 63118.

Table 20-18

2-Amino-2-methyl-1,3-propanediol Buffer [7]   constant ionic strength
($pK'_b = 5.22$)

2-Amino-2-methyl-1,3-propanediol,[a] *1.0M.* Dissolve 105.14 g of the base in CO_2-free water to make 1 liter.
Hydrochloric acid, 1.0M.

Add to a 1 liter volumetric flask the designated volume of the 1.0M solution of the base or the designated weight of the base. Add 50 ml of HCl for ionic strength = 0.05, 100 ml of HCl for ionic strength = 0.10, or 200 ml of HCl for ionic strength = 0.20, add water if necessary to dissolve the base, and dilute to 1 liter.

pH	Ionic Strength = 0.05 Amount of Base ml	(g)	Ionic Strength = 0.10 Amount of Base ml	(g)	Ionic Strength = 0.20 Amount of Base ml	(g)
8.2	61.7	(6.487)	120.9	(12.711)	239	(25.13)
8.4	68.2	(7.171)	133.7	(14.057)	260	(27.34)
8.6	79.0	(8.306)	154	(16.19)	294	(30.93)
8.8	95.9	(10.083)	183	(19.24)	348	(36.59)
9.0	122	(12.83)	229	(24.08)	441	(46.37)
9.2	164	(17.24)	303	(31.86)	606	(63.72)
9.4	232	(24.39)	424	(44.58)	—	—
9.6	338	(35.54)	626	(65.82)	—	—
9.8	511	(53.73)	—	—	—	—
10.0	790	(83.06)	—	—	—	—

[a] Obtainable from Sigma Chemical Co., St. Louis, Mo. 63118.

Table 20-19

Borate Buffer [15, 20]   0.05*M*

Boric acid, 0.2M, + potassium chloride, 0.2M. Dissolve 12.37 g of reagent grade boric acid and 14.91 g of reagent grade KCl in water to make 1 liter.
Sodium hydroxide, 0.1M.

To a 100 ml volumetric flask add either 25.00 ml of the boric acid-KCl solution or 0.309 g of boric acid and 0.373 g of KCl, and the designated amount of sodium hydroxide solution. Add water if necessary to dissolve the salts and dilute to 100 ml.

pH	NaOH ml	pH	NaOH ml
7.8	2.65	9.0	21.40
8.0	4.00	9.2	26.70
8.2	5.90	9.4	32.00
8.4	8.55	9.6	36.85
8.6	12.00	9.8	40.80
8.8	16.40	10.0	43.90

These buffers are about 0.1 pH unit lower at 37°C than at 25°C.

Table 20-20

Glycine-Sodium Hydroxide Buffer [11]   0.1*M*

Glycine, 0.1M, + sodium chloride, 0.1M. Dissolve 7.505 g of reagent grade glycine and 5.85 g of reagent grade sodium chloride in water to make 1 liter.
Sodium hydroxide, 0.1M.

Mix together the designated volumes of the two solutions.

pH	Glycine ml	NaOH ml	pH	Glycine ml	NaOH ml
8.4	95.0	5.0	10.6	52.5	47.5
8.6	92.0	8.0	10.8	51.0	49.0
8.8	88.0	12.0	11.1	50.0	50.0
9.0	83.0	17.0	11.4	49.0	51.0
9.2	77.0	23.0	11.6	47.5	52.5
9.4	72.0	28.0	11.8	45.0	55.0
9.6	67.0	33.0	12.0	43.0	57.0
9.8	62.5	37.5	12.2	40.0	60.0
10.0	58.5	41.5	12.4	30.0	70.0
10.2	56.0	44.0	12.6	20.0	80.0
10.4	54.0	46.0	12.8	10.0	90.0

The data as given were interpolated from the data of Sorensen [11]. Note that at pH near 11, the pH changes quite rapidly with small changes in added solution, hence the values given may not be very accurate and should always be checked with a pH meter for accurate work.
At 37°C the pH values will be approximately 0.3 unit lower than at 25°C.

Table 20-21

Glycine-Sodium Hydroxide Buffer [7, 19]   constant ionic strength

Glycine, 2.0M. Dissolve 150.14 g of reagent grade glycine in water to make 1 liter.
Sodium hydroxide, 1.0M.

A. For ionic strength = 0.05, to 50 ml of the sodium hydroxide solution add the designated amount of glycine solution or solid glycine given in column A and dilute to 1 liter.
B. For ionic strength = 0.10, use 100 ml of the sodium hydroxide solution and the volume or weight of glycine given in column B and dilute to 1 liter.
C. For ionic strength = 0.20, use 200 ml of the sodium hydroxide solution and the amount of glycine given in column C and dilute to 1 liter.

pH	A Glycine ml	(g)	B Glycine ml	(g)	C Glycine ml	(g)
8.8	234	(35.13)	467	(70.12)	—	—
9.0	157	(23.57)	313	(46.99)	633	(95.04)
9.2	108	(16.22)	216	(32.43)	437	(65.61)
9.4	77.4	(11.62)	155	(23.27)	312	(46.84)
9.6	58.1	(8.72)	116	(17.42)	234	(35.13)
9.8	45.8	(6.88)	91.7	(13.77)	185	(27.78)
10.0	38.1	(5.72)	76.3	(11.46)	153	(22.97)
10.2	33.2	(4.99)	66.5	(9.98)	134	(20.12)
10.4	30.1	(4.52)	60.3	(9.05)	121	(18.17)
10.6	28.0	(4.20)	56.4	(8.47)	113	(16.97)
10.8	26.7	(4.01)	53.8	(8.08)	108	(16.22)

Table 20-22

Carbonate Buffer [18, 25]   0.01M
(pK$'_a$ = 9.9)

Sodium carbonate, 0.1M. Dissolve 10.60 g of reagent grade anhydrous Na_2CO_3 in water and dilute to 1 liter. The salt is hygroscopic; if it has been exposed to air, it should be dried for several hours at 250°C and cooled in a desiccator before use.
Sodium bicarbonate, 0.1M. Dissolve 8.40 g of reagent grade $NaHCO_3$ in water to make 1 liter.

Mix together the designated quantities of the two solutions.

pH	Na_2CO_3 ml	$NaHCO_3$ ml	pH	Na_2CO_3 ml	$NaHCO_3$ ml
9.2	14.0	86.0	10.0	58.0	42.0
9.4	22.0	78.0	10.2	69.5	30.5
9.6	32.5	67.5	10.4	79.0	21.0
9.8	45.0	55.0	10.6	88.0	12.0

At 37°C these buffers are about 0.1 pH unit lower than at 25°C.
The data are a combination of that cited by Henry [18] and experiments in our own laboratories. These buffers are the same as those of Delory and King as cited in Tietz [25].

741

Table 20-23

Michaelis "Universal Buffer" [21]
(pK$'_a$ = 4.7, 8.0)

ionic strength = 0.16

Sodium acetate-sodium barbital buffer. Dissolve 19.428 g of sodium acetate trihydrate, reagent grade, and 28.428 g of sodium barbital, U.S.P., in water to make 1 liter.
Hydrochloric acid, 0.1M.
Sodium chloride, 8.5 g/100 ml. Dissolve 8.50 g of reagent grade sodium chloride in water to make 100 ml.

For each solution add 20 ml of the acetate-barbital solution, 8 ml of the sodium chloride solution, and the designated amount of HCl solution and dilute to 100 ml with water.

pH	HCl ml	pH	HCl ml	pH	HCl ml
2.6	60.1	5.0	30.8	7.4	20.2
2.8	59.0	5.2	28.8	7.6	17.2
3.0	57.6	5.4	27.6	7.8	13.8
3.2	56.0	5.6	26.8	8.0	10.4
3.4	54.1	5.8	26.4	8.2	7.7
3.6	52.1	6.0	26.1	8.4	5.3
3.8	49.4	6.2	25.9	8.6	3.6
4.0	45.9	6.4	25.6	8.8	2.4
4.2	43.2	6.6	25.3	9.0	1.6
4.4	40.8	6.8	24.8	9.2	0.9
4.6	36.7	7.0	23.9	9.4	0.5
4.8	34.0	7.2	22.5		

These values are obtained by interpolation from the data of Michaelis [21].
Note that at several places the difference of only a fraction of a milliliter of the HCl will cause a considerable change in pH; thus the values may not be very accurate and should always be checked with a pH meter for careful work.

Table 20-24

McIlvaine Standard Buffer [22–24]
($pK'_a = 2.1, 3.1, 4.7, 6.4, 6.7$)

Disodium phosphate, 0.2M. Dissolve 28.41 g of reagent grade anhydrous Na_2HPO_4 in water to make 1 liter. The salt is somewhat hygroscopic; if it has been exposed to air, it should be dried for a few hours at 105°C and then cooled in a desiccator before use.
Citric acid, 0.1M. Dissolve 21.01 g of citric acid monohydrate in water to make 1 liter. This acid tends to lose water on contact with air; a fresh bottle should be used or the acid may be dried to constant weight at 100°C and 19.21 g of the anhydrous acid used.

Mix together the designated quantities of the two solutions.

pH	Na_2HPO_4 ml	Citric Acid ml	pH	Na_2HPO_4 ml	Citric Acid ml
2.2	2.00	98.00	5.2	53.60	46.40
2.4	6.20	93.80	5.4	55.75	44.25
2.6	10.90	89.10	5.6	58.00	42.00
2.8	15.85	84.15	5.8	60.45	39.55
3.0	20.55	79.45	6.0	63.15	36.85
3.2	24.70	75.30	6.2	66.10	33.90
3.4	28.50	71.50	6.4	69.25	30.75
3.6	32.20	67.80	6.6	72.75	27.25
3.8	35.50	64.50	6.8	77.25	22.75
4.0	38.55	61.45	7.0	82.35	17.65
4.2	41.40	58.60	7.2	86.95	13.05
4.4	44.10	55.90	7.4	90.85	9.15
4.6	46.75	53.25	7.6	93.65	6.35
4.8	49.30	50.70	7.8	95.75	4.25
5.0	51.50	48.50	8.0	97.25	2.75

Special Buffers

Potassium Tetraoxalate [26] $0.05M$

Dissolve 12.71 g of potassium tetraoxalate, $KH_3(C_2O_4)_2 \cdot 2\ H_2O$ (SRM-189)[*] in water to make 1 liter. This solution will have a pH of 1.68 at 25°C and may be used as a low pH standard for determining gastric acidity (Chap. 17).

Potassium Bitartrate [26]   saturated solution, approx. $0.034M$

Mix about 1 g of the salt with 50 ml of water, and stir well or shake for several minutes. Pour off supernatant liquid and use. This will have a pH of 3.56 at 25°C. It is a convenient way of obtaining a fairly rapid low pH buffer for standardization of a pH meter.

Potassium Biphthalate [26] $0.05M$

Dissolve 10.21 g of potassium acid phthalate (SRM-1852), which has been previously dried at 105°C for 1 hr, in ammonia-free water and dilute to 1 liter. This solution will have a pH of 4.01 at 25°C. When the solution is accurately prepared, the pH is exact but the solution has little buffering power.

Borax [20] $0.01M$

Dissolve 3.81 g of $Na_2B_4O_7 \cdot 10\ H_2O$ (SRM-187b) in water and dilute to 1 liter. This solution will have a pH of 9.18 at 25°C.

Phosphate Buffers for Blood pH Standardization

For accurate work these must be made from NBS salts KH_2PO_4 (SRM-186 Ic) and NaH_2PO_4 (SRM-186 IIc) which have been dried at 110°C for 2 hr and cooled in a desiccator.

1. Phosphate buffer [13] $0.025M$, pH 6.84 at 37°C

Dissolve 3.40 g of KH_2PO_4 and 3.55 g of Na_2HPO_4 in CO_2- and ammonia-free water and dilute to 1 liter at 25°C. This buffer will have a pH of 6.86 at 25°C.

2. Phosphate buffer [27]   ionic strength = 0.1, pH 7.384 at 37°C

Dissolve 1.179 g of KH_2PO_4 and 4.303 g of Na_2HPO_4 in CO_2- and ammonia-free water and dilute to 1 liter at 25°C. This solution will have a pH of 7.413 at 25°C.

---

[*] Salts obtainable from the National Bureau of Standards, Washington, D.C. 20234, as Standard Reference Material carry a designation such as (SRM-189).

Table 20-25. Approximate Molarity of Concentrated Acids and Bases and Dilutions to Make 1.0M Solutions

Acid or Base	Approx. Molarity of Conc. Acid or Base	Amount of Conc. Acid Diluted to 1 Liter for Preparation of 1.0M (except as noted)
Hydrochloric acid (36%)	12M	82 ml to make 1.0M
Sulfuric acid (98%)	18M	28 ml to make 0.5M (1N)
Glacial acetic acid (100%)	17.5M	57 ml to make 1.0M
Nitric acid (70%)	16M	63 ml to make 1.0M
Phosphoric acid (85%)	15M	67 ml to make 1.0M
Perchloric acid (70%)	16.5M	62 ml to make 1.0M
Alkali		
ammonium hydroxide (30%)	15.5M	65 ml to make 1.0M
Sodium hydroxide (50% by wt)	12.5M	80 ml to make 1.0M

Preparation of Standard Acids and Bases

For the preparation of a standard acid, an aliquot of the concentrated reagent grade acid is diluted as calculated from the data of Table 20-25, using about 2% more acid than designated to give a solution that is slightly over strength. The solution is mixed, allowed to cool to room temperature, diluted to the final volume and mixed well. It is then standardized by titration as outlined below. The acids usually used for acidimetry and for the accurate preparation of buffers are hydrochloric and acetic. For these purposes the solutions must be accurately standardized. For ordinary reagents it is usually sufficient to dilute the acids carefully using amounts as calculated from the table, and use as reagents without further standardization.

Hydrochloric acid may be standardized by the use of tris-(hydroxymethyl) aminomethane. For many purposes a good reagent grade is satisfactory, such as Trizma.† For very accurate work a primary standard may be obtained from the National Bureau of Standards (SRM-723). The material is dried at 105°C for 1 hr and then stored in a desiccator. To standardize a 1M solution of acid, accurately weigh out close to 3 g of the tris and transfer quantitatively to a 250 ml flask with about 50 ml of distilled water. Swirl to dissolve most of the base, add 2 to 3 drops of indicator [4-(4-dimethylamino-1-naphthylazo)-3-methoxybenzene sulfonic acid,‡ 0.1 g in 100 ml of 60% ethanol], and

† Sigma Chemical Co., St. Louis, Mo. 63118.
‡ #1954, Eastman Kodak Co., Rochester, N.Y. 14650.

titrate with the acid using a high-quality buret. The endpoint is a change from orange-yellow to violet, the titration being carried out to the intermediate pink color. The color change is not very sharp and an excess of indicator makes the change more difficult to see. The molarity of the acid is then calculated as follows:

$$\frac{\text{weight of tris used (g)}}{\text{ml titration} \times 0.12112} = \text{molarity of acid}$$

One could prepare a solution of exactly $1M$ tris and use aliquots of this for titrations, but the suggested method is probably more accurate. Several titrations are carried out and the average value taken. If desired, one can dilute the solution to an exact molarity. If made up according to the suggestions, it will be only a few percentage points higher than $1.00M$, and the dilution can be made as follows: If the molarity as determined is M_1 and the desired exact molarity is M_2, then to A ml of the acid $(M_1 - M_2)A/M_2$ ml of water is added. If the first few titrations indicate a molarity of, say, 1.05 and one desires exactly $1.000M$, it is best to add somewhat less than the calculated amount of water; in this example, add about 45 ml of water instead of the calculated 50 ml/liter. If accurate standardization then yields a molarity of, say, 1.005, one can add 5 ml of water/liter and assume the mixture to be $1.000M$ without further standardization.

Acetic acid solutions cannot be standardized against tris and are best standardized against a sodium hydroxide solution of known molarity using phenolphthalein (1% in ethanol) as indicator.

For the preparation of accurate solutions of sodium hydroxide, it is best to use carbonate-free NaOH. This may be prepared by mixing 500 g of reagent grade NaOH and 500 ml of water. This should be done in a Pyrex flask. After cooling, the solution is transferred to a polyethylene bottle and allowed to stand for several weeks until all the carbonate has settled out. The carbonate-free solution is then decanted off. It is usually more convenient to purchase the solution as a 50% by weight carbonate-free NaOH solution. Slightly over 80 g of this solution is weighed out on a rough balance, dissolved in CO_2 free distilled water, and diluted to 1 liter. This is then standardized against potassium biphthalate. Here also, for most purposes a reagent primary standard grade is sufficiently pure. For very accurate work, that obtained from the Bureau of Standards (SRM-84h) may be used. The salt is dried at 110°C for several hours and then cooled in a desiccator. The procedure is similar to that for the standardization of acid. Between 4 and 5 g of the salt is accurately weighed out and transferred quantitatively to a 250 ml Erlenmeyer flask with the aid of about 50 ml of water. The mixture is swirled to dissolve some of the salt, a few drops of 1% phenolphthalein are added, and the solution is titrated with the sodium hydroxide to the faint pink endpoint. The biphthalate is not very soluble and will not all dissolve at the beginning. It dissolves as the titration proceeds, and one must make certain that all the salt is dissolved at the endpoint. The calculation is similar to that for acids:

$$\frac{\text{weight of biphthalate taken (g)}}{\text{ml titration} \times 0.20422} = \text{molarity of NaOH}$$

The sodium hydroxide solution may be diluted to an exact molarity similarly to the procedure outlined for HCl.

References

1 Clark, W. M., and Lubs, H. A. *J. Biol. Chem.* 25:479, 1916.
2. Homer, W. J., Pinching, G. D., and Acree, S. F. *J. Res. Natl. Bur. Stds.* 35:539, 1945.
3. Lillie, R. D. *Histological Technique.* Philadelphia: Blakiston, 1948.
4. Gomori, G. In S. P. Colowick and N. O. Kaplan [Eds.], *Methods in Enzymology*, vol. 1. New York: Academic, 1955, P. 138.
5. Bates, R. G., Siegel, G. L., and Acree, S. F. *J. Res. Natl. Bur. Stds.* 30:347, 1943.
6. Walpole, G. S. *J. Chem. Soc.* 105:2105, 1914.
7. Datta, S. P., and Grzybowski, A. K. In C. Long [Ed.], *Biochemists Handbook.* New York: Van Nostrand, 1961. P. 19.
8. Temple, J. W. *J. Am. Chem. Soc.* 51:1754, 1929.
9. Sorensen, S. P. *Biochem. Z.* 21:131, 1909.
10. Sorensen, S. P. *Biochem. Z.* 22:352, 1909.
11. Sorensen, S. P. *Ergeb. Physiol.* 12:393, 1912.
12. Bates, R. G. *J. Res. Natl. Bur. Stds.* 39:411, 1947.
13. Bates, R. G., and Acree, S. F. *J. Res. Natl. Bur. Stds.* 34:372, 1945.
14. Mertz, E. T., and Owen, C. A. *Proc. Soc. Exp. Biol. Med.* 43:204, 1940.
15. Gomori, G. *Proc. Soc. Exp. Biol. Med.* 62:33, 1946.
16. Manov, G., Chuette, K. E., and Kirk, S. F. *J. Res. Natl. Bur. Stds.* 48:84, 1952.
17. Michaelis, L. *J. Biol. Chem.* 87:33, 1930.
18. Henry, R. J. *Clinical Chemistry, Principles and Technics.* New York: Harper & Row, 1964. P. 966.
19. Richterich, R. *Klinische Chemie.* Basel: Karger, 1965.
20. Manov, G. G., DeLollis, N. J., Lindvall, R. W., and Acree, S. F. *J. Res. Natl. Bur. Stds.* 36:543, 1946.
21. Michaelis, L. *Biochem. Z.* 234:139, 1931.
22. McIlvaine, T. C. *J. Biol. Chem.* 49:183, 1921.
23. Bates, R. G. *J. Res. Natl. Bur. Stds.* 47:127, 1951.
24. Elving, R. J., Markowitz, J. M., and Rosenthal, I. *Anal. Chem.* 28:1179, 1956.

25. Bermes, E. W., Jr., and Foreman D. T.   In N. W. Tietz [Ed.], *Fundamentals of Clinical Chemistry*. Philadelphia: Saunders, 1970. P. 33.
26. Bates, R. G., Pinching, G. D., and Smith, E. R.   *J. Res. Natl. Bur. Stds.* **45:418,** 1950.
27. Bower, V. E., Paabo, M., and Bates, R. G.   *Clin. Chem.* 7:292, 1961.

Index